In Situ Monitoring of Aquatic Systems
Chemical Analysis and Speciation

IUPAC SERIES ON ANALYTICAL AND PHYSICAL CHEMISTRY OF ENVIRONMENTAL SYSTEMS

Series Editors

Jacques Buffle, *University of Geneva, Geneva, Switzerland*
Herman P. van Leeuwen, *Wageningen University, The Netherlands*

Series published within the framework of the activities of the IUPAC Commission on Fundamental Environmental Chemistry, Division of Chemistry and the Environment.

INTERNATIONAL UNION OF PURE AND APPLIED CHEMISTRY (IUPAC)
Secretariat, PO Box 13757, 104 T.W. Alexander Drive, Building 19,
Research Triangle Park, NC 27709-3757, USA

Previously published volumes (Lewis Publishers):
Environmental Particles Vol. 1 (1992) ISBN 0-87371-589-6
Edited by Jacques Buffle and Herman P. van Leeuwen
Environmental Particles Vol. 2 (1993) ISBN 0-87371-895-X
Edited by Jacques Buffle and Herman P. van Leeuwen

Previously published volumes (John Wiley & Sons, Ltd):
Metal Speciation and Bioavailability in Aquatic Systems Vol. 3 (1995)
ISBN 0-471-95830-1
Edited by André Tessier and David R. Turner
Structure and Surface Reactions of Soil Particles Vol. 4 (1998)
ISBN 0-471-95936-7
Edited by P. M. Huang, N. Senesi and J. Buffle
Atmospheric Particles Vol. 5 (1998)
ISBN 0-471-95935-9
Edited by Roy M. Harrison and René E. van Grieken

IUPAC Series on Analytical and Physical Chemistry
of Environmental Systems. Volume 6

In Situ Monitoring of Aquatic Systems

Chemical Analysis and Speciation

Edited by

JACQUES BUFFLE
University of Geneva, Geneva, Switzerland

GEORGE HORVAI
Technical University of Budapest, Budapest, Hungary

JOHN WILEY & SONS, LTD
Chichester · New York · Weinheim · Brisbane · Singapore · Toronto

Published in 2000 by John Wiley & Sons Ltd,
Baffins Lane, Chichester,
West Sussex PO19 1UD, England

National 01243 779777
International (+44) 1243 779777

e-mail (for orders and customer service enquiries): cs- books@wiley.co.uk
Visit our Home Page on http://www.wiley.co.uk
or http://www.wiley.com

Other Wiley Editorial Offices

John Wiley & Sons, Inc., 605 Third Avenue,
New York, NY 10158–0012, USA

WILEY-VCH Verlag GmbH, Pappelallee 3,
D-69469 Weinheim, Germany

Jacaranda Wiley Ltd, 33 Park Road, Milton,
Queensland 4064, Australia

John Wiley & Sons (Asia) Pte Ltd, 2 Clementi Loop #02–01,
Jin Xing Distripark, Singapore 129809

John Wiley & Sons (Canada) Ltd, 22 Worcester Road,
Rexdale, Ontario M9W 1L1, Canada

Library of Congress Cataloging-in-Publication Data

In-situ monitoring of aquatic systems: chemical analysis and speciation / edited by Jacques Buffle, George Horvai.
p. cm.—(IUPAC series on analytical and physical chemistry of environmental systems; v.6)
Includes bibliographical references and index.
ISBN 0–471–48979–4 (alk. paper)
1. Water quality bioassay. 2. Chemistry, Analytic. 3. Speciation (Chemistry) I. Buffle, J. (Jacques), 1943– II. Horvai, George. III. Series.

QH90.57.B5 15 2000
577.6′028′7—dc21

99–089123

British Library Cataloguing in Publication Data

A catalogue record for this book is available from the British Library

ISBN 0 471 48979 4

Typeset in 10/12pt Times by Kolam Information Services Pvt Ltd, Pondicherry, India
Printed and bound in Great Britain by Biddles Ltd, Guildford and King's Lynn
This book is printed on acid-free paper responsibly manufactured from sustainable forestry, in which at least two trees are planted for each one used for paper production.

Contents

Jacques Buffle is professor of analytical and environmental chemistry. He joined the staff of the Department of Inorganic, Analytical and Applied Chemistry of the University of Geneva, Switzerland, in 1969. He is also an Invited Professor at INRS-Eau, since 1982, and has been Associate Professor of Electroanalytical Chemistry at the University of Lausanne, Switzerland. He was the first chairman of the IUPAC Commission on Fundamental Environmental Chemistry, and a member of the Committee of the IUPAC Division on Chemistry and the Environment. He is a member of the Research Council of the Swiss National Foundation. He has authored more than 150 research papers, 2 books, and is a co-editor of 5 other monographs. He teaches under-and postgraduate courses in analytical and environmental chemistry.

Dr. Buffle has received diploma in biological chemistry, chemical engineering and computer sciences, and a PhD degree in analytical chemistry. His research interests are the understanding of microstructures of aquatic environmental compartments (water, sediment, soils), the physiochemical processes regulating the circulation of chemical compounds in these compartments, the roles of these processes on biota, and the development of *in situ* sensors and integrated analytical systems for the measurement of the corresponding key physico-chemical parameters. He is particularly interested in contributing to the understanding of macroscale effects, at the level of whole soils, lakes etc.., by microscale phenomena, at the molecular, colloidal and microbial level, by considering multidisciplinary approaches.

George Horvai is professor of chemistry at the Budapest University of Technology and Economics (the former Technical University of Budapest) in Hungary. He graduated and received his PhD degree from the same university. He worked for two decades in the Institute for General and Analytical Chemistry and became head of the newly founded Division of Chemical Information Technology in 1994. He was visiting researcher for one year each at the ETH Zurich and the University of Illinois, respectively. He is secretary of the IUPAC Commission on Fundamental Environmental Chemistry.

Professor Horvai's main research interest lies with the physical chemistry and applications of analytical sensors, particularly electroanalytical sensors. He has also extensively worked with the automation of analytical instrumentation. He is fascinated by complicated trace analysis problems. Recently he started working with molecularly imprinted polymers as selective analyte enrichment tools in environmental analysis. He has about 70 publications in the fields mentioned.

List of Contributors

Jacques Buffle
CABE, Sciences II, 30 Quai E Ansermet, 1211 Genève 4, Switzerland, jacques.buffle@cabe.unige.ch

Wei-Jun Cai
Department of Marine Sciences, The University of Georgia, Athens, Georgia 30602, USA, wcai@arches.uga.edu

Kenneth H. Coale
Moss Landing Marine Laboratories and Monterey Bay Aquarium Research Institute, 7700 Sandholdt Road, Moss Landing, CA 95039, USA

Lars R. Damgaard
Institute of Biological Sciences, Department of Microbial Ecology, University of Aarhus, Ny Munkegade, DK-8000 Aarhus C, Denmark

William Davison
Environmental Sciences Division, LEBS, Lancaster University, Lancaster, LA1 4YQ, UK, w.davison@lancaster.ac.uk

Dick De Beer
Max Planck Institute for Marine Microbiology, Celsiusstrasse 1, D-28359 Bremen, Germany, dbeer@mpi-bremen.de

Nico F. de Rooij
Institute of Microtechnology, University of Neuchâtel, rue Jacquet-Droz 1, 2007 Neuchâtel, Switzerland, nico.derooij@imt.unine.ch

N.-K. Djane
CABE, Sciences II, 30 Quai E Ansermet, 1211 Genève 4, Switzerland

Virginia A. Elrod
Moss Landing Marine Laboratories and Monterey Bay Aquarium Research Institute, 7700 Sandholdt Road, Moss Landing, CA 95039, USA

Jean-Charles Fiaccabrino
Institute of Microtechnology, University of Neuchâtel, rue Jacquet-Droz 1, 2007 Neuchâtel, Switzerland, jean-charles.fiaccabrino@imt.unine.ch

G. Fones
Environmental Sciences Division, LEBS, Lancaster University, Lancaster, LA1 4YQ, UK

Ronnie Glud
Marine Biological Laboratory, University of Copenhagen, Strandpromenaden 5, DK-3000 Helsingør, Denmark, mblrg@mail.centrum.dk

J.K. Gundersen
Marine Biological Laboratory, University of Copenhagen, Strandpromenaden 5, DK-3000 Helsingør, Denmark, jens.gundersen@biology.aau.dk

M. Harper
Environmental Sciences Division, LEBS, Lancaster University, Lancaster, LA1 4YQ, UK

J. Hendrikse
MESA Research Institute, University of Twente, Enschede, The Netherlands

George Horvai
Technical University Budapest, Division of Chemical Information Technology, Gellért tér 4, H-1111 Budapest, Hungary, horvai@ch.bme.hu

Kenneth S. Johnson
Moss Landing Marine Laboratories and Monterey Bay Aquarium Research Institute, 7700 Sandholdt Road, Moss Landing, CA 95039, USA, johnson@mbari.org

Thomas Kjær
Institute of Biological Sciences, Department of Microbial Ecology, University of Aarhus, Ny Munkegade, DK-8000 Aarhus C, Denmark

Milena Koudelka-Hep
Institute of Microtechnology, University of Neuchâtel, rue Jacquet-Droz 1, 2007 Neuchâtel, Switzerland, milena.koudelka@imt.unine.ch

Michael Kühl
Marine Biological Laboratory, University of Copenhagen, Strandpromenaden 5, DK-3000 Helsingør, Denmark, mblmik@inet.uniz.dk

Lars Hauer Larsen
Unisense ApS, Science Park Aarhus, Gustav Wieds Vej 10, DK-8000 Aarhus C, Denmark

Jan Lorenzen
Institute of Biological Sciences, Department of Microbial Ecology, University of Aarhus, Ny Munkegade, DK-8000 Aarhus C, Denmark

L. Matthiasson
CABE, Sciences II, 30 Quai E Ansermet, 1211 Genève 4, Switzerland

Jocelyn L. Nowicki
Moss Landing Marine Laboratories and Monterey Bay Aquarium Research Institute, 7700 Sandholdt Road, Moss Landing, CA 95039, USA

N. Parthasarathy
CABE, Sciences II, 30 Quai E Ansermet, 1211 Genève 4, Switzerland, nalini.parthasarathy@cabe.unige.ch

N.B. Ramsin
Marine Biological Laboratory, University of Copenhagen, Strandpromenaden 5, DK-3000 Helsingør, Denmark

Clare E. Reimers
College of Oceanic and Atmospheric Studies, Oregon State University, Corvallis, OR 97331, USA

Niels Peter Revsbech
Institute of Biological Sciences, Department of Microbial Ecology, University of Aarhus, Ny Munkegade, DK-8000 Aarhus C, Denmark, revsbech@biology.aau.dk

Carsten Steuckart
Max Planck Institute for Marine Microbiology, Celsiusstrasse 1, D-28359 Bremen, Germany

P. Teasdale
Environmental Sciences Division, LEBS, Lancaster University, Lancaster, LA1 4YQ, UK

M.-L. Tercier-Waeber
CABE, Sciences II, 30 Quai E Ansermet, 1211 Genève 4, Switzerland, marylou.tercier@cabe.unige.ch

Albert van den Berg
MESA Research Institute, University of Twente, Enschede, The Netherlands, a.vandenberg@el.utwente.nl

Herman P. van Leeuwen
Agricultural University of Wageningen, Laboratory of Physical and Colloid Chemistry, Postbus 8083, NL-6700 EK Wageningen, The Netherlands, herman@fenk.wau.nl

Heidi Zamzow
Moss Landing Marine Laboratories and Monterey Bay Aquarium Research Institute, 7700 Sandholdt Road, Moss Landing, CA 95039, USA

M. Zhang
Environmental Sciences Division, LEBS, Lancaster University, Lancaster, LA1 4YQ, UK

Series Preface

The main purpose of the IUPAC Series on Analytical and Physical Chemistry of Environmental Systems is to make chemists and other scientists aware of the most important bio-physico-chemical conditions and processes which define the behavior of environmental systems. Thus the various volumes of the series emphasize the fundamental theoretical concepts of environmental and bio-environmental processes, taking into account their specific aspects such as physical and chemical heterogeneity. Another major goal of the series is to discuss the analytical tools which exist or should be developed to study those processes. Indeed, there is a great need for methodology developed specifically for the field of analytical/physical chemistry of the environment, in close connection with the corresponding process studies.

The present volume of the series focuses on the *in situ* chemical analysis and speciation in aquatic systems. It critically discusses the various techniques available for the *in situ* determination of the local chemical and physicochemical properties of the system. The volume was realized within the frame of our activities for the IUPAC Commission on Fundamental Environmental Chemistry in the Division of Chemistry and the Environment. We thank the IUPAC officers responsible, especially the executive director Dr. John Jost, for their support and assistance. We also thank the International Council of Scientific Unions (ICSU) for financial support of the work of the Commission. This enabled us to materialize the plenary discussion meeting (Geneva, 1998) which formed such an essential step in the preparation and harmonization of the various chapters of this book.

The series is growing prosperously, and a number of future volumes are on their way. Volume 7, on the Biogeochemistry of Iron in Seawater, is in press, and volumes 8 and 9, on topics that analyze the interplay between environmental chemistry and (micro)biology, are in preparation. As with the earlier books in the Series, these volumes collect critical reviews that characterize the current state of the art and provide guidelines for future research.

Jacques Buffle and Herman P. van Leeuwen
Series Editors

Preface

The impacts of human activities on environmental systems, and in particular chemical pollution, are becoming increasingly important issues. To overcome these problems, both understanding of environmental processes and the development of monitoring systems are necessary. Natural ecosystems are complicated regulated systems, including a large number of physical, chemical and biological actions and reactions, which altogether ensure the functioning of the system, and its homeostasis. Correct functioning requires that all these processes, in particular the chemical ones, are well tuned and balanced. In this context, chemical pollution may just result from concentration changes of major (e.g. CO_2 in atmosphere, or O_2 in waters) or minor constituents (e.g. trace metals in waters), which may shift equilibria from one state to another. But they may also result from the introduction of man-made chemical products (e.g. pesticides) which may completely block or modify chemical pathways.

The above presentation of ecosystems shows that they are functioning very much like living organisms. Maintaining ecosystems in good 'health' thus requires the same type of approach as for human beings. It should be based on the fundamental understanding of ecosystem functioning, and in particular, as far as chemistry is concerned, on environmental physical chemistry (i.e. the equivalent of biochemistry or biophysics); networks of ecosystem monitoring (the equivalent of medical analysis) should also be developed to assess any disfunctioning of the test ecosystem as quickly as possible. Finally, diagnosis and possible treatment should be established, by comparing the recorded analytical data with those of correctly functioning systems.

To enable efficient interpretation of the functioning of ecosystems such as lakes, oceans, or ground water, the recording of large data sets is essential, in order to correctly take into account natural spatial and temporal variations. This requires the use of a network of *in situ* or on field sensors or analytical devices, for continuous, real-time monitoring of major, minor and trace components, simultaneously at a large number of locations in the ecosystem, and at various depths in the water columns or ground water. Such a huge number of analyses are not feasible by using the classical approach based on sample collection, storage and transportation, followed by sample handling in laboratory. Robust sensors and instruments for automatic *in situ* or on site measurements should thus be developed. This approach is required not only for the above-mentioned reason of cost effectiveness, but also for scientific reasons. Indeed, determinations of minor and trace inorganic and organic compounds have become more and more important for water quality assessment. Classical analyses of these compounds, however, are often prone to many artefacts which can only be overcome by *in situ* measurements, avoiding the sampling step.

The development of more or less sophisticated electrodes, optodes, sensors and analytical systems (see Chapter 1 for definition) for *in situ* measurements in water and sediments, is still at its beginning. This is because it is based on other young disciplines, such as (i) the understanding of microbiological and physicochemical environmental processes, (ii) the organic synthesis of new sophisticated reagents and materials for sensor construction, and (iii) the necessary microtechniques for machining robust microanalytical systems. As a consequence, a very large development can be expected in the field of *in situ* sensors and analytical systems, in the next 10–20 years.

Pioneering work, however, has already been done in the last 15 years, to develop *in situ* analytical devices, in particular in water and sediments, and critical evaluation of the corresponding results can serve as a good basis for determining guidelines for future developments. This is a major goal of this book. Although some sensors (e.g. pH, O_2) are more robust than others, and routine continuous measurements are already feasible with several systems, improvements are needed with all of them. The reader thus should not expect to find in the book recipes working in all conditions, irrespective of the test sample. All chapters critically discuss the state of the art of the corresponding analytical device, its capabilities, the major possible improvements, and the best way(s) to do it.

Sensor development not only requires technical considerations, but also theoretical knowledge. As discussed in Chapter 1 and in some other specific chapters, diffusional transport of the analyte and in some cases of reagents is a key aspect in this type of development. In addition, for minor and trace compounds, chemical speciation (i.e. discrimination between the nature and properties of different chemical forms of the analyte, corresponding to different environmental impacts) is also essential. Thus optimal functioning of sensors often requires mathematical modelling of transport coupled to chemical reactions. Finally, laboratory studies are usually far from sufficient to lead to the development of reliable and robust *in situ* sensors measuring environmentally relevant parameters. Field work has to be done (i) at the first stage of development to define the sensor concepts, and (ii) at later stages to test its analytical performance, to improve its robustness *in situ* and to develop the appropriate signal acquisition and transmission equipment. Thus development of *in situ* sensors and analytical systems requires a complex combination of expertises, in analytical chemistry, physical chemistry, environmental chemistry, microtechnology and mechanical and electronic engineering. This book covers all these aspects, except the mechanical and engineering ones. It describes relatively well-established sensors and analytical methods, but also the more recent principles and techniques, in order to promote new ideas in the field. In all cases, however, only analytical devices based on rigorous scientific grounds, and for which the feasibility of field and in most cases *in situ* applications has been demonstrated, are described. Optodes are discussed in Chapter 2 for oxygen measurements. A similar discussion could unfortunately not be included for the measurement of trace metals or organics. This field is still in its infancy but is likely to progress quickly in the future.

The book includes the most important existing *in situ* sensors and analytical systems. Chapter 1 discusses general concepts which should be considered for the development of any type of sensor, in order to get reliable and environmentally relevant information. It also helps the reader to place the various chapters in perspective to each other, inside a common frame. The next four chapters deal with sensors for *in situ* measurements of major components: O_2 (Chapter 2), pH and CO_2 (Chapter 3), S(–II) (Chapter 4), Ca^{2+} and N species (Chapter 5). Chapters 6, 7 and 9–11 deal with sensors and analytical systems for minor or trace organic or inorganic components. In these cases, the signal most often depends on the speciation of the test analyte. All these chapters thus discuss speciation aspects relevant to each technique. Chapter 8 is specifically devoted to the physico-chemical principles needed to understand how dynamic chemical equilibria, such as metal complexation, affect the signal of analytical devices based on flux measurements. Because most trace compound determinations are based on such flux measurements, we have found it important that a rigorous formulation of these general physico-chemical concepts, and some examples of their applications to a few sensor types, be described in a specific chapter. Finally, the book ends with the existing microtechniques which could be used for the fabrication of *in situ* sensors or microanalytical systems. Although very few complete analytical systems have yet been built based on this technology, and none of them for environmental application, it is clear that the fabrication of at least key components (such as microelectrodes, microreactors, etc.) of *in situ* analytical systems should be performed at the microscale. The development of new *in situ* analytical devices should thus greatly profit from microtechnologies. The main purpose of this chapter is to stimulate ideas for new microsensor or micro-analytical system construction, by using the concepts of microsensors described in Chapters 2–6, which were built with more classical technologies.

This book should provide researchers interested in the development of *in situ* sensors and analytical systems with the appropriate updated literature and critically evaluated information. However, we hope that it will be even more helpful to laboratories in charge of water quality assessment, by providing them with updated information on existing sensors and analytical systems, their present capabilities and the expected future developments. In most cases, either detailed technical information is given or the corresponding literature is cited, which should help any interested scientist to start using these analytical devices in an appropriate manner. Thanks to the theoretical background discussed in particular for methods related to speciation, correct interpretation of the data should also be made easier even for the non-specialist.

We gratefully acknowledge the cooperation of all the authors who have done their best to make the chapters accessible to non-specialists, while, simultaneously, giving enough information to provide in depth critical evaluation of the existing analytical systems, and much information on the corresponding literature. A financial contribution of the International Council of Scientific Unions is also acknowledged. Last but not least, we would like to thank particularly the help of Montserrat Filella who has participated in reviewing the chapters, with respect to editing, IUPAC terminology and SI units and symbols. It is our hope that the critical synthesis of information presented here will promote easier use and further developments of *in situ* sensors and analytical systems.

Jacques Buffle and George Horvai

1 General Concepts

J. BUFFLE
University of Geneva, Switzerland

G. HORVAI
Technical University of Bupapest, Hungary

1 IMPORTANCE OF *IN SITU* MEASUREMENTS

Basically three approaches can be used for water quality analysis and monitoring. The most often used is water sampling followed by sample storage, handling and analysis in laboratory. This approach, however, is costly and does not provide large data sets of spatial distributions or temporal evolutions of the parameters of interest. In addition, delicate analyses such as the determination of dissolved gases or trace compounds are often prone to significant artefacts. The second approach consists of performing the analysis directly on the lake, river or sea shore, or on board a ship, after manual or automatic sampling. This approach is

In Situ Monitoring of Aquatic Systems: Chemical Analysis and Speciation Edited by J. Buffle and G. Horvai.

called, throughout this book, *on site* analysis. It often uses laboratory procedures and instruments, although they are sometimes adapted to these specific conditions. On site analyses approach the ideal of real-time measurements and minimize some artefacts associated with sample storage. In the third approach, measurements are made exactly at the location of interest (e.g. at depth in water, or within sediment or soil). This approach will be called *in situ* analysis in this book. It minimizes most of the artefacts, not only those due to storage, but also those due to sampling such as changes in pressure (e.g. gas evolution) or oxic/anoxic conditions. It also allows automatic real-time measurements, sometimes in locations difficult to access, such as great depths, or in environmental systems which should not be perturbed. Instruments and analytical procedures should, however, be specially developed for this purpose. This book deals particularly with such specific developments for on site or *in situ* measurements (collectively denoted as *in the field*). They are expected to have a significant impact in the future on (i) our understanding of ecosystem functioning, and (ii) our ability to monitor, model and therefore to efficiently protect these ecosystems.

The major advantages of *in situ* and on site measurements, compared with laboratory analysis, are summarized in Table 1. The last item is crucial in studies dealing with the understanding of ecosystem functioning. In most cases, the determination of fluxes at interfaces is required (e.g. at air–water, water–sediment, water–biofilm), which are partly controlled by the physical structure of the interface, sometimes at the micrometer to millimeter level. In such cases, the measurements of concentration gradients, *in situ*, using non-perturbing probes, is essential. The first item of Table 1 is of key importance both for studies of ecosystem functioning and for water, sediment or soil quality monitoring. Indeed changes of conditions during sampling (temperature, pressure, oxygen concentration) and sample transformation during storage and handling (see below), may affect most of the chemical compounds of interest, including gases and all trace compounds, either organics or inorganics. For the

Table 1. Advantages of *in situ* probes for natural water monitoring (adapted from ref [1])

- Elimination of many of the artefacts due to sample handling, *i.e.* no or minimum sample transformation (see below)
- Minimization of the overall cost of data collection (in particular, due to a reduction of analysis time)
- Possibility of real-time analysis, allowing the rapid detection of pollutant inputs (e.g. monitoring of industrial wastes or water quality in water treatment plants)
- Ability to accumulate detailed spatial and temporal data banks of complete ecosystems (lakes, aquifers, etc.)
- Possibility to perform measurements in locations which are difficult to access (boreholes, deep lakes or oceans)
- Possibility of measuring concentration gradients and fluxes at environmental interfaces (sediment–water; air–water), at high (sub-mm) spatial resolution

latter, understanding the real impact of a particular compound in its environment requires a knowledge of its distribution amongst various types of species, such as those which are bioavailable, freely mobile by diffusion, transported by colloids or fixed on solid phases (see e.g. Figure 16 in Chapter 10). These species distributions are very sensitive to physico-chemical changes and measurements of these species distributions (referred to below as *speciation*) are often perturbed by sampling, storage and handling. *In situ* measurement is probably the best way to overcome a major problem of aquatic analytical chemists, i.e. how to maintain the sample in its original state during sampling and handling, without disturbing any of the very dynamic chemical, biological and physical processes which may continue after sample collection. As examples, some of these processes which may modify compound speciation, in particular that of trace compounds, include [1,2]:

(a) Losses of compound or compound complexants (especially macromolecules and colloids) by adsorption to any surface used during sample handling (polymer/glassware, filtration apparatus, etc.).
(b) Release of either the test or interfering compounds from the polymer/glassware, filters and analytical apparatus. (During *in situ* analysis in water, the sensor or analytical system used is usually small enough to be readily equilibrated with the test sample, thus minimizing problems (a) and (b).)
(c) Gaseous re-equilibration of the sample with the atmosphere due to pressure change. Re-equilibration of gases with acid–base properties (e.g. CO_2) may cause significant variations in pH (and thus modify compound speciation). When anoxic samples are equilibrated with the atmosphere, oxidation of some of the inorganic species ($Mn^{2+}, Fe^{2+}, S(-II)$) may produce colloidal particles ($MnO_2, Fe(OH)_3, S^0$) which may drastically change the species distribution of many trace compounds, owing to their strong redox or adsorption reactions with these colloids.
(d) Coagulation of colloidal matter, followed by sedimentation of the aggregates and the associated trace compounds (Colloids are ubiquitous in aquatic samples and include a large fraction of trace compounds.)
(e) Microbial activity. The continued metabolism of microbes during sample storage may significantly alter the chemical composition of the sample. For example, the pH may vary because of continued respiration (pH decrease) or photosynthesis (pH increase). Concentrations of trace compounds may also be changed as a result of their continued uptake or release by living microorganisms. Complexation or enzymatic properties may also change owing to the release of biomolecules.

Some of these problems may be minimized by special (often tedious) precautions, but problems (d) and (e) are natural processes which cannot be stopped without drastically perturbing the sample. Indeed, in their natural environment, aquatic samples are not at thermodynamic equilibrium; at best, they may be in steady state conditions due to the continuous input (e.g. soil leaching, atmospheric inputs, cell growth) and output (e.g. coagulation/sedimentation, cell death) of colloids and microbes [2,3]. While the sampling process stops most of the inputs, coagulation and microbial turnover may continue and any anticoagulant or antibiotic may either induce drastic changes in the chemical speciation of the test compounds or cause analytical problems [3].

A number of criteria should be fulfilled for the development of *in situ* probes [1] (Table 2). Measurements should, of course, be reliable and sensitive enough to enable the accurate measurement of concentrations in the ranges of 10^{-2} to 10^{-6} mol L^{-1} for *major* components (such as O_2, CO_2, alkalinity, NO_3^-, phosphate) and 10^{-6} to 10^{-15} mol L^{-1} for *minor or trace* compounds (e.g. most elements of the periodic table as well as organic pollutants) [4]. Reliability and robustness are extremely important criteria for *in situ* probes. An analytical instrument which is 80–90 % reliable in laboratory conditions may be acceptable provided technical staff are close at hand. However, in the field, probe reliability must be much higher, especially for measurements at depth in which no visual control is possible and repairs are both time consuming and difficult. Unfortunately, improving reliability from 80 to 100 % can be as difficult as developing the basic analytical principle to the point at which 80 % reliability is attained.

The capability of *multicompound analysis* is an important requirement for minor or trace compounds, in particular metals and organics, since the environmental impact usually depends on the combined effect of several compounds with similar but not identical properties (same type of metals, or organics with similar functional groups). In such cases, the use of one probe per compound would lead to very large unwieldy multiprobe systems. Multicompound analysers usually based on voltammetric (Chapter 9) or spectrometric measurements are therefore needed. In the following chapters, the term *sensor* will be applied to any device which enables the measurements of a single specific species. A sensor may be significantly more complicated than a classical metallic or ion-selective *electrode*, or an *optode*, as it may include (bio)chemical transformation of the analyte inside a reaction chamber (e.g. Chapter 6). Throughout this book, the

Table 2. Criteria for the design of *in situ* probes for natural water monitoring

- Reliable, automatic measurements
- Simple, compact, low cost apparatus
- No or minimum sample transformation (minimization of artifacts)
- High sensitivity for minor and trace compounds (10^{-7}–10^{-15} mol L^{-1})
- Multicompound analysis capability for minor or trace compounds
- Speciation capabilities
- Physically and chemically non-perturbing for the system tested
- Preferably measurement time faster than the time scale of the process studied

term *analytical system* (Figure 1) is used for instruments which enable the analysis of several (sometimes many) compounds, by combining a sensor (e.g. spectrometric or voltammetric sensors; Chapter 9), with the separation and possible chemical transformation of analytes by added reagents (Chapter 7). Contrary to sensors, in which transport processes are based only on diffusion, analytical systems often include pumps, valves, flow-through cells, and electronics

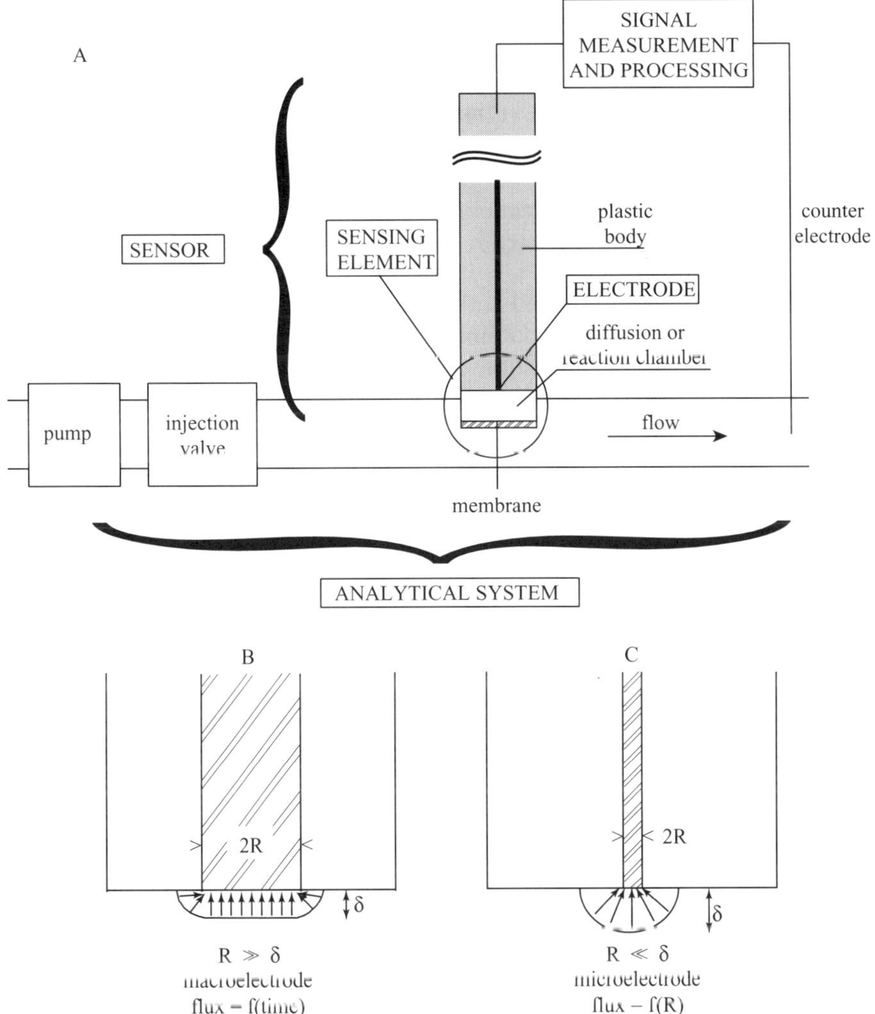

Figure 1. (A) Schematic representation explaining the terminology related to 'electrode', 'sensing element', 'sensor', and 'analytical system' used in this book. In this representation the electrode might be replaced by an optode. (B, C) Schematic representation of diffusion layers at (B) a macroelectrode and (C) a microelectrode

(sometimes optics) for signal processing, which make them more complicated, but also more versatile. It must be noted that the division between electrodes (or optodes), sensors and analytical systems is somewhat arbitrary. For example, the sensors of Chapter 6 can be seen as small analytical systems. This terminology mainly expresses the existence of an increasing complexity when passing from electrodes (or optodes) to sensors and to analytical systems (see Figure 1 and Section 2.1 for more details).

The intrinsic *response time* (also called below measurement time) requirement for sensors or analytical systems depends on the time evolution of the studied process. This latter may vary from seconds (e.g. changes in biofilms) to hours (e.g. processes linked to photosynthetic activities or tides) to weeks (e.g. lake eutrophication). Some sensors or analytical systems may have very fast measurement time (e.g. fractions of seconds for O_2 or pH electrodes; Chapters 2 and 3) up to longer measurement times (e.g. minutes for an NO_3^- electrode, Chapter 5, or *in situ* voltammetry, Chapter 9, and hours for DGT techniques (diffusive gradient in thin films, Chapter 11). In all cases the important point is that the measurement time should be well known and either much shorter than the time scale of the process studied (enabling fluctuation recording) or much longer, then enabling statistical integration of fluctuations, to provide a well-defined time average value (e.g. see discussion in Chapter 11).

Speciation capability depends on the test compound. Species distributions of major elements (see above for definition) usually include only a few species in natural conditions, e.g. 1 for O_2, 3 for S(−II) (i.e. H_2S, HS^-, S^{2-}), 1 for NO_3^-, etc. The development of specific sensors for each of these particular species is then feasible and useful. This approach is discussed in Chapters 2 (O_2), 3 (pH and CO_2), 4 (S(−II), 5 (N species and Ca^{2+}) and 6 (NO_3^- and CH_4). In contrast, many minor or trace compounds (metals or organics) are often distributed between a very large number of species (free, complexed to many types of complexants etc.), so that analytical systems or devices responding to several environmentally relevant groups of species are more appropriate. The properties of the species of each group should then be well established. The conceptual and practical aspects of this approach are discussed in Chapters 8–11 mostly for trace metals, as well as in Chapter 6 for the BOD sensor.

2 MICROELECTRODES, MICROSENSORS AND MICROANALYTICAL SYSTEMS FOR *IN SITU* MEASUREMENTS: CONCEPTS AND TERMINOLOGY

2.1 CHARACTERISTIC DIMENSIONS OF SENSORS AND ENVIRONMENTAL SYSTEMS

The words microsensors, ultramicroelectrodes, micrototal analytical system (μ-TAS), microtechnology are often used in the literature, even though there is no real consensus on the size limit between macro-, micro- or ultramicro-dimensions. In the field of electro-analytical chemistry, the last two terms are sometimes used to denote dimensions of 10–100 μm and 1–10 μm respectively. In studies of fluxes at the sediment–water interface and in biomedical applications, the word micro(bio)sensor often denotes dimensions from 100 μm down to 1 μm or below. The lack of consensus is because various users employ different criteria. For instance, sedimentologists or biophysicists usually link the sensor size to the perturbation created by its insertion in the system tested or to the spatial resolution they would like to reach; clearly here, the whole sensor body is the important factor. In contrast, electrochemists often use, as a criterion, the mode of transport of the chemical species towards the sensing part of the electrode (e.g. Pt disc or Hg drop electrode); in this case, only the size of this sensing element is important, and not the thickness of its embodiment.

Realizing in which size scale a sensor or analytical system is working is not just a question of wording and arbitrary definition. As discussed below, for the measurement of well-defined parameters, *in situ*, in complicated, microstructured, environmental media, a clear link must be made between the dimensions related to the environmental process tested and those of the sensor-related processes. As explained below, these dimensions are also directly related to the time scales of these processes (see Chapter 8 for a detailed discussion on electrode/sensor size). To get an overall picture in mind, not one size criterion, but several criteria must be used and compared with each other, as discussed below. This is of utmost importance when chemical speciation plays a significant role.

In that respect, it is convenient to discuss separately the following analytical devices and environmental systems (Figures 1 and 2). In each case, the important dimension is that of the rate limiting part of the system. As far as analytical devices are concerned it is useful to discriminate between electrodes (or optodes), sensors (or sensing element) and analytical systems (see above and Figure 1).

An electrode is typically a metal surface (for redox amperometric, voltammetric and potentiometric electrodes) or a solid or liquid membrane made of ionic material (for ion-selective electrodes). An optode is typically an optical fiber whose tip is covered with an optically active reagent. Electrodes and optodes are the parts of the system which enable the transformation of the chemical information of interest (related to the test compound), into the measured current, potential or light intensity. An electrode or optode can be used directly *in situ* or combined into a sensor or sensing element.

A sensing element is a significantly more complicated device than an electrode or optode. It usually includes one (sometimes more) diffusion or reaction chambers (Figure 1) separated from the sample by a semi-permeable membrane. A sensor is an electrode-type device in which the sensing element is the most important part. It can be inserted directly *in situ* in the test medium.

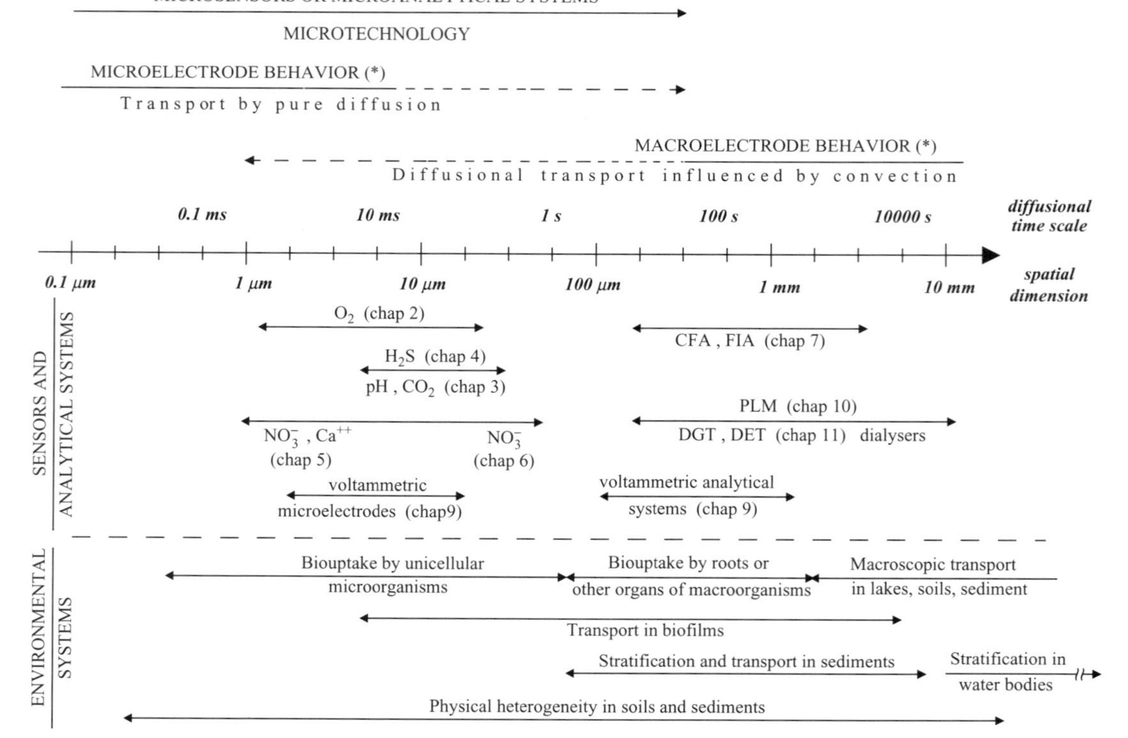

Figure 2. Comparative spatial dimensions of a few environmental systems, analytical systems, sensors, and electrodes. * Rigorous discrimination between micro- and macroelectrodes should be based on the ratio R/δ (see text and Figure 1), where the diffusion layer thickness δ may depend on time and experimental conditions. The diffusional time scale and spatial scale in the figure are related together by equation (1). The time scale is intended to give an order of magnitude of diffusional transport in the various systems

However, a sensing element can also be a part of a more complicated analytical system.

The heart of an analytical system is either an electrode (or optode) or a sensing element, but it also includes other components (e.g. pumps, injection valve, heating or mixing systems) and the whole electronics necessary to record and process the signal *in situ*. Contrary to sensors, analytical systems are built to measure simultaneously a large number of different components.

In all cases, the transport of the analyte to the electrode or optode is a key part of the analytical process, and it has to be considered carefully to choose the size of the electrode or sensor (see below). Perturbation of the medium by insertion of the analytical device, as well as understanding the role of the heterogeneity of the test medium on the measured signal are two other important factors linked to the test environmental systems to be considered. These various aspects are briefly discussed below.

- For any analytical principle based on *electrode or optode* on which the analyte is consumed (e.g. amperometric or voltammetric electrode), the flux of analyte at the electrode surface is the key process to measure, even if the consumption of analyte is negligible with respect to the environmental medium studied. This flux may depend on the concentration of the analyte (through diffusional transport), its chemical reactions (related to speciation), and convection. Because convection is usually an ill-controlled process, transport by pure diffusion is preferable, particularly when information on speciation of the analyte is of interest (Chapters 8–11). For this reason, the limit between *microelectrodes* and *macroelectrodes* [5] is fixed by the ratio R/δ, where R is the electrode radius (assuming a sphere or a disc) and δ is the diffusion layer thickness at the electrode surface. δ depends on time (see below and Chapters 8 and 9), electrode geometry and convection. The important point is that:
  — For $R/\delta \gg 1$ (Figure 1B; corresponding to macroelectrodes), the measured flux depends on δ (but not R) and therefore on ill-controlled natural convection.
  — For $R/\delta \ll 1$ (Figure 1C; corresponding to microelectrodes), the measured flux depends only on R and is therefore independent of natural mixing. In particular it does not depend on the liquid flow of the test water which is essential when these electrodes are used directly in the test medium. In addition, contrary to the previous case, so-called spherical diffusion occurs at microelectrodes, which ensures the maintenance of steady-state conditions and constant flux at the electrode surface, even in unstirred solution as in a closed reaction chamber (Figure 1). This is of major importance for development of sensors and sensing elements for analytical systems.

Since δ depends on time and on the electrochemical conditions used, there is no unique size limit between micro- and macroelectrode. In typical conditions, however, electrodes behave as microelectrodes for $R < 1\ \mu m$, and as macroelectrodes for $R > 100\ \mu m$ (Figure 2).

- As far as the analytical signal of *sensors or sensing elements* (Chapters 2–6, 10 and 11) is concerned, the key dimensions are those of the part of the device which limits the transport rate. It may be the dimensions of the outer membrane area, its thickness or the size of the internal chamber (Figure 1), where diffusion-controlled (and reaction) processes occur before the detection. These dimensions control the *measurement time* of the whole sensor or sensing element, since transport of analyte (and sometimes reagents) is a limiting step for most sensors. Small dimensions of the sensing element (particularly below 10–100 μm, Figure 2) favor both a quick response time and a better reproducibility, as transport is less or not affected by the usually ill-controlled hydrodynamic processes (convection).
- The *external dimensions* of the body of the sensor (or analytical system) are linked to the perturbation of the test medium. Therefore it is also a key factor, but only for *in situ* measurements in structured (non-homogeneous) media, in particular soil and sediment. The size of the body close to the very location where analyte is measured is the most important factor. This dimension is also directly linked to the *effective response time of the sensor* in *the tested medium*. Indeed, a realistic value of the test parameter is obtained only after re-equilibration of the medium, after the perturbation due to the insertion of the sensor. The time required is directly related to the fluxes created by the perturbation and thus to the size of the perturbing sensor body (see Chapter 11 for a quantitative discussion). Since the external dimensions of the sensor body are always larger than those of its sensing element, the effective response time of the complete sensor is usually longer than the measurement time of the sensing element (see above), particularly in highly stratified media where perturbation may be important.
- The above dimensions of the analytical device must be compared with the *characteristic dimension linked to the heterogeneity* of the environmental system studied. This is, for example, the distance between the stratified chemical layers in a eutrophic lake (meter) or sediment (0.1 to 1 mm), or the distances between microniches of a sediment or soil which control its horizontal heterogeneity. Comparison of this 'heterogeneity' dimension with both sensor external dimensions and sensing element dimensions is important in understanding the spatial resolution limit of the sensor. Ideally, realistic stratification or heterogeneity data are obtained only when the dimensions of the sensing element of the sensor are much smaller than the heterogeneity dimensions of the medium (provided the sensor body is not too

large). In principle, recording a large number of horizontal and vertical measurements can then provide a full picture of the overall spatial structure [6]. However grain and pore sizes in soil and sediments usually follow very broad distributions [7]. Therefore it may be expected that after re-equilibration of the perturbation, most sensors, even the smallest, still measure a spatially weighted average value of the test parameter. The weighting factor will depend directly on the size of the sensing element which defines the spatial and time domain over which the measured flux is integrated. Detailed computations have shown, for instance, how diffusion inside the DGT membrane determines the lower limit of the measurable spatial resolution. [8] (Chapter 11).

Transient phenomena in soils or sediments can only be studied by following re-equilibration after the perturbation. After that re-equilibration, the time resolution is fixed by the measurement time of the sensor (see above). For direct measurements without preconcentration (e.g. O_2, pH, ISE, biosensors, some voltammetries, Chapters 2–6 and 9), this is directly linked to the size of the sensing element or the internal size of the analytical system (Chapter 7). With sensor sizes less than tens of micrometers, time resolution better than seconds (Figure 2) can thus be obtained. For techniques with preconcentration (in particular for trace compounds), the time resolution is limited by the preconcentration time (minutes for voltammetry, Chapter 9, fractions of an hour for PLM, Chapter 10, hours to days for DGT, Chapter 11). When time average parameters are measured, exact knowledge of the flux at the sensor during the measurement or preconcentration time (Chapter 8) is a key requirement to relate the measured signal to the real-time fluctuations in the environmental system.

2.2 TIME AND SPATIAL SCALES: DEFINITION OF MICRO- AND MACROSENSORS AND ANALYTICAL SYSTEMS

Both at the sensing element surface and in the studied environment, the flux of an analyte depends on the corresponding concentration gradient which results from (i) diffusional transport, (ii) (bio)chemical reactions linked to the production or decomposition of the compounds and (iii) hydrodynamic transport due to all kinds of controlled or uncontrolled convection. The last two components may or may not exist depending on the conditions. Diffusional transport, however, always exists in a non-homogeneous system. Therefore it is particularly relevant to compare any spatial or time scales with those of this process. This is easily done by using the Stokes–Einstein law, which relates the average distance, l, traveled through by a compound with diffusion coefficient, D, during the time, t:

$$l = \sqrt{2Dt} \tag{1}$$

Replacing l by (i) the diffusion layer thickness, δ, at an electrode surface, (ii) the size of the sensing element, (iii) the external sensor dimensions, or (iv) the characteristic dimension of heterogeneity, provides orders of magnitudes of, respectively, the measurement time of the electrode or sensor (for measurements without preconcentration), the effective response time due to perturbation and the time scale of natural processes in the absence of convection and chemical reaction.

Comparison of these various time scales enables estimation of the best sizes for the electrode, sensing element or the external dimension of the sensor, with respect to the process studied. Rigorous mathematical equations should of course be used for precise calculations [8,9] (Chapters 2, 8 and 11).

Figure 2 compares the diffusional times and spatial scales of typical environmental processes, with the typical dimensions of electrodes, sensors and analytical systems described in this book. The time and size scales are related by equation (1), assuming $D = 10^{-9}\,\mathrm{m^2\,s^{-1}}$, which corresponds to the upper limit of diffusion coefficients of small ions and molecules in water. Although the domain of pure diffusive transport depends on conditions (see discussion on electrode size in section 2.1), typically, in environmental or sensing systems below about 1 μm (time scale below 1 ms), transport occurs solely by pure diffusion. Above a few hundreds of micrometers, natural or man-driven convection is usually dominant. Between these limits, both convection and diffusion may play a significant role.

As discussed in section 2.1, the practical limit between micro- and macroelectrodes depends on R/δ and is usually between 1 and 100 μm. Most sensors (Chapters 2–6) and all analytical systems (Chapters 7, 9–11) are significantly more complex than microelectrodes. In the fields of environmental and biomedical applications, the word *microsensor* is traditionally used for sensors for which the size of the sensing element is typically below 100 μm. Similarly, the words *micro-(analytical) systems*, and *microtechniques* are currently used in the field of microtechnology for systems with sizes below 100 μm (Figure 2). This terminology is used in this book.

The figure exemplifies how the sizes of the various analytical devices compare with those of the environmental processes of interest. To get high spatial resolution, the size of the sensing element of the sensor should be much smaller than the environmental heterogeneity studied. However, other considerations may be important. For example, the signal of a microsensor of 1–10 μm is more informative for understanding the uptake of compounds by unicellular microorganisms with similar sizes, than those of either much smaller or much larger sensors or analytical systems such as the DGT or PLM systems with sizes of 0.3–10 mm. However, the latter may be more appropriate for understanding biouptake by roots or transport in 0.1–1 mm thick biofilms, since they provide parameters corresponding to fluxes averaged over similar spatial dimensions and time scales.

3 TRENDS IN THE DEVELOPMENT OF ANALYTICAL SYSTEMS AND SENSORS FOR ROUTINE MONITORING

3.1 NEED FOR ANALYTICAL SYSTEMS/SENSORS IN WATER MONITORING

The development of new *in situ* chemical probes will be a challenging analytical problem for the next tens of years. Time and spatial requirements will force new ideas and technologies into play. Only a few *in situ* chemical probes (pH and oxygen, NO_3^-, Ca^{2+}, S(−II)) are presently reliable enough for routine use, yet there is a very large need for chemical probes and sensors (Table 3), especially for use at depths greater than 2–3 m. This is true not only for electrochemical sensors which are presently the most developed, but also for optical sensors which are undergoing rapid development. The lack of robust sensors is partly due to the analytical difficulties which must be overcome for their development, but also because new analytical and (bio)physical concepts are required. A major goal of this book is to stimulate such ideas.

3.2 CONCEPTUAL AND TECHNICAL GUIDELINES FOR NEW DEVELOPMENTS

Although the field of *in situ* environmental sensors and analytical systems is still in its infancy, a few guidelines and words of caution can be given for future developments.

3.2.1 Micro- Versus Macro-analytical Systems or Sensors

It has been emphasized above that microsystems have a number of advantages, and Chapter 12 of this book shows that our present knowledge in microtechnology enables us to envisage the fabrication of complete analytical instruments with volumes of a few milliliters or even less. It should be realized, however, that such microtechnological means are also at the beginning of their development

Table 3. Compounds for which submersible chemical probes are required by water treatment companies

- *Cations*: calcium, magnesium, aluminum, ammonium, iron, manganese, copper, zinc, cadmium, chromium, mercury, nickel, lead
- *Anions*: chloride, nitrite, nitrate, orthophosphate, fluoride, cyanide, selenium, carbonates, arsenate, silicate
- *Xenobiotic organics*: pesticides, surfactants, phenols, amines, polycyclic hydrocarbons, complexants
- *Heterogeneous natural complexants*: natural organic matter, suspended particles and colloids

and that considerable effort must still be made to test each component before the robust systems required for *in situ* measurements can be built. In addition, some components (such as power supplies) are still sometimes at the macro-scale.

When developing an *in situ* sensor or analytical system, a careful analysis must be undertaken to evaluate which part really needs to be fabricated at the microscale. As discussed in Chapter 12, it is still often preferable to build hybrid systems or sensors, with the micropart (size $< 100\,\mu m$) restricted to that required by the analytical problem at hand. The rest can be built at the milli- to centimeter level by traditional mechanic or electronic means. Microfabrication is usually important in the following cases:

- Fabrication of the *sensing element of the sensor or analytical system* (electrode, reaction chamber) at the microscale level (preferably $< 100\,\mu m$ for reaction chambers and less than a few micrometers for amperometric/voltammetric electrodes) has several advantages, in particular short measurement time, better signal/noise ratio for voltammetric microelectrodes and fast and reproducible transport of analyte and reagents to or in the sensing element. It must be mentioned, however, that potentiometric microelectrodes may have more problems than advantages (see Chapter 5)
- The *external body* of the sensor/analytical system should be of microsize only when perturbation of the measured system is important (e.g. in sediments, biofilms, etc.). Even in such cases, it may be sufficient that the sensor tip is microsized; the rest of the body can have millimeter size (see Chapter 2 for discussion). Building the whole sensor body at microscale is not recommended when it is not useful, as it may induce several practical problems (such as fragility).
- Even when macroscopic probes are used, e.g. in the water column, *the overall internal volume* of the system should be as small as possible to minimize reagent consumption, and usually also energy consumption, as, for example, larger pumps are needed for larger systems. This requirement is valid mostly for probes devoted to continuous long-term unattended measurements. Using typical values of 1 L for the total volume of reagent stock in the probe, and the minimum values of two reagents, 20 measurements per day during 15 d require that the internal volume is not larger than 1–3 mL. For higher frequency and/or measurements at longer term, maximum internal volumes should be a few hundreds of microliters (leading to channel/chamber sizes smaller than a few hundreds of micrometers).
- *Integration of the electronics* of even a complicated instrument on a chip is feasible and tempting (Chapters 9 and 12). It must be realized, however, that the performance of such 'microinstruments' (in terms of signal/noise ratio) is still significantly lower than those of instruments based on classical electronics. Therefore such types of miniaturization are preferably used only when

absolutely needed and when high sensitivity or accuracy is not a requirement. Often, integration of a preamplifier onto the sensing chip is sufficient and gives much better performances (Chapter 12).

3.2.2 Other Aspects

A few additional key factors need to be considered in the development of *in situ* sensors and analytical systems:

- The developments of sensors specific to a single chemical species, or of analytical systems (such as voltammetric probes, Chapter 9, or CFA/FIA with spectrometric detection, Chapter 7) are complementary. The choice depends on the problem at hand, and above all on the nature of the compound of interest, sensors being better adapted to major compounds, while analytical systems are usually preferable for minor or trace components. In particular, analytical systems are necessary for multielement determination and when detailed information on complicated speciation is required.
- Both sensors and analytical systems must be reliable and robust. This implies that traditional academic development of new concepts must be complemented by substantial technical developments in order to obtain a routine apparatus working *in situ*. Industry should be strongly encouraged to invest in this aspect of the research.
- It should also be clearly realized that *in situ* sensor and analytical system developments are not limited to technical aspects. The measured signal largely depends on the chemical reactivity of the analyte in the test medium. Therefore environmentally relevant (and interpretable) parameters will be obtained only if sensor or analytical system developments are based on detailed (bio)physicochemical studies of the interactions of the analyte with its medium and the corresponding understanding of environmental processes.
- For the analysis of a minor or trace compound, it is important to develop new *in situ* techniques to measure specific groups of species (e.g. complexes or redox states; speciation measurements). Different *in situ* techniques may measure different, often complementary, types of species, and very few of the newly developed *in situ* techniques will measure total concentrations. This point is worth emphasizing. Data obtained by *in situ* probes will necessarily be of a different nature to data obtained by traditional laboratory techniques, which are often determined subsequent to complicated handling and chemical treatment. Since such treatments must be avoided or largely minimized *in situ*, analytical principles which are technically less complicated but intrinsically selective to particular species will be favored. Because of their novel nature, it will be necessary to relate the data obtained by *in situ* probes to those presently obtained routinely by classical methods. This implies that

water treatment or environmental monitoring criteria will have to be adapted to the newly available (and hopefully more relevant) parameters. As much as possible, water managers should be involved in the development of these new techniques, as cooperation between the utilization and development is the best guarantee for their success.

GLOSSARY

In situ (Measurements) performed without sampling, at the exact location of interest (at depth in the water column, sediment or soil).

On site (Measurements) performed immediately after sampling, close to the location of interest.

In the field General expression to include on site and *in situ* measurements, as opposed to laboratory measurements.

Microelectrode An electrode in which transport is controlled by diffusion only and for which $R/\delta \ll 1$ with R = electrode radius and δ = diffusion layer thickness (typical size is less than a few micrometers) (Figure 1).

Microsensor A sensor with size smaller than a few hundreds of micrometers (Figure 1).

(Micro)analytical system An analytical instrument composed, in a fully or semi-integrated manner, of a detector, pumps, valves and flow-through cell, capable, in appropriate combinations to determine a large number of different chemical species, as opposed to sensors.

Sensor An analytical device developed to measure a single or a few specific species, and composed, in an integrated body, of an electrode or optode combined with one or several diffusion/reaction chambers, separated from the test medium by a membrane.

ACKNOWLEDGEMENTS

D. De Beer, W. Davison, H.P. van Leeuwen, N. Revsbech, and M.-L. Tercier are greatfully acknowledged for their inputs to the manuscript and René Menghetti for drawing the figures.

REFERENCES

1. Buffle J., Wilkinson K. J., Tercier M.-L. and Parthasarathy N. (1997). *In situ* monitoring and speciation of trace metals in natural waters, *Annali di Chimica*, **87**, 67–83.
2. Buffle, J. and Leppard, G. G. (1995). Characterization of aquatic colloids and macromolecules: Part I, Structure and behaviour of colloidal material, *Environ. Sci. Technol.*, **29**, 2169–2175.

3. Chen Y. and Buffle J. (1996). Physicochemical and microbial preservation of colloid characteristics of natural water samples II. Physico-chemical and microbial evolution *Water Res.*, **30**, 2185–2192.
4. Buffle, J. and Stumm, W. (1994). General chemistry of aquatic systems, in *Chemical and Biological Regulations of Aquatic Systems*, eds Buffle, J. and De Vitre, R. R. Lewis, Boca Raton, FL, Chapter 1, p.1.
5. Stulik, K., Amatore, C., Holub, K., Marecek, V. and Kutner, W. (2000). Microelectrodes: definitions, characterization and application, *Pure Appl. Chem.*, in press.
6. Jorgensen, B. B.. and Des Marais, D. J. (1990). *Limnol. Oceanogr*, **35**, 1343.
7. Lerman, A. (1979). *Geochemical Processes*, Wiley-Interscience, New York.
8. Harper, M. P., Davison, W. and Tych, W. (1999). Estimation of pore water concentrations from DGT profiles: a modelling approach, *Aq. Geochem.*, **5**, 337–335.
9. Crank, J. (1964). *The Mathematics of Diffusion*, Clarendon Press, Oxford.

2 Electrochemical and Optical Oxygen Microsensors for *In Situ* Measurements

R. N. GLUD
University of Copenhagen, Denmark
J. K. GUNDERSEN AND N. B. RAMSING
University of Aarhus, Denmark

In Situ Monitoring of Aquatic Systems: Chemical Analysis and Speciation Edited by J. Buffle and G. Horvai.

1 INTRODUCTION

1.1 TECHNIQUES FOR MEASURING DISSOLVED O_2

1.1.1 Laboratory Techniques

Oxygen (O_2) is an absolutely critical key molecule in nature. Oxygen is produced by photosynthesis, and it is the ultimate electron acceptor for degradation of organic material. It is thus an excellent tracer for biological activity in the pelagic environment. Adequate availability of molecular oxygen is a prerequisite for most fauna and flora, and microbial processes are largely governed by the availability of oxygen as a terminal electron acceptor. In an evolutionary perspective, the advent of oxygenic photosynthesis and the gradual oxidation of the biosphere was thought to be essential to the subsequent evolution of multicellular organisms and all higher life forms. Because of its universal importance numerous techniques for measuring dissolved O_2 have been developed and have been applied with variable success to the aquatic environment.

The iodometric titration technique [1] probably represents the oldest and most widely applied method for O_2 determination and it is still used with only minor modifications since the original description. Techniques such as gasometry [2], mass spectrometry [3] and gas chromatography [4] have been employed, but have never become widely used, and they were not well suited for *in situ* applications. Various polarographic and voltammetric measuring principles have been applied for O_2 sensing especially within the fields of medicine and physiology [5–11]. Voltammetric O_2 measurements will be discussed in Chapter 9. However, amperometric O_2 sensors have probably been the most successful within the aquatic sciences and most certainly the most widely applied principle for *in situ* studies. The working principle of these sensors is based on the measurement of a current that is proportional to the rate at which O_2 is reduced on a metal surface kept at a fixed negative potential (around -0.8 V). Numerous amperometric macrosensor designs for pelagic work and *in situ* respirometry have been presented in recent decades and the field has been extensively reviewed [12–15]. The aim of this review is thus not to provide a comprehensive overview of all O_2 sensors applied in aquatic science. Rather we would like to present insights into the possibilities of performing O_2 analysis using O_2 microsensors, with emphasis on *in situ* measurements, that have evolved since Revsbech and Jorgensen [16] reviewed the use of microelectrodes in microbial ecology.

Basically two types of O_2 microelectrodes have been applied *in situ*; cathode- and Clark-type sensors. Recently, microsensors based on optochemical measuring principles have successfully been introduced to aquatic science and have demonstrated great potential for *in situ* applications in the water column as well as for benthic studies.

1.1.2 Water Column and Sediment–Water Interface Studies

Most *in situ* work has been performed in the pelagic environment. Here constraints on sensor size, power supply, data transmission etc. are less than for studies across the sediment–water interface. This is simply because in the water column the spatial and temporal scales for O_2 variability are several orders of magnitudes larger than the variability found across the sediment–water interface. In this chapter we apply the term interface for the boundary between sediment and water. The vertical O_2 gradient in the water column typically extends over depth scales of meters (or hundreds of meters) and O_2 concentrations may be stable for hours or days [17]. The size of sensors and associated electronics is therefore of minor importance and often constant cable connections between ship and equipment put few constraints on the power requirements.

At the interface O_2 gradients are very steep and the characteristic depth scales are on the order of mm or μm [18]. Additionally, the O_2 concentration across such interfaces may change drastically within seconds as a result of changes in hydrodynamics, the faunal activity, or photosynthesis. Meaningful measurements at interfaces therefore put severe constraints on the applied sensors in terms of size, response time and stirring sensitivity. Operation at such a small scale also makes a constant cable connection to a ship problematic since mechanical disturbance via the cable can affect the position of the sensors.

The introduction of microsensors to aquatic science in the early eighties opened many new research fields, especially for benthic research [19–21]. Additionally, the miniaturization offered several advantages over the more commonly applied macro-(or mini-) sensors (see below), although these advantages have not yet been widely appreciated and exploited. Oxygen microsensors are therefore only gradually being introduced to studies in the water column and for *in situ* respirometry [22,23].

Various definitions for a microsensor have been used (see Chapter 1). In the following we have chosen to use a rather broad definition stating that an O_2 microsensor is any sensor that is capable of measuring the O_2 distribution with a spatial resolution better than 100 μm.

1.2 O_2 MICROSENSORS

The main advantage of microsensors is the very small size of the sensing tip. This allows measurements at the same scale as the structures relevant for production or consumption of O_2 in the aquatic environment [16,24–28]. However, the miniaturization also makes it possible to exploit working principles that would not work at macroscale. This is because on a small scale diffusion is a very efficient transport mechanism and thus scaling down improves signal stability, and reduces the response time and the analyte consumption rates.

The first O_2 microsensors were simple bare platinum cathodes. However, in order to provide a chemical and physical shielding towards the environment most cathode sensors that have been applied were coated with membranes [29–31] (Figure 1A). The membrane must allow O_2 to penetrate, but at the same time it must be electrically conducting. This means that at least some ions must be able to pass the membrane, this can cause interference problems with other solutes in the environment (see section 2.6). Various membrane materials have been utilized within different fields of research: Collodion, polystyrene, Zapon Lacquer, silicone, acrylic polymers, cellulose acetate, DPX resin etc. [32–34]. The different materials have their advantages and disadvantages in relation to adhesion, mechanical stability, permeability towards O_2 and various electrolytes, etc. The optimal membrane material for a given application has to be carefully evaluated. For applications in benthic research DPX-membranes have proven to be successful [33,35]. The polarogram of gold-plated cathodes has a broader plateau around −0.8 V than that of platinum cathodes [32]. This gives rise to a more stable current, which is less affected by small physical or biological fluctuations during application. Most microelectrodes therefore operate with a gold-plated cathode. The tip diameter of cathode-type sensors can be made very small (<1 μm) and they have proven very useful for physiological studies [32]. The very first field and *in situ* deep-sea measurements were performed with such sensors and they are still occasionally being applied *in situ* [35–37]. In order to increase the signal size and the robustness of such sensors the so-called needle-electrodes with tip diameters in the order of 600–800 μm have been developed and applied in situ [38,39]. Figure 1 schematically present the outlines of cathode-type sensors. A major improvement in O_2 microsensing was achieved by miniaturizing the Clark-type O_2 sensor [40,41]. As compared with the cathode-type sensor the Clark-type sensor is more complicated to manufacture, but has some major advantages (see below).

Some typical sensor characteristics for micro- and mini-sensors that have been applied for benthic *in situ* studies are compiled in Table 1. However, the sensor properties of individual electrodes vary, so the tabulated values can only

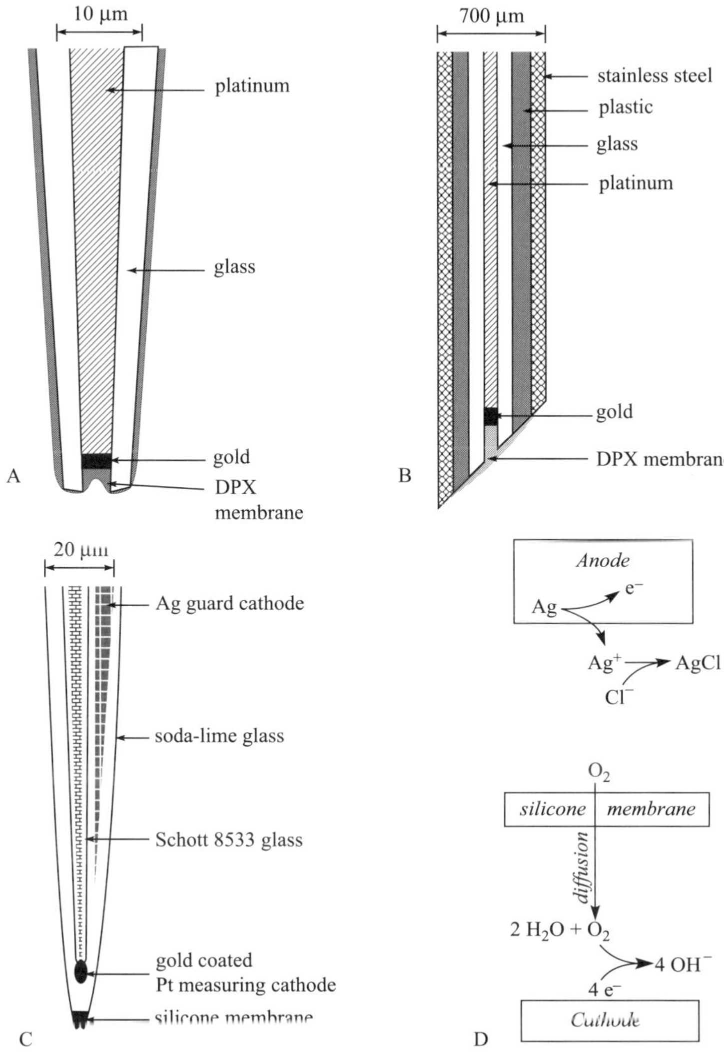

Figure 1. (A) The basic structure of a gold-plated and DPX-coated cathode sensor. (B) The tip zone of a needle-type DPX-coated cathode sensor. (C) The tip of a Clark-type O_2 microelectrode. The anode which is usually fixed further up in the casing is not shown. (D) Schematic drawing of the measurement principle. The upper panel shows the reaction at the anode placed in the bulk electrolyte, while the lower panel illustrates the reaction at the negatively polarized cathode (DPX: see glossary)

Table 1. Characteristics of three types of amperometric O_2 sensors applied for *in situ* interface studies

	Sensor		
	Micro-cathode*	Micro-Clark	Needle-type
Tip size (μm)	0.2–5	1–10	600–800
Stirring sensitivity (%)	2–50	0–2	5–50
90 % Response time (s)	0.1–2	0.5–2	60–120
Detection limit ($\mu mol\ L^{-1}$)	0.1	0.1	1
Signal size at air saturation (pA)	100–200	50–200	5000
Main interference	Mg^{2+}, Ca^{2+}, H_2S	H_2S	Mg^{2+}, Ca^{2+}, H_2S
Key references	32–34	45	38, 39

* Membrane coated.

be taken as typical electrode performance. It should also be stressed that the importance of various sensor characteristics must be evaluated in the context of the scientific question addressed. In some instances, stirring sensitivity may be crucial for the interpretation of recorded data while it is irrelevant in others. The most appropriate sensor properties therefore vary with a given task. In the following section the various sensor characteristics are discussed in detail and examples of applications where sensor characteristics are of major importance are presented and discussed.

For water column studies instrument packages (so-called conductivity–temperature–depth instruments or CTDs) capable of carrying O_2 macrosensors have been commercially available for many years. Most available systems are modular and the configurations are custom designed, and optimized for a given application. For *in situ* microsensor studies at the benthic interface various *in situ* vehicles have been constructed and applied. Independently operating platforms working directly at the sea floor for interface studies are termed 'landers'. A review of technical solutions and designs of benthic landers was recently presented [42]. Further, instrumental concerns in relation to *in situ* measurements in the benthic boundary layer have also been discussed recently [43] and special attention to *in situ* measurements with microsensors was given by another recent review [44]. In the present context we will therefore refrain from further discussion on *in situ* instrumentation and will refer to the above-mentioned reviews.

2 CLARK-TYPE O_2 MICROELECTRODE

2.1 CONSTRUCTION AND FUNCTION

The measuring cathode of the Clark-type O_2 microsensor is immersed in an aqueous electrolyte chamber which contains the internal Ag/AgCl reference anode (Figure 1C). The alkaline electrolyte (pH 10) of a standard microelectrode contains 0.5 mol L^{-1} KCl in 0.5 mol L^{-1} carbonate buffer. The internal solution and electrodes are separated from the exterior by a glass casing and an O_2 permeable silicone membrane (Figure 1C). A silver anode within the electrolyte acts as the reference anode. An additional improvement of the sensor was made by placing a guard cathode behind the sensing cathode [45] (Figure 1C). This prevented interference from O_2 diffusing toward the sensor tip from the bulk electrolyte chamber within the electrode and thereby ensured a constant low zero-current. A review of the construction of the Clark-type O_2 microsensor can be found elsewhere [16].

In the working mode the measuring and the guard cathode are both polarized at −0.8 V versus the Ag/AgCl reference electrode. Lower polarization may result in an inefficient reduction of O_2 (low unstable signals), while higher polarization may lead to reduction of water (high unstable signals) followed by gas formation and insensitivity. The measuring cathode consists of a gold-plated platinum wire. The reaction stoichiometry at the cathode is shown in Figure 1D. The redox reaction of O_2 is very complex and not completely understood, but H_2O_2 is known to be an important intermediate [46]. A porous structure of the gold surface making up the measuring cathode gives a very large catalytic area, ensuring a complete reduction of oxygen and thus a stable signal. During measurements, there is a net diffusion of O_2 through the highly permeable silicone membrane. The oxygen is dissolved in the electrolyte and is subsequently reduced at the gold-coated platinum cathode. The Clark-type O_2 microsensor thus responds to the O_2 partial pressure of the medium which is determined by the O_2 concentration and other environmental parameters such as temperature, salinity etc. (see below). At the reference anode solid Ag is oxidized and subsequently precipitates as AgCl (Figure 1D).

2.2 CALIBRATION (TEMPERATURE AND SALINITY EFFECTS)

In the case of constant temperature, salinity and pressure, the Clark-type microsensor has a perfectly linear response to O_2. In most *in situ* applications it is a simple matter to perform a two-point calibration by measuring the electrode signal in two different water samples whose O_2 concentration is subsequently determined by Winkler titration. Alternatively, for benthic microprofile measurements two calibration points are typically inherent in the measurements: the constant reading in the bottom water and a zero reading in the deep anoxic sediment [44]. All that is required is thus a Winkler determination of the oxygen concentration in the bottom water.

Often O_2 measurements at variable salinity and/or temperature are required [22,47]. Electrode calibration in such complex cases demands a quantitative understanding of signal dependence on sensor dimension, salinity and temperature. A multivariate analysis of electrode signals as a function of various external variables has been applied in order to construct a calibration model for Clark-type sensors [48]. Even though such an empirical approach may be useful in special cases, it is not generally applicable as it is not based on a full understanding of the function of the sensor. A mathematical description of the essential transport processes within a Clark-type O_2 microsensor was recently developed [49]. The model assumes that the internal geometry of the tip zone is conical and that the sensor tip can be described by three radii, the length of the

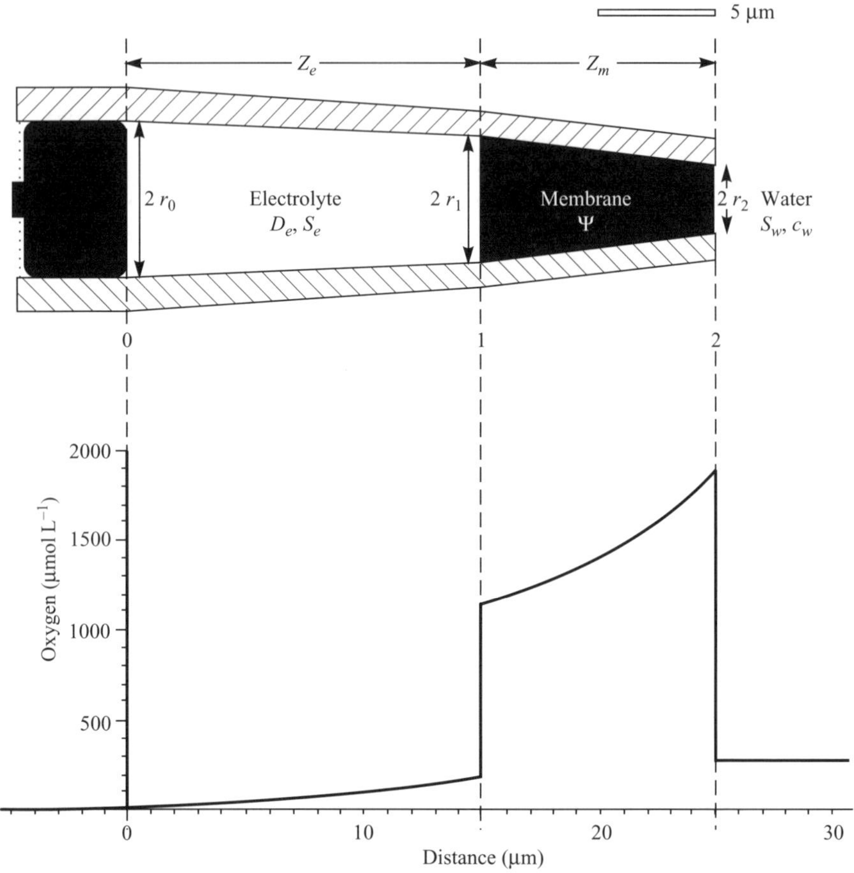

Figure 2. The tip of a Clark-type O_2 microelectrode as described by Equation (1). In case of the following physical conditions: $c_W = 273\ \mu\text{mol L}^{-1}$, $T = 295$ K, salinity $= 34\,‰$, $Z_e = 15.0\ \mu$m, $Z_m = 10.0\ \mu$m, $r_0 = 3.4\ \mu$m, $r_1 = 2.8\ \mu$m, $r_2 = 1.5\ \mu$m, the sensor signal can be calculated from Equation (1) to be 259 pA. The resulting O_2 concentration profile through the tip zone is indicated. (Redrawn from Gundersen, J. K. *et al.*, *Limnol. Oceanogr.*, **43**, 1932 (1998). Reproduced by permission of American Society of Limnology & Oceanography.)

silicone membrane, and the distance between membrane and cathode surface (Figure 2). Additionally, two other assumptions are made: (i) the O_2 concentration at the cathode surface is zero (as is the case after a short polarization period) and (ii) the O_2 concentration at the outer silicone membrane surface equals the concentration in the bulk water phase. The latter assumption is only valid for sensors with low stirring sensitivity, yet the model can be extended to include the diffusive processes outside the sensor tip (this extension is discussed in section 2.3). In such a case the electrode signal, Si, given by the O_2 reduction current (i.e. sensor output minus zero current) can be described by applying simple Fickian diffusion theory and mass conservation [49,50]:

$$\mathrm{Si} = \Phi \pi r_1 p_\mathrm{w} \left(\frac{Z_\mathrm{m}}{r_2 \Psi} + \frac{Z_\mathrm{e}}{r_0 D_\mathrm{e} S_\mathrm{e}} \right)^{-1} \quad (1)$$

where (for an explanation of symbols please also refer to Figure 2), Φ = current generated per mole O_2 reduced (3.86×10^{-5} A s mol^{-1}), p_w = partial O_2 pressure in the ambient water (Pa), Z_m and Z_e = lengths of silicone membrane and electrolyte distance, respectively (m), Ψ = O_2 permeability of the silicone membrane (mol m^{-1} Pa^{-1} s^{-1}), D_e = O_2 diffusion coefficient in the electrolyte (m^2 s^{-1}), S_e, S_w = O_2 solubility in the electrolyte or in water sample (mol L^{-1} Pa^{-1}), and r_0, r_1, r_2 = the internal sensor radii at the cathode, inner and outer position of the silicone membrane, respectively (m).

The geometry of a given sensor (r_0, r_1, r_2, Z_m and Z_e) can be determined with a microscope. The solubility, S_e, and the diffusion coefficients, D_e, can be found in the literature [51,52] and can be recalculated to the correct temperature and salinity [53]. The O_2 permeability, Ψ, (defined as the product of the diffusion coefficient of O_2 in the membrane material and the distribution coefficient between membrane and solution) of the silicone used for the sensor membranes has been measured to be 11.2×10^{-15} mol s^{-1} m^{-1} Pa^{-1} [49]. However, the permeability of an electrode membrane at a given temperature is dependent on the degree of hydration and the age of the membrane. For most practical applications, it may be simpler to perform a one-point calibration for the determination of Ψ by applying equation (1). The model was experimentally verified by comparing the signal from 23 different O_2 microsensors to signals predicted by equation (1) (Figure 3). For this figure we used a fixed silicone permeability of 11.2×10^{-15} mol s^{-1} m^{-1} Pa^{-1}. The presented equation differs from previous calibration equations [54] in two ways. (i) it is a theoretical description of the molecular diffusion processes within the electrode, which does not rely on empirically determined relationships and (ii) it predicts the O_2 consumption of a given electrode.

If the value Z_e is 0 (i.e. the silicone membrane is coated directly on the cathode, as on a cathode microsensor), the signal becomes directly proportional

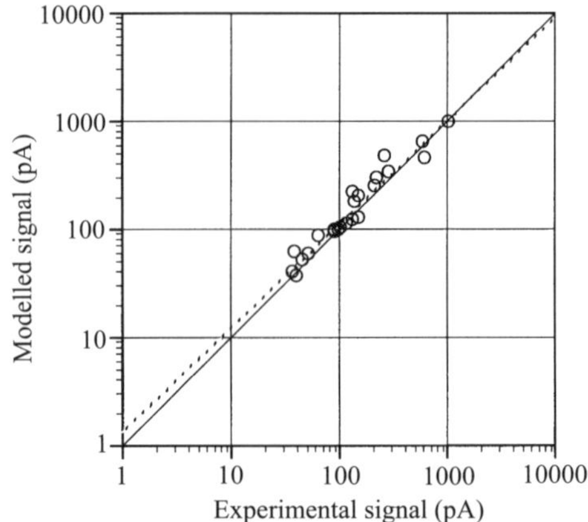

Figure 3. The signals of 23 Clark-type O_2 microelectrodes as calculated using equation (1) versus experimentally determined values. The solid line is the theoretical line with slope = 1 and ordinate at origin = 0, while the dashed line is the best linear fit. (From Gundersen, J. K. *et al.*, *Limnol. Oceanogr.* **43**, 1932 (1998). Reproduced by permission of American Society of Limnology & Oceanography.)

to the O_2 partial pressure multiplied by the membrane permeability and divided by the membrane length. This is in accordance with previous calibration equations for oxygen macro electrodes [54]. In Clark-type O_2 microsensors the distance Z_e typically accounts for 30–90 % of the internal diffusion distance and thus has to be included in a model of the electrode signal [49,55]. This can be demonstrated by model predictions of the sensor signal as a function of temperature and salinity. In the case of a constant O_2 concentration but an increased salinity of the bulk water surrounding the sensor, the partial pressure (p) will increase correspondingly. As the partial pressure is the only parameter in equation (1) affected by salinity, the sensor signal will increase irrespective of the membrane thickness (Figure 4A). The relative membrane thickness is defined as $(Z_m/Z_e) + Z_m$. If we instead increase the temperature, then the partial pressure (p), the permeability (Ψ), and the transport coefficient (D_e) will increase while the O_2 solubility (S_e) in the electrolyte will decrease; the combined effect will be a signal increase that will be more pronounced in case of thicker membranes (Figure 4B). Increasing the relative membrane thickness will result in relatively higher sensor signals and the effect will increase with temperature (Figure 4C) [49]. An earlier presented model described the sensor signal from a silver cathode, polyethylene-coated Clark-type macroelectrode [55]. The resulting equations bear resemblance to the model presented above.

The useful microelectrode signal is the sensor output minus the 'zero-current', i.e. the current at zero O_2 concentration. The zero-current is typically in the order of 1–2 pA (or approximately 1 % of the sensor signal in air-saturated

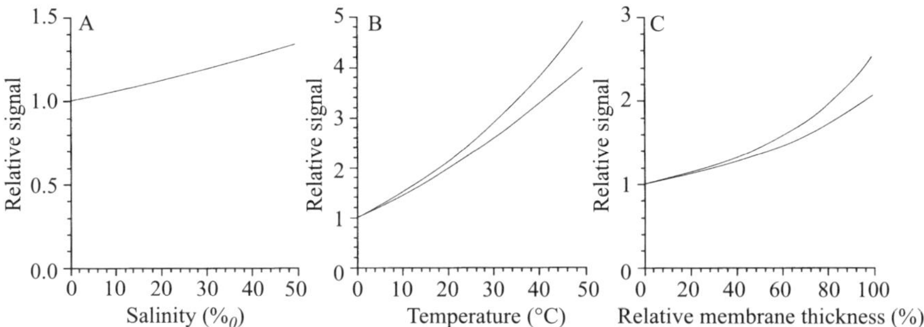

Figure 4. Relative signal change as a function of salinity (A) and temperature (B). The salinity effect is independent of the relative membrane thickness. However, the temperature-induced change is membrane thickness dependent as indicated by the upper (relative thickness 99.9 %) and the lower lines (relative thickness 0.1 %) in panel (B). Panel (C) presents the relative signal increase as a function of the relative membrane thickness at 50 °C (upper line) and 0 °C (lower line). (From Gundersen, J. K. *et al.*, *Limnol. Oceanogr.*, **43**, 1932 (1998). Reproduced by permission of American Society of Limnology & Oceanography.)

water). The zero-current is ascribed to conductivity and reducible constituents in the glass of the sensor [56], but minimal zero currents can be achieved by using highly insulating glasses. Measurements have shown that the size of the gold cathode has relatively little effect on the zero-current [56]. During long-term use of a microelectrode the zero-current typically decreases; this is partly due to a gradual consumption of O_2 dissolved in the electrolyte but may also be due to a gradual reduction of impurities in the glass. At high temperatures the zero-current may contribute significantly to the sensor signal (Figure 5) and it is therefore important to determine the zero-current at the working temperature. At elevated temperatures, for instance, it can be advantageous to reduce the polarization voltage in order to reduce the zero-current.

The model above contains only a single variable, the membrane permeability (Ψ), which has to be estimated at one known temperature and salinity (i.e. a one-point calibration). This one-point calibration allows for O_2 concentration determinations based on sensor signals obtained at any temperature and salinity. This makes calibration of microsensors for their application in water columns or in vent sites possible. The model also makes it possible to optimize sensor construction with respect to the signal to noise ratio, response time, and stirring sensitivity (see below).

2.2.1 Case Study: Measurements at a Vent Site

An extreme case of temperature effects on *in situ* measurements of O_2 is experienced during studies around hydrothermal vents, where temperatures

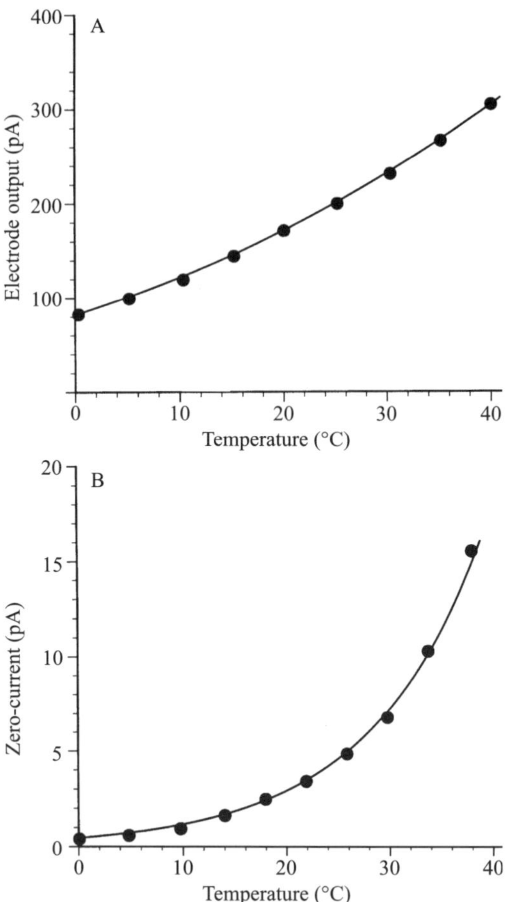

Figure 5. The output (A) and the zero current (B) as a function of temperature. (From Gundersen, J. K. *et al.*, *Limnol. Oceanogr.*, **43**, 1932 (1998). Reproduced by permission of American Society of Limnology & Oceanography.)

may vary from 1 to > 100 °C. In such cases a reliable calibration of the O_2 signals requires a good quantitative understanding of the temperature effects on the O_2 microsensor. During a study at 10 m water depth in the Aegean sea, O_2 and temperature microprofiles were recorded simultaneously at a variable distance from a vent center (Figure 6). It was observed that O_2 penetration depths and O_2 fluctuations at a given depth horizon gradually increased towards the vent center [57]. The simultaneously obtained temperature measurements expressed an extreme horizontal and vertical temperature gradient that had to be accounted for during sensor calibration (Figure 6).

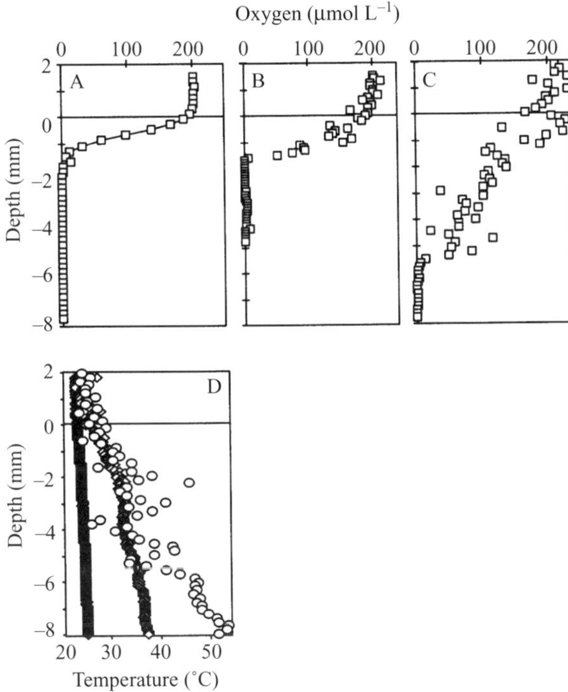

Figure 6. Panels (A), (B) and (C) show three microprofiles obtained *in situ* by Clark-type O_2 microelectrodes at distances of 4.0 m (A), 1.0 m (B) and 0.1 m (C) from a hydrothermal vent center. (D) illustrates the simultaneously obtained temperature microprofiles at the three respective sites (closed diamonds 4 m away; open diamonds 1 m away and open circles 0.1 m away from the vent center). (Modified from *Mar. Chem.*, **69**, 43–54 (2000), Wenzhöfer *et al.*, In situ microsensor studies of a hydrothermal vent at Milso (Greece), with permission from Elsevier Science.)

From microprofiles and continuous sensor readings at given positions a microcirculation pattern driven by the centrally out-flowing seep-water was resolved [57].

2.3 SENSITIVITY TO ANALYTE CONSUMPTION AND STIRRING

The measuring principle of an amperometric O_2 sensor is based on analyte consumption. Commonly applied macrosensors have current outputs in the order of 0.1–10 μA [13], which corresponds to an O_2 consumption rate by the sensors of 0.2–22 μ mol d^{-1}. For a Clark-type microsensor this value is in the order of 20 pmol O_2 d^{-1} which is equivalent to the O_2 consumption of an individual dinoflagellate. The analyte consumption of O_2 sensors in itself rarely

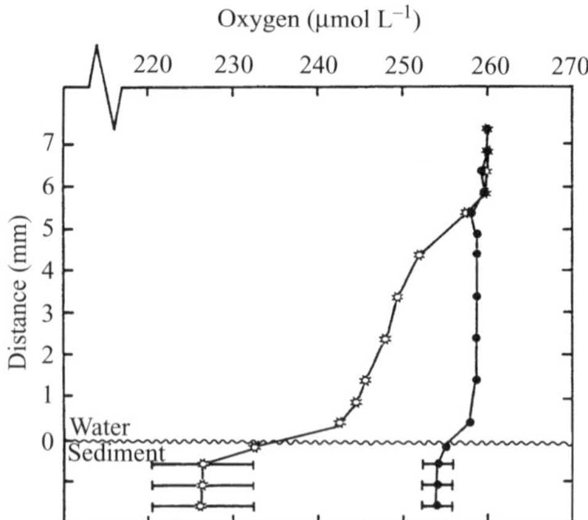

Figure 7. Apparent O_2 profiles as measured with two Clark-type O_2 microelectrodes having stirring sensitivities of 6% (✲) and 1%(●) respectively. The gradients are measured through a stirred 8 mm thick water film covering a layer of inert glass beads (no O_2 consumption). The observed gradients are solely due to a gradual decrease in flow velocities. (From Revsbech, N. P. R., *Limnol. Oceanogr.*, **34**, 474 (1989). Reproduced by permission of American Society of Limnology & Oceanography.)

poses a problem to measurements, and for respirometry the cathode size is usually scaled down below the expected O_2 consumption rate of the investigated object [58].

However, the fact that the sensor consumes the analyte, and thus requires a steady flux of analyte into the sensing tip, implies that the sensor signal will be affected by the transport of the analyte in the surrounding medium—the so-called 'stirring sensitivity' [59]. For interface studies the stirring sensitivity is a very important parameter, since the moving of a sensor tip from the free flowing water to the stagnant intersititial water of the sediments will cause a signal decrease solely due to the decrease in mass transport of oxygen (Figure 7). Stirring sensitivity is usually quantified as the percentage increase in sensor signal following the onset of vigorous agitation around a submerged sensor tip. This principle has recently been applied for the construction of a diffusivity sensor [60].

The stirring sensitivity of a Clark-type O_2 microsensor is largely determined by the sensor geometry and the diffusive path that O_2 must travel from the medium to reach the polarized catalytic surface (see also chapter 8). The path can be divided into two parts (Figure 8): (i) an internal part within the electrode (i.e. the distance from the sensor tip to the catalytic surface) and which is insensitive to stirring and (ii) an external part within the diffusion layer (DL) surrounding the sensor tip (Figure 8). The thickness of the DL will be a function of the water flow, as an increasing flow velocity will gradually erode the DL

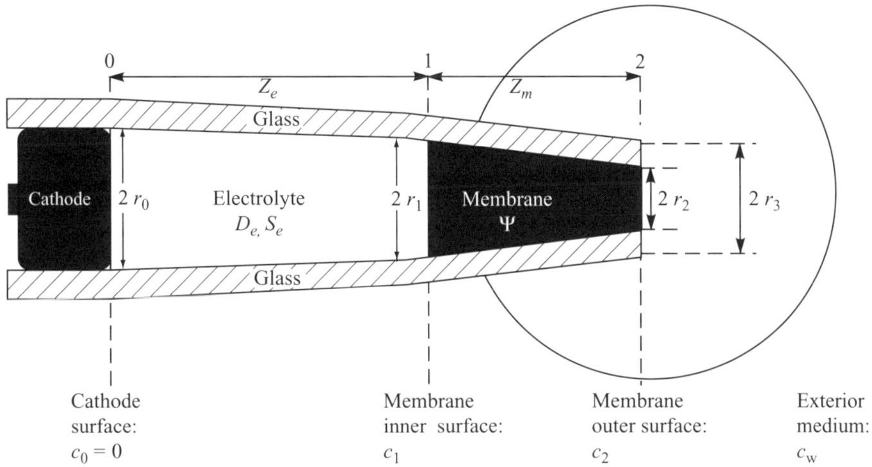

Figure 8. The tip of a Clark-type O_2 microelectrode including a hypothetical diffusion sphere (DL) outside the sensor tip

away. The analyte supply through the external medium is thus sensitive to stirring. The overall stirring sensitivity of the sensor is determined by the relative importance of the external and the internal paths in limiting O_2 flux to the sensing cathode. If the main restriction to O_2 flux is imposed within the sensor, then the electrode will be largely unaffected by stirring.

The model presented in section 2.2 describes the diffusive processes within the sensor. However, it assumes that the sensor is insensitive to stirring, i.e. the O_2 concentration at the sensor tip surface is equal to the concentration of the bulk water. The stirring effect can be included if we assume spherical oxygen gradients outside the sensor tip which replenish the O_2 consumed by the sensor (Figure 8). In stagnant water the O_2 distribution outside the sensor can be calculated by rearranging equation (1) into:

$$c_{\mathrm{w}} = \mathrm{Si}S_{\mathrm{w}}\left(\frac{Z_{\mathrm{m}}}{r_2\Psi} + \frac{Z_{\mathrm{e}}}{r_0 D_{\mathrm{e}} S_{\mathrm{e}}}\right)(\Phi\pi r_1)^{-1} \tag{2}$$

where c_{w} is the O_2 concentration. The outside gradient can in theory be completely eroded away and the stirring sensitivity thereby becomes equivalent to the relative difference between O_2 concentration at the sensor surface and infinitely far away from the tip. Applying the O_2 concentration calculated from this relation and the oxygen flux as calculated from the electrode signal (Si), we can estimate the stirring sensitivity (ST) by using spherical diffusion equations [50]:

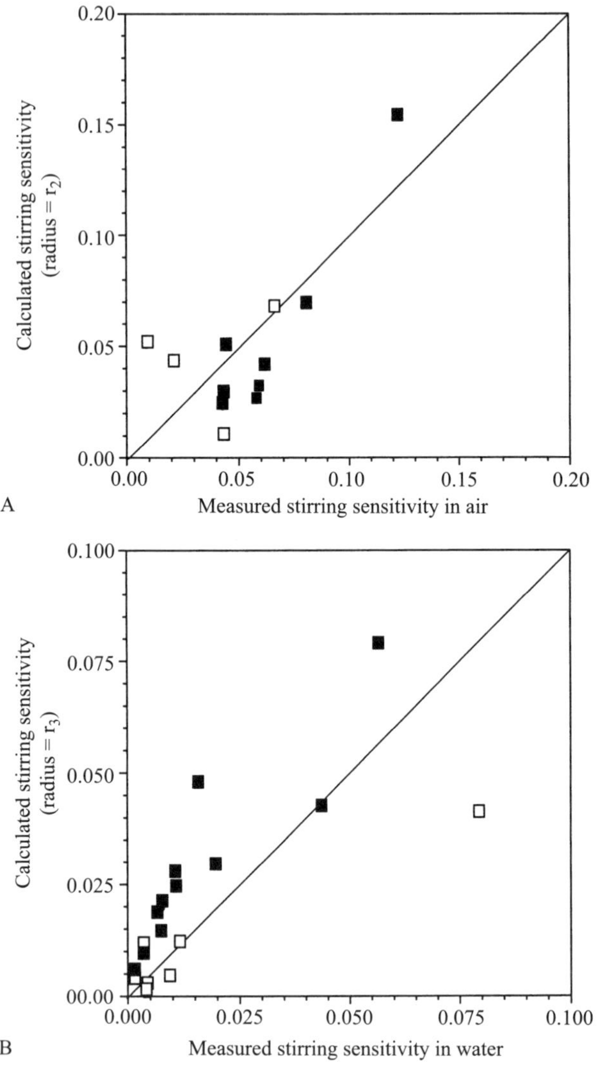

Figure 9. Calculated stirring sensitivity as a function of measured stirring sensitivity for 17 different Clark-type O_2 microelectrodes, in air (A) and water (B). Filled symbols represent sensors with a low r_3/r_2 ratio < 3 (please refer to Figure 8) and open symbols represent sensors with a high r_3/r_2 ratio > 3. r_2 and r_3 have been used for calculation of stirring sensitivity in (A) and (B), respectively.

$$\mathrm{ST} = \frac{(c_{\mathrm{w}} - c_2)}{c_{\mathrm{w}}} = \frac{\mathrm{Si}}{4r\pi D\Psi}\left(c_2 + \frac{\mathrm{Si}}{4r\pi D\Psi}\right)^{-1} \tag{3}$$

where r equals the radius of the diffusing sphere (which is approximately equal to electrode opening) and c_{w} and c_2 are the O_2 concentrations in the exterior medium far away from the sensor and at the sensor surface, respectively.

It is not experimentally possible to erode the DL of a submerged electrode completely. However, by using the signal reading in air as being equivalent to recording under vigorously stirred conditions it was possible to compare theoretically and experimentally determined stirring sensitivities (Figure 9A). The comparison shows some scatter, but the points fall evenly around the 1:1 line. The poor correlation is probably a result of difficulties in determining the exact geometry of the various sensor tips under a microscope, and in measuring the correct temperature of the air-exposed electrode tip. If we thus assume that the stirring sensitivity corresponds to the erodible spherical concentration gradient outside the outer electrode diameter (instead of the inner electrode opening used in the previous model), we get a reasonable correlation between the calculated stirring sensitivity and the experimentally determined value (Figure 9B). The calculations can be used to identify the geometric parameters of importance for the stirring sensitivity and thereby for optimization of sensor design.

From the insight gained above it is clear that bare cathode-type O_2 microsensors or cathode sensors with a relatively thin membrane will have an extremely high stirring sensitivity. Cathode-type microsensors with stirring sensitivity as high as 50% have been applied in the laboratory [34]. However, by constructing microsensors with appropriate sensor geometry, membrane thickness, aspect ratio etc., cathode-type microsensors with stirring sensitivities $< 5\%$ have been made routinely [33,37]. The stirring sensitivities of Clark-type O_2 microsensors are usually around 1–2% but they can be made without any measurable stirring sensitivity if needed [40]. There is a trade-off between low stirring sensitivity and temporal resolution. A low stirring sensitivity is achieved by pulling the cathode back from the sensor opening but this will, however, also increase the response time of the sensor (see section 2.4.2).

Clark-type macroelectrodes being applied for measurements in water columns or within an enclosed water phase usually have significant stirring sensitivity in the order of 10–20% [13,61]. This is the result of large cathodes being applied with relatively thin membranes. In order to reduce this problem double membrane electrodes have been constructed [62]. However, in most commercial macro-probes, flow-through systems or stirred measuring chambers are mounted around the sensor tips in order to ensure insensitivity to the ambient water flow [15,63].

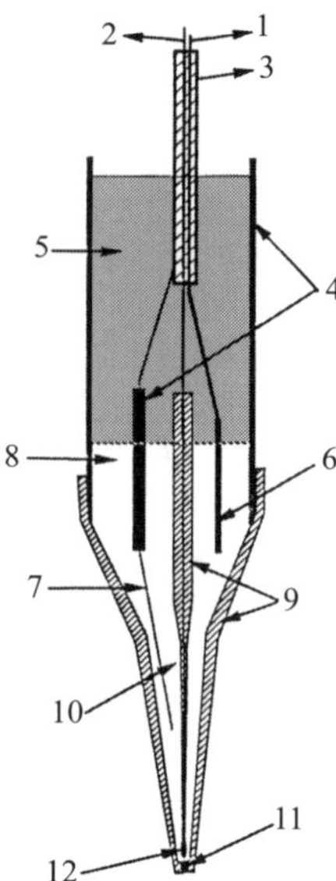

Figure 10. A Clark-type O_2 'mini'-electrode specially developed for fast, stirring-insensitive, measurements in water columns. Tip diameter is 30 μm. 1 = anode, 2 = cathode, 3 = ground, 4 = soda glass, 5 = epoxy resin, 6 = silver wire (anode), 7 = silver wire (guard cathode), 8 = 1.5 mol L^{-1} KCl, 9 = Schott 8533 glass, 10 = platinum wire (measuring cathode), 11 = silastic membrane, 12 = gold-plated cathode. (From Oldham, C., *Limnol. Oceanogr.*, **39**, 1959 (1994). Reproduced by permission of American Society of Limnology & Oceanography.)

The experience gained from O_2 microsensing at interfaces was recently adapted to a mini-CTD developed for studying fine-scale O_2 patches in water columns [22]. The O_2 sensor developed was in principle a Clark-type microelectrode, with a reinforced outer glass casing leading to an outer sensor tip diameter of 30 μm. The internal sensor hole, however, was only a few μm (Figure 10). Thereby the advantages of low stirring sensitivity (a few %) and fast response time (around 0.15 s for 90 % response) characteristic of microsensors was combined with a rugged design [22]. Owing to the fast response and no requirement for a constant water flow over the sensor surface the O_2 'mini'-sensor could be incorporated into a free – ascending CTD instrument, measuring high resolution water column profiles of temperature, and O_2 [22]. A similar sensor, with slight modifications resulting in better long-term stability, was developed for O_2 measurements during deep-sea deployments of benthic chamber landers [64].

2.4 SPATIAL AND TEMPORAL RESOLUTION

As already mentioned in section 1.1.2 discussions of spatial and temporal resolution of microsensors are mostly relevant for studies at interfaces or within the sediment matrix. Necessary sensor characteristics for water column measurements can practically always be met by standard O_2 microsensors, because of their small size, low stirring sensitivity and fast response.

2.4.1 Spatial Resolution

Microsensors do not change the O_2 concentration significantly at the measuring site because of their very limited O_2 consumption rate. Therefore, the spatial resolution by which a steady state microgradient can be resolved is in principle set by the diameter of the sensor tip. As a general rule we assume that the spatial resolution of a sensor is in the order of two times the outside tip diameter. Cathode-type microsensors can be made extremely small with tip diameters below 1 μm [32], while the tip diameter of Clark-type microsensors is usually around 1–10 μm [40]. For most practical purposes such spatial resolution is more than sufficient to resolve the concentration gradients across interfaces within aquatic environments [16]. In most cases mechanical problems related to electrode positioning (using manual or motor-controlled micromanipulators) or to physical disturbances within the environment are the factors limiting the spatial resolution, rather than the tip size of the microsensors. Needle-like cathode sensors with tip diameters of 600–800 μm have a correspondingly lower resolution and their high O_2 consumption rates may cause local distortion of the O_2 gradient.

A practical problem related to sensor size that should be mentioned is that very thin sensors are fragile. In many sediments (and especially on solid surfaces) this has to be taken into account. Years of *in situ* application of microsensors has taught us that tip sizes in the order of 10–25 μm are sufficiently rigid to survive microprofiling through the oxic zone of normal coastal sediments. Another practical concern is the extent to which microsensors mechanically disturb the structure of the sediment as they penetrate. This is still an open question, although repeated measurements at the same location with microsensors usually reveal only slight differences between the first and the subsequent profiles, indicating that any mechanical disturbance is of minor importance. Another study concluded that microprofiles resolved by needle electrodes (tip diameter 0.7 mm) and cathode microelectrodes (tip diameter 2–10 μm) were very 'similar' and that no 'significant difference' between such profiles could be observed at a spatial resolution of 0.5 mm [38]. The study was performed in a silty sediment with O_2 penetration depths of approximately 10 mm. Any potential effect will probably depend on the sediment characteristics and the slope of the O_2 gradients, and it still remains to be shown to what extent sensor dimensions affect measured porewater profiles.

The high spatial resolution that can be obtained with O_2 microelectrodes has allowed studies on O_2 concentration gradients towards or within meiofauna individuals and marine aggregates (marine snow) [27,65,66]. For 'on site' or laboratory-based studies within microbiota or in symbiosis it has to be recognized that microsensor techniques are invasive. Phobic responses to mechanical disturbances or related to penetration of tissue cannot always be neglected [27], and are a matter of concern in physiological studies [67].

2.4.2 Temporal Resolution

After experiencing an abrupt change in the O_2 concentration, the signal of a microelectrode follows an exponential response curve [32]. It is therefore difficult to evaluate the exact time period required for a 100 % sensor response. The response times given for various microsensors are therefore usually the 90 % response time, i.e. the time required to reach 90 % of the total signal change, after an abrupt change in the O_2 concentration. The response time of a sensor is due to the time it takes diffusion to establish a new steady-state gradient through the sensor tip following a change in partial pressure. The response time (t_{90}) is thereby a function of the transport coefficient (D and/or Ψ) and the distance from the bulk water to the sensing cathode. As the transport both in the electrolyte and in the membrane material may present the rate limiting transport, the response time can be estimated by $t_{90} = (K_e Z_e^2 D^{-1}) + (K_m Z_m^2 s_m \Psi^{-1})$ where K_e and K_m are constants and s_m the solubility in the membranes [49]. Most membrane materials have O_2 permeabilities of the same order of magnitude as that of stagnant water and seldom pose a limit to the response time of a sensor. However, the O_2 solubility in some membranes, e.g. the silicone frequently used for Clark-type O_2 sensors, is much higher than the solubility in water. The membrane thus has to be filled up before a steady state gradient can be reached, which will increase the response time, while improving the selectivity for the many hydrophilic ions/compounds of the medium which may interfere with measurements. In all cases the response time is proportional to Z^2, which makes the distance between the cathode and the exterior medium the most critical parameter. Bare cathode-type sensors are therefore extremely fast (a few μs), while membrane-coated cathodes or Clark-type sensors are somewhat slower (see Table 1). As mentioned above, there is a trade-off between stirring sensitivity and response time of microsensors, and the optimal sensor for a given application has to be evaluated in that context. The geometry of a Clark-type O_2 microsensor can be optimized so that stirring sensitivities are kept below 1 % and the 90 % response time is below 0.5 s. This, however, requires a very skilful constructor.

2.4.3 Case Study 1: Thickness and Stability of the Benthic Diffusive Boundary Layer (DBL)

Although the existence of the DBL had been known for many years [68], the physical reality and its significance first became apparent when microelectrodes were applied to studies of benthic interfaces [69]. *In situ* application of membrane-coated cathode sensors combined with high resolution video recording demonstrated a signal decrease, significantly larger than the 3–5% stirring sensitivity of the applied sensors, just above the sediment surface [70]. This was used to indicate the existence of a DBL covering marine sediments, and the thickness above deep sea sediments was estimated to be of the order of 0.5–1.5 mm [70]. These values were later refined by high resolution microsensor measurements both in coastal and deep sea sediments [71,72]. In laboratory studies it was shown that the thickness of the DBL was primarily a function of the water flow velocity above the sediment and the sediment roughness (Figure 11) [26,69], as might be theoretically expected.

At the scale of a microsensor tip the sediment–water interface is usually not a planar surface but a landscape with 'mountains' and 'valleys'. A detailed 3D mapping demonstrated how the DBL covered the sediment surface like a

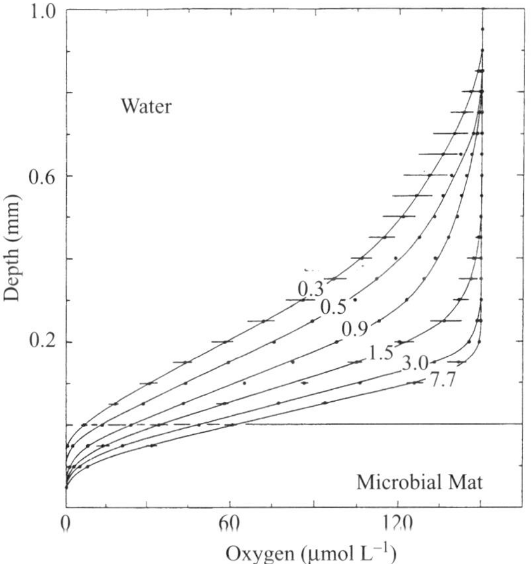

Figure 11. Oxygen microprofiles measured above a cyanobacterial mat at six different free-flow velocities ranging from 0.3 to 7.7 cm s^{-1} (indicated in figure). Depth 0.0 is equivalent to the mat surface and small horizontal bars indicate the standard deviation of four measurements (only shown for velocities of 0.3, 1.5 and 7.7 cm s^{-1}). (From Jorgensen, B. B., and Des Marias. D. J., *Limnol. Oceanogr.*, **35**, 1343 (1990). Reproduced by permission of American Society of Limnology & Oceanography.)

'carpet' being thinner on the upstream sides and thicker on the lee side of small mounts [26] (Figure 12). This shows that in the case of rough microtopography a geometric correction term has to be included to calculate accurate fluxes from vertical profiles, as the assumption of planar diffusion is invalid and since the profiles usually are not perpendicular to the DBL and sediment surfaces [26]. Continuous recordings of the O_2 concentration at given depths within the DBL have also demonstrated a fluctuating signal with a characteristic pattern [71]. The amplitude of the fluctuations increased in the upper part of the DBL only to decrease together with the frequency of the O_2 fluctuations as the sediment surface was approached. Increasing flow velocities of the overlying water reduced the amplitude of the oscillations but increased their frequency (Figure 13). These phenomena were interpreted as eddy intrusions from the turbulent phase above the DBL which penetrate down into the linear flow of the DBL [71,73]. At high flow velocities, the fluctuation frequency was as fast as the response time of the applied sensors [71]. The frequency and amplitudes of these fluctuations measured at high flow velocities are therefore probably underestimated in these studies. For more detailed studies on the O_2 fluctuations, faster electrodes (e.g. bare cathode sensors) could be applied, although their higher stirring sensitivity could lead to ambiguous results.

The linear O_2 concentration gradients within the DBL (Figure 11) are often used to calculate the net O_2 uptake (or release) of sediments [74]. The calculations are based on Fick's first law of diffusion: $J = D\,\mathrm{d}c/\mathrm{d}z$, where J is the flux, D the diffusion coefficient, c the O_2 concentration and z the depth [50]. An advantage of this approach is a well-established transport coefficient, the molecular diffusion coefficient for oxygen, within the DBL. It is, however, still unclear to what extent intrusions from the overlying water into the DBL affects the mass-transport within the DBL. Flux estimates within the sediment require an empirical determination of the transport coefficient which incorporates the molecular diffusion coefficient, porosity, and tortuosity factors [75,76]. The main problem for *in situ* studies of the DBL is to determine the exact position of the sediment surface. Recently, a reflection-based fiber optical sensor was used to determine the exact position of the sediment surface [77]. Such a sensor could easily be fixed on a microelectrode for simultaneous surface detection. Other commonly applied techniques for determining the position of the sediment surface are resistivity probes [75,78] or changes in the slope of the concentration profiles which indicate the reduced transport coefficients of sediments [79].

In section 2.4.1 we discussed the physical disturbance which could be associated with measurements in sediments or tissue. It was previously anticipated that the small tip diameter of O_2 microsensors allowed measurements to be made within the DBL without any disturbance [26,71]. However, detailed 3D mapping of the DBL structure, with a microsensor coming from below, up through the sediment matrix, revealed a major distortion of the DBL structure

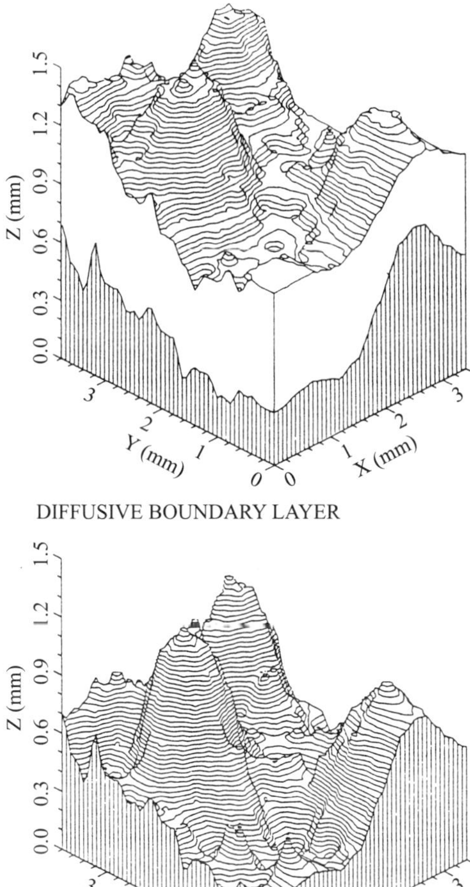

Figure 12. Three-dimensional plot of the microtopography of a microbial mat and the overlying DBL, measured by multiple insertions of a microelectrode. A structure of 0.5–1.0 mm protruding from the investigated area (3.6 mm along the *X*-axis and 3.8 mm along the *Y*-axis) caused smoothened deflections of the DBL. The Z-axis has a maximum value of 1.5 mm in both panels. The flow was in the direction of the positive *Y*-axis and the free flow velocity was 0.4 cm s^{-1}. (From Jorgensen, B. B., and Des Marias. D. J., *Limnol. Oceanogr.*, **35**, 1343 (1990). Reproduced by permission of American Society of Limnology & Oceanography.)

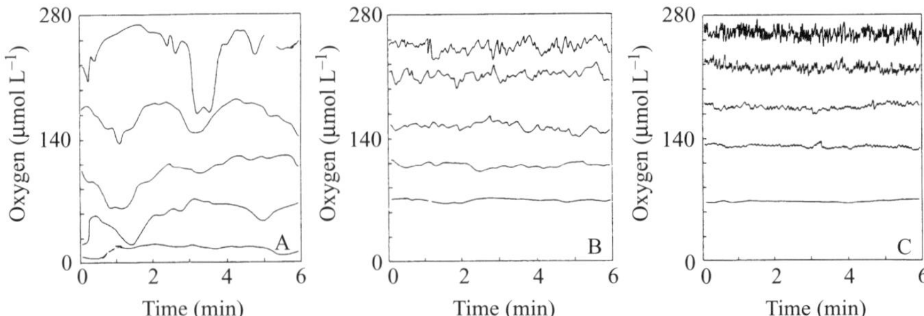

Figure 13. Continuous recordings of the O_2 concentrations at various distances above and below a microbial mat surface at three different flow velocities: (A) 0.3, (B) 1.3 and (C) 9.3 cm s^{-1}. The data were recorded at the following distances in relation to the mat surface: (A) 0.50, 0.30, 0.10, 0.00 , and −0.10 mm; (negative below mat surface); (B) 0.30, 0.20, 0.10, 0.00, and −0.10 mm; (C) 0.10, 0.05, 0.00, −0.05 , and −0.20 mm. The fluctuations mirror the effect of eddies in the flowing water that interact with the O_2 concentrations within the DBL. (Redrawn from Gundersen, J. K., and Jorgensen, B.B., *Nature*, **345**, 604 (1990). With permission.)

around a microsensor tip coming from above (Figure 14) [80]. The DBL immediately below the sensor tip was compressed by 25–45 % and previous estimates of the DBL thickness with microsensors were most likely underestimated correspondingly. The mechanism behind this effect is still not clear, but is most likely associated with pressure differences due to acceleration and deceleration of water flow around the electrode shaft, which is placed in a vertical velocity gradient [80]. The quantitative importance of the effect is mainly a function of the water flow velocity and the sensor geometry. Thick and stubby sensors impose a larger problem, however, even very thin and elongated sensors (< 10 μm thick) show clear evidence of DBL depression. Over very irregular surfaces the effect may to some extent be masked by interaction between flowing water, the sensor, and the microtopography [81].

2.4.4 Case Study 2: Gross Photosynthesis Measurements

Studies on benthic photosynthesis require microsensors with high temporal and spatial resolution. Steady state microprofiles with a spatial resolution of 50–100 μm have often been used to estimate the net O_2 export out of photosynthetic communities [82,83]. However, using the co-called 'light/dark shift technique' it is also possible to quantify the gross photosynthetic activity at a relatively high spatial resolution [84,85]. Using this technique the gross photosynthesis is calculated from the initial decrease in oxygen concentration at a given position after eclipse of the light source, assuming: (I) a steady-state O_2 distribution in the community before darkening, (II) an identical

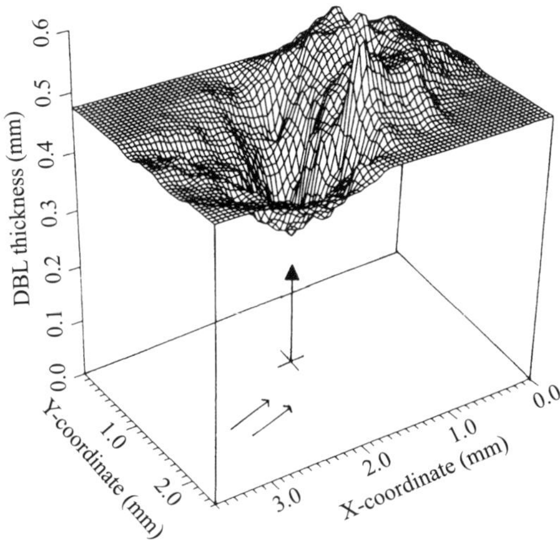

Figure 14. The disturbed DBL structure following the placement of a microelectrode (tip size 8 μm) at a depth position of 0.3 mm (indicated by large arrowhead). Prior to insertion the DBL was flat as was the benthic interface itself. Small arrows indicate the flow direction (flow velocity was approximately $3\,cm\,s^{-1}$). The sediment surface is at a depth position of 0.0 mm. (From Glud *et al.*, *Limnol. Oceanogr.*, **39**, 462 (1994). Reproduced by permission of American Society of Limnology & Oceanography.)

O_2 consumption before and during the light–dark shift, and (III) constant O_2 gradients within the community during the dark incubation. It has been demonstrated that assumptions (I) and (II) can be fulfilled in benthic photosynthetic communities but that (III) requires fast sampling during a short eclipse period so that the absolutely initial decrease in O_2 concentration is used for the calculation [85,86].

Within a cyanobacteria-dominated system, O_2 concentrations during light/dark cycles were recorded at a total of 22 depth horizons at a temporal resolution of 100 ms. From these data the O_2 microprofiles in light and after 1, 2, and 3.8 s of darkness were calculated (Figure 15A). The continuous recordings at four sediment depths are shown in Figure 15B. They clearly demonstrate that the rate of O_2 decrease changes during the dark incubation. In other words the O_2 concentration gradients changed after the eclipse of the light source (assumption (III) is not fulfilled). The gross photosynthetic rate measured after 1.1 and 2.6 s of darkness (Figure 16) show that the depth distribution of the measured activity widens out, so that peak activities were underestimated and the low activities are overestimated with longer incubation time. The depth-integrated activity, however, remained constant within a dark incubation period of 4 s (Figure 16B) [85,86]. The observed effect is the result of a net O_2 diffusion along the concentration gradients during the dark incubation, and

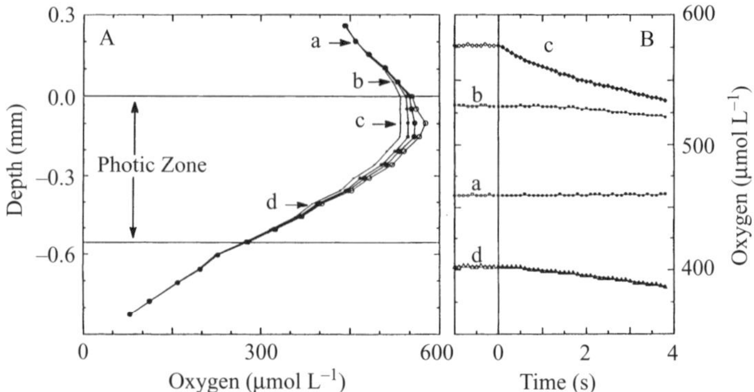

Figure 15. Oxygen distributions immediately above and within a cyanobacteria-dominated microbial mat during a light–dark shift. The depth of 0.00 mm indicates the mat surface, while position −0.55 mm indicates the lower boundary of the photic zone. The constructed profiles were obtained in light (large open circles), 1 s after darkening (large filled circles), 2 s after darkening (medium filled circles) and 3.8 s after darkening (small filled circles). Letters in the left panel indicate the positions of the continuous recordings presented in the right panel. (From Glud, R.N., *et al.*, *J. Phycol.*, **28**, 51 (1992). With permission.)

demonstrate that, in order to measure the true distribution of photosynthesis, the O_2 decrease immediately after onset of darkness has to be obtained.

A rough estimate of the spatial resolution obtainable in gross photosynthetic measurements can be deduced from the following equation: $\delta = (2Dt)^{0.5}$; where δ is the distance O_2 molecules move away from a given depth horizon in the time interval t, given a diffusion coefficient of D [87]. This means that within a period of 1 s at 20 °C ($D = 2.0 \times 10^{-9}\, m^2\, s^{-1}$), 32 % of the O_2 molecules would move more than 63 μm [84]. Gross photosynthetic measurements in two adjacent layers are therefore not independent if the applied dark period is too long compared with the specified spatial resolution. For measurements of gross photosynthesis it is therefore crucial to apply fast sensors (and measuring instruments) [85,86].

2.5 HYDROSTATIC PRESSURE EFFECTS

2.5.1 General Discussion

Interface studies or *in situ* respirometry are usually performed at constant but possibly elevated hydrostatic pressures, while water column measurements are conducted through a hydrostatic pressure gradient. Proper calibration of the sensor signals requires a thorough understanding of how hydrostatic pressure affects the electrode signal.

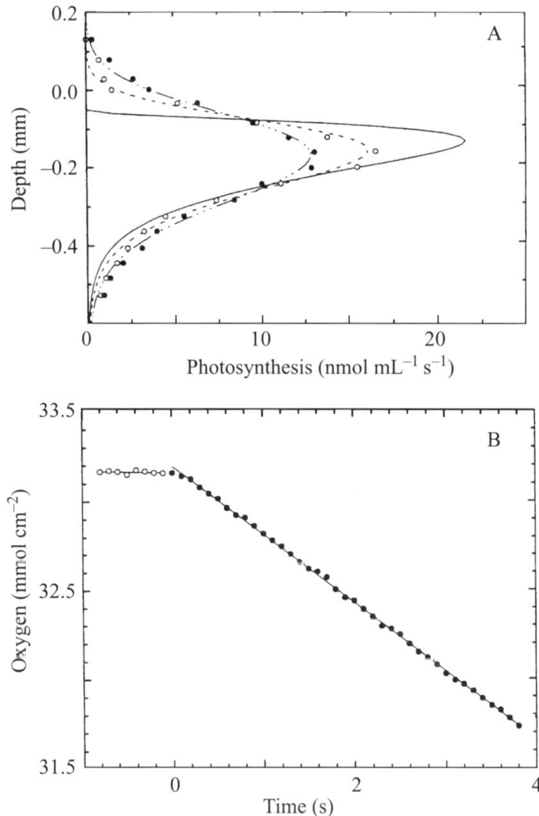

Figure 16. (A) Measured and simulated microprofiles of gross photosynthesis in a cyanobacterial microbial mat. Data recorded after 1.1 (open circles) and 2.6 s (filled circles) of darkness are displayed together with the equivalent simulated profiles, the simulated activity profile after 0.0 s is shown as a continuous curve (for the simulation procedure, refer to the original manuscript). (B) The depth-integrated O_2 concentration within the photic zone during a light–dark shift (time 0.0 s). The linear decrease indicates a constant O_2 consumption rate from the onset of darkness. Open symbols were recorded in light, while the filled symbols were recorded in darkness. (From Glud, R.N., *et al.*, *J. Phycol.*, **28**, 51 (1992). With permission.)

For understanding the effects of pressure on electrodes it is important to realize that membrane-coated sensors respond to the partial pressures (actually the fugacity of O_2), rather than O_2 concentrations. Various parameters of importance for the signal of an O_2 electrode will change with increasing hydrostatic pressure: (I) the O_2 partial pressure (fugacity), (II) the membrane permeability, (III) diffusion coefficient of O_2, (IV) the activation volumes at the cathode and anode.

Since the dissolved O_2 is generally not in direct equilibrium with air at any given water depth the partial pressure of O_2 is not defined. It is therefore more appropriate to use the term O_2 fugacity. The O_2 fugacity change (f) as a function of hydrostatic pressure is described by the following equation:

$$f = f_0 \exp(v\Delta P/RT) \tag{4}$$

where f_0 is the fugacity at the water column surface, v is the partial molar volume of O_2, ΔP is the hydrostatic pressure change, R is the gas constant, and T is the absolute temperature [88]. A laboratory study obtained an excellent match between the output of a DPX-coated cathode microelectrode exposed to hydrostatic pressures between 0.1 and 60 MPa (1–600 bar) and equation (4) (Figure 17) [89]. This indicates that for that sensor type the O_2 permeability of the DPX membrane was insensitive to hydrostatic pressure.

For Clark-type macrosensors it is a common observation that the sensor signal decreases with increasing hydrostatic pressure, and the effect is ascribed to decreasing permeability of the sensor membrane [90–92]. In order to avoid disruption of Clark-type O_2 sensors during deep-sea application (or pressure testing) the electrolyte is pressure compensated over a flexible membrane [44]. The Clark-type microsensor exhibits a close to linear signal decrease with increasing hydrostatic pressure (Figure 18A), as was found for macrosensors

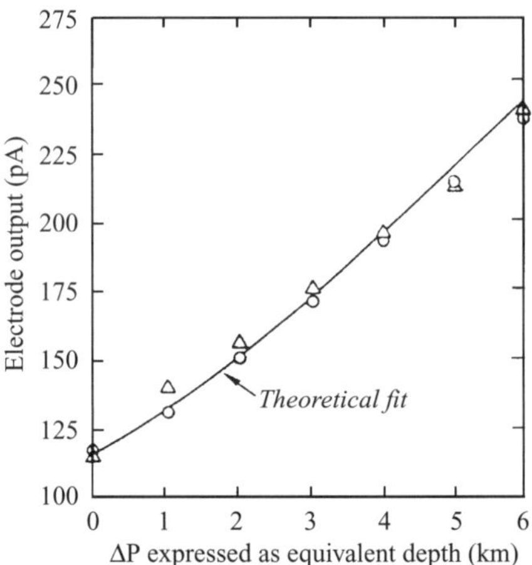

Figure 17. The hydrostatic pressure response of a cathode-type O_2 microelectrode equipped with a DPX membrane. Two replicate measurements were performed in two subsequent pressure cycles (circles and triangles). The output signal is in picoamperes (p_A). The theoretical response as predicted by Equation (2) is indicated by a solid line. (An erratum has later made a slight correction to the theoretical fit, but without importance in the present context.) (Reproduced from *Deep-Sea Res.*, **34**, Reimers, C.E., 2019 (1987), with permission from Elsevier Science.)

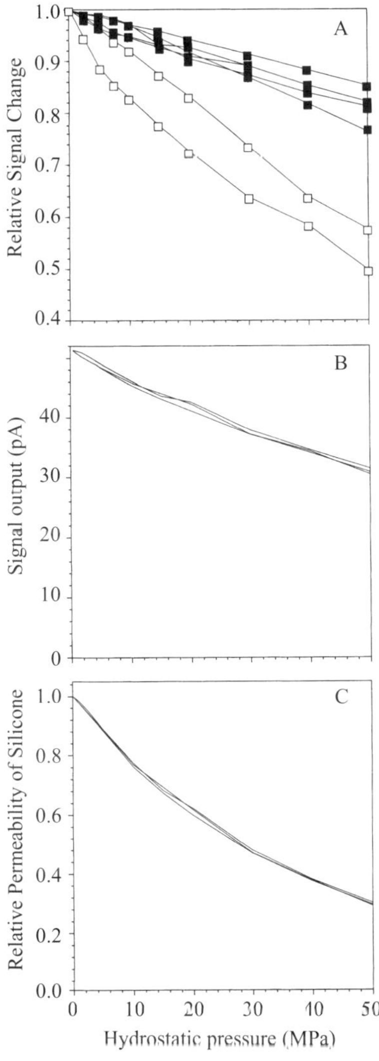

Figure 18. (A) Changes in the relative signal of six different Clark-type O_2 microelectrodes with increasing hydrostatic pressure. The two most pressure sensitive sensors also had the longest silicone membrane. (B) The response of a microelectrode (in picoamperes during three replicate pressure cycles and (C) the decrease in silicone permeability that was required in order to explain the observed sensor response solely by permeability changes

Generally, we observe that microelectrodes with a relatively thick silicone membrane show a stronger signal decrease with pressure. This suggests that the effect is related to a decrease in membrane permeability, which exceeds the effect of the partial pressure increase. Hysteresis effects are seldom seen and repeated pressure cycles reveal very reproducible sensor responses (Figure 18B). The change in membrane permeability that is required to give the observed sensor response can be calculated by applying the model described in section 2.2 (Figure 18C). The estimated permeability decreases exponentially with increasing pressure. The pressure at which the permeability is 50 % of the value at atmospheric pressure is around 40 MPa (400 bar) for all sensors tested, although the absolute permeability differs amongst electrodes. This is likely to be related to variability in the physical characteristics of the silicone membranes (aging, hydration etc.) and the fact that the geometry of a given sensor is difficult to measure exactly. When applying a Clark-type O_2 microsensor under variable hydrostatic pressure, it is therefore recommended to perform an empirical determination of pressure response of each sensor and to include the results in the calibration procedure. For interface studies or respirometry at elevated hydrostatic pressures the sensor calibration should preferably be performed *in situ* [44,72]. Measurements in pressure chambers have shown that the zero current of the Clark-type O_2 microsensors is independent of hydrostatic pressure (data not shown). For commercially available macrosensors, standardized production procedures ensure that the empirical relationship describing hydrostatic pressure effects on the sensor signal can be incorporated in the calibration software which is usually purchased along with the sensors [92,93].

The diffusion coefficients decrease along with the viscosity at elevated hydrostatic pressures. However, the effect is minimal (4 % for ΔP of 60 MPa), and does not have a major effect on the sensor output. For a commercially available O_2 macroelectrode the hydrostatic pressure effects on activation volumes were calculated to be of minor importance [91], and a study concluded that change in membrane permeability was the controlling factor for the electrode signal at hydrostatic pressures up to at least 100 MPa [91].

2.5.2 Case Study 3: Deep-sea Measurements (*In Situ* and On Site)

Deep-sea interface studies are performed at constant but very high hydrostatic pressures. However, if the electrodes are protected by a pressure compensating system between electrolyte and ambient water, they remain fully functional [44]. Figure 19A presents O_2 profiles measured over the sediment water interface at stations along a depth transect in the SE Atlantic [72]. The profiles reflect the gradual increase in O_2 penetration depth as a result of the reduced sedimentation of organic material in deep-sea environments (and to much lesser extent an increase in the O_2 concentration of the bottom water). Owing to the extreme O_2 penetration depth at the deepest investigated site the microelectrode

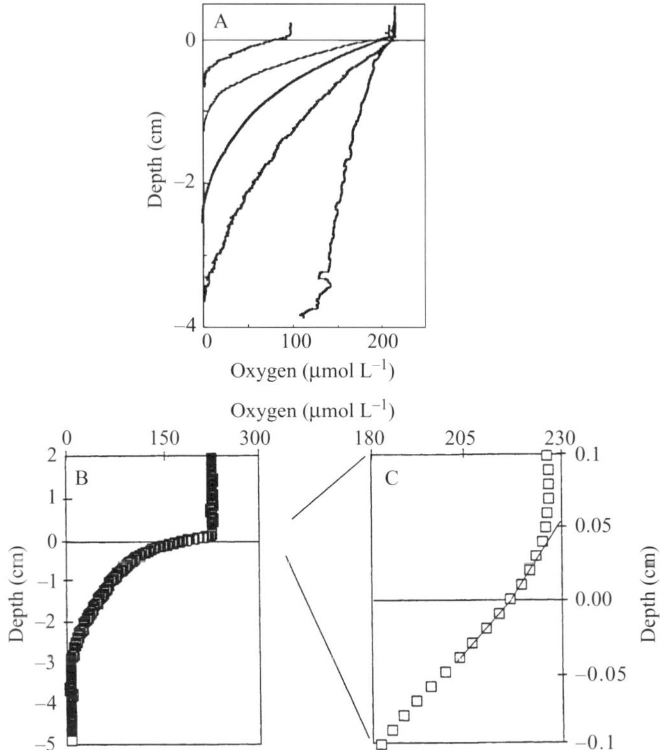

Figure 19. (A) *In situ* O_2 microprofiles obtained at a spatial resolution of 100 µm along a depth transect in the SE Atlantic. Profiles show a gradual increase in O_2 penetration with the water depth: 600, 1743, 1960, 3100 and 4986 m. (B) The profile obtained at 3100 m with an enlargement of the sediment water interface (horizontal line). (Reprinted from *Deep-Sea Res.*, **41**, Glud *et al.*, 1767 (1994), with permission from Elsevier Science.)

broke before reaching the anoxic horizon. Figure 19B presents a magnified view of the interface measurements performed at 3100 m water depth and demonstrates the high spatial resolution required to resolve the O_2 gradients within the DBL. In order to obtain complete microprofiles and high resolution DBL measurements simultaneously, it may be necessary to equip the instrument package with a combination of large robust sensors (for deep profiles) and thin sensors for the DBL measurements [94]. In order to resolve the very small concentration change across the DBL in the deep sea, it is necessary that the applied sensors have practically no stirring sensitivity (see section 2.3).

Parallel *in situ* and on site measurements have demonstrated a significant change in the surficial O_2 distribution after recovery of deep-sea sediment cores.

For instance, the 'on site' O_2 penetration depth was only 20% of the *in situ* value while the calculated sediment O_2 uptake was 3.5 times higher in the laboratory as compared with *in situ* [72]. The effects are probably related to transient heating during core recovery and pressure-release effects on specially adapted meio- and microfauna [72,95]. These findings emphasize the need for *in situ* O_2 measurements to obtain realistic estimates of the benthic distribution and exchange of O_2 in the deep sea.

2.6 STABILITY, LIFETIME, AND INTERFERENCES

It is difficult to give any absolute values for the stability and lifetime of various O_2 microelectrodes, since they are very much dependent upon the environmental conditions. Under laboratory conditions [32] it has been found that for cathode-type sensors the signal at air saturation and in anoxia varied by 9.3% and 1.3%, respectively ($n = 7$) over a 30 d period. Also under laboratory conditions, we have measured the average drift in the air saturation signal of Clark-type O_2 microsensors to be in the order of 2% d^{-1} ($n = 25$). However, this drift may to some extent be due to small changes in temperature and air pressure. As estimated from equation (1) a signal change of 2% could be the result of a minor temperature change of <1 °C. It is often observed that after polarization the sensor signal as well as the zero-current show an initial decrease. This is associated with an initial reduction of O_2 dissolved in the electrolyte, and it is therefore recommended to keep Clark-type O_2 sensors polarized for a few hours prior to application. The most common reason for a significant signal drift during measurements is malfunctioning or leaky membranes [59]. The lifetime of both cathode-type and Clark-type O_2 microelectrodes is usually in the order of months to years unless they are mechanically damaged or 'poisoned' by interfering compounds [96].

The ion-permeable membrane-coated O_2 cathode sensor, which has an external reference is an open system, and solutes in the environment can therefore potentially interfere with the measuring circuit. Cathode sensors can be poisoned by ions, in particular, Ca^{2+} and Mg^{2+}, presumably due to precipitation of hydroxides and carbonates at the cathode surface and in the membrane [16,32]. This process is stimulated by the alkaline microenvironment generated by the reduction of O_2 to OH^-. Dissolved organic material may in a similar manner change the cathode surface and the structure of the membrane. The result is a non-linear and a drifting calibration curve. For these reasons cathode sensors with hydrophilic membrane coatings are not optimal for *in situ* O_2 measurements. To our knowledge no detailed studies on the interference of various agents on hydrophilic-coated cathode sensors have been performed. However, for *in situ* measurements such sensors are gradually being replaced by Clark-type microelectrodes (or microoptodes—see below).

For microsensors with hydrophobic membranes the only major interfering agent is H_2S, which can pass the gas-permeable membrane and interfere with the electrochemical reactions of the sensor. The result is usually an increase and drift of the signal. Precipitation of various sulfur–metal complexes within the sensor tip, which change the catalytic surface as well as the membrane structure, can be observed [59]. In the worst case H_2S-poisoned sensors lose their sensitivity to O_2. The mechanism of H_2S poisoning of microelectrodes, however, is very complex and not fully understood [59]. It is noticeable that for cathode-type sensors it has been observed that partly poisoned electrodes, with a reduced sensitivity to O_2, become insensitive to H_2S and can subsequently be used for O_2 measurements within an H_2S gradient [20]. The reason for this behavior is still unclear. It is, however, important to note that, for most relevant habitats, the H_2S and O_2 containing horizons are well separated spatially.

A sulfide-resistant macro-O_2 electrode based on a gold cathode, and with an alkaline sodium sulfide solution has been developed [59]. The anode is Ag/Ag_2S, whose potential is approximately −0.7 V. The cathode is kept at a fixed potential of −0.1 V versus this reference electrode. The principle has been tested on a microscale; the lifetime of such sensors is limited (approximately 1 week) owing to precipitation of S^0 at the sensor tip [56]. In general, interfering agents are a minor problem for *in situ* application of Clark-type O_2 microelectrodes.

3 OPTICAL MICROSENSORS (MICROOPTODES)

Recently a new generation of microsensors for *in situ* O_2 measurements has been introduced to aquatic science. The sensors are not based on any electrochemical principle, but take advantage of an optochemical technique. The so-called microoptodes were developed by minaturizing and optimizing sensors that previously had been applied in other fields of science [97–99]. Such microsensors were recently tested for their potential for *in situ* applications [100]. Needle-supported minioptodes (diameter 1–3 mm), and macrooptodes applying the same principles, have been developed for applications in water columns, coarse sediments and for measuring very deep O_2 profiles [94,101,102]. The development of optodes, and associated electronics for *in situ* applications in aquatic environments, is still in its early stage. So far only measuring systems where the fluorescence intensity is used as analyte-dependent parameter have been realized. However, fluorescent lifetime-based sensing schemes have great potential and will probably lead to further advances in coming years (see Section 3.4).

3.1 CONSTRUCTION AND FUNCTION

The basic principle of O_2 microoptodes is dynamic fluorescence quenching [103]. In the absence of O_2 the fluorophore absorbs light at a given wavelength and releases the absorbed energy by emitting fluorescence with a defined intensity and lifetime. However, in the presence of O_2, quenching of the fluorophore results in a decrease of the fluorescence light intensity as well as in the fluorescence lifetime.

The first microoptode was based on a ruthenium complex (ruthenium(II) tris-4–7–diphenyl-1,10–phenanthroline perchlorate) Ru(diph)$_3$, which was dissolved in polystyrene (PS) [104] (Figure 20). Titanium dioxide particles (diameter 1 μm) were added to the indicator–polymer mixture in order to enhance the mechanical stability and to provide more efficient scattering of the emitted light. An optical silica fiber with an outer diameter of 140 μm was tapered to a final diameter of 30 μm by heating. Subsequently the fiber tip was dip-coated with the indicator–matrix cocktail. After evaporation of all solvent (approximately 24 h), an additional layer of black silicone was coated onto the sensor tip in order to shield out ambient and backscattered light during measurements. A shielding is also required in order to avoid stimulation of any potential phototrophic activity in the investigated communities. The applied fluorophore has a large Stoke shift with an optimal absorption around 450 nm and an emission wavelength of 610 nm. This ensures that relatively cheap glass filters can be applied in the measuring set-up. The shaft was mechanically supported by a needle or by a glass capillary, and the tip diameter was approximately 40 μm (Figure 20). Physically the tip diameter could be made smaller, but this resulted in unacceptably low signal levels [104].

Besides the Ru(diph)$_3$–PS reactive layer, numerous other combinations of fluorophore and immobilization materials have been tested for their potential use in macro- and microoptodes [105–108]. Different chemically reactive layers have various advantages in relation to stability, response time, and sensitivity range, and selecting the optimal combination for a given task is often a trade-off between the various characteristics of the fluorophore and the

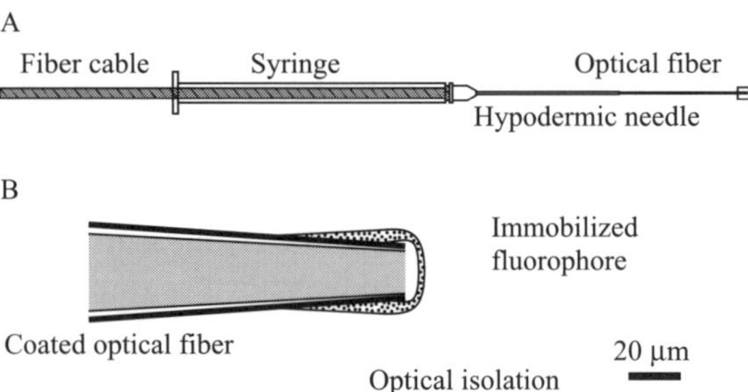

Figure 20. (A) Schematic presentation of an intensity based microoptode with an optical insulation layer. (B) Enlargement of the sensor tip (From Klimant *et al*., *Limnol. Oceanogr*., **40**, 1159 (1995). Reproduced by permission of American Society of Limnology & Oceanography.)

immobilization material [108]. The following criteria should in general be met by fluorophores and immobilization materials applied for intensity-based microoptodes: (I) the fluorophore should be excitable by a light emitting diode (LED), (II) the fluorophore must be sufficiently photostable, (III) the fluorophore should have a relatively large Stoke shift, (IV) the fluorophore should be non-soluble in water to ensure minimal dye leaching, (V) the fluorophore should be soluble in the hydrophobic polymer used for immobilization, (VI) O_2 should be highly soluble in the immobilization material, (VII) the immobilization material should be mechanically stable under the measuring conditions and adhere well to the fiber. Soft polymers such as silicone, plasticized poly(vinyl cloride) (PVC), or cellulose derivates do not adhere very well and are easily mechanically damaged during sensor application. Mechanically stable hydrophobic materials such as PS or poly(methyl methacylate) (PMMA) appear to be better suited as immobilization materials [108]. Recently sol–gels (ormosils) have proven to be robust and stable as immobilization materials [109]. Sol–gels are condensed silica alkoxides, which form a highly permeable non-crystalline structure. The addition of organic-substituted precursors may increase the flexibility and lower the brittleness of the material [110]. Table 2 compiles the characteristics of some chemically reactive layers that have been tested for use in microoptodes.

Table 2. Composition and selected properties for sensing material, which have been tested for microoptodes (modified from Klimant *et al.*, 1997)

Fluorophore (luminophore)/ Immobilization material	Signal	Response time (t_{90})	Sensitivity*	Excitation/ Emission light (nm)	Photostability/ Mechanical stability
$Ru(diph)_3$ / PS†	very high	< 2 s	22%	450/600	very good/ very good
$Ru(diph)_3$ / PVC	very high	< 200 ms	59%	450/600	poor /poor
Pt-OEP‡/PS	high	< 2 s	80%	400 and 535/640	moderate/very good
Pt-OEP/PMMA§	high	< 5 s	35%	400 and 535/640	moderate/good
Pt-OEKP¶/PS	moderate	< 2 s	75%	400 and 592/760	good/very good
Pd-OEP**/PS	moderate	< 2 s	98%	400 and 545/670	poor/very good

* Signal decrease following transfer from nitrogen to air.
† Polystyrene.
‡ Platinum(II)-octaethylporphyrin.
§ Poly (methyl-methacrylate).
¶ Platinum(II)-octaethylketoporphyrin.
** Palladium(II)-octaethylporphyrin.

$Ru(diph)_3$ is a fluorophore which meets all of the criteria mentioned above and has been selected by various research groups as the optimal indicator for *in situ* application [94,100,101,111]. As immobilization materials, both PS and sol-gels have proven to be reliable, and have been applied for *in situ* optode measurements [100,101].

3.2 CALIBRATION

The signal of optodes decreases non-linearly with increasing oxygen concentration, in contrast to oxygen electrodes (Figure 21). For an ideal dissolved fluorophore the calibration curve can be linearized using the Stern–Volmer equation [112]:

$$\frac{I_0}{I} = 1 + k_{sv}c \qquad (5)$$

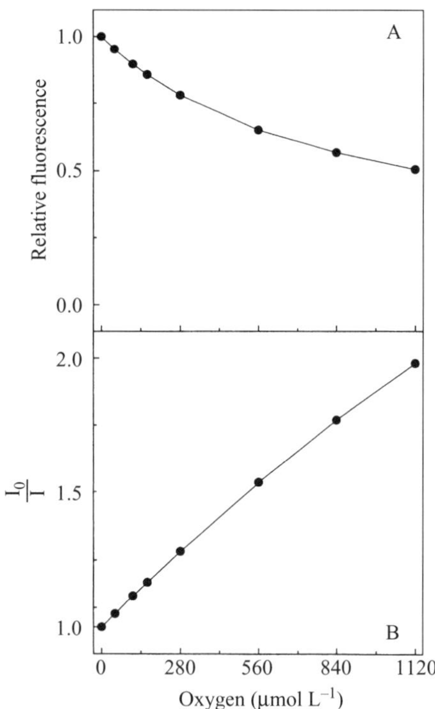

Figure 21. The calibration curve of an $Ru(diph)_3$ – PS – based microoptode (A) and the corresponding Stern–Volmer plot (B). (From Klimant *et al.*, *Limnol. Oceanogr.*, **40**, 1159 (1995). Reproduced by permission of American Society of Limnology & Oceanography.)

where I_0 and I are the fluorescence intensity in the absence and presence of oxygen, respectively, c is the oxygen concentration in the solution and k_{SV} the quenching coefficient. However, in reality O_2-sensitive fluorophores, which are dissolved in a solid matrix, exhibit non-linear Stern–Volmer calibration curves. A general calibration equation based on a two-site O_2 quenching model, which accounts for the non-ideal behavior of immobilized fluorophores, has been suggested [113]. The equation is, however, relatively complex and requires multiple calibration points, so for most practical applications a modified Stern–Volmer equation (which adequately describes the calibration of micro-optodes) has been suggested [104]:

$$I = I_0\left(\alpha + (1-\alpha)\frac{1}{1+k_{sv}c}\right) \tag{6}$$

In practical applications it has been shown that by applying an α value of around 0.15 this empirically derived equation describes the calibration curve of the Ru(diph)$_3$–based microoptode adequately [104]. The sensor signal is proportional to the O_2 partial pressure of the ambient water and proper sensor calibration requires only two calibration points, typically the readings obtained in anoxic and air-saturation conditions. Empirical relations based on polynomial regressions have also been applied for calibration of Ru(diph)$_3$–based mini- and microoptodes [101, 111]. Microoptodes have, as a result of the non-linear calibration curve, increased O_2 sensitivity in the range close to anoxia. Additionally, as a result of higher signal at low O_2 concentrations the signal to noise ratio improves, allowing for a better signal resolution. These advantages have been used for microoptode-based investigation of microaerophilic activity [111].

The signals of Ru(diph)$_3$–based optodes are not affected by changes in salinity [101, 104]. However, the fluorescence signal is dependent upon temperature, and ruthenium(II)-based complexes embedded in O_2-impermeable materials have actually been used for temperature sensing [114,115]. For PVC-immobilized fluorophores the I_0 value decreases almost linearly with temperature (0.5% $°K^{-1}$), while k_{SV} increases by approximately 1% $°K^{-1}$ of temperature increase [116]. The net result is a decrease in O_2 sensitivity with increasing temperature which has to be accounted for during sensor calibration [101]. Preliminary data suggest that the temperature response is very sensor specific and that the temperature effect has to be quantified for each individual sensor [101,117]. In the case where optodes are used at constant temperatures the calibration should preferably be performed at the *in situ* temperature [100]. Temperature sensitivity is, however, very dependent on the immobilization material, and temperature-compensated microprobes applying combinations of fluorophore sensing chemistries can potentially be used for *in situ* work [108, 115].

3.3 CAPABILITIES AND LIMITATIONS

We still have very limited experience with *in situ* application of microoptodes and do not fully understand the details of their functioning under *in situ* conditions. However, their potential is promising and their main advantages seem to be a much simpler manufacturing procedure, no analyte consumption, and a superior stability to that of the Clark-type O_2 microelectrodes. Their main disadvantages are limited experience in their use, slower response times and a relatively high noise to signal ratio for the first prototypes. Still being a relatively new technique for the aquatic environment further instrumental optimization is probably required before the technique will be more generally applied.

3.3.1 Spatial and Temporal Resolution

In order to ensure sufficient fluorescence the minimal tip diameter of micro-optodes is approximately 25 μm (alternatively very elaborate amplification instrumentation is required) [104]. The spatial resolution of microoptodes is consequently no better than around 50 μm (twice the outside tip diameter). Comparisons of microprofiles obtained with Clark-type O_2 microelectrodes (tip diameter < 10 μm) and microoptodes (tip diameter > 25 μm did not reveal any significant difference at a spatial resolution of 50 μm [104]. The larger tip size of microoptodes probably does not affect porewater measurements, but invasive measurements in, for example, aggregates, symbionts, or around individual microbiota may be problematic.

Microoptodes do not consume the analyte and consequently they have no stirring sensitivity. As discussed in section 2.3, this can be an advantage. The sensor signal is dependent upon a thermodynamic equilibrium, and the response time is significantly longer than for standard O_2 microelectrodes (Table 2). This is a consequence of the relatively thick sensing layers (which includes the optical isolation) and of the low O_2 permeability of mechanically stable immobilization materials. Microoptodes can be made with a faster response time by using materials with a higher permeability and a lower O_2 solubility (e.g. PVC) and by using thinner sensing layers. This, however, will be a trade-off with lower sensor stability and lower fluorescent signals (Table 2). In general, microoptodes are therefore not as suitable as microelectrodes for gross photosynthesis measurements or for studies on DBL O_2 dynamics.

3.3.2 Hydrostatic Pressure Effects

The relationship between hydrostatic pressure and the optode signal was invest-igated for $Ru(diph)_3$–polystyrene-based microoptodes [100]. The fluorescent

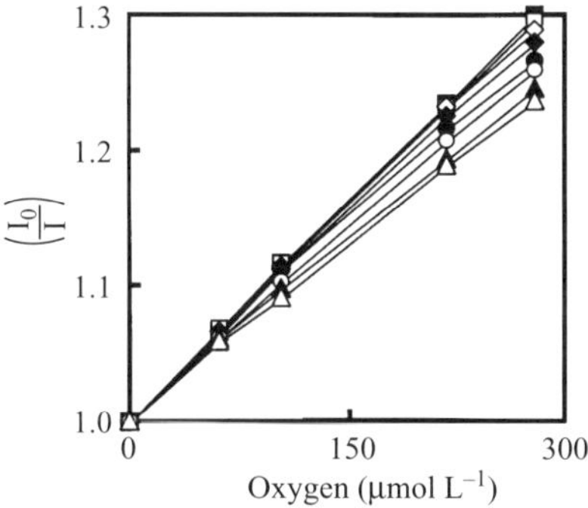

Figure 22. The effect of hydrostatic pressure on the Stern–Volmer plot of a Ru(Diph)$_3$ – PS – based microoptode. Measurements were performed at: 0.1 MPa (1 atm) (■–■); (5 MPa (50 atm) (□–□); 10 MPa (100 atm). (◇–◇); 20 MPa (200 atm.); (◆–◆); 30 MPa (300 atm) (●–●); 40 MPa (400 atm). (○–○); 50 MPa (500 atm) (▲–▲);60 MPa (600 atm). (△–△). A gradual decrease in O_2 sensitivity is apparent. Reprinted from *Deep-sea Res.*, **46**, 171 (1999) Glud *et al.*, with permission from Elsevier Science

signal in the absence of O_2 (I_0) showed a slight linear increase in the range of 0.1 – 60 MPa. The effect was ascribed to mechanical stress on the light guidance from the fluorophore to the measuring equipment. However, as the O_2 concentration of the ambient water was increased the effect of hydrostatic pressure became more pronounced and reached a maximum of 0.02 % atm^{-1} at surface air-saturation (301 μmol L^{-1}). The decreased quenching at elevated pressures caused a 20 % reduction in the O_2 sensitivity at a hydrostatic pressure equivalent to full ocean depth (60 MPa) (Figure 22). The sensors remained fully operational, and for *in situ* work at constant hydrostatic pressure, the effect is of minor importance, if calibration is performed *in situ* [100].

The decreased sensitivity is most likely related to a decrease in the O_2 permeability of the PS, but effects on the quenching process itself cannot be excluded. Microoptodes based on other immobilization materials (PVC and sol–gels) also express reduced sensitivity with increasing pressures [118]. Future applications of optodes for water column studies require a quantitative understanding of how the hydrostatic pressure affects the fluorescence signal.

3.3.3 *In Situ* Tests on a Benthic Lander

The first *in situ* measurements with microoptodes were recently performed [100] on a profiling lander and on a benthic chamber lander. Measurements were performed in a coastal sediment. The porewater profiles of the microoptodes were very similar to the profiles simultaneously obtained by Clark-type O_2 microelectrodes (Figure 23A). The optode measurements performed by the benthic chamber, however, appeared to be more 'noisy' than the comparable

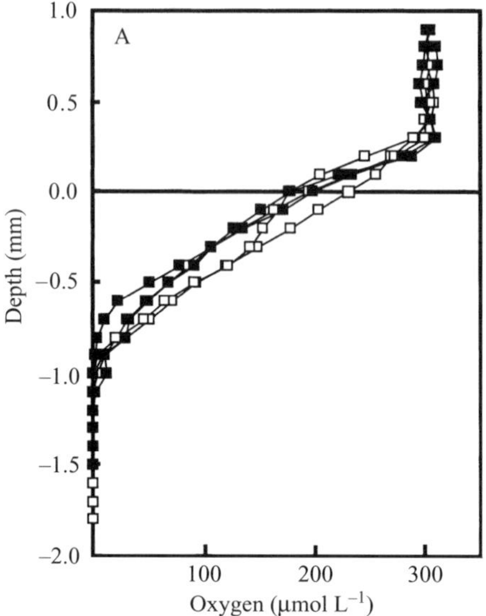

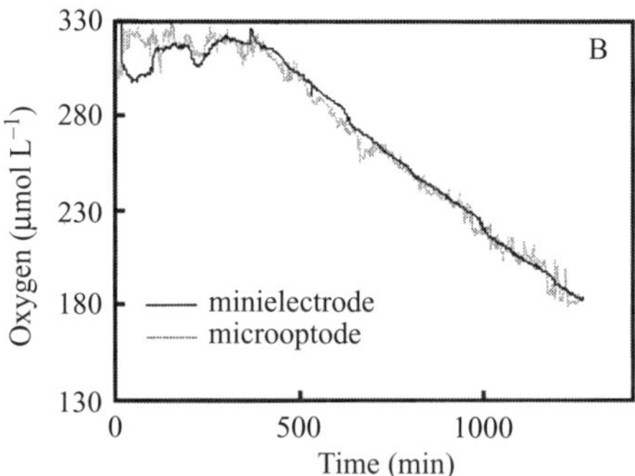

Figure 23. (A) *In situ* microprofiles measured simultaneously by Clark-type O_2 microelectrodes (open squares) and O_2 microoptodes (filled squares). Horizontal line indicates the estimated sediment surface. (B) Continuous recordings of a Clark-type O_2 mini-electrode and O_2 microoptode during an *in situ* incubation of a benthic chamber. The lid was closed approximately 450 min after deployment. Reprinted from *Deep-sea Res.*, **46**, 171, (1999) Glud *et al.*, , with permission from Elsevier Science

electrode measurements (Figure 23B). For the benthic chamber it was necessary to have a 1.5 m long optical fiber between the sensors and the measuring electronics [100]. The fluctuations in signal were most likely caused by bending of the fiber cable induced by water movements at the seafloor. This effect can be reduced by mechanical improvements, and be eliminated by lifetime-based measurements (see below).

3.3.4 Stability and Interference

Laboratory investigations have shown that after an initial phase (1–2 d) in which the sensitivity decreased (probably related to evaporation of solvents and material curing) the calibration characteristics of microoptodes remained constant for $>$ 1 year. No significant drift was observed during continuous measurements with $Ru(diph)_3$–based optodes for periods up to 50 h [104]. Any signal drift is likely to be associated with photodegradation of the fluorophore. This can be reduced by exciting the fluorophore only at the frequency required for the measurements [100]. The excellent long-term stability of microoptodes can be further improved by using the fluorescence lifetime rather than the fluorescence intensity as a signal carrier [119]. The O_2 quenching is extremely specific for $Ru(diph)_3$–PS-based microoptodes, and interference by solutes of relevance for benthic *in situ* studies, i.e. pH, CO_2, H_2S, and heavy metals, has not been detected [104].

3.4 INTENSITY VERSUS LIFETIME-BASED MEASUREMENTS

The collisional quenching of molecular O_2 has two effects on the excited immobilized fluorophore: the fluorescent intensity is decreased and the lifetime of the fluorescence is shortened [120]. So far only the intensity-based measuring scheme has been developed for *in situ* O_2 measurements. The main problems associated with intensity-based measurements are the demand for a constant fluorescent signal, whereas microbending of the fibers will lead to noisy signals (see Figure 23B), and photodegradation and dye leaching will result in a loss of O_2 sensitivity. Further, in order to block out scattering and reflection effects, optical isolation of the sensor tip is required, even without any phototrophs in the surroundings.

The fluorescence lifetime signal is independent of the absolute light intensity, and thus avoids all of the potential problems related to intensity-based measurements. Lifetime-based O_2 microsensing has been realized in the laboratory [121] and for macrosensors used *in situ* [122]. The relationship between fluorescence lifetime and the partial pressure of O_2 is similar to equations (5) and (6) since:

$$\frac{\tau}{\tau_0} = \frac{I}{I_0} \tag{7}$$

where τ and τ_0 are the lifetimes of the fluorescence in the presence and absence of O_2, respectively [123]. For $Ru(diph)_3$-PS-based sensors τ_0 is around 4.5 μs while τ in 100 % O_2 is approximately 2 μs. These values are in a range still accessible by standard techniques for resolving lifetimes [121,123]. Other indicator molecules with longer lifetimes such as phosphorescent palladium and platinum porphyrins with lifetimes that are 20–100 times longer may, however, be considered for future lifetime-based *in situ* applications [108,124].

There are three principal techniques for determining luminescence lifetimes: (i) frequency domain method, (ii) ratioing method, (iii) and time domain method. Below we briefly present the various techniques but refer to the references mentioned below for a more thorough discussion of their various applications.

In the frequency domain method the fluorophore is excited with sinusoidally modulated light. The emitted light will reflect this modulation with a phase angle (φ), which is relatively easy to measure for a properly selected modulation frequency (f_{mod}). It is related to the fluorescence lifetime by: $\tan \varphi = \tau 2\pi f_{mod}$. This approach has been successfully applied in the laboratory for $Ru(diph)_3$–PS-based microoptodes [121, 123] and for water column, *in situ* measurements by applying a macrooptode based on a modified ruthenium complex immobilized in a sol–gel [122].

The ratioing method is based on a rectangular modulation of the excitation light (on/off), and the subsequent rise and decay of luminescence intensity contain information about the luminescence lifetime and can be quantified through simple ratioing of the signals in light and darkness. The method has been combined with imaging for medical applications [125].

In the time-domain method the immobilized fluorophore is exposed to an excitation pulse. Subsequently, the fluorescence intensity is quantified in well-defined time windows after the eclipse of light, and the fluorescence lifetime can be calculated from the exponential decay in fluorescence intensity. This approach has very recently been applied in combination with planar sensing for resolving two-dimensional O_2 distribution at benthic interfaces [126] (see section 3.5).

Lifetime-based methods have a promising potential for *in situ* O_2 measurements. However, the necessary measuring electronics are still relatively sophisticated and expensive [123], and further optimization and miniaturization is required before lifetime-based microoptodes can be applied *in situ*.

3.5 TWO-DIMENSIONAL OPTICAL SENSING

Microsensors measure the O_2 concentration at a single point and depth profiles are obtained by stepwise movement of the sensors. The oxygen distribution in many benthic communities is characterized by significant variations in time and space [127, 128]. Applying microsensors in order to describe or to overcome such variability may be an overwhelming task, and in bioturbated sediments, rhizospheres, biofilms, microbial mats etc. it can be virtually impossible. This has led into the development of sensor arrays that can apply up to eight microoptodes simultaneously within an area of approximately 10 cm^2 [123]. However, the horizontal resolution of such approaches is rather coarse and the number of microsensors that can be operated simultaneously is very limited. To overcome these problems, techniques based on medical applications of planar sensors combined with imaging were recently introduced in aquatic biology [116].

Instead of fixing the $Ru(diph)_3$–PS and the necessary optical insulation on the tip of a fiber (Figure 20) a transparent support foil was coated with the active gel layer. The 200 μm thick foils were glued onto one of the glass plates making up the side of a flume aquarium. The flume was then filled with sediment or microbial mats in such a way that the foil was partly covered (Figure 24). The excitation light was supplied by a halogen lamp equipped with an appropriate excitation filter. A CCD camera equipped with a 50 mm macro-lens and an emission filter recorded the two-dimensional distribution of fluorescent light (Figure 24). The images are calibrated using equation (6), but owing to inhomogeneities in the sensor and the applied light field each of the 3.4×10^5 measuring points making up the images was individually calibrated (the CCD chip contained 1317 × 1035 pixels, but images were obtained with a binning factor of 2). In the configuration described the calibrated oxygen images covered 13 × 17 mm and the oxygen concentration is expressed on an 8 bit scale at a spatial resolution of 26 × 26 μm. The planar optodes can resolve the O_2 distribution better than any traditional microsensor approach and allow for more detailed studies of the benthic O_2 dynamics (Figure 25A). The spatial resolution and the area covered by an image can be regulated by attaching appropriate optical lenses to the CCD camera. Sensor characteristics are the same as for the equivalent microoptodes described above. To what extent the placement of planar optodes affects the O_2 distribution at the investigated interface still remains to be studied in greater detail [116, 129]. Since the first demonstration of the measuring approach the technique has been successfully applied in other biological systems [129, 130].

Recently an optimized camera system applying lifetime-based imaging has been successfully combined with planar oxygen sensors for benthic interface studies [126]. The approach has interesting potential and offers many advantages compared with the intensity-based approach. The optical isolation layer

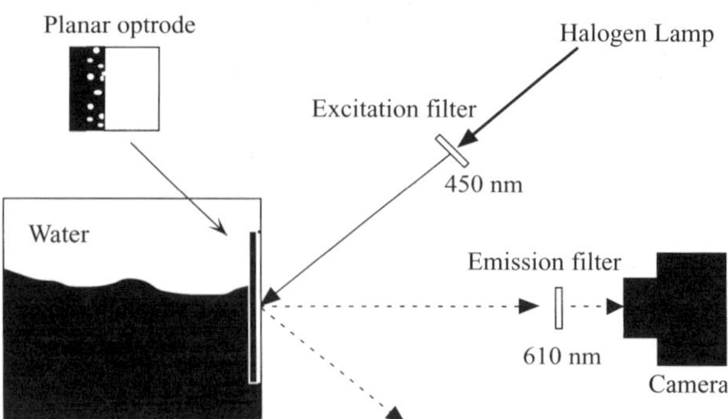

Figure 24. Basic set-up for the first planar optode measurements. An enlargement of the three-layered planar optode (support foil, sensing chemistry with Ti-granules and black silicone) is included. (Redrawn from Glud *et al.*, *Mar Ecol. Prog. Ser.*, **140**, 217 (1996))

can be omitted, which allows direct visual inspection of the sample during oxygen measurements [110].

So far the planar optode technique has been used only in the laboratory and requires further optimization before *in situ* applications can be realized. However, camera systems, based on the principle of inverted periscopes, have for some years been applied *in situ* for obtaining vertical images of biogenic structures in surficial sediments [131, 132]. Combining such camera systems with planar optodes should be possible in the coming years. This would represent a big step forward in studying the *in situ* oxygen dynamics at benthic interfaces.

4 CONCLUSIONS

A range of applicable O_2 microsensors is currently at our disposal for *in situ* investigations, and the lack of appropiate sensors is no longer a limiting factor for *in situ* investigations of O_2 dynamics. Clark-type O_2 microelectrodes probably represent the current 'state of the art'. However, microoptodes have potential advantages in areas of future *in situ* research. In general it is important to define the required sensor characteristics for a given task— and then to choose the proper sensor design. In some instances the optimal choice is to perform complementary measurements with different sensor types. Table 3 compiles the characteristics of O_2 micro- and mini-sensors discussed in this chapter with comments on prominent advantages and disadvantages.

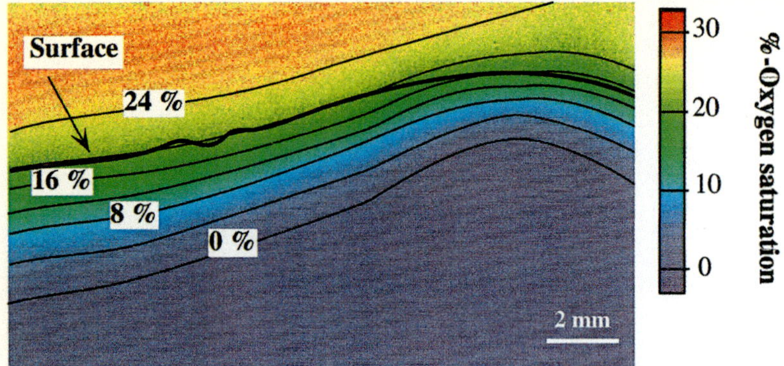

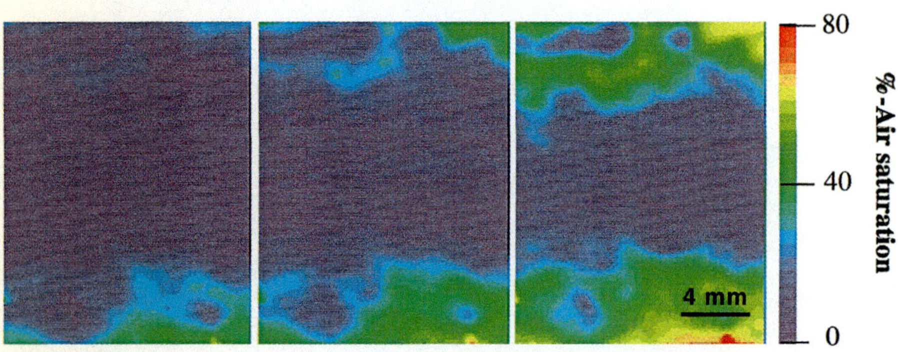

Plate 1 (Top) The O_2 distribution across the interface of an intertidal sediment as measured by an $Ru(diph)_3$–PS planar optode. Thin horizontal lines indicate the O_2 isopletes of 0, 4, 8, 12, 16, 20 and 24% air saturation. The thick line indicates the estimated position of the sediment surface. (Bottom) The O_2 distribution at the base of a 400μm thick biofilm, which was grown directly on the sensor at three different flow velocities: 20.5, 33.1 and 35.1cm s^{-1}. The better ventilation of the base film is apparent as the flow velocity is increased. (From Glud *et al., Mar. Ecol. Prog. Ser.*, **140**, 217 (1996) and Glud *et al., Aqua. Microb. Ecol.*, **14**, 223 (1998). Reproduced by permission of Inter-Research.

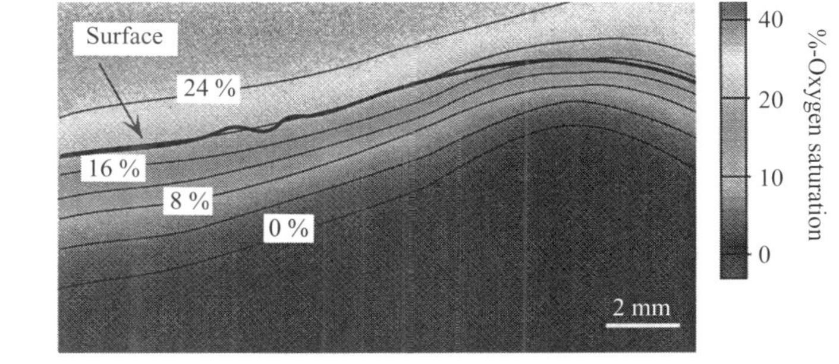

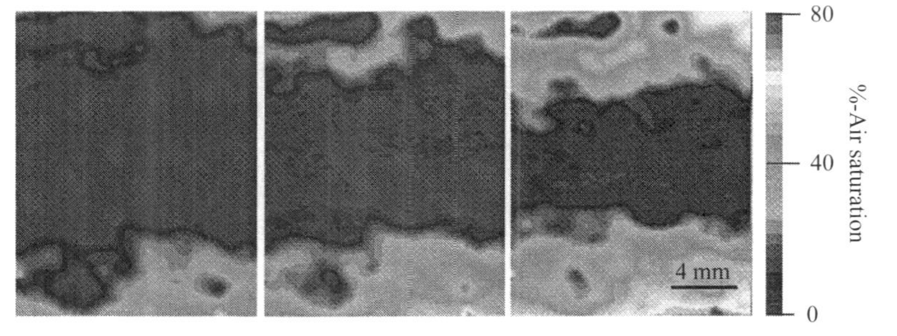

Figure 25. (A) The O_2 distribution across the interface of an intertidal sediment as measured by an Ru(diph)$_3$–PS planar optode. Thin horizontal lines indicate the O_2 isopletes of 0, 4,8,12,16,20 and 24% air saturation. The thick line indicates the estimated position of the sediment surface. (B) The O_2 distribution at the base of a 400 μm thick biofilm, which was grown directly on the sensor at three different flow velocities: 20.5, 33.1 and 35.1 cm s^{-1}. The better ventilation of the base film is apparent as the flow velocity is increased. (From Glud *et al*., *Mar. Ecol. Prog. Ser*., **140**, 217 (1996) and Glud *et al*., *Aqua. Microb. Ecol*., **14**, 223 (1998). Reproduced by permission of Inter-Research.)

Table 3. Micro- and mini-sensors relevant for aquatic *in situ* investigations

Sensor type	Outside tip-diameter (μm)	Stirring sensitivity (%)	Response time t_{90} (s)	Long-term stability	Primary application areas	Prime references
DPX-coated cathode sensor	0.2–2	2–50	0.1–2	poor	Sediments/mats/biofilms Invasive approaches	16,32, 33
Needle DPX-coated cathode sensor	700	5–50	60–120	poor	Coarse or very oxic sediments	38
Clark-type microelectrode	1–10	0–2	0.5–2	good	Sediments/mats/biofilms Photosynthesis/water column/ respirometry / invasive approaches	40,45
Clark-type minielectrode*	1000–3000	2–5	0.2	good	Water column/ respirometry	22
$Ru(diph)_3$–PS microoptode	25–50	none	2–5	good†	Sediments/mats/biofilms/ water column/ respirometry	104,108
$Ru(diph)_3$–PS mini-needle optode‡	1000	none	250	good†	Coarse or very oxic sediments	94
$Ru(diph)_3$–PS planar optode	–	none	2–5	good†	Sediments/mats/biofilms Photosynthesis	116,126

* A comparable sensor for respirometry has been developed. That sensor has a longer response time but a superior stability[23].
† For optode measurements *in situ* lifetime-based sensing will further improve the sensor stability.
‡ A comparable sensor with a sol–gel matrix has been developed. That sensor has a tip diameter of 3 mm[101].

ACKNOWLEGDEMENTS

Following funding agencies are gratefully acknowledged for their financial support: The Danish National Science Research Council, the European Commission, and the Carlsberg Foundation. Rodney Roberts is thanked for his help with improving the language.

GLOSSARY

CCD camera Charged coupled device camera.
CTD Conductivity, temperature, and depth instrument.
DPX resin resin for microelectrode (Fig 1) = n-(2-methyl-4-amino-5-pyrimidyl)-n-(1-methyl-2-thiopbut(1)en-4-diphosphatidyl) amide
Benthic In relation to the sea bottom.
Diffusion layer (DL) The region around a sensor tip where diffusion is the main transport mode.
Diffusive boundary layer (DBL) The horizon above a substratum where diffusion is the main transport mode.
Interface Boundary between sediment (or any other substratum) and water phase.
Lander Independent vehicle operating at the sea floor.
Pelagic In relation to the water column far from sea shore.
Sol–gel (ormosil) Condensed silica alkoxides.
Stirring sensitivity Change in sensor signal caused by change in ambient water flow

LIST OF SYMBOLS

(the point of first use is given in parentheses)

c Concentration
c_w Concentration in water, infinitely far away from the sensor
c_2 Concentration at the sensor surface
D Diffusion coefficient
D_e Diffusion coefficient in electrolyte
f Fugacity
f_0 Fugacity at water surface
f_{mod} Modulation frequency
I Fluorescence intensity
I_0 Fluorescence intensity in absence of O_2
J Diffusive flux
k_{SV} Quenching coefficient

p_w	Partial pressure in ambient water
ΔP	Hydrostatic pressure
R	Gas constant
r	Radius
S_e	Solubility in electrolyte
Si	Electrode signal
ST	Stirring sensitivity
T	Absolute temperature
t_{90}	90% response time
v	partial molar volume
Z_m	Membrane length
Z_e	Distance within the electrolyte
Ψ	Permeability
Φ	Current generated by O_2 reduction
φ	Phase angle
τ	Fluorescence lifetime
τ_0	Fluorescence lifetime in absence of O_2

REFERENCES

1. Winkler, L. W. (1888). Die Bestimmung des im Wasser gelösten Sauerstoffes, *Ber. Dtsch. Chem. Ges. Berlin*, **21**, 2843.
2. Scholander, P. F., van Dam, L., Claff, C. L. and Kanwisher, J. W. (1955). Microgasometric determination of dissolved oxygen and nitrogen *Biol. Bull.*, **109**, 328.
3. Benson, B. B. and Parker, P. D. M. (1961). Relations among the solubilities of nitrogen, argon and oxygen in distilled water and sea water, *J. Phys. Chem.*, **65**, 1489.
4. Swinnerton, J. W., Linnebom V. J. and Cheek C. H. (1962). Determination of dissolved gases in aqueous solutions by gas chromatography *Anal. Chem.*, **34**, 483.
5. Davies, P. W. and Brink, F. (1942). Microelectrodes for measuring local oxygen tension in animal tissue, *Rev. Sci. Instrum.* **13**, 524.
6. Sproule, B. J., Miller, W. F., Cushing, I.E. and Chapman, C. B. (1957). An improved polarographic method of measuring oxygen tension in whole blood, *J. Appl. Physiol.*, **30**, 272.
7. Kanwisher, J. (1959). A polarographic oxygen electrode, *Limnol. Oceanogr.*, **4**, 210.
8. Davison, W. and Whitfield, M. (1977). Modulated polarographic and voltametric techniques in the study of natural water chemistry, *J. Electroanal. Chem.*, **75**, 763.
9. Yim, H. S. and Meyerhoff, M. E. (1992). Reversible potentiometric oxygen sensors based on polymeric and metallic film electrodes, *Anal. Chem.*, **64**, 1777.
10. Brendel P. J. and Luther, III G. W. (1995). Development of a gold amalgam voltametric microelectrode for the determination of dissolved Fe, Mn, O_2, and S^{2-} of marine and freshwater sediments, *Environ. Sci. Technol.*, **29**, 751.
11. Atkinson, M. J., Thomas, F. I. M., Terrill, E., Morita, K. and Liu, C. C. (1995). A micro-hole potentiostactic oxygen sensor for oceanic CTD's, *Deep-sea Res.*, **42**, 761.

12. Fatt, I. (1976). *Polarographic Oxygen Sensor*, CRC Press, Cleveland, OH, 279 pp.
13. Gnaiger E. and Forstner, H. (1983). *Polarographic Oxygen Sensors*, Springer, Berlin, 370 pp.
14. Smith, K. L., Jr. and Baldwin, R. J. (1983). Deep-sea respirometry: In situ techniques. In *Polarographic Oxygen Sensors*, ed. Gnaiger, E. and Forstner, H., Springer, p. 37.
15. Thomas, F. I. M., McCarthy, S. A., Bower, J., Krothapalli, S., Atkinson, M. J. and Flament, P. (1995). Response characteristics of two oxygen sensors for oceanic CTD's, *J. Atmos. Ocean. Technol* **12**, 687.
16. Revsbech, N. P. and Jorgensen, B. B. (1986). Microelectrodes: their use in microbial ecology, *Adv. Microb. Ecol.*, **9**, 293.
17. Warren, B. A. (1973). The transpacific hydrographic sections at Lats. 43 °S and 28 °S: the SCORPIO Expedition—II. Deep water. *Deep-Sea Res.*, **20**, 9.
18. Archer, D. and Devol., A. (1992). Benthic oxygen fluxes on the Washington shelf and slope: A comparison of in situ microelectrode and chamber measurements. *Limnol. Oceanogr.*, **37**, 614.
19. Revsbech, N. P., Jørgensen, B. B. and Blackburn, T. H., (1980). Oxygen in the seabottom measured with a microelectrode, *Science*, **207**, 1355.
20. Revsbech, N. P., Sorensen, J., Blackburn, T. H. and Lomholt, J. P. (1980). Distribution of oxygen in marine sediments measured with microelectrodes, *Limnol. Oceanogr.*, **25**, 403.
21. Chen, Y.S. and Bungay, H. R. (1981). Microelectrode studies of oxygen transfer in trickling filter slimes, *Biotech. Bioeng.*, **23**, 781.
22. Oldham, C. (1994). A fast-response oxygen sensor for use on fine scale and microstructure CTD-profiles, *Limnol Oceanogr.*, **39**, 1959.
23. Glud, R. N., Gundersen, J. K., Revsbech, N. P., Jorgensen, B. B. and Hüttel, M. (1995). Calibration and performance of the stirred flux chamber from the benthic lander Elinor, *Deep-sea Res.*, **42**, 1029.
24. Kuenen, J. G., Jørgensen B. B. and Revsbech N. P. (1986). Oxygen measurements in trickling filter biofilms, *Wat. Res.*, **20**, 1589.
25. Sweerts J. P. R. A, Louis V. and Cappenberg T. E. (1989). Oxygen concentration profiles and exchange in sediment cores with circulating overlying water, *Fresh. Biol.*, **21**, 401.
26. Jorgensen, B. B. and Des Marais, D. J. (1990). The diffusive boundary layer of sediments: Oxygen microgradients over a microbial mat, *Limnol. Oceanogr.*, **35**, 1343.
27. Kühl, M., Cohen, Y., Dalsgaard, T., Jørgensen, B. B. and Revsbech N. P. (1995). Microenvironment and photosynthesis of zooxanthellae in scleratinian corals studied with microsensors for O_2, pH, and light, *Mar. Ecol. Prog. Ser.*, **117**, 159.
28. Ploug, H., Kühl, M., Buchholz, B. and Jørgensen, B. B. (1997). Anoxic aggregates–an emphemeral phenomenon in the ocean, *Aquat. Microb. Ecol.*, **13**, 285.
29. Whalen, W. J., Riley J. and Nair P. (1967). A microelectrode for measuring intracellular pO_2, *J. Appl. Physiol.*, **23**, 798.
30. Bungay H. R. (1969). Microprobe technique for determining diffusivities and respiration rates in microbial slime systems, *Biotech. Bioeng.*, **11**, 765.
31. Baumgärtl H. and Lübbers D. W. (1973). Platinum needle electrodes for polarographic measurement of oxygen and hydrogen. In *Oxygen Supply*, ed. Kessler, M. Urban and Schwarzenberg, München, p. 130.
32. Baumgärtl, H. and Lübbers D. W. (1983). Microcoaxial needle sensor for polarographic measurements of local O_2 pressure in the cellular range of living tissue. Its

construction and properties. In *Polarographic Oxygen Sensors*, (ed. Gnaiger, E. and Forstner, H., Springer, Berlin, p. 37.

33. Revsbech, N. P. (1983). *In situ* measurements of oxygen profiles of sediments by use of oxygen microelectrodes. In *Polarographic Oxygen Sensors*, (ed. Gnaiger, E. and Forstner, H., Springer, Berlin, p. 265.
34. Gust G., Booij K., Helder W. and Sundby B., (1987). On the velocity sensitivity (stirring effect) of polarographic oxygen microelectrodes, *Neth. J. Sea Res.*, **21**, 255.
35. Epping, E., and Helder , W. (1997). Oxygen budgets for Northern Adriatic sediments calculated from in situ oxygen microprofiles, *Contin. Shelf Res.*, **17**, 1737.
36. Revsbech, N. P., Sørensen, J., Blackburn, T. H. and Lomholt, J. P. (1980). Distribution of oxygen in marine sediments measured with microelectrodes, *Limnol. Oceanogr.*, **25**, 403.
37. Reimers, C. E., Kalhorn S., Emerson, S. R and Nealson, K. H. (1984). Oxygen consumptiom rates in pelagic sediments from the Central Pacific: First estimates from microelectrode profiles, *Geochim. Cosmochim. Acta.*, **48**, 903.
38. Helder, W. and Bakker, J.F. (1985). Shipboard comparison of micro and minielectrodes for measuring oxygen distribution in marine sediments, *Limnol. Oceanogr.*, **30**, 1106.
39. Visscher, P. T., Beukema, J. and van Germerden, H. (1991). In situ measurements of oxygen and sulfide profiles with a novel combined needle-electrode, *Limnol. Oceanogr.*, **36**, 1476.
40. Revsbech, N. P. and Ward, D. (1983). Oxygen microelectrode that is insensitive to medium chemical composition: Use in an acid microbial mat dominated by *Cyanidium caldarium*, *Appl. Environ. Microbiol.*, **45**, 755.
41. Clark, L. C., Wolf, R., Granger, D. and Taylor A. (1953). Continuous recording of blood oxygen tension by polarography, *J. Appl. Physiol.*, **6**, 183.
42. Tengberg A., De Bovee, F., Hall, P., Berelson, W., Cicceri, G., Crassous, P., Devol, A., Emerson, S., Glud, R. N., Graziottin, F., Gundersen, J. K., Hammond, D., Helder, W., Holby, O., Jahnke, R., Khripounoff, A., Nuppenau, V., Pfannkuche, O., Reimers, C., Rowe, G., Sahami, A., Sayles, F., Schuster, M., Wehrli, B. and De Wilde, P. (1995). Benthic chamber and profile landers in oceanography—A review of design, technical solutions and functioning, *Prog. Oceanogr.*, **35**, 253.
43. Reimers, C. E., Jahnke, R. A. and Thomsen, L. (1999). In situ boundary layer measurements. In *The Benthic Boundary Layer: Transport and Biogeochemical Processes*, (ed. Boudreau, B. and Jorgensen, B. B., Oxford University Press, Oxford.
44. Reimers, C. E. and Glud, R. N. (2000). In situ chemical measurements at the sediment–water interface. In *Chemical Sensors in Oceanography*, (ed. Varney, M., Gordon and Breach.
45. Revsbech, N. P., (1989). An oxygen microelectrode with a guard cathode, *Limnol. Oceanogr.*, **34**, 474.
46. Bucher, R., (1983). Electrolytes. In *Polarographic Oxygen Sensors*, (ed. Gnaiger, E. and Forstner, H., Springer, Berlin, p. 66.
47. Gundersen, J. K., Jorgensen, B. B., Larsen, E. and Jannasch, H. W. (1992). Mats of giant sulphur bacteria on deep-sea sediments due to fluctuating hydrothermal flow, *Nature*, **360**, 454.
48. Berntsson, M. A., Tengberg, A., Hall, P. O. J. and Josefson, M. (1998). Multivariate experimental methodology applied to the calibration of a Clark-type oxygen sensor, *Anal. Chim. Acta.*, **355**, 43.

49. Gundersen, J. K., Ramsing, N. B. and Glud, R. N. (1998). Predicting the signal of O_2 microsensors from physical dimensions, salinity and O_2 concentration, *Limnol. Oceanogr.*, **43**, 1932.
50. Crank, J. (1989). *The Mathematics of Diffusion*, Clarendon Press, Oxford, 414 pp.
51. Garcia, H. E. and Gordon, L. I. (1992). Oxygen solubility in seawater: Better fitting equations, *Limnol. Oceanogr.* **37**, 1307.
52. Broecker, W. S. and Peng, T. H. (1974). Gas exchange rates between air and sea, *Tellus.*, **26**, 21.
53. Li, Y. H. and Gregory, S. (1974). Diffusion of ions in deep-sea sediments, *Geochim. Cosmochim. Acta*, **38**, 703.
54. Millard, R. C. (1982). CTD calibration and data processing techniques at WHOI using 1978 practical salinity scale. In *Proc. Intl STD Conf. Workshop*, Mar. Rech. Soc., 19 pp.
55. Mancy, K. H., Okun, D. A. and Reilley, C. N. (1962). A galvanic cell oxygen analyzer, *J. Electroanal. Chem.*, **4**, 65.
56. Revsbech, N. P., personal communication.
57. Wenzhöfer, F., Holby, O., Glud R. N., Nielsen, H. K. and Gundersen, J. K. (2000). In situ microsensor studies of a hydrothermal vent at Milso (Greece), *Mar. Chem.*, **69**, 43.
58. Quetin, L. B. and Mickel, T. J. (1983). Sealed respirometers for small invertebrates. In *Polarographic Oxygen Sensors* (ed. Gnaiger, E. and Forstner, H., Springer, Berlin, p. 184.
59. Hale, J. M. (1983). Factors influencing the stability of polarographic oxygen sensors. In *Polarographic Oxygen Sensors*, (ed. Gnaiger, E. and Forstner, H., Springer, Berlin, p. 3.
60. Revsbech, N. P., Nielsen, L. P. and Ramsing, N. B. (1998). A novel microsensor principle for determination of apparent diffusivity in sediments, *Limnol. Oceanogr.*, **43**, 986.
61. Jirka, G. H. and Ho, A. H. W. (1990). Measurements of gas concentration fluctuations at the water surface, *J. Hydraul. Eng.*, **116**, 835.
62. Bühler, H. (1983). A double-membrane sterilizable oxygen sensor. In *Polarographic Oxygen Sensors*, (ed. Gnaiger, E. and Forstner, H., Springer, Berlin p. 76.
63. Gnaiger, E. (1983). In situ measurement of oxygen profiles in lakes: Microstratifications, oscillations and the limits of comparison with chemical methods. In *Polarographic Oxygen Sensors* ed. Gnaiger, E. and Forstner, H., Springer, Berlin, p. 245.
64. Glud, R. N., Gundersen, J. K., Revsbech, N. P., Jorgensen, B. B. and Hüttel, M. (1995). Calibration and performance of the stirred flux chamber from the benthic lander Elinor, *Deep-sea Res.*, **42**, 1029.
65. Rink, S., Kühl, M., Bijma, J. and Spero, H. J. (1998). Microsensor studies of photosynthesis and respiration in the symbiotic foraminifer, *Orbulina universa*, *Mar. Biol.*, **131**, 583.
66. Plough, H. and Jorgensen, B. B. (1999). A net-jet flow system for mass transfer and microsensor studies of sinking aggregates, *Mar. Ecol. Prog. Ser.*, **176**, 279.
67. Albanese, R. A. (1973). On microelectrode distortion of tissue oxygen tensions, *J. Theor. Biol.*, **38**, 143.
68. Boudreau, B. P. and Guinasso, N. L. (1982). The influence of a diffusive sublayer on accretion, dissolution, and diagenesis at the sea floor. In *The Dynamic Environment of the Ocean Floor*, ed. Fanning, K. A. and Manheim, F. T., Lexington, pp. 115–145.
69. Jorgensen, B. B. and Revsbech, N. P. (1985). Diffusive boundary layers and the oxygen uptake of sediments and detritus, *Limnol. Oceanogr.*, **30**, 111.

70. Archer, D., Emerson, S. and Smith, C. R. (1989). Direct measurement of the diffusive sublayer at the deep sea floor using oxygen microelectrodes, *Nature* **340**, 623.
71. Gundersen J. K. and Jorgensen, B. B. (1990). Microstructure of diffusive boundary layers and the oxygen uptake of the sea floor, *Nature* **345**, 604.
72. Glud, R. N, Gundersen, J .K., Jorgensen, B. B., Revsbech, N. P. and Schulz, H. D. (1994). Diffusive and total oxygen uptake of deep-sea sediments in the eastern South Atlantic Ocean: *In Situ* and laboratory measurements, *Deep Sea Res.*, **41**, 1767.
73. Jorgensen, B. B. (1994). Diffusion processes and boundary layers in microbial mats. In *Microbial Mats*, (ed. Stal, L. J. and Caumette, P., NATO ASI Series, Vol. G35.
74. Rasmussen, H. and Jorgensen, B.B. (1992). Microelectrode studies of seasonal oxygen uptake in a coastal sediment: role of molecular diffusion, *Mar Ecol. Prog. Ser.*, **81**, 289.
75. Ullman, W. J. and Aller R. C. (1982). Diffusion coefficients in nearshore marine sediments, *Limnol. Oceanogr.*, **27**, 552.
76. Iversen, N. and Jorgensen, B.B. (1993). Diffusion coefficients of sulfate and methane in marine sediments: Influence of porosity, *Geochim Cosmochim Acta.*, **75**, 571.
77. Klimant, I., Holst, G. and Kühl, M. (1997). A simple fiber-optic sensor to detect the penetration of microsensors into sediments and other biological materials, *Limnol. Oceanogr.*, **42**, 1638.
78. Reimers, C. E. (1992). Carbon fluxes and burial rates over the continental slope and rise off central California with implications for the global carbon cycle, *Glob. Biogeochem.* **6**, 199.
79. Sweertz J. P. R. A., Lois, V. and Cappenberg, T. E. (1989). Oxygen concentration profiles and exchange in sediment cores with circulating overlying water, *Fresh. Biol.*, **21**, 401.
80. Glud, R. N., Gundersen, J. K., Revsbech, N. P. and Jorgensen B. B. (1994). Effects on the benthic diffusive boundary layer imposed by microelectrodes, *Limnol. Oceanogr.*, **39**, 462.
81. Lorenzen, J., Glud, R. N. and Revsbech, N. P. (1995). Impact of microsensor caused changes in diffusive boundary layer thickness on O_2 profiles and photosynthetic rates in benthic communities of microorganisms, *Mar. Ecol. Prog. Ser.*, **119**, 237.
82. Epping E. H. G. and Jorgensen, B. B. (1995). Light enhanced oxygen respiration in benthic phototrophic communities, *Mar. Ecol. Prog. Ser.*, **139**, 193.
83. Kühl, M., Glud, R. N., Ploug, H. and Ramsing N. B. (1996). Microenvironmental control of photosynthesis and photosynthesis-coupled respiration in an epilithic cyanobacterial biofilm, *J. Phycol.*, **32**, 799.
84. Revsbech N. P. and Jorgensen, B. B. (1983). Photosynthesis of benthic microflora measured with high spatial resolution by the microprofile method: capabilities and limitations of the method, *Limnol. Oceanogr.*, **28**, 749.
85. Glud, R. N., Ramsing, N. B. and Revsbech, N. P. (1992). Photosynthesis and photosynthesis-coupled respiration in natural bio-films quantified with microsensors, *J. Phycol.*, **28**, 51.
86. Lassen, C., Glud, R. N., Ramsing, N. B. and Revsbech N. P. (1998). A method to improve the spatial resolution of photosynthetic rates obtained by oxygen microsensors, *J. Phycol.* **34**, 89.
87. Jost, W. (1964). Fundamental aspects of diffusion processes, *Angew. Chem. Int. Ed. Engl.*, **3**, 713.

88. Enns T., Scholander, P. F. and Bradstreet, E. D. (1965). Effect of hydrostatic pressure on gases dissolved in water, *J. Phys. Chem.*, **69**, 389.
89. Reimers, C. E. (1987). An in situ microprofiling instrument for measuring interfacial pore water gradients: methods and oxygen profiles from North Pacific Ocean, *Deep-sea Res.*, **34**, 2019.
90. Greene, M. W., Gafford, R. D. and Rohrbaugh, D. G. (1970). A continuous profiling deep-submersible, dissolved monitor, *Mar. Technol. Soc.*, **2**, 1485.
91. Morild, E. and Olmheim, J. E. (1980). Pressure dependence of the oxygen electrode, *J. Electrochem. Soc.*, **127**, 2356.
92. Owens, W. B. and Millard, R. C., Jr. (1985). A new algorithm for CETD oxygen calibration, *J. Phys. Oceanogr.*, **15**, 621.
93. Atkinson, M. J., Thomas, F. I. M. and Larson, N. (1996). Effect of pressure on oxygen sensors. *J. Atm. Oceanic Technol..*, **13**, 1267.
94. Wenzhöfer, F., Holby, O. and Kohls, O. (1999). Deep penetrating oxygen profiles measured in situ by oxygen optodes, *Deep-sea Res.*, in press.
95. Glud, R. N., Gundersen, J. K. and Holby, O. (1999). Benthic in situ respiration in the upwelling area off central Chile, *Mar. Ecol. Prog. Ser.*, **186**, 9.
96. Kühl, M. and Revsbech, N. P. (2000). Microsensors for the study of interfacial biogeochemical processes. In *The Benthic Boundary Layer: Transport and Biogeochemical Processes*, (ed. Boudreau, B. and Jorgensen, B. B., Oxford University Press, Oxford.
97. Kroneis, H. W. and Marsoner, H. J. (1983). A fluorescence-based sterilizable oxygen probe for use in bioreactors, *Sens. Actuators*, **4**, 587.
98. Peterson, J. I., Fitzgerald, R. V. and Buckhold, D. K. (1984). Fiber-optic probe for in vivo measurement of oxygen partial pressure, *Anal. Chem.*, **56**, 474.
99. Leiner, M. J. P. (1991). Luminescence chemical sensors for biomedical applications: Scope and limitations, *Anal. Chim. Acta*, **225**, 209.
100. Glud, R. N., Klimant, I., Holst, G., Kohls, O., Meyer, V., Kühl, M. and Gundersen, J. K. (1999). Adaptation, test and *in situ* measurements with O_2 microoptodes on benthic landers, *Deep-sea Res.*, **46**, 171.
101. Wang, W., Reimers, C. E., Shahriari, M. and Wainright, S. C. (1999). A fiber optic sensor for monitoring the variability of dissolved oxygen in the coastal zone. In Proceedings of The Marine Technology Society, Annual Conference, 1998.
102. Kohls, O. (1995). Optische Sauerstoffsensoren. Ph.D. thesis, University Hannover.
103. Kautsky, H. (1939). Quenching of luminescence by oxygen, *Trans. Faraday Soc.*, **35**, 216.
104. Klimant, I., Meyer, V. and Kühl, M. (1995). Fiber-optic oxygen microsensors, a new tool in aquatic biology, *Limnol. Oceanogr.*, **40**, 1159.
105. Wolfbeis, O. S., Leiner, M. J. P. and Posch, H. E. (1986). A new sensing material for optical oxygen measurement with the indicator embedded in an aqueous phase, *Microchim. Act*, **3**, 359
106. Bacon, J. R. and Demas, J. N. (1987). Determination of oxygen concentration by luminescence quenching of a polymer immobilized transition-metal complex, *Anal. Chem.*, **59**, 2780.
107. Klimant, I. and Wolbeis, O. S. (1995). Oxygen sensitive luminescent materials based on silicone-soluble ruthenium diimine complexes, *Anal. Chem.*, **34**, 3160.
108. Klimant, I., Kühl, M., Glud, R. N. and Holst, G. (1997). Optical measurement of oxygen and temperature in microscale: strategies and biological applications, *Sens. Actuators*, **38–39**, 29.

109. McEvoy, A. K., McDonagh, C. M. and MacCraith, B. D. (1996). Dissolved oxygen sensor based on fluorescence quenching of oxygen sensitive ruthenium complexes immobilised in Sol–Gel derived porous silica coatings, *Anal.*, **121**, 785.
110. Koenig, B., Holst, G., Glud, R. N. and Kühl, M. (2000). Imaging of oxygen distribution at benthic interfaces: A mini-review . In *Organism–Sediment Interactions*, (ed. J. Y. Aller, Woodin, S. A., and Aller, R. C.). The Belle W. Baruch Library in Marine Science, in press.
111. Riess, W., Giere, O., Kohls, O. and Sarbu, S. M. (1999). Anoxic thermomineral waters and bacterial mats a habitat for freshwater nematodes, *Aquat. Microb. Biol.*, **18**, 157.
112. Stern, O., and Volmer, M., (1919). Über die Abklingzeit der Fluoreszenz, Phys. Z. **20**, 183.
113. Carraway, E. R., Demas, J. N. and DeGraff, B. A. (1991). Luminescence quenching mechanisms for microheterogenous systems, *Anal. Chem.*, **63**, 332.
114. Demas, J. N. and DeGraff, B. A. (1992). On the design of luminescence based temperature sensors, *Proc. SPIE*, **1796**, 71.
115. Holst, G., Kühl, M. and Klimant, I. (1996). New temperature micro-optodes. In *Abstracts 3rd Eur. Conf. Optical Chemical Sensors and Biosensors (EUROPT(R)ODE III)*, Zurich, Switzerland.
116. Glud, R. N., Ramsing, N. B., Gundersen, J. K. and Klimant, I. (1996). Planar optrodes, a new tool for fine scale measurements of two dimensional O_2 distribution in benthic communities, *Mar. Ecol. Prog. Ser.*, **140**, 217.
117. Klimant, I, personal communication.
118. Kohls, O. and colleagues, unpublished results.
119. Lippitsch, M. E., Pusterhoffer, J., Leiner, M. J. P. and Wolfbeis, O. (1988). Fiber-optic oxygen sensor with the fluorescence decay time as the information carrier, *Anal. Chim. Acta.*, **205**, 16.
120. Lübbers D. W. (1992). Fluorescence based chemical sensors, *Adv. Biosensors*, **2**, 215.
121. Holst, G., Kühl, M. and Klimant, I. (1995). A novel measuring system for oxygen microoptodes based on a phase modulation technique, *SPIE Proc.*, **2508**, 387.
122. Kohls, O. and colleagues (1999). Work in progress.
123. Holst, G., Glud, R. N., Kühl, M. and Klimant, I. (1997). A microoptode array for fine-scale measurement of oxygen distribution, *Sens. Actuators*, **38–39**, 122.
124. Papkovsky, D. B. (1995). New oxygen sensors and their application to biosensing, *Sens. Actuators*, **29**, 213.
125. Hartman, P., Ziegler, W., Holst, G. and Lübbers, D. W. (1997). Oxygen flux fluorescence lifetime imaging, *Sens. Actuators* **38–39**, 110.
126. Holst, G., Kohls, O., Klimant, I., König, B., Kühl, M. and Richter, T. (1998). A modular luminescence lifetime imaging system for mapping oxygen distribution in biological samples, *Sens. Actuators*, **51**, 163.
127. Jorgensen, B. B., Revsbech, N. P. and Cohen, Y. (1983). Photosynthesis and structure of benthic microbial mats: Microelectrode and SEM studies of four cyanobacterial communities, *Limnol. Oceanogr.*, **28**, 1075.
128. Glud, R. N., Holby,O., Hofmann, F. and Canfield, D. (1998). Benthic mineralization in Arctic sediments (Svalbard), *Mar. Ecol. Prog. Ser.*, **173**, 237.
129. Glud, R. N., Santegoeds, C. M., de Beer, D., Kohls, O. and Ramsing, N. B. (1998). Oxygen dynamics at the base of a biofilm studied with planar optrodes, *Aquat. Microb. Ecol.*, **14**, 223.

130. Glud, R. N., Kühl, M., Kohls, O. and Ramsing N. B. (1999). Heterogeneity of oxygen production and consumption in a photosynthetic microbial mat as studied by planar optodes, *J. Phycol.*, **35**, 270.
131. Rhoads, D. C. and Germano, J. D. (1982). Characterization of organism–sediment relations using sediment profile imaging: An efficient method of remote ecological monitoring of the seafloor (Remots system), *Mar. Ecol. Prog. Ser.*, **8**, 115.
132. Grehan, A. J. Keegan, B. F., Bhaud, M. and Guille, A. (1992). Sediment profile imaging of soft substrates in the western Mediterranean: The extent and importance of faunal reworking, *C. R. Acad. Sci. Paris*, **309**, 315.

3 Sensors for *In Situ* pH and pCO_2 Measurements in Seawater and at the Sediment–Water Interface

WEI-JUN CAI AND
The University of Georgia, USA
CLARE E. REIMERS
Oregon State University, USA

In Situ Monitoring of Aquatic Systems: Chemical Analysis and Speciation Edited by J. Buffle and G. Horvai.

1 INTRODUCTION

Carbon dioxide is the dominant end product of organic carbon degradation in almost all marine environments and its variation is often a measure of net ecosystem metabolism [1–3]. Therefore, in marine biogeochemical studies, it is desirable to measure parameters that define the carbon dioxide system. There are four readily measurable parameters of the marine carbon dioxide system: pH, pCO_2, total dissolved inorganic carbon (DIC) and total alkalinity (TA) (see section 2 and definitions in the List of Symbols and Acronyms). Among them, pH 'reflects the thermodynamic state of all the various acid–base systems present in seawater, particularly the geochemically important carbon dioxide system and is indicative of the processes involved in biological production and respiration' [4]. The partial pressure of carbon dioxide, pCO_2, is closely linked to pH, but it is also affected by the physical and chemical factors that control the solubility and transport of a non-ideal gas. For example, the pCO_2 of surface seawater is most often shifting towards equilibrium with the atmosphere because of gas exchange between these two reservoirs. Direct measurements of pH profiles in the ocean were first accomplished with sensors beginning in the 1960s [5,6]. Direct sensor measurements of pCO_2 have only been accomplished very recently.

Interestingly, when the first *in situ* oceanographic pH measurements were made, a few point measurements were also reported in the sediments at the base of extended water column profiles [5,7]. These sediment pH values were considerably lower than the water column observations, reflecting the processes of organic carbon and $CaCO_3$ cycling near the sediment–water interface. Variable amounts of calcium carbonate precipitation induced by changes in pressure and temperature when bringing a core from the deep-sea floor to the sea surface greatly hampered early attempts to describe in detail the inorganic carbon chemistry of sediment porewaters [8,9]. Therefore several *in situ* methods were developed for benthic studies including chambers [10–12], *in situ* porewater samplers [13–15], and microelectrodes [16–19]. Microelectrodes or other forms of microsensors represent the only measurement technique that can provide millimeter to sub-mm scale *in situ* profiles of porewater parameters that are free of artifacts that may be caused by sample disturbance and the separation of porewaters. This chapter will focus on sensors that have been developed for pH and $p\mathrm{CO_2}$ *in situ* measurements. Since many of the most successful of these sensors are microsensors, the operation principles, preparation procedures and performance limitations of these types of sensors will be emphasized. As background, the solution chemistry of carbon dioxide is first reviewed and an error analysis presented to evaluate the uncertainties in concentrations of DIC, carbon alkalinity (CA), HCO_3^- or CO_3^{2-} that may result from calculation from *in situ* pH and $p\mathrm{CO_2}$ measurements. Finally, examples of *in situ* pH and $p\mathrm{CO_2}$ measurements that meet requirements of high precision, accuracy and low cost will be given, leading to suggestions for future applications.

2 THE SOLUTION CHEMISTRY OF CARBON DIOXIDE

2.1 THE DISSOLUTION, HYDRATION AND DISSOCIATION OF CARBON DIOXIDE

2.1.1 Solubility

Carbon dioxide gas has a strong tendency to dissolve in water with a solubility of about 28 times that of O_2 and Ar gases [20–22]. Equilibrium between the gas phase and water is governed to a close approximation by Henry's law, which states that for an ideally dilute solution, the gas vapor pressure of a volatile solute is proportional to its mole fraction in the solution. For aquatic applications, it is a common practice to use a modified version of Henry's law that relates a gravimetric or molar concentration (mol kg^{-1} or mol L^{-1}) of the dissolved gas $[CO_2]$ (rather than its mole fraction), to a solubility coefficient K_H, the fugacity of carbon dioxide, $f\mathrm{CO_2}$, and an exponential term that depends on the total pressure P and temperature T of the system [21]

$$[CO_2] = K_H f\mathrm{CO_2} \exp[-(P-1)\bar{V}_{CO_2}/RT] \tag{1}$$

In equation (1), $\bar{V}_{CO_2}$ represents the partial molar volume of CO_2 in solution. Equations to predict the solubility coefficient, K_H, as a function of temperature and salinity have been fitted to experimental measurements of CO_2 solubility by

Table 1. pCO_2 levels in natural aquatic environments and in physiological solutions. Note that Pascal (Pa) is the unit adopted by the International System of Units (SI) for gas pressures although various other units are still used in the literature. $1 atm = 1.01325 \times 10^5$ Pa and is close to 1bar($= 10^5$Pa). A gas mixture containing 0.1 % CO_2 has a partial pressure of 1000 μatm (or 1 matm) at 1 atm total pressure. Present-day (1990s) atmospheric pCO_2 is about 35.8 Pa or 358 μatm.

System	pCO_2 (100 Pa or 10^{-3} atm)	References
Atmosphere	0.358 ± 0.01	[23]
Open ocean surface water	0.360 ± 0.08	[24–27]
Coastal ocean surface water	0.1–0.6	[28]
Deep open ocean water and sediment	0.800–3	[29,30]
Near shore sediments	1–200	[31,32]
Photosynthetic mats	<0.3	[33]
Physiological solutions	40–50	[34–36]

Weiss [21]. The use of fugacity corrects for deviations of the gas from ideal behavior. However, in most sensor applications the fugacity of carbon dioxide in equation (1) is taken as equal to the partial pressure because the difference amounts to less than 1 %. Furthermore, equation (1) is often simplified to $[CO_2] = K_H pCO_2$ if surface waters are being considered. When describing the inorganic carbon chemistry of deep waters, fCO_2 or pCO_2 is defined thermodynamically by the equality expressed in equation 1, even though there is usually no gas phase physically in contact with the solution. Thus, pCO_2 levels vary widely in natural environments (Table 1).

2.1.2 Hydration and Dehydration

Dissolved CO_2 in water exists in two forms: one as free or non-hydrated CO_2 and another as hydrated CO_2 (i.e., H_2CO_3) [37]. The non-hydrated species is most often denoted as CO_{2aq} indicating that it is dissolved in aqueous solution [38,39], although, CO_{2f} is sometimes used, see ref. [37]. Since it is not easy to differentiate CO_{2aq} and H_2CO_3, they are often considered together as $[CO_2^*] = [CO_{2aq}] + [H_2CO_3]$. In this chapter, we will use $[CO_2^*]$ or simply $[CO_2]$ (as in equation (1)) to indicate $[CO_{2aq}] + [H_2CO_3]$, the total concentration of dissolved CO_2 molecules (carbonic acid). The kinetics of hydration and dehydration will be discussed.

Most of the dissolved CO_2 molecules remain as free, unassociated CO_{2aq} (>99.5 %). Only a small fraction (<0.5 %) combines with water.

$$CO_{2aq} + H_2O \underset{k_{H_2CO_3}}{\overset{k_{CO_2}}{\rightleftharpoons}} H_2CO_3 \tag{2}$$

The rate of hydration is pseudo first order since the concentration of water does not normally change. The overall rate expression is [37,38]:

Table 2. Rate constants and half-life values of CO_2 hydration, k_{CO_2}, and $k_{H_2CO_3}$, dehydration. Half life, $t_{1/2} = ln2/k$

T (°C)	$k_{CO_2}(s^{-1})$	$t_{1/2}(s)$	$k_{H_2CO_3}(s^{-1})$	$t_{1/2}(s)$
0	0.0021	330	2	0.35
15	0.014	49.5	11	0.063
25	0.032	21.6	26.6	0.026
38	0.12	5.8	89	0.008

$$\text{rate} = \frac{-d[CO_{2aq}]}{dt} = k_{CO_2}[CO_{2aq}] - k_{H_2CO_3}[H_2CO_3] \tag{3}$$

Values of k_{CO_2} and $k_{H_2CO_3}$ are functions of temperature and are listed in Table 2 together with half-life values. It is clear from Table 2 that CO_2 hydration at room temperature is relatively slow but tolerable with respect to how this process affects a sensor's response time ($t_{1/2} = 22\,s$ at 25°C). The slower hydration rates at low temperatures (1–2°C) pose an obstacle to obtaining fast CO_2 measurements with pCO_2 sensors in deep-sea or high-latitude ocean environments. In certain living cells (for example, in the human lung), a carbonic anhydrase enzyme is produced to catalyze the CO_2 hydration process. The effect of slow hydration on a CO_2 sensor and the applications of enzyme to catalyze the hydration process will be discussed in section 6.3.2.

The hydration constant is defined as

$$K_0 = [H_2CO_3]/[CO_{2aq}] = k_{CO_2}/k_{H_2CO_3} \tag{4}$$

This constant at 25 °C $\approx 0.03/27 = 1/830$ which serves to show that CO_{2aq} is the dominant species in water.

2.1.3 Equilibrium in Solution

In water, carbonic acid dissociates as an acid in two steps

$$\begin{array}{ccccc} CO_{2aq} + H_2O & \rightleftharpoons & H_2CO_3 & \rightleftharpoons & HCO_3^- + H^+ \\ & & & & \uparrow\downarrow \\ & & & & CO_3^{2-} \quad + H^+ \end{array} \tag{5}$$

These dissociation reactions are very fast (at 10^{-6} second level [38]) and their equilibrium can be defined by so-called mixed acidity constants [39]:

$$K_1 = \frac{\{H\}^+[HCO_3^-]}{[CO_2*]} \tag{6}$$

and

$$K_2 = \frac{\{H^+\}[CO_3^{2-}]}{[HCO_3^-]} \tag{7}$$

where $\{H^+\}$ is the activity of hydrogen ion, but the other species are expressed as concentrations. (In marine chemistry, it is now also common to express H^+ with a concentration scale [4].) The inclusion of $[CO_2^*]$ in the K_1 expression implies that K_1 is a composite constant which incorporates K_0. K_1 and K_2 vary with temperature and ionic strength. K_1 is about $10^{-5.86}$ while K_2 is $10^{-8.95}$ in $0.7\,\text{mol}L^{-1}$ ionic media (seawater) at 25°C [40]. Several sets of precisely determined K_1 and K_2 at various temperatures and salinities are available and have been evaluated extensively in [26,41,42].

2.2 EQUILIBRIUM CALCULATIONS AND ERROR ANALYSIS FOR THE CARBON DIOXIDE SYSTEM

Parameters of the marine carbonate system we have introduced so far are pH, pCO_2, $[CO_2^*]$, $[HCO_3^-]$, and $[CO_3^{2-}]$. Remaining are DIC, CA and TA. DIC is the total dissolved inorganic carbon defined as:

$$\text{DIC} = [CO_2^*] + [HCO_3^-] + [CO_3^{2-}] \tag{8}$$

Oceanographers measure DIC most often by acidification of water samples and subsequent quantification of the extracted CO_2 gas by a coulometer or by an infrared CO_2 analyzer [43]. DIC can be measured very precisely (0.05–0.1 %) and accurately (0.1–0.2 %) when sample volume is not limiting (more than 10 mL for the coulometric method) and Certified Standard Reference Materials (CRMs) are used for calibration purposes.

TA is the deficiency of H^+ or the excess base with respect to the zero proton level at the CO_2 equivalence point (about pH = 4.5) [44,45]. Alkalinity by this definition can be determined by HCl titration of the water sample to the CO_2 equivalence point (the Gran titration [46]) or by a curve fitting method (Dickson, personal communication). Mathematically, it is defined as

$$\begin{aligned}\text{TA} = {} & [HCO_3^-] + 2[CO_3^{2-}] + [OH^-] - [H^+] + [B(OH)_4^-] + 2[PO_4^{3-}] \\ & + [HPO_4^{2-}] - [H_3PO_4] + [SiO(OH)_3^-] + \ldots \text{contributions of} \\ & \text{other minor acid or base species}\end{aligned} \tag{9}$$

With water sample volumes greater than 50 mL, TA in seawater can be measured to a precision of 0.1 %. CRMs for alkalinity have been in use since 1996 [27]. With these standards, the reported accuracy of analyses is better than ± 0.2 % [47].

The other two measurable parameters of the carbon dioxide system, pH and pCO_2, are the foci of this chapter. The equations needed to calculate DIC, CA and $[CO_3^{2-}]$ from measured pH and pCO_2, and the error functions of such

Table 3. Definitions and equations for the carbon dioxide system

$$CO_{2(g)} \rightleftharpoons CO_{2(aq)} \qquad K_H = \frac{[CO_2^*]}{pCO_2} \tag{i}$$

where $[CO_2^*] = [CO_{2aq}] + [H_2CO_3]$

$$H_2CO_3 \rightleftharpoons H^+ + HCO_3^- \qquad K_1 = \frac{\{H\}^+[HCO_3^-]}{[CO_2^*]} \tag{ii}$$

$$HCO_3^- \rightleftharpoons H^+ + CO_3^{2-} \qquad K_2 = \frac{\{H^+\}[CO_3^{2-}]}{[HCO_3^-]} \tag{iii}$$

definition

$$DIC = [CO_2^*] + [HCO_3^-] + [CO_3^{2-}] \tag{iv}$$

$$CA = [HCO_3^-] + 2[CO_3^{2-}] \tag{v}$$

Calculation from pH and pCO_2 measurements

$$[CO_2^*] = k_H pCO_2 \tag{vi}$$

$$[HCO_3^-] = k_H K_1 pCO_2 10^{pH} \tag{vii}$$

$$[CO_3^{2-}] = k_H K_1 K_2 pCO_2 / 10^{2pH} \tag{viii}$$

Error analysis in % (concentrations or pressures) and logarithmic unit (pH),

$$\sigma_{DIC} = \left[\left(\frac{K_1 10^{pH} + 2K_1K_2 10^{2pH}}{1 + K_1 10^{pH} + K_1 K_2 10^{2pH}} \sigma_{pH}\right)^2 + \left(\sigma_{pCO_2}\right)^2\right]^{1/2} \tag{ix}$$

$$\sigma_{CA} = \left[\left(\frac{10^{pH} + 4K_2 10^{2pH}}{10^{pH} + 2K_2 10^{2pH}} \sigma_{pH}\right)^2 + \left(\sigma_{pCO_2}\right)^2\right]^{1/2} \tag{x}$$

$$\sigma_{[CO_3^{2-}]} = \left[\left(2\sigma_{pH} \ln 10\right)^2 + \left(\sigma_{pCO_2}\right)^2\right]^{1/2} \tag{xi}$$

calculations are presented in Table 3 [26,48–51]. The maximum error of a quantity derived from two other measured quantities at near-seawater conditions (and assuming realizable measurement errors) is given in Table 4. In general, very high accuracy is needed in determining pH and pCO_2 for the purpose of accurately describing the marine carbon dioxide system. Some calculated quantities have small errors (e.g. pCO_2 and $[CO_3^{2-}]$ from DIC and pH) while others may have large errors (e.g. pCO_2 and $[CO_3^{2-}]$ from DIC and CA).

Figure 1 presents the error functions of DIC, CA, $[CO_3^{2-}]$ calculated from pH and pCO_2 as described by equations in Table 3. A 0.01 pH unit error is assumed for pH ($\sigma_{pH} = 0.01$). A 2.3 % error is assumed for pCO_2 ($\sigma_{pCO_2} = 2.3\,\%$). First, we notice that if pH measurement is error-free ($\sigma_{pH} = 0$ in equation (ix) in Table 3), then errors in the calculated DIC, CA and $[CO_3^{2-}]$ depend only on the error in pCO_2 measurement and are independent of the pH value of the solution. We also notice that the errors in the calculated quantities caused by the error in

Table 4. The maximum error estimation in calculating CO_2 system parameters from a certain combination of measurements. The combinations produce the largest possible errors in the derived properties. For example, the combination of +1 % error in CA and +1 % error in DIC only produces an error of 0.9 % in the $[CO_3^{2-}]$ while the combination of +1 % error in CA and −1 % error in DIC produces an error of +39 % in the $[CO_3^{2-}]$. Errors are given in % except for pH, which is in pH units. It was based on a model porewater defined as TA = 2500 mmol kg^{-1}, DIC = 2450 mmol kg^{-1}, T = 2 °C, $P = 400 \times 10^5$ Pa and S = 34.7. The TA consists of CA and $B(OH)_4^-$ contribution only

Error in measured parameters		Error in derived parameters		
CA	DIC	pH	pCO$_2$	$[CO_3^{2-}]$
±1	∓1	±0.16	−30 to +48	+39 to -31
±0.5	∓0.5	±0.08	−17 to +21	+19 to -17
pH	pCO$_2$	DIC	CA	$[CO_3^{2-}]$
±0.02	±4.6	±9	±10	±14
±0.01	±2.3	±5	±5	±7
DIC	pH	CA	pCO$_2$	$[CO_3^{2-}]$
±1	±0.02	±1	∓3.6	±5.6
±0.5	±0.01	±0.6	∓1.9	±2.8

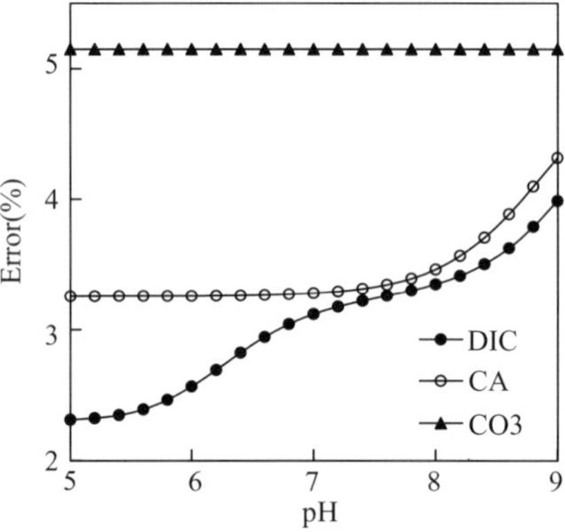

Figure 1. Errors that can result in values of DIC, CA, and CO_3^{2-} calculated from pH and pCO_2. It is assumed that $\sigma_{pH} = 0.01$ and $\sigma_{pCO_2} = 2.3\,\%$

pH measurement (0.01 unit) increase as the pH value of the system increases. $[CO_2]$ is only proportional to pCO_2, while $[HCO_3^-]$ has a first order relationship with $\{H^+\}$, and $[CO_3^{2-}]$ has a second order relationship with $\{H^+\}$ (equations (vi)–(viii) in Table 3). A calculation of CA will have a larger error than a calculation of DIC due to the $2[CO_3^{2-}]$ term. Calculations of $[CO_3^{2-}]$ are the most sensitive to errors in the pH measurement. Therefore, direct determinations of $[CO_3^{2-}]$ would be preferable. A PVC liquid membrane $[CO_3^{2-}]$ sensor was used recently to measure [CO_3^{2-}] profiles in cores of lake sediment [52]. Its detection limit was approximately 1 μmolL^{-1}, and its selectivity for [CO_3^{2-}] was assessed to be high enough for applications in artificial seawater without interference from other ions. Further evaluation of this sensor for marine applications is needed.

TA and DIC are the properties measured the most routinely in hydrocast samples of deep waters and in extracted sediment porewaters. This is because these properties are not affected by changes in temperature and pressure. There are, however, a few potential problems in using these two properties to constrain the carbonate system that are especially acute in sediment porewater applications.

In porewaters from organic-rich sediments, dissolved silicate and phosphate concentrations can be very high. For example, at 10 cm depth in California Borderland Basin sediments, total $PO_4 = 100$ μmol L^{-1}, total Si = 250 μmol L^{-1} [53], TA = 2.7 mmol L^{-1}, and DIC = 2.7 mmol L^{-1} [54]. Parameters derived from TA and DIC with/without considering the contribution of PO_4 and Si are as follows: pH, 7.28/ 7.55; pCO_2: 298/161 Pa; calcite saturation state, 0.41/0.76. These differences are very significant. Additionally, the contribution of organic alkalinity associated with dissolved organic matter in porewaters can be substantial. This is especially true in freshwater sediments that have low TA, but this problem is not well recognized [45,55]. For highly reduced sediments, another potentially serious problem with alkalinity measurement is the oxidation of inorganic species (such as Fe^{2+}) during sample storage or titration, which generates acid (if not in N_2 atmosphere).

These problems can be avoided by performing *in situ* pH and pCO_2 measurements or by the combination of *in situ* pH and pCO_2 measurements with DIC measurements if sampling artifacts are minimized for the latter. Calculation of carbonate species from pH and pCO_2 or pH (or pCO_2) and DIC are unambiguous (i.e. these calculations do not need to account for P, Si and organic species).

At present, most porewater DIC and TA determinations have not been reported with accuracies better than ±0.5 %, largely because of limited sample volumes. These uncertainties can cause 0.08 unit uncertainty in the calculated pH value, and ±18 % uncertainty in the [CO_3^{2-}] (Table 4). If various sampling artifacts are involved, the errors can be even larger.

According to Table 4, traditional porewater parameters, DIC and TA, provide the least accurate prediction of carbon dioxide equilibrium calculation even if no artifacts are introduced during core retrieval and porewater extraction. Direct measurements of porewater pH and pCO_2 have the advantage of avoiding ambiguities involving other acid–base species and allow us to calculate individual CO_2 species and the carbonate saturation state more precisely. As far as calculating DIC from measured pH and pCO_2 is concerned, the result may not be as precise as direct measurement. This is true both in the water column and porewater. The determination of *in situ* pH and pCO_2 together with DIC measurements appears to be the best combination of parameters for the study of the carbon dioxide system.

3 PRINCIPLES OF ELECTROCHEMICAL pH AND pCO_2 MEASUREMENTS

3.1 ELECTROCHEMICAL pH MEASUREMENT

There are numerous discussions in the literature on the theory behind electrochemical measurement of pH in aquatic environments. The following brief description is adopted from Dickson [4]. The potentiometric measurement of pH is based on the use of the cell

ref. elec. | concentrated KCl || test solution | electrode reversible to $H^+(aq)$

where, typically, the electrode reversible to hydrogen ion is a glass electrode that is assumed to exhibit Nernstian behavior. The emf of this cell may be represented as

$$E = E^{0'} + k_N \log\{H^+\} + E_J \quad (10)$$

where k_N is the Nernst constant ($RT \ln 10/F$), E_J is the liquid junction potential that arises because of ionic strength differences between the electrolyte solution of the reference electrode and the test solution [56], and $E^{0'}$ is the 'conditional potential' of the cell which depends on conditions (such as the filling solution) in the pH electrode [57] and the contribution of the reference electrode potential. During routine measurements, electrode readings in one or more reference solution(s) are compared with that of the test solution. By assuming the values of E_J in both the test and the reference solution are the same ($\Delta E_J = 0$), the pH of the test solution is therefore, operationally defined [58,59] as

$$\mathrm{pH(X)} = \mathrm{pH(S)} + (E_s - E_x)/k_N \quad (11)$$

where X and S indicate the test and reference solution, respectively.

One problem with applying standard pH electrodes in the ocean is that $E^{0'}$ tends to increase with increasing pressure and such pressure effects will be different for each individual electrode [60]. Such potential changes are negligible on electrodes with flat membranes but more serious with bulb-type glass electrodes because of the asymmetry of the pH sensitive membrane [7]. However, when pH is measured in porewaters at the seafloor, pressure is constant. Therefore, in these applications it is convenient to let E_S equal the electrode potential in bottom waters just above the sediment and to calculate pH(S) from DIC, TA and concurrent nutrient data acquired at the study site [17,61]. The slope (k_N) of the electrode is assumed not to vary with pressure as is defined by the Nernst equation and was verified by Distéche and Distéche [60,62] during laboratory experiments with pH and reference electrodes at high hydrostatic pressures.

Problems of pH measurement in saline media arise most acutely when the ionic strength and composition is variable, for example in surveys of estuarine waters. These problems are caused by changes in liquid junction potential, and they have no ready remedy [63–66]. It is important to emphasize that the reproducibility of liquid junction potentials, not their absolute value, is most critical to pH measurements involving concentrated KCl salt bridges. Liquid junction potential varies with the physical design of the junction and is not reproducible with standard reference electrodes (i.e. even with the 'identical' ceramic or asbestos junctions [65,66]). Liquid junction potential cannot be predicted precisely on a theoretical basis either. Uncertainties of 0.1 pH units or larger are expected in low salinity water [22,64] owing to differences in liquid junction potential between samples and standards. Special cells, such as a flowing cell [64,67,68] or a series of seawater TRIS buffers with different salinities [63] have been proposed to eliminate this problem, but have been used with only limited success. The trouble with special cells is that they are difficult to maintain, while buffers with different salinities do not completely eliminate the reproducibility problem. High concentrations of dissolved organic matter, including organic acids, may be a second factor complicating pH measurements in estuarine waters [22]. Readers interested in in-depth discussions of the theory and problems of pH measurement in marine, estuarine and lacustrine environments, should refer to refs [63–66].

An added problem that has been considered when pH is measured in sediments is the 'suspension/charge effect'. The suspension/charge effect is a shift in the potential difference between a pH electrode and a reference electrode that is caused by transferring the electrodes from a pure solution to a suspension or sediment [69]. There is no consensus on the theoretical explanation of the suspension effect. It has been documented in reference to acidic suspensions (such as strong acid ion exchange resins or acid soils), but is generally not significant in suspensions where the aqueous phase is seawater [17,70–73]. For example, Siever *et al.* (see ref. [71]) showed that, if CO_2 loss during porewater expression is prevented, pH values measured in the expressed water and in the sediment are the same. Furthermore, Cai and Reimers [17] measured pH in various buffered slurries of seawater and marine sediments with pH microelectrodes and commercial pH electrodes and found no significant potential difference (<0.2 mV) compared with the buffered seawater alone.

3.2 POTENTIOMETRIC $p\mathrm{CO_2}$ SENSORS

A $p\mathrm{CO_2}$ sensor operates by measuring the pH in a thin layer of $NaHCO_3$ solution [17] (or inner NaOH solution [74]) trapped between the tip of a pH

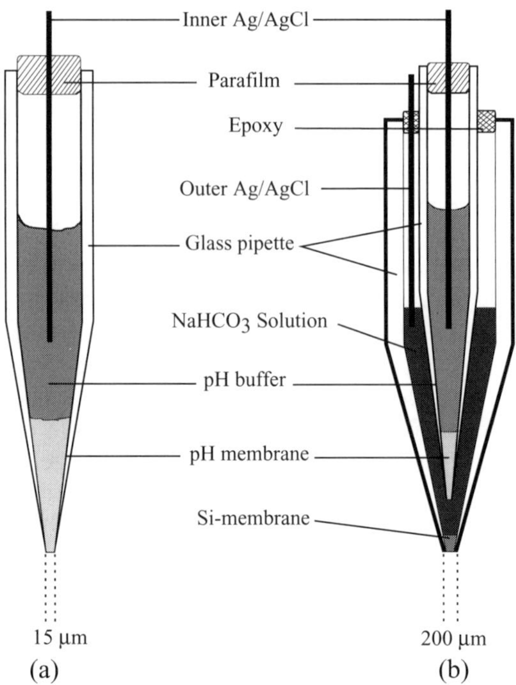

Figure 2. Schematic representation of a liquid membrane pH microelectrode (a) and a pCO_2 microelectrode (b). Parts are not drawn proportionately.

sensor and a CO_2 permeable hydrophobic membrane (Figure 2). The $\{H^+\}$ or $[H^+]$ in the $NaHCO_3$ inner solution is linearly related to the inner solution's $[CO_2]$ or pCO_2, and the latter, at equilibrium, equals the pCO_2 in the solution outside of the CO_2 permeable hydrophobic membrane. When the pH of the $NaHCO_3$ solution is measured by a pH electrode, the emf of a pCO_2 electrode is

$$E = E' + k_N \log\{H^+\} = E' + k_N \log K_1 - k_N \log[HCO_3^-] + k_N \log[CO_2] \quad (12)$$

where $[HCO_3^-]$ is the hydrogen carbonate concentration in the inner solution. Assuming $[HCO_3^-]$ does not change during measurement, equation (12) can be simplified to

$$E = E'' + k_N \log[CO_2] = E^* + k_N \log pCO_2 \quad (13)$$

where $E'' = E' + k_N \log K_1 - k_N \log[HCO_3^-]$ and E^* includes the logarithm of Henry's law constant, K_H. An effectively constant E'' or E^* can be guaranteed for CO_2 measurements in physiological solutions (Table 1) by using moderately high $[HCO_3^-]$ (10^{-3} – $10^{-2}\,\mathrm{mol\,L^{-1}}$) in the internal solution of the sensor [75,76]. Under such conditions, $\log pCO_2$ is linearly related to the EMF or pH.

In natural aquatic environments, pCO_2 ranges from very low to very high (Table 1), which at the extremes can cause some of the simplifying assumptions which led to equation (13) to be invalid. A more accurate representation of the equilibrium pH value in the $NaHCO_3$ film can be described with the following equation derived from a charge balance [75]:

$$[H^+]^3 + [Na^+][H^+]^2 - (K_1 K_H\, pCO_2 + K_w)[H^+] - 2K_1 K_2 K_H pCO_2 = 0 \quad (14)$$

where K_w is the water dissociation constant. A similar equation that includes indicator dye can be found in DeGrandpre [74]. The difference of pH, ΔpH, calculated from the $[H^+]$ given in equation (14) versus the $[H^+]$ in equation (13), is a function of sample pCO_2 and the concentration of the internal $NaHCO_3$ solution, $c(T)$. Table 5 illustrates the dependence of ΔpH (expressed as $\Delta E = -59\ \Delta pH$) on pCO_2 for different inner $c(T)$ values. It can be seen from Table 5 that considerable deviation from assumed linearity can occur at low inner $c(T)$ and high sample pCO_2 or at high inner $c(T)$ and low sample pCO_2. Therefore, the high $c(T)$ should be used for high sample pCO_2 measurement (e.g. in blood and some sediments) and low $c(T)$ for low sample pCO_2 measurements, in particular, if a linear calibration is assumed.

An $NaHCO_3$ concentration of 1–3 mmol L^{-1} has been recommended for marine sediment analysis [17,32]. It can be seen that within the pCO_2 range of interest (~30 Pa in surface waters to over 2×10^4 Pa in organic rich sediments, Table 1), the deviation ΔE is less than 1 mV (see the bold part of Table 5)

Table 5. The theoretical deviation from linearity, ΔE(mV), of the potential of a pCO_2 microelectrode as a function of sample pCO_2 and the concentration of the internal sodium bicarbonate solution, $c(T)$ (modified from Zhao and Cai [32]). 100 Pa is approximately equal to 10^{-3} atm and 0.1 Pa is about equal to 1 μatm (see Table 1).

$c(T)$ (mmol L^{-1})	0.01	0.1	0.5	1	2	10
Inner pCO_2 (100 Pa)*	0.014	0.054	0.18	0.334	0.637	3.062
Sample pCO_2 (100 Pa)						
0.01	−1.61	−2.64	−6.06	−9.09	−13.29	−27.01
0.1	0.24	−0.28	−0.79	−1.38	−2.45	−8.28
0.2	0.71	−0.14	**−0.40**	**−0.71**	**−1.31**	−3.63
0.3	1.11	−0.08	**−0.27**	**−0.48**	**−0.89**	−3.63
1	3.38	0.01	**−0.08**	**−0.15**	**−0.28**	−1.25
10	15.99	0.40	**0.01**	**−0.01**	**−0.03**	−0.13
100	38.84	3.40	**0.16**	**0.04**	**0.01**	−0.01
$(pCO_2)_L$ (0.1 Pa)†	16	25	77	150	251	1200

* $[CO_2]$ here is the initial $[CO_2]$ of the $NaHCO_3$ solution and is calculated from the equation $[CO_2] = \frac{c(T)}{K_1}\sqrt{\frac{K_2 c(T)+K_w}{1+c(T)/K_1}}$. Values of K_1, K_2 and K_w in infinitely dilute solution and at 25 °C are used (for 0 ionic strength taken from ref [39]).

† The lowest sample pCO_2 the electrodes can detect when $\Delta E = 1$ mV

for a 0.5–2 mmol L^{-1} HCO_3^- inner solution. Decreasing the $[HCO_3^-]$ can significantly improve the detection limit. However, low concentration filling solutions are less stable and easier to contaminate.

The process of determining pCO_2 with a macro-sensor that uses a flat tip pH glass electrode involves the following steps: (a) equilibrium between sample CO_{2aq} and CO_{2g} in the hydrophobic membrane, (b) diffusion of CO_{2g} in the membrane, (c) equilibrium between CO_{2g} in the hydrophobic membrane and dissolved CO_{2aq} in the inner solution, (d) hydration of CO_{2aq} in the inner solution; (e) diffusion of CO_{2aq}, H_2CO_3, HCO_3^- and CO_3^{2-} in the inner solution, and (f) H^+-equilibrium with the surface of the pH electrode (Figure 3) [77]. Steps (d) and (e) can proceed simultaneously. In other words, CO_2 does not have to be hydrated near the hydrophobic membrane. It can diffuse towards the pH electrode while being hydrated and dissociated.

In general the equilibration time of a pCO_2 sensor is determined by the rate of hydration of CO_{2aq}, or diffusion of CO_2 through the silicone hydrophobic membrane [17,32,75,76,78]. Cases where hydration is not likely to be a rate determining step are sensors with thick membranes (over 100 μm), sensors where the carbonic anhydrase enzyme is used in the internal solution [17], or applications at high pCO_2 (more than 10 Pa in physiological systems). The membrane limitation has been assumed in most of the literature describing pCO_2 macro-sensors because of membrane thickness [75–77]. Assuming the concentration gradient in the membrane is linear (instead of curved as in Figure 3), an equation describing the dynamic response of a membrane limited sensor was derived by Ross *et al* [76] and examined in detail by Jensen and Rechnitz [75]. In the Ross equation, the response time is directly proportional to the thickness of the CO_2 permeable hydrophobic membrane (m) and thickness of the HCO_3^- thin layer (l). Although other response time relationships have been proposed [75], all suggest that in making pCO_2 microelectrodes for rapid-response marine applications, m and l should be as thin as possible. Conversely,

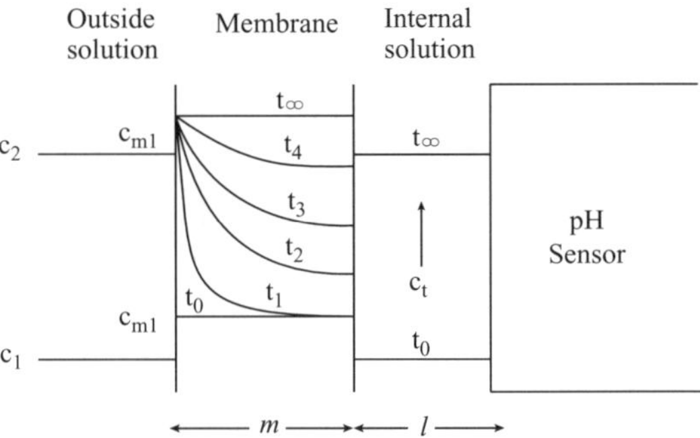

Figure 3. Schematic representative of how a pCO_2 electrode responds to a change in $[CO_2]$ from c_1 to c_2 in a sample solution (from ref. [77]). Symbols are explained in the text

we have found for deep-sea use the hydrophobic membrane must be thick enough to prevent it from collapsing onto the internal pH sensor when a temporary pressure difference occurs between the inside and the outside of the pCO_2 electrode. In the final analysis, the optimum thickness depends on the geometry of the electrode tip and the conditions under which the sensor will be used.

The thickness of the HCO_3^- layer (l) is critical to the response time for a reason that may be explained with an analogy. When a large group of people are rushing into a theatre, whether the slowest are moving at the gate (diffusion) or they are those delayed from taking their seats because of the distribution of 3–D viewing glasses (hydration), the time it takes to fill the threatre is always positively correlated (in the Ross equation it is proportional) to the size of the threatre (the thickness of the HCO_3^- layer).

The Ross equation also indicates that the response rate will be slower when CO_2 concentration is decreasing than when CO_2 concentration is increasing. Another cause for a hysteresis effect is that in most pCO_2 sensors there are two compartments of the internal hydrogen carbonate solution, a thin film between the sensor membrane and the tip of the internal pH sensor and a reservoir of hydrogen carbonate [17]. Between these two compartments there is exchange of CO_2 and other carbon species if there is a concentration difference [75]. Such transfers are probably negligible for most pCO_2 macro-sensors since their geometry helps restrict exchange. For pCO_2 microelectrodes based on pH microelectrodes, serious hysteresis effects may result from this leakage. However, an advantage of this exchange is that it can provide a long-term stability to the microelectrode by constantly renewing the thin HCO_3^- solution at the very tip.

Changes in the ionic composition of an isolated thin layer of $NaHCO_3$ solution will affect the response behavior of pCO_2 sensors [32]. These changes often result from ions leaching into or out of the solution (i.e. due to reactions with the sensor materials such as the glass surface). Such interference can be eliminated by continuously (or intermittently) renewing the internal solution being sensed by the pCO_2 sensor. The extremely precise and stable pCO_2 measurements achieved with fiber optic methods by DeGrandpre and co-workers [28,74,79] relied on such a mechanism. So far, this novel idea has not been successfully applied to the pCO_2 microelectrode.

4 OPTICAL pH AND pCO_2 MEASUREMENTS

4.1 PHOTOMETRIC pH MEASUREMENT

A second approach used for determining pH in aquatic environments involves the use of an indicator dye that is added directly to a test solution or is confined within a sensing cell or matrix that may be illuminated and monitored by one or more optical fiber(s). The most common indicators used for seawater analyses belong to the family of sulfonephthalein dyes. These dyes are weak diprotic electrolytes that have different absorption properties in their weak acid (HD^z) and conjugate base (D^{z-1}) forms, where z is a charge number (0 or -1).

The equilibrium reaction that is applied to the sulfonephthalein dyes at the pH of seawater is a second dissociation such that $z = -1$ and

$$HD^z = H^+ + D^{z-1}, \tag{15}$$

which can be characterized with a concentration-based constant

$$K_c = [H^+][D^{z-1}]/[HD^z]. \tag{16}$$

pH is defined using either the free hydrogen ion scale [80,81] or the total hydrogen ion scale [4,82]. A detailed discussion on the relationship between different pH scales can be found in Dickson [65] and Culberson [66]. The pH of an unknown solution is determined from the pK_c of the indicator where p$K_c = -\log K_c$, and by detecting the concentration ratio of D^{z-1} and HD^z by means of optical measurements. The equilibrium-based relationship is

$$pH = pK_c + \log[D^{z-1}]/[HD^z]. \tag{17}$$

Most researchers who have pursued measuring pH photometrically in discrete seawater samples have used spectrophotometric absorbance measurements at two absorbing wavelengths, λ_1 and λ_2, and one non-absorbing wavelength, λ_3, at a controlled temperature (e.g. 25°C) to estimate the ratio $[D^{z-1}]/[HD^z]$. They assume that since the absorbance, A_i, of the sample plus dye at any one wavelength, λ_i, is determined by the Beer–Lambert law as

$$A_i = (\varepsilon_{i,\mathrm{HD}}\,[HD^z] + \varepsilon_{i,\mathrm{D}}\,[D^{z-1}])l \tag{18}$$

where l is the optical path length and $\varepsilon_{i,\mathrm{D}}$ and $\varepsilon_{i,\mathrm{D}}$ are the molar absorptivies of the protonated and unprotonated forms of the dye [43], absorbances at two wavelengths, λ_1 and λ_2, can be ratioed (after subtracting out any background absorbance measured in the sample without dye) and rearranged to give

$$\frac{[D^{z-1}]}{[HD^z]} = \frac{A_1/A_2 - \varepsilon_{1,\mathrm{HD}}/\varepsilon_{2,\mathrm{HD}}}{\varepsilon_{1,\mathrm{D}}/\varepsilon_{2,\mathrm{HD}} - (A_1/A_2)(\varepsilon_{2,\mathrm{D}}/\varepsilon_{2,\mathrm{HD}})} \tag{19}$$

The absorbances of each sample with and without dye at the non-absorbing wavelength are used to correct for any instrumental drift between measurements.

Combining equations (17) and (19) gives

$$pH = pK_c + \log\left(\frac{R_A - e_1}{e_2 - R_A e_3}\right) \tag{20}$$

which enables the calculation of pH provided molar absorption coefficient ratios, $e_1 = \varepsilon_{1,HD}/\varepsilon_{2,HD}$; $e_2 = \varepsilon_{1,D}/\varepsilon_{2,HD}$; $e_3 = \varepsilon_{2,D}/\varepsilon_{2,HD}$, and values of the equilibrium constant, K_c, are determined in the laboratory as a function of temperature and salinity. In equation (20), $R_A = A_1/A_2$. Clayton and Byrne [82] and Zhang and Byrne [83] report empirical expressions for the variation of these parameters with temperature and salinity for the dyes *m*-cresol purple and thymol blue. These expressions are reproduced in Table 6.

The most significant advantage of spectrophotometric methods is their high precision (usually better than 0.001 pH units). There are no liquid junction or high impedance problems with optical pH measurements, and calibration is inherent in prior knowledge of the thermodynamic properties of the indicator dye. Therefore, the method offers freedom from frequent calibration and is ideal for continuous measurements of both pH and total alkalinity (after acid addition) using automated shipboard analysers that can intake samples and operate while a ship is under way [84–86]. The only disadvantage is that in many applications the sample's measurement temperature and pressure on ship do not match *in situ* conditions. When this occurs, *in situ* pH may be calculated from the measured shipboard pH and DIC or TA, using thermodynamic constants of the carbon dioxide system at shipboard and *in situ* conditions, respectively. An additional problem may arise in application of spectrophotometric methods in some fresh and brackish waters that have very high humic substances, because of the strong background color.

pH fiber optic sensor or 'optrode' measurements are based on most of the same principles as spectrophotometric measurements except that usually the indicator dye and the sample are either separated by a proton permeable membrane or in different proton-permeable phases. Sol–gel films are one example of a matrix material that can easily incorporate pH sensitive dyes, added at the sol stage, while remaining accessible to the diffusion of H+ via the pores of the gel phase [87]. Absorbances, or in many cases reflectance, from a membrane or film impregnated with the absorbing reagent (e.g. [88]) are again

Table 6. Empirical expressions for the acid–base dissociation constant and extinction coefficient ratios for two sulfonephthalein indicators used to determine seawater pH

Parameter	For *m*-cresol purple [82]* $\lambda_1 = 578$, $\lambda_2 = 434$	For thymol blue [83] $\lambda_1 = 596$, $\lambda_2 = 435$
pK_c	$= 1245.69/T + 3.8275 + 0.00211(35 - S)$	$= 4.706S/T + 26.3300 - 7.17218 \log T - 0.017316S$
e_1	$= 0.0069$	$= -0.00132 + 1.600 \times 10^{-5} T$
e_2	$= 2.222$	$= 7.2326 - 0.0299717T + 4.600 \times 10^{-5} T^2$
e_3	$= 0.133$	$= 0.0223 + 0.0003917T$

* 293 K < T < 303 K and 30 < S < 37.

generally monitored at two or more wavelengths but with the aid of filter and photodiode systems. The fundamentals and many designs of pH optrodes are described by Leiner and Wolfbeis [89]. Surprisingly, pH optrodes have been used most widely in the ocean as components of pCO_2 sensors or analysers. For this reason, the issues that have limited use of fiber optic pH sensors will be discussed below in the context of pCO_2 sensors.

4.2 FIBER OPTIC pCO_2 MEASUREMENTS

The pCO_2 system developed by DeGrandpre and co-workers [28,74,79] is a complete analyzer. It consists of a fiber-optic pH sensor in a flow cell connected to a CO_2–permeable membrane equilibrator, a light source and detection hardware, and a pumping system that delivers blank solutions and renews the pH indicator solution (5×10^{-5} mol L^{-1} bromothymol blue degassed and titrated to pH 9.0 with 1.05 mol L^{-1} NaOH) (Figure 4). DeGrandpre has named this instrument the Submersible Autonomous Moored Instrument for CO_2 (SAMI-CO_2). The indicator solution is stored in polyethylene-coated aluminum bags to ensure constant alkalinity, and the periodic renewal of this

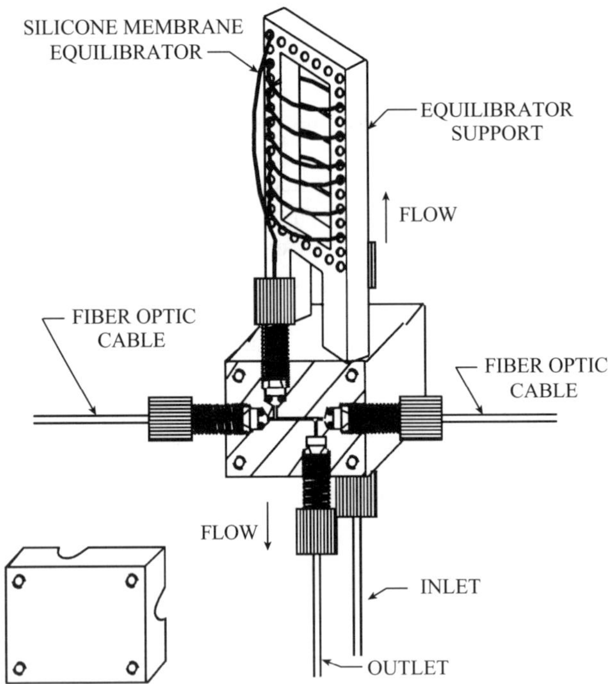

Figure 4. A cutaway view of the SAMI-CO_2 fiber-optic flow cell with the tubular gas-permeable membrane equilibrator [28]

Table 7. SAMI-CO_2 operating characteristics, from DeGrandpre *et al.* [28]. 0.1 Pa is roughly 1 μatm

Tested operating range	20–60 Pa
Measurement precision	$\pm$0.1 Pa
Measurement accuracy	$\pm$0.2 Pa
Response time	5 min
Temperature coefficient	0.4 Pa °K^{-1}
Reagent consumption	4 ml d^{-1}
Power consumption	14 A h*

*About 2 months operation.

solution enhances the stability and sensitivity of the instrument. Table 7 summarizes the SAMI-CO_2 operating characteristics. It has been used on many ocean moorings and has provided reliable pCO_2 measurements for periods up to 6 months. Leaking reagent bags and biofouling have been the only significant problems that have compromised some data collections (DeGrandpre, personal communication). The hardware components used with fiber optic sensors (e.g. light source, optical couplers, amplifiers and photo diode components) are very reliable and rarely limit sensor resolution and performance [90].

Most other fiber optic analytical systems for the measurement of pCO_2 have not progressed beyond the prototype stage. Goyet *et al.* [91] developed a pCO_2 sensor for seawater based upon the fluorescence of HPTS (hydrooxypyrenetrisulfonic acid) combined with two absorbing dyes that increase the fluorescence intensity change over a narrow pH range. These dyes were encapsulated behind a molded, gas-permeable, silicone membrane that was held in place at the end of a 220 μm diameter silica fiber. The sensor was used at sea to measure pCO_2 in seawater samples collected by hydrographic casts. By comparisons with gas chromatographic analyses, the sensor was accurate to $\leq 2\%$ (0.9 Pa). However, Goyet *et al.* report that the sensor response time was approximately 40 min, the temperature dependence of the dye mixture was complex, and after 2 weeks of experiments, photobleaching of the encapsulated fluorescent dye degraded the sensor performance.

Another fluorescence-based fiber optic sensor has been developed by Tabacco *et al.* [92] and deployed off the side of a buoy $\sim$ 2 m below the sea surface in Vineyard Sound, off Massachusetts, USA, for a period of 7 weeks. This sensor relies on a new pH sensitive dye, 5,6–carboxyseminaphthofluorescein, which is held within an internal membrane of poly(*N*-vinyl)pyrrolidone at the face of a 400 μm diameter single-core fiber. A 10 μm thick Teflon membrane serves as the CO_2-permeable barrier to the outside solution and the disk of poly(*N*-vinyl)pyrrolidone polymer is backed by a reservoir of $\sim$ 50 μL of indicator solution that is held in a cavity around the fiber. At constant temperature, the ratio of dye emission at 540/630 nm can be linearly related to

pCO_2 over the 20–100 Pa range. This is the range that is most commonly encountered in surface seawater. The sensor, however, has a very long equilibrium time (about 100 min), and not surprisingly it requires an independent measurement of temperature for accurate determination of pCO_2. When deployed in the field it clearly tracked diurnal variations in pCO_2 but showed an upward drift in the signal ratio over time that was attributed to biofouling. The authors report they are addressing the biofouling problem by coating all supporting tubing with antifoulant paint.

In contrast to surface seawater, porewaters may have much higher pCO_2 values that must be resolved rapidly over small spatial scales. The most successful prototype fiber-optic pCO_2 sensor for *in situ* porewater profiling is described by Hales *et al.* [30]. This absorbance-based sensor is bulky compared with microelectrodes, and in field trials it suffered from leakage problems under high pressure as a result of the differing compressibilities of silicone-based and glass components. However, it has a response time of only a few minutes and is capable of resolving differences in pCO_2 of less than 8%. Hales *et al.* [30] have suggested that the sensor is better suited for freshwater measurements than marine because the internal solution of bromothymol blue may be made up to higher concentrations in a low ionic-strength aqueous solution without the problem of an osmotic gradient across the sensor membrane.

5 pH MICROELECTRODES

5.1 VARIETIES OF pH MICRO-SENSORS

As early as 1927, Taylor and Whitaker [93] developed a platinum-based pH microelectrode for physiological research. However, major advances in pH microsensor applications began only after glass pH microelectrodes were developed in the 1960s [94,95] and in the 1980s when pH liquid membrane microelectrodes based on neutral carriers became available [96–98]. Beginning in 1983, pH microelectrodes were applied by marine scientists to characterize biofilms, microbial mats [16,78,99–102], sediments [17,18,29,78,103] and microenvironments in and around marine snow and planktonic organisms [104,105]. This work is being continued actively today.

Three types of pH microelectrodes are currently available: glass, metal and liquid membrane (including polymeric membrane). Glass microelectrodes have excellent selectivity and cover a wide dynamic range (pH 0–14), but they require special equipment and skill to fabricate. They also have a very high resistance (more than 10 GΩ [17,95]). Metal oxide electrodes have very low resistance (note that a high impedance electrometer, such as a pH meter, is still required to limit the current drawn to pA level and to ensure a stable potential reading), but they may be subject to severe interference by redox reactions [106]. Therefore their applications have been limited. Ion-selective liquid membrane microelectrodes (LMM) based on neutral carriers are now a powerful technology [96]. They have dynamic response characteristics similar to those of glass electrodes, but unlike glass microelectrodes, they can be constructed easily and their inner resistance is not very high.

An additional type of pH sensor that may have a great potential for *in situ* analysis is based on the ion-sensitive field-effect transistor (ISFET). The ISFET is a new integrated device composed of an ion-selective electrode and an insulated-gate FET [107] (also see Chapter 12 of this book). ISFETs are advantageous in their rapid response, rugged structure, small size and low impedance. ISFETs are not expected to be sensitive to organic contaminants and redox species in natural environments. An interesting device has been developed that places the ISFET sensor next to an actuator that generates H^+ or OH^- coulometrically to determine base or acid concentrations rapidly and *in situ* [108,109]. This device is intriguing to marine chemists because it could be developed into an *in situ* alkalinity analyzer for deep-sea use. We, however, have not seen any reports of application of ISFET-based sensors in seawater.

5.2 PRINCIPLES GOVERNING POLYMERIC LIQUID MEMBRANE pH MICROELECTRODES BASED ON NEUTRAL CARRIERS

In this and the next two sections the principles and properties that are responsible for the characteristics of polymeric liquid membrane pH electrodes are described, because these sensors have clear advantages over other types of pH sensors and because their aquatic applications show great promise. The key part of a polymeric liquid membrane microelectrode is a pH-sensitive membrane that consists of a pH-selective material composed of a neutral proton carrier dissolved in a membrane solvent. The membrane solvent is not miscible with water and so forms an organic phase that separates the aqueous sample solution from the aqueous internal filling solution. The neutral carriers are capable of selectively extracting ions from aqueous solution into the membrane phase and transporting them across the organic phase by carrier translocation. A membrane potential is established during the process and can be measured against a reference electrode. It is important to realize that, similar to pH measurements with glass membrane electrodes, the proton flux across the polymeric liquid membrane and the resultant current (at pA level) should be extremely low to ensure a stable membrane potential which is the parameter to be measured. This is different in principle from the permeation liquid membrane (PLM) (see chapter 10 of this book) where the flux through the membrane is the measured parameter and should be high to ensure a good sensitivity. Furthermore, unlike liquid membranes that contain ion exchangers, neutral carriers act to complex analyte ions and this selectivity is governed by the binding properties of the neutral carrier, instead of the properties of the solvent. Therefore, the membrane containing a neutral carrier usually has a better selectivity than the membrane containing an ion exchanger [96].

The purpose of the membrane matrix polymers (usually PVC) in polymeric liquid membranes with neutral carriers is only to support the membrane solution. They are added to the pH membrane solution with other additives that may act as lipophilic anionic sites to maintain electroneutrality in the membrane [110]. The lipophilic anionic additives also play a very important role in membrane selectivity by excluding lipophilic anions in the test solution (especially blood samples) from entering the membrane. The matrix polymers do not alter the microelectrode selectivity or slope, but may increase the lifetime of the microelectrodes by reducing the loss of the membrane solution. However, membrane resistance will increase by adding polymer to the membrane.

Microelectrodes usually have higher resistance than macroelectrodes. It is therefore, important to reduce the membrane resistance of microelectrodes [98]. Polar solvents are used in membrane solutions for this purpose. Polar solvents, however, are more soluble in aqueous solution and therefore reduce the lifetime of membrane electrodes. Increasing the percentage of the neutral carriers and additives in the membrane solution can also reduce the membrane resistance since the concentration of charged particles will increase as the concentration of neutral carrier increases. The amount of additives must be much lower than the neutral carriers (Table 8); otherwise it leads to a dramatic decrease in membrane

Table 8. Compositions of some polymeric membranes for pH microelectrodes (after Zhao and Cai [33]).*

Membrane components	Membrane contents		
	A	B	C
Liquid membrane	66.4%	70.1%	86.6%
PVC support	33.6%	29.9%	–
PVC-COOH support			13.4%
Liquid membrane composition			
TDDA	9.9%	–	9.8%
ETH 1907	–	6.2%	
KT_4ClPB	1.1%	1.1%	1.2%
2-NPOE	90.0%	92.7%	88.9%

* The polymeric membrane composition is shown in the first half of the table where liquid membrane as a whole is viewed as one component. The detailed composition of liquid membrane is shown in second half of the table. Composition A is used most often in Cai's laboratory. Other polymeric membranes, such as membranes containing 5%, 14% and 42% PVC, 14% PVC-COOH, or mixed membranes containing 15% PVC and 15% PVC-COOH have been tested with TDDA. TDDA from 5% to 10% has also been tested.

selectivity and may yield a membrane with properties comparable with those of a classical cation-exchanger membrane microelectrode [111]. Therefore the amount of additive that can be added is limited. The upper limit of the carrier concentration is usually restricted by the solubility of the carrier molecules in the membrane phase.

Lastly when conducting pH measurements in sediments, the sensing membrane in the tip of pH microelectrodes should be able to withstand abrasion by the sediment particles. Liquid membrane pH microelectrodes without matrix polymers (PVC) will not satisfy this requirement. Addition of less than 42% of PVC does not affect the response characteristics of a pH microelectrode. This issue is discussed in detail in ref. [33].

5.3 PREPARATION OF THE LIQUID MEMBRANE pH MICROELECTRODE

5.3.1 Reagents and Materials

The two most frequently used neutral carriers are tridodecylamine (TDDA) and 4–nonadecylpyridine (ETH1907). A new type of neutral carrier is also available from Fluka Chemical Inc. and was used by de Beer *et al.* [78]. The other reagents most frequently used for liquid membrane solutions are: 2–nitrophenyl octyl ether (2–NPOE) as the solvent, potassium tetrakis(4–chlorophenyl)borate (KT_4ClPB) as lipophilic anion, and poly(vinyl chloride) (PVC) as polymer. A silanization reagent is *N*, *N*-dimethyltrimethylsilylamine. These reagents are all available from Fluka Chemical Inc. Commercially available Pyrex glass capillary tubing is most convenient for the body of pH microelectrodes. Capillaries can also be pulled from larger borosilicate or aluminosilicate glass tubing. Table 8 lists various components that have been used successfully in constructing liquid membrane microelectrodes.

5.3.2 Methods

The preparation of liquid membrane pH microelectrodes involves three steps.

(1) Preparation of glass micropipettes. Glass micropipettes are drawn by a micropipette puller (an inexpensive vertical puller from Sutter Instrument Co. works fine). The tips of the newly pulled micropipettes are then broken to a diameter of about 20 μm for sediment applications or smaller for biofilm applications. This may be accomplished by lowering the tip until it touches a polished plexiglass surface (or a quartz surface). When the tip and its mirror image in the plexiglass are in contact (seen under a microscope), a slight tap on the holder of the plexiglass usually causes the tip to break. If the tip diameter is to be larger than 20 μm, it can be readily made with a fine glass tubing cutter.

(2) Silanization. The inner surface of a glass micropipette has to be made hydrophobic in order for the membrane solution to be held in the tip [112]. To do so, a batch of micropipettes with appropriate tip diameters are put in an oven on a watch glass covered with a glass bottle. After pre-drying at 150°C for 30 min, the temperature is raised to 200 °C and then about 50 μL of silanization reagent (*N*, *N*-dimethyltrimethylsilylamine) is injected into the bottle and is allowed to react with the glass surface in the vapor phase at 200 °C for 30 min. This procedure is the same as described in ref. [97] except that a bottle is used instead of a beaker to prevent the loss of the silanizing reagent. The silanized micropipettes can be stored in a desiccator with silica gel for about 1 week.

(3) Filling. The liquid membrane solution for the pH microelectrode typically consists of a neutral carrier, membrane additive (KT_4ClPB), membrane solvent (2–NPOE) and membrane matrix polymers [32,78]. Since the polymer cannot dissolve in the solvent directly, newly distilled tetrahydrofuran (THF) is added to the mixture of components at a volume ratio of about three to four times, and then the membrane solution is vibrated to accelerate the dissolution process. When a homogeneous membrane solution forms, the THF is allowed to evaporate at room temperature until the membrane solution has a suitable viscosity for filling (when tilting the vial, the solution can still flow along the inner wall slowly; some practice is needed to obtain a 'suitable viscosity'). The silanized micropipettes are filled with a pH 7 phosphate buffer from the shaft-end by using a fine plastic capillary (pulled from a plastic 100 μL pipette tip). A pressure to the top of the micropipette can be applied to force the internal solution to flow into the tip. Then the micropipette is immersed into the polymeric membrane solution while observing the process under a microscope. A vacuum is applied to the shaft-end of the micropipette to draw a column of membrane solution into the tip of the micropipette (filling height between 500 and 1000 *μm*). No air space should be left between the membrane solution and the internal filling solution. Finally, an Ag/AgCl wire is inserted into the shaft of the micropipette as an inner reference electrode. The microelectrodes are allowed to age and stabilize (conditioning) in pH 7 buffer (or air) overnight before use. If conditioned in air, a microelectrode will need to be placed in buffer for at least half an hour before use. After conditioning, the final membrane height decreases to 200–500 μm because of the evaporation of THF. Figure 2 illustrates the whole electrode.

5.4 PERFORMANCE OF LIQUID MEMBRANE pH MICROELECTRODES

5.4.1 Response Time and Resistance

Response time and resistance are associated with each other. High resistance usually causes long response time. The PVC membrane microelectrode based

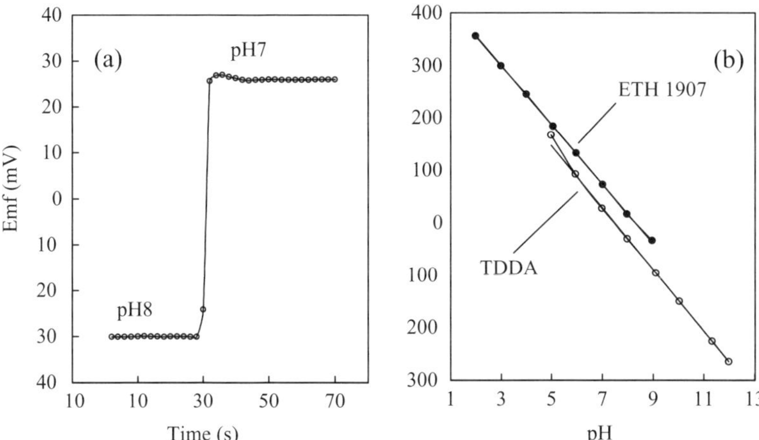

Figure 5. Emf versus response time and Emf versus pH of a liquid membrane pH microelectrode (tip diameter =15 μm for TDDA and 25 μm for ETH1907) (from ref. [33])

on TDDA or ETH1907 responds rapidly in buffers (98 % of response, $t_{98} < 4$ s) (Figure 5). The corresponding membrane has a resistance of 1–4 GΩ for microelectrodes with tip diameters of 10–30 μm. Glass pH microelectrodes normally have a higher resistance and slower response (over 10 GΩ, with tip diameter of 100–200 μm [17]) than the liquid membrane pH microelectrode. The resistance of a liquid membrane pH microelectrode decreases as the tip diameter increases. Therefore, larger microelectrodes tend to have shorter response times and are more stable. However, if the tip diameter is larger than 60 μm, the membrane solution can run out easily. Large tips may also disturb microenvironments. We recommend using liquid membrane pH microelectrodes with tip diameters of 10–30 μm for studies of fine-grained sediments.

5.4.2 Slope and Dynamic Response Range

The dynamic response range is from pH 5.5 to12 and from pH 2 to 9 for TDDA- and ETH 1907-based pH microelectrodes, respectively (Figure 5). These ranges cover the usual pH conditions found in seawater and marine sediments (pH 5.5–8.5). TDDA-based pH microelectrodes often have a super-Nernst response from pH = 5 to 5.5 [97,98], but are Nernstian above pH 5.5. Occasionally, the pH microelectrodes based on neutral carriers may have a super-Nernstian (60–65 mV pH^{-1} at 25°C) or sub-Nernstian slope (55–58 mV pH^{-1} at 25°C) in the normal dynamic response range. This abnormality may be linked to the quality of the reagents. We have found that the electrodes exhibit ideal slopes when all the membrane reagents are replaced with new ones. Newly bought reagents can be used for more than half a year without problems developing [33].

5.4.3 Lifetime and Performance

The lifetime and performance of pH polymeric membrane microelectrodes based on neutral carriers depends on several factors including the quality of the silanized glass pipettes, the height of membrane column and tip diameter of the glass pipettes. If a glass pipette with a low quality of silanization is used, the pH microelectrode may not respond properly to pH changes due to charge leakage. The silanized glass pipettes should be stored in a dust-free and dry location (such as a desiccator) and should not be used if they have been stored for more than 1 week. If the initial filling height is too short, membrane shrinkage and leakage can occur after several days. In addition, if the tip of the micropipette is too large, the membrane will be lost easily. Under normal conditions, the lifetime of the pH microelectrode is 2–5 weeks.

The pH microelectrodes without polymer can be used immediately, but the pH microelectrodes with polymer should be conditioned overnight. Additionally, the TDDA membrane is very unstable in the certified pH 4 buffer and contact with this buffer should be avoided. It is not clear whether phthalate, the main component of the pH 4 buffer, interferes with the TDDA membrane as an anion [33].

5.4.4 Comparison of Liquid Membrane pH Electrodes and Glass Electrodes

Glass pH electrodes are known for their stability in many environments including sediments [16,17,95,113] (see also Chapter 5). Although we have not systematically compared glass microelectrodes with liquid membrane microelectrodes, our experience suggests that the glass microelectrodes are more durable and have a longer lifetime than the liquid membrane microelectrodes when these sensors are used for profiling sediments. Because of the very high impedance of the glass membrane, glass microelectrodes also have a longer response time than liquid membrane microelectrodes (e.g. most glass pH microelectrodes have reported 98 % response times of a few minutes, whereas the liquid membrane microelectrodes have 98 % response times of only a few seconds). While glass microelectrodes of various tip sizes can be made [16,17,95,113], glass mini-pH electrodes (tip diameter 2–3 mm, from Mettler Toledo/Ingold or 1 mm from Microelectrode Inc. with separate reference electrodes) are commercially available and convenient for comparison with the polymeric microelectrode. The glass mini-pH electrodes (from Mettler Toledo/Ingold) have also been used for *in situ* porewater pH measurements in deep-sea sediments [114,115]. Figure 6 shows pH profiles obtained by a glass mini-electrode (tip diameter 2–3 mm) and two PVC liquid membrane

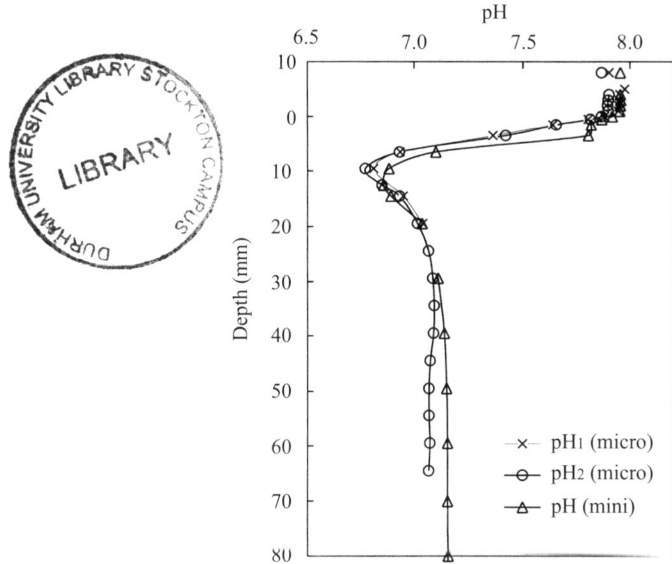

Figure 6. Comparison of porewater pH profiles measured by two PVC liquid membrane microelectrodes, pH_1 (micro) and pH_2 (micro), and a glass mini-electrode. The microelectrodes are PVC liquid membrane electrodes (tip diameter ~15 μm) while the glass mini-electrode has a tip diameter about 2–3 mm. The sediment was collected on June 22, 1998 from Buzzards Bay (Cape Cod, MA, a site with numerous previous studies [117]). The sediment was homogenized and then incubated in the laboratory for 10 d. Large benthic animals were excluded from the incubated sediment. Overlying water was replaced with seawater collected from the same site every other day during the incubation. Profiles measured in intact cores have similar features (Cai, unpublished).

microelectrodes (tip diameter≈ 15 μm) in a marine sediment (mud). The pH profiles measured by both types of electrodes agreed in general. This comparison supports the reliability of pH measurements by polymeric membrane microelectrodes in aquatic sediments. However, the mini-electrode severely underestimated the pH change near the sediment–water interface. Probably, the mini-electrode pushed mud from the uppermost sediment into lower layers, altering the pH in the lower layers [116]. We suggest that the commercially available glass mini-electrode may be used to estimate the general shape of pH profiles in marine sediments for its availability, convenience and mechanical strength. However, if these sensors are used to determine sharp pH gradients near the sediment–water interface, these gradients may be seriously underestimated.

6 pCO_2 MICROELECTRODES

6.1 GENERAL REVIEW OF THE DEVELOPMENT OF pCO_2 MICROELECTRODES

There are three types of potentiometric pCO_2 microelectrodes including the classical Severinghaus type [17,118,119]. In one of the newly designed pCO_2 microelectrodes, the gas-permeable hydrophobic membrane is replaced by an air gap (Figure 7) [120]. This design achieves a fast response by eliminating the need of gas transfer through a hydrophobic membrane and is excellent for measurements of pCO_2 in small volume aqueous samples in the laboratory. However, it is not suitable for underwater or sediment measurements since the air gap will disappear under high pressure and will be intruded by sediment particles.

In another design, a pH-sensing PVC membrane also functions as a gas-permeable membrane (Figure 8) [121]. Two nearly identical pH micro-pipettes can be pulled together side by side. One of them (A) is filled with pH buffer and the other (B) with $NaHCO_3$ solution. This design is simple to construct and allows simultaneous measurements of pH (by A) and pCO_2 (by the difference of

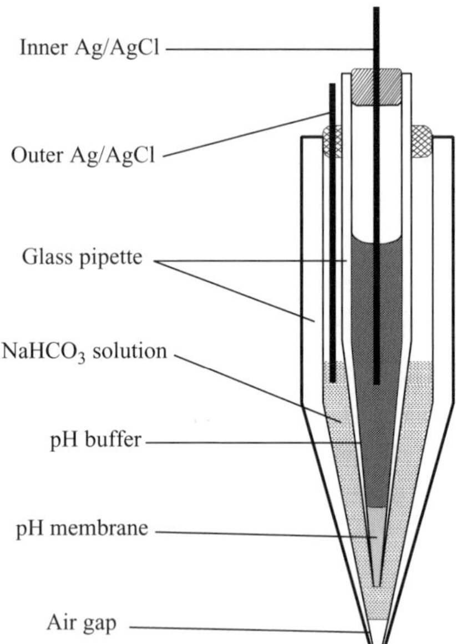

Figure 7. An air gap pCO_2 microelectrode similar to that of Ma [120]. The air gap (length is 125–250 μm) was made hydrophobic by repeatedly dipping the pipette tip into dimethyldichlorosilane /carbon tetrachloride solution and heating. Parts are not drawn proportionately. Such an effort was also made earlier by Pui *et al.* [36].

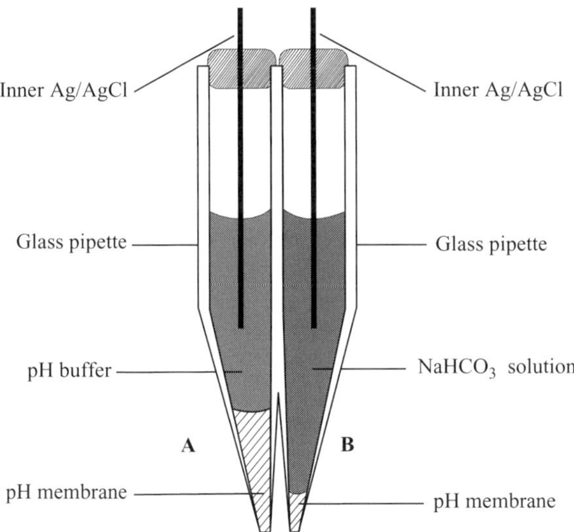

Figure 8. A pH and pCO_2 combination microelectrode (modified after Bomsztyk and Calalb [121]). Parts are not drawn proportionately

A and B) at the same location. However, there is a serious drawback with the design. CO_2 will not come to equilibrium quickly near the pH-sensitive membrane (in B), since there is no internal barrier (such as the pH electrode in the Severinghaus design) to block the exchange of CO_2 with the entire reservoir of bicarbonate solution. This design problem may be modified by inserting a barrier inside electrode B. Another serious problem may be described as a dilemma between having a thin film (a fast response) and a physically strong and durable membrane. Unlike a silicone membrane, most thin PVC membranes do not adhere to the glass tightly, which results in the leakage of the internal solution and therefore very short lifetime (Cai, unpublished). Therefore, it seems that Severinghaus's design is still the best design for pCO_2 measurements, especially for the purpose of low pCO_2 measurements in natural environments. Nevertheless, the design in Figure 8 is very attractive since it is a combination electrode of pH and pCO_2.

The early use of pCO_2 electrodes and microelectrodes was confined to studies of media with high $[CO_2]$ such as blood [118,119]. As indicated earlier, the $[CO_2]$ or pCO_2 of most natural waters (10–100 μmol kg^{-1} or 30–300 Pa, Table 1) is well below the detection limit of commercial pCO_2 electrodes while pCO_2 values in sediment porewater can be much higher (over 10^4 Pa). Although there are several commercial suppliers of pCO_2 microelectrodes, it should be noted that these electrodes are not designed for pCO_2 measurement in common aquatic environments. In addition, while these electrodes are called 'microelectrodes', their diameters are normally a few mm. The first pCO_2 microelectrode to be applied in a sediment for biogeochemical study was prepared by Cai and Reimers [17] using a glass pH microelectrode as the internal sensing unit. The electrode had a tip diameter of about 250 μm and had a response time of 5–10 min when it was newly made, but the performance degraded quickly (longer response time, less sensitive and more serious hysteresis). Because of this deterioration problem, *in situ* pCO_2 measurements at the sea floor have been only moderately successful [17,29]. However, using similar pH and pCO_2 microelectrodes and taking care to let the sensors respond fully, Komada *et al.* [31] successfully conducted a study of the diffusive behavior of carbon dioxide species in sediments incubated in the laboratory.

It is hoped that greater pCO_2 microelectrode use may improve our understanding of several areas of sediment biogeochemistry, in particular the rate of $CaCO_3$ dissolution in deep-sea sediments. Much effort has been given to the development of a more stable and faster or a smaller pCO_2 microelectrode [31,32,78]. Cai (unpublished) built a pCO_2 microelectrode with the Ir-oxide pH sensor and achieved a similar response as in Cai and Reimers [17] but a much longer lifetime and better stability. It has also been found that the most important factor that limits the lifetime and performance of all previous pCO_2 microelectrodes is the release of interference materials (see section 6.3.1) from the glass tubing used as the outside housing [32].

6.2 PREPARATION OF THE SEVERINGHAUS-TYPE pCO_2 MICROELECTRODE

A Severinghaus-type pCO_2 microelectrode after the design of Zhao and Cai [32] is presented in Figure 2. The outer pipette of aluminosilicate glass is pulled by a micropipette puller in two steps, first to a diameter of ~ 2 mm and then to ~ 1 μm. The tip of the outer pipette is then cut to 100–250 μm. The tip size can be further reduced if needed [78]. A silicone membrane is formed in the tip by dipping into a silicone adhesive while observing under a microscope. If the original silicone adhesive is too viscous for dipping, it can be diluted by THF to a suitable viscosity in order to prepare a very thin membrane (10–30 μm [32]). The silicone membrane is cured in the air for at least 24 h. The micropipette, with the silicone membrane, is then filled with internal solution (2 mmol L^{-1} $NaHCO_3$, 0.5 mol L^{-1} NaCl) by using a plastic tube with a very fine capillary tip so that the tip can be placed as close as possible to the silicone membrane. Any air gap formed between the silicone membrane and the hydrogen carbonate solution can be dispelled by several methods. If the volume of the air gap is very small, a pressure can be applied to the other end to push the air in the tip through the membrane. Air, trapped at the tip, can also be dispelled by tapping the larger end gently or by placing the sensor under a vacuum.

The next step is to match the inner pH microelectrode and the outer pipette while observing under a microscope. To do this, the outer pipette is fixed on a microscope stage and the pH microelectrode on a 3–D manipulator. The outer pipette and the pH electrode must be kept in the same horizontal plane. The relative position of the outer pipette and the pH electrode is adjusted carefully in order to set the pH microelectrode in the center of the outer micropipette to a distance of ~ 5–30 μm from the silicone membrane. UV-sensitive epoxy glue is used to fix the inner and outer pipettes together. Then, a two-part epoxy glue can be used to strengthen the bond. Finally, a Ag/AgCl reference electrode is inserted into the internal solution in the outer pipette, and the pCO_2 microelectrode is ready for use.

6.3 IMPORTANT CONSTRAINTS ON THE PERFORMANCE OF pCO_2 MICROELECTRODES

6.3.1 Interference from Glass Material

Morf *et al.*[122] have discussed the response time of a CO_2 sensor to an interfering gas that acts as an acid or base (for example, H_2S and NH_3). Acids and bases can also form from the dissolution of glass materials species (i.e. those of Si, Pb, Al, B, etc.) into the $NaHCO_3$ solution. The chemical durability of a glass tubing mainly depends on the composition of the glass [123,124]. Table 9 shows the composition and some properties of the four glasses (Al, B, Pb, and soda lime) that have been used for pCO_2 microelectrodes. The chemical durability of the glasses decreases in the sequence: Al > B > Pb~soda lime. Exact behavior depends on the content of alkali, which can easily dissolve in aqueous solution, and the content of Al_2O_3 and B_2O_3, which can reduce the solubility of alkali (Al_2O_3 is more effective) [123]. The initial step of the corrosion of alkali–silicate glass is the release of alkali ions into the $NaHCO_3$ solution, and it can increase the alkalinity of the solution ($Na^+_{(s)} + H_2O = H^+_{(s)} + Na^+ + OH^-$, here s indicates solid surface). After this initial step, the Si–O network will be broken-up: $SiO_{2(s)} + H_2O = H_2SiO_3$. This step releases a weak acid into the solution. pCO_2 microelectrodes made from Pb-glass showed great drift and very short lifetimes because of serious corrosion of alkali–silicate glass (Cai, unpublished). In almost all the previous pCO_2 microelectrodes, Pyrex was used to construct the outer pipettes [35,120,121,125–127] Unfortunately, Pyrex also dissolves in aqueous solution (Table 9). This may be the main reason why the pCO_2 microelectrodes reported previously showed considerable drift and short lifetime. Alumina glass has a very small amount of alkali and shows the best chemical durability [95,123,124,128]. Cai and Reimers [17] and de Beers *et al.* [78] used soda lime glass for the outer pipettes of their pCO_2 microelectrodes. Therefore, these microelectrodes also had serious drift and a short lifetime.

Table 9. The composition and properties of glasses used for pCO_2 microelectrodes*

Glass	Corning number	Softening point (°C)†	Weight loss ($mg\ cm^{-2}$)‡	Composition (wt %)							
				SiO_2	B_2O_3	Al_2O_3	CaO	MgO	PbO	Na_2O	K_2O
Alumina	1710§	915	0.35	64	4.5	10.4	8.9	10.2		1.3	0.2
Pyrex	7740	820	1.4	81	13	2				4	
Lead	8870§	630	3.6	62		2			22	7	7
Soda lime	0080	696	1.1	72.6	0.8	1.7	4.6	3.6		15.2	

* Data are from Bansal and Doremus [124].
† The softening point data are from Houde Glass Catalogue for 1720,7740,0120 and 0080.
‡ Weight loss in 5% NaOH solution at 100 °C for 6 h.
§ Zhao and Cai [32] used Corning 1720 Al glass and Corning 0120 Pb glass, which have similar composition to Corning glasses 1710 and 8870.Cai and Reimers [17] also used 0120 Pb glass for the pH microelectrode.

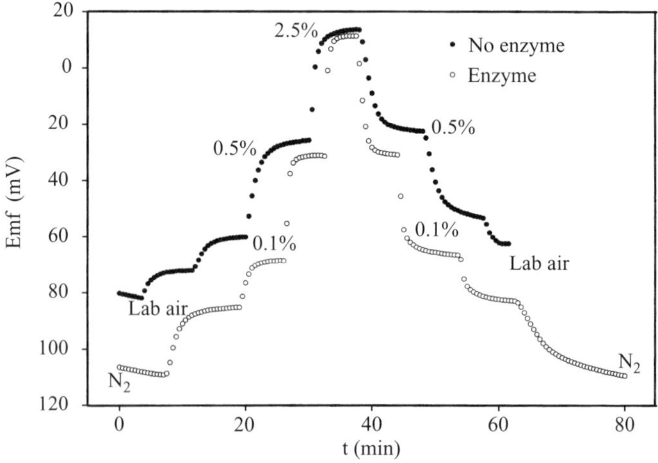

Figure 9. Effect of carbonic anhydrase enzyme on the signals of pCO_2 microelectrodes subjected to gases with variable CO_2 content. The electrode without the enzyme had an outer tip of 170 μm, a silicone membrane with a thickness of 30 μm, and an inner pH electrode with a tip diameter of 25 μm. The electrode with the carbonic anahydrase had an outer tip of 130 μm, a silicone membrane with a thickness of 35 μm, and an inner pH electrode with a tip diameter of 25 μm. The inner solutions were 2 mmol L^{-1} $NaHCO_3$ + 0.5 mol L^{-1} NaCl. The carbonic anhydrase concentration was 0.75 g L^{-1}.

6.3.2 Kinetics of CO_2 Hydration and the Use of Enzyme

It is well known that the CO_2 hydration reaction is accelerated by the enzyme carbonic anhydrase [129]. Some studies have shown that adding carbonic anhydrase to the $NaHCO_3$ solution in pCO_2 electrodes can cause a pronounced reduction of response time and hysteresis (keeping in mind that the response curve for an increase in pCO_2 is significantly different from that for a decrease in pCO_2). Severinghaus initially found that the enzyme had no effect on response rate [118], but later reported that the response might be accelerated if $1\,mg\,mL^{-1}$ carbonic anhydrase were added to the filling solution [119]. Donaldson [130,131] confirmed that carbonic anhydrase could improve the response and hysteresis of Severinghaus's pCO_2 electrodes. Carbonic anhydrase has been used also in an optical fiber pCO_2 sensor [132].

Figure 9 demonstrates that the response of a pCO_2 microelectrode can be improved greatly after adding the enzyme, that is, the response time and hysteresis are reduced and the response slope is enhanced. The detection limit is also lowered after the addition of enzyme. The effect of enzyme is not obvious for high pCO_2 measurement but is very clear during low pCO_2 measurement [32]. This might be the reason there has been no consensus on the role of enzyme in the past when high pCO_2 situations were studied.

6.4 PERFORMANCE OF NEW pCO_2 MICROELECTRODE DESIGNS

By using glass tubing with high chemical stability (i.e. aluminosilicate glass), a neutral carrier-based PVC liquid membrane pH microelectrode (tip diameter 15 μm) and a CO_2 transfer enzyme (carbonic anhydrase), the new pCO_2 microelectrode outperforms all previous ones in its response time (2 min for 98% response at pCO_2 above 80 Pa), stability ($0.4\,mV\,h^{-1}$) and shelf life (1 month) [32]. A typical response and calibration result of such a pCO_2 microelectrode is

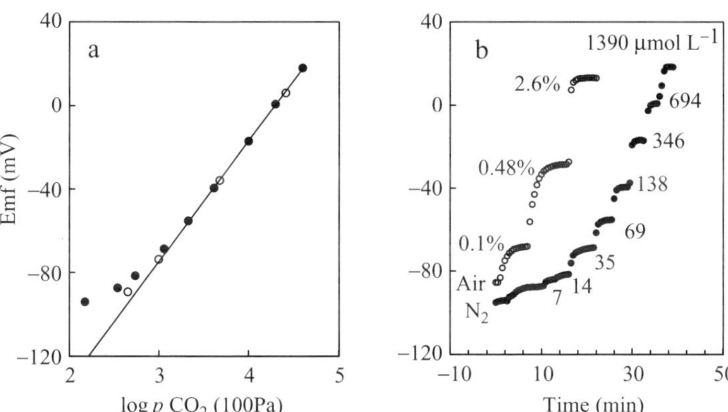

Figure 10. The calibration (part a, Emf versus log pCO_2) and response (part b, Emf versus time) curves of a typical pCO_2 microelectrode. The internal solution contains $2\,mmol\,L^{-1}$ $NaHCO_3$, $0.5\,mol\,L^{-1}$ NaCl and $0.75\,g\,L^{-1}$ enzyme (carbonic anhydrase). Curves with open circles are calibration and response in standard CO_2 gases. Curves with closed circles are calibration and response in HCl solution with added known amount of $NaHCO_3$ solution. In part b, the numbers by the curve indicate partial pressure (%) or concentration ($\mu mol\,L^{-1}$) of CO_2 for curve (o) and curve (•) respectively. Laboratory air contains ~40 Pa (~400 μatm) CO_2

given in Figure10. Nearly perfect Nernstain response is achieved with the new pCO_2 microelectrode. Accurate sensor measurements have also been confirmed by pCO_2 calculations based on sediment porewaters whose DIC and alkalinity values were measured independently (Cai, unpublished). Calculated DIC from pH and pCO_2 data can also be accurate ([31], Cai, unpublished and this chapter).

Another pCO_2 microelectrode based on neutral carrier liquid PVC membranes has been developed by de Beer *et al.* [78]. De Beer *et al.*'s microelectrode is smaller in size (10 μm diameter) and may have a faster response (10 s for 90 % response). This sensor is particularly useful for applications in bio-film and microbial mats where fine resolution may be critical. However, it is reported to have a serious drift problem and a short lifetime. Zhao and Cai's [32] pCO_2 microelectrode has a much longer lifetime and a higher stability which are preferred for field applications.

7 APPLICATION OF pH AND pCO_2 SENSORS

Extensive evaluation of the pH and pCO_2 microelectrodes for *in situ* measurements in the field is clearly needed if we are to use them with full confidence to study organic carbon degradation and $CaCO_3$ dissolution in the water column or near sediment–water interfaces. In this section, applications of the pH and pCO_2 sensors for *in situ* measurements in the water column and in sediments are discussed in more detail.

7.1 APPLICATIONS IN THE WATER COLUMN

The early efforts to use glass electrodes to measure water column pH profiles [5,6] have been largely replaced by more precise spectrophotometry-based methods. Most notably, Byrne *et al.* [133] have developed a deployable analyzer that measures conductivity, temperature depth, and pH via the absorption of seawater mixed with thymol blue indicator in two cells at 435 and 536 nm. This instrument can return one pH measurement per second with a precision of ±0.001 pH units and has been deployed to 500 m. Several other groups are now routinely measuring the pH of a pumped stream of surface water during underway surveys either using rugged diode array spectrophotometers [84] or reverse flow injection analysis [134].

pCO_2 may also be measured with automated shipboard-based equilibrator systems [24,25], but as discussed in section 2.3, the use of compact pCO_2 sensors has been largely confined to moorings. Figure 11 illustrates time-series data from a moored SAMI system [135] which indicates very large pCO_2 changes in short time periods (<1 d) in ocean surface waters. These changes may be explained by the superimposition of diurnal biological processes,

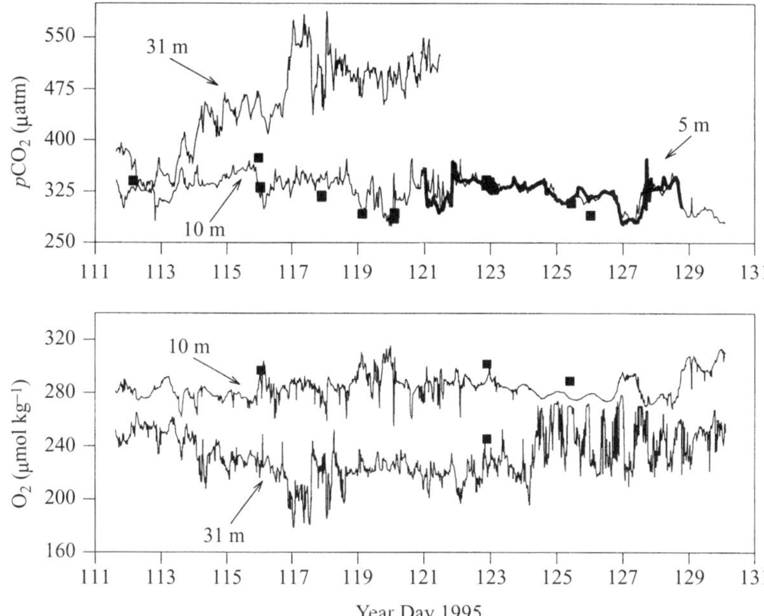

Figure 11. A 19 d time series of seawater partial pressure of carbon dioxide (pCO_2) (measured by the SAMI pCO_2 analyzer) and dissolved O_2 (measured by pulsed O_2 electrodes made by YSI Inc., Yellow Springs, Ohio, USA) at different ocean depths during April–May 1995 off Monterey, California. The squares represent discrete measurements from a surface vessel. The 5 m pCO_2 data constitute the thick line that overlays the 10 m data. In the original figure, (~0.1 Pa) is used.

physical advection, gas exchange and temperature effects. The only comparable measurements are those of Friederich *et al.* [136], who measured the difference between sea-surface and atmospheric pCO_2 by means of a moored infrared analyzer (LI-COR, Inc) operating in a differential mode with inlets from a compact equilibrator and the atmosphere.

7.2 APPLICATIONS IN SEDIMENTS

The first application of a pH microelectrode *in situ* under deep sea conditions was conducted by Archer *et al.* [18]. pH was observed to decrease by less than 0.1 units from bottom water to depths of 1–2 cm in calcite-rich sediments. Archer *et al.* [18] modeled the processes responsible for these pH profiles as well as for O_2 profiles collected simultaneously. They inferred that the calcite dissolution rate constant for deep-sea sediment was much lower than that previously measured in laboratory experiments [137]. *In situ* pH profiles have also supported the suggestion by Emerson and Bender [138] that $CaCO_3$ is dissolved in sediments by the CO_2 introduced to porewaters by organic matter degradation [18,29,114,115]. This work has led to the hypothesis that changes in the relative rain rates of organic carbon and $CaCO_3$ to the sea floor will alter pCO_2 levels at the sea surface and in the atmosphere on glacial–interglacial time scales (thousands of years) [139].

The first *in situ* porewater pCO_2 micro-profile was collected in San Diego Bay at a 4 m water depth [17]. In only a few mm depth in sediment, pH (seawater scale) decreased sharply from 7.9 to 7.1 while pCO_2 increased from 58 Pa to nearly 300 Pa. Such a large decrease in pH is a result of oxidation of reduced species (i.e. HS^-, Mn^{2+}, and Fe^{2+}) near the sediment–water interface. The large pCO_2 increase with depth was thus caused in part by changes in carbonate equilibria associated with the pH shift. At a continental rise site (4100 m), pH and pCO_2 changes were observed to be much smaller and a reflection of the oxic degradation of organic carbon [29]. Anoxic sediments with significant amounts of manganese and iron reduction can show large decreases in porewater pH [103] with coincident increases in pCO_2 and DIC (Cai and Luther, unpublished).

As mentioned earlier it is often a goal to calculate fine scale DIC profiles from pH and pCO_2 microelectrode measurements [31,61]. While it is still a very daunting task, Figure 12 shows that smooth and credible increases in DIC may

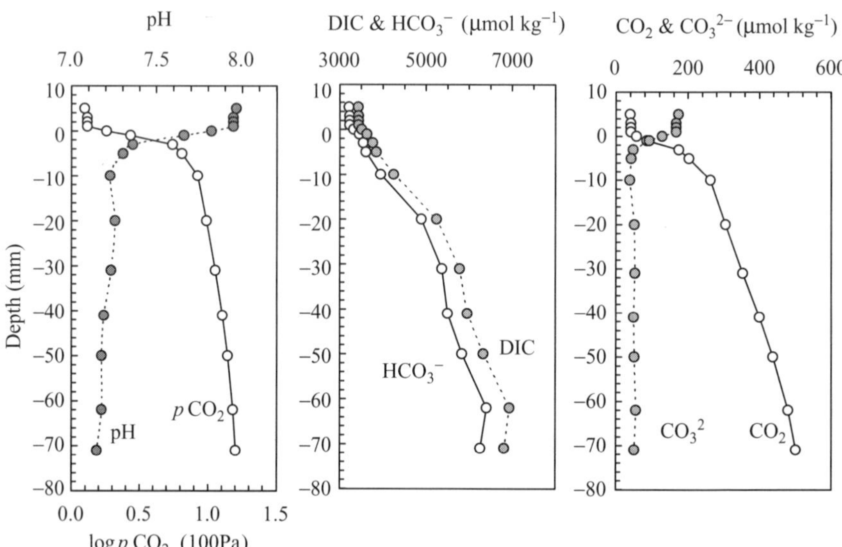

Figure 12. Porewater pH and pCO_2 profiles measured in an incubated sediment from the Satilla River estuary, GA. The incubation time was 6 weeks. pH values are in NBS scale. The CO_2, HCO_3^-, CO_3^{2-} concentration profiles are calculated from the pH and pCO_2 data

be obtained with high resolution profiling. We believe this approach will provide accurate DIC profiles and fluxes for the deep sea once the liquid membrane-based sensors are applied *in situ* more widely. The derivation of DIC profiles from pH and pCO_2 measurements in sediments, however, can be misleading even when both pH and pCO_2 profiles are measured accurately. First, the gradients of these profiles in coastal sediments or microbial mats are very sharp. A depth mismatching of a few hundred microns of the two sensors can generate erroneous predictions of DIC gradients. Second, surface sediments may be highly heterogeneous for various reasons. If pH and pCO_2 electrodes are fixed together, one can reduce mismatching problems but not problems of very fine scale heterogeneity. Measuring a large number of profiles should help define the limits of variability caused by heterogeneity.

8 CONCLUSIONS AND FUTURE WORK

Over the last 15 years, designs and applications of pH and pCO_2 sensors for *in situ* measurements at sea have developed greatly. Successful long-term mooring of pCO_2 by an autonomous fiber-optic-based sensor system is a great achievement [28], and probably represents the direction of many future *in situ* measurements at the sea surface.

Spectrophotometric fiber-optic-based sensors have superior stability and precision to potentiometric electrodes and so are recommended for seawater monitoring. Further work on the temperature and pressure dependence of many pH-sensitive dyes is needed, however. Microelectrodes have been proven as tools capable of characterizing sediments and biofilms with minimal interference to the system under study. Potentiometric measurements also have the advantage of requiring only the equivalent of a high-quality pH-meter in the way of instrumentation and so should see further development as a means for water column profiling.

The lifetime of a pCO_2 microelectrode is still a serious hindrance which prevents its wide application. Although relatively long shelf life has been reported for the newest designs [32], no field test during a long cruise has been conducted. Measuring pH and pCO_2 simultaneously within the same microenvironment for accurate characterization of the carbonate system is still a daunting task. Development of a better combined pH and pCO_2 microelectrode is needed for the study of respiration, photosynthesis, and $CaCO_3$ formation and dissolution in many marine environments.

ACKNOWLEDGMENT

We would like to thank Mr. Pingsan Zhao for assistance. Prof. Buffle and Prof. Horvai have provided many constructive suggestions to improve the presentation of this chapter. This work is partly supported by NSF grants OCE-9596143 and DEB-9612291 to WJC and OCE-9872042 to CER.

LIST OF SYMBOLS AND ACRONYMS

A	Absorbance
e	Ratio of molar absorptivities
E_J:	Liquid junction potential
$\{H^+\}$	Activity of proton
CA	Carbonate alkalinity, $CA = [HCO_3^-] + 2[CO_3^{2-}]$
$[CO_2*]$	$[CO_{2aq}] + [H_2CO_3]$, where CO_{2aq} indicates a free species dissolved in aqueous solution but not hydrated while H_2CO_3 is the hydrated species
$c(T)$:	Total (analytical) concentration of $NaHCO_3$ in the internal solution of the pCO$_2$ electrode. It is equivalent to $[Na^+]$ or DIC inside the electrode
DIC:	Total dissolved inorganic carbon, DIC $= [CO_2*] + [HCO_3^-] + [CO_3^{2-}]$.
Emf	Electromotive force
fCO$_2$	Fugacity of carbon dioxide
k_N	Nernst constant $= RT \ln 10/F$, where R is the gas constant and F the Faraday constant
pH	$-\log H^+$, roman p indicate $-\log$
pCO$_2$:	Partial pressure of CO_2, italic p indicate partial pressure
R_A	Ratio of the absorbances at different wavelengths
TA	Total alkalinity, which is defined as $TA = [HCO_3^-] + 2[CO_3^{2-}] + [OH^-] - [H^+] + [B(OH)_4^-] + 2[PO_4^{3-}] + [HPO_4^{2-}] - [H_3PO_4] + [SiO(OH)_3^-] + \ldots$ contributions of other minor acid or base species
V_{CO_2}	Partial molar volume of CO_2
ε	Molar absorptivity
λ	Wavelength

Pressure unit used in this chapter is Pascal (Pa). Various units have been used in the literature. 1 atm $= 1.01325 \times 10^5$ Pa. 1 bar $= 10^5$ Pa. A 0.1% of CO_2 in atmosphere is 1000μatm (or 1 matm) or 1000 ppm.

REFERENCES

1. Smith, S. V. and Hollibaugh, J. T. (1997). Annual cycle and interannual variability of ecosystem metabolism in a temperate climate embayment, *Ecology/Ecological Monographs*, **67**, 509.

2. Smith, S. V. and Hollibaugh, J. T. (1993). Coastal metabolism and the oceanic organic carbon balance, *Rev. Geophys.*, **31**, 75.
3. Hopkinson, C. S. (1985). Shallow-water and pelagic metabolism: Evidence of heterotrophy in the near-shore Georgia Bight, *Mar. Biol.*, **87**, 19.
4. Dickson, A. G. (1993). pH buffers for sea water media based on the total hydrogen ion concentration scale, *Deep-Sea Res.*, **40**, 107.
5. Ben-Yaakov, S. and Kaplan, I. R. (1968). pH–Temperature profiles in ocean and lakes using an in situ probe, *Limnol. Oceanogr.*, **13**, 688.
6. Park, K. (1966). Deep sea pH, *Science*, **154**, 1540.
7. Whitfield, M. (1971). A compact potentiometric sensor of novel design. In situ determination of pH, pS^{2-}, and E_h, *Limnol. Oceanogr.*, **16**, 829.
8. Emerson, S., Grundmanis, V. and Graham, D. (1982). Carbonate chemistry in marine pore waters: MANOP sites C and S, *Earth Planet. Sci. Lett.*, **61**, 220.
9. Murray, J. W., Emerson, S. and Jahnke, R. (1980). Pressure effect on the alkalinity of deep-sea interstitial water samples, *Geochim. Cosmochim. Acta*, **44**, 963.
10. Berelson, W. M., Hammond, D.E., Smith, K. L., Jr., Jahnke, R. A., Devol, A. H., Hinga, K. R., Rower, G. T. and Sayles, F. (1987). In situ benthic flux measurement devices: bottom lander technology, *MTS J.*, **21**, 26.
11. Jahnke, R. A. and Christiansen, M. B. (1989). A free-vehicle benthic chamber instrument for sea floor studies, *Deep-Sea Res.*, **36**, 625.
12. Sayles, F. and Dickinson, W. (1991). The ROLAID lander: a benthic lander for the study of exchange across sediment-water interface, *Deep-Sea Res.*, **38**, 505.
13. Sayles, F. L., Mangelsdorf, P. C. J., Wilson, T. R. S. and Hume, D. N. (1976). A sampler for the in situ collection of marine sedimentary pore waters, *Deep-Sea Res.*, **23**, 259.
14. Martin, W. R. and Sayles, F. L. (1996). $CaCO_3$ dissolution in sediments of the Ceara Rise, western equatorial Atlantic, *Geochim. Cosmochim. Acta*, **60**, 243.
15. Bender, M., Jahnke, R., Weiss, R., Matin, W. *et al.* (1989). Organic carbon oxidation and benthic nitrogen and silica dynamics in San Clemente Basin, a continental borderland site. *Geochim. Cosmochim. Acta*, **53**, 685.
16. Revsbech, N. P., Jorgensen, B. B. and Blackburn, T. H. (1983). Microelectrode studies of the photosynthesis and O_2, H_2S and pH profiles of a microbial mat, *Limnol. Oceanogr.*, **28**, 1062.
17. Cai, W.-J. and Reimers, C. E. (1993). The development of pH and pCO_2 microelectrodes for studying the carbonate chemistry of pore waters near the sediment–water interface, *Limnol. Oceanogr.*, **38**, 1776.
18. Archer, D., Emerson, S. and Reimers, C. E. (1989). Dissolution of calcite in deep-sea sediments: pH and O_2 microelectrode results, *Geochim. Cosmochim. Acta*, **53**, 2831.
19. Reimers, C. E. (1987). An in situ microprofiling instrument for measuring interfacial pore water gradients: methods and oxygen profiles from North Pacific Ocean, *Deep-Sea Res.*, **34**, 2019.
20. Weiss, R. F. and Craig, H. (1973). Precise shipboard determination of dissolved nitrogen, oxygen, argon, and total inorganic carbon by gas chromatography, *Deep-Sea Res.*, **20**, 291.
21. Weiss, R. F. (1974). Carbon dioxide in water and seawater: The solution of a non-ideal gas, *Mar. Chem.*, **2**, 203.
22. Cai, W.-J. and Wang, Y. (1998). The chemistry, fluxes and sources of carbon dioxide in the estuarine waters of the Satilla and Altamaha Rivers, Georgia, *Limnol. Oceanogr.*, **43**, 657.

23. Takahashi, T., Feely, R. A., Weiss, R.F., Wanninkhof, R. H., Chipman, D. N., Sutherland, S. C., Takahashi, T. T. (1997). Global air-sea Flux CO_2: An estimate based on measurements of sea-air pCO_2 difference. *Proc. NaH. Acad. Sci.*, **94**, 8292–8299.
24. Goyet, C. and Peltzer, E. T. (1997). Variation of CO_2 partial pressure in surface seawater in the equatorial Pacific Ocean, *Deep-Sea Res. Part I*, **44**, 1611.
25. Goyet, C. A. R and Eischeid, G. (1998). Observations of the CO_2 system properties in the tropical Atlantic Ocean, *Mar. Chem.*, **60**, 49.
26. Millero, F. J. (1995). Thermodynamics of the carbon dioxide system in the oceans, *Geochim. Cosmochim. Acta*, **59**, 661.
27. Millero, F. J., Lee, K. and Roche, M. P. (1998). Distribution of alkalinity in the surface water of the major oceans, *Mar. Chem.*, **60**, 109.
28. DeGrandpre, M., Hammar, T. R., Smith, S. P. and Sayles, F. L. (1995). In-situ measurements of seawater pCO_2, *Limnol. Oceanogr.*, **40**, 969.
29. Cai, W.-J., Reimers, C. E. and Shaw, T. (1995). Microelectrode studies of organic carbon degradation and calcite dissolution at a California continental rise site, *Geochim. Cosmochim. Acta*, **59**, 497.
30. Hales, B., Burgess, L. and Emerson, S. (1997). An absorbance-based fiber-optic sensor for CO_2aq measurement in porewaters of sea floor sediments, *Mar. Chem.*, **59**, 51.
31. Komada, T., Reimers, C. E. and Boehme, S. E. (1998). Dissolved inorganic carbon profiles and fluxes determination using pH and pCO_2 microelectrodes., *Limnol. Oceanogr.*, **43**, 769.
32. Zhao, P. and Cai, W.-J. (1997). An improved pCO_2 microelectrode, *Anal. Chem.*, **69**, 5052.
33. Zhao, P. and Cai, W.-J. (1999). pH polymeric membrane microelectrodes based on neutral carriers and their application in aquatic environments, *Anal. Chim. Acta*, **395**, 285–291.
34. Siesjo, B. K. (1961). A method for continuous measurement of the carbon dioxide tension on the cerebral cortex. *Acta Physiol. Scand.* **51**, 297.
35. Pucacco, L. R. and Carter, N. W. (1978). An improved pCO_2 microelectrode, *Anal. Biochem.*, **90**, 427.
36. Pui, C. P., Rechnitz, G. A. and Miller, R. F. (1978). Micro-size potentiometric probes for gas and substrate sensing, *Anal. Chem.*, **50**, 330.
37. Pilson, M. E. Q. (1998). *An Introduction to the Chemistry of the Sea*, Prentice Hall, Upper Saddle River, NJ.
38. Morel, F. M. M. and Hering, J. G. (1992). *Principles and Applications of Aquatic Chemistry*. 2nd edn, John Wiley and Sons, New York.
39. Stumm, W. and Morgan, J. J. (1981). *Aquatic Chemistry*, 2nd edn, John Wiley and Sons, New York.
40. Butler, J. N. (1982). *Carbon Dioxide Equilibria and Their Applications*, Addison-Wesley, Reading, MA.
41. Dickson, A. G. and Millero, F. (1987). A comparison of the equilibrium constants for the dissolution of carbonic acid in seawater media, *Deep-Sea Res.*, **34**, 1733.
42. Lee, K. and Millero, F. J. (1995). Thermodynamic studies of the carbonate system in seawater, *Deep-Sea Res. I*, **42**, 2035.
43. Dickson, A. and Goyet, C. (1994). *DOE Handbook of Methods for the Analysis of the Various Parameters of the Carbon Dioxide System in Sea Water*, Version 2.
44. Dickson, A. G. (1981). An exact definition of total alkalinity and a procedure for the estimation of alkalinity and total CO_2 from titration data, *Deep-Sea Res.*, **28**, 15–21.

45. Cai, W.-J., Wang, Y. and Hodson, R. E. (1998). Acid–base properties of dissolved organic matter in the estuarine waters of Georgia, *Geochim. Cosmochim. Acta*, **62**, 473.
46. Gran, G. (1952). Determination of the equivalence point in potentiometric titrations. Part II., *Analyst*, **77**, 661.
47. Steinberg, P. A., Millero, F. J. and Zhu, X. (1998). Carbonate system response to iron enrichment, *Mar. Chem.*, **62**, 31.
48. Dickson, A. G. and Riley, J. P. (1978). The effect of analytical error on the evaluation of the components of the aquatic carbon-dioxide system, *Mar. Chem.*, **6**, 77.
49. Gieskes, J. M., ed. (1974). The CO_2 system and alkalinity. In *The Sea; Ideas and Observations on Progress in the Study of the Seas*, Vol. 5, *Marine Chemistry*, ed. Goldbert, D. D., Wiley-Interscience, New York, p. 123.
50. Park, P. K. (1969). Oceanic CO_2 system: an evaluation of ten methods of investigation, *Limnol. Oceanogr.*, **14**, 179.
51. Skirrow, G. (1975) The dissolved gases–carbon dioxide. 2nd edn *Chemical Oceanography*, ed. Riley, J. P. and Skirrow, G., Vol. 2, Academic Press, London.
52. Mueller, B., Buis, K., Stierli, R. and Wehrli, B. (1998). High spatial resolution measurements in lake sediments with PVC based liquid membrane ion-selective electrodes, *Limnol. Oceanogr.*, **43**, 1728.
53. Shaw, T. (1988).The early diagenesis of transition metals in nearshore sediments, Ph. D. Dissertation, Scripps Institute of Oceanography, University of California, San Diego, p. 164.
54. McCorkle, D. C. and Emerson, S. R. (1988). The relationship between pore water carbon isotopic composition and bottom water oxygen concentration, *Geochim. Cosmochim. Acta*, **52**, 1169.
55. Cantrell, K. J., Serkiz, S. M. and Perdue, E. M. (1990). Evaluation of acid neutralizing capacity data for solutions containing natural organic acids, *Geochim. Cosmochim. Acta*, **54**, 1247.
56. Sawyer, D. T., Sobkowiak, A. and Roberts, J. L., Jr. (1995). *Electrochemistry for Chemists*, 2nd edn, John Wiley & Sons, New York.
57. Oldham, H. B., Myland, J. C. (1994). *Fundamentals of Electrochemical Science*, Academic Press, San Diego, CA.
58. Bates, R. G. (1974). *Determination of pH Theory and Practice*, John Wiley & Sons, New York.
59. IUPAC (1979). Manual of symbols and terminology for physicochemical quantities and units, *Pure and Appl. Chem.*, **51**, 1.
60. Distéche, A. (1959). pH measurements with a glass electrode withstanding 1500 kg/cm^2 hydrostatic pressure, *Rev. Sci. Instrum.*, **30**, 474.
61. Reimers, C. E. and Glud, R. N. (1999). In situ chemical sensor measurements at the sediment–water interface, in *Chemical Sensors in Oceanography*, ed. Varney, M., Gordon and Breach.
62. Distéche, A. and Distéche, S. (1965). The effect of pressure on pH and dissociation constants from measurements with buffered and unbuffered glass electrode cells, *J. Electrochem. Soc.*, **112**, 350.
63. Millero, F. (1986). The pH of estuarine waters, *Limnol. Oceanogr.*, **31**, 839.
64. Whitfield, M., Butler, R. A. and Covington, A. K. (1985). The determination of pH in estuarine waters. Part I. Definition of pH scales and the selection of buffers, *Oceanol. Acta*, **8**, 423.
65. Dickson, A. G. (1984). pH scale and proton-transfer reactions in saline media such as sea water, *Geochim. Cosmochim. Acta*, **48**, 2299

66. Culberson, C. H. (1981). Direct potentiometry, In *Marine Electrochemistry*, ed. Whitedfield M. and Jagner, D. John Wiley & Sons, Chichester, p. 187.
67. Butler, R. A., Covington, A. K. and Whitfield, M. (1985). The determination of pH in estuarine waters. I. Practical considerations, *Oceanologica Acta*, **8,** 433.
68. Covington, A. K., Whalley, P. D. and Davison, W. (1983). Procedures for the measurement of pH in low ionic strength solutions including freshwater, *Analyst*, **108**, 1528.
69. Overbeek, J. T. G. (1953). Donnan-E. M. F. and suspension effect, *J. Colloid Sci.*, **8,** 593.
70. Yang, S. X., Cheng, K. L., Kurtz, L. T. and Peck, T. R. (1993). Suspension effect in potentiometry, *Particle Sci. Technol.*, **7,** 139.
71. Garrels, R. M. and Christ, C. L. (1965). *Solution, Minerals and Equilibria*, Harper & Row, New York.
72. Brezinski, D. P. (1983). Influence of colloidal charge on response of pH and reference electrodes: the suspension effect, *Talanta*, **30**, 347.
73. Jenny, H., Nielsen, T. R., Coleman, N. T. and Williams, D. E. (1950). Concerning the measurement of pH, ion activities, and membrance potentails in colloidal systems, *Science*, **112**, 164.
74. DeGrandpre, M. D. (1993). Measurement of seawater pCO_2 using a renewable-reagent fiber optic sensor with colorimetric detection, *Anal. Chem.*, **65**, 331.
75. Jensen, M. A. and Rechnitz, G. A. (1979). Response time characteristics of the pCO_2 electrode, *Anal. Chem.*, **51**, 1972.
76. Ross, J. W., Riseman, J. H. and Krueger, J. A. (1973). Potentiometric gas sensing electrodes, *Pure Appl. Chem.*, **36**, 473.
77. Buffle, J. and Spoerri, M. (1981). Role of the diffusion process on the response time of CO_2 electrodes, *J. Electroanal. Chem.*, **129**, 67.
78. de Beer, D., Glud, A., Epping, E. and Kühl, M. (1997). A fast-responding CO_2 microelectrode for profiling sediments, microbial mats, and biofilms., *Limnol. Oceanogr.*, **42**, 1590.
79. DeGrandpre, M. D., Hammar, T. R., Wallace, D. W. R. and Wirick, C. D. (1997). Simultaneous mooring based measurements of seawater CO_2 and O_2 off Cape Hatteras, North Carolina, *Limnol. Oceanogr.*, **42**, 21–28, .
80. King, D. W. and Kester, D. R. (1989). Determination of seawater pH from 1.5 to 8.5 using colorimetric indicators, *Mar. Chem.*, **26**, 5.
81. Byrne, R. H., Robert-Baldo, G., Thompson, S. W. and Chen, C. T. A. (1988). Seawater pH measurements: an at-sea comparison of spectrophotometric and potentiometric methods, *Deep-Sea Res.*, **35**, 1405.
82. Clayton, T. D. and Byrne, R. H. (1993). Spectrophotometric seawater pH measurements: total hydrogen ion concentration scale calibration of *m*-cresol purple and at-sea results, *Deep-Sea Res.*, **40**, 2315.
83. Zhang, H. and Byrne, R. H. (1996). Spectrophotometric pH measurements of surface seawater at in-situ conditions: absorbance and protonation behavior of thymol blue, *Mar. Chem.*, **52**, 17.
84. Dickson, A. G. (1997). Underway measurement of pH; method and meaning, in *Marc'h Mor Workshop Proceedings*, IUEM Brest, France.
85. Roche, M. P. and Millero, F. J. (1998). Measurement of total alkalinity of surface waters using a continuous flowing spectrophotometric technique, *Mar. Chem.*, **60**, 85.
86. Yao, W. and Byrne, R. H. (1998). Simplified seawater alkalinity analysis: Use of linear array spectrometers, *Deep-Sea Res. I*, **45**, 1383.

87. Butler, T. M., MacCraith, B. D. and McDonagh, C. (1995). Development of an extended range fiber optic pH sensor using evanescent wave absorption of sol–gel entrapped pH indicators, *Proc. SPIE*, **2508**, 168.
88. Monici, M., Boniforti, R., Buzzigoli, G., DeRossi, D. *et al.* (1987). Fibre-optic pH sensor for seawater monitoring, *Proc. SPIE*, **798**, 294.
89. Leiner, M. J. P. and Wolfbeis, O. S. (1991). Fiber optic pH sensors, in *Fiber Optic Chemical Sensors and Biosensors*, ed. Wolfbeis, O. S., CRC Press, New York, p. 359.
90. Modlin, D. N. and Milanovich, F. P. (1991). Instrumentation for fiber optic chemical sensors, in *Fiber Optic Chemical Sensors and Biosensors*, ed. Wolfbeis, O. S., CRC Press, New York, p. 237.
91. Goyet, C., Walt, D. R. and Brewer, P. G. (1992). Development of a fiber optic sensor for the measurement of pCO_2 in sea water: design criteria and sea trials, *Deep-Sea Res.*, **39**, 1015.
92. Tabacco, M. B., Uttamlalt, M., McAllister, M. and Walt, D. R. (1998). An autonomous sensor and telemetry system for low-level pCO_2 measurements in seawater, *Anal. Chem.*, **71**, 154.
93. Taylor, C. V. and Whitaker, D. M. (1927). Potential determination in the protoplasm and cell-sap of Ntella, *Protoplasma*, **3,** 1.
94. Kostyuk, P. G. and Sorokina, Z. A. (1961). On the mechanism of hydrogen ion distribution between cell protoplasm and the medium, in *Membrane Transport Metabolism*, ed. Kleineller, A. and Kotyk, A., Academic Press, New York. p. 3.
95. Thomas, R. C. (1982). pH microelectrodes: Tips on making the recessed-tip type for intracellullar use. In *Intracellullar pH: Its Measurement, Regulation, and Utilization in Cellullar Functions*, ed. Nuccitelli, R. Alan R. Liss, p. 1.
96. Ammann, D. (1986). *Ion-Selective Microelectrodes Principles, Design and Application*, Springer, New York.
97. Ammann, D., Lanter, F., Steiner, R. A., Schulthess, P., *et al.* (1981). Neutral carrier based hydrogen ion selective microelectrode for extra- and intracellular studies, *Anal. Chem.*, **53**, 2267.
98. Chao (Zhao), P., Ammann, D., Oesch, U., Simon, W. *et al.* (1988). Extra- and intracellular hydrogen ion-selective microelectrode based on neutral carriers with extended pH response range in acid media, *Pflugers Arch.*, **411**, 216.
99. Revsbech, N. P. (1994). Analysis of microbial mats by use of electrochemical microsensors: recent advances. In *Microbial Mats: Structure, Development and Environmental Significance*, ed. Stal, L. J. and Caumette, P., Springer, Berlin, p. 135.
100. Van Den Heuvel, J. C., De Beer, D. and Cronenberg, C. C. H. (1992). Microelectrodes: A versatile tool in biofilm research. In *Biofilms—Science and Technology*, ed. Melo, L. F., Kluwer, Dordrecht, p. 631.
101. Revsbech, N. P. and Jorgensen, B. B. (1986). Microelectrodes: their use in microbial ecology. In *Advances in Microbial Ecology, Vol. 9*, ed. Marshall, K. C., Plenum, New York.
102. Hartley, A. M., House, W. A., Leadbeater, B. S. C. and Callow, M. E. (1996). The use of microelectrodes to study the precipitation of calcite upon algal biofilms, *J. Colloid. Interface. Sci.*, **183**, 498.
103. Reimers, C. E., Ruttenberg, K. C., Canfield, D. E., Christiansen, M. B. *et al.* (1996). Porewater pH and authigenic phases formed in the uppermost sediments of the Santa-Barbara Basin, *Geochim. Cosmochim. Acta*, **60**, 4037.
104. Alldredge, A. L. and Cohen, Y. (1987). Can microscale chemical patches persist in the sea? Microelectrode study of marine snow, fecal pellets, *Science*, **235**, 689.

105. Jorgensen, B. B., Erez, J., Revsbech, N. P. and Cohen, Y. (1985). Symbiotic photosynthesis in a planktonic foraminiferan, Globigerinoides sacculifer (Brady), studies with microelectrodes, *Limnol. Oceanogr.*, **30**, 1253.
106. Papeschi, G., Bordi, S., Beni, C. and Ventura, L. (1976). Use of an iridium electrode for direct measurement of pH of proteins after isoelectric focusing in polyacrylamide, *Biochim. Biophys. Acta*, **453**, 192.
107. Matsuo, T. and Esashi, M. (1981). Methods of ISFET fabrication, *Sens. Actuators*, **1**, 77.
108. Van der Schoot, B. H. and Bergveld, P. (1988). Coulometric sensors, the application of a sensor–actuator system for long-term stability in chemical sensing, *Sens. Actuators*, **13**, 251.
109. Olthuis, W., Luo, J., Van der Schoot, B. H. and Bergveld, P. (1990). Modelling of non-steady-state concentration profiles at ISFET-based coulometric sensor–actuator systems, *Anal. Chim. Acta*, **229**, 71.
110. Morf, W. E. (1981). *The Principles of Ion-Selective Electrodes and of Membrane Transport*, Elsevier, Amsterdam.
111. Ammann, D., Morf, W. E., Anker, P., Meier, P. C. *et al.* (1983). Neutral carrrier based ion-selective electrodes, *Ion-selective Electrode Rev.*, **5**, 3.
112. Walker, J. L. J. (1971). Ion specific ion exchanger microelectrode, *J. Neuronsci. Methods*, **11**, 187.
113. Hinke, J. A. M. (1967). Cation-selective microelectrodes for intracellullar use, in *Glass Electrodes for Hydrogen and Other Cations, ed.* Eiseman, G., Marcel Dekker, New York, p. 598.
114. Hales, B. and S. E. Emerson (1997). Calcite dissolution in sediments of the Ceara Rise: In situ measurements of porewater O_2, pH, and CO_2(aq), *Geochim. Cosmochim. Acta*, **61**, 501.
115. Hales, B., Emerson, S. E. and Archer, D. (1994). Respiration and dissolution in the sediments of the western North Atlantic: estimates from models of in situ microelectrode measurements of porewater oxygen and pH, *Deep-Sea Res.*, **41**, 695.
116. Zhao, P. (1998).Development of new generation of pH and pCO_2 microelectrodes for marine sediment studies, M. S. Thesis, Department of Marine Sciences, University of Georgia, Athens, GA, p. 67.
117. McNichol, A. P., Lee, C. L. and Druffel, E. R. M. (1988). Carbon cycling in coastal sediments: 1. A quantitative estimate of the remineralization of organic carbon in the sediments of Buzzards Bay, MA, *Geochim. Cosmochim. Acta*, **52**, 1531.
118. Severinghaus, J. W. and Bradley., A. F. (1958). Electrodes for blood pO_2 and pCO_2 determination, *J. Appl. Physiol.*, **13**, 515.
119. Severinghaus, J. W. (1968). Measurements of blood gases: pO_2 and pCO_2, *Ann. New York Acad. Sci.*, **14**8, 115.
120. Ma, Y. (1990). A carbon dioxide air-gap microelectrode based on a neutral hydrogen ion exchanger pH microelectrode, *Anal. Biochem.*, **18**6, 74.
121. Bomsztyk, K. and Calalb, M. K. (1986). A new microelectrode method for simultaneous measurement of pH and pCO_2, *Am. J. Physiol.*, **25**1, F933.
122. Morf, W. E., Mostert, I. A. and Simon, W. (1985). The response of potentiometric gas sensors to primary and interfering species, *Anal. Chem.*, **57**, 1122.
123. Paul, A. (1990). Chemistry of Glasses. 2nd edn, Chapman and Hall, New York.
124. Bansal, N. P. and Doremus, R. H. (1986). *Handbook of Glass Properties*, Academic Press, Orlando, FL, pp. 31–45 and pp. 647–656.
125. Voipio, J. and Kaila, K. (1993). Intestial pCO_2 and pH in rat hippocampal slices measured by means of a novel fast CO_2/H^+-sensitive microeletrode base on a PVC-gelled membrane, *Pflügers Arch.*, **423**, 193.

126. Sohtell, M. and Karlmark, B. (1976). In vivo micropunture pCO_2 measurements, *Pflügers Arch.*, **363**, 179.
127. Collany, H. T., Schumacher, T. E., Rue, R. R. and Liu, S. Y. (1993). A carbon dioxide microelectrode for in situ pCO_2 measurement, *Microchem. J.*, **48**, 42.
128. Coon, R. L., Lai, N. C. J. and Kampine, J. P. Evaluation of dual-function pH and pCO_2 in vivo sensor, *J. Appl. Physiol.*, **640**, 625.
129. Bauer, C., Gros, G. and Bartels, H. (1980). *Biophysics and Physiology of Carbon Dioxide*, Springer, New York, pp. 3–63 and pp. 133–150.
130. Donaldson, T. L. and Palmer, H. J. (1979). Dynamic response of the carbon dioxide electrode, *AIChE J.*, **25**, 43.
131. Donaldson, T. L. and Ho, S. P. (1985). Electrokinetic effects: Carbonic anhydrase and CO_2 electrodes dynamics, *Chem. Eng. Commun*, **37**, 223.
132. Marazuela, M. D. and Bondi, M. C. M. a. O., G. (1995). Enhanced performance of a fibre-optic luminescence CO_2 sensor using carbonic anhydrase, *Sens. Actuators B*, **29**, 126.
133. Waterbury, R. D., Byrne, R. H., Kelly, J., Leader, B. *et al.* (1996). Development of an underwater in-situ spectrophotometric sensor for seawater pH. In *Chemical, Biological, and Environmental Fiber Sensor VIII*, SPIE—The International Society for Optical Engineering, Denver, CO.
134. Bellerby, R. G. J., Turner, D. R., Millward, G. E. and Worsfold, P. J. (1995). Shipboard flow-injection determination of sea-water pH with spectrophotometric detection, *Anal. Chim. Acta*, **309**, 259.
135. DeGrandpre, M. D., Hammar, T. R. and Wirick, C. D. (1998). Short-term pCO_2 and O_2 dynamics in California coastal waters, *Deep-Sea Res.*, **45**, 1557.
136. Friederich, G. E., Brewer, P. G., Herlien, R. and Chavez, F. P. (1995). Measurement of sea surface partial pressure of CO_2 from a moored buoy, *Deep-Sea Res. I*, **42**, 1175.
137. Keir, R. S. (1980). The dissolution kinetics of biogenic calcium carbonates in seawater, *Geochim. Cosmochim. Acta*, **44**, 241.
138. Emerson, S. and Bender, M. L. (1981). Carbon fluxes at the sediment–water interface of the deep sea: calcium carbonate preservation, *J. Mar. Res.*, **39**, 139.
139. Archer, D. and Maier-Reimer, E. (1994). Effect of deep-sea sediment calcite preservation on atmospheric CO_2 concentration, *Nature*, **367**, 260.

4 Sensors for *In situ* Analysis of Sulfide in Aquatic Systems

M. KÜHL
University of Copenhagen Denmark

C. STEUCKART
Max-Planck-Institute for Marine Microbiology, Bremen, Germany

1 INTRODUCTION

Sulfur cycling is of major importance for the biogeochemistry of ecosystems, and here microbial and chemically catalyzed sulfide conversions play a key role [1]. In the environment, sulfide is not only a strong poison to all aerobic organisms through its high affinity to metal containing enzymes [2,3] but is also an important product of anaerobic microbial activity. Under oxic conditions microorganisms are preferentially utilizing oxygen as an electron acceptor, whereas under anoxic conditions other substances such as metal ions (Fe(III), Mn(IV)), nitrate, sulfate or even carbon dioxide are used. In the marine environment sulfate is usually the most available electron acceptor owing to its high

In Situ Monitoring of Aquatic Systems: Chemical Analysis and Speciation Edited by J. Buffle and G. Horvai.

concentration in seawater (ca 25 mmol L^{-1}) and more than 50 % of the carbon mineralization in marine sediments can go via sulfate respiration [4]. This leads to a high H_2S production by microbial sulfate reduction below the oxic/anoxic interface in sediments [5,6], biofilms [7], and in stratified water masses [8,9], where oxygen concentration is low and sulfate availability is high. The sulfide produced reacts with heavy metal ions in the environment and forms precipitates such as FeS, and FeS_2, which in the latter case is preserved in the geological record. A significant amount of sulfide is oxidized via various chemical and microbial pathways involving O_2, NO_3^-, Fe(III) and Mn(IV) [5]. Sulfide oxidation thus binds the sulfur cycle together with the cycling of other key elements in nature [1,5].

In certain areas of industry and waste water treatment, sulfide poses problems. In sewage treatment plants, excessive sulfide production can lead to inhibition of efficient nutrient removal and clogging of activated sludge basins. In sewers, sulfide containing waste water can lead to unpleasant odors and, when oxidized to sulfate, severe corrosion of sewer pipes [10]. Sulfur removal from solids, liquids and gases is of major industrial interest, e.g. in the paper industry [11,12] and the mining and fossil fuel industry [13,14]. Also, corrosion processes due to biofilm formation on immersed surfaces are in part induced by the presence of sulfide-producing bacteria [15,16]. A better understanding of sulfide conversion processes is thus both of general biogeochemical interest and has important socioeconomic implications. This calls for suitable analytical techniques for detecting sulfide in the environment.

In this chapter, we review the current methodology for measuring sulfide in aquatic systems. While we give an overview of currently available analytical techniques, we will focus on direct measurements of sulfide with sensors, that is, devices that can be used directly in natural waters without previous conditioning steps, and that respond specifically and reversibly to sulfide. Special attention is given to sulfide micro-sensors that allow measurements at high spatial (< 0.1 mm) and temporal ($t_{90} = 90\,\%$ of response time < 1–10 s) resolution, with minimal consumption of the analyte, and, therefore, without significant effects on the sulfide equilibria and gradients present in the aquatic enviroment. More general reviews on microsensors and their application in environmental analysis appear elsewhere [17–24] and in other chapters of this book.

2 SULFIDE IN AQUATIC SYSTEMS

Here, we only briefly summarize some important characteristics of sulfide in natural waters. More detailed accounts can be found e.g. in the work of Millero and co-workers [25–27]. In aqueous solution hydrogen sulfide is found to be a weak acid and, neglecting metal complexes and solid phases, the total sulfide concentration, $[S(-II)]_t$, consists of H_2S (dissolved hydrogen sulfide), HS^-

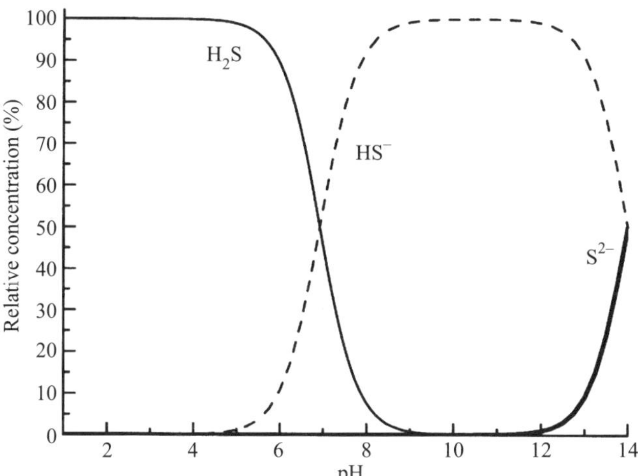

Figure 1. Relative concentrations of the dissociation products of H_2S at different pH ($T = 298\,^\circ K$, $I \leq 1\,\text{mol}\,L^{-1}$) calculated from equations (3)–(5). For the calculations we used $pK_1 = 6.921$ (298 °K) as calculated from equation (6) [28] and $pK_2 \approx 14$ [29]

(hydrogeno sulfide ion), and S^{2-} (sulfide ion). The relation of the actual concentrations of these species is determined by the dissociation constants K_1 and K_2 (equations (1) and (2); Figure 1), and $[S(-II)]_t$ can be calculated from the measured species concentrations, pH, ionic strength and temperature using equations (3)–(5):

$$H_2S \overset{H_2O}{\rightleftharpoons} H_3O^+ + HS^- \qquad K_1 = \frac{[H_3O^+][HS^-]}{[H_2S]} \tag{1}$$

$$HS^- \overset{H_2O}{\rightleftharpoons} H_3O^+ + S^{2-} \qquad K_2 = \frac{[H_3O^+][S^{2-}]}{[HS^-]} \tag{2}$$

$$[H_2S] = \frac{[S(-II)]_t}{1 + \dfrac{K_1}{[H_3O^+]} + \dfrac{K_1K_2}{[H_3O^+]^2}} \tag{3}$$

$$[HS^-] = \frac{[S(-II)]_t}{1 + \dfrac{[H_3O^+]}{K_1} + \dfrac{K_2}{[H_3O^+]}} \tag{4}$$

$$[S^{2-}] = \frac{[S(-II)]_t}{1 + \dfrac{[H_3O^+]}{K_2} + \dfrac{[H_3O^+]^2}{K_1K_2}} \tag{5}$$

The determination of the exact value of the dissociation constants has been a subject of scientific debate. For the first constant, Broderius and Smith[30] performed a direct photometric determination of H_2S in the gas phase and found the empirical formula, at infinite dilution:

$$pK_1 = 3.122 + \frac{1132}{T} \qquad \text{for } 283\,^\circ\text{K} \leq \text{T} \leq 298\,^\circ\text{K} \tag{6}$$

giving a value of $pK_1 = 6.921$ at 298 °K which is in good agreement with the constant found by Barbero *et al.*[31] using the empirical expression:

$$pK_1 = 19.840 + \frac{930.8}{T} - 2.800\ln T \tag{7a}$$

Other empirical relations for the first dissociation constant as a function of temperature and salinity in natural waters are given by Millero *et al.* [26]:

$$\begin{aligned} pK_1 &= -98.080 + \frac{5765.4}{T} + 15.0455\ln T \quad \text{at infinite dilution} \\ pK_1^* &= pK_1 - 0.1498\sqrt{S} + 0.0119S \qquad \text{at salinity, } S, \text{(in ppt)} \end{aligned} \tag{7b}$$

The main problem in the determination of the second dissociation constant is to determine very precisely one of the components HS^- or S^{2-}. Licht and co-workers [32–34] tried to measure sulfide ion using different methods and found a very low value for K_2 of around 10^{-17}. This would mean for most natural conditions (pH 7–9 and 0–10 mmol L^{-1} total sulfide) that the sulfide ion could not be detected by any currently available analytical method except ISE in S(−II) buffered samples.

Licht and co-workers [32–34] used highly concentrated sulfide solutions of 3–6 mol L^{-1}. Physico-chemical parameters in aqueous solutions can, however, only be readily calculated from experimental data when the ionic strength of the solution is located inside the Debye–Hückel region, which describes properties of aqueous solutions up to a maximum ionic strength of about 1 mol L^{-1} (provided empirical corrections are included). Ionic strength outside this region, resulting for instance from sulfide concentrations of 3 mol L^{-1}, leads to a destruction of the water structure and hence to undefined conditions. In addition, there is no way to extrapolate values of Licht and co-workers to infinitely diluted solutions or even to conditions where the ionic strength is inside the Debye–Hückel region.

A more reliable method to get information about the acidity of HS^- was reported by Widmer and Schwarzenbach [29]. They investigated the complex formation of $[HgS_2]^{2-}$ at a mercury electrode and found at an ionic strength of 1 mol L^{-1} that

$$pK_1(20\,^\circ\text{C}) = 6.88 \pm 0.02 \qquad pK_2(20\,^\circ\text{C}) = 14.15 \pm 0.05 \tag{8}$$

The existence of dissociation constants pK_0 and pK_{00} corresponding to the equilibria:

$$H_3S^+ \rightleftharpoons H^+ + H_2S \qquad K_0 = \frac{[H^+][H_2S]}{[H_3S^+]} \tag{9}$$

$$H_4S^{2+} \rightleftharpoons H^+ + H_3S^+ \qquad K_{00} = \frac{[H^+][H_3S^+]}{[H_4S^{2+}]} \tag{10}$$

as proposed by Su *et al.* [35] is not very likely since a decrease of H_2S concentration at lower pH is generally not observed. In addition, the authors used a sulfide ion selective electrode (sulfide ISE) at $pH < 5$, which is not adequate as these electrodes are responding specifically to the sulfide ion. Even with the amplified technique of Su *et al.*, there is no way to overcome the limitations of the detection principle. Their potential readings must therefore be caused by other phenomena such as chemical destruction of the membrane and formation of potentials at the silver electrode due to the high solubility of Ag_2S in acidic media (see Section 4.2).

Aqueous sulfide solutions are easily oxidized by oxygen, peroxides (e.g. H_2O_2), halogens, nitric acid, lead dioxide and other oxidants [26,36,37]. Thereby, the final products (see Table 1) as well as the reaction rates are highly determined by the pH. At $pH < 6$ the reaction rate is very slow, shows maxima at pH 8 and 11 and decreases at $pH > 11$ [38]. On the overall pH range, heavy metal ions (mainly Fe^{3+}, Fe^{2+} and Ni^{2+}) are increasing the reaction rate [41]. In the case of iron, this is due to the local formation of H_2O_2:

$$Fe^{2+} + O_2 \xrightarrow{H_3O^+} Fe^{3+} + HO_2 \tag{11}$$

$$2HO_2 \rightarrow H_2O_2 + O_2 \tag{12}$$

The sensitivity of sulfide solutions to oxidation plays an important role in the accuracy of analytical determination. Some established analytical procedures require sampling, and sample stabilization, e.g. by the addition of zinc acetate to form zinc sulfide, before the actual analysis is performed. Zinc sulfide is stable for several weeks against oxidation by oxygen [42]. The most important problems involved with sampling are the losses of the analyte by evaporation, adsorption and oxidation prior to the stabilization step. This requires careful handling of samples and specialized procedures for the calibration of all analytical methods, for the determination of sulfide. This is just as important when *in situ*-methods are used and calibrated.

Table 1. Primary oxidation products of sulfide species at different pH

pH range	Main sulfide species	Primary oxidation products	References
acidic	H_2S	S^0, SO_4^{2-}	[38]
neutral	HS^-	$S_2O_3^{2-}, S_nO_6^{2-}, SO_3^{2-}, SO_4^{2-}, S_n^{2-}$	[38,39]
alkaline	HS^- and S^{2-}	$SO_4^{2-}, S_2O_3^{2-}, S^0$	[40]

Calibration of analytical methods for sulfide determination in aqueous samples should involve the following steps:

- A relatively concentrated sulfide solution of about 0.1 mol L^{-1} [S(−II)] is prepared by weighing a certain amount of Na_2S (7–9)H_2O and adding deaerated, deionized water. The determination of the exact sulfide content of this stock solution can be done by iodometric titration [43]. Such concentrated sulfide solutions are stable for months when protected against the impact of oxygen, light and heavy metals, e.g. by storing the solution under argon in brown gas-tight glass flasks. It is not optimal to use flasks with rubber washers or other parts made of silicone, rubber, PVC, Teflon etc. for such long time storage, as these materials are permeable for oxygen and can contain unknown amounts of heavy metals. Furthermore, some rubber materials tend to adsorb sulfide.
- The stock solution is used to make a diluted working solution. The sulfide content can be determined by the methylene blue method [43,44] (see Section 3.1). This diluted working solution is then used to calibrate the analytical method by exactly the same procedure, which is applied on the samples, i.e. stabilization with zinc acetate, dilution steps etc. The working solution cannot be stored and is stable only for some hours.

Another more accurate and easy way to perform calibration of analytical methods for sulfide determination was proposed by Jeroschewski and Schmuhl [45,46]. The method is based on the use of a novel sulfide generator (Figure 2), which performs an electrochemical reduction of HgS in a flow-through apparatus, whereby H_2S is produced in a deaerated, acidic carrier solution. The exact H_2S concentration [μmol L^{-1}] is determined by the flow velocity and the applied current according to the Faraday law:

$$c_{H_2S} = \frac{6 \times 10^4 I}{Fnv} \tag{13}$$

where I is the applied generator current in μA, F is the Faraday constant in c.mol^{-1}, n is the number of exchanged electrons (2), and v is the flow velocity [mL min^{-1}] of the carrier solution through the generator cell.

The sulfide generator thus performs a coulometrically controlled formation of hydrogen sulfide. The method can only be used in a flow-through configuration, but it is in principle possible to prepare stabilized sulfide solutions with an exactly known content by adding the sulfide-containing carrier solution to a zinc acetate solution. While the sulfide generator is ideal for calibration of e.g. amperometric H_2S sensors, it cannot be used for calibration of sulfide ion-selective electrodes as small amounts of mercury are released, which will interact with the Ag/Ag_2S membrane of such electrodes (see section 4).

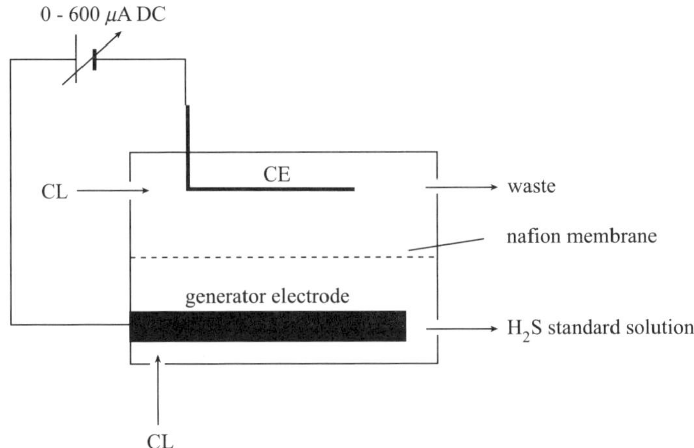

Figure 2. Scheme of a coulometric H_2S generator. CL, carrier solution of 5×10^{-3} mol L^{-1} H_2SO_4; CE, counter electrode; the generator electrode contains HgS; DC, direct current

3 MEASURING TECHNIQUES FOR SULFIDE

A large variety of techniques are available for measuring either total sulfide, sulfide ion or dissolved hydrogen sulfide. The techniques for the determination of $[S(-II)]_t$ usually are *ex situ* methods, i.e. sampling is required. Here the most important techniques are using the methylene blue reaction, other spectrophotometric methods, or chromatographic methods (see Table 2 and 3).

3.1 SPECTROPHOTOMETRIC METHODS

Spectrophotometric methods usually involve conditioning and treatment of the analyte sample with chemical reactions prior to the spectroscopic analysis. There exists a large variety of such analytical methods (Table 2). Only a few authors have proposed direct spectrophotometric determination of H_2S, mostly in the gas phase [76–83]. Although the analytical procedure in this case is rather simple, the direct measurement of H_2S demands a significant amount of (expensive) technical equipment, it suffers from serious interferences, especially from SO_2, and it requires great care to avoid analyte losses through evaporation, adsorption and rapid oxidation. The method is, therefore, seldom used for environmental analysis of sulfide. This is also true for kinetic methods, i.e. sulfide-catalyzed or -inhibited reactions [84–87,103,104] because under most conditions the parameter time cannot be controlled accurately. Furthermore, in natural waters interference with the kinetic reactions, e.g. by metal ions, is a problem.

Table 2. Spectrophotometric methods for sulfide determination in aqueous systems (Fl., Fluorescence; Ph., Photometry; FPh., Flame Photometry; k.Ph., kinetic photometry; MECA, molecular emission cavity analysis; GP-MAS, gas phase molecular absorption spectrometry; LOD, limit of detection)

Method	Procedure	Dynamic range ($mol\,L^{-1}$)	LOD ($mol\,L^{-1}$)	Interference	References
Fl.	fluorescence quenching of fluorescein mercury acetate (FMA)	$< 10^{-5}$	10^{-7}	NO_2^-, SO_3^{2-}, COS, $(CH_3)_2S$, CS_2	[47–49]
Fl.	fluorescence quenching of FMA incorporated in ethylcellulose*	no data	56×10^{-9}	cystein	[50]
Fl.	fluorescence quenching of thionine*				[51, 52]
Fl.	reduction of potassium-1,2-naphtoquinone-4-sulfonate	$(2.5–15) \times 10^{-6}$	No data	$S_2O_3^{2-}$, SO_3^{2-}, NO_2^-	[53]
Ph.	formation of methylene blue	$<1 \times 10^{-3}$†	$(0.3–1) \times 10^{-6}$	I^-, NO_2^-, CS_2, $S_2O_3^{2-}$, CN^-	[28, 43, 44, 54–56]
Ph.	formation of methylene blue (FIA)	$<5 \times 10^{-3}$†	$(0.1–1) \times 10^{-6}$	NH_4^+, CO_3^{2-} (at high conc.), CS_2, $S_2O_3^{2-}$	[51–63]
Ph.	formation of methylene blue (FIA)	$(6–600) \times 10^{-6}$†	$(0.3–20) \times 10^{-6}$		[64–66]
Ph.	addition of HS^- to Brilliant Green	$<62.3 \times 10^{-6}$	0.6×10^{-6}	SO_3^{2-}	[67]
Ph.	indirect determination with formation of SCN^-	$(10–600) \times 10^{-6}$	No data	Br^-, I^-, S(0), $S_nO_4^{2-}$	[68–71]
Ph.	complexation with organic mercury compounds (FIA)	$(0.3–1.1) \times 10^{-3}$	$(10–80) \times 10^{-6}$	oxalate, acetate	[72,73]
Ph.	reaction with cacothelin (FIA)	$(1–1100) \times 10^{-6}$	No data	SO_3^{2-}	[73]
Ph.	exchange with SCN^- and detection with Fe(III)	$(0.3–1.1) \times 10^{-3}$	No data	many other ions	[73]
Ph.	reaction with $[Fe(CN)_5NO]^{2-}$	$<25 \times 10^{-3}$	0.5×10^{-3}	selective	[74]
Ph.	reaction with Fe(III)/nitrilotriacetic acid	$(0.6–3.1) \times 10^{-3}$	No data	selective	[75]
MECA	after evaporation	$<1.4 \times 10^{-3}$	0.6×10^{-6}	all S-compounds	[76]

FPh.	previous separation with IC	$(0.4–1.2) \times 10^{-3}$	No data	selective	[77]
GP-MAS	after evaporation	$<18 \times 10^{-3}$†	$(0.8–22) \times 10^{-6}$	SO_3^{2-}, CN^-, NO_2^-	[78–83]
k.Ph.	reduction of toluidin blue by S^{2-} catalyzed by Se(IV)	$(8.8–53.1) \times 10^{-6}$	1.5×10^{-6}	SO_3^{2-}	[84]
k.Ph.	reaction of methylorange with BrO_3^- catalyzed by S^{2-}	$<156.3 \times 10^{-6}$	5.6×10^{-6}	no data	[85]
k.Ph.	utilizes the iodine–azide reaction	$(0.6–15.6) \times 10^{-6}$	0.3×10^{-6}	Cr(IV), V(V), Ce(IV), $S_2O_3^{2-}$, SO_3^{2-}	[86]
k.Ph.	reaction of Pyronine-G with hypophosphite catalyzed by Pd(II) and inhibited by S^{2-}	$(0.3–6.2) \times 10^{-6}$	0.12×10^{-6}	I^-, Br^-, IO_3^-, SO_3^{2-}, NO_2^-	[87]

* irreversible under anaerobic conditions.
† depending on experimental parameters.

Table 3. Chromatographic methods for sulfide determination in aqueous systems (PID, photoionization detector; FPD, flame photometric detector; SCD, sulfur chemoluminescence detector; amp., amperometric detector; photom., photometric detector; RP-HPLC, reverse phase high pressure liquid chromatography; LOD, limit of detection).

Method	Procedure	Dynamic range ($mol\,L^{-1}$)	LOD ($mol\,L^{-1}$)	References
GC/PID	previous enrichment with cryofocussing	$< 2.2 \times 10^{-3}$	12.7×10^{-9}	[88]
HS-GC/MS	cryofocussing of headspace sample (HS)	$< 156 \times 10^{-9}$	29×10^{-9}	[89]
HS-GC/FPD	cryofocussing of headspace sample	$< 0.5 \times 10^{-9}$	0.2×10^{-12}	[90]
GC/SCD	direct aqueous injection	$< 58.8 \times 10^{-6}$	0.25×10^{-6}	[91]
IC/amp.	gas dialysis	$< 2.5 \times 10^{-6}$	0.06×10^{-6}	[92]
IC/amp.	previous enrichment by zinc acetate fixation	$< 1.2 \times 10^{-6}$	0.16×10^{-6}	[93]
IC/amp.	direct injection of dialysed samples (see Chapter 11)	no data	$< 0.31 \times 10^{-6}$*	[94,95]
IC/photom.	post-column reaction of $KBrO_3$+ methylorange catalyzed by sulfide	no data	0.3×10^{-6}	[96]
IC/photom.	post-column reaction with I_2	$< 1.2 \times 10^{-6}$	0.1×10^{-6}	[97]
IC/photom.	post-column reaction with I_2	$< 0.5 \times 10^{-3}$	1.8×10^{-6}	[98]
RP-HPLC	pre-column derivatization to methylene blue	$< 1 \times 10^{-3}$	0.5×10^{-6}	[99]
RP-HPLC	pre-column derivatization to methylene blue and enrichment on silica gel	no data	3.1×10^{-9}	[100]
RP-HPLC	gas dialysis and pre-column derivatization with monobromobimane	$< 10 \times 10^{-6}$	0.04×10^{-6}	[101]
RP-HPLC	pre-column derivatization to 1-methyl-2-thiopyridone	$< 156 \times 10^{-6}$	0.06×10^{-6}	[102]

* depending on column age

The reaction of *N,N*-dimethyl-1,4-phenylenediamine with H_2S, which is catalyzed by Fe(III), via addition of $FeCl_3$, and leads to formation of methylene blue (3,7-*bis*(dimethylamino)-phenothiazin-5-ium chloride), was originally introduced by Fischer in 1883 for the determination of total sulfide in aqueous solution [44]. During the last century, the methylene blue method was modified and adapted to specific analytical problems. The method was applied in various fields of scientific research on sulfide-containing systems, and has become a standard method in analytical chemistry [28,43,54,55]. An important recent improvement was the development of an *in situ* analyzer (Scanner = submersible chemical analyzer) by Sakamoto-Arnold and Johnson [57–59] (see also Chapter 7 in this book), which uses a flow injection analysis (FIA) variation of the methylene blue method. This analyzer has been applied for *in situ* determination of sulfide in hydrothermal vent fields on the deep-sea floor at several kilometers water depth, yielding important and new knowledge about these extreme sulfidic habitats.

Another modification of the original methylene blue method was introduced by Cline [56], and is widely applied in limnological and marine research. Instead of the standardized method, which uses separate solutions of *N,N*-dimethyl-1,4–phenylenediamine and $FeCl_3$, Cline proposed a mixed solution of the two reagents to overcome problems arising with salinity, temperature and pH. We recently compared this approach with the standard method and were not able to realize any advantage (Steuckart *et al.*, unpublished data). Furthermore, the method of Cline shows an important disadvantage if the sample is pH buffered (which is the case in many natural waters). Reliable quantitative results can only be obtained when the pH of the reaction mixture (reagents + sample) is kept very low, as the absorption spectrum of methylene blue depends on its state of protonation (Figure 3A). Also, the sensitivity characteristics of the method vary when mixed reagent solutions are used. Under the same conditions, i.e. same sulfide and reagent concentration and same protonation state of

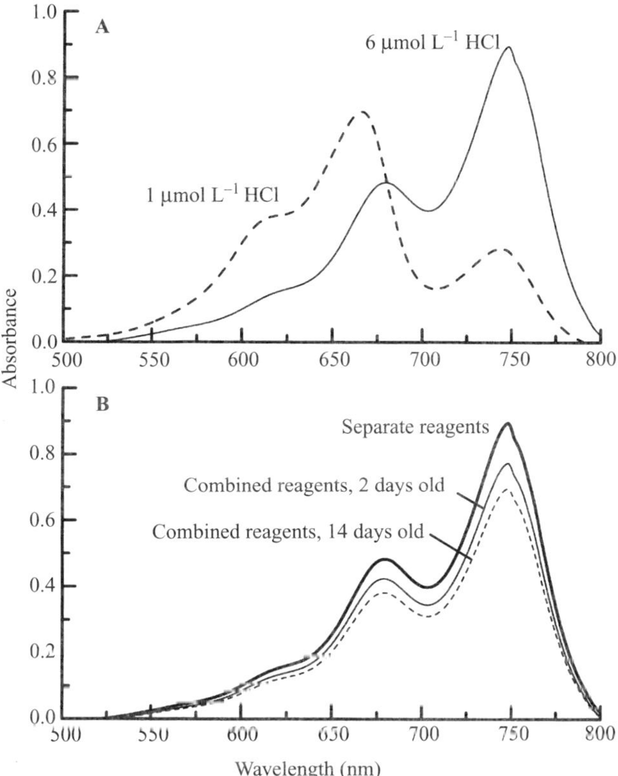

Figure 3. Absorbance spectra of methylene blue: (A) at different pH; (B) obtained with different types of sample/reagent treatment, i.e. with separate reagents, and with 2 d and 14 d old solutions of mixed reagents, respectively (see text for details)

the methylene blue, the mixing of the reagents some hours prior to the analytical procedure leads to a significant drop in sensitivity, which increases with the age of the mixed solution (Figure 3B). The experiments were not performed at the upper border of the dynamic range, i.e. the concentrations of the reagents were always high enough to convert all sulfide to methylene blue, hence resulting in the same theoretical methylene blue concentration. Therefore, the drop in sensitivity can only be explained by a deterioration of the reagent mixture and/or the formation of interfering substances over time. This is consistent with a reaction mechanism proposed by Kubàñ *et al.* [105], where the first step of the methylene blue reaction is the oxidation of the diaminoaniline by Fe(III) leading to a slowly established equilibrium between a cation radical and a quinone diimine, where the latter is not involved in the further reaction with H_2S. In conclusion, we recommend the use of separate reaction solutions when using the methylene blue method.

The formation of ethylene blue instead of methylene blue, by using *p*-diethylaminoaniline as the reactive substrate, has been successfully established for the determination of sulfide in Kraft liquors [64–66]. This method is well suited for sulfide determination even in complex wastewater from paper mills (white, green and black liquors consisting of highly concentrated mixtures of hydroxide and sulfide).

An alternative to the classical methylene blue method for environmental sulfide analysis was proposed by Koh *et al.* [68–71] and is based on several variations of the oxidative reaction of sulfide with cyanide to thiocyanide:

$$S^{2-} + I_2 + CN^- \rightarrow SCN^- + 2I^- \tag{14}$$

Though the methodological parameters (e.g. linear dynamic range, LOD, precision) are similar to those of the methylene blue method, the method is significantly more time consuming and complicated. We conclude that the classical methylene blue method is still the spectrophotometric method of choice for sulfide determination in natural waters.

3.2 CHROMATOGRAPHIC METHODS

Chromatographic methods find many applications in environmental chemistry. They combine powerful separation methods and sensitive detection techniques. The main disadvantage of chromatography is the need for sampling prior to analysis, and *in situ* measurements using chromatographic methods are very complicated to realize. In addition, sample treatment (pre-column or post-column) as well as the separation procedure itself change the chemical and/or biological state of the sample. Therefore, information about the actual situation with respect to sulfide speciation, pH, redox equilibria etc. cannot easily be obtained. In spite of this, gas chromatography (GC), ion chromatography (IC), and high performance liquid chromatography (HPLC) are used for the separation and determination of sulfide in waters (Table 3). Gas chromatography requires the removal of the analyte (H_2S) from the aqueous matrix prior to the analysis followed by enrichment procedures, such as cryofocusing of the head-space [89,90]. Tang and Heaton [91] have shown that it is possible to inject aqueous samples into the gas chromatograph. The detectors normally used for the gas chromatographic determination of H_2S are the photo-ionization detector (PID), the flame photometric detector (FPD), and the sulfur chemoluminescence detector (SCD), which is extremely selective towards sulfur compounds.

Ion chromatographic methods can be performed directly with filtered aqueous samples but problems such as (i) enrichment due to the reaction of sulfide with heavy metal contaminations of the reagents, or (ii) a strong adsorption of sulfide onto the analytical column can be observed [106]. Because of the low dissociation of H_2S, the widespread conductivity detectors cannot be used for sulfide determination. In most cases, ion chromatographic methods are combined with amperometric (Ag versus SCE) [92–95] or photometric (e.g. post-column oxidation with iodine) [97,98] detection principles. More sensitive HPLC methods involve a pre-column derivatization of sulfide mostly to methylene blue and these techniques have detection limits in the nmol L^{-1} range [100–102].

3.3 ELECTROCHEMICAL METHODS

Electrochemistry offers a variety of measuring principles for determining specific sulfur species directly in natural waters (see Table 4). Currently, three types of electrochemical techniques are mostly used for *in situ* environmental analysis of sulfide speciation: (i) the potentiometric sulfide ion-selective electrode, (ii) the amperometric H_2S sensor, and (iii) methods based on voltammetry with either amalgamated gold or Hg-coated iridium microelectrodes. These methods will be discussed in more detail in the following section.

4 SENSORS FOR MEASURING SULFIDE IN AQUATIC SYSTEMS

The quantitative determination of chemical variables in the environment requires analytical methods that are minimally invasive in terms of disturbance of local chemical or redox equilibria and in terms of mechanical disturbance. Electrochemical sensors are suitable analytical tools for this purpose because no sampling and/or sample treatment is required. Such sensors can be constructed with geometric parameters of the sensor tip in the micrometer range, which minimizes mechanical disturbance and, in the case of electrochemical methods, offers many other advantages compared with macroelectrodes. For instance, amperometric and voltammetric methods are characterized by a consumption

Table 4. Electrochemical methods for sulfide determination in aqueous systems (RDE, rotating disc electrode; SMDE, static mercury drop electrode; ISE, ion-selective electrode; LOD, limit of detection)

Method	Procedure	Dynamic range (μmol L^{-1})	LOD (μmol L^{-1})	References
Polarography	cathodic stripping polarography	no data	0.09	[107]
Polarography	differential pulse polarography	< 1500	0.1	[108–110]
Voltammetry (Ag-RDE)	cathodic stripping voltammetry	0.01–10	0.01	[111,112]
Voltammetry (SMDE)	alternating current voltammetry	0.01–900	no data	[113]
Potentiometry	S^{2-} – ISE(Ag/Ag_2S)	< 10^6	< 10^{-17*}	[114–117]
Potentiometry	S^{2-} – ISE(Ag/Ag_2S/pH)	< 10^6	$\leqslant$10; $[S(-II)]_t$	[118–126]
Potentiometry	S^{2-} – ISE (Ag/Ag_2S in FIA with/without gas dialysis)	15–1500 $[S(-II)]_t$	1 – 6	[127–130]
Potentiometry	S^{2-} – ISE (air-gap variation of Ag/Ag_2S)	> 10; $[S(-II)]_t$	0.01; $[S(-II)]_t$	[131]
Potentiometry	CN^- – ISE (Ag/AgCN with acidic evaporation)	> 12; $[S(-II)]_t$	0.3; $[S(-II)]_t$	[132]
Potentiometry	S^{2-}/Cl^- – ISE (Ag/Agx (S,Cl) with FIA)	no data	< 10; $[S(-II)]_t$	[133]
Potentiometry	potentiometric titration with Pb(II) with Pb-ISE	no data	no data	[134]
Potentiometry	S^{2-} – ISE (Co(II)-phtalocyanines and -porphyrines as electrocatalysts)†	depends on pH	depends on pH	[135–137]
Potentiometry	HS^- – ISE (carrier system)	0.2–20; $[S(-II)]_t$	0.06; $[S(-II)]_t$	[138]
Potentiometry	HS^- – ISE (carbon paste electrode)	< 1000; HS^-	no data	[139]
Amperometry	porous Au electrode with pneumato-amperometry	3 – 3300	no data	[140]
Amperometry	H_2S specific biosensor with immobilized *Thiobacillus thiooxidans*	20–400	no data	[141]
Amperometry	H_2S specific sensor with redox mediator	3–90	~ 1	[142–146]
Amperometry	H_2S specific sensor with redox mediator (FIA)	1–750	~ 1	[147,148]
Amperometry	H_2S microsensor	1–750	< 1	[6,149,150]

* Free concentration in buffered medium; value stated by manufacturer (see Section 4.2 for details).
† blocked by I^-, SCN^- and CN^-.

of the analyte through reduction or oxidation. The analytical parameter is the corresponding current which is time dependent in the case of macroelectrodes and is described by the Cottrell equation [151] :

$$i_l = \frac{nFA\sqrt{D}c^*}{\sqrt{\pi t}} \tag{15}$$

where, i_l is the limiting current, t is the time, n is the number of exchanged electrons, F is the Faraday constant, A is the electrode surface area, D is the analyte diffusion coefficient, and c^* is the bulk concentration of the analyte.

In the case of microelectrodes the limiting current is time independent [151] :

$$i_l = 4r_0\pi nFDc^* \tag{16}$$

where i_l is the limiting current, r_0 is the electrode radius, and the other parameters have the same meaning as in equation (15). This important measuring characteristic makes microelectrodes very suitable for the analysis of dynamic processes and solute distribution at high spatial and temporal resolution. In addition, since the radius of the electrode, r, is very small, the limiting current, i_l, is also small and analysis in highly resistive solutions is possible since the $i.R$ drop is small. The formation of spherical, hemispherical and cylindrical diffusion layers enhances the mass transport of the analyte to the microelectrode leading to short response times and low dependence of the sensor signal on stirring rate of the sensor tip environment. A more detailed discussion of macro-versus microscale electrochemical sensors can be found in Chapters 8 and 9 of this book.

The different measuring characteristics of macro- and microelectrodes are illustrated in Figure 4. Cyclic voltammetry of a reversible redox system (here: hexacyanoferrate(III)/(II) in alkaline solution) at macroelectrodes shows a typical asymetrical peak of the I/E curve due to a time-dependent diffusion layer thickness, whereas at a microelectrode spherical diffusion leads to a time-independent flux of analyte at the electrode surface, and thus a characteristic S-shape of the I/E curve.

Most sensor principles available for the determination of sulfide species in the environment are based on potentiometry (section 4.2), voltammetry (section 4.3), and amperometry (section 4.4). The available sensors respond either specifically to the sulfide ion or to H_2S, or to the whole of labile S(−II) species. Some attempts to develop HS^- specific sensors have been reported [138,139]. However, a reliable analytical method for HS^- is still to be developed. Also, reversible optical sensors for the determination of S(−II) species have not yet been developed [50–52,152].

4.1 OPTICAL AND BIOSENSORS FOR SULFIDE

Choi *et al.* [50] proposed an optical sensor based on the fluorescence quenching of fluoresceine by mercury(II) acetate (FMA), which is reacting with sulfide. The measuring principle is reversible only in the presence of oxygen, which is used to regenerate the sulfide sensitivity of the sensor after analytical measurement. A similar behavior was found for the fluorescent dye thionine, which is also irreversibly quenched by sulfide in the absence of oxygen [51,52]. Evidently this imposes severe problems for the practical use of these sensor principles as S(−II) is unstable in the presence of oxygen. Therefore, the currently available

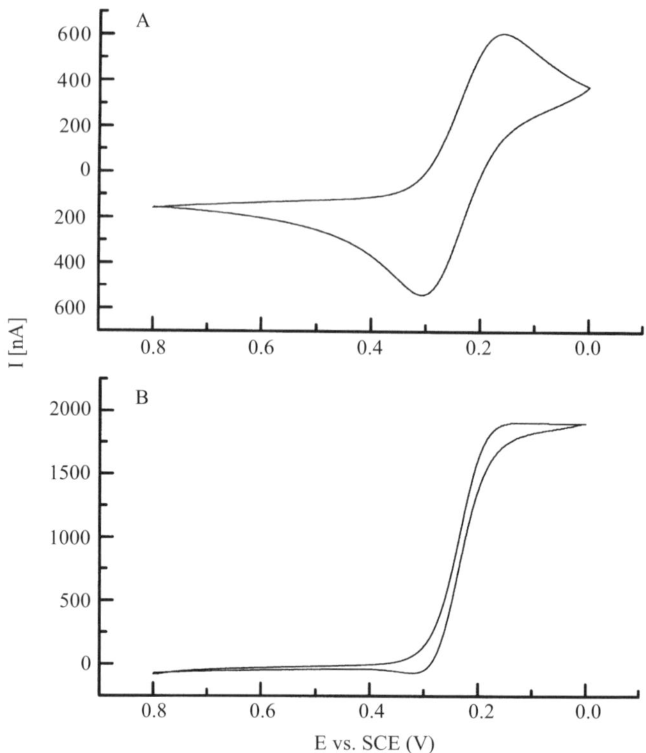

Figure 4. Cyclic voltammogram of 0.05 mol L^{-1} $K_3Fe(CN)_6$ in 0.5 mol L^{-1} carbonate buffer solution measured at a Pt disc electrode. (A) Electrode diameter = 200 μm; scan rate = 0.05 V s^{-1}. (B) Electrode diameter = 10 μm; scan rate 0.1 V s^{-1}

optical sensors do not allow for the continuous determination of sulfide under anaerobic conditions in natural waters. Cardoso *et al.* [152] presented a reversible sensor principle using FMA for the determination of atmospheric hydrogen sulfide, which cannot, however, be used in natural waters.

An interesting possibility for future sensor developments was proposed by Kurosawa *et al.* [141]. Their H_2S specific sensor is based on the oxidation of sulfide by *Thiobacillus thiooxidans* (a colourless sulfur bacterium) with oxygen, which is monitored by a Clark-type oxygen sensor as an internal transducer in the biosensor. More details on various biosensors are presented in Chapter 6 in this book.

4.2 POTENTIOMETRIC SULFIDE SENSORS

Potentiometric methods are based on the measurement of equilibrium potentials without electrochemical consumption of the analyte and are, therefore, well suited for *in situ* determination. The most important tools used in potentiometry are ion-selective electrodes (ISE) consisting of a solid state or (semi-) liquid membrane and a suitable transducer [153]. The sensor membrane, which has to be electrically conductive and practically insoluble in the medium to be investigated, is the sensitive part of such electrodes. Although this type of electrode is referred to as ion selective, the analytical practice shows interferences by many other species. Problems arise when species are present that may complex one of the components of the membrane or incorporate into the solid membrane (e.g. Cl^-, Br^-, I^- in the case of Ag/Ag_2S membranes) [154]. A more general discussion of potentiometric techniques and ion-selective microsensors can be found in Chapter 5 of this book.

The Ag/Ag_2S electrode, which was originally used as a reference electrode, was first introduced as a potentiometric sulfide ISE by Berner in 1963 [114]. In 1983, Revsbech *et al.* [155] introduced an Ag/Ag_2S microelectrode for use in environmental analysis. More robust needle-type Ag/Ag_2S sensors have also been developed [117].

Silver sulfide is a semiconductor with a high ionic conductance similar to its electrical conductance, i.e. $\sim 5.4 \times 10^{-4}\,\mathrm{S\,cm^{-1}}$, with Ag^+ as the mobile ion in an S^{2-} network [156]. The detection principle is based on the formation of equilibrium potentials in the electrochemical chain $Ag/Ag_2S/S^{2-}$ described by the Nernst equation:

$$E = E^0 - \frac{RT}{2F}\ln a(\mathrm{S}^{2-}) \tag{17}$$

where $a(S^{2-})$ is the activity of the ion S^{2-}. This implies a strong influence of temperature and ionic strength mainly due to a change in the activity coefficient of the sulfide ion. With the help of equation (5) the calibration curve for sulfide ion (E versus $\ln(a(\mathrm{S}^{2-}))$) obtained with the ISE at a given pH and ionic strength can be converted to a calibration curve for total sulfide (E versus ln $[\mathrm{S(-II)}]_t$, which has a theoretical slope of $-29\,\mathrm{mV\,decade^{-1}}$ at 298 K. As the potentiometric Ag/Ag_2S electrode responds only to the sulfide ion, and not to HS^- or H_2S, the E versus $\ln[\mathrm{S(-II)}]_t$ calibration graph exhibits a parallel shift by $29\,\mathrm{mV\,pH^{-1}}$ at pH 7–13, while at pH < 7 the shift amounts to $59\,\mathrm{mV\,pH^{-1}}$. The slope of the E versus $\ln[\mathrm{S}^{2-}]$ calibration curve is theoretically independent of pH. The theoretical limit of detection can be calculated from the solubility product of Ag_2S[157]:

$$[\mathrm{Ag}^+]^2[\mathrm{S}^{2-}] = 10^{-51} \tag{18}$$

giving a value of $[\mathrm{S}^{2-}] = 6.3 \times 10^{-18}\,\mathrm{mol\,L^{-1}}$, which corresponds to $2\,\mathrm{pmol\,L^{-1}}$ of $[\mathrm{S(-II)}]_t$ at pH 8.5). In practice, the limit of detection is far higher (about $0.1\,\mu\mathrm{mol\,L^{-1}}$ of $[\mathrm{S(-II)}]_t$) owing to the formation of mixed potentials, silver complexes, elemental silver and other factors leading to the so-called super-Nernstian behavior (see below).

There are several practical problems with the use of Ag/Ag_2S electrodes in complex environmental samples. Dissolved silver ions, heavy metals incorporated (as metal sulfides) in the membrane, halides (as silver halides), and pseudohalides (cyanide) can build new electrochemical chains of the general form $Ag/(Ag, Me)_x(S, Hal)/(Ag^+, Me^{n+}, S^{2-}, Hal^-)$ and hence lead to unpredictable mixed potentials and irreproducible quantitative results [154,158].

In particular, the presence of Hg(II) in solution leads to an irreversible destruction of the Ag_2S membrane by formation of HgS precipitates at the membrane surface. Because of the lower solubility product of HgS as compared with Ag_2S and the unpredictable amount of the HgS formed, uncontrolled mixed potentials occur, which can result in large analytical errors. For that reason, Dobcnik *et al.* [116] proposed to coat the Ag_2S membrane with mercury sulfide in order to improve the sensitivity, the reproducibility and the response time. We tested this modification but were not able to obtain reliable quantitative results because of the non-linearity of the calibration graph [150]. This is consistent with the investigations of De Marco *et al.* [154]. Yu *et al.* [159] presented a modified sulfide microelectrode which was pretreated with $HgCl_2$ solution similarly to the procedure proposed by Dobcnik *et al.* [116]. The calibration of their microelectrode, as well as that of a commercial sulfide ISE used in their study, showed increasing slopes of the calibration curves with decreasing pH up to about 50 mV $(pS^{2-})^{-1}$ at pH 7.2. The authors interpret this behavior as a response of the sulfide ISE to HS^- and S^{2-}, which is in contrast to the well-known theory for the Ag/Ag_2S electrode.

Another problem with the potentiometric sulfide ISE is the so-called super-Nernstian behavior. At low sulfide concentrations ($< 1\ \mu mol\ L^{-1}$ $[S(-II)]_t$ at pH 12.7), reducing conditions and high pH (e.g. by use of so-called sulfide antioxidant buffer, SAOB, with commercial macroelectrodes [115]) the slope of the E versus $\ln[S(-II)]_t$ calibration graph can show a significant deviation from $-29\ mV\ decade^{-1}$ [119,160,161]. This is due to the reduction of Ag_2S under these conditions, which leads to the formation of elemental silver at the electrode surface [156,158].

Frevert and coworkers developed a so-called pH_2S sensor based on the Ag/Ag_2S electrode in conjunction with a pH glass electrode, thus avoiding a liquid junction reference [118–122,125]. Although, it was stated that the influence of ionic strength was reduced by 50 % and that the sensor signal was pH independent at pH < 6, the basis of Frevert's sensor remains the Ag/Ag_2S electrode including all the above-mentioned problems and disadvantages.

In conclusion, Ag/Ag_2S-based sulfide sensors can be used for environmental analysis under near neutral to alkaline conditions in the water column as well as in sediments and biofilms with detection limits of ca. $1\ \mu mol\ L^{-1}$ for $[S(-II)]_t$. The construction of well-functioning Ag/Ag_2S-based sulfide sensors and calibration of such sensors is, however, complicated by the above-mentioned factors and this puts a limitation on the practical use of such sensors for *in situ* analysis. Nevertheless, such sensors have been used for laboratory [7,155,162] as well as *in situ* applications [117,163,164] in various aquatic systems (see section 5).

4.3 VOLTAMMETRIC SULFIDE SENSORS

Voltammetry is an electroanalytical method, which enables multi-species analysis by measuring oxidation or reduction currents of chemical species as function of the potential imposed to the electrode. The potential at which the electrode reaction occurs is primarily determined by the redox potential of the electron transfer, but it is additionally influenced by pH (when H_3O^+ is involved in the redox reaction), complexation of the test species and its diffusion properties. A more detailed account of voltammetric techniques in environmental analysis in water and sediment is given by Buffle and Tercier–Waeber in Chapter 9 of this book. In the following, we only address some aspects relating to voltammetric determination of sulfide.

Various voltammetric techniques have been used with Hg electrodes to measure S(−II) in natural waters [110]. Recently, Brendel and Luther [165] used differential pulse polarography at an amalgamated gold electrode (tip diameter ~ 100 μm) for the direct measurement of S(−II) in sediment cores. First applications of this technique for the combined measurement of concentration gradients of sulfide, oxygen, I^-, Fe(II), and Mn(II) in porewater of sediments have been reported [166,167]. Also, data from *in situ* profiling of sulfide and other chemical variables in sediments have been obtained by Reimers and Luther (cited in ref. [23]). These studies demonstrate that new information on the complex porewater chemistry of aquatic sediments can be obtained. However, voltammetry with bare electrodes in complex media can be problematic. In the following, we list a few concerns on the limitations of voltammetric techniques with respect to sulfide analysis and give some suggestions for improvement. Nevertheless, besides a need for technical optimizations the approach of performing *in situ* voltammetry with microelectrodes seems very promising and we regard the technique as having a large potential for the quantification of various redox species in natural systems, especially for fine scale analysis of iron and manganese species.

Electrodes with a diameter of ~ 100 μm or larger are not microelectrodes (see also Chapter 9) and, therefore, the well-known advantages of microelectrodes such as measurement in low conductivity freshwaters and independence of stirring rate will not be fully realized. Such relatively large electrodes exhibit currents in the nA range, related to a large local analyte consumption that may disturb local gradients and equilibria around the sensor tip. Microsensors in μm size range have a much smaller consumption of analyte (typical measuring currents in the pA range) and are the only ones to which spherical diffusion occurs. The advantageous measuring characteristics of microelectrodes are obtained only when the dimensions of the sensor tip, where the reactions take place, are smaller than the thickness of the diffusion layer at the electrode surface, leading to a spherical or hemispherical diffusion field. [150]

One of the main advantages of using mercury as an electrode is its high overpotential for the reduction of water to form hydrogen. When it is coated onto substrates of high solubility such as Au or Ag, however, this advantage is decreased. Glassy carbon electrodes do not suffer from this problem, but the deposition of mercury is not very reproducible on this substrate [168,169]. Buffle and coworkers [170,171] have shown that iridium is by far the best substrate for mercury because of its low solubility in mercury ($< 10^{-6}$ wt%) and its good wettability by Hg.

Another aspect to consider is the measurement of sulfide by oxidation at metal electrodes with pulse techniques. Shimizu *et al.* [172] wrote in 1981: "It may be dangerous to extend the frequently employed pulse techniques to an electrochemical system forming a deposit or film at a solid electrode. A variation of the thickness of the deposited film with time may cause a change in double-layer capacity, and thus capacitive currents may still remain at the current sampling time". In addition, in all cases double peaks are observed at high sulfide concentrations (>10–100 μmol L^{-1}) when Hg electrodes were used with DPP [108,109]. Canterford attributed this phenomenon to the formation of dense HgS films. Davison and Gabbutt [173] observed similar problems with DPP when sulfide concentrations in natural waters exceeded 2 μmol L^{-1}, but at lower concentrations good linear calibration curves were obtained. Normal pulse polarography, in conditions which prevent surface accumulation, can be recommended for higher concentrations, while the reliable determination of sulfide at a very low concentration level is possible by the use of cathodic stripping techniques [107,111,112] or a.c. voltammetry [113] provided the accumulation time is short enough so that a multilayer film is never formed. The important point is that at high concentration the electrode must always be placed at $E < E_{\text{pic}}$ (to avoid oxidation of Hg) except during the analysis, where a fast positive scan must be used followed immediately by a return of the electrode potential to $E < E_{\text{pic}}$ (The reaction is: $S(-II) + Hg \rightarrow HgS + 2e^-$). At low sulfide concentration, the electrode can be put at $E > E_{\text{p}}$ for a given time to accumulate HgS, but not too long in order to avoid a multilayer film formation, and then a negative scan is used (the reaction is: $HgS + 2e^- \rightarrow Hg + S(-II)$) (see Chapter 9, section 5 for more details).

Electrochemical measurements at an unprotected electrode under complex environmental conditions are problematic because of fouling of the electrode (see Chapter 9) by natural waters, colloidal and particulate forms of organic and inorganic matter that can be reduced, oxidized or simply adsorbed onto the electrode surface. This often interferes with the electrode reaction of interest and drastically perturbs the voltammetric peaks, which can result in large analytical errors, hard to quantify. This is especially true in sediments exhibiting complex porewater composition and high concentrations and gradients of both inorganic and organic compounds. In some cases electrochemical conditioning between scans can help alleviate such interferences but reproducibility is rarely good and the presence of memory effects should be carefully checked.

The problems associated with *in situ* voltammetry with bare electrodes may also partly apply to bare potentiometric electrodes. Thus, for *in situ* applications a protection of the electrode surface by a membrane, permeable to the test analyte, is usually necessary (a detailed discussion of voltammetry on bare and membrane-covered electrodes is given in Chapter 9).

4.4 AMPEROMETRIC HYDROGEN SULFIDE SENSORS

Most electrochemical gas sensors have a gas permeable membrane, through which the analyte can diffuse into an inner electrolyte compartment, where the electrochemical reactions take place under well-defined conditions. Consequently, gas sensors often exhibit much better measuring characteristics in terms of stability and, especially, selectivity as compared with potentiometric and voltammetric sensors. Gas sensors are well suited for *in situ* analysis and especially Clark-type oxygen microelectrodes have proven to be excellent tools for environmental analysis (see Chapter 1) [174, 175].

An amperometric detection principle for the determination of dissolved hydrogen sulfide in aquatic systems was developed by Jeroschewski and coworkers. [143–148] and several macrosensors for H_2S were realized. In collaboration with our group, a new H_2S microsensor based on this amperometric measuring principle was developed [6,149,150]. As the sensor principle is relatively new in comparison with potentiometric and voltammetric sulfide sensing, we describe the new H_2S microsensor in some more detail below.

The sensor design is based on the Clark principle, i.e. the sensor consists of electrodes in an electrolyte filling an inner compartment, which itself is separated from the analyte solution by a gas permeable membrane (silicone). All electrodes (working electrode, guard electrode, and counter-electrode; Figure 5A) are made of platinum and are placed in a glass casing made of a Pasteur pipette, that is tapered to a tip diameter of a few micrometers and sealed with a thin silicone membrane. The working and guard electrode have a tip diameter of a few micrometers and are prepared by electrochemical etching in concentrated KCN solution. For the sake of mechanical stabilization, electrical insulation towards the guard electrode and minimization of the active electrode surface, the working electrode is coated with a highly resistive glass except for the very tip of a few micrometers. The analyte, H_2S, diffuses through the gas permeable membrane of the casing and is electrochemically determined inside the sensor.

The direct oxidation of sulfide to elemental sulfur at platinum electrodes leads to an inactivation of the electrode surface [176,177]. Hence, the Clark

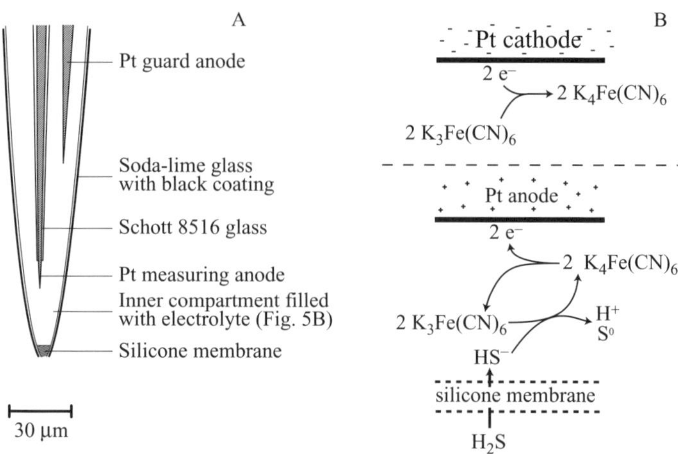

Figure 5. Schematic drawing of an H_2S microsensor tip (A), and the measuring principle of the microsensor (B) (redrawn from Kühl *et al.* [6] by permission of Inter-Research). Note that only the working and the guard electrodes are shown in (A). The Pt counter-electrode is situated further away from the sensor tip up in the bulk part of the electrolyte-filled outer casing. The electrolyte is shielded from photodegradation by painting the outer casing with a black paint. Any optically dense paint which shows a good adhesion to glass can be used. We have good experience with black enamel paint containing xylene as solvent

principle was improved by using a redox mediator (hexacyanoferrate(III)), which oxidizes sulfide to sulfur before it reaches the charged Pt surface. This reaction forms hexacyanoferrate(II) that diffuses to the Pt microanode, where it is reoxidized to hexacyanoferrate(III) (Figure 5B). A more detailed reaction scheme is as follows [149]:
Anode reactions:

$$H_2S + 2[Fe(CN)_6]^{3-} \rightleftarrows 2[Fe(CN)_6]^{4-} + S^0 + 2H^+ \qquad \text{(chemical step)} \quad (19)$$

$$2[Fe(CN)_6]^{4-} \rightleftarrows 2[Fe(CN)_6]^{3-} + 2e^- \qquad \text{(electrochemical step)} \quad (20)$$

Cathode reaction:

$$2[Fe(CN)_6]^{3-} + 2e^- \rightleftarrows 2[Fe(CN)_6]^{4-} \qquad \text{(electrochemical step)} \quad (21)$$

Brutto reaction:

$$H_2S + 2[Fe(CN)_6]^{3-} \rightleftarrows 2[Fe(CN)_6]^{4-} + S^0 + 2H^+ \qquad (22)$$

A potential difference, ΔE_{pol} (Figure 6), between +80 and +150 mV is imposed between the working or guard electrode and the counter-electrode. The guard electrode serves to shield the measuring electrode from reduced components,

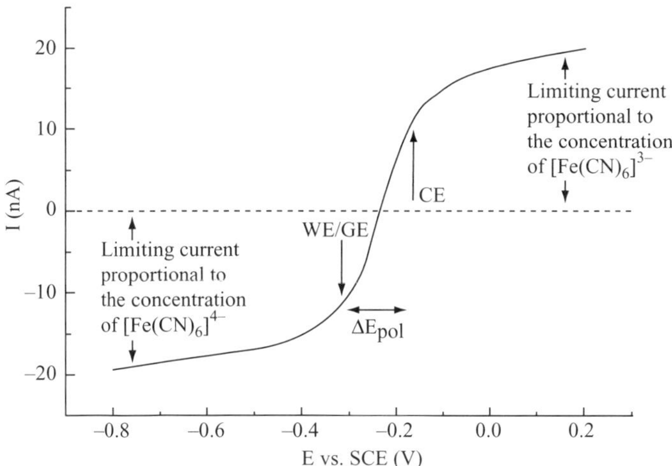

Figure 6. A single-sweep voltammogram of 0.05 mol L^{-1} hexacyanoferrate(II)/(III) in 0.5 mol L^{-1} carbonate buffer measured with a Pt disc electrode (diameter = 5 μm) versus SCE at a scan rate of 0.5 V s^{-1}. The potentials imposed on each individual electrode in a H_2S microsensor are indicated on the figure. WE, working electrode; GE, guard electrode; CE, counter-electrode; ΔE_{pol} = polarization voltage imposed on the sensor

e.g. hexacyanoferrate(II) produced at the counter-electrode, that diffuse towards the sensor tip. The guard electrode thus helps to keep a constant high ratio of hexacyanoferrate(III) to hexacyanoferrate(II) in the microsensor tip compartment, which is a prerequisite for a low zero-current and good signal stability.

The individual potentials of the electrodes are determined by the relative concentrations of hexacyanoferrate(II) and hexacyanoferrate(III) in the inner compartment. This is very important for the lifetime of the sensor. Figure 6 illustrates the potentials of the electrodes with respect to each other. When the sensor is filled with electrolyte, the sensor electrodes are at a potential of ~ 0.27 V versus SCE, i.e. close to the standard potential of the redox system hexacyanoferrate (II)/(III). When the sensor is polarized, the potential of the counter-electrode, CE, is shifted by ΔE_{pol} versus WE/GE towards more reducing potentials and a current can flow between counter-electrode and the measuring and guard electrode, respectively. Over time, equation (22) leads to a slow change of the ratio of hexacyanoferrate(II) to hexacyanoferrate(III), which results in a slow vertical shift of the i–E curve in Figure 6, from top to bottom. This results in a slow drift towards more negative potentials of the working, guard and counter-electrodes, respectively. When the working and guard electrode potential gets close to the jump of the i–E curve (Figure 6), small changes in the potential, e.g. by electrical noise, lead to high changes of the current, resulting in a lower signal to noise ratio. This can be overcome by a stepwise increase of the polarization voltage. When the concentration of hexacyanoferrate(II) is so high that the potential of the counter-electrode is shifted to very negative potentials, the sensor lifetime ends and the inner solution must be changed. The latter is, however, seldom practicable with H_2S microsensors. This is only one of several parameters determining the lifetime of the microsensor, which is typically 2–8 weeks. More details of sensor characteristics are discussed elsewhere [6,149,150].

The H_2S microsensor exhibits a linear response from $\sim 0.1\ \mu mol\ L^{-1}$ to $>$1–2 $mmol\ L^{-1}$ H_2S, with a sensitivity of 0.2–3 pA $(\mu mol\ L^{-1})^{-1}$. The detection limit is $\sim$0.1–1 $\mu mol\ L^{-1}$ H_2S, depending on the actual sensitivity of the sensor and the accuracy of the ampere meter, which is used in connection with the H_2S microsensor. We recommend the use of picoampere meters with a resolution of 0.1–1 pA, with the possibility for polarizing both measuring and guard electrode. Such meters are commercially available (e.g. from Unisense Aps, Denmark).

The small consumption of H_2S by the microsensor in the test water minimizes the corresponding concentration gradients and, therefore, the effect of external convection at the sensor tip. The sensor signal in stagnant water exhibits a decrease by $<$1–2 % of the value measured in vigorously stirred water. In addition, there is no disturbance of sulfide equilibria at the microsensor tip. Therefore, the sensor signal, for a given total sulfide concentration, at various pH, can be predicted from the protonation equilibiria of the sulfide system (equation (3)). The only siginificant interfering agent is SO_2, which is not important in most natural waters. As the electrolyte of the H_2S microsensor is photodegraded, we recommend shielding the inner electrolyte-filled compartment from bright light by painting the outer sensor casing with an optically dense paint with a good adhesion to glass, e.g. a black enamel paint containing xylene as the solvent.

5 SULFIDE AND H_2S MICROSENSOR APPLICATIONS IN AQUATIC SYSTEMS

Most studies of sulfide in the water column are still based on spectrophotometric methods (see Section 3.1) and relatively few measuring devices equipped with sulfide sensors have been used for profiling the pelagic environment. As a representative example, Eckert *et al.* [125] described a measuring device equipped with a potentiometric Ag/Ag_2S electrode, that was used to monitor the vertical sulfide distribution in the hypolimnion of a stratified lake.

Water sampling and subsequent sulfide analysis or monitoring with large sulfide sensors can in principle provide sufficient resolution for applications in the water column. The consumption of H_2S may, however, be high and stirring effects may therefore be important. Furthermore, with water sampling the risk of degassing or oxidation by O_2 during sampling is very high. *In situ* measurements are therefore important. It is even more so in the porewaters of sediments and biofilms, where sulfide cycling is of major importance, steep concentration profiles of sulfide and other redox species prevail and must be determined [5–7]. Nevertheless, porewater extraction from defined sediment strata followed by spectrophotometric (Section 3.1) or chromatographic (Section 3.2) analysis are still the most widely used methods for sulfide analysis of sediments in biogeochemistry, despite the inherent limitations in spatial and temporal resolution, and the necessity for sample destruction by this method.

During the last decade a slowly increasing number of studies have used microsensors for measuring sulfide species in benthic systems. Besides the advantages of microsensors over macrosensors in terms of e.g. signal stability, response time and low stirring sensitivity, they are also ideal tools for monitoring steep sulfide gradients owing to their negligible analyte consumption and small mechanical disturbance of the sediment or biofilm matrix. Hence relatively fast and repetitive measurements in practically undisturbed environmental samples become possible, from which detailed information about sulfide producing and consuming processes can be obtained. Below, we list some examples of such sulfide microsensor applications in natural biofilms and sediments, both in the laboratory and *in situ* (Figures 7 and 8)

In spite of the many practical problems with Ag/Ag_2S microsensors discussed in section 4.2, they have been used successfully in many different systems ranging from wastewater biofilms [7,159], sediments and cultures of microorganisms [117,162,179] to microbial mats growing under extreme environmental conditions in hot springs [180] and hypersaline lakes [155,181]. Ag/Ag_2S microsensors were also the first sulfide microsensors to be deployed on benthic lander intruments for autonomous *in situ* profiling of sediments and microbial mats near hydrothermal vents in the deep sea [23,164].

Voltammetric S(−II) microsensors (section 4.3) have been applied in several studies of marine sediment, both in the laboratory and *in situ* [23,166,167]. Although promising, these sensors still need improvement and further development (see Chapter 9). The most recently developed amperometric H_2S microsensor (section 4.4) has already found several laboratory applications in waste water, freshwater and marine biofilms and sediments [6,178,182,183], and the H_2S microsensor now seems to be used more frequently than the Ag_2S microsensor. First *in situ* applications were recently performed in coastal sediments and near shallow water hydrothermal vents [184,185] The H_2S microsensor seems a favorable alternative to the Ag/Ag_2S microsensor in most applications where $pH < 8.5$, whereas for studies of more alkaline waters and sediments the Ag/Ag_2S microsensor is still the best choice. Above all, with the H_2S microsensor it is now possible to study sulfide distribution and dynamics at fine scale in acid environments [6]. The fast response time of this sensor has also allowed the first reliable estimates of anoxygenic photosynthesis from measurements of sulfide dynamics around experimental light–dark shift events [186].

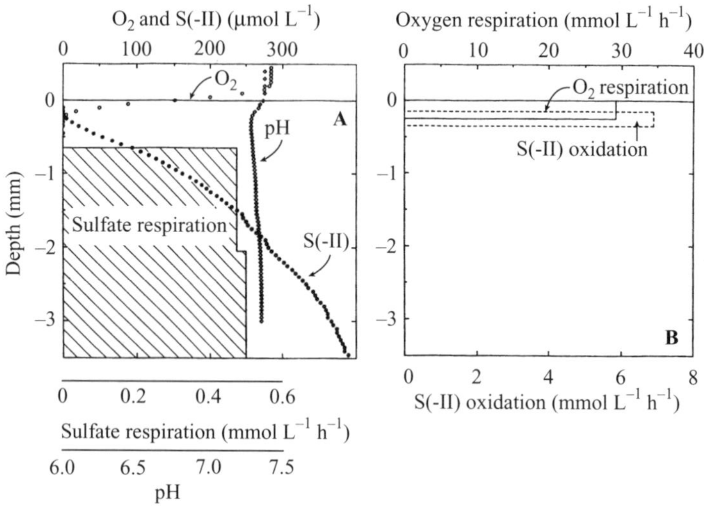

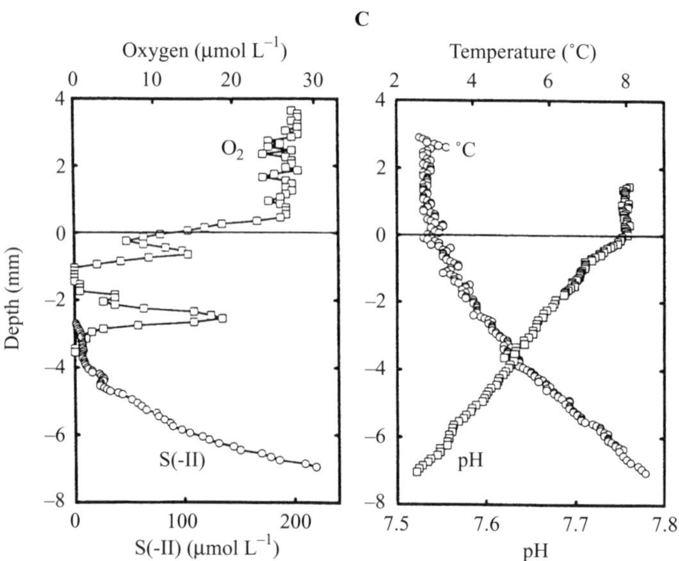

Figure 7. Examples of laboratory (A,B) and *in situ* (C) applications of Ag/Ag_2S microelectrodes. (A,B) Measurements of oxygen and sulfide in a biofilm and modeled reaction rates for sulfide production and consumption (redrawn from Kühl and Jørgenson [7] by permission of American Society for Microbiology). (C) Measurements of sulfide, oxygen, pH and temperature in a microbial mat subject to advective porewater transport; data were obtained *in situ* at a hydrothermal vent at 2 km water depth with a small measuring module controlled by the deep-sea research submersible Alvin. Reprinted with permission from *Nature* [164]. Copyright (1992) Macmillan Magazines Limited

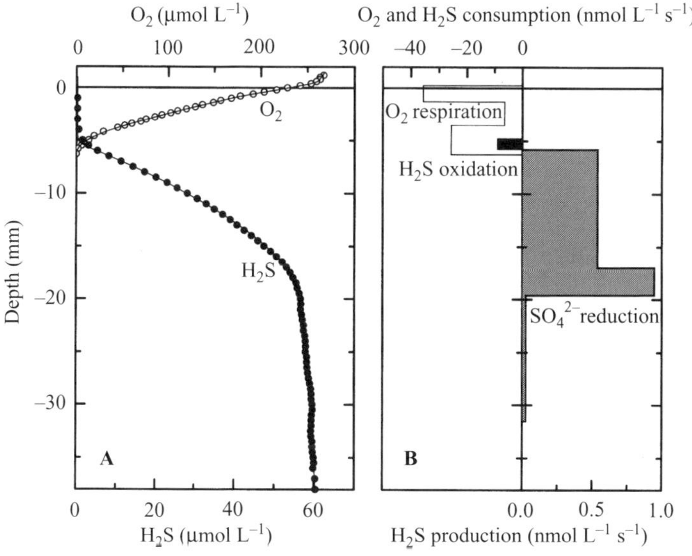

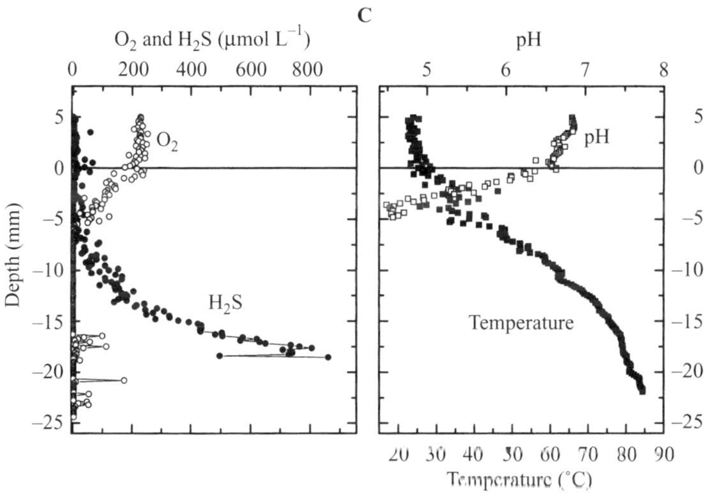

Figure 8. Examples of laboratory (A,B) and *in situ* (C) applications of H_2S microsensors. (A) Oxygen and sulfide measurements in acid (pH < 5) lake sediment. (B) Oxygen and sulfide turnover rates modeled from the microprofiles. (C) *In situ* measurements of oxygen, sulfide, pH and temperature in a sediment near a shallow-water (7 m) hydrothermal vent (Milos, Greece). Redrawn from Kühl *et al.* [6] by permission of Inter-Research, and from ref. [178]

6 SUMMARY AND DIRECTIONS FOR FUTURE RESEARCH

Several sensor techniques are available for direct sulfide analysis in the aquatic environment. Especially microsensors are well-suited for this purpose owing to their minimal consumption of the analyte, fast response times and their ability to measure with minimal invasion the distribution and dynamics of sulfide at high spatial and temporal resolution. Every analytical technique, however, has its advantages and drawbacks, which should be carefully considered for each application. Therefore, we see a need for a detailed experimental comparison of currently available sulfide microsensor techniques. Preferentially, such a comparison should involve measurements not only in defined sulfide solutions but also in defined gradients of sulfide and pH. The latter, would give a much better idea about how the various microsensors affect the analyte gradient itself, and, therefore, how applicable the various techniques are for fine scale measurements in environmental analysis.

Another complicating factor when working with *in situ* sulfide analysis is the instability of dilute sulfide solutions, which requires careful and strictly anaerobic calibration procedures under environmental conditions. For this, more detailed studies of sensor performance as a function of environmental variables such as temperature, salinity, and hydrostatic pressure are required. Besides the use of sulfide standards in traditional dosimetric calibration procedures, new devices may also simplify calibration of sulfide sensors [45].

At the moment, no reliable optical sulfide sensors are available for environmental analysis [20]. However, with the rapid development of optical sensor technology in recent years [19,187,188] optical sulfide sensors will probably become available. Also, advanced gel sampling techniques are becoming available for fine scale sulfide measurements [189] (see also Chapter 11). Last but not least, sulfide measurement with voltammetric microsensors seems a promising direction for future research, provided suitable sensors can be designed to limit interference, when measuring in complex natural systems (Chapter 9). Biosensors for sulfide species have already been developed [141], and with the recent development of new measuring and construction principles for microbiosensors by Revsbech and coworkers (Chapter 6), much better sulfide biosensors can now be realized.

Even if suitable and well-characterized sensors exist, it is a major undertaking to transfer the technology from the laboratory to the field in order to perform *in situ* environmental analysis. An adaptation of the sensors for special measuring platforms needs to be realized and first sucessful deployments of such measuring systems for *in situ* measurements of sulfide have been reported in the literature [92,144,156,159].

In conclusion, with the present array of sulfide sensors, detailed studies of the sulfur cycle can be performed in waters, biofilms and sediments. It must be emphasized, however, that the most suitable techniques have to be carefully chosen and optimized for each application. The largest gap in our ability to characterize the sulfide equilibria in nature is presently the lack of suitable sensors for measuring HS^-, which is the most abundant sulfide species in most natural aquatic environments.

ACKNOWLEDGEMENTS

The authors acknowledge the financial support by the Max Planck Gesellschaft, the Danish Natural Science Research Council (project 9700549), and the European Commission (projects MICROMARE, MAS3–CT950029, and MICROFLOW, MAS3–CT970078). We thank J. Buffle and W. Davison for their help with writing the parts on voltammetry in this chapter. The editors and other contributors to this book, and especially Jaques Buffle are thanked for their patience and for providing constructive criticism of earlier versions of this chapter.

REFERENCES

1. Howarth, R. W., Stewart, J. W. B. and Ivanov, M. V. (1992). *Sulfur Cycling on the Continents: Wetlands, Terrestrial Ecosystems and Associated Water Bodies*, John Wiley & Sons, Chichester.
2. Grieshaber, M. K. and Völkel, S. (1998). Animal adaptations for tolerance and exploitation of poisonous sulfide, *Ann. Rev. of Physiol.*, **60**, 33–53.
3. Vismann, B.(1991). Sulfide tolerance: physiological mechanisms and ecological implications, *Ophelia*, **34**, 1–27.
4. Jørgensen, B. B. (1982). Mineralization of organic matter in the sea bed—the role of sulphate reduction, *Nature*, **296**, 643–645.
5. Jørgensen, B. B. (1987). Ecology of the sulphur cycle: oxidative pathways in sediments. In *The Nitrogen and Sulphur Cycles, Soc. for General Microbiology Symposium 42*, ed. Cole, J. A. and Ferguson, S., Cambridge University Press, Cambridge, pp. 31–63.
6. Kühl, M., Steuckart, C., Eickert, G. and Jeroschewski, P. (1998). A H_2S microsensor for profiling biofilms and sediments: application in an acidic lake sediment, *Aq. Microb. Ecol.*, **15**, 201–209.
7. Kühl, M. and Jørgensen, B. B. (1992). Microsensor measurements of sulfate reduction and sulfide oxidation in compact microbial communities of aerobic biofilms, *Appl. Environ. Microbiol*, **58**, 1164–1174
8. Jørgensen, B. B., Kuenen, J. G. and Cohen, Y. (1979). Microbial transformations of sulfur compounds in a stratified lake (Solar Lake, Sinai), *Limnol. Oceanogr.*, **24**, 799–822.
9. Fenchel, T., Bernard, C., Esteban, G., Finlay, B., Hansen, P. J. and Iversen, N. (1995). Microbial diversity and activity in a Danish fjord with anoxic deep water, *Ophelia*, **43**, 45–100.
10. Thistlethwayte, D. K. B. (1972). *The Control of Sulphides in Sewerage Systems*, Butterworth, Sydney.

11. Rinzema, A. and Lettinga, G. (1988). Anaerobic treatment of sulfate containing wastewater, In *Biotreatment Systems*, Vol. III, ed. Wise, D. L., CRC Press, Boca Raton, FL, pp. 107–121.
12. Rintala, J. and Puhakka, J. (1994). Anaerobic treatment in pulp and paper mill waste management: a review, *Biores. Technol.*, **47**, 1–18.
13. Tichy, R., Lens, P., Grotenhuis, J. T. C. and Bos, P. (1998). Solid state reduced sulfur compounds: Environmental aspects and bio-remediation, *Crit. Rev. Environ. Sci. Technol.*, **28**, 1–40.
14. Monticello, D. J. and Finnerty, W. R. (1985). Microbial desulfurization of fossil fuels, *Annu. Rev. Microbiol.*, **1985**, 371–389.
15. Hamilton, W. A. and Lee, W. (1995). Sulfate-reducing bacteria. In *Biocorrosion*, ed. Barton, L. L., Plenum, New York, pp. 243–264.
16. Lee, W., Lewandowski, Z., Nielsen, P. H. and Hamilton, W. A. (1995). Role of sulfate-reducing bacteria in corrosion of mild steel: a review, *Biofouling*, **8**, 165–194.
17. Kühl, M. and Revsbech, N. P. (2000). Microsensors for BBL studies. In *The Benthic Boundary Layer*, ed. Jørgensen, B. B. and Boudreau, B. P., Oxford University Press, Oxford, *in press*.
18. Revsbech, N. P. and Jørgensen, B. B. (1986). Microelectrodes: their use in microbial ecology, *Adv. Microb. Ecol.*, **9**, 293–352.
19. Dakin, J. and Culshaw, B. (1997). *Optical Fiber Sensors. 4. Applications, Analysis and Future Trends*, Artech House, Norwood.
20. Holst, G., Kohls, O., Klimant, I. and Kühl, M. (2000). Optical microsensors. In *Chemical Sensors in Oceanography*, ed. Varney, M., Gordon & Breach, pp. 143–188.
21. Amann, R. and Kühl, M. (1998). *In-situ* methods for assessment of microorganisms and their activities, *Cur. Opin. Microbiol.*, **1**, 352–358.
22. Varney, M., ed. (2000). *Chemical Sensors in Oceanography*. Gordon & Breach.
23. Reimers, C. and Glud, R. N. (2000). *In-situ* chemical measurements at the sediment–water interface. In *Chemical Sensors in Oceanography*, ed. Varney, M., Gordon and Breach.
24. Santegoeds, C. M., Schramm, A. and de Beer, D. (1998). Microsensors as a tool to determine chemical microgradients and bacterial activity in wastewater biofilms and flocs, *Biodegradation*, **9**, 159–167.
25. Millero, F. J. (1986). The thermodynamics and kinetics of the hydrogen sulfide system in natural waters, *Mar. Chem.*, **18**, 121–147.
26. Millero, F. J. and Hershey, J. P. (1989). Thermodynamics and kinetics of hydrogen sulfide in natural waters, In *Biogenic Sulfur in the Environment*, ed. Saltzman, E. S. and Cooper, W. J., American Chemical Society, Washington, DC, pp. 282–313.
27. Millero, F. J., Plese, T. and Fernandez, M. (1988). The dissociation of hydrogen sulfide in seawater, *Limnol. Oceanogr.*, **33**, 269–274.
28. Deutsche Einheitsverfahren zur Wasser- und Schlammuntersuchung (1975). *D7*, 7. Lieferung.
29. Widmer, M. and Schwarzenbach, G. (1964). Die Acidität des Hydrogensulfidions HS^-, *Helv. Chim. Acta*, **47**, 266–271.
30. Broderius, S. J. and Smith, Jr., L. L. (1977). Direct determination and calculation of aqueous hydrogen sulfide, *Anal. Chem.*, **49**, 424–428.
31. Barbero, J. A., McCurdy, K. G. and Tremaine, P. R. (1982). Apparent molal heat capacities and volumes of aqueous hydrogen sulfide near 25 °C; the temperature dependence of H_2S ionization, *Can. J. Chem.*, **60**, 1872–1880.
32. Licht, S., Forouzan, F. and Longo, K. (1990). Differential densometric analysis of equilibria in highly concentrated media: determination of the aqueous second acid dissociation constant of H_2S, *Anal. Chem.*, **62**, 1356–1360.

33. Licht, S. (1988). Aqueous solubilities, solubility products and standard oxidation–reduction potentials of the metal sulfides, *J. Electrochem. Soc.*, **135**, 2971–2975.
34. Peramunage, D., Forouzan, F. and Licht, S. (1994). Activity and spectroscopic analysis of concentrated solutions of K_2S, *Anal. Chem.*, **66**, 378–383.
35. Su, Y. S., Cheng, K. L. and Jean, Y. C. (1997). Amplified potentiometric determination of pK_{00}, pK_0, pK_1, and pK_2 of hydrogen sulfides with Ag_2S ISE, *Talanta*, **44**, 1757–1763.
36. Hollemann, A. F. and Wiberg, E. (1985). *Lehrbuch der Anorganischen Chemie*, Walter deGruyter, Berlin, pp. 91–100.
37. Gmelin-Institut Clausthal-Zellerfeld (ed.) (1953). *Gmelins Handbuch der anorganischen Chemie*, edn 8, Part 9 B(1), Verlag Chemie, Weinheim.
38. Chen, K. Y. and Morris, J. C. (1972). Kinetics of oxidation of aqueous sulfide by O_2, *Environ. Sci. Technol.*, **6**, 529–537.
39. Cline, J. D. and Richards, F. A. (1969). Oxygenation of hydrogen sulfide in seawater at constant salinity, temperature and pH, *Environ. Sci. Technol.*, **3**, 838–843.
40. Avrahami, M. and Golding, R. M. (1968). The oxidation of the sulphide ion at very low concentrations in aqueous solutions, *J. Chem. Soc. (A)*, **16**, 647–651.
41. Zhang, J. -Z. and Millero, F. J. (1994). Kinetics of oxidation of hydrogen sulfide in natural waters. In *Environmental Geochemistry of Sulfide Oxidation*, ed. Alpers, C. N. and Blowers, D. W., *ACS Symp. Ser. 550*, American Chemical Society, Washington, DC, Chapter 26.
42. Pomeroy, R. (1953). Auxiliary pretreatment by zinc acetate in sulfide analyses, *Anal. Chem.*, **25**, 571–572.
43. Fonselius, S. H. (1983). Determination of hydrogen sulphide. In *Methods in Seawater Analysis*, ed. Grasshoff, M., Erhardt, K. and Kremling, K., Verlag Chemie, Weinheim.
44. Fischer, E. (1883). Bildung von Methylenblau als Reaktion auf Schwefelwasserstoff, *Chem. Ber.*, **16**, 2234–2236.
45. Jeroschewski, P. and Schmuhl, A. (1993). Coulometrische Herstellung von Schwefelwasserstof-Standardlösungen, *Rostock Meeresbiol. Beitr.*, **1**, 117–122.
46. Schmuhl, A. (1995). Untersuchungen zur elektrochemischen Herstellung von Schwefelwasserstoff-/Sulfid-Standard-Lösungen im Spurenbereich, Ph. D. Thesis, Universität Rostock, Germany.
47. Grünert, A., Ballschmiter, K. and Tölg, G. (1968). Fluoreszenzanalytische Bestimmung von Sulfidionen im Nanogrammbereich, *Talanta*, **15**, 451–457.
48. Bottrell, S., Banks, D., Bird, D. and Raiswell, R. (1991). A simple method for field determination of some reduced sulphur species in soil gases, *Environ. Technol.*, **12**, 393–398.
49. Jaeschke, W., Schunn, H. and Haunold, W. (1995). A continuous flow instrument with flow regulation and automatic sensitivity range adaption for the determination of atmospheric H_2S, *Fresensius' J. Anal. Chem.*, **351**, 27–32.
50. Choi, M. F. and Hawkins, P. (1997). Development of sulphide-selective optode membranes based on fluorescence quenching, *Anal. Chim. Acta*, **344**, 105–110.
51. Shahriari, M. R. and Ding, J. (1994). Active silica-gel films for hydrogen sulfide optical sensor application, *Opt. Lett.*, **19**, 1085–1087.
52. Kohls, O., Klimant, I., Holst, G. and Kühl, M. (1996). Thionine as an indicator for use as a hydrogen sulfide optode, *Proc. SPIE*, **2836**, 311–321.
53. Punta, A., Barragàn, F. J., Ternero, M. and Guiraum, A. (1990). Spectrofluorimetric determination of sulphide in natural and wastewaters with 1,2–naphtoquinone-4–sulphonate, *Analyst*, **115**, 1499–1503.
54. DIN 38405, Teil D26 (1989).

55. DIN 38405, Teil D27 (1992).
56. Cline, J. D. (1969). Spectrophotometric determination of hydrogen sulfide in natural waters, *Limnol. Oceanogr.*, 14, 454–458.
57. Sakamoto-Arnold, C. M., Johnson, K. S. and Beehler, C. L. (1986). Determination of hydrogen sulfide in seawater using flow injection analysis and flow analysis, *Limnol. Oceanogr.* , **31**, 894–900.
58. Johnson, K. S., Beehler, C. L. and Sakamoto-Arnold, C. M. (1986). A submersible flow analysis system, *Anal. Chim. Acta*, **179**, 245–257.
59. Johnson, K. S., Beehler, C. L., Sakamoto-Arnold, C. M. and Childress, J. J. (1986). In-situ measurements of chemical distributions in a deep-sea hydrothermal vent field, *Science*, **231**, 1139–1141.
60. Leggett, D. J., Chen, N. H. and Mahadevappa, D. S. (1981). Flow injection method for sulfide determination by the methylene blue method, *Anal. Chim. Acta*, **128**, 163–168.
61. Francom, D., Goodwin, L. R. and Dieken, F. P. (1989). Determination of low level sulfides in environmental waters by automated gas dialysis/methylene blue colorimetry, *Anal. Lett.*, **22**, 2587–2600.
62. Lei, W. and Dasgupta, P. K. (1989). Determination of sulfide and mercaptans in caustic scrubbing liquor, *Anal. Chim. Acta*, **226**, 165–170.
63. Gamo, T., Sakai, H., Nakayama, E., Ishida, K. and Kimoto, H. (1994). A submersible flow-through analyzer for *in-situ* colorimetric measurement down to 2000 m depth in the ocean, *Anal. Sci.*, **10**, 843–848.
64. Babiker, M. O. and Dalziel, J. A. W. (1983). Studies on the determination of sulphide using N,N-Diethyl-p-phenylenediamine, *Anal. Proc.*, **20**, 609–611.
65. Papaefstathiou, I. and Luque de Castro, M. D. (1995). Approaches for improving the precision and sensitivity of continuous pervaporation processes, *Anal. Lett.*, **28**, 2063–2076.
66. Papaefstathiou, I., Luque de Castro, M. D. and Valcàrcel, M. (1996). Flow-injection/pervaporation coupling for the determination of sulphide in Kraft liquors, *Fresenius' J. Anal. Chem.*, **354**, 442–446.
67. Ensafi, A. A. (1992). Flow-injection determination of traces of sulfide by the brilliant green–sulfide reaction with spectrophotometric detection, *Anal. Lett.*, **25**, 1525–1543.
68. Koh, T., Miura, Y., Yamamuro, N. and Takaki, T. (1990). Spectrophotometric determination of trace amounts of sulphide and hydrogen sulphide by formation of thiocyanate, *Analyst*, **115**, 1133–1137.
69. Koh, T. and Okabe, K. (1992). Separation and spectrophotometric determination of thiosulfate, sulfite and sulfide in their mixtures, *Anal. Sci.*, **8**, 285–291.
70. Koh, T., Takahashi, N., Yamamuro, N. and Miura, Y. (1993). Spectrophotometric determination of sulfide at the 10^{-6} mol l^{-1} level by formation of thiocyanate and its solvent extraction with methylene blue, *Anal. Sci.*, **9**, 487–492.
71. Koh, T., Okabe, K. and Miura, Y. (1993). Separation and determination of thiosulfate, sulfite and sulfide in mixtures, *Analyst*, **118**, 669–672.
72. Yaqoob, M., Anwar, M., Masood, A. S. and Masoom, M. (1991). Flow injection method for sulphide determination using an organic mercury compound, *Anal. Lett.*, **24**, 581–588.
73. Yaqoob, M., Rishi, L. and Masoom, M. (1991). Comparative study of different methods for sulphide determination when adopted to flow system, *J. Chem. Soc. Pak.*, **13**, 32–37.
74. Sonne, K. and Dasgupta, P. K. (1991). Simultaneous photometric flow injection determination of sulfide, polysulfide, sulfite, thiosulfate, and sulfate, *Anal. Chem.*, **63**, 427–432.

75. Kester, M. D., Shiundu, P. M. and Wade, A. P. (1992). Spectrophotometric method for determination of sulfide with iron(III) and nitrilotriacetic acid by flow injection, *Talanta*, **39**, 299–312.
76. Burguera, J. L. and Burguera, M. (1984). Determination of sulphur anions by flow injection with a molecular emission cavity detector, *Anal. Chim. Acta*, **157**, 177–181.
77. Hauge, S., Marøy, K. and Thorlacius, A. (1991). Detection of some sulphur anions and colloidal sulphur by flame molecular emission spectrometry, *Anal. Chim. Acta*, **251**, 197–203.
78. Arowolo, T. A. and Cresser, M. S. (1991). Automated determination of sulphide by gas-phase molecular absorption spectrometry, *Analyst*, **116**, 595–599.
79. Jin, Q., Zhang, H., Duan, Y., Yu, A., Liu, X. and Wang, L. (1992). Trace determination of sulphide and sulphur dioxide by vapor molecular absorption spectrometry using magnesium and tellurium hollow cathode lamps, *Talanta*, **39**, 967–970.
80. Sanz, J., Cabredo, S. and Galbàn, J. (1992). Sulphide determination in water by gas-phase molecular absorption spectrometry, *Anal. Lett.*, **25**, 2095–2105.
81. Parvinen, P. and Lajunen, L. H. -J. (1994). Determination of sulfide as hydrogen sulfide in water and sludge samples by gas phase molecular AS, *At. Spectrosc.*, **15**, 83–86.
82. Cabredo, S., Sanz, I., Sanz, J. and Galbàn, J. (1995). Simultaneous determination of sulphide and sulphite by gas-phase molecular absorption spectrometry. Comparative study of different calculation methods, *Talanta*, **42**, 937–943.
83. Ebdon, L., Hill, S. J., Jameel, M., Corns, W. T. and Stockwell, P. B. (1997). Automated determination of sulfide as hydrogen sulfide in waste streams by gas-phase molecular absorption spectrometry, *Analyst*, **122**, 689–693.
84. Mousavi, M. F. and Shamsipur, M. (1992). Spectrophotometric determination of trace amounts of sulfide ion based on its catalytic reduction of toluidine blue, *Bull. Chem. Soc. Jpn*, **65**, 2770–2772.
85. Kawakubo, S., Iwatsuki, M. and Fukasawa, T. (1993). Flow system based on sequential delivery of air-sandwiched solutions into a micro cell for spectrophotometric catalytic analysis, *Anal. Chim. Acta*, **282**, 389–395.
86. Ensafi, A. A. and Samimifar, M. (1994). Spectrophotometric reaction-rate method for the determination of sulfide by catalytic action on the oxidation of sodium azide by iodine, *Anal. Lett.*, **27**, 153–167.
87. Sànchez-Pedreño, C., Garcìa, M. S., Albero, M. I. and Tobal, L. (1996). Kinetic determination of sulphide in water and air samples, *Int. J. Environ. Anal. Chem.*, **62**, 273–279.
88. Cutter, G. A. and Oatts, T. J. (1987). Determination of dissolved sulfide and sedimentary sulfur speciation using gas chromatography–photoionization detection, *Anal. Chem.*, **59**, 717–721.
89. Jacobsson, S. and Falk, O. (1989). Determination of hydrogen sulphide by porous-layer open-tubular column gas chromatography–mass spectroscopy, *J. Chromatogr.*, **479**, 194–199.
90. Radford-Knœry, J. and Cutter, G. A. (1993). Determination of carbonyl sulfide and hydrogen sulfide species in natural waters using specialized collection procedures and gas chromatography with flame photometric detection, *Anal. Chem.*, **65**, 976–982.
91. Tang, H. and Heaton, P. (1996). Analysis of volatile trace sulfur compounds in water with GC-SCD by direct aqueous injection, *Int. J. Environ. Anal. Chem.*, **62**, 263–271.
92. Goodwin, L. R., Francom, D., Urso, A. and Dieken, F. P. (1988). Determination of trace sulfides in turbid waters by gas dialysis/ion chromatography, *Anal. Chem.*, **60**, 216–219.

93. Okutani, T., Yamakawa, K., Sakuragawa, A. and Gotoh, R. (1993). Determination of a micro amount of sulfide by ion chromatography with amperometric detection after coprecipitation with basic zinc carbonate, *Anal. Sci.*, **9**, 731–734.
94. Steinmann, P. and Shotyk, W. (1995). Ion chromatography of organic-rich natural waters from peatlands IV. Dissolved free sulfide and acid-volatile sulfur, *J. Chromatogr. A*, **706**, 287–292.
95. Steinmann, P. and Shotyk, W. (1996). Sampling anoxic pore waters in peatlands using "peepers" for in-situ-filtration, *Fresenius' J. Anal. Chem.*, **354**, 709–713.
96. Obrezkov, O. N., Shlyamin, V. I., Shpigun, O. A. and Zolotov, Y. A. (1991). Use of kinetic detection with ion chromatography for the determination of some anions, *Mendeleev Commun.*, **1991**, 38–39.
97. Miura, Y., Tsubamoto, M. and Koh, T. (1994). Ion chromatographic determination of sulfide, sulfite and thiosulfate in mixtures by means of their postcolumn reactions with iodine, *Anal. Sci.*, **10**, 595–600.
98. Miura, Y., Maruyama, T. and Koh, T. (1995). Ion chromatographic determination of L-Ascorbic acid, sulfite, sulfide and thiosulfate using a cation-exchanger of low crosslinking, *Anal. Sci.*, **11**, 617–621.
99. Savage, J. C. and Gould, D. H. (1990). Determination of sulfide in brain tissue and rumen fluid by ion-interaction reversed-phase high-performance liquid chromatography, *J. Chromatogr. Biomed. Appl.*, **526**, 540–545.
100. do Nascimento, P., Hinkamp, S. and Schwedt, G. (1992). Verfahren zur Bestimmung von Sulfidspuren mit Anreicherung, HPLC und FIA, *Vom Wasser*, **78**, 21–31.
101. Togawa, T., Ogawa, M., Nawata, M., Ogasawara, Y., Kawanabe, K. and Tanabe, S. (1992). High performance liquid chromatographic determination of bound sulfide and sulfite and thiosulfate at their low levels in human serum by pre-column fluorescence derivatization with monobromobimane, *Chem. Pharm. Bull.*, **40**, 3000–3004.
102. Bald, E. and Sypniewski, S. (1993). Reversed-phase high-performance liquid chromatographic determination of sulphide in an aqueous matrix using 2–iodo-1–methylpyridinium chloride as a precolumn ultraviolet derivatization reagent, *J. Chromatogr.*, **641**, 184–188.
103. Safavi, A., Rahmani, A. and Hosseini, V. N. (1996). Kinetic–spectrophotometric determination of sulfide by its reaction with resazurin, *Fresenius' J. Anal. Chem.*, **354**, 502–504.
104. Safavi, A. and Ramezani, Z. (1997). Kinetic spectrophotometric determination of traces of sulfide, *Talanta*, **44**, 1225–1230.
105. Kubàñ, V., Dasgupta, P. K. and Marx, J. N. (1992). Nitroprusside and methylene blue methods for silicone membrane differentiated flow injection determination of sulfide in water and wastewater, *Anal. Chem.*, **64**, 36–43.
106. Han, K. and Koch, W. F. (1987). Determination of sulfide at the parts-per-billion level by ion chromatography with electrochemical detection, *Anal. Chem.*, **59**, 1016–1020.
107. Berge, H. and Jeroschewski, P. (1964). Quantitative Bestimmung geringer Sulfidmengen mit Hilfe des hängenden Quecksilbertropfens als Elektrode, *Z. Anal. Chem.*, **207**, 110–113.
108. Canterford, D. R. and Buchanan, A. S. (1973). Application of differential pulse polarography to anodic electrode processes involving mercury compound formation, *J. Electroanal. Chem.*, **44**, 291–298.
109. Canterford, D. R. (1974). Effect of film formation on the normal pulse polarographic behaviour of sulphide, *J. Electroanal. Chem.*, **52**, 144–147.

110. Davison, W., Buffle, J. and DeVitre, R. (1988). Direct polarographic determination of O_2, Fe(II), Mn(II), S(−II) and related species in anoxic waters, *Pure Appl. Chem.*, **6**, 1535–1548.
111. Shimizu, K. and Osteryoung, R. A. (1981). Determination of sulfide by cathodic stripping voltammetry of silver sulfide films at a rotating silver disk electrode, *Anal. Chem.*, **53**, 584–588.
112. Tanaka, T., Ishiyama, T. and Mizuike, A. (1992). Cathodic stripping voltammetry of sulfide with a silver disk electrode, *Anal. Sci.*, **8**, 281–284.
113. Batina, N., Ciglenecki, I. and Cosovic, B. (1992). Determination of elemental sulphur, sulphide and their mixtures in electrolyte solutions by a. c. voltammetry, *Anal. Chim. Acta*, **267**, 157–164.
114. Berner, R. A. (1963). Electrode studies of hydrogen sulfide in marine sediments, *Geochim. Cosmochim. Acta*, **27**, 563–575.
115. *Instruction Manual Sulfide Ion Activity Electrode Model 94–16*, Orion Research, 1st edn (1966).
116. Dobcnik, D., Gomiscek, S. and Stergulec, J. (1990). Preparation of a sulphide ion-selective microelectrode with chemical pretreatment of silver wire in Hg(II) solution, *Fresenius' J. Anal. Chem.*, **337**, 369–371.
117. Visscher, P. T., Beukema, J. and van Gemerden, H. (1991). In-situ characterization of sediments: Measurements of oxygen and sulfide profiles with a novel combined needle electrode, *Limnol. Oceanogr.*, **36**, 1476–1480.
118. Frevert, T. and Galster, H. (1978). Schnelle und einfache Methode zur In-situ-Bestimmung von Schwefelwasserstoff in Gewässern und Sedimenten, *Schweiz. Z. Hydrol.*, **40**, 199–208.
119. Frevert, T. (1980). Determination of hydrogen sulfide in saline solutions, *Schweiz. Z. Hydrol.*, **42**, 255–268.
120. Gulens, J., Herrington, H. D., Thorpe, J. W., Mainprize, G., Cooke, M. G., Dal Bello, P. and Macdougall, S. (1982). Comparison of sulfide-selective electrode and gas-stripping monitors for hydrogen sulfide in effluents, *Anal. Chim. Acta*, **138**, 55–63.
121. Dan, T. B., Frevert, T. and Cavari, B. (1985). The sulfide electrode in bacterial studies, *Wat. Res.*, **19**, 983–985.
122. Peiffer, S. and Frevert, T. (1987). Potentiometric determination of heavy metal sulphide solubilities using a pH_2S (glass $|Ag^0, Ag_2S$) electrode cell, *Analyst*, **112**, 951–954.
123. Tóth, I., Solymosi, P. and Szabò, Z. (1988). Application of a sulphide-selective electrode in the absence of a pH-buffer, *Talanta*, **35**, 783–788.
124. Guterman, H., Ben-Yaakov, S. and Abeliovich, A. (1983). Determination of total dissolved sulfide in the pH range 7. 5 to 11. 5 by ion selective electrodes, *Anal. Chem.*, **55**, 1731–1734.
125. Eckert, W., Frevert, T. and Trüper, H. G. (1990). A new liquid-junction free probe for the *in-situ* determination of pH, pH_2S and redox values, *Wat. Res.*, **24**, 1341–1346.
126. Schmidt, E., Marton, A. and Hlavay, I. (1994). Determination of the total dissolved sulphide in the pH range 3–11. 4 with sulphide selective ISE and Ag/Ag_2S electrodes, *Talanta*, **41**, 1219–1224.
127. Brunt, K. (1984). Rapid determination of sulfide in waste waters by continuous flow analysis and gas diffusion and a potentiometric detector, *Anal. Chim. Acta*, **163**, 293–297.
128. Thomas, J. D. R. (1985). Electrochemical analysis of sulphur compounds of environmental interest, *Int. J. Environ. Anal. Chem.*, **20**, 167–177.

129. Glaister, M. G., Moody, G. J. and Thomas, J. D. R. (1985). Studies on flow injection analysis with sulphide ion-selective electrodes, *Analyst*, **110**, 113–119.
130. Lee, H. L., Bae, Z. U. and Oh, S. -H. (1992). Gas-sensing membrane electrodes for the determination of dissolved gases (IV). Continuous-automated determination of sulfide ion using tubular PVC membrane type pH electrode, *J. Korean Chem. Soc.*, **36**, 638.
131. Fligier, J., Czichon, P., Gregorowicz, Z. and Gratzl, M. (1988). Air-gap sulfide sensor with super-Nernstian response in the low concentration range, *Anal. Chim. Acta*, **204**, 213–221.
132. Lindell, H. and Jäppinen, P. and Savolainen, H. (1988). Determination of sulphide in blood with an ion-selective electrode by pre-concentration of trapped sulphide in sodium hydroxide solution, *Analyst*, **113**, 839–840.
133. Lima, J. L. F. C. and Rocha, L. S. M. (1990). FIA tubular potentiometric detectors based on homogeneous crystalline membranes. Their use in the determination of chloride and sulphide ions in water, *Int. J. Environ. Anal. Chem.*, **38**, 127–133.
134. Raba, J., Mallea, M. A., Quintar, S. and Cortinez, V. A. (1992). Determination of sulfide with chloranilic acid by biamperometric and automated potentiometric end-point detection with a lead chloranilate selective electrode, *Talanta*, **39**, 1007–1011.
135. Tse, Y. -H., Janda, P. and Lever, A. B. P. (1994). Electrode with electrochemically deposited N,N′,N″,N‴-Tetramethyltetra-3,4-pyridinoporphyrazinocobalt(I) for detection of sulfide ion, *Anal. Chem.*, **66**, 384–390.
136. Tse, Y. -H., Janda, P., Lam, H. and Lever, A. B. P. (1995). Electrode with electropolymerized tetraaminophtalocyanatocobalt(II) for detection of sulfide ion, *Anal. Chem.*, **67**, 981–985.
137. Zhang, J., Lever, A. B. P. and Pietro, W. J. (1995). Electrocatalytic activity of a graphite electrode coated with hexadecachlorophtalocyanatoiron(II) toward sulfide oxidation, and its possible application in electroanalysis, *Can. J. Chem.*, **73**, 1072–1077.
138. Ma, Y. L., Galal, A., Zimmer, H., Mark, Jr., H. B., Huang, Z. F. and Bishop, P. L. (1994). Potentiometric selective determination of hydrogen sulfide by an electropolymerized membrane electrode based on binaphtyl-20–crown-6, *Anal. Chim. Acta*, **289**, 21–26.
139. Hu, X. and Leng, Z. (1996). Determination of trace-levels of sulfide by high-sensitivity potentiometry with a carbon paste electrode, *Anal. Commun.*, **33**, 297–298.
140. Opekar, F. and Bruckenstein, S. (1985). Pneumatoamperometric determination of cyanide, sulfide and their mixtures, *Anal. Chim. Acta*, **169**, 407–412.
141. Kurosawa, H., Hirano, T., Nakamura, K. and Amano, Y. (1994). Microbial sensor for selective determination of sulphide, *Appl. Microbiol. Biotechnol.*, **41**, 556–559.
142. Söllig, M. (1986). Untersuchungen zur elektrochemischen Bestimmung von Schwefelwasserstoff, Ph. D. Thesis, Universität Rostock, Germany.
143. Jeroschewski, P., Söllig, M. and Berge, H. (1988). Amperometrische Bestimmung von Schwefelwasserstoff, *Z. Chem.*, **28**, 75.
144. Haase, K. and Trommer, A. (1992). Amperometrischer Sensor zur Bestimmung von Schwefelwasserstoff/Sulfid im Spurenbereich, Ph. D. Thesis, Universität Rostock, Germany.
145. Jeroschewski, P., Haase, K., Trommer, A. and Gründler, P. (1993). Galvanic sensor for the determination of hydrogen sulphide/sulphide in aqueous media, *Fresenius' J. Anal. Chem.*, **346**, 930–933.
146. Jeroschewski, P., Haase, K., Trommer, A. and Gründler, P. (1994). Galvanic sensor for determination of hydrogen sulfide, *Electroanalysis*, **6**, 769–772.

147. Braun, S. (1994). Fliessanalysensystem für die H_2S/Sulfid-Analytik mit einem amperometrischen Detektor, Ph. D. Thesis, Universität Rostock, Germany.
148. Jeroschewski, P. and Braun, S. (1996). A flow analysis system with an amperometric detector for the determination of hydrogen sulphide in waters, *Fresenius' J. Anal. Chem.*, **354**, 169–172.
149. Jeroschewski, P., Steuckart, C. and Kühl, M. (1996). An amperometric microsensor for the determination of H_2S in aquatic environments, *Anal. Chem.*, **68**, 4351–4357.
150. Steuckart, C. (1997). Amperometrischer Mikrosensor für Schwefelwasserstoff in wässrigen Systemen, Ph. D. Thesis, Universität Rostock, Germany.
151. Heinze, J. (1993). Elektrochemie mit Ultramikroelektroden, *Angew. Chem.*, **105**, 1327–1349.
152. Cardoso, A. A., Liu, H. and Dasgupta, P. K. (1997). Fluorometric fiber optic drop sensor for atmospheric hydrogen sulfide, *Talanta*, **44**, 1099–1106.
153. Camman, K. and Galster, H. (1996). *Das Arbeiten mit ionenselektiven Elektroden*, 3rd edn, Springer, Berlin.
154. De Marco, R., Cattrall, R. W., Liesegang, J., Nyberg, G. L. and Hamilton, I. C. (1990). Surface studies of the silver sulfide ion selective electrode membrane, *Anal. Chem.*, **62**, 2339–2346.
155. Revsbech, N. P., Jørgensen, B. B., Blackburn, T. H. and Cohen, Y. (1983). Microelectrode studies of the photosynthesis and O_2, H_2S, and pH profiles of a microbial mat, *Limnol. Oceanogr.*, **28**, 1062–1074.
156. Hepel, M., Bruckenstein, S. and Tang, G. C. (1989). The formation and electroreduction of silver sulfide films at a silver metal electrode, *J. Electroanal. Chem.*, **261**, 389–400.
157. Kaltofen, R. (1986). *Tabellenbuch Chemie*, 10th ed., Verlag Harri Deutsch, Thun.
158. van Staden, J. F. (1988). Determination of sulphide using flow injection analysis with a coated tubular solid-state silver sulphide ion-selective electrode, *Analyst*, **113**, 885–889.
159. Yu, T., Bishop, P. L., Galal, A. and Mark, Jr., H. B. (1997). Fabrication and evaluation of a sulfide microelectrode for biofilm studies. In *Chemistry and Technology of Chemical and Biosensors*, ed. Akmel, N. and Usmani, A. M., American Chemical Society, Washington, DC.
160. Baumann, E. W. (1974). Determination of parts per billion sulfide in water with the sulfide-selective electrode, *Anal. Chem.*, **46**, 1345–1347.
161. Harsànyi, E. G., Tòth, K. and Pungor, E. (1984). The behaviour of the silver sulphide precipitate-based ion-selective electrode in the low concentration range, *Anal. Chim. Acta*, **161**, 333–341.
162. Jørgensen, B. B. and Revsbech, N. P. (1983). Colorless sulfur bacteria, *Beggiatoa* spp. and *Thiovulum* spp. in O_2 and H_2S microgradients, *Appl. Environ. Microbiol.*, **45**, 1261–1270.
163. Komada, T., Reimers, C. E. and Boehme, S. E. (1998). Dissolved inorganic carbon profiles and fluxes determined by using pH and pCO_2 microelectrodes, *Limnol. Oceanogr.*, **43**, 769–781
164. Gundersen, J. K. and Jørgensen, B. B. (1992). Mats of giant sulphur bacteria on deep-sea sediments due to fluctuating hydrothermal flow, *Nature*, **360**, 454–455.
165. Brendel, P. J. and Luther III, G. W. (1995). Development of a gold amalgam voltammetric microelectrode for the determination of dissolved Fe, Mn, O_2, and S(−II) in porewaters of marine and freshwater sediments, *Environ. Sci. Technol.*, **29**, 751–761.
166. Luther III, G. W, Brendel, P. J., Lewis, B. L., Sundby, B., Lefrancois, L., Silverberg, N. and Nuzzio, D. B. (1998). Simultaneous measurement of O_2, Mn, Fe, I^-,

and S(−II) in marine porewaters with a solid-state voltammetric microelectrode, *Limnol. Oceanogr.*, **43**, 325–333.
167. Huettel, M., Ziebis, W., Forster, S. and Luther III, G. W. (1998). Advective transport affecting metal and nutrient distributions and interfacial fluxes in permeable sediments, *Geochim. Cosmochim. Acta*, **62**, 613–631.
168. Baranski, A. S. (1987). Rapid anodic stripping analysis with ultramicroelectrodes, *Anal. Chem.*, **59**, 662–666.
169. Wechter, C. and Osteryoung, J. (1990). Voltammetric charaterization of small platinum–iridium-based mercury film electrodes, *Anal. Chim. Acta*, **234**, 275–284.
170. Kounaves, S. P. and Buffle, J. (1986). Deposition and stripping properties of mercury on iridium electrodes, *J. Electrochem. Soc.*, **133**, 2495–2498.
171. Tercier, M. -L., Parthasarathy, N. and Buffle, J. (1995). Reproducible, reliable and rugged Hg-plated Ir-based microelectrode for in-situ measurements in natural waters, *Electroanalysis*, **7**, 55–63.
172. Shimizu, K., Aoki, K. and Osteryoung, R. A. (1981). Electrochemical behavior of sulfide at the silver rotating disc electrode Part I. Polarization behavior of silver sulfide films, *J. Electroanal. Chem.*, **129**, 159–169.
173. Davison, W. and Gabbutt, C. D. (1979). Polarographic methods for measuring uncomplexed sulphide ions in natural waters, *J. Electroanal. Chem.*, **99**, 311–320.
174. Revsbech, N. P (1989). An oxygen microelectrode with a guard cathode, *Limnol. Oceanogr.*, **34**, 474–478.
175. Gundersen, J. K., Ramsing, N. B. and Glud, R. N. (1998). Predicting the signal of O_2 microsensors from physical dimensions, temperature, salinity, and O_2 concentration, *Limnol. Oceanogr.*, **43**, 1932–1937.
176. Ramasubramanian, N. (1975). Anodic behavior of platinum electrodes in sulfide solutions and the formation of platinum sulfide, *J. Electroanal. Chem.*, **64**, 21–37.
177. Kapusta, S., Viehbeck, A., Wilhelm, S. M. and Hackerman, N. (1983). The anodic oxidation of sulfide on platinum electrodes, *J. Electroanal. Chem.*, **153**, 157–174.
178. Santegoeds, C. M., Ferdelman, T. G., Muyzer, G. and de Beer, D. (1998). Structural and functional dynamics of sulfate-reducing populations in bacterial biofilms, *Appl. Environ. Microbiol.*, **64**, 3731–3739.
179. Nelson, D. C., Revsbech, N. P. and Jørgensen, B. B. (1986). Microoxic–anoxic niche of *Beggiatoa* spp.: microelectrode survey of marine and freshwater strains, *Appl. Environ. Microbiol.*, **52**, 161–168.
180. Castenholz, R. W., Jørgensen, B. B., D'Amelio, E. and Bauld, J. (1991). Photosynthetic and behavioral versatility of the cyanobacterium *Oscillatoria boryana* in a sulfide-rich microbial mat, *FEMS Microbiol. Ecol.* , **86**, 43–58.
181. Jørgensen, B. B. and Des Marais, D. J. (1986). Competition for sulfide among colorless and purple sulfur bacteria in a cyanobacterial mat, *FEMS Microbiol. Ecol.*, **38**, 179–186.
182. Pringault, O., Kühl, M., de Wit and Caumette, P. (1998). Growth of green sulphur bacteria in experimental benthic oxygen, sulphide, pH and light gradients, *Microbiology*, **144**, 1051–1061.
183. Wieland, A. and Kühl, M. (2000). Short term temperature effects on oxygen and sulfide cycling in a hypersaline microbial mat (Solar Lake, Egypt), *Mar. Ecol. Progr. Ser.*, in press.
184. Wenzhöfer, F. (1999). In-situ microsensor studies of a hydrothermal vent at Milos (Greece), *Mar. Chem.*, **69**, 43–54.
185. Ziebes, W. unpublished results.

186. Pringault, O., Epping, E., Guyoneaud, R., Khalili, A. and Kühl, M. (1999). Dynamic of anoxygenic photosynthesis in an experimental green sulfur bacterial biofilm, *Environ. Microbiol.* , **1**, 295–305.
187. Wolfbeis, O. S. (1991). *Fiber Optic Sensors and Biosensors*, CRC Press, Ann Arbor, MI.
188. Wolfbeis, O. S. (1997). Chemical sensing using indicator dyes. In *Optical Fiber Sensors. 4. Applications, Analysis and Future Trends*, ed. Dakin, J. and Culshaw, B., Artech House, Norwood, pp. 53–107.
189. Davison, W. personal communication.

5 Potentiometric Microsensors for *In Situ* Measurements in Aquatic Environments

D. DE BEER
Max Planck Institute for Marine Microbiology, Bremen, Germany

1 INTRODUCTION

The objective of this chapter is to relate a short overview of potentiometric principles to measurements with ion-selective microsensors. Potentiometric analyses are based upon measurement of the potential difference of an electrochemical cell in the absence of current. With this method, ion concentrations are measured using ion selective membrane electrodes. These electrodes provide a rapid and convenient way for measuring a range of anions and cations (see Table 1 for examples). Miniaturization of some of these sensors to microelectrodes is a recent development. Although the principles are the same,

In Situ Monitoring of Aquatic Systems: Chemical Analysis and Speciation Edited by J. Buffle and G. Horvai.

Table 1. Overview of compounds measurable with potentiometry. Selectivity coefficients are determined with the separate solution method. The values in bold are estimated from the maximum ratio ($c_{interferent}/c_{analyte}$) for which interference is negligible [1]. The underlined compounds must be absent for reliable determination of the analyte.

Analyte	Detection limit (mol L^{-1})	Interfering compounds ($\log(K_{i,j}$ / L mol^{-1}))	References
Br^-	5×10^{-6}	CN^- (**4**), I^- (**4**), Cl^- (**−2**), OH^- (**−4**), $\underline{S^{2-}}$	1
BF_4^-	10^{-5}	ClO_4^- (3), I^- (2), ClO_3^- (1), CN^- (0), Br^- (−1), NO_2^- (−1), NO_3^- (−2), HCO_3^- (−2), Cl^- (−2), SO_4^{2-} (−4)	1
Ca^{2+}	10^{-8}	H^+ (−2.5), Na^+ (−5.8), K^+ (−7.2), Mg^{2+} (−6.7), Ba^{2+} (−2.5), NH_4^+ (−3.6), Mn^{2+} (−1.6), Co^{2+} (−4.6)	2
Cd^{2+}	10^{-7}	$\underline{Hg^{2+}, Ag^+, Cu^{2+}}$	1, 3–5
Cl^-	5×10^{-5}	CN^- (**−7**), I^- (**−7**), Br^- (**−3**), $S_2O_3^{2-}$ (**−2**), OH^- (**2**), $\underline{S^{2-}}$	1
CN^-	10^{-6}	I^- (**1**), Br^- (**−3**), Cl^- (**6**), $\underline{S^{2-}}$	1
Cu^{2+}	10^{-8}	$\underline{Hg^{2+}, Ag^+, Cu^+}$	1, 4, 5
CO_2	10^{-6}	H_2S (0.3)	6, 7
F^-	10^{-6}	OH^- (**−3**)	1
H^+	10^{-12}	Na^+ (−13, − 14)	8
K^+	10^{-6}	H^+ (−3.4), Na^+ (−4.1), Mg^{2+} (−5.7), Ca^{2+} (−5.2)	9
Mg^{2+}	10^{-6}	H^+ (1.7), Na^+ (−3.8), Li^+ (−3.1), Ca^{2+} (0)	10
Na^+	3×10^{-5}	H^+ (0.5), Li^+ (−1.6), K^+ (−0.4), NH_4^+ (−0.9), Mg^{2+} (−3.6), Ca^{2+} (−3.3)	11
NH_4^+	10^{-6}	H^+ (−1.8), Li^+ (−2.4), Na^+ (−2.7), K^+ (−0.9), Mg^{2+} (−5.5), Ca^{2+} (−3.8)	12
NO_2^-	10^{-8}	SCN^- (0.3), I^- (−2), HCO_3^- (−2.2), ClO_4^- (−2.5), Br^- (−3), NO_3^- (−3.2), Cl^- (−3.2), SO_4^- (−3.5), $\underline{S_2O_3^{2-}}$, $\underline{S^{2-}}$	13
NO_3^-	10^{-5}	I^- (1), NO_2^- (−1), Cl^- (−2), HCO_3^- (−3), acetate(−3)	14
HPO_4^{2-}	10^{-6}	Cl^- (−2.3), NO_3^- (−2.8), SO_4^{2-} (−3), lactate(−3), SCN^- (−2.3), acetate(−3.2)	15
S^{2-}	10^{-6} (total sulfide, pH 7)	$\underline{Hg^{2+}}$, $\underline{O_2}$	16, 17

preparation and use of potentiometric microsensors requires special equipment, training and patience of the researcher.

As potentiometric microsensors are much more fragile, noise sensitive and difficult to prepare than macrosensors, they are not the instruments of choice for routine water analysis. For that purpose it is better to use either macrosensors or alternative analytical techniques. Microsensors should be used when spatial resolution is needed, i.e. for measurements of steep gradients at interfaces. As steep gradients are usually caused by microbial processes, potentiometric microsensors for aquatic environments are still typical research tools, almost exclusively used for microbiological studies.

2 POTENTIOMETRIC SENSORS

The potentiometric sensors relevant for environmental monitoring are generally ion-selective membrane electrodes. The following discussion will be limited, therefore, to this class of potentiometric sensors.

2.1 PRINCIPLES OF POTENTIOMETRY

Potentiometric determinations are based on the measurement of an electrical potential difference across a selective membrane. An example of a measuring circuit is:

ref 1 | 3 mol L^{-1} KCl | filling electrolyte | ion-selective membrane | sample | **3 mol L^{-1} KCl | ref 2**

where the underlined part is the ion-selective sensor and the bold part the external reference electrode. In Figure 1 the measuring circuit for potentiometric sensors is sketched. Ref 1 and ref 2 indicate connections between the electrolytes and the wiring, usually an Ag/AgCl or calomel electrode. Ref 1 is the internal reference element located inside the measuring electrode; ref 2 is part of the external reference electrode.

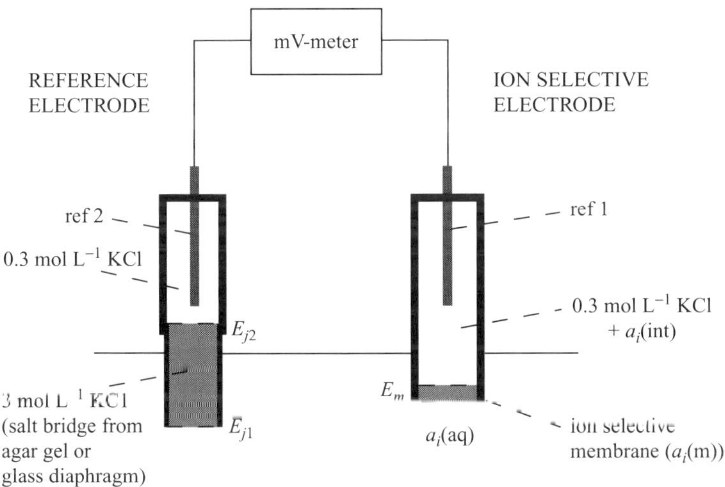

Figure 1. Measuring circuit for potentiometric sensors. E_m = membrane potential, E_{j1}, E_{j2} = liquid junction potentials, a_i(aq), a_i(m), a_i(int) = activitiy of ion i in the test aqueous solution, in the membrane, and in the internal solution, respectively

The reference electrodes, which have a constant potential relative to the solution they contact, can be either calomel or Ag/AgCl half-cells. The Ag/AgCl is most widely used as it equilibrates faster than calomel electrodes after a temperature change [8] and is easy to integrate in sensors. The external reference electrode is connected to the sample by a salt bridge, a glass diaphragm or an agar gel with 3 mol L^{-1} KCl.

The electrical potential drop across this circuit is the sum of all individual contributions, of which most are sample independent and thus constant. The measured potential is then

$$\Delta E = E_{\text{const}} + E_{\text{j1}} + E_{\text{m}} \tag{1}$$

where E_{j1} is the junction potential between sample and reference, and E_{m} is the membrane potential. E_{j1} is small and is almost constant if 3 mol L^{-1} KCl is used in the salt bridge, therefore, E_{m} determines the response of the circuit. While previously E_{m} was described as the sum of the two interface potentials and the diffusion potential inside the membrane, it is now recognized that the inner phase boundary potential can be considered as constant and the diffusion potential in the membrane as negligible. Thus, in most cases, the phase boundary potential at the sample side of the membrane is the main determinant of E_{m} and thus of the sensor response, ΔE [18]. In ref. [18], the authors derived from thermodynamic considerations an expression for the phase boundary potential, E_{PB}:

$$E_{\text{PB}} = -\frac{\mu_i^{\text{o}}(\text{org}) - \mu_i^{\text{o}}(\text{aq})}{zF} + \frac{RT}{zF}\ln\frac{a_i(\text{aq})}{a_i(\text{m})} \tag{2}$$

where μ is the chemical potential, a_i the activity of the ion i in the sample (aq) and membrane (m), R is the gas constant, T the absolute temperature, z the algebraic charge number of the ion, F the Faraday constant. Under the further condition that $a_i(\text{m})$ remains constant, the Nernst equation is obtained

$$\Delta E = E^\circ + \frac{RT}{zF}\ln a_i(\text{aq}) \quad \text{or} \quad \Delta E = E^\circ + k \log a_i(\text{aq}) \tag{3}$$

where E° is the offset potential, i.e. the potential difference across the cell when $a_i = 1$, and k is the slope factor equal to 59.16 z^{-1} mV at 25 °C.

The essential condition for the phase boundary potential model, that the membrane composition remains constant, is met by a series of conditions. The membrane must have ion-exchange or complexing properties, and the main determinant for $a_i(\text{m})$ is the concentration of the corresponding sites in the membrane. The membrane must also be hydrophobic, to prevent co-extraction of the counter-ions. If these conditions are not met, $a_i(\text{m})$ will be proportional to $a_i(\text{aq})$. Incorporation of an ionophore that selectively binds the analyte ion ensures that a_i(m) remains constant in the presence of interfering ions. The ionophore thus determines the selectivity of the sensor.

The phase boundary model as described above explains and predicts the behavior of ion-selective membrane sensors better than the ion-translocation model [19]. For example, the short response time of liquid membrane sensors is incompatible with the transport of ions through thick membranes, and the need for lipophilic ions of opposite charge as membrane components.

In practice, ideal Nernst behavior (equation (3)) is not always observed. Particularly, deviations from the Nernst equation become significant at low activities. The deviations are explained by additional contributions to ΔE, in particular liquid junction potentials, E_{j1} and E_{j2}, phase boundary potentials due to interfering ions, and potential leaks (Figures 1 and 2) [19]. E_{j1} is dependent on the sample composition, but this dependence can be made almost negligible by using a high concentration salt bridge, such as 3 mol L^{-1} KCl which is normally used. The junction potential between the salt bridge and the sample is dominated by the diffusivity of K^+ and Cl^-, and since these are approximately equal, E_{j1} becomes close to zero. E_{j2} is constant since the composition of the salt bridge and the internal solution of the reference electrodes are constant. The contributions to ΔE by interfering ions are expressed by the Nicolskii–Eisenman equation:

$$\Delta E = E^{\circ\prime} + k \log \left(a_i + \sum K_{i,j} (a_j)^{z_i/z_j} \right) \qquad (4)$$

with $E^{\circ\prime}$ the observed offset potential, K_{ij} the selectivity factor of ion i over j and a_j the activities of interfering ions. As pointed out by Bakker *et al.* (1997) the Nicolskii–Eisenman formalism is inconsistent when $z_i \neq z_j$. They solved this problem using the phase boundary potential model, for which the reader is referred to the relevant literature [18].

The reference electrode is usually a commercial Ag/AgCl or a calomel half-cell. It is a mistake to position the external reference very close to the measuring electrode. First of all, the level of the internal electrolyte of the reference electrode should be above the sample solution to avoid contamination of the electrolyte by analyte solution and dilution of the salt bridge. In the case of *in situ* measurements at large depth the reference should be slightly pressurized to avoid this effect, especially in freshwater environments [8]. Secondly, for measurements in sediments, biofilms or microbial mats, the reference electrode should not be pushed into the matrix. Then the 'suspension effect' can change the junction potential between the salt bridge and the sample, especially when colloids are present [8]. This poorly understood phenomenon can result in a considerable change in offset. The suspension effect changes only the potential of the reference electrode, i.e. the junction potential at the salt bridge, but not the potential of the ion-selective electrode.

2.2 OVERVIEW OF EXISTING MICROSENSORS

In many aquatic systems conversions mainly take place in the sediments, microbial mats and biofilms covering the sediments and solid surfaces, instead of in the water phase. Because of mass transfer limitations inside these structures, gradients of substrates and products are present, the concentrations of substrates being lower and the concentrations of products higher than in the water phase. The slope of the concentration gradients depends on the conversion rates and mass transfer rates. Significant changes can occur within 10 μm in highly active biofilms while in less active deep-sea sediments significant changes occur typically within millimeters to centimeters. In such systems porewater analysis has limitations. The extraction of porewater may influence the concentration profiles and in the case of biofilms its spatial resolution is insufficient. Thus *in situ* measurements are needed with high spatial resolution, of at least 50 μm. The best technique available nowadays is the use of microsensors, needle-shaped devices with a tip size of 1–20 μm which can measure the concentration of a specific compound. Owing to the small sensing tip, highly localized measurements are possible, since the spatial resolution should be approximately equal to the tip size of the sensor. There are indications, however, from theoretical and experimental studies that microsensors can influence the concentration profiles. The evidence is conflicting: while the theory predicts underestimation especially by sensors larger than 10 μm [20], experiments with O_2 microsensors in a biofilm showed an overestimation of local concentrations by sensors larger than 16 μm [21]. Microsensors may change the local concentrations by the consumption of substrate (in the case of amperometric sensors), by compression of the local matrix [22], by changing the diffusion field (blocking diffusion by the sensor body) [20], or by compressing the boundary layer [23]. Although the microelectrode technique is invasive and the tips have a small influence on structures and processes, it is the best choice for direct concentration measurements inside biofilms, mats and sediments.

Microsensors were introduced in microbial ecology by Bungay *et al.* (1969) [24] who measured O_2 profiles in biofilms. The technique was strongly improved by Revsbech who constructed reliable O_2 microsensors for profiling sediments and biofilms [25,26]. More microsensors relevant for microbial ecology were developed and used, such as for N_2O [27], pH [28], NH_4^+ [29], NO_3^- [30–32], S^{2-} [33], H_2S [34], NO_2^- [35], CH_4 [36], Ca^{2+} [2] and CO_2 [6].

A survey of the compounds which can be measured with membrane electrodes (see Table 1 and list of microsensors for microbial ecology above) showed that they can be grouped in three, rather distinct, clusters corresponding to the three groups of elements in the periodic system: (I) elements from the alkali and alkaline earth metal groups, (II) b-metals, (III) elements in the top right corner of the periodic system (with the exclusion of the noble gases, but including the halogens, C, N, O and S). Important elements such as iron, manganese and silicon are placed far from these clusters in the periodic system, which makes the prospects of measuring these elements with potentiometric methods rather poor. Iron and manganese can be measured by stripping voltammetry [37], but the technology has not yet been applied to needle-type microsensors.

There are several different types of potentiometric microsensors. Their miniaturization is described below.

(1) Crystalline membrane. These membranes consist, e.g. of metal halides and respond to the concentration of the halogen and metal ion in the sample. Lists of sensor types based on this principle can be found in reviews [38]. Only the Ag_2S membrane sensor has been miniaturized. The sensor responds to Ag^+ and S^{2-}. This microelectrode has been useful in studies of the sulfur cycle in microbial mats and biofilms. The application and peculiarities of this sensor are discussed in Chapter 4.

(2) Full glass. Glass membrane sensors are described for Na^+, K^+ and H^+ [39] and all have been miniaturized, mainly for use in animal physiology [39]. As Na^+ or K^+ microgradients do not normally occur in sediments, mats and biofilms, only the pH microsensor is useful for environmental application. The full glass pH microsensor can be used for many purposes [40], but because of its rather large (50–100 μm long) sensing tip, it has a rather low spatial resolution. The long tip responds to the pH gradient along the length of the sensing surface, but it averages in an unknown way. Thus it is impossible to resolve pH differences within a distance of less than the tip length. Using the recessed tip principle, for which the sensor tip is placed in a casing with a small opening at the tip, can considerably increase the spatial resolution [39]. However, this increases the response time considerably. The pH sensor is discussed in detail in Chapter 3.

(3) Liquid membranes. These can be divided into classical ion-exchanging membranes and carrier-based membranes. The ion-exchanging membranes contain hydrophobic ions (ion-exchanger ions) and the analyte counter-ion. Carrier-based membranes contain complexes between specifically binding hydrophobic agents (also called ionophores) and the analyte ion. The ion specific behavior of liquid membranes with dissolved carrier was discovered in 1967 [41], and later these membranes were incorporated in micro-pipettes [42]. In the past, liquid membrane macrosensors for water analysis have been sold commercially. The liquid membrane microsensor technique was developed by cell physiologists for intracellular measurements (mostly H^+, CO_3^{2-}, Mg^{2+}, Ca^{2+}, Li^+, Na^+ and K^+) [39, 43, 44] These sensors can be very small, with a tip diameter of less than 1 μm, i.e. the size of a bacterial cell. Relevant liquid membrane microsensors for use in microbial ecology are NH_4^+ [29], NO_3^- [30], NO_2^- [35, 45], H^+ [46], Ca^{2+} [2], and CO_3^- [47]. The CO_2 microsensor [6] is a special electrode, with a liquid membrane pH sensor as transducer. As H^+ and CO_2 microsensors are treated in Chapter 2, this chapter will focus on liquid membrane sensors for determination of N-compounds and Ca^{2+}.

2.3 LIQUID MEMBRANE MICROSENSORS

Potentiometric microsensors function in the same way as macrosensors, although they deviate more often from ideal behavior, because of their small tip size. Liquid membrane microelectrodes consist of a glass micropipette with ion-exchanging liquid in the tip acting as the functional membrane. The potential build up by the membrane may be dissipated through the glass wall of the micropipette and through the shunt between the ion-exchanging membrane and the glass wall (see Figure 2) [48]. Obviously, miniaturization increases the occurrence of leaks as the ratio of the tip surface to the total capillary surface decreases. Consequently, for some microsensors a relatively large tip is required. For example, NO_2^- sensors are only functional if the tip is at least 10 μm [35]; smaller sensors only respond to concentrations above 100 μmol L^{-1} [45]. However, Ca^{2+} microsensors with 1 μm tips function excellently, having fast response ($t_{90} < 1$ s) at low concentrations (< 1 μmol L^{-1}) [2]. The selectivity is partially an intrinsic property of the membrane and partially determined by the size and shape of the sensor. Generally, the smaller the size of a sensor, the worse the selectivities, as non-specific ion exchange in the glass wall and between the liquid membrane and the glass wall more strongly contributes to the signal. Also, the quality of the silanization may vary, as it is a poorly controllable process, influenced (in a way that is not fully understood) by local temperature, humidity, and shape of the tip [48]. In the author's experience, the best results (with respect to the ease of filling with membrane,

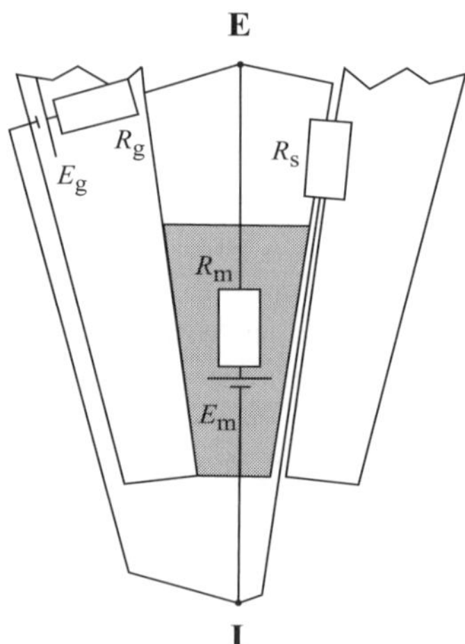

Figure 2. The electrical components of a potentiometric microsensor. E_m = membrane potential, R_m = membrane resistance, E_g = glass potential (potential by ion exchange through glass), R_g = resistance of glass wall, R_s = resistance shunt (seal between membrane and glass wall). **I** denotes sample, **E** sensor electrolyte

Table 2. Selectivity coefficients of liquid membrane microsensors (approximate values)

Cation sensors	$\log(K_{i,Na^+})$	$\log(K_{i,K^+})$	$\log(K_{i,H^+})$	$\log(K_{i,Ca^{2+}})$	$\log(K_{i,Li^+})$	$\log(K_{i,Mg^{2+}})$	References
H^+	−10	−10	–	−11	−10	?	46, 49, 50
Ca^{2+}	−6	−5	−4	–	?	−7	2, 51
NH_4^+	−2	−1	−4	−5	−4	−6	29, 52
Anion sensors	$\log(K_{i,Cl^-})$	$\log(K_{i,HCO_3^-})$	$\log(K_{i,HS^-})$	$\log(K_{i,NO_2^-})$	$\log(K_{i,NO_3^-})$	$\log(K_{i,SO_4^{2-}})$	
NO_3^-	−2	−3	?	−2	–	−3	14, 30, 53
NO_2^-	−5	−4	poisoned	–	−5	−6	13, 35

signal stability, detection limit, sensor lifetime and physical strength of tip) are obtained if the tip is tapered between 5 and 10°. In Table 2 a list of selectivity coefficients for microsensors is given. The values are approximate, and may vary slightly between individual microsensors. In actual practice, with Ca^{2+} and N– compound microelectrodes, determination of the selectivity factors for each microsensor is not needed. A calibration performed in a solution with the same composition as the sample (but with adjusted concentrations of the analyte) is sufficient, thereby determining $E^{o\prime}$, k and the concentration range within which the electrode can be used. Ca^{2+} sensors used in seawater can be calibrated in water with the main salts present in seawater. pH microsensors used in seawater should be calibrated in special seawater buffers. NH_4^+, NO_3^- and NO_2^- sensors used for waste water biofilm research can be calibrated in artificial waste water containing the main constituents of the actual waste water in which the measurements are performed. This way of calibrating is reliable, if the interfering ions are constant in background, i.e. if there are no gradients of interfering species. Then their contribution to the signal ($\sum K_{i,j}(a_j)^{z_i/z_j}$ in equation (4)) is constant at each point in the profile. The strongest interfering species for NH_4^+ are Na^+ and K^+, which are not consumed or produced in microbial processes, and therefore constant over the distance over which the profiles are measured (typically a few mm). The strongest interfering species for NO_3^- are ClO_4^-, SCN^-, I^- and Cl^- that are normally not present in significant quantities in the freshwater environments where nitrate sensors are used. Except in biofilms converting halogenated substrates Cl^- is not usually a biological product, so no gradients can be expected. HCO_3^- has been reported as interfering with the measurements of NO_3^-, and being a major product of microbial conversion it can lead to errors. The error is strongly dependent on the selectivity of the ion-exchanging membrane, as will be discussed later.

Although the electrodes respond to the activity of the ion, electrodes are usually calibrated for concentration as under life-sustaining (low ionic strength, pressure and temperature) conditions activity coefficients are approximately constant. The calibration is log–linear over a concentration range depending on

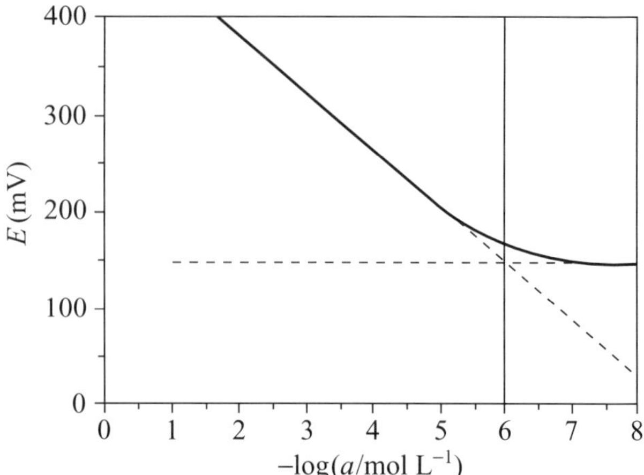

Figure 3. Typical response curve of a potentiometric sensor. The detection limit is given by the intercept of the dashed lines

the composition of the medium. At lower concentrations the calibration tapers off and finally the sensor becomes insensitive to concentration change (Figure 3). The intercept of the linear parts of the log–linear response curve is the detection limit. However, the curved part of the calibration line can be used below this concentration thereby extending the actual measuring range 1 decade below the detection limit (albeit with reduced accuracy). In medium where the analyte is complexed however, the use of the curved part of the calibration plot is not recommended.

It should be kept in mind that potentiometric sensors respond to the activity of the free ions. This is important for Ca^{2+} which easily binds to various organic molecules. Consequently, Ca^{2+} profiles measured with microsensors may not represent the total mobile concentration. This should be considered for correct estimation of the fluxes. This may especially occur in organic-rich sediments and biofilms in reactors with high organic loading. Also NH_4^+ can bind to sediments, particularly to clay and humic substances.

2.3.1 Membrane Chemistry

Cations. For cations, carrier-based liquid membrane sensors are used, in which the selectivity of the membrane is determined by a lipophilic ligand that complexes reversibly and specifically with the measured ion. The carrier–ion complex is so strong that the concentration of free ions in the membrane is negligible compared with the concentration of ions bound in the complex. This is an important difference with the classical ion-exchanging membranes (see below). Most relevant carrier-based liquid membrane sensors for microbial ecology measure H^+, NH_4^+ and Ca^{2+}. The response time (t_{90}) is expressed as the time needed to reach 90% of the signal change after a step change in concentration. According to the phase boundary model, transport in the sensor should not play a role. Since the membrane phase boundary has a thickness of a few nm, transport through the aqueous diffusion layer surrounding the microsensor should limit the response time [18]. Then sensors with a tip size of 1–3 μm should have a t_{90} of the order of 1–100 ms. However, usually the response time is of the order of seconds, so what actually affects the response time of ion-selective sensors is not well understood. The response time is influenced by the tip size of the sensor (which determines the electrical resistance of the measuring system) and the capacitance of the measuring circuit. However, this effect occurs if the tip size is 1 μm or less, while for environmental studies, sensors have tip sizes of 3 μm, and relatively low tip resistance. The response time of the nitrite sensor depends on the concentration and varies from 15 s in the μmol L^{-1} range to 1 s in the mmol L^{-1} range; for other microsensors it is of the order of 1 s over the whole concentration range.

Anions. For anions, usually classical ion-exchanging membranes are used based on lipophilic salts (e.g. quaternary ammonium salt) that act as exchanger sites. In these membranes complexation between the anions and cationic sites in the membrane phase is negligible, so that all anions are freely dissolved in the membrane. They all exhibit the same selectivity sequence with a preference for lipophilic over hydrophilic ions, following the Hofmeister series:

$$ClO_4^- > SCN^- > I^- > NO_3^- > NO_2^- > Cl^- > HCO_3^- > SO_4^{2-} > HPO_4^{2-}$$

The selectivity is determined by the distribution coefficient of the anions between sample solution and membrane, which is mainly determined by the standard Gibbs energy for transfer of the ion from the water phase into the membrane (for full treatment of this issue see Koryta (1983), pp. 30–39) [54]. Although the membrane composition has an effect on the selectivity coefficients, the selectivity sequence cannot be changed. This is demonstrated in Figure 4, which shows the selectivity of various membrane chemistries for a variety of ions. As a consequence of this fixed sequence of selectivities, the only sensor based on this principle, useful for environmental application, is the NO_3 sensor, often applied in freshwater sediments and biofilms. The strongest interference can be expected from ClO_4^-, SCN^- and I^- species which are not common in freshwater environments. Also carbonate sensors and their use in animal physiology, have been described [47]. Because of their sensitivity to NO_3^- they cannot be used in aquatic environments. Innovations for anion sensors (deviations from the Hofmeister series) can only be expected from development of anion-selective carriers. The use of tri-*n*-octyltin chloride

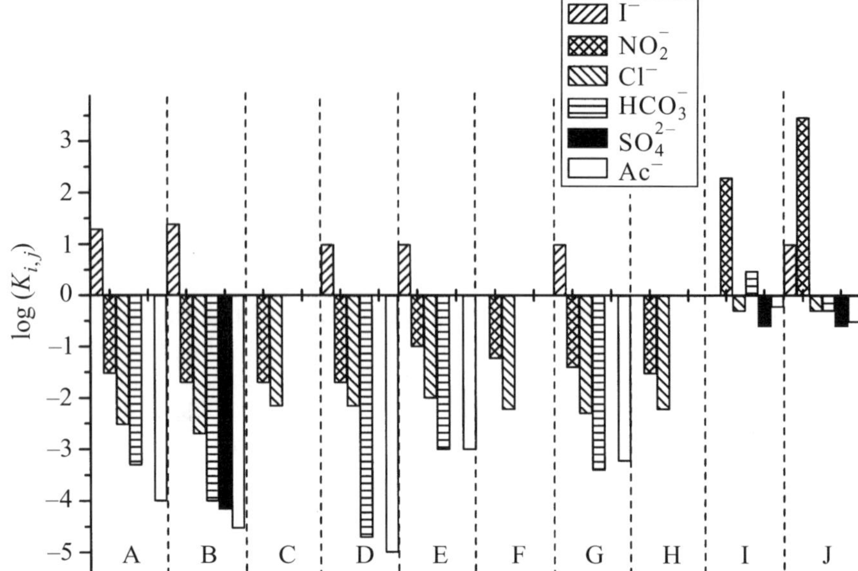

Figure 4. Selectivity coefficients of various liquid ion exchange membranes (LIX), measured in microelectrodes (except for G and J) with the separate solution method. (A) 10 % Co(III)-iodo-salocden in nitrophenyl octyl ether (NPOE) [55], (B) as for (A) but 20 % [55], (C) 20 % Co(III)-triphenylphosphine-bromo-salocden in NPOE [55], (D) 20 % Co(III)-methoxo-salocden in NPOE [55], (E) Orion exchanger [55], (F) Orion exchanger [30], (G) Crytur [55], (H) Radiometer [53], (I) 9 % aquocyanocobalt(III)-hepta(2-phenylethyl)cobyrinate in NPOE [45], (J) 1 % aquocyanocobalt(III)-hepta(2-phenylethyl)cobyrinate in NPOE [45]. (I) and (J) are carrier-based membranes.

increased the selectivity towards HCO_3^- ions [9], although no application of this carrier has been reported. A lipophilic vitamin B12 derivative proved to be a highly selective carrier for NO_2^- (Figures 4 and 5) [13]. The developments of phosphate ionophores [15, 56] and a promising sulfate carrier [44] have been reported; however, sensor applications based on these compounds have not yet been published.

Carrier-based liquid membranes consist of an inert, viscous, non-volatile and hydrophobic solvent, such as 2,3–dimethylnitrobenzene, *o*-nitrophenyl *n*-octyl ether (most widely used), bis(1–butylpentyl)adipate or bis(2-ethylhexyl)sebacate in which the lipophilic complexing agents are dissolved. Addition of lipophilic anions (such as sodium tetraphenylborate or potassium tetrakis(*p*-chlorophenyl)borate) has been shown to be essential for the carrier-based sensor characteristics by increasing the selectivity and reducing the response times [19]. As is now understood, the lipophilic ions act as ion-exchangers needed for the functioning of the membrane [18]. Finally, addition of poly(vinyl

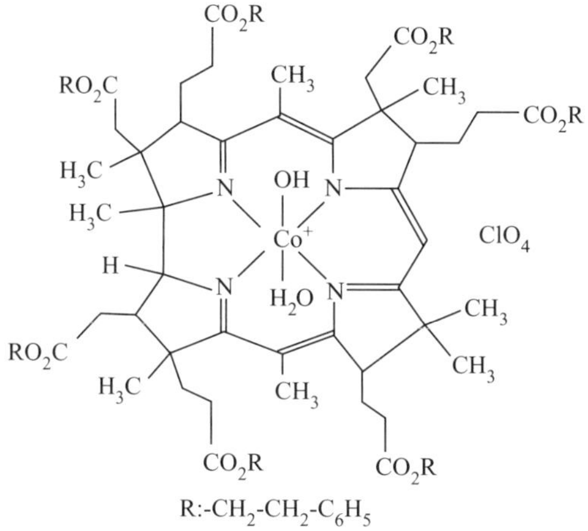

R:-CH_2-CH_2-C_6H_5

Figure 5. Nitrite ionophore used in liquid membranes. It is a vitamin B12 derivative to which alkyl groups are attached to make the complex hydrophobic. The nitrite ion is thought to complex by exchange with an axial ligand of the Co^{3+} [13]

chloride) (PVC) improves the signal stability, lifetime and sensitivity of the sensor. Solidification of the liquid membrane physically stabilizes the membrane in the tip. Furthermore, the tip of the microcapillary gets coated by liquid membrane which decreases the ion-exchange through the thin glass wall. The PVC is dissolved in tetrahydrofuran (THF) and mixed with the liquid membrane. This cocktail is brought into the tip of the microsensor and after evaporation of the THF a solid membrane remains. Without addition of PVC the tip size cannot be larger than 1 μm, as in larger tips the capillary force is too weak to maintain the membrane. Solidification allows preparation of larger sensors (3–20 μm), that are stronger than 'real' microsensors with 1 μm tips.

2.3.2 Preparation of Liquid Membrane Microsensors

The preparation of liquid membrane microsensors involves pulling of the capillaries, silanization of the glass, shielding, filling with electrolyte and liquid membrane, and application of a coating. The liquid membrane microsensors were originally developed for intracellular measurements. These sensors were of simple design, consisting of a silanized capillary with a tip size of less than 1 μm, filled with electrolyte and with a droplet of ion-exchanger in the tip. Such sensors are extremely fragile and noise sensitive. Thus experiments have to be performed in a Faraday cage for noise protection, and measurements must be done in a quiet working environment. Moreover, these sensors will break often in sediments. For experiments in environmental samples the sensor manufacturing has been modified to make them more sturdy and less noise sensitive. The following procedure can be used for all types of liquid membrane sensors, that differ only in the filling electrolyte and membrane.

Glass tubing (e.g. 3.5 mm OD, 8516 Schott) is heated in a flame and pulled to a thickness of ca. 1 mm over a length of ca 20 cm. The capillary is further pulled in an electrical heating loop to a thickness of ca. 200 μm. Then, using a thin platinum heating loop mounted on a micromanipulator, the glass is elongated until a thickness of ca. 30 μm is reached. Then the heat of the loop is reduced, the loop is moved slowly closer to the glass until it falls off. The capillary should gradually taper to a 1–3-μm tip.

To stabilize the hydrophobic membrane in the tip of the microcapillary the glass must be made hydrophobic by a silanization procedure. The capillaries are baked in a glass container (2–3 L) at 150 °C for 3 h, to remove bound water. Then 250 μL *N*,*N*-dimethyltrimethylsilylamine (Fluka) is added and the closed container is left overnight at 200 °C. An alternative procedure is dipping the freshly pulled microcapillaries in trichlorosilane, followed by baking at 125 °C for 15 min. This procedure is faster, but with inferior results [48]. The silanized capillaries are glued with silicone glue in a glass shielding made from Pasteur pipettes, with the tips protruding 1–2 cm [53]. The finished pipettes can be stored dry and dust-free for an unlimited time.

Before filling, the tips can be broken to the desired size. For most sensors 1–3 μm tips are ideal, for NO_2^- sensors the tip should be 10–15 μm. Breaking is done under microscopic guidance, by moving gently against a clean Pasteur pipette. The sensor tips are filled with the appropriate electrolyte over a length of ca. 4 cm. The electrolyte contains usually 5–50 mmol L^{-1} of the ion measured, or for pH sensors a pH buffer with 300 mmol L^{-1} KCl. Then some liquid membrane without PVC is introduced in the tip (ca. 300 μm). Additionally, 100–200 μm of the PVC-containing liquid membrane is introduced. After ca. 2 h the THF is evaporated and a solidified ion-selective membrane is left in the tip. In the author's experience, liquid membranes completely gelled with PVC are not stable in capillaries larger than 1 μm. The membrane often loosens from the glass wall, or is upon immersion partially extruded through the tip, for unknown reasons. Only the combination of PVC gelled and non-gelled liquid membrane results in functional sensors.

Initially good working electrodes can be destroyed by contact with environmental samples [57]. Sudden potential changes, drift and the absence of response to concentration changes are the typical symptoms. This is probably caused by dissolution of hydrophobic substances from the biomass into the liquid membrane. If this occurs, often in cyanobacterial mats and dense environmental biofilms, a coating should be applied to protect the sensor from direct contact with the sample. A good separation is obtained by a thin (< 1 μm) layer of protein cross-linked by glutaraldehyde [35]. This coating is impermeable to the contaminants from the biomass, but smaller hydrophilic molecules can pass.

The selectivity is not affected. NO_2^- sensors become slower, while the response time of NO_3^-, pH and NH_4^+ sensors remains the same.

Finally, the shielding surrounding the sensor is filled with $0.3\,mol\,L^{-1}$ KCl and connected with a silver wire to the reference. The filling electrolyte is connected with a chlorinated silver wire, by a coaxial cable to the voltmeter input. A voltmeter with high input impedance (10^{15} **Ω**) is needed. The resistance of the voltmeter must be at least 1000 times that of the microsensors, which have a resistance of 10^{10}–10^{12} **Ω**; therefore, a voltmeter used for measurements with macrosensors (input impedance $< 10^{13}$ **Ω**) is unsuitable. The outer shielding of the coaxial cable must be positioned below the liquid shielding. The shielding is essentially a Faraday cage, but much more effective and convenient than a cage enveloping the whole equipment. A finished liquid membrane sensor is depicted in Figure 6.

The basic equipment for calibration and measurements consists of appropriate amplifiers and micromanipulators. Use of motorized micromanipulators is recommended to increase spatial resolution and to avoid vibrations by manual operation. A computer can be used for data collection and control of the motorized micromanipulator. A recorder should be used in parallel as it is helpful in recognizing occurrence of drift and slow response times.

2.3.3 Liquid Membrane Sensors for N Compounds

NH_4^+, NO_3^- and NO_2^- are the main compounds from the nitrogen cycle that can be measured by liquid membrane microsensors. NH_4^+ is released during

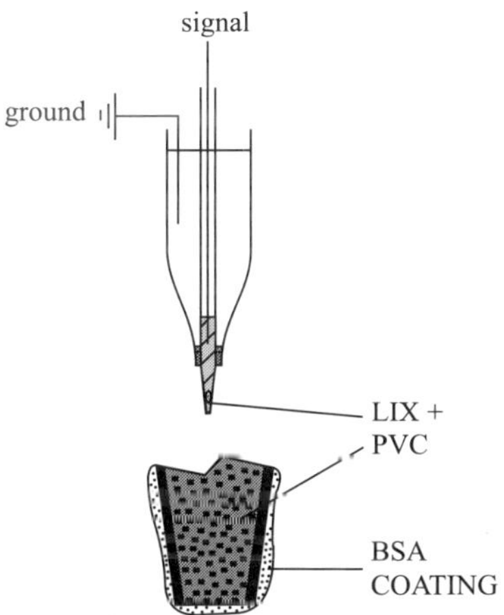

Figure 6. Scheme of a liquid membrane microsensor, with magnified tip. The BSA coating is a thin layer of cross-linked bovine serum albumin, the liquid membrane (LIX) is solidified with PVC

mineralization and oxidized by nitrification. NO_3^- is consumed mainly by denitrification and NO_2^- is an intermediate of both nitrification and denitrification. The low selectivity of liquid membrane sensors for NH_4^+ towards Na^+ and K^+ and of those for NO_3^- and NO_2^- towards Cl^-, prohibits measurements in seawater. Fortunately, NO_3^- can be measured in seawater with the recently developed biosensor [58]. However, for studies in freshwater, liquid membrane sensors may be preferred because of the ease of preparation.

The NH_4^+ electrode was developed first [29] and tested in studies of diffusion and reaction in enzymatic model systems. Since the measured profiles correlated well with those calculated with a diffusion–reaction model the electrode was considered reliable for profile measurements. The response time is ca. 1 s. The lifetime is approximately 1–2 d. The detection limit in distilled water is 1 μmol L^{-1}. The selectivity towards K^+ and Na^+ is low, so the sensor cannot be used in seawater. In most media the response is not log–linear below 10 μmol L^{-1}, but in freshwater systems, such as lake and river sediments and biofilms from waste water plants, the sensor can be used for measurements down to 1 μmol L^{-1}.

The later developed NO_3^- microsensor was also first tested in a model system, containing denitrifying bacteria [30]. The sensor is fast, with a t_{90} less than 1 s. Various liquid membrane chemistries have been used. The first NO_3^- sensor was based on the ion-exchanger from the Orion macroelectrode (1,10–phenanthroline nickel nitrate), and later other sensor chemistries were used in microsensors, such as methyltridodecylammonium nitrate, tricaprylylmethylammonium nitrate and tridodecylhexadecylammonium nitrate (Corning) [14]. Comparison of different microsensors showed that the nickel-based Orion liquid membrane had the best selectivity towards Cl^-, while the tridodecylhexadecylammonium-based sensor had the lowest detection limit. This comparison was made for plant physiology where Cl^- is the main interferent, although for microbial ecology studies the sensitivity towards hydrogen carbonate is more important [53]. Interfering substances are not very problematic when these are homogeneously distributed and thus present a constant background. Then they only increase the detection limit of the measurement. However, when the interfering compound is varying with depth one cannot subtract it as a constant background from the signal. In sediments and biofilms hydrogeno carbonate gradients are present, due to pH gradients and degradation of organic matter. Because of this problem NO_3^- did not seem to deplete in the denitrifying zones of sediments [53]. A comparison between different sensor chemistries showed that the Orion liquid membrane was much less sensitive towards hydrogeno carbonate than the liquid membrane used by Jensen (decyltrioctylammonium bromide and PVC in dibutyl phthalate, Radiometer) [55]. Indeed with microsensors based on the Orion liquid membrane the hydrogen carbonate interference is not a problem, and measurements showed that in denitrifying biofilms and sediments NO_3^- is depleted [30, 35, 59]. Therefore, the nickel-based Orion compound is recommended for use in denitrifying systems, where determination of NO_3^- depletion is important. Regrettably, Orion has stopped this production, but a recipe is available [60]. Recently, a new sensor chemistry for NO_3^- microsensors has been developed [61]. The reported selectivities and detection limits exceed those of the Orion liquid membrane, but it has not yet been used in ecological studies. A disadvantage of this sensor is its sensitivity towards sulfide, which irreversibly poisons the sensor. Sulfide also affects the signal of sensors with the nickel-based liquid membrane, although not irreversibly. Most NO_3^- liquid membranes have a detection limit in distilled water of 10 μmol L^{-1}, but measurements can be made down to 1 μmol L^{-1} (using the Orion liquid membrane). Surprisingly, NO_3^- microsensors coated with cross-linked BSA have a lower detection limit (1 μmol L^{-1} in interferent-free water) than uncoated sensors. The lifetime of the sensors is ca. 5 h, and 8–12 h when the liquid membrane is gelled with PVC.

The most recent development was the NO_2^- microsensor [35], which was applied in nitrifying and denitrifying biofilms. This carrier-based sensor is highly selective for NO_2^-, as can be appreciated from Table 1. The response of this sensor is strongly size dependent: the performance of microsensors with 1–2 μm tips is too poor, with respect to selectivity and detection limit, for practical use [45]. This is probably a wall effect as excellent sensors are obtained when the tip size is 10–20 μm. When stored dry, this sensor can be used for weeks. The response time is concentration dependent: in the range of 100–1000 μmol L^{-1} it is 1 s, from 1–100 μmol L^{-1} it is ca. 10–15 s, and below 1 μmol L^{-1} it increases significantly to minutes. The detection limit in distilled water is in the sub μmol L^{-1} range, and a log–linear response can be obtained down to 10^{-8} mol L^{-1}. In seawater the detection limit is 10 μmol L^{-1}, below which the response curve tapers off sharply. A serious problem with this sensor is its sensitivity towards sulfide and other reduced sulfur compounds, that reduce the Co(III) in the carrier molecule. The poisoning is irreversible, and results in a continuous drift. This phenomenon is a major drawback, as it may prohibit measurements in anoxic zones. Poisoning was observed in sulfate reducing biofilms, at depths where sulfide was detected (data not shown). However, also in anoxic zones of freshwater sediments, where no sulfide was detected, poisoning occurred, probably by reduced sulfide species not detected by the sulfide microsensors (data not shown). The recently developed new NO_2^- carrier may not have this problem.

2.3.4 Ca^{2+} Microsensors

Ca^{2+} sensors were developed for animal physiology and as ultra-microelectrodes (< 1 μm) are often used for intracellular studies. They are useful in this field because of their extremely low detection limit. Several carriers are available, with different detection limits (0.1 nmol L^{-1} – 0.1 μmol L^{-1}), selectivities and response times (1–5 s). The actual response times are probably less than 200 ms [61], but cannot be determined accurately because of the slow mixing of calibration solutions [2]. For environmental studies all are suitable, as the concentrations are measured in the $\mu mol\,L^{-1}$ to $mmol\,L^{-1}$ range instead of the $nmol\,L^{-1}$ range. The sensors for Ca^{2+} are, together with those for H^+, the best liquid membrane sensors available. The selectivity allows measurements in marine and hypersaline conditions.

2.3.5 Comparison of Liquid Membrane with Other Microsensors

Except for pH, liquid membrane sensors measure a unique range of compounds, making true comparison difficult. In general, both the signal stability and lifetime of liquid membrane microsensors are poor compared with those of full glass, amperometric O_2 and H_2S, and optical microsensors. Typically, liquid membrane sensors can be used for a few days, after which the detection limit is too high or the calibration levels off (slope factor k decreases). Especially when a low detection limit is important, freshly-prepared sensors should be used. Also the signal drifts, usually 0.5–2 $mV\,h^{-1}$, resulting in an error of 2–7.5% h^{-1} for monovalent ions. Drift is caused by hydration of the microcapillaries and degradation of the liquid membrane. NO_3^- sensors last less than 1 d, because the nickel complex is slightly water soluble and disappears from the membrane. NH_4^+ and proton carriers are much more hydrophobic and these microsensors last for a few days to 1 week, when stored dry between measurements. NO_2^- sensors can be stored for weeks. For comparison, drift of amperometric O_2 [26] and H_2S [34] microsensors is a few % per day, and for microoptrodes, it is in the order of a few % per month; pH glass microsensors do not significantly drift during 1 week. The lifetime of microsensors is ca. 1 month (amperometric H_2S), half a year (amperometric O_2) or more than a year (optrodes and full glass pH sensors).

This comparison is unfavorable for liquid membrane sensors, but they do have their positive sides. For NH_4^+, NO_2^- and Ca^{2+} no alternative sensors exist. The biosensor [58] (see also Chapter 4) does not distinguish between NO_2^- and NO_3^-. Drift can be determined between experiments by calibrations or by determining the signal in the bulk liquid, which usually has a constant composition. Since drift is constant, the signal can simply be corrected. Liquid membrane sensors are easy and fast to prepare, which partially compensates for their short lifetime. The reported damage by contact with biomass [57] can be prevented by a hydrophilic coating. Their small size and absence of substrate consumption may be crucial for some applications. For example, the small size of pH liquid membrane sensors is essential for the functioning of fast, small ($< 10\,\mu m$ tip) and sensitive CO_2 microsensors [6]. Full glass pH sensors are too large for this purpose and no functioning microsensors could be made. Also the full glass pH sensors may have insufficient spatial resolution in highly active systems. The NO_3^- biosensor has a tip size of 20 μm, which may disturb the matrix structure and profiles in highly active systems with steep gradients. Thus for highly active biofilms, liquid membrane microsensors may be a better choice. In conclusion it can be stated that, despite their disadvantages, liquid membrane microsensors are highly useful tools for microprofiling of nitrogen species in freshwater systems and for measurements of pH and Ca^{2+} in all aquatic environments.

3 ECO-PHYSIOLOGICAL STUDIES WITH LIQUID MEMBRANE SENSORS

Microsensors are ideal tools for fundamental studies on microbial processes in sediments and biofilms. Most commonly used is the O_2 microsensor, while only a limited number of research groups have worked with liquid membrane electrodes. Examples of studies using liquid membrane sensors on the nitrogen cycle, with emphasis on nitrification and denitrification, will be given (see Table 3 for an extensive overview). NO_3^- and NH_4^+ liquid membrane sensors have also been used for plant physiological studies, i.e. in uptake in roots, but this application will not be reviewed here. Secondly, some applications of the Ca^{2+} microsensor in environmental research will be shown.

Concentration profiles are a result of conversions and mass transport. In most sediments and biofilms transport of substrate and product takes place by diffusion. Consequently, if the diffusion coefficients are known, local conversion rates can be derived. This is demonstrated in Figure 7, which schematically shows substrate and product profiles occurring during nitrification and denitrification. As a result of nitrification NH_4^+ will decrease in the nitrifying zone, usually determined by the O_2 penetration depth, and in the same zone NO_3^- will increase. Because of denitrification NO_3^- will decrease in the anoxic zones. In layers where consumption and production occurs the profiles of the respective compounds are curved. From the profiles, NH_4^+ and NO_3^- fluxes can be determined using Fick's law ($J = D\,\mathrm{d}c/\mathrm{d}x$), at different depths in the sediment. Then the volumetric conversion rates can be determined in a layer of interest, by subtracting the fluxes at the top and the bottom of that layer and dividing this number through the thickness of the layer (see Figure 13). If the concentration profiles are recorded at sufficient spatial resolution, a fine scale distribution of activity can be determined [71]. Alternatively, one can use a reaction–diffusion model to calculate the profiles and by an iterative procedure fit the modeled profiles to those measured. In such a procedure the activity profile is changed until a good fit is obtained [29,30,70,76].

Table 3. Overview of *in situ* studies with microsensors

Subject	Microsensors	Metabolic activity	Goal	References
aggregates	NO_3^-, NH_4^+, O_2, pH	nitrification	distribution activity	62
aggregates	NO_3^-, NO_2^-, NH_4^+, O_2, pH	nitrification	distribution activity and microbial species	63
biofilm	NO_3^-, NH_4^+, O_2	nitrification, aerobic mineralization	distribution activity; effect of conditions	64
biofilm	NO_3^-, NO_2^-, NH_4^+, O_2, H_2S	denitrification, sulfate reduction, sulfide oxidation, aerobic mineralization	distribution activities; anaerobic sulfide oxidation	65
biofilm	NO_3^-, NO_2^-, NH_4^+, O_2, flow	nitrification, denitrification, flow, aerobic mineralization	distribution activity	35
biofilm	NO_3^-, NO_2^-, NH_4^+, O_2	nitrification, denitrification, flow, aerobic mineralization	distribution activity; distribution of microbial species; development of biofilm	63, 66
activated sludge	NO_3^-, NO_2^-, NH_4^+, O_2	nitrification, denitrification, aerobic mineralization	anoxic processes in aerated sludge	67
Beggiatoa mat	NO_3^-, O_2	denitrification	interfacial fluxes	59, 68, 69
sediment	NO_3^-, NH_4^+, O_2	nitrification, denitrification, aerobic mineralization	distribution activities	17, 30, 59, 69
sediment	NO_3^-, O_2	nitrification, denitrification, aerobic mineralization	regulation and distribution of activities	53, 70
sediment	NO_3^-, NH_4^+, O_2	nitrification, denitrification, aerobic mineralization, photosynthesis	distribution of activities; effect of photosynthesis of benthic algae on nitrification and denitrification	71
sediment	NO_3^-, NH_4^+, O_2	nitrification, denitrification	distribution activities; effect of plants roots on nitrification and denitrification	72
sediment	Ca^{2+}	calcium precipitation	effect of photosynthesis by benthic algae	73
Chara corallina	Ca^{2+}, pH	calcification	Ca^{2+}, H^+ uptake	74, 75

3.1 STUDIES ON THE N-CYCLE IN BIOFILMS

Most information on a process is obtained if all reactants are determined; therefore, results from a combination of sensors are most informative. Such a study was first performed on nitrifying aggregates from a fluidized bed reactor [62]. The reactor was fed with a mineral medium; therefore, nitrification was the main process (besides some heterotrophic growth on decaying nitrifiers). At that time the NO_2^- microsensor was not yet developed, so only NH_4^+, NO_3^-

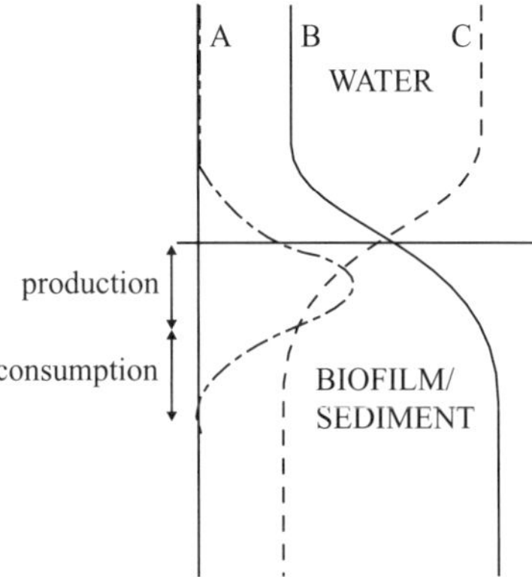

Figure 7. Schematic representation of substrate (C), end product (B) and intermediate product (A) profiles. The intermediate is produced in the top of the biofilm or sediment and consumed in the deeper zone. Such profiles can be found in nitrifying/denitrifying systems, where nitrate is formed in the upper oxic zone and consumed in the anoxic zone. The zones of production and consumption of the intermediate are indicated by arrows

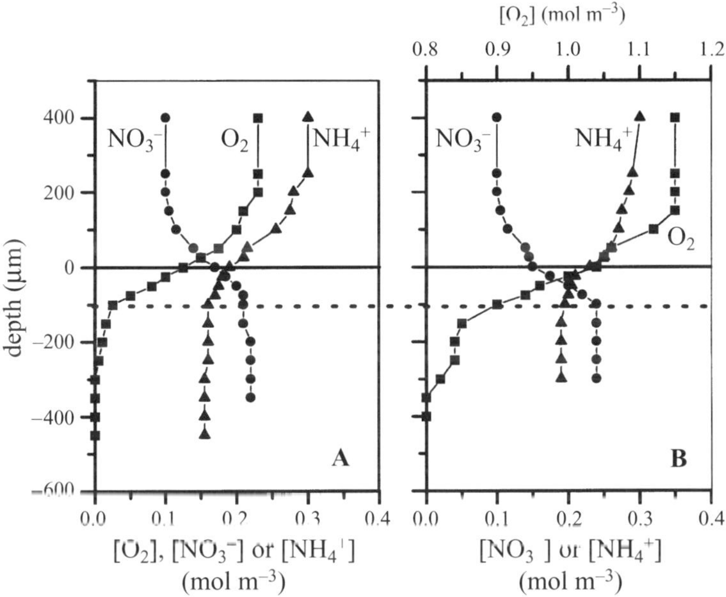

Figure 8. NH_4^+ (▲), NO_3^- (●) and O_2(■) profiles in a nitrifying aggregate under aeration (A) and during sparging with pure oxygen (B). Depth = 0 indicates the aggregate surface, negative values indicate positions inside the aggregate. The dotted line indicates the boundary of the nitrifying zone [62]

and O_2 profiles were measured (see Figure 8). If incubated under normally aerated conditions (A), O_2 penetrated the outer 100 μm of the aggregates, and NH_4^+, NO_3^- and O_2 fluxes were close to the stoichiometry of NH_4^+ oxidation to NO_3^-. At higher concentrations (B), O_2 penetrated the whole aggregate, but the NO_3^- and NH_4^+ profiles remained unchanged. It was concluded that the nitrifying organisms were predominantly located in the outer layer of 100 μm thickness, the center of the 2 mm sized aggregates was inactive. For the NO_3^- microsensors the Orion liquid ion exchanger (LIX) was used. In these measurements hydrogen carbonate interference could not become a problem as in this autotrophic system no hydrogen carbonate production occurred.

A later study on the same reactor showed similar NH_4^+, NO_3^- and O_2 profiles. Now also NO_2^- was measured, and found to be present in insignificant amounts ($< 10\,\mu mol\,L^{-1}$) under reactor conditions [63]. This study was combined with molecular analysis of the nitrifying population, to determine the dominant species and their distribution. Interestingly, no *Nitrobacter* or *Nitrosomonas* could be detected, the species most commonly found in nitrifying systems with microbial cultivation techniques. A more detailed analysis, by cloning and sequencing 16S rDNA, showed the presence of a consortium of new *Nitrospira* and *Nitrosospira* species. *In situ* hybridization with 16S rRNA probes, designed using the sequences found, showed that the two types of cells formed separate dense clusters that were in close contact with each other. It was hypothesized that *Nitrobacter* and *Nitrosomonas* strains are adapted to high substrate concentrations, which explains their occurrence in enrichment cultures. *Nitrospira* and *Nitrosospira* can compete under low NH_4^+ and NO_2^- concentrations as present in most natural environments. These species may well be more relevant nitrifiers than *Nitrobacter* and *Nitrosomonas*, and kinetic parameters derived from *Nitrobacter* and *Nitrosomonas* may not be useful to describe nitrification in nature and waste water treatment. Very recently activity profiles measured with microsensors were combined with specific cell counts by *in situ* hybridization, resulting in specific activities (i.e. conversion rates per cell). The results confirmed that *Nitrospira* and *Nitrosospira* have a much higher substrate affinity than *Nitrobacter* and *Nitrosomonas* [60].

Similar microsensor studies were performed on flat biofilms from a laboratory reactor [64]. Also these researchers showed that in a nitrifying biofilm most of the nitrifying activity was present in the outer 100–200 μm (see Figure 9). In this autotrophic biofilm no short term effects of glucose addition were observed, demonstrating the absence of a heterotrophic population. In a biofilm grown in a medium amended with organics both nitrification and aerobic mineralization of organics occurred. The NH_4^+ and NO_3^- profiles indicated that in this mixed heterotrophic–autotrophic biofilm nitrification was inhibited by addition of glucose, owing to competition for O_2 by the heterotrophic population. Strangely, no denitrification was observed, as NO_3^- concentrations in the anoxic zone were equal to that in the oxic zone. The NO_3^- measurements

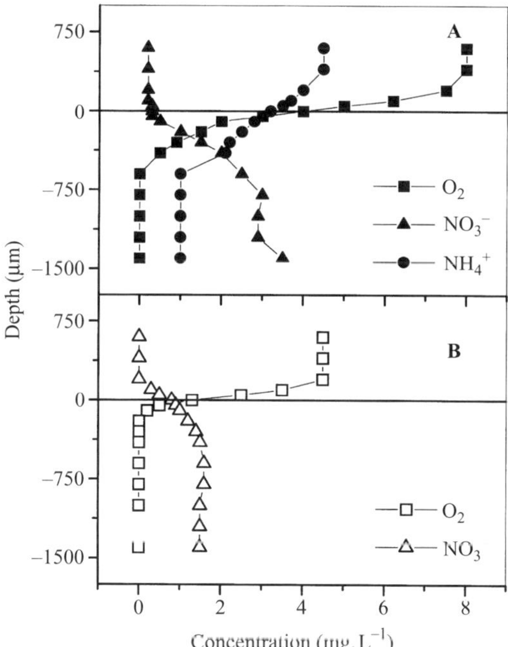

Figure 9. (A) NH_4^+, NO_3^- and O_2 profiles in a nitrifying biofilm without addition of glucose; (B) NO_3^- and O_2 profiles in the same biofilm, but with addition of glucose. Addition of glucose (80 mg L^{-1}) reduced the nitrification, probably because of competition for oxygen between heterotrophic and nitrifying populations. Glucose addition did not lead to denitrification, which would have resulted in nitrate decrease or depletion in the anoxic zone. Depth = 0 indicates the aggregate surface, negative values indicate positions outside the aggregate. Reproduced from ref. [64]

were performed with microsensors based on the Orion LIX. These researchers did not mention any technical problems with microsensors.

The effect of NO_3^- addition on sulfide profiles was investigated in a 2 cm thick biofilm obtained from the wall of a waste water plant (Figure 10) [35, 65]. In absence of NO_3^- the O_2 profile overlapped with the sulfide profile, indicating the aerobic oxidation of sulfide. Addition of nitrate resulted in a separation of the sulfide and O_2 profiles as shown previously [16], indicating anaerobic sulfide oxidation. NO_3^- (measured with an Orion LIX microsensor) penetrated the biofilm deeper (to 250 µm) than O_2 (to 150 µm) and overlapped with the sulfide profile. No HCO_3^- interference was noticed as measurements showed complete NO_3^- depletion. In the denitrifying zone a significant NO_2^- peak was observed, so that both NO_2^- and NO_3^- could be terminal electron acceptors for sulfide oxidation. In contrast with hydrogeno carbonate, sulfide did cause

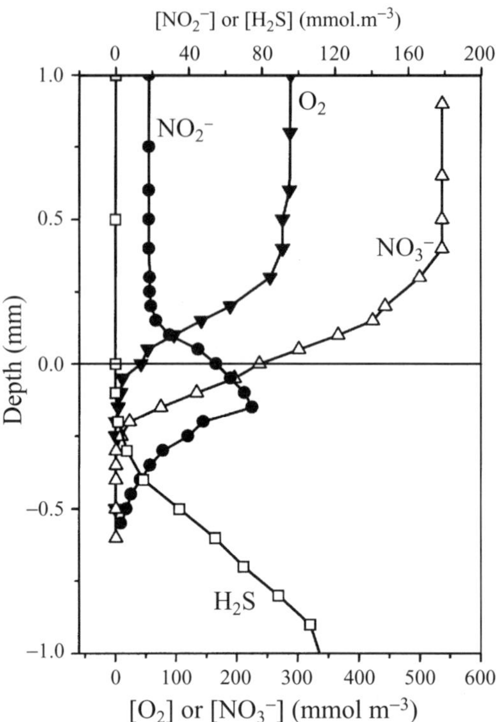

Figure 10. NO_3^-(Δ), NO_2^-(●), H_2S(□) and O_2 (▲) profiles in a thick biofilm from a wastewater treatment plant. In absence of NO_3^-, H_2S and O_2 profiles overlapped, indicating aerobic sulfide oxidation (not shown). After addition of nitrate the H_2S and O_2 profiles separated, indicative of anaerobic sulfide oxidation, with either NO_3^- or NO_2^- as e-acceptor

serious interference. At a concentration of more than 50 μmol L^{-1} sulfide, drift of both NO_3^- and NO_2^- sensors made further profiling impossible. However, meaningful profiles could be recorded as both NO_2^- and NO_3^- were depleted before the interference became a problem.

The studies mentioned above were performed on biofilms and aggregates placed in flowcells specially designed for microsensor studies, i.e. in samples removed from the reactor in which they were grown. With the shielded sensors *in situ* measurements can be done, inside an operating reactor. An example is a study on a biofilm from a membrane reactor, in which both nitrification and denitrification occurred simultaneously [35]. The biofilm was growing on a silicon tubing pressurized with 3 atm air. O_2 diffused through the silicon tube ca. 300 μm deep into the base of the 2.5 mm thick biofilm. NH_4^+ and organics were supplied from the bulk phase. NH_4^+ was consumed in the oxic zone, and converted to NO_3^-. In this nitrifying zone a peak of NO_2^- was observed. NO_2^- and NO_3^- were consumed in the anoxic zone, owing to denitrification. In this reactor the development of a nitrifying biofilm occurred in ca. 2 months. Nitrification was not complete and NO_2^- was the main product. After switching to heterotrophic conditions denitrification started within a few hours. Initially NO_2^- consumption was located in patches but after 2 weeks a homogeneous anoxic denitrifying layer covered the nitrifying base film [66].

3.2 STUDIES ON THE N-CYCLE IN SEDIMENTS

The first studies with liquid membrane sensors for NH_4^+ and NO_3^-, combined with O_2 profiles, were performed in sandy and silty sediments from a meso-eutrophic lake (see Figures 11 and 12) [17, 59]. In both types of sediment NH_4^+ oxidation occurred in the aerobic zone, and denitrification in the anaerobic zone. Conversion rates were calculated from the profiles, showing that in organic-rich, silty sediments both nitrification and denitrification were more intense than in sandy sediments that were low in organics. The NO_3^- measurements were performed with the Orion LIX, and no hydrogen carbonate interference was found. A problem in the profiling of sandy sediments was irregularities in the signal due to collisions with sand grains. Therefore, the profiles with liquid membrane sensors were recorded while retracting the microsensor out of the sediment.

At the surface of sulfide-rich sediments sulfur bacteria from the genus *Beggiatoa* often grow abundantly in white mats. Most *Beggiatoa* species cannot be cultivated, so little is known about their metabolic activities. A study on semi-purified mats was performed with NO_3^- microsensors [68]. *Beggiatoa* mats were

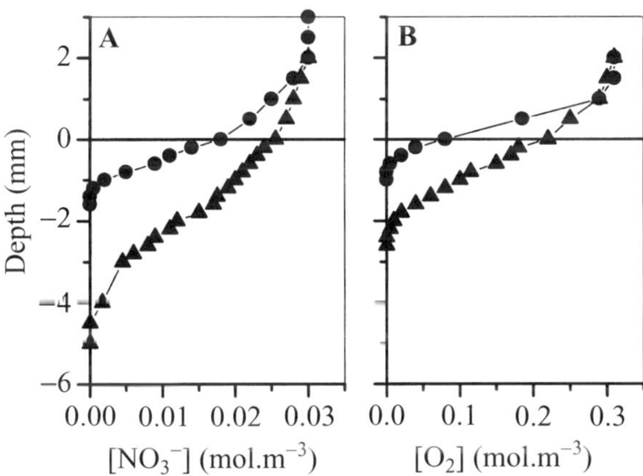

Figure 11. NO_3^- and O_2 profiles in sediments from a lake. Profiles in organic rich silt (●) and sand (▲). In both sediments, nitrate penetrates deeper than oxygen

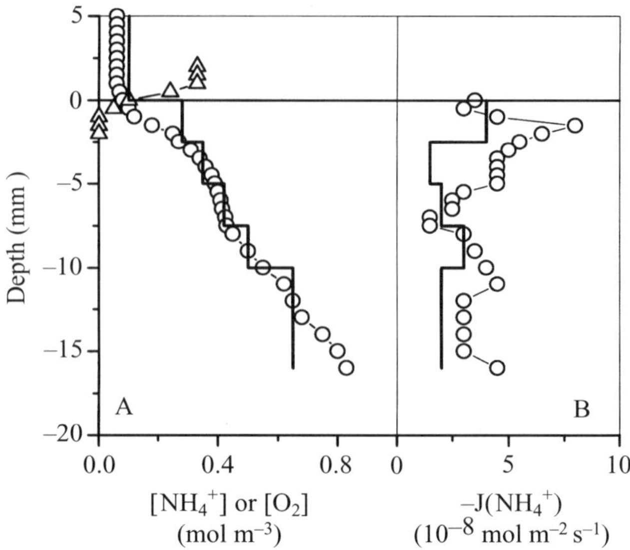

Figure 12. NH_4^+(○) and O_2(△) profiles (A), and the corresponding fluxes, *J* (B), in the same silty sediments as in Figure 11. The block diagrams indicate ammonium concentrations measured by porewater extraction

removed from a sediment surface and washed carefully with sterile water until no bacteria other than *Beggiatoa* were present. After a 1 h incubation in natural lake water these mats were profiled and found to consume nitrate at very high rates. Oxygen did not penetrate more than 10 μm in the mats. This led to the conclusion that these organisms are efficient denitrifiers and because of their abundance may have significant effects on the nitrogen budgets in aquatic systems. Recent insights make it much more likely that these organisms do not denitrify, but ammonify (conversion of NO_3^- to NH_4^+ instead of N_2).

Two later studies in model sediment, from which the burrowing animals were removed, were done with NO_3^- and O_2 microsensors [53, 70]. The effect of O_2 and NO_3^- concentration in the overlying water on the localization of nitrification and denitrification was determined, by fitting the NO_3^- profiles with a diffusion–reaction model. Also in this research denitrification was exclusively found in the anaerobic zone. At increasing O_2 concentrations, nitrification in the sediment became a more important NO_3^- source for denitrification than the bulk liquid, owing to increased path length for NO_3^- diffusion from the bulk to the anoxic zone. Secondly, at high O_2 concentrations nitrification was localized in the lower parts of the aerobic zones, adjacent to the anaerobic zones. It was hypothesized that NH_4^+ then became limiting for nitrification. Unfortunately, NH_4^+ was not measured in these studies. The serious problems due to hydrogen carbonate interference, as reported in these studies, can be attributed to the choice of LIX chemistry (Radiometer). These researchers introduced the shielding surrounding the liquid membrane sensor to protect the signal from noise. This was an important improvement, that made the use of a Faraday cage obsolete.

Photosynthesizing benthic algae can have a strong effect on the nitrogen cycle, as shown in a study with NH_4^+, NO_3^- and O_2 microsensors (Figure 13) [71]. In a freshwater sediment containing benthic algae, illumination increased the O_2 penetration depth. This stimulated nitrification, and thus the NH_4^+ profile developed a dip and NO_3^- formed a peak in the aerobic zone. NO_3^- diffused both downwards into the anaerobic zone, where it was used by denitrification, and upwards to the sediment surface where it was consumed, most likely by the benthic algae. Thus it was possible to visualize with microsensors part of the N-cycle in sediments: NH_4^+ was formed in the deeper sediments by mineralization, diffused upwards into the nitrification zone, was converted to NO_3^-, which was partially used in denitrification and partially incorporated in algal biomass, which after mineralization will again contribute to the NH_4^+ pool.

Recently it was shown, using a combination of ^{15}N isotope pairing and microsensor techniques (O_2, NO_3^- and NH_4^+), how plants enhance nitrification and denitrification activity in sediments [72]. Microprofiles showed that O_2

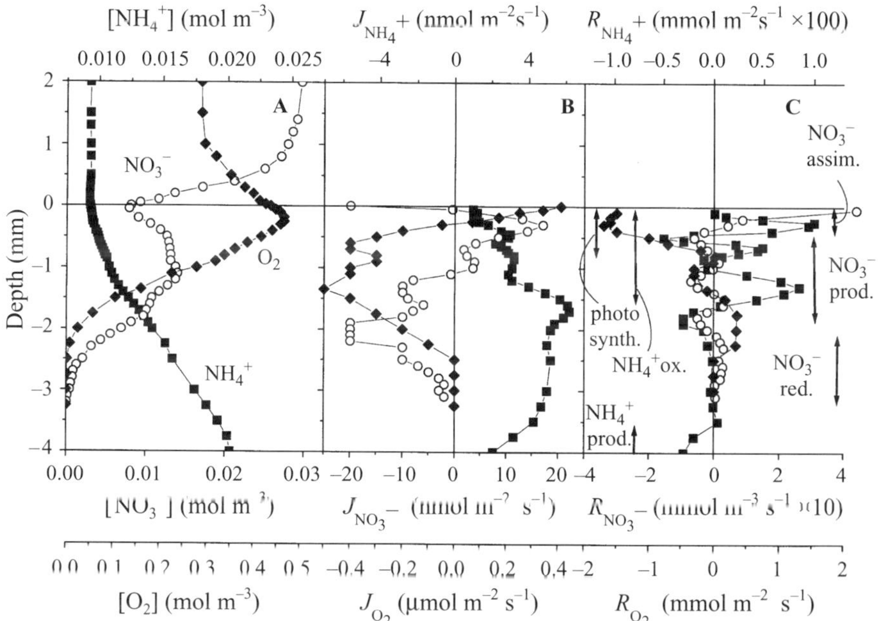

Figure 13. The distribution of NO_3^-(○), O_2(◆) and NH_4^+(■) in a lake sediment during light incubation. From the concentration profiles (A), the local fluxes, *J*, were calculated (B) and from these the local conversion rates, *R*, (C)

diffused out of the roots, and that in a nitrifying zone adjacent to the roots NH_4^+ was consumed and NO_3^- was formed. Illumination increased the O_2 content of the sediment and stimulated nitrification. The root-associated nitrifying zone was surrounded by zones of denitrification, leading to a close coupling between nitrification and denitrification. No problems with microsensors were reported in this study.

3.3 ECO-PHYSIOLOGICAL MEASUREMENTS WITH Ca^{2+} SENSORS

The Ca^{2+} sensor is widely used in the field of animal physiology. Since calcium is involved in a large number of biological and biogeochemical processes, it is surprising that this sensor has been used so rarely in aquatic studies.

Photosynthesis is an important regulating mechanism of calcification. The CO_2 fixation by photosynthesis and concomitant pH increase can lead to oversaturation of $CaCO_3$ and thus stimulate its precipitation. If precipitation rates equal photosynthesis no net CO_2 is consumed and the pH remains constant. Ca^{2+} microprofiles were recorded in a freshwater algal film [73] and Ca^{2+} fluxes were calculated from $0.5–1.3 \times 10^{-6}$ mol m^{-2} s^{-1}. In this study photosynthesis was not determined. Other studies [33,77] have reported photosynthesis rates of ca. 3×10^{-6} mol m^{-2} s^{-1} for sediments with benthic algae, which is in the same order of magnitude as the Ca^{2+} fluxes. More detailed measurements are needed to clarify the relation between photosynthesis, water chemistry and calcification in algal mats.

Elegant studies have been done on the giant cell alga *Chara corallina* [74,75]. This alga has a calcification to photosynthesis ratio of 1. Calcification is confined to distinct bands around the stem. Protons generated by calcification are transported to the cell surface in between calcifying bands. Locally the pH is lowered and the carbonate system shifted towards CO_2 which is taken up for photosynthesis. Microsensors for pH and Ca^{2+} were positioned at the cell surface and the cells were subjected to light–dark shifts. These measurements indicated that Ca^{2+} is excreted at calcifying sites in exchange for H^+, while Ca^{2+} is taken up in between calcifying bands. By this mechanism calcification is regulated by the alga, and can be fine-tuned to the photosynthesis, independent of the Ca^{2+} concentration in the water.

Due to the signal stability, the Ca^{2+} sensor can be used to detect small changes in environments with a large background concentration. For example, it is possible to measure concentration changes of the order of a few μmol L^{-1} in seawater, which has a concentration of ca. 8 mmol L^{-1}. In the sea, at depths of less than ca. 4000 m, calcium carbonate is oversaturated, and thus can precipitate. Marine organisms have to regulate the calcium precipitation carefully, in order not to get encrusted. Corals are probably the most spectacular calcifying systems. Despite this, the regulation of calcification in corals is poorly understood. With a combination of microsensors (H^+, O_2, CO_2 and Ca^{2+}) the

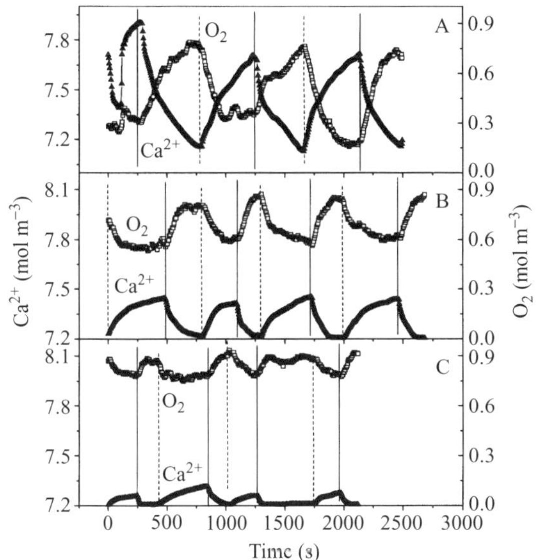

Figure 14. Ca^{2+}(▲) and O_2(□) dynamics at the surface of a coral induced by light–dark shifts, measured at 510 (A), 230 (B) and 70 (C) μmol photons $m^{-2} s^{-1}$. The simultaneous response, similar initial rates and amplitude of the dynamics indicate a close coupling

relation between calcification and photosynthesis in corals [78] and foraminifera [79] is currently being studied in our laboratory. In Figure 14 the effect of light–dark shifts in the O_2 and Ca^{2+} dynamics at the surface of coral tissue is shown. Since both the amplitude and the initial rates of the dynamics are similar, processes regulating O_2 and Ca^{2+} concentrations at the surface seem to be coupled. Indeed, a direct relation between calcium uptake and the activity of the photosynthetic apparatus could be demonstrated. Such a coupling could not be shown between the dynamics of CO_2 or H^+ and Ca^{2+}. Results of the measurements are promising and publications can be expected soon.

4 OUTLOOK

Although potentiometric analysis has been an established technique for decades, new developments also continue in this field. These developments sooner or later reach the level of microsensors, where these inventions can be used for eco-physiological, environmental and diagenetic studies. Interesting is the finding of new ionophores that may lead to improved or new sensors. Hopefully, the new nitrite ionophore will not be poisoned by sulfide. The reported finding of a sulfate ionophore may lead to new possibilities in detection of sulfate reduction.

Furthermore new detection techniques are under development for measurement of trace metals. Ion-sensitive electrodes have detection limits in the $\mu mol\,L^{-1}$ range, unsuitably high for trace metals. It was shown that this was determined by leaking of analyte ions from the internal electrolyte through the membrane [80]. By choosing an electrolyte that buffers the analyte ion at a low concentration, detection limits could be reduced down to the $pmol\,L^{-1}$ range [81]. This, together with the development of new highly selective ionophores for heavy metals [3,44] may open perspectives for the analysis of trace metals by potentiometric methods.

REFERENCES

1. Skoog, D. A. and Leary, J. (1992). *Principles of Instrumental Analysis*, 4th edn, Saunders College Publishing, Fort Worth.
2. Ammann, D., Bührer, T., Schefer, U., Müller, M. and Simon, W. (1987). Intracellular neutral carrier based Ca^{2+} microelectrode with sub-nanomolar detection limit, *Pflügers Arch.*, **409**, 223.
3. Pineros, M. A., Shaff, J. E. and Kochian, L. V. (1998). Development, characterization, and application of a cadmium-selective microelectrode for the measurement of cadmium fluxes in roots of *Thlaspi* species and wheat, *Plant Physiol.*, **116**, 1393.
4. Midorikawa, T., Tanoue, E. and Sugimura, Y. (1990). Determination of complexing ability of natural ligands in seawater for various metal ions using ion selective electrodes, *Anal. Chem.*, **62**, 1737.
5. Lake, D. L., Kirk, P. W. W. and Lester, J. N. (1989). Heavy metal solids association in sewage sludges, *Wat. Res.*, **23**, 285.
6. De Beer, D., Glud, A., Epping, E. and Kühl, M. (1997). A fast responding CO_2 micro-electrode for profiling sediments, microbial mats and biofilms, *Limnol. Oceanogr.*, **42**, 1590.
7. Cai, W. J. and Reimers, C. E. (1993). The development of pH and pCO_2 microelectrodes for studying the carbonate of pore waters near the sediment–water interface, *Limnol. Oceanogr.*, **38**, 1762.
8. Westcott, C. C. (1978). *pH Measurements*, Academic Press, San Diego.
9. Oesch, U., Ammann, D. and Simon, W. (1986). Ion-selective membrane electrodes for clinical use, *Clin. Chem.*, **32**, 1448.
10. Müller, M., Rouilly, M., Rustenholz, B., Maj-Zurawska, M. Z. H. and Simon, W. (1988). Magnesium selective electrodes for blood serum studies and water hardness measurement, *Mikrochim. Acta*, **3**, 283.
11. Kessler, M. (1976). *Ion and Enzyme Electrodes in Biology and Medicine*, Urban & Schwarzenberg, Munich.
12. Scholer, R. P. and Simon, W. (1970). Antibiotika-Membranelektrode zur selektiven Erfassung von Ammoniumionenaktivitäten, *Chimia*, **24**, 372.
13. Stepanek, R., Krautler, B., Schulthess, P., Lindemann, B., Ammann, D. and Simon, W. (1986). Aquocyanocobalt(III)-hepta(2–phenylethyl)-cobyrinate as a cationic carrier for nitrite-selective liquid-membrane electrodes, *Anal. Chim. Acta*, **182**, 83.

14. Zhen, R.-G., Smith, S. J. and Miller, A. J. (1992). A comparison of nitrate-selective microelectrodes made with different nitrate sensors and the measuremets of intracellular nitrate activities in cells of excised barley roots, *J. Exp. Botany*, **43**, 131.
15. Carey, C. M. and Riggan, W. B. (1994). Cyclic polyamine ionophore for use in a dibasic phosphate-selective electrode, *Anal. Chem.*, **66**, 3587.
16. Kühl, M. and Jørgensen, B. B. (1992). Microsensor measurements of sulfate reduction and sulfide oxidation in compact microbial communities of aerobic biofilms, *Appl. Environ. Microbiol.*, **58**, 1164.
17. De Beer, D., Sweerts, J.-P. R. A. and van den Heuvel, J. C. (1991). Microelectrode measurement of ammonium profiles in freshwater sediments, *FEMS Microbiol. Ecol.*, **86**, 1.
18. Bakker, E., Bühlmann, P. and Pretsch, E. (1997). Carrier based ion-selective electrodes and bulk optodes. 1. General characteristics, *Chem. Rev.*, **97**, 3083.
19. Ammann, D. (1986). *Ion-selective Microelectrodes: Principles, Design and Applications*, Springer, Berlin.
20. Albanese, R. A. (1973). On microelectrode distortion of tissue oxygen tensions, *J. Theor. Biol.*, **38**, 143.
21. Zhang, T. C. and Bishop, P. L. (1994). Evaluation of tortuosity factors and effective diffusivities in biofilms, *Wat. Res.*, **5**, 2279.
22. Muller, W., Winnefeld, A., Kohls, O., Scheper, T., Zimelda, W. and Baumgartl, H. (1994). Real and pseudo oxygen gradients in Ca-alginate beads monitored during polarographic pO_2–measurements using Pt-needle microelectrodes, *Biotechnol. Bioeng*, **44**, 617.
23. Glud, R. N., Gundersen, J. K., Revsbech, N. P. and Jørgensen, B. B. (1994). Effects on the benthic diffusive boundary layer imposed by microelectrodes, *Limnol. Oceanogr.*, **39**, 462.
24. Bungay, H. R., Whalen, W. J. and Sanders, W. M. (1969). Microprobe techniques for determining diffusivities and respiration rates in microbial slimes, *Biotechnol. and Bioeng.*, **11**, 765.
25. Revsbech, N. P. and Ward, D. M. (1983). Oxygen microelectrode that is insensitive to medium chemical composition: Use in an acid microbial mat dominated by *Cyanidium caldarum, Appl. Environ. Microbiol.*, **45**, 755.
26. Revsbech, N. P. (1989). An oxygen microelectrode with a guard cathode, *Limnol. Oceanogr.*, **55**, 1907.
27. Revsbech, N. P., Nielsen, L. P., Christensen, P. B. and Sorensen, J. (1988). A combined oxygen and nitrous oxide microsensor for denitrification studies, *Appl. Environ. Microbiol.*, **45**, 2245.
28. Hinke, J. (1969). Glass microelectrodes for the study of binding and compartmentalisation of intracellular ions. In *Glass Microelectrodes*, ed. Lavallee, M., Schanne, O. F. and Herbert, N. C., John Wiley & Sons, New York, p. 349.
29. De Beer, D. and Van den Heuvel, J. C. (1988). Response of ammonium-selective microelectrodes based on the neutral carrier nonactin, *Talanta*, **35**, 728.
30. De Beer, D. and Sweerts, J. P. R. A. (1989). Measurements of nitrate gradients with an ion-selective microelectrode, *Anal. Chim. Acta*, **219**, 351.
31. Larsen, L. H., Revsbech, N. and Binnerup, S. J. (1996). A microsensor for nitrate based on immobilized denitrifying bacteria, *Appl. Environ. Microbiol.*, **62**, 148.
32. Larsen, L. H., Kjaer, T. and Revsbech, N. P. (1997). A microscale NO_3^- biosensor for environmental applications, *Anal. Chem.*, **69**, 3527.
33. Revsbech, N. P., Jørgensen, B. B., Blackburn, T. H. and Cohen, Y. (1983). Microelectrode studies of the photosynthesis and O_2, H_2S and pH profiles of a microbial mat, *Limnol. Oceanogr.*, **28**, 1062.

34. Jeroschewski, P., Steukart, C. and Kühl, M. (1996). An amperometric microsensor for the determination of H_2S in aquatic environments, *Anal. Chem.*, **68**, 4351.
35. De Beer, D., Schramm, A., Santegoeds, C. M. and Kühl, M. (1997). A nitrite microsensor for profiling environmental biofilms, *Appl. Environ. Microbiol.*, **63**, 973.
36. Damgaard, L. R. and Revsbech, N. P. (1997). A microscale biosensor for methane, *Anal. Chem.*, **69**, 2262.
37. Tercier, M. L., Parthasarathy, N. and Buffle, J. (1995). Reproducible, reliable and rugged Hg-plated Ir-based microelectrode for in situ measurement in natural waters, *Electroanalysis*, **7**, 55.
38. Awasthi, S. P. (1990). *Ion-selective Electrodes—a Review*, World Scientific, Singapore.
39. Thomas, R. C. (1978). *Ion-sensitive Intracellular Microelectrodes, How to Make and Use Them*, Academic Press, London.
40. Revsbech, N. P. and Jørgensen, B. B. (1986). Microelectrodes: their use in microbial ecology, *Adv.Microbial Ecol.*, **9**, 293.
41. Stefanac, Z. and Simon, W. (1967). Ion specific electrochemical behaviour of macrotetrolides in membranes, *Microchem. J.*, **12**, 125.
42. Walker, J. L. (1971). Ion specific liquid ion exchanger microelectrodes, *Anal. Chem.*, **43**, 89.
43. Morf, W. E. and Simon, W. (1978). *Ion-selective Electrodes Based on Neutral Carriers*, Plenum, New York.
44. Bühlmann, P., Pretsch, E. and Bakker, E. (1998). Carrier based ion-selective electrodes and bulk optodes, *Chem. Rev.*, **98**, 1593.
45. Schaller, U., Bakker, E., Spichiger, U. and Pretsch, E. (1994). Nitrite selective microelectrodes, *Talanta*, **41**, 1001.
46. Schulthess, P., Shijo, Y., Pham, H. V., Pretsch, E., Ammann, D. and Simon, W. (1981). A hydrogen ion-selective liquid-membrane electrode based on tri-n-dodecylamine as neutral carrier, *Anal. Chim. Acta*, **131**, 111.
47. Khuri, R. N., Bogharian, K. K. and Agulian, S. K. (1974). Intracellular bicarbonate in single skeletal muscle fibers, *Pflügers Arch.*, **349**, 285.
48. Munoz, J. L., Deymhimi, F. and Coles, J. A. (1983). Silanization of glass in the making of ion-selective microelectrodes:, *J. Neurosci. Methods*, **8**, 231.
49. Ammann, D., Lanter, F., Steiner, R. A., Schulthess, P., Shio, Y. and Simon, W. (1981). Neutral carrier based hydrogen-ion selective microsensor for extra- and intracellular studies, *Anal. Chem.*, **53**, 2267.
50. Chao, P., Ammann, D., Oesch, U., Simon, W. and Lang, F. (1988). Extracellular and intracellular hydrogen ion-selective microelectrode based on neutral carriers, *Pflügers Arch.*, **411**, 216.
51. Tsien, R. Y. and Rink, T. J. (1981). Ca^{2+}-selective electrodes: a novel PVC-gelled neutral carrier mixture compared with other currently available sensors, *J. Neurosci. Methods*, **4**, 73.
52. Buhrer, T., Peter, H. and Simon, W. (1988). Ammonium ion-selective microelectrode based on the antibiotics nonactin-monactin, *Pflügers Arch.*, **412**, 359.
53. Jensen, K., Revsbech, N. P. and Nielsen, L. P. (1993). Microscale distribution of nitrification activity in sediment determined with a shielded microsensor for nitrate, *Appl. Environ. Microbiol.*, **59**, 3287.
54. Koryta, J. and Stulik, K. (1983). *Ion-selective Electrodes*, 2nd edn, Cambridge University Press, Cambridge.
55. Verschuren, P. G., van der Baan, J. L., de Beer, D. and van den Heuvel, J. C. (1998). A nitrate-selective microelectrode based on a lipophilic derivative of iodocobalt (III) (salen), *Fres. J. Anal. Chem.*, **363**, 595–598.

56. Rudkevich, D. M., Stauthamer, W. P. R. V., Verboom, W., Engbersen, J. F. J., Harkema, S. and Reinhoudt, D. N. (1992). UO_2–salenes: neutral receptors for anions with a high selectivity for dihydrogen phosphate, *J. Am. Chem. Soc.*, **114**, 9671.
57. Revsbech, N. P. (1994). Analysis of microbial mats by use of electrochemical microsensors: recent advances. In *Microbial Mats*, ed. Stal, L. and Caumette, P., Springer, Berlin, Vol. **35**, p. 135.
58. Damgaard, L. R., Larsen, L. H. and Revsbech, N. P. (1995). Microscale biosensors for environmental monitoring, *Trends Anal. Chem.*, **14**, 300.
59. Sweerts, J.-P. R. A. and de Beer, D. (1989). Microelectrode measurements of nitrate gradients in the littoral and profundal sediments of a meso-eutrophic lake (Lake Vechten, The Netherlands), *Appl. Environ. Microbiol.*, **55**, 754.
60. Ross, J. W. (1969). Anion selective microelectrode. In *United States Patent Office*, Orion Research Incorporated, Cambridge, MA, USA, p. 6.
61. Meyer, T., Wensel, T. and Stryer, L. (1990). Kinetics of calcium channel opening by inositol 1,4,5-triphosphate, *Biochemistry*, **29**, 32.
62. De Beer, D., van den Heuvel, J. C. and Ottengraf, S. P. P. (1993). Microelectrode measurements of the activity distribution in nitrifying bacterial aggregates, *Appl. Environ. Microbiol.*, **59**, 573.
63. Schramm, A., de Beer, D., Wagner, M. and Amann, R. (1998). *Nitrosospira* and *Nitrospira* sp. as dominant populations in a nitrifying fluidized bed reactor: identification and activity in situ. *Appl. Environ. Microbiol.*, **64**, 3480.
64. Zhang, T. C., Fu, Y. C. and Bishop, P. L. (1994). Competition in biofilms. *Wat. Environ. Res.*, **67**, 992.
65. Santegoeds, C. M., Muyzer, G. and de Beer, D. (1998). Biofilm dynamics studied with microsensors and molecular techniques, *Wat. Sci. Technol.*, **3**, 125.
66. Schramm, A. (1998). In situ structure and function analysis of nitrifying/denitrifying biofilms, Bremen University.
67. de Beer, D., Schramm, A., Santegoeds, C. M. and Nielsen, H. K. (1998). Anaerobic processes in activated sludge, *Wat. Sci. Technol.*, **37**, 605.
68. Sweerts, J.-P. R. A., de Beer, D., Nielsen, L. P., Verdouw, H., van den Heuvel, J. C., Cohen, Y. and Cappenberg, T. E. (1990). Denitrification by sulphur oxidizing *Beggiatoa* spp. mats on freshwater sediments, *Nature*, **344**, 762.
69. Sweerts, J.-P. R. A., Bar-Gillisen, M. J., Cornelise, A. A. and Cappenberg, T. E. (1991). Oxygen consuming processes at the profundal and littoral sediment-water interface of a small meso-eutrophic lake (Lake Vechten, the Netherlands), *Limnol. Oceanogr.*, **36**, 1124.
70. Jensen, K., Sloth, N. P., Risgaard-Petersen, N., Rysgaard, S. and Revsbech, N. P. (1994). Estimation of nitrification and denitrification from microprofiles of oxygen and nitrate in model sediment systems, *Appl. Environ. Microbiol.*, **60**, 2064.
71. De Beer, D. (1998). Use of microelectrodes to measure in situ microbial activities in biofilms, sediments and microbial mats. In *Molecular Microbial Ecology Manual*, ed. Akkermans, A. D. L., van Elsas, J. D. and de Bruyn, F. J., Kluwer, Dordrecht, Vol. **8.1.3**, in press.
72. Risgaard-Petersen, N. and Jensen, K. (1997). Nitrification and denitrification in the rhizosphere of the aquatic macrophyte *Lobelia dortmanna* L., *Limnol. Oceanogr.*, **42**, 529.
73. Hartley, A. M., House, W. A., Leadbeater, B. S. C. and Callow, M. E. (1996). The use of microelectrodes to study the precipitation of calcite upon algal biofilms, *J. Colloid Interface Sci.*, **183**, 498.

74. McConnaughey, T., D. and Falk, R. H. (1991). Calcium proton exchange during algal calcification, *Biol. Bull.*, **180**, 185.
75. McConnaughey, T. D. (1991). Calcification in *Chara corallina*: CO_2 hydroxylation generates protons for bicarbonate assimilation, *Limnol. Oceanogr.*, **36**, 619.
76. Revsbech, N. P., Madsen, B. and Jørgensen, B. B. (1986). Oxygen production and consumption in sediments determined at high spatial resolution by computer simulation of oxygen microelectrode data, *Limnol. Oceanogr.*, **31**, 293.
77. Revsbech, N. P., Jørgensen, B. B. and Brix, O. (1981). Primary production of microalgae in sediments measured by oxygen microprofile, $H^{14}CO_3^-$ fixation, and oxygen exchange methods, *Limnol.Oceanogr.*, **26**, 717.
78. de Beer, D., Kühl, M., Stambler, N., and Vaki, L. (2000). A microsensor study of light enhanced CA^{2+} uptake and photosynthesis in the reef-building hermatynic coral *Favin sp. Mul. Ecol. Ploy. Series* **194**, 75
79 Köhler-Rink, S., and Kühl, M., (2000). Microsensor studies of photosynthesis and respiration in larger symbiotic foraminifera. I The physico-chemical microenvironment of *Marginopora vertebralis, Amphistegnia lobifera and Amphisoris hemprechii. Marine Biol.* In press.
80. Mathison, S. and Bakker, E. (1998). Effect of transmembrane electrolyte diffusion on the detection limit of carrier based potentiometric ion sensors, *Anal. Chem.*, **70**, 303.
81. Sokaliski, T., Ceresa, A., Zwickl, T. and Pretsch, E. (1997). Large improvement of the lower detection limit of ion-selective polymer membrane electrodes, *J. Am. Chem. Soc.*, **119**, 11347.

6 Biosensors for Analysis of Water, Sludge and Sediments with Emphasis on Microscale Biosensors

N. P. REVSBECH, T. KJÆR, L. R. DAMGAARD, AND J. LORENZEN
University of Aarhus, Denmark

L. H. LARSEN
Unisense ApS, Aarhus, Denmark

In Situ Monitoring of Aquatic Systems: Chemical Analysis and Speciation Edited by J. Buffle and G. Horvai.

1 INTRODUCTION

1.1 PRESENT AVAILABILITY OF SENSORS FOR LONG-TERM ENVIRONMENTAL MONITORING

Large resources have been allocated to the development of sensors for environmental monitoring. For some areas the work has been very successful, such as the development of sensors for automobile exhaust. For other areas the success with invention of sensors that have resulted in extensive use has been more limited. Currently the only sensors for analysis of chemical parameters in natural aquatic environments that are commercially available and have resulted in widespread use are electrochemical and optical O_2 and pH sensors as described elsewhere (Chapters 2 and 3) in this volume. It is apparent from the content of this volume that sensors may be used for many other analyses in the aquatic environment, but none of these sensors have until now gained widespread use, mostly because of unsatisfactory long-term stability, but also often because of the presence of interfering species in the natural environment. For scientific use many of these sensors are, however, highly interesting, as they can be used to collect essential information about the natural environment that could not be obtained by other means. In addition to the 'real' sensors, miniaturized analytical systems for continuous and long-term *in situ* use are under development (Chapter 12 this volume) [1,2,3], and UV absorption [4] or UV-caused fluorescence [5] are also used extensively.

1.2 ADVANTAGES OF MICROSCALE SENSOR DESIGN

The development of microscale sensors has been a key research field for the authors, and our description of environmentally relevant biosensors in this chapter is therefore heavily biased towards microscale biosensors. We thus do not attempt to give a full description of all types of environmentally relevant biosensors, but a brief overview of the various biosensor types is presented. We started to develop microscale sensors as we—being microbial ecologists—needed such tools to elucidate the chemistry and the transformation rates at a scale relevant to the world of bacteria. It turned out, however, that by making sensors small we gained more than just increased spatial resolution of our measurements. First of all the signals from microscale sensors are only marginally affected by changes in flow rate or diffusional characteristics of the medium. The small-scale spherical diffusion field around microsensor tips described in detail in Chapters 8 and 9 can explain this. Microscale sensors may also be characterized by very rapid responses to changes in analyte concentration. According to the Einstein–Smoluchowski equation [6] diffusion can be described by $l = (2Dt)^{1/2}$, where l is diffusional path length covered during time t and D is the diffusion coefficient. According to this equation, a small molecule such as oxygen ($D = 2 \times 10^{-9}\,m^2 s^{-1}$ in water) will, on average, migrate about 0.06 mm in 1 s and 0.6 mm in 100 s. Sensors with external or internal concentration gradients extending < 0.1 mm will thus exhibit 90% response times of a few seconds or less, while sensors with longer diffusion distances may be slow. Examples of actual response times of small sensors are given for the nitrate biosensor treated below. Finally the small size of microsensors makes it possible to use completely new principles of detection as an effective supply of reactants can be mediated by diffusion from internal reservoirs. Both the NO_3^- and the CH_4 biosensors described below are thus dependent on such efficient supply of reactants from internal reservoirs. The external dimensions of a sensor do, however, not need to be microscale even when the sensing elements inside the sensor are. Clark-type O_2 sensors may thus be made with a 1 mm tip consisting of almost solid glass, but with a membrane-filled pore of a few micrometers in diameter in the center and containing a cathode also being a few micrometers in diameter [7]. Such a sensor will still have the desired characteristics of low stirring sensitivity and rapid response, although the external diameter has been increased to improve physical sturdiness.

1.3 DEFINITION OF A BIOSENSOR

Much of the effort in sensor technology has been devoted to the development of biosensors, and a substantial part of this effort has been on the development of sensors for environmental use. The interest for biologically mediated reactions is, of course, caused by the extreme specificity of enzymatic and immunological reactions. The term 'biosensor' has been used for any biological component used to sense any parameter. Usually this biological component is attached to some piece of hardware such as an electrochemical detection system or an optical fiber, but the term biosensor has, as an example, also been used about genetically modified bacteria that emit light or show some other measurable gene expression when exposed to some environmental variable. The principles of a 'classical' biosensor can be exemplified by the glucose biosensor [8], where a platinum anode is coated with a layer of the enzyme glucose oxidase. The glucose oxidase catalyzes the reaction between glucose and O_2, whereby hydrogen peroxide is formed. This hydrogen peroxide is oxidized at the anode, and the current generated by this oxidation is then a measure for the glucose concentration in the medium. The enzyme may be shielded from the sample by some kind of semi-permeable membrane.

1.4 (MACROSCALE) BIOSENSORS OF RELEVANCE FOR ENVIRONMENTAL MONITORING

Although large resources have been allocated to biosensor development, the only biosensor (not counting immunological disposable strips for xenobiotics) regularly used for monitoring of aquatic environments is the so-called BOD (biological oxygen demand) sensor [9], which gives semi-quantitative estimates of the content of dissolved, easily degradable organic matter in waste water. It is basically just a layer of immobilized heterotrophic microorganisms placed in front of a conventional O_2 sensor. Because of the respiratory activity of the microorganisms, governed by the concentration of dissolved organics in the analyzed medium, they regulate the amount of O_2 reaching the O_2 sensor by diffusion from the medium. For interpretation of the signal from this sensor it is thus critical to know the O_2 concentration in the medium. It is apparent from the BOD sensor review of Praet *et al.* [10] that the signal from a BOD sensor may be quite difficult to calibrate to some well-defined expression of dissolved organic concentration, as the responses to different dissolved organic species differ, and the response to each individual species is furthermore dependent on the recent history of the sensor. Particulate organic matter is not included in the reading. BOD sensors are usually exposed to a pulse of the water to be analyzed which results in a peak in respiratory activity (i.e. lower reading by the O_2 sensor), and after this peak a considerable period of time is needed to approach some kind of baseline respiration. The lifetime of a macroscopic BOD sensor may be up to months [9] without replacement of the immobilized microbial cells. The same design as that of the BOD sensor has been used extensively with other microorganisms for detection of a wide range of chemical species [10], but apparently the characteristics of these sensors have not allowed for extensive practical use.

The field of biosensors for environmental use was recently reviewed in a special issue of *Trends in Analytical Chemistry* [11], and Table 1 is largely a summary of this issue. A lot of work has been devoted to the development of enzyme-based sensors for pesticides [16] or other pollutants such as phenolic compounds. [19] The pesticide sensors may be based on enzyme inhibition, where some enzyme-catalyzed reaction is inhibited owing to the presence of a pollutant, and they therefore cannot be used for real *in situ* monitoring, as the

Table 1. Examples of biosensors that could possibly be used for environmental monitoring (cons. = consumption)

Chemical parameter	Bio-component	Detection principle	Lifetime	Analytical range (mol L^{-1})	Interferences / comments	Continuous operation	References
Ammonium	Bacteria	Oxygen cons.	?	?	?	+	12, 13
BOD	Bacteria/yeasts	Oxygen cons.	months	Ill-defined	Ill-defined	–	9, 10
CO_2	Bacteria	Oxygen cons.	1 month	3–12% sat.	None	+	14, 15
Herbicides	Algae	Photosyn. inh.	?	Very variable	?	–	16
Methane	Bacteria	Oxygen cons.	months	10^{-6}–10^{-3}	Sulfide	+	17
Nitrite	Bacteria	Oxygen cons.	?	?	?	+	12
Nitrate + nitrite	Bacteria	N_2O evol.	> 2 months	10^{-7}–10^{-2}	N_2O	+	18
Pesticides	Enzymes	Enzyme inh.	?	$\geqslant 10^{-8}$	?	–	16
Phenols	Enzyme(s)	Oxygen cons.	?	$\geqslant 10^{-8}$	Pre-treatment is necessary	–	19
Surfactants	Bacteria	Oxygen cons.	?	?	?	–	20
Toxicity	Bacteria	Respiration inh.	?	?	?	+	13, 21
Xenobiotics	Antibody	Many detection principles	?	Very variable	Highly specific	–	22

primary substrates for the reaction must be added to the test medium. Immunological reactions have also been used for analysis of such xenobiotic compounds [22]. It has, however, turned out to be extremely difficult to make reversible enzyme or immunological sensors that can be used for direct long-term monitoring of natural aquatic environments, either because the sensing reaction itself results in inactivation or because humic substances etc. interfere or gradually inactivate the sensor. The use of enzyme or antibody preparations for use in sensors for environmental analysis thus seems to be highly problematic when long-term stable biosensors are essential, but more sophisticated approaches where highly selective membranes protect the biological components [23] may in the future result in more stable enzyme sensors. The glucose biosensor has at present, to our knowledge, the best long-term stability among the enzyme-based sensors [24], as glucose oxidase is extremely stable. A microscale version of this sensor has also been described [8]. Glucose is, however, not an important freely dissolved chemical species of most natural environments. It should be stressed that enzyme sensors may be used for detection of many environmentally relevant chemical species if long-term stability is of minor importance (Table 1), especially if pretreatment of the sample to remove interfering agents, to adjust pH, and to add essential chemicals for the reaction (and also often to extract the analyte) is possible.

By using actively growing microorganisms instead of enzyme preparations the essential enzyme(s) can be efficiently shielded against inactivation, and the pool of enzymes is also continuously replenished by growth of the microorganisms. It should be stressed that the idea of using bacteria or eucaryotic cells instead of enzymes in biosensors was invented long ago. Examples of such whole cell biosensors are the BOD sensor and similar designs sensing, among other things, SO_3^{2-}, NO_2^-, CO_2, and NH_4^+ [12]. Most whole-cell biosensors used until now are almost identical to the BOD sensor in terms of design, but any microbiological reaction giving rise to a measurable product or consuming a measurable reactant within the sensor may form the basis of a biosensor. Compared with sensors based on purified enzyme preparations, whole cell biosensors may seem less specific, as whole cells contain a large number of different enzymes, but many bacteria are lithoautotrophic (cannot assimilate or oxidize organic matter) and may rely on oxidation of only one or a few inorganic species while assimilating CO_2. The O_2 consumption inside sensors based on these organisms can thus be a measure of the concentration of such inorganic species. Others use oxidized inorganic compounds as electron acceptors and reduce these to chemical species, which can be detected. An example of this is the NO_2^- / NO_3^- biosensor described below where NO_2^- and NO_3^- are reduced to N_2O. The final option with whole cells is not to use the enzymes themselves but rather gene expression (see section about bioluminescence below) as a measure of the inducer concentration. Such gene expression is often extremely specific for the chemical species in question. While metabolic activity as such will only work in sensors specific to a few chemical species, gene expression may be used for virtually all types of both inorganic and organic species. As outlined below gene expression is, however, difficult to handle owing to long response times and irreversibility.

2 MICROSCALE WHOLE CELL BIOSENSORS—GENERAL PRINCIPLES

It is possible to use bacteria or yeasts in macroscale biosensors as mentioned in section 1. However, most macroscale designs impose some fundamental restrictions on the nutrient supply to the microorganisms, as all necessary chemical species must be supplied through the membrane tip; otherwise, the sensor no longer acts as a real sensor but rather as a microbiological assay in a stirred liquid medium [25]. These limitations may be overcome by making the sensor tip very small and conical. Then, all necessary growth components, except for the one to be quantified by the sensor, can be readily supplied to the microorganisms in the tip by diffusion from an internal reservoir (Figure 1). The microsensor then works very much like a continuous culture vessel where the growth-limiting factor is the supply of an electron donor or acceptor through the tip membrane. The tip membrane may be more or less restrictive in terms of permeability, one example being a silicone membrane only allowing relatively small, uncharged molecules to pass, another example being ion-permeable membranes made from cellulose acetate or other materials.

The design illustrated in Figure 1 makes it theoretically possible to make biosensors with an extremely long lifetime and stability, as the flux through the tip membrane and thus also the need for nutrients from the internal reservoir is extremely small. Often biosensors made this way have a 'reaction chamber' in the tip which is only 10^{-7} times the volume of the 1 mL large medium reservoir,

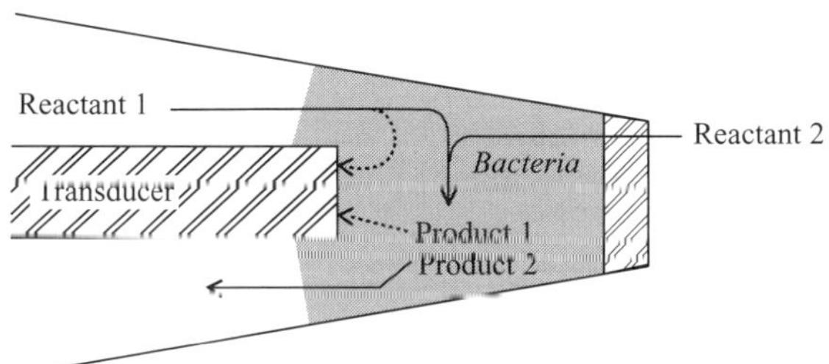

Figure 1. Microscale continuous-culture principle of microscale biosensors based on living microorganisms. The growth of microorganisms is limited to the tip region as one or more growth-limiting chemical species enter through the tip membrane

and the reservoir may last for years. Another advantage is that the volume of bacteria mediating the reaction is extremely well defined, as the relative positions of the tip membrane and the internal sensing element are fixed. The metabolic status of the bacteria may, of course, vary as a result of their recent life history, but compared with macroscale analog the metabolic status is also more defined as only actively growing microorganisms are found in the tip, and excess microbial biomass is squeezed behind the tip where the microorganisms eventually perish. For the $NO_2^- + NO_3^-$ and CH_4 biosensors described in this chapter, the metabolic status of the microorganisms is actually irrelevant as long as the total population between the tip membrane and the sensing element is sufficient to mediate a full conversion. The length of the reaction chamber essentially governs the response time of the sensor, and it should thus be kept as short as possible. Not all microbial reactions are equally rapid, however, and the length needed for a full conversion of chemical species entering through the tip may thus vary. For the CH_4 and $NO_2^- + NO_3^-$ biosensors described below, the reaction chambers are 100–300 μm long.

One fundamental advantage of the sensor design shown in Figure 1 is that it is possible to change the supply of ions through sensors equipped with ion-permeable membranes by applying a voltage between environment and medium reservoir. For the $NO_2^- + NO_3^-$ biosensors it is thereby possible to increase the sensitivity by a factor of more than 10, and it is then possible to analyze $NO_2^- + NO_3^-$ down to 0.1 $\mu mol\,L^{-1}$ [26]. The use of such electrophoretic migration of ions has, to our knowledge, not been used successfully in macroscopic sensors, as it is essential to have a well-defined region where the electrical resistance of the sensor is located (i.e. our microsensor tip), and both internal electrolyte reservoir and electrode capacities must be large to minimize the effects of polarization.

The design shown in Figure 1 is not limited to analyses of growth limiting substrates entering the microsensor through the tip membrane. A fundamental condition for its performance is that growth is restricted to the tip, and at least one growth limiting substance must therefore enter through the tip. However, this growth limiting substance, typically O_2, may serve only to keep a microbial population dense and active in the tip, and a signal to be detected may then originate from a minor constituent also entering through the sensor tip. Such a signal could typically be bioluminescence, and bacteria emitting light by exposure to a large variety of environmental parameters have been engineered (e.g. ref. [46]) The approximately 1 h response time of bioluminescence, which is usually coupled to gene expression, has, however, limited the applicability of such techniques, but even with slow response there is no doubt that bioluminescence-based sensors will find widespread application in the future. Microorganisms may be engineered to emit light when exposed to various xenobiotics or to heavy metals, and sensors containing such bacteria may serve as warning systems for industrial effluents etc.

3 MICROSCALE BOD BIOSENSOR

A simple microscale, whole cell BOD sensor (Figure 2) not utilizing the continuous culture principle illustrated in Figure 1 has been described by Neudörfer and Meyer-Reil [27]. It is based on an O_2 microsensor situated behind poly(vinyl alcohol) immobilized yeast cells (*Rhodotorula mucilaginosa*). The glass surfaces were treated with a silane to facilitate adhesion of the polyvinyl alcohol to the glass. The sensor was shown to respond to glucose concentrations in a reproducible manner (Figure 3), and was subsequently used to quantify depth profiles of dissolved organic matter in a sediment. As mentioned in section 1, BOD sensors are difficult to calibrate to some well-defined expression of organic matter concentration as the response to different organic compounds differ.

A microscale design of a BOD sensor as shown in Figure 2 may result in faster response than those of macroscale analogs, but it is also clear that a design without a tip membrane will result in a limited lifetime (days) owing to growth of contaminating microorganisms in the tip. The greatest advantage of a microscale BOD sensor is the ability to measure the microdistribution of dissolved organic matter in sediments and similar stratified communities, but unfortunately this ability is restricted to the oxic zone where such concentrations are

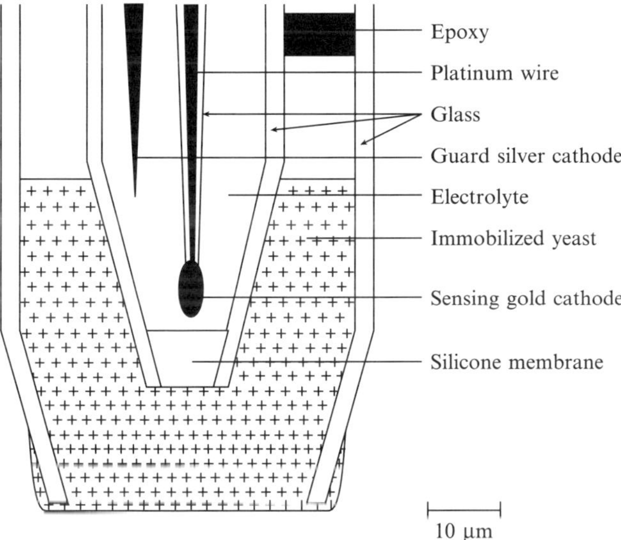

Figure 2. Microscale BOD (biological oxygen demand) sensor based on immobilized yeast cells. The respiration of the yeast cells limits the amount of O_2 reaching the internal O_2 microsensor, and the respiration is governed by the concentration of dissolved organic matter in the surrounding medium. Redrawn from Neudörfer and Meyer-Reil [27] by permission of Inter-Research

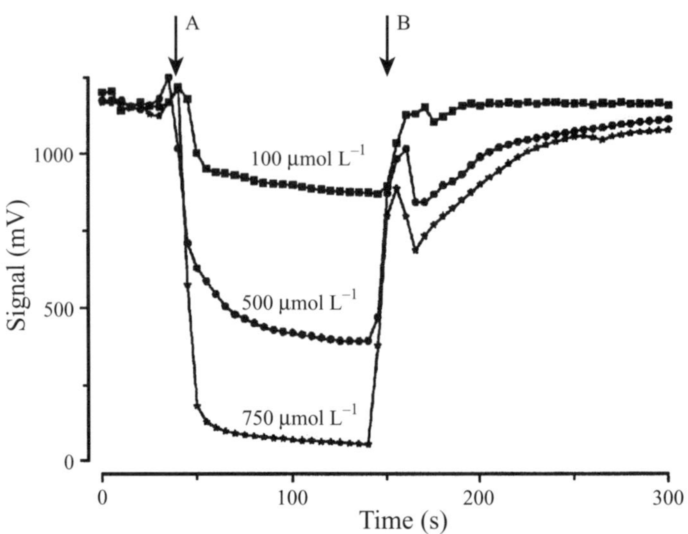

Figure 3. Response to various glucose concentrations of the microscale BOD sensor shown in Figure 2. The sensor was exposed to glucose at 'A' and reintroduced into water without dissolved organics at 'B'. Exposure to: ■, 100 μmol L^{-1}; ●, 500 μmol L^{-1}; *, 750 μmol L^{-1}. Redrawn from Neudörfer and Meyer-Reil [27].

low. The reading from the sensor is dependent on the concentration of O_2, so parallel readings of O_2 must always be performed.

4 CONTINUOUS CULTURE, WHOLE CELL MICROSCALE BIOSENSOR FOR NITRATE AND A SEMI-MICRO SENSOR FOR CONTROL OF WASTE WATER TREATMENT

4.1 GENERAL DESCRIPTION OF NITRATE BIOSENSOR

Nitrate can be reduced by heterotrophic bacteria to either NO_2^-, N_2O, N_2, or NH_4^+. Electrochemical microsensors for NO_2^-, N_2O, and NH_4^+ have been described [28–30], and a biosensor for NO_3^- can thus be made by a bacterial reduction of NO_3^- to one of these species. Nitrite will be measured as well if biosensors are based on electrochemical N_2O or NH_4^+ microsensors. We have constructed [18] such a $NO_3^- + NO_2^-$ biosensor (Figure 4) by using a N_2O microsensor as a transducer, but the NO_2^- microsensor may actually be sufficiently good for use as a transducer in an alternative biosensor design. We usually refer to the $NO_3^- + NO_2^-$ biosensor as a NO_3^- biosensor, as most environments contain little NO_2^- compared with NO_3^-. The liquid medium inside the sensor contains a high concentration (0.5 wt %) of tryptic soy broth,

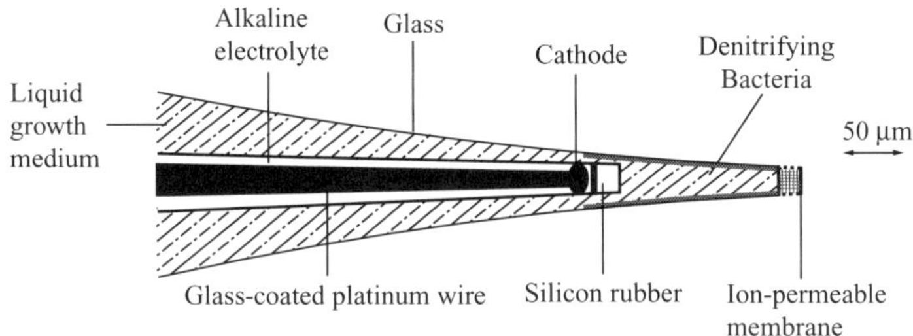

Figure 4. Tip of biosensor for NO_3^- based on bacterial conversion of NO_3^- to N_2O and subsequent electrochemical detection of the N_2O

which supports rapid metabolism of the applied strain of the denitrifying bacterium *Agrobacterium radiobacter*. This strain has no N_2O reductase, and therefore the reduction of NO_3^- and NO_2^- stops at the N_2O stage. The medium is kept at relatively high salinity (about 1 %) to facilitate electrophoretic attraction or repulsion (see section 2) of $NO_3^- + NO_2^-$. The voltages used to mediate electrophoretic attraction are usually from +0.1 to +1.0 V versus an external standard calomel electrode, while a potential of –0.8 V practically excludes any entry of NO_3^- and NO_2^- through the ion-permeable tip membrane. Glass tips with inserted ion-permeable membranes are available commercially.

It should be stressed that the sensor actually senses $NO_3^- + NO_2^-$, and any N_2O present will interfere. The interference of N_2O can, however, be compensated for by measuring the N_2O concentration while entry of NO_3^- and NO_2^- into the sensor is prevented by a high negative potential (–0.8V) applied versus an external calomel electrode [26]. The concentration of N_2O in the environment is, however, rarely so high that it causes pronounced interference.

The NO_3^- biosensor is shown in Figure 4 with a tip diameter of about 25 μm, which is the smallest possible tip diameter if concentrations below 1 μmol L^{-1} should be measured. It is still not known how large the diameter of the sensor tips can be made before the supply of electron donors from the internal reservoir starts to be limiting. The distance between membrane and transducer should be kept below 150 μm to obtain 90 % response times of about 15–30 s, and at a maximum of 300 μm if a response time of 2 min is acceptable.

The response of the NO_3^- biosensor is linear as long as the bacteria can mediate a full conversion of NO_3^- in the reaction space between internal transducer tip and the tip membrane. Figure 5 shows calibration curves for a sensor measuring in air-saturated water and in O_2 saturated water. A significant proportion of the bacteria respires aerobically when the analyzed water is O_2 saturated, so the capacity to reduce NO_3^- is lowered, resulting in a lower range for linear response. The sensitivity in the linear range is, however, independent of O_2 concentration. This independence of O_2 concentration may not appear logical, but it can be shown both experimentally and by mathematical modeling

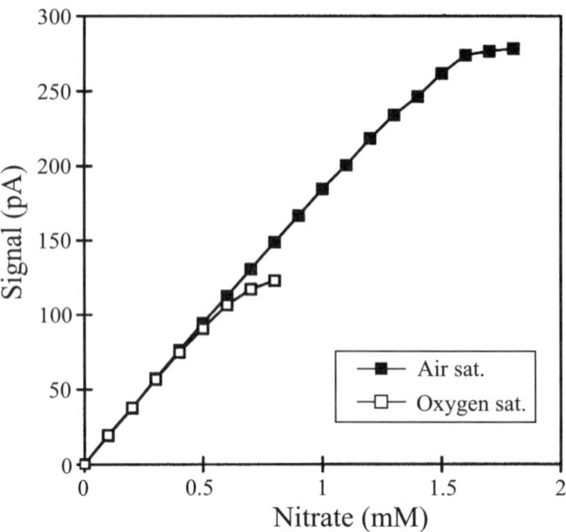

Figure 5. Calibration curves for NO_3^- biosensor in air- and O_2 saturated water. Within the linear range the calibration curves are identical, but the maximum NO_3^- concentration for linear response is lower at high O_2 as the bacteria prefer O_2 to NO_3^-. Reprinted with permission from Larsen *et al.* [31]. Copyright American Chemical Society

of the concentration gradients within the sensor (T. Kjær *et al.*, unpublished results). The linearity is even unaffected by the shape of the tip region (parallel sided or conical). There are no interfering substances except for the N_2O mentioned above. In the first publication on the NO_3^- biosensor [31] sulfide was mentioned as an irreversible inhibitor, but a change in the cathode material plated onto the platinum electrode from silver to palladium has removed the sulfide interference [26].

The bacteria (*Agrobacterium radiobacter*) used in the sensor are remarkably active under a wide range of environmental conditions. The NO_3^- biosensor has thus been used successfully in a temperature range from 3 to 42 °C, but the dynamic range was very low at 3 °C, and for routine operation the temperature should be above 5 °C. The bacteria are also active under a wide range of salinities, and the sensor has thus been used in freshwater as well as oceanic strength seawater. Toxic chemical species seem also to have relatively little effect on the bacteria, probably because they experience relatively low concentrations owing to an efficient diffusional exchange between the sensor tip region and the large internal medium reservoir. Phenol, which goes readily through the membrane, was thus in one instance used to stop bacterial activity in a model wastewater treatment plant while a NO_3^- biosensor was inserted. All biological activity in the plant ceased immediately after the phenol addition, but the biosensor still functioned. The tolerable extremes of external pH have still not been determined for the NO_3^- biosensor, but strong buffering of the internal medium is possible so that the sensor should operate over a wide range of pH values. Other types of bacteria are, however, very sensitive to toxic compounds, and biosensors containing nitrifying bacteria [13] have been constructed to serve as warning devices for toxic emissions.

Nitrate biosensors should be stored in NO_3^--containing (e.g. $1\,mmol\,L^{-1}$) water when not in use, and a couple of days with continuous polarization is needed before a stable baseline for zero NO_3^- is obtained. Continuous polarization is actually recommended for maximum lifetime and performance. The lifetimes of real microscale NO_3^- biosensors have until now only been a maximum of a few weeks, but work is in progress to improve this. The problems causing a limited lifetime have apparently not been microbiological, but rather membrane and N_2O transducer stability.

4.2 MONITORING IN WASTE WATER TREATMENT PLANTS

There is a large demand for a stable and fast-responding NO_3^- sensor for regulation and emission control of wastewater treatment plants, and a robust version of the NO_3^- biosensor is therefore being developed for this purpose. A result of a test run in an alternating oxic/anoxic waste water treatment plant is shown in Figure 6 together with simultaneous readings of O_2 made by a robust version of an O_2 microsensor. It can be observed that nitrification, i.e. formation of $NO_3^- + NO_2^-$, starts immediately after onset of aeration and that

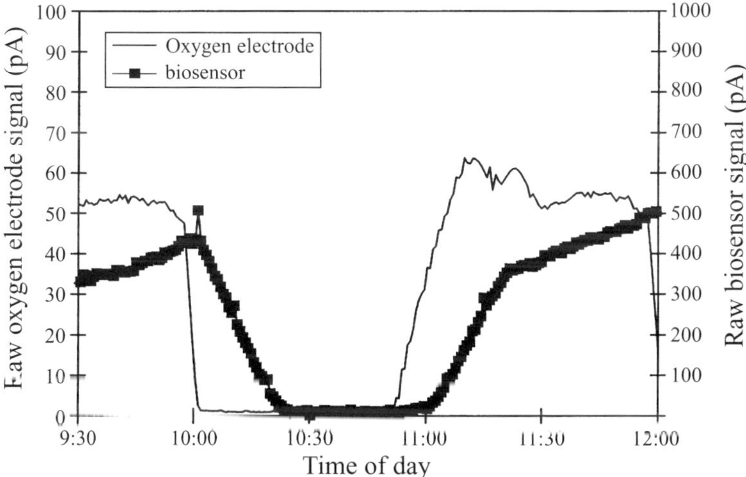

Figure 6. O_2 and NO_3^- concentrations during an aeration cycle of an activated sludge tank in a waste water treatment plant. Nitrification results in a rapid increase in NO_3^- after start of aeration, and denitrification consumes the NO_3^- after onset of anoxic conditions

denitrification starts immediately after onset of anoxic conditions. It is also remarkable that nitrification was very intense during the initial 10–15 min of the oxic period owing to the presence of NH_4^+ formed during the preceding anoxic period, and that nitrification thereafter had to be based on the continuously liberated NH_4^+ resulting in a lower rate. From a manager's point of view it is evident that the anoxic periods could have been reduced considerably with resulting increase in efficiency, as denitrification was complete within the first 30 min of the 60 min period without aeration. Use of a fast-responding NO_3^- sensor in wastewater treatment plants could thus be of great value. At present it is possible to make fast-responding semi-micro NO_3^- biosensors which exhibit < 20 % drift in signal over the initial 2 months of continuous operation in a waste water treatment plant. The present dynamic range of these semi-micro sensors is from 5 to 3.500 μmol L^{-1} at 20 °C, but work is being conducted to insert the sensors in constant-temperature units so that the measuring range is independent of ambient temperature.

4.3 *IN SITU* MONITORING OF NATURAL AQUATIC ENVIRONMENTS

The only field experiment conducted until now with a microscale NO_3^- biosensor was performed in a Danish lake while the water temperature was 8 °C (L.H. Larsen *et al.*, unpublished results). The biosensor was mounted on a benthic lander [32] and profiles of NO_3^- in the sediment were measured at intervals during a 10 d period. The experiment was very successful, and detailed NO_3^- profiles were obtained although the water phase NO_3^- concentration was only 8 μmol L^{-1}. The signal drift of the NO_3^- sensor during the period was less than 1 % d^{-1}.

4.4 EFFECTS OF CHANGES IN DIFFUSIVITY AND FLOW ON MEASUREMENTS

The obvious advantage of microsensors is that they may be used to analyze the spatial distribution of chemical or physical parameters. Much work of this type has, however, been done without taking into account that the sensor should be characterized by very low stirring effect (< 2 %) if reliable results should be obtained [33]. A sensor with a 2 % difference between stirred and stagnant water with identical concentration of the species being sensed will typically exhibit a 3–6 % difference between a reading in a sediment matrix and in stirred water because of the low transport coefficients in a stagnant sediment matrix, and even a 2 % stirring effect may thus be critical for calculations based on concentration gradients near the sediment–water interface. When used without electrophoretic transport of ions into the NO_3^- biosensor, there is only a small sensitivity to stirring. The same is not always the case if the sensitivity is improved by applying a positive potential to the sensor (see section 2), but for 25 μm thick sensors experiments showed relatively small effects (<5% change from stirred to stagnant water). There is, however, one additional problem which should not be neglected: when relatively thick and conical microsensors are approaching a sediment they affect the water flow in the immediate vicinity of the sensor, and the readings are thus made under another flow regime than found in the absence of the sensor [34]. The diffusive boundary layer above the sediment may be eroded down from, for example, 200 to 100 μm, and the concentration of NO_3^- at the sediment surface is therefore increased. Such an effect is negligible if the NO_3^- penetration is several millimeters, but the effect can be pronounced in very active systems where large concentration changes occur over less than 1 mm [35], as a significant proportion of the decrease in NO_3^- concentration then occurs in the diffusive boundary layer. The effect on the local flow conditions is smaller when very thin sensors are used, but introduction of even very thin O_2 sensors with tip diameters < 10 μm did result in significant effects on the thickness of the diffusive boundary layer [34].

4.5 MEASUREMENT IN MARINE ENVIRONMENTS WITH LOW NITRATE CONCENTRATIONS

An example of a NO_3^- profile in a marine sediment as obtained by a 25 μm thick biosensor is shown in Figure 7. The readings were performed while sensitivity was improved by applying a potential of +0.6 V to the biosensor

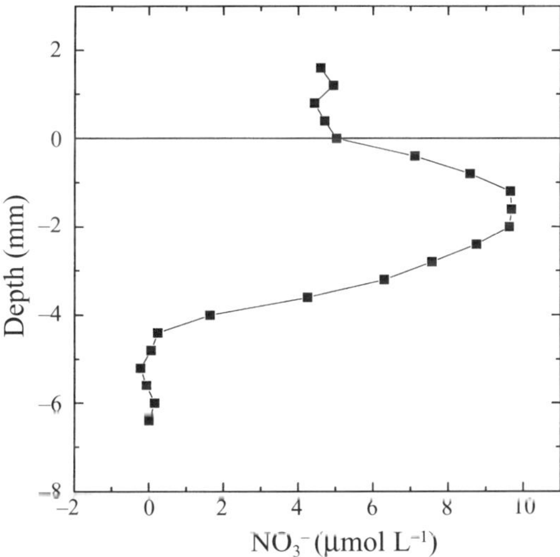

Figure 7. Profile of NO_3^- in a marine sediment at 16 °C. Nitrification results in a peak of NO_3^- up to 10 μmol L^{-1} in the (oxic) 0–3.5 mm surface layer while denitrification causes NO_3^- depletion in the (anoxic) 3.5–4.5 mm layer.

versus an external calomel reference electrode. It should be noticed that the NO_3^- concentration in the overlying water was only 4 μmol L^{-1}, and that the resolution of the readings was about 0.1 μmol L^{-1}. There was a peak in NO_3^- (+NO_2^-) caused by NO_3^- (+NO_2^-) production (nitrification) in the upper 3–3.5 mm of the sediment followed by NO_3^- (+NO_2^-) consumption (denitrification) below ca. 3.5 mm depth.

4.6 DETAILED MAPPING OF MICROSCALE DISTRIBUTION OF NITRIFICATION AND DENITRIFICATION IN A SEDIMENT

It is obvious from the data presented in Figure 7 that NO_3^- distribution in sediments can be analyzed at great accuracy by the use of NO_3^- biosensors, and that these sensors can thus be used to study nitrification and denitrification. The data presented in Figure 8 illustrate this in more detail (see Lorenzen

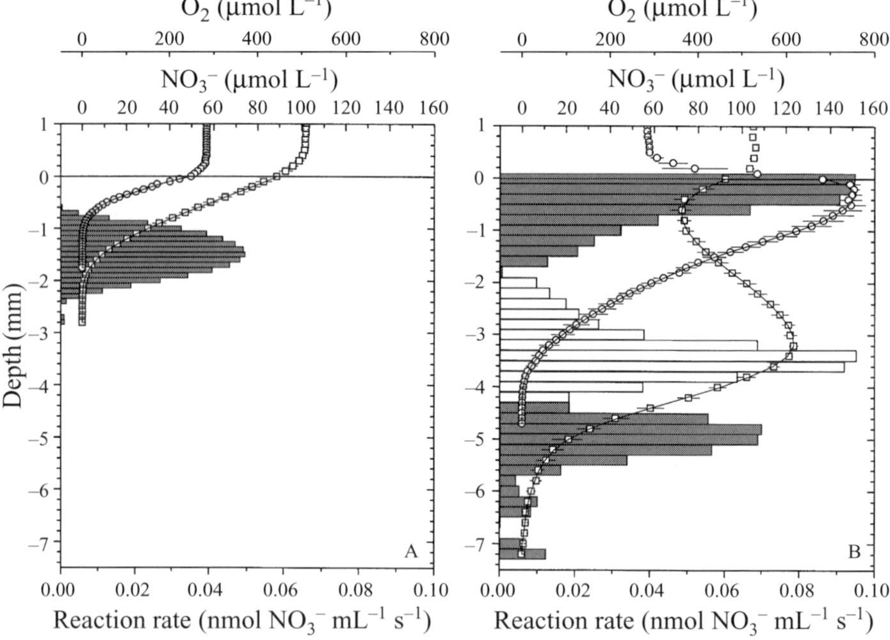

Figure 8. Profiles of O_2 (○, mean values with bars indicating SD, $n = 6$), NO_3^- (□, mean values with bars indicating SD, $n = 6$), rates of NO_3^- assimilation (grey bars in the 0–2 mm layer of panel B), rates of nitrification (light bars), and rates of denitrification (grey bars in panel A and the 4.4–7.2 mm layer of panel B) in a diatom-covered sediment during darkness (A) and during illumination (B). All profiles were measured at different sites in the sediment core, but as shown by the standard deviations, the different profiles were very similar

et al. [36] for a thorough discussion of similar data). The data were recorded in a diatom-covered sediment core from a freshwater lake that was exposed to 12 h light and 12 h dark diurnal cycles. The NO_3^- and O_2 profiles in Figure 8A represent steady-state conditions during the night, whereas the data of Figure 8B represent steady-state light conditions. During the night, O_2 penetrated to only 1 mm depth. The NO_3^- profile through the oxic layer was almost linear, indicating no net transformation of NO_3^-, whereas denitrification in the anoxic layers below 1 mm depth caused depletion of NO_3^- at a depth of about 2 mm. In the light, the diatoms in the top 1.5 mm produced O_2, and the maximum O_2 concentration was about three times air saturation. The O_2 penetration was increased from 1 mm in the dark to 5 mm in the light. The NO_3^- profile was also heavily affected by the light and associated microphytobenthic photosynthesis. There was thus a minimum in NO_3^- in the diatom layer caused by assimilation. In the oxic zone below the diatom layer there was a peak in NO_3^- caused by nitrifying bacteria oxidizing NH_4^+ to NO_3^-, followed by NO_3^- depletion in the anoxic layers below 4.3 mm depth due to denitrification.

Metabolic rates (bars in Figure 8) were calculated from the concentration profiles by a computer-implemented diffusion-reaction model [37]. To do this it is, however, necessary to know the depth profiles of diffusivity, but this is now a relatively simple task as a microsensor for the determination of microscale water flow or sediment diffusivity has been developed [38] (see also section 6) It should be stressed that the modeled rates in Figure 8 are net rates, so in principle a rate of zero at some depth could be due to identical production and consumption rates at that depth.

4.7 COMPARISON OF NITRATE BIOSENSOR WITH ION-EXCHANGER BASED SENSORS

As compared to the liquid ion-exchanger (LIX) type NO_3^- [39] and NO_2^- [28] micro- and macrosensors (see Chapter 5), the NO_3^- biosensor has both advantages and limitations. The LIX electrodes are relatively easy to make, whereas the biosensors require great skill. The LIX electrodes can also be made with extremely small tips (at least for the NO_3^- electrodes down to sub-micrometer diameter), whereas the biosensors lose sensitivity if made with tip diameters below about 25 μm. The biosensors are, however, able to measure accurately in water containing interfering ions, including seawater. When operated with an applied positive tip potential the biosensors may also be much more sensitive than the ion-exchanger electrodes, where the practical detection limit in environmental waters is very dependent on the concentration of interfering ions. Finally it is possible to make biosensors which are extremely long-term stable. We still have not found procedures that reproducibly result in long-term stable microscale biosensors, but as described above the semi-microscale NO_3^- biosensor may operate continuously, even in wastewater, for periods of months.

Ion-exchanger-type NO_3^- electrodes are marketed [4], but the stability in waste water is apparently too poor for widespread use in waste water treatment.

5 MICROSCALE BIOSENSOR FOR METHANE

5.1 PREVIOUS METHODS FOR RESOLVING METHANE GRADIENTS

Usable electrochemical sensors for CH_4 have not been described, as CH_4 is very inert, so the spatial resolution of CH_4 in sediments has been determined by gas sampling through membrane-equipped capillaries with subsequent GC analysis of the collected gas [40], or by membrane-inlet mass spectrometry [41]. These methods suffer, however, from the need for a relatively large and highly permeable membrane-covered window to ensure a sufficient gas flux for the analysis, and the probes are therefore characterized by a high stirring sensitivity of about 100–200 %. As described for the NO_3^- biosensors above, high stirring sensitivities lead to inaccurate readings when the diffusive properties of analyzed stagnant media change, and changes in the reading may then be due both to real changes in concentration and to local changes in diffusivity. A so-called biosensor for CH_4 has also been described [25], but it was based on addition of large samples to a stirred culture of methane-oxidizing bacteria with subsequent monitoring of the decrease in O_2 concentration. Methane is, however, mostly present in anaerobic environments, so the ideal sensor would be one that could measure without the need for external oxygen.

5.2 GENERAL DESCRIPTION OF METHANE BIOSENSOR

A biosensor for the determination of CH_4 under anoxic conditions was developed by using the design illustrated in Figure 9, where CH_4-oxidizing bacteria are cultured in the thin microsensor tip [17]. The principle is basically the same microscale continuous culture vessel as illustrated in Figure 1, but the CH_4 sensor is made a little more complicated by the supply of O_2 to the tip via an internal gas-filled capillary containing an O_2 microsensor with its tip permanently positioned near the surface of the membrane covering the gas-filled capillary. The O_2 microsensor monitors the O_2 gradient within the biosensor as illustrated in Figure 10. The current in the measuring circuit is high for zero CH_4 and decreases with increasing methane concentration. The calibration curve may be linear over the full range from 0 to 100 % CH_4 saturation (Figure 11), or it may be linear only at relatively low CH_4 concentrations. The O_2 partial pressure is always constant at the inner surface of the silicone rubber (Figure 10), as the diffusion coefficient of O_2 in air is about 10^4 times higher than the diffusion coefficient in water or silicone rubber. The response time is determined by the relatively long distance from the air reservoir to the biosensor

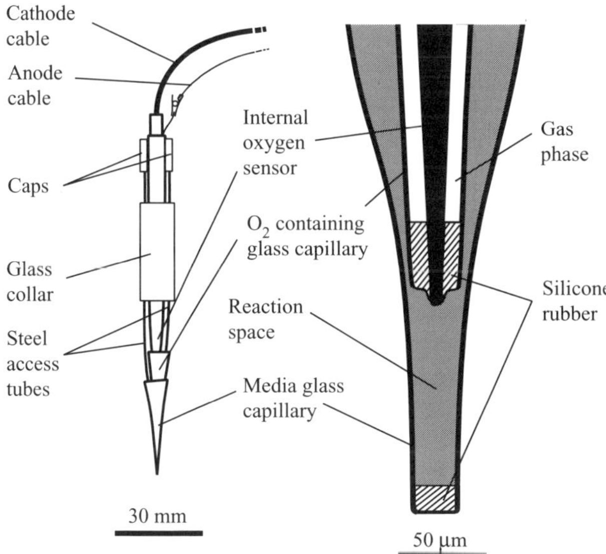

Figure 9. Microscale biosensor for CH_4 based on CH_4 – oxidizing bacteria living in a gradient of CH_4 from the analyzed medium and O_2 from an internal reservoir. An internal O_2 microsensor monitors the O_2 gradient within the sensor. Left: entire sensor. Right: enlarged section through the tip region. Reprinted with permission from Damgaard and Revsbech [17]. Copyright (1997) American Chemical Society

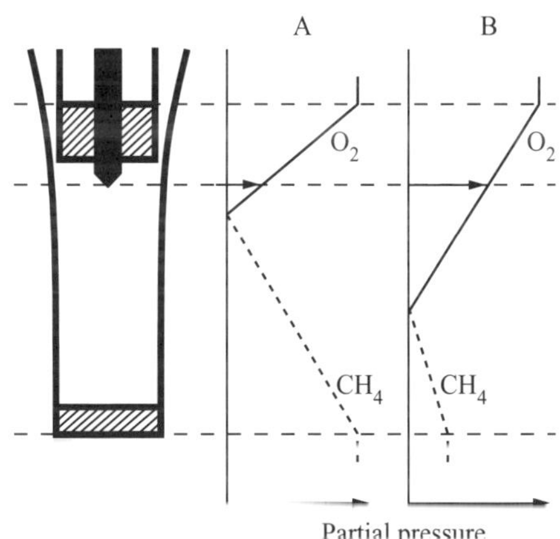

Figure 10. Functioning of the CH_4 biosensor. The sensor tip is shown schematically to the left. The two diagrams (A) and (B) illustrate how changes in CH_4 concentration affect the O_2 gradient and thereby the signal (illustrated with an arrow) from the internal O_2 microsensor. Reprinted with permission from Damgaard and Revsbech [17]. Copyright (1997) American Chemical Society

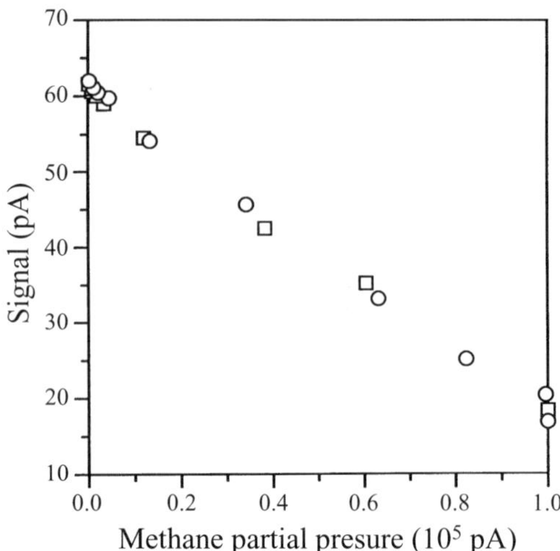

Figure 11. Calibration of a CH_4 biosensor performed twice with an intervening 18 h interval: ○, calibration at start of experiment; □, calibration after 18 h. Reprinted with permission from Damgaard and Revsbech [17]. Copyright American Chemical Society

tip, and the 90 % response time for a sensor with dimensions as shown in Figure 9 is about 30 s.

The signal from the CH_4 biosensor is not as ideal as the signal from the NO_3^- biosensor. First of all, there is a high current from the internal O_2 microsensor at low CH_4 concentration and a low current at high CH_4. This results in lower accuracy at low CH_4 concentrations, as a temperature change of only 1°C affects the current in the circuit by about 3 %. The 'simple' CH_4 biosensor shown in Figure 9 is not as insensitive to stirring as is the case with the NO_3^- biosensor, and the stirring effect is an offset of the whole calibration curve corresponding to 2–4 μmol L^{-1} CH_4, so this creates problems in the quantification of very low methane concentrations. The problems with stirring effects described here are, however, negligible as compared with the > 100 % stirring effect of the alternative membrane probe sampling procedures. Sulfide interferes. The sensitivity to sulfide is about 25 % of that to CH_4 at pH = 7. Long-term exposure to sulfide may, lead to depositions of elemental sulfur inside the sensor, but such long-term exposure has not yet been tested. Hydrogen may also interfere, but this interference is apparently due to a non-axenic (i.e. contaminated) methanotrophic culture in the microsensor tip and might be alleviated by sterilization of the sensor before the bacteria are added. The signal for hydrogen has, however, been much lower than that for CH_4 in the sensors investigated, and as hydrogen in methanogenic environments is usually present at concentrations < 1 μmol L^{-1} this interference is in most cases irrelevant.

The CH_4 biosensor may function for months, but calibration should be performed at regular intervals. The bacteria in the tip respond rapidly by increased respiration rate when they are exposed to CH_4, but there will always be some residual metabolism even in the absence of CH_4, and such a residual metabolism may change as a function of the life history of the biosensor. It is actually strange that such a residual metabolism does not result in pronounced baseline problems as are known for BOD sensors [10], where the usual practice is to incubate the sensor in nutrient-free medium for a considerable period, so that the metabolism can stabilize at a low level before each exposure to a new sample.

5.3 METHANE BIOSENSOR WITH OXYGEN GUARD AND ITS USE IN RICE PADDY SOIL

Methanogenic environments may be investigated in great detail by use of the CH_4 biosensor shown in Figure 9, but it only works under anoxic conditions as all O_2 must be supplied from the internal reservoir. It is, however, possible to add an O_2 scavenging system to the sensor tip as shown in Figure 12. The sensor

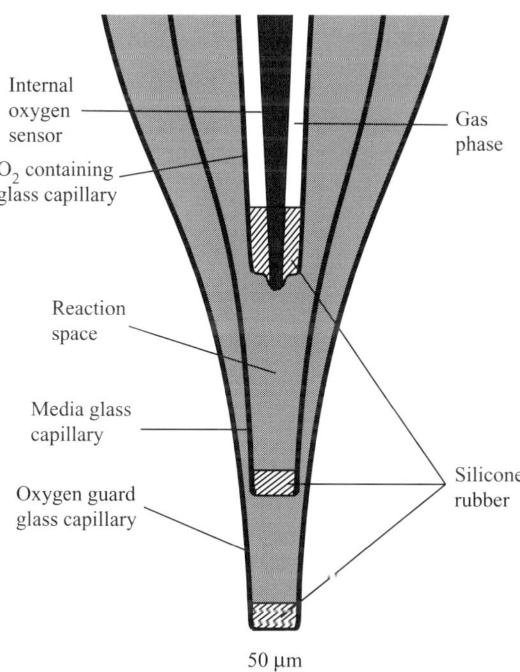

Figure 12. Tip of CH_4 biosensor equipped with an O_2 guard capillary containing the heterotrophic bacterium *Agrobacterium radiobacter* in a 1% tryptic soy broth medium. Reproduced from Damgaard *et al.* [42] by permission of American Society for Microbiology

shown in Figure 12 used heterotrophic bacteria immobilized in front of the capillary with the CH_4-oxidizing bacteria to remove the O_2 [42], but a higher efficiency may theoretically be obtained by using a $0.5\,mol\,L^{-1}$ solution of ascorbate at pH 13, and it should thereby be possible to reduce the distance between the two membranes to 30 μm. By adding this O_2 guard the CH_4 sensor is made insensitive to external O_2, and the (small) stirring effect seen by the 'simple' CH_4 biosensor is practically removed, so the modification could seem to be ideal. There are, however, also negative aspects of the O_2 guard. First of all, the level of complexity is increased, and the construction of a complete sensor is quite difficult and tedious. The addition of a guard does, however, also lead to lower signal and to slower response. The distance between the exterior and the internal O_2 microsensor is increased, and as the response time increases with the square of the distance (twice the distance gives four times longer response time) this is in itself a problem. What is worse, however, is that CH_4 may accumulate in the O_2 guard behind the tip of the CH_4 biosensor, and this gives a very slow response to large changes in CH_4 concentration. It can thus only be recommended to use an O_2 guard when absolutely necessary. When aerobic CH_4 oxidation is studied there is, however, no choice. Overlapping CH_4 and O_2 profiles from a rice paddy as measured with an O_2 microsensor and a CH_4 biosensor with O_2 guard are shown in Figure 13. During darkness

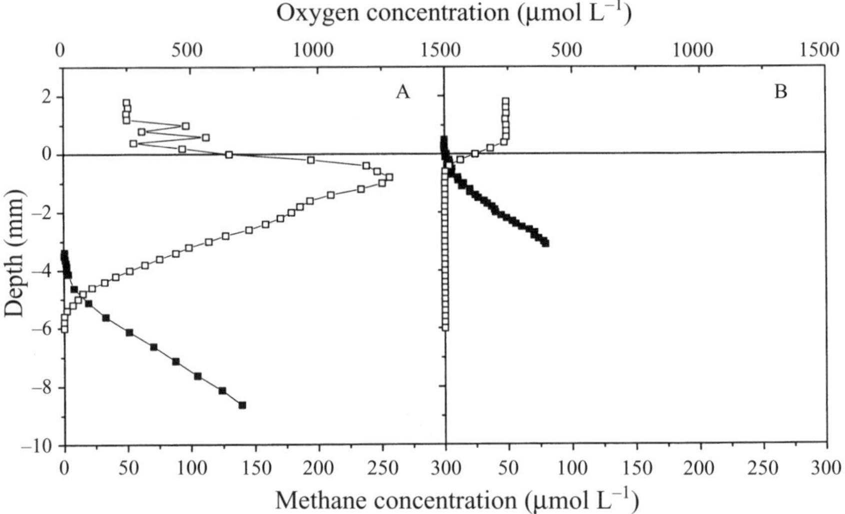

Figure 13. Oxygen (□) and CH_4 (■) profiles in a rice paddy soil as measured by microsensors. (A) Profiles during the day with deep O_2 penetration caused by cyanobacterial photosynthesis. (B) Profiles at night. The O_2 and CH_4 profiles were measured with different sensors, so they may not be perfectly aligned. Reproduced from Damgaard *et al.* [42] by permission of American Society for Microbiology

(Figure 13B) the O_2 penetration was less than 1 mm, but extensive CH_4 oxidation in this 1 mm led to almost full CH_4 depletion below the sediment surface. During the day (Figure 13A) illumination caused O_2 production by cyanobacteria living in the top soil layers so that the CH_4 oxidation horizon was now found at 4–6 mm depth.

6 CALCULATION OF METABOLIC RATES BASED ON DIFFUSIVITY SENSORS AND DEPTH PROFILES OF CHEMICAL SPECIES MEASURED WITH MICROSCALE BIOSENSORS

The results shown in Figure 8 illustrate how detailed data obtained by microscale (bio)sensors can be, but they also illustrate how calculations of depth profiles of metabolic rates can be performed based on the chemical profiles. The diffusivity profiles necessary for performing such calculations based on Fick's first and second laws of diffusion can now be measured with a diffusivity sensor [38]. This diffusivity sensor contains a reservoir of tracer gas which diffuses out into the surrounding medium through a membrane in the sensor tip while a built-in microsensor for the gas in question monitors the gas concentration at the membrane surface. A low diffusivity in the surrounding medium will result in impeded diffusion of the tracer gas away from the sensor tip and thus in a high reading from the built-in sensor, whereas the opposite is the case for a high diffusivity. The same sensor can also be used to quantify flow rates down to very low values ($< 10\,\mu m\,s^{-1}$). Flow/diffusivity sensors based on O_2 as a tracer are commercially available, but for environmental applications more inert tracers such as acetylene (L.R. Damgaard *et al.*, unpublished results) should be used. An alternative optical determination of microscale diffusivity distribution has also been described [43].

7 FUTURE DEVELOPMENTS IN MICROSCALE BIOSENSORS FOR ENVIRONMENTAL MONITORING

As mentioned in the section about NO_3^- biosensors a long-term stable NO_3^- biosensor for control of waste water treatment has been developed. At present this sensor and the O_2 and pH sensors are, to our knowledge, the only real chemical sensors (i.e. not counting miniaturized flow injection and spectroscopic devices) that will function continuously on-line for periods of months while immersed in complex media such as waste water. There are, however, several other possibilities for new types of microscale biosensors for long-term environmental monitoring, and we expect that such biosensors will be based on whole cells, as enzyme-based sensors most probably cannot be made sufficiently long-term stable. Biosensors with a short lifetime (for measuring, e.g.

xenobiotic compounds) may, however, be used extensively in the future, and there may be advantages of applying microscale designs here also. In the beginning of the chapter the possibility of making bioluminescence-based biosensors has already been mentioned, and this is probably the most extensive open area for new developments as such sensors may detect very low concentrations.

It may be possible to make new microscale biosensors based on whole cells for chemical species such as NH_4^+ [12] and SO_4^{2-}. Ammonium-oxidizing bacteria (*Nitrosomonas* sp.) that might be used in a possible micro-biosensor are, however, very sensitive to variations in environmental parameters, and as quite good electrochemical NH_4^+ sensors exist, the niche for use of an NH_4^+ biosensor will be relatively narrow. Analysis of NH_4^+ in marine sediments could be done with such sensors, but a more interesting possibility is long-term monitoring of NH_4^+ in waste water, where a biosensor might outperform purely electrochemical sensors in terms of lifetime and long-term stability if based on a microscale design. A microscale SO_4^{2-} biosensor would be of great scientific interest, as no reliable electrochemical sensor for SO_4^{2-} exists, and the principle of bacterial SO_4^{2-} reduction followed by electrochemical detection of the sulfide evolved should be tested in microscale biosensors.

By applying the proper microorganisms and membranes, the CH_4 biosensor design shown in Figure 9 can be used for analysis of many different organic or inorganic compounds. The main problem is, however, that except for CH_4, HS^-, and NH_4^+, most oxidizable low-molecular weight chemical species do not build up as large dissolved pools in natural sediments and biofilms. An exception to this is acetate (and other short-chain carboxylic acids), as many methanogenic environments contain freely dissolved acetate in appreciable concentrations ($10^{-5} - 10^{-2}$ mol L^{-1}). A modified CH_4 biosensor containing acetate-oxidizing bacteria such as a *Pseudomonas* sp. may thus be used for analysis of acetate. Other easily degradable organic species are usually found in concentrations below the few micromolar level necessary for detection by the O_2 consumption within the sensor. Iron and manganese may build up to appreciable concentrations, but various aspects of solubility, diffusivity, and possible interferences do not make the construction of biosensors for these species feasible. The tip of a biosensor based on oxidation of Fe^{2+} or M_n^{2+} would rapidly be filled with insoluble oxides and hydroxides, and the very slow diffusion into the sensor (the diffusivity of these ions is much lower than for NO_3^-, CH_4, O_2, etc.) [44] would also give a poor sensitivity. At present voltammetry (Buffle, Chapter 9) and dialysis methods (Davison, Chapter 11) seem to be the best *in situ* detection principles for Fe^{2+} and Mn^{2+}.

One sensor that would be extremely valuable and where a satisfactory detection scheme still has to be devised is a phosphate sensor. Many biological reactions and transport systems are highly specific for phosphate, but although attempts have been made it has until now not been possible to couple this specificity with a satisfactory detection principle.

8 CONCLUSION

Taken as a whole, the combination of available microscale electrochemical sensors, optodes, and biosensors (a short review was presented by Kühl and Revsbech [45]) now makes it possible to analyze the microscale chemistry of aquatic environments in great detail, although especially a phosphate sensor is still missing. A considerable amount of information about our environment has already been gained by use of these sensors, but the potential for considerable expansion of our knowledge is still there. The development of microscale sensors does, however, also have a broader scope. The analytical schemes utilized in the microscale NO_3^- and CH_4 biosensors also work in sensors with diameters up to about 0.5 mm, and such semi-macro biosensors may in the future contribute significantly to environmental monitoring and to efficient control of waste water treatment.

ACKNOWLEDGEMENTS

We thank the Commission of the European Communities (Mast III Programme MICROMARE, Project no. 950029) and the Danish Biotechnology Programme for support.

GLOSSARY

Axenic Culture of organisms with only one type being present

Benthic lander Instrument made for *in situ* investigation of the sediment–water interface.

Bioluminescence Biological emission of light based on enzymatic oxidation of an aldehyde.

Biosensor Often used for sensors based on any biological component that can be used to obtain a signal for a chemical parameter. In this chapter, biosensor is used in a more restricted sense, i.e. it is a physical device based on the combination of microorganisms and a detection system that enables the measurement of chemical species.

Cyanobacteria Photosynthetic microorganisms also often referred to as blue-green algae.

Denitrification Bacterial respiration with NO_3^- and NO_2^- whereby NO_3^- and NO_2^- are reduced to N_2 or N_2O.

Diatoms Eucaryotic photosynthetic microorganisms with a silica shell.

Diffusive boundary layer The thin layer just above a surface where diffusional transport of dissolved species perpendicular to the surface dominates over transport by flow.

Gene expression Translation of the genetic code in DNA to RNA and often further to protein.

Inducer In molecular biology, this term is used for some chemical species that causes gene expression (see above).

Eucaryotic Organisms having a nuclear membrane as opposed to bacteria. Eucaryotic cells are usually larger than bacterial cells.

Heterotrophic Organisms assimilating organic species as opposed to the autotrophic ones assimilating CO_2.

LIX Liquid ion exchanger, i.e. some ion exchanger for a specific ion dissolved into a hydrophobic liquid.

Methanogenic Methane-producing.

Microphytobenthos Photosynthetic microorganisms living on the sediment surface.

Nitrification Oxidation of NH_4^+ to NO_2^- by one type of bacteria followed by further oxidation of NO_2^- to NO_3^- by another type of bacteria.

Strain Bacterial species are difficult to define, and it is therefore common to refer to specific isolates (or mutants), also called strains.

Xenobiotic Non-biological, man-made chemical species.

REFERENCES

1. Lynggaard-Jensen, A., Eisum, N. H., Rasmussen, I., Svankjær-Jacobsen, H. and Stenstrøm, T. (1996). Description and test of a new generation of nutrient sensors, *Wat. Sci. Technol.*, **33**, 25.
2. Carlsson, K., Jacobsen, H. S., Lynggaard Jensen, A., Stenstrøm, T. and Karlberg, B. (1997). Micro-continuous flow system for wet chemical analysis, *Anal. Chim. Acta*, **354**, 35.
3. Jannasch, H. W., Johnson, K. S. and Sakamoto, C. M. (1994). Submersible, osmotically pumped analyzers for continuous determination of nitrate *in situ*, *Anal. Chem.*, **66**, 3352.
4. Wacheux, H., Da Silva, S. and Lesavre, J. (1993). Inventory and assessment of automatic nitrate analyzers for urban sewage works, *Wat. Sci. Technol.*, **28**, 489.
5. Isaacs, S. and Henze, M. (1994). Fluorescense monitoring of an alternating activated sludge process, *Wat. Sci. Technol.*, **30**, 229.
6. Crank, J. (1983). *The Mathematics of Diffusion*, Oxford University Press, London.
7. Revsbech, N. P. (1989). An oxygen microelectrode with a guard cathode, *Limnol. Oceanogr.*, **34**, 472.
8. Cronenberg, C. C. H., Van Groen, H., De Beer, D. and Van den Heuvel, J. C. (1991). Oxygen-independent glucose microsensor based on glucose oxidase, *Anal. Chim. Acta*, **242**, 275.
9. Riedel, K., Lange, K. P., Stein, H.-J., Kühn, M., Ott, P. and Scheller, F. (1990). A microbial sensor for BOD, *Wat. Res.*, **24**, 883.
10. Praet, E., Reuter, V., Gaillard, T. and Vasel, J.-L. (1995). Bioreactors and biomembranes for biochemical oxygen demand estimation, *Trends Anal. Chem.*, **14**, 371.

11. Barcelo D., ed. (1995). Special issue: Biosensors for environmental monitoring, *Trends Anal. Chem.*, **14**(7).
12. Karube, I. and Tamiya, E. (1987). Biosensors for environmental control, *Pure Appl. Chem.*, **59**, 545.
13. König, A., Secker, J., Riedel, K. and Metzger, J. W. (1998). A microbial sensor for measuring inhibitors and substrates for nitrification in wastewater, *Int. Lab.*, July 1998, 15.
14. Suzuki, H., Kojima, N., Sugama, A., Takei, F, Ikegami, K., Tamiya E., and Karube, I. (1989). Fabrication of a microbial carbon dioxide sensor using semiconductor fabrication techniques, *Electroanalysis*, **1**, 305.
15. Suzuki, H., Tamiya, E. and Karube, I. (1991). Disposable amperometric CO_2 sensor employing bacteria and a miniature oxygen electrode, *Electroanalysis*, **3**, 53.
16. Marty, J.-L., Garcia, D. and Rouillon, R. (1995). Biosensors: potential in pesticide detection, *Trends Anal. Chem*, **14**, 329.
17. Damgaard, L. R. and Revsbech, N. P. (1997). A microscale biosensor for methane, *Anal. Chem.*, **69**, 2262.
18. Larsen, L. H., Kjær, T. and Revsbech, N. P. (1997). A microscale NO_3^- biosensor for environmental applications, *Anal. Chem.*, **69**, 3527.
19. Marko-Varga, G., Emnéus, J., Gorton, L., and Ruzgas, T. (1995). Development of enzyme-based amperometric sensors for the determination of phenolic compounds, *Trends in Anal. Chem.*, **14**, 319.
20. Nomura, Y., Ikeburo, K., Yokoama, K., Takeuchi, T., Arikawa, Y., Ohono, S. and Karube, I. (1994). A novel microbial sensor for anionic surfactant determination, *Anal. Lett.*, **27**, 3095.
21. Rogers, K. R. (1995) Biosensors for environmental applications, *Biosens. Bioelectron*, **10**, 535.
22. Marco, M.-P., Gee, S. and Hammock, B. D. (1995). Immunological techniques for environmental analysis. I. Immunosensors, *Trends Anal. Chem.*, **14**, 341.
23. Moretto, L. M., Ugo, P., Zanata, M., Guerriero, P., Martin, C. R. (1998). Nitrate biosensor based on the ultrathin-film composite membrane concept, *Anal. Chem.*, **70**, 2163.
24. Eggins, B. R. (1996). *Biosensors: An Introduction*, Wiley, Chichester.
25. Karube, I., Okada, T. and Suzuki, S. (1982). A methane gas sensor based on oxidizing bacteria, *Anal. Chim. Acta*, **135**, 61.
26. Kjær, T., Larsen, L. H., and Revsbech, N. P. (1999). Electrophoretic sensitivity control of microscale biosensors, *Anal. Chim. Acta*, **391**, 57.
27. Neudörfer, F. and Meyer-Reil, L.-A. (1997). A microbial biosensor for the microscale measurement of bioavailable organic carbon in oxic sediments, *Mar. Ecol. Prog. Ser.*, **147**, 295.
28. Beer, D., Schramm, A., Santegoeds, C. M. and Kühl, M. (1997). A nitrite microsensor for profiling environmental biofilms, *Appl. Environ. Microbiol.*, **63**, 973.
29. Revsbech, N. P., Nielsen, L. P., Christensen, P. B. and Sørensen, J. (1988). Combined oxygen and nitrous oxide microsensor for denitrification studies, *Appl. Environ. Microbiol.*, **54**, 2245–2249.
30. Beer, D. and van den Heuvel, J. C. (1988). Response of ammonium-selective microelectrodes based on the neutral carrier nonactin, *Talanta*, **35**, 728.
31. Larsen, L. H., Revsbech, N. P. and Binnerup, S. J.. 199 A microsensor for nitrate based on immobilized denitrifying bacteria, *Appl. Environ. Microbiol.* **62**, 1248.
32. Glud, R. N., Gundersen, J. K., and Jørgensen, B. B., Revsbech, N. P. and Schulz, H. D. (1994). Diffusive and total oxygen uptake of deep-sea sediments in the eastern

and South Atlantic Ocean: *in situ* and laboratory measurements, *Deep-Sea Res.*, **41**, 1767.
33. Gust, G., Booij, K., Helder, W. and Sundby, B. (1987). On the velocity sensitivity (stirring effect) of polarographic oxygen microelectrodes, *Neth. J. Sea Res.*, **21**, 255.
34. Glud, R. N., Gundersen, J. K., Revsbech, N. P. and Jørgensen, B. B. (1994). Effects on the benthic diffusive boundary layer imposed by microelectrodes, *Limnol. Oceanogr.*, **39**, 462.
35. Schramm. A., Larsen, L. H., Revsbech, N. P., Ramsing, N. B., Amann, R. and Schleifer, K.-H. (1996). Structure and function of a nitrifying biofilm as determined by in situ hybridization and microelectrodes, *Appl. Environ. Microbiol.*, **62**, 4641.
36. Lorenzen, J., Larsen, L. H., Kjær, T. and Revsbech. N. P. (1998). Biosensor determination of the microscale distribution of nitrate, nitrate assimilation, nitrification, and denitrification in a diatom inhabited freshwater sediment, *Appl. Environ. Microbiol.*, **64**, 3264.
37. Revsbech, N. P., Madsen, B. and Jørgensen, B. B. (1986). Oxygen production and consumption in sediments determined at high spatial resolution by computer simulation of oxygen microelectrode data, *Limnol. Oceanogr.*, **31**, 293.
38. Revsbech, N. P., Nielsen, L. P. and Ramsing, N. B. (1998). A novel microsensor for the determination of diffusivity in sediments and biofilms, *Limnol. Oceanogr.*, **43**, 986.
39. de Beer, D. and Sweerts, J.-P. R. A. (1989). Measurement of nitrate gradients with an ion-selective microelectrode, *Anal. Chim. Acta*, **219**, 351.
40. Rothfuss, F. and Conrad, R. (1994). Development of gas diffusion probe for the determination of methane concentrations and diffusion characteristics in flooded paddy soil, *FEMS Microbiol. Ecol.*, **14**, 307.
41. Lloyd, D, Thomas, K., Price, D., O'Neil, B., Oliver, K. and Williams, T. N. (1996). A membrane-inlet mass spectrometer miniprobe for the direct simultaneous measurement of multiple gas species with spatial resolution of 1 mm, *J. Microbiol. Methods*, **25**, 145.
42. Damgaard, L. R., Revsbech, N. P. and Reichardt, W. (1998). An oxygen insensitive microscale biosensor for methane used to measure methane concentration profiles in a rice paddy, *Appl. Environ. Microbiol.*, **64**, 867–870.
43. De Beer, D., Stoodley, P. and Lewandowski, Z. (1997). Measurement of local diffusion coefficients in biofilms by microinjection and confocal microscopy, *Biotechnol. Bioeng.*, **53**, 151.
44. Li, Y. H. and Gregory, S. (1974). Diffusion of ions in sea water and in deep-sea sediments, *Geochim. Cosmochim. Acta*, **38**, 703.
45. Kühl, M. and Revsbech, N. P. (2000). Microsensors for studies of interfacial biogeochemical processes, In *The Benthic Boundary Layer: Transport Processes and Biogeochemistry*, ed. Boudreau, B. and Jørgensen, B. B., Oxford University Press, Oxford, in press.
46. Lee, S. M., Suzuki, M., Kumagai, M., Ikeda, H., Tamiya, E. and Karube, I. (1992). Bioluminescense detection system of mutagen using firefly luciferase genes induced in Escherichia coli lysogenic strain, *Anal. Chem.*, **64**, 1755.

7 Continuous Flow Techniques for On Site and *In Situ* Measurements of Metals and Nutrients in Sea Water

K. S. JOHNSON, V. A. ELROD, J. L. NOWICKI, K. H. COALE AND H. ZAMZOW

Moss Landing Marine Laboratories and Monterey Bay Aquarium Research Institute, USA

1 INTRODUCTION

Aquatic systems are sufficiently complicated that laboratory experiments cannot provide complete insights into the processes that control chemical distributions or the impacts of chemicals on ecosystem structure [1]. Scientists must

In Situ Monitoring of Aquatic Systems: Chemical Analysis and Speciation Edited by J. Buffle and G. Horvai.

monitor concentration changes in natural systems to understand the biogeochemical processes that are operating. A few observations in natural systems cannot adequately represent the distribution of chemicals in such systems or their influence on ecosystems. Undersampling the chemical environment may lead to a model of environmental interaction that is either oversimplified or inaccurate [2]. Whenever possible, environmental studies are greatly strengthened by using on site or *in situ* measurement techniques that allow large numbers of measurements, rather than collecting samples and transporting them to the laboratory for analysis. These near real time measurements can identify important trends in the data. On site analysis can also minimize the probability of chemical changes in the sample due to reactions during storage. Artifacts due to sample contamination can be identified, which allows the analyst to correct potential problems.

In the laboratory, the analytical challenges that are presented by natural waters can be overcome through the use of very sophisticated instrumentation, such as graphite furnace atomic absorption spectrophotometry, gas chromatography–mass spectrometry or inductively coupled plasma–mass spectrometry (ICPMS). While a number of laboratories have developed portable instrumentation based on these sophisticated methods [3], they do not always operate well under the harsh conditions found at remote sites or on ships at sea. This mandates alternative approaches for on site and *in situ* analyses. There are two general approaches that could be used for the determination of dissolved chemical concentrations on site or *in situ*: chemical sensors and chemical analyzers [4,5]. We define chemical sensors as devices in which passive diffusion transports the chemical species to be determined to the detector. Sensors ultimately offer the simplest solution to on site and *in situ* chemical analysis. However, the analytical challenges involved in measuring chemical concentrations in natural systems can be formidable [6]. For example, only seven ions are present in seawater at concentrations greater than 1 mmol L^{-1}. Hidden within this matrix of major ions are vanishingly small quantities of the remaining elements (Figure 1). Determination of these trace chemicals is often complicated by the major ion matrix and severe problems due to contamination [7]. However, tremendous developments are being achieved in this field. For example, a single electrochemical detector can now detect a variety of chemical species [8]. These devices, which include most electrochemical analyzers with chemically treated electrodes, are generally treated in other chapters of this volume.

Chemical analyzers move a sample by mass transport through the instrument. This confers several advantages, but at the cost of additional complexity [9]. Mass transport of the sample makes it feasible to add a valve at the sample inlet which can be switched to a standard solution with a known concentration of the analyte. Calibration *in situ* is, therefore, possible in such an analyzer. This means that instrument stability is less critical than in a sensor

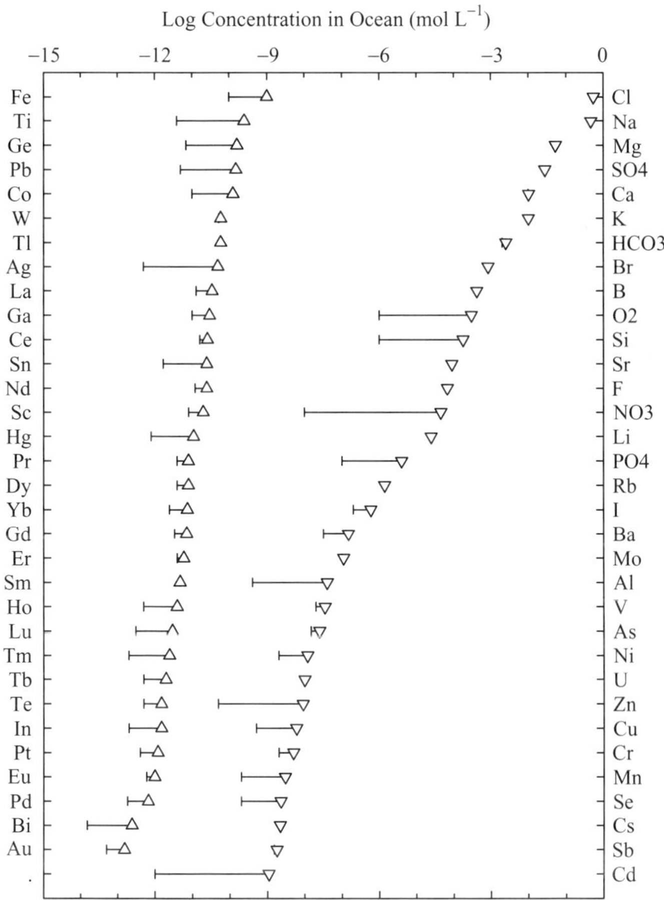

Figure 1. Element concentrations in seawater [6]

that must maintain its calibration for the duration of a deployment. It is also relatively easy to perform multistep reactions in an analyzer by merging the flowing sample stream with reagents at various points along the flow path. This is a significant advantage because it allows more complicated sample processing that can compensate for a lack of selectivity or sensitivity in an analysis. For example, in seawater many divalent cations cannot be discriminated at a sensor surface because the large concentrations of Mg^{2+} and Ca^{2+} overwhelm the sensor signal. However, multistep reactions make it possible to develop much more selective detection schemes. This chapter will discuss on site and *in situ* applications of unsegmented continuous flow analytical methods using optical methods of detection, including spectrophotometry, fluorescence and chemiluminescence. The focus of this chapter is primarily on *in situ* chemical analysis. Methods that have been developed for on site studies, and which have lead to *in situ* analysis, are also considered. The examples presented come primarily from oceanographic studies with which the authors are most familiar.

Although this chapter focuses primarily on analyses of seawater, many of the challenges are similar in freshwater studies. It has often been assumed that metal concentrations in freshwater systems are much higher owing to the greater contact with sediments and anthropogenic activities. The exhaustive anticontamination procedures needed to measure dissolved trace elements accurately in marine environments have not been widely used in freshwater systems. However, several recent studies have shown that many of the analytical techniques used routinely for freshwater studies result in severe sample contamination and the levels of metals are much lower than previously thought [10–12]. These studies indicate that trace element concentrations in lakes and rivers may be as low as the mean ocean values. Benoit *et al.* [13] recently assessed the sources of the contamination that occur during metal measurements in freshwater systems and they found that stringent anticontamination procedures were necessary at every step.

2 CONTINUOUS FLOW ANALYSIS

Continuous flow analyzers were developed to automate many of the steps performed in a typical chemical analysis. A continuous flow analyzer uses a pump to propel sample through a tubing manifold to a chemical detector. In most cases, the sample is treated chemically by merging the sample stream with various reagent streams to derivatize the analyte species before it enters the detector. However, sensor systems may be incorporated into a continuous flow analyzer to detect the analyte species directly. Reagents can be added to buffer or to derivatize a sample. Solid phase reactors [14] may be used in such systems to concentrate trace elements or to separate them from the background matrix and to separate dissolved gases from the aqueous matrix. Standard and blank solutions can be substituted for the sample to provide automatic calibration of the system.

The earliest applications of continuous flow analysis (CFA) were based on segmented systems, which use gas bubbles to segment the liquid sample stream. The bubbles limit dispersion of the sample in the reaction manifold. Gas-segmented continuous flow analyzers, primarily based on Technicon Autoanalyzers or their derivatives, have been widely used in oceanographic research for chemical analyses onboard ship [15,16]. In the mid 1970s, scientists in several laboratories realized that gas bubbles were not necessary to segment flow in narrow tubing. Ruzicka and Hansen [17] coined the term flow injection analysis (FIA) to describe this method in 1975. The development of FIA demonstrated that continuous flow analyzers could operate equally well without air bubbles to segment the sample stream if small diameter tubing (≈.05 mm i.d.) was used to carry the sample. Flow is laminar in these systems and the dimensions are small enough that radial diffusion can act to limit dispersion of the sample [18].

FIA was rapidly adapted for environmental analyses [19]. These applications include analysis of gases, anions, cations, metals and organics. The technique has also been adapted for *in situ* applications by several laboratories. In the following sections, we review applications of CFA for on site and for *in situ* applications. We begin by reviewing applications to measurements of the major plant nutrients (nitrate, ammonia, phosphate and silicate). Applications to the determination of metal concentration and speciation are then considered. Finally, applications to other trace substances, including non-metals and organics, are considered.

A continuous flow or FIA system consists of one or more pumps used to propel sample and reagents, a hydraulic manifold system that carries the liquids and acts to mix them, a set of reagent reservoirs, and a flow-through detector that monitors the concentration of the analyte species [18,19]. The FIA system will also include a set of valves that are used to inject samples, reagents or standards. Complete systems for laboratory and on site applications are available from a variety of sources and a recent review [20] examines the various systems. Several manufacturers also maintain Internet sites with extensive information on FIA components and systems (e.g., Global FIA—http://www.globalfia.com/ and Alitea Instruments—http://www.flowinjection.com/). Virtually all of the *in situ* systems have been custom built, although at least one company (Chelsea Instruments, http://www.chelsea.co.uk/ci/index.htm) has begun to offer them commercially. It is beyond the scope of this chapter to review the various components that constitute these systems in detail. However, we will comment on some of the major choices that may be made where appropriate.

3 NUTRIENTS

3.1 ON SITE APPLICATIONS

The storage of samples prior to analysis at the very high levels of accuracy and precision required in oceanographic studies remains a contentious issue that is best resolved by analyzing samples on site [21]. On site analysis of the major nutrients has been a requirement of virtually all of the major oceanographic programs in recent years. The initial applications of CFA to on site analyses were made using gas-segmented systems [15,16]. However, gas-segmented CFA has little potential for *in situ* use, despite one early report of a prototype system designed to operate underwater [22].

FIA was introduced to oceanographic research by Andersen [23], who developed a system for nitrite and nitrate analyses. Early work with FIA identified several potential difficulties with the technique. Sensitivity was not as good as that obtained with segmented CFA systems because of dilution of the sample slug in the reaction manifold. The technique of reversed FIA was developed to overcome this difficulty [24]. In conventional FIA, a small volume of sample is injected into a carrier stream and the sample becomes diluted as it is dispersed into the carrier. If the sample is used as the carrier and an essential reagent is injected, then dispersion dilutes the reagent. The concentration of sample in the zone of the injected reagent increases over time. This can improve sensitivity several fold. A second problem can arise when the sample and carrier have different salinities. This creates a difference in refractive index between the sample and the carrier, which produces a signal in spectrophotometric detectors. The careful design of the detector, including the use of a reference wavelength where the analyte species does not absorb light, can minimize this effect [25]. These issues are considered in a review on optical absorption detector design [26]. It is also possible to design the reaction manifold to eliminate the refractive index signal [27].

FIA has come to be widely used in oceanographic studies, where it is used for a variety of analyses. The reviews by Atienza *et al.* [28,29] detail many of the applications to analyses of seawater. Analytical methods for major plant nutrients, inorganic carbon, organic carbon species and oxygen have been developed. Early applications of FIA in oceanography included interfacing the systems to pumps deployed over the side of ships to obtain high resolution vertical profiles of dissolved chemicals. These experiments permitted measurements of nutrients [30] and trace chemicals such as hydrogen peroxide [31,32] at high spatial resolution in the upper ocean. For example, Figure 2 shows the concentrations of nitrate measured over time with an FIA system [33] operating on board ship and interfaced to a flow-through pumping system. The large variability in nitrate is produced by the presence of a frontal system that is advecting water of different temperatures and chlorophyll concentrations past the stationary pump. Such variability cannot be observed by collecting a limited number of samples and returning them to a shore-based laboratory.

FIA systems can be designed to operate with relatively little maintenance. This has made it feasible to deploy them for long term monitoring. Early applications of such systems were primarily for industrial process monitoring [34]. However, several systems have also been developed for long term environmental studies, primarily in freshwater [35] or waste water [36]. These ruggedized systems are deployed above water and operate unattended for periods of time up to 1 month (Figure 3). They have proven to be useful for monitoring waste water streams for compliance with regulatory limits. Another important application for FIA systems is in studies where the volume of sample that is available for analysis is small. This includes measurements of chemical distributions in the interstitial waters of sediments [37].

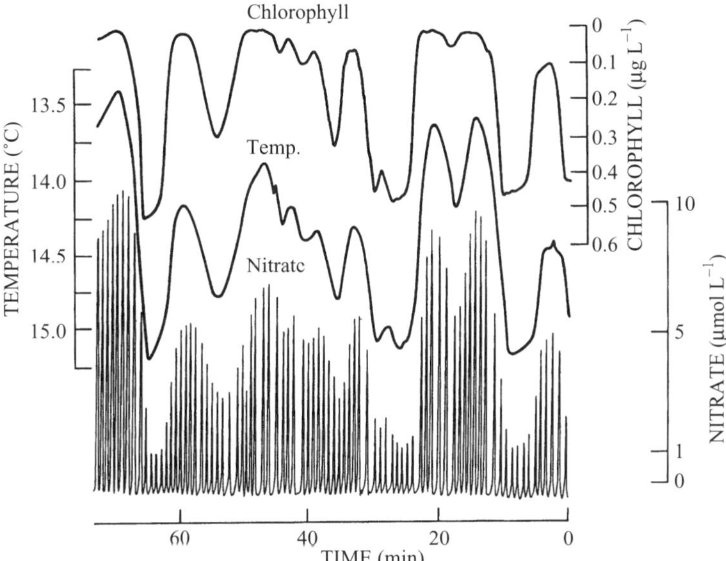

Figure 2. Values of nitrate (μmol L^{-1}), chlorophyll fluorescence (converted to units of μg L^{-1}) and temperature measured simultaneously in seawater pumped on board ship while the vessel drifted across a frontal zone in the Santa Barbara Channel, California [30]. These data illustrate the high temporal resolution that can be obtained for chemical measurements made with FIA systems

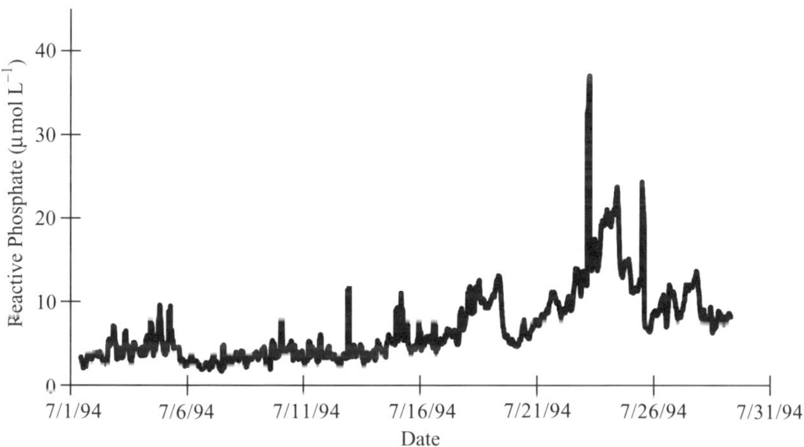

Figure 3. Measurements of phosphate in a wastewater stream made over a 1 month period (7 to 31 July 1994) with an unattended FIA analyzer [36]. Phosphate was measured spectrophotometrically as the reduced phosphomolybdate dye

3.2 *IN SITU* APPLICATIONS

3.2.1 Continuous Flow Analysis versus Flow Injection Analysis

The versatility of CFA, particularly FIA systems, suggested that it would be feasible to develop systems that could be operated under water. This would be particularly beneficial for deep-sea studies, where the great water depths make it very difficult to return samples to a ship with a pump [38]. A continuous flow analyzer, based upon the principles of unsegmented CFA, was developed in 1984 for *in situ* measurements in deep sea hydrothermal systems [39]. This Submersible chemical analyzer (commonly called a Scanner) was capable of simultaneously measuring two chemicals by colorimetry, as well as monitoring dissolved O_2 in the sample stream and temperature at the sample inlet. Submersible chemical analyzers are one of the most versatile means available today that allow remote monitoring in the ocean of the concentrations of dissolved nutrients. Work with *in situ* analyzers is ongoing in several laboratories. Recent advances have made it possible to produce systems with greatly reduced complexity and increased reliability for long term deployments. It is now possible to monitor concentrations of nutrient species for periods of time greater than 1 month. Below, we summarize the work that has been done to develop chemical analyzers that will operate under water.

A schematic layout of a Scanner system configured for the analysis of dissolved NO_3^- is shown in Figure 4 [40]. The determination of NO_3^- requires three reaction steps, each of which is essentially irreversible (Figure 5). Nitrate is reduced to NO_2^- by buffering the sample to a neutral pH and pumping the sample/buffer mixture through a column of copperized cadmium. The nitrite is then reacted with sulfanilamide in acid to form a diazonium ion. Finally, the diazonium ion is reacted with *N*-(1-naphthyl)ethylenediamine to form a brightly colored azo dye molecule, which can be quantified by spectrophotometry. The sample and reagents are propelled through the reaction manifold by a multi-channel pump and delivered to a colorimeter. The sample (or standard) is pumped continuously in this system, unlike FIA where the sample is introduced only periodically. The system can be calibrated by switching the valve on the inlet to select the sample, a blank or a standard solution with a known NO_3^- concentration (Figure 6).

These systems technically are not FIA. They are part of the more general class of analyses termed unsegmented CFA. They are derived directly from the principles of FIA, however, and lack only the injection valve commonly found in FIA. True FIA systems for *in situ* applications have now been developed by Taylor *et al.* [41], Daniel *et al.* [25] and David *et al.* [42]. The injection valve that is used in true FIA systems to periodically inject samples allows the detector baseline to be fully resolved between each injection. This contributes greatly to the stability of laboratory-based analytical systems because variations in detector baseline signal can be corrected for each sample. We have found that the greatest source of baseline variability in spectrophotometric systems is due to trapping of small air bubbles in the flow cell. Formation of air bubbles is not a problem when systems are operated at high pressures *in situ*. Baseline stability is quite high and variations can be adequately controlled for by analyses of

NITRATE ANALYTICAL MANIFOLD

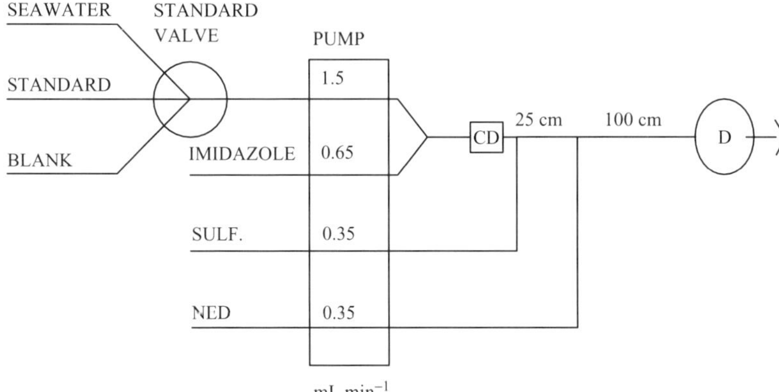

Figure 4. Reaction manifold used for the determination of nitrate *in situ* with a CFA system [40]. D is a photometer using a light emitting diode for the light source [26], CD is a granular cadmium column used to reduce nitrate to nitrite, SULF. is the sulfanilamide reagent and NED is the *N*-(1–napthyl)ethylenediamine reagent. The pump is a peristaltic pump with the motor encased in oil and the pump head in the water

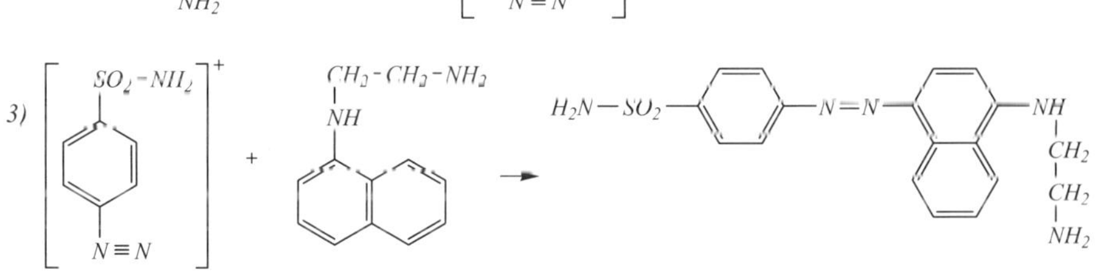

1) $NO_3^- + Cd + 2H^+ \rightarrow NO_2^- + Cd^{2+} + H_2O$

2) NO_2^- + (SO_2-NH_2 / NH_2) $+ 2H^+ \rightarrow$ [SO_2-NH_2 / $N\equiv N$]$^+$ $+ 2H_2O$

3) [SO_2-NH_2 / $N\equiv N$]$^+$ + ($CH_2-CH_2-NH_2$ / NH) $\rightarrow$ H_2N-SO_2- $-N=N-$ $-NH-CH_2-CH_2-NH_2$

Figure 5. The sequence of reactions used for the determination of nitrate

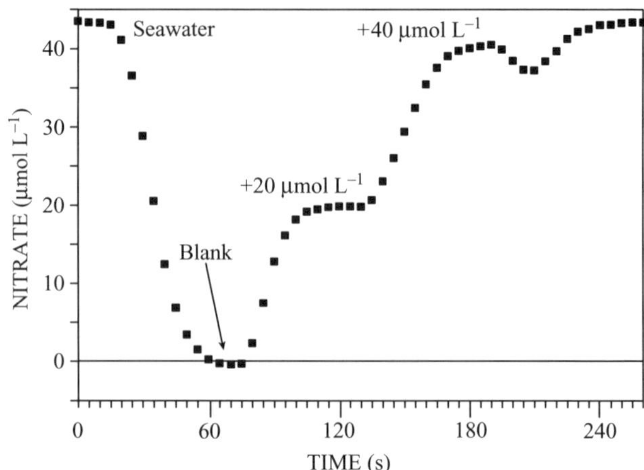

Figure 6. An *in situ* nitrate calibration at 2000 m depth off the California coast [40]. The analytical system was similar to that shown in Figure 4, except the standard valve allowed three standards (blank, 20 μmol L^{-1}, 40 μmol L^{-1}), as well as seawater to be analyzed. The figure shows the detector response changing from seawater with 43 μmol L^{-1} nitrate, to blank, to intermediate standard to high standard, followed by seawater. A small bolus of blank preceeds the last transition to seawater, causing the signal transition to lower nitrate concentrations

blank solutions at longer time intervals than used in FIA. The greater mechanical simplicity of systems without injection valves seems to outweigh the advantages of more baseline resolution.

3.2.2 Applications of CFA Systems

The first applications of a Scanner system were to studies of chemical distributions around chemoautotrophic animal communities in deep-sea hydrothermal systems [43]. These communities are typically located at depths greater than 2000 m. The animals in these communities are hosts to symbiotic, sulfur oxidizing bacteria [44]. The animals take up hydrogen sulfide from the reducing hydrothermal solutions and oxygen from the surrounding seawater, as well as carbon dioxide, and deliver the chemicals to their internal bacteria population. The bacteria then use the energy derived from oxidizing the sulfide to fix the inorganic carbon into reduced organic molecules via the Calvin–Benson cycle. These animals can only survive where they can obtain both the reduced (sulfide) and oxidized (oxygen) chemicals. High resolution measurements of the distribution of sulfide and oxygen around animals are an essential component for ecological studies of the distributions of these animals.

The original Scanner was configured to measure sulfide in the hydrothermal environment by the methylene blue colorimetric method [39]. It was also designed to simultaneously measure dissolved silicate (orthosilic acid) by the reduced silicomolybdate dye method. Silicate is a conservative tracer in vent environments, whose concentration changes are due primarily to dilution of hydrothermal solutions with seawater. It can be used to normalize sulfide concentrations so that sulfide changes due to uptake by the animal community can be separated from changes due to dilution with seawater. Figure 7 shows measurements of dissolved sulfide, silicate, oxygen and temperature as the inlet to the Scanner system is pushed from ambient seawater at a depth of 2200 m into a hydrothermal vent [43]. These measurements span a distance of less than 1 m and demonstrate the tremendous spatial variability that can occur in these systems. The Scanner system was mounted on the Deep Submergence Vehicle ALVIN to obtain these observations.

A variety of instruments based on CFA have now been developed at oceanographic laboratories around the world. Scanners have been used to determine the vertical distribution of chemicals in the ocean [40,41,45]. They have been used to monitor chemical variability in estuaries [25] and over the continental shelf [42]. Submersible chemical analyzers designed for observing hydrothermal vent systems have also been developed and applied to a variety of studies [46].

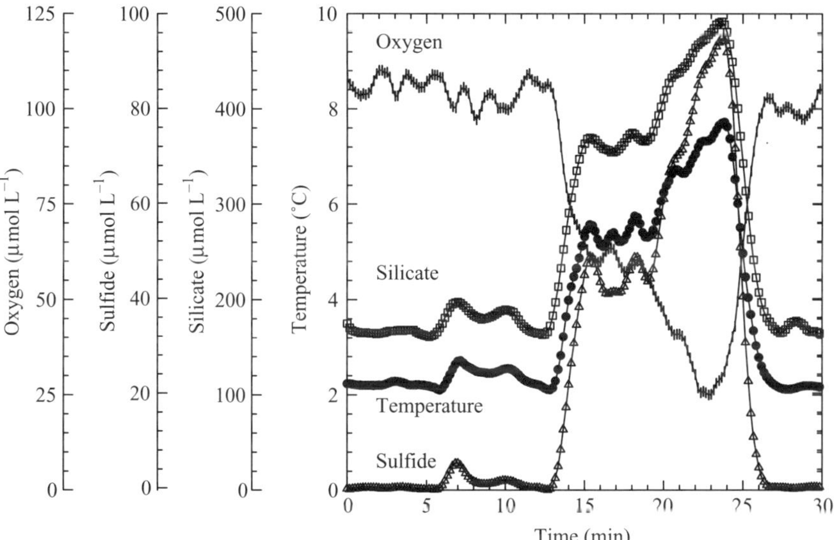

Figure 7. Measurements of silicate by the reduced silico-molybdate method and sulfide by the methylene blue method [39] made *in situ* with a CFA system at a hydrothermal vent system in the Galapagos Rift [43]. Corresponding measurements of temperature and oxygen are also shown. The measurements were made while the sample inlet was moved from ambient seawater into the vent by the Deep Submersible Vehicle ALVIN.

3.2.3 Pumping Systems

Long term observations of chemical variability using chemical analyzers placed on oceanographic moorings remain an important goal. The early generation of Scanners were not practical for such an application because, with few exceptions [41], they used peristaltic pumps to propel the sample and reagents. These pumps require large amounts of power to compress the pump tubing and they require frequent servicing to replace pump tubes. Improved pumping systems are required to meet long term *in situ* monitoring objectives. A significant step towards long term observations was made with the development of Scanner systems based on osmotically powered pumps [47]. These pumps utilize the osmotic pressure differential between seawater and saturated salt solutions and they require no electrical power. They have been deployed for periods up to 3 months. Instruments based on osmotic pumps have been used to collect very interesting data on nitrate variability in open ocean environments, which demonstrate the role of eddies in augmenting the vertical flux of nutrients into the euphotic zone [4,48].

Osmotic pumps cannot be easily turned on and off, however. This can limit their versatility, particularly in complex manifold systems. We have explored the application of micro-solenoid pumps in systems for the analysis of seawater nutrients and trace metals. Solenoid pumps are relatively robust and low powered, as they require only brief pulses of electricity to operate them. A prototype system for the analysis of nitrite using solenoid pumps has been described [49]. This system has been used to monitor dissolved chemical variability while submerged in coastal systems (Figure 8). Solenoid pumps may offer an efficient alternative for future *in situ* CFA systems.

3.2.4 Stability and Consumption of Reagents

There are several challenges in addition to those presented in developing mechanical systems that will operate *in situ* with no physical intervention from the operator. Standards and reagents must be stable for the duration of an instrument deployment. This presents one of the greatest challenges to the chemist when designing systems for long term deployments. Recent work has shown that it is possible to prepare standard solutions for many chemical species that are stable for long time periods [50]. However, the stability of some reagents, especially redox sensitive ones such as hydrogen peroxide, can be a limitation. The challenge to the chemist in operating these systems is to find suitable sets of reagents that remain stable for sufficiently long time periods. Over long time periods the volume of reagents consumed by chemical analyzer systems must be minimized. It is hard to envision deploying systems that contain more than about 1 L of each reagent without the physical size becoming an obstacle. This implies quite low flow rates for deployments longer than a few

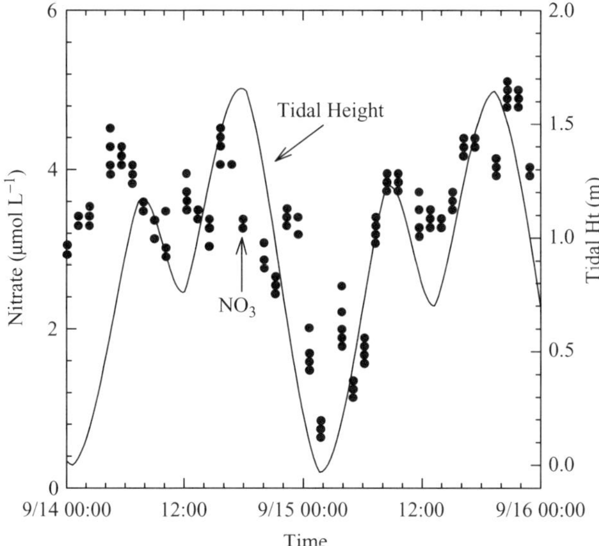

Figure 8. Measurements of nitrate made in the Monterey, California harbor at 2 m depth with a submersible CFA system that uses individual solenoid pumps [49] for each sample and reagent line. The instrument was programmed to collect six nitrate measurements at 30 s intervals every 70 min. Every 280 min the instrument recalibrates itself and nitrate was not determined. Note the strong correlation with local tides. This correlation is produced by oscillations of a front between high and low nitrate waters that is driven by tidal currents

days. At a flow rate of 0.5 mL min^{-1}, 720 mL would be pumped in 1 d. Flow rates must be substantially less than 0.5 mL min^{-1} over a 30 d deployment, or an analyzer must operate at a duty cycle on the order of 1/30, i.e. approximately 1 h d^{-1}. On the bright side, the relatively low flow rates make sample filtration a feasibility. Further, the low, linear flow rates in a FIA system (approx. 100 cm min^{-1} at a flow rate of 0.5 mL min^{-1} in 0.8 mm i.d. tubing) are small compared with settling rates of particles larger than 100 μm.

4 METALS

4.1 ON SITE APPLICATIONS

Trace metals can play an extremely important role in regulating the structure of aquatic communities. In some ocean ecosystems, concentrations of dissolved iron in the low picomolar range (<50 pmol L^{-1}) limit the growth rates of phytoplankton [51] and uptake of carbon dioxide [52]. This creates a mechanism that may allow iron to influence climate by regulating ocean phytoplankton primary production rates and the concentration of carbon dioxide in the atmosphere [51]. In the other extreme, high concentrations of copper in small embayments can regulate the structure of the phytoplankton community by limiting the growth of picoplankton, which have a low tolerance for copper [53].

Two major advances have made it possible to determine total concentrations of dissolved trace elements in seawater. First was the realization of the extraordinary efforts required to carry a seawater sample through each step of the sampling and analytical process without contamination [7]. Second was the adaptation of analytical methodologies with sufficient sensitivity and selectivity to determine trace metals at extremely low levels. The analytical methods most commonly used in oceanography involve a preconcentration and separation step using ion exchange resins or chelation and solvent extraction in shore-based laboratories. Even after concentration by factors from 100 to 1000, most trace elements in a seawater sample can be detected using only the most sensitive instrumentation. Organic extraction or solid-phase separations followed by graphite furnace atomic absorption spectrometry or ICPMS have been most widely used. Electrochemical methods such as anodic stripping voltammetry (ASV) or adsorptive cathodic stripping voltammetry (CSV), which involve preconcentration of the trace element on an electrode surface, are also popular for laboratory analyses and have been applied to *in situ* trace element analysis (see Chapter 9).

Much of the focus in the analytical chemistry of trace metals in seawater is now on the development of methods that can be used at sea to produce large data sets [e.g. 1, 54, 55]. Shipboard methods for the determination of trace metals have, in general, precluded atomic absorption and mass spectrometric detection as these instruments are sensitive to the constant vibration and accelerations experienced on board ships. Instead, a variety of methods have been developed for use at sea, including atomic fluorescence (Hg) [56], gas chromatography (Al, Be, As and Se) [e.g. 57], and electrochemical techniques such as differential pulsed (DP) ASV and DPCSV (Zn, Cd, Cu, Pb, Co, Ni and Fe) [58].

Recently, flow injection methods have been developed which have suitable detection limits for a suite of metals present in seawater, including Cu, Fe, Mn, Co, Al and Zn (Table 1). These techniques have used chemiluminescence, kinetic spectrophotometric and fluorescence methods of detection. Most FIA methods based on direct detection of a charge transfer complex between a metal ion and ligand do not provide sufficient sensitivity to determine the very low concentrations of dissolved metals found in seawater, unless there is a very large preconcentration factor [59]. Except in cases where metal concentrations are strongly elevated above natural levels, spectrophotometric detection methods in FIA are only practical using reactions where the metal has a catalytic effect that allows each metal atom to produce many dye molecules [60]. Chemiluminescence [61], fluorescence [62] and kinetic spectrophotometric [60] methods for analysis have been known for many years. These methods all have potentially

Table 1. Flow injection methods for on site analysis of trace metals in seawater that have been demonstrated to operate at ambient concentrations

Metal	FIA method of detection		
	Chemiluminescence	Kinetic–spectrophotometry	Fluorescence
Iron	66, 74, 77, 91, 92, 93	94	
Manganese	95, 96, 97	98, 99	
Cobalt	64		
Copper	73, 70		
Zinc			100
Aluminum			101

greater sensitivity than spectrophotometry based on formation of charge transfer complexes. However, chemiluminescence and fluorescence quantum yields are strongly dependent on environmental conditions. Small changes in sample properties can alter the yield of light sufficiently to bias the results. Kinetic spectrophotometric and chemiluminescence measurements are both proportional to the rate of a reaction. Variations in properties such as pH can easily affect the results. As a result of this sensitivity to environmental conditions, these detection methods have not previously been widely used for analysis of metals in environmental samples.

The development of FIA has made these methods much more practical. Flow injection systems allow analyses to be performed in a highly reproducible manner. If a column with an immobilized ligand is used in the analytical system, then the analyte species can be separated from interfering compounds, such as the major ions in seawater, and eluted into a medium with a uniform composition before analysis. In addition, a column allows the sample to be concentrated before analysis. Sensitivity can also be enhanced by adding micelles to many of the systems, which reduces the quenching of fluorescence and chemiluminescence that occurs in aqueous media [63]. Most of the methods used to measure metals at ambient concentrations in seawater employ all of these techniques to enhance sensitivity. The column can also be used to separate trace elements from the background matrix of major ions found in high salinity samples. Separation of the metals bound on the column from matrix elements in the solution that fills the interstices of the column may require a second valve in the system (Figure 9). The second valve can switch the sample stream to ultrapure water that is buffered at an appropriate pH [64]. The ultrapure water is used to flush the matrix elements from the interstitial spaces in the column and, if necessary, to elute weakly bound alkali and alkaline earth elements retained by the column.

FIA systems based on these principles can be quite compact and rugged, and they are well suited for deployment in the field. The total dissolved (filtered

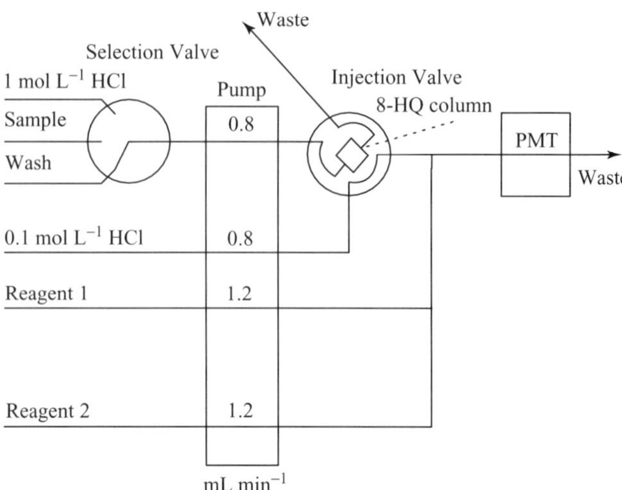

Figure 9. Two valve manifold for determination of metals in seawater by chemiluminescence or fluorescence. Manifolds similar to that shown have been used for determinations of Co [64], Mn [96], Cu [73], Fe [74] and Zn [100]. A column of immobilized 8-hydroxyquinoline is placed in the sample loop of the injection valve. A stream of 1 mol L^{-1} HCl is directed through the column, while the injection valve is in the load position, to clean it. The stream selection valve is then switched to the sample position and metal is loaded onto the column. After sufficient metal is loaded on the column, the stream selection valve is switched to a wash solution (ultrapure water buffered to an appropriate pH) to rinse seawater from the column interstices. The injection valve is then switched to the inject position, causing a weak HCl solution to pass through the column and elute the bound metals. The stream of eluent acid merges with several reagent streams and chemiluminescence or fluroescence is monitored at the detector

through 0.4 μm) metal concentrations determined with these methods compare very favorably with measurements by the classical methods using sample preconcentration followed by graphite furnace atomic absorption spectrophotometry. Vertical profiles of the concentrations of Co, Mn, Zn and Cu determined at open ocean stations in the Pacific Ocean are shown in Figure 10. The results are compared with conventional measurements obtained by sample preconcentration and GFAAS. These measurements performed at sea are generally indistinguishable from measurements made in shore-based laboratories for a suite of metals.

Organic ligands present in seawater bind much of the dissolved fraction of many metals, including copper, zinc and iron [65]. The organically bound fraction of these metals may be undetectable in systems that use a column because equilibrium would not be reached with the chelating column in the short time that sample passes through the system [66]. This problem can be avoided by acidifying samples to a pH near 3 before analysis. The results shown

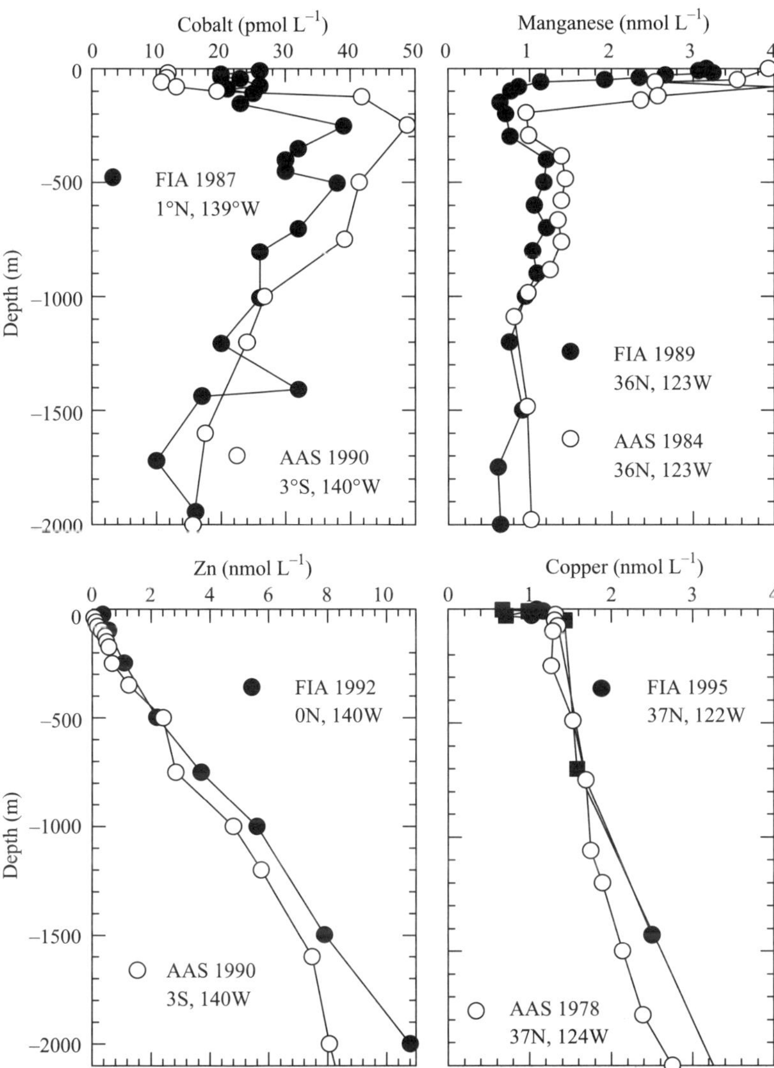

Figure 10. Comparison of dissolved metal concentrations determined at sea using FIA with chemiluminescence (Co, [64]; Mn, [96], and Cu [70]) or fluorescence (Zn, [100]) detection with metal concentrations determined by electrothermal atomization atomic absorption spectroscopy after preconcentration of the samples by organic extraction in shore-based laboratories at nearby stations. AAS data: Co, Moss Landing Marine Laboratories (MLML) unpublished; Mn, [102]; Zn, MLML unpublished; Cu, [103]

for zinc and copper in Figure 10 are for acidified samples. The agreement with conventional analyses demonstrates that accurate results can be obtained even for the strongly bound metals.

CFA techniques have been particularly useful for mapping metal distributions in the marine environment. Measurements of iron made on board of a ship during a deliberate iron addition experiment in the open ocean [51], are shown in Figure 11. The ability to measure iron in real time has been an important component of the success of these open ocean iron fertilization experiments.

4.2 METAL SPECIATION

4.2.1 Metal Complexation

Measurements of total element concentrations are generally insufficient to define chemical behavior in the environment. The chemical speciation of an element plays a major role in regulating its biological availability, toxicity and geochemical reactivity in seawater [65]. For example, copper is present in surface seawater at concentrations of less than 1 nmol L^{-1} [7]. Most of the copper is bound by an organic ligand that appears to be produced by organisms in the upper ocean [67]. This ligand has a high specificity for the copper. The ligand is

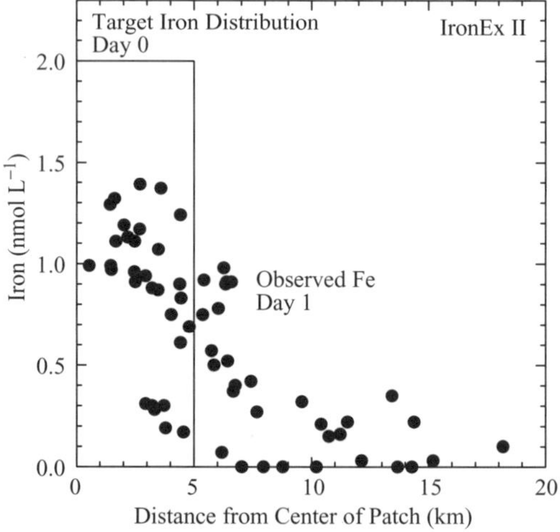

Figure 11. Dissolvable iron concentrations determined in the equatorial Pacific on the first day after fertilization of a 10 km × 10 km area with 200 kg of Fe (as $FeSO_4$ dissolved in acidified seawater) to reach a nominal concentration of 2 nmol L^{-1} [51]. Concentrations are plotted versus distance from the center of the patch. Iron was measured at sea by FIA with chemiluminescence detection using luminol as a reagent [66]. Difference between the observed iron distribution and the target distribution are due to iron adsorption onto sinking particles and due to horizontal diffusion

also present at concentrations on the order of 1 nmol L^{-1} and it regulates the copper(II) aquo-ion concentration. Concentrations of the copper aquo-ion are calculated to be on the order of 10^{-14} mol L^{-1} in solutions with 10^{-9} mol L^{-1} total copper [67]. These concentrations are undetectable by most analytical methods although some electrochemical methods that incorporate preconcentration at electrode surfaces can measure them and give speciation information (see Chapter 9). Studies of copper speciation have traditionally been performed using electrochemical methods such as DPASV at a thin mercury film/glassy carbon electrode, or cathodic stripping methods (DPCSV) using a hanging mercury drop electrode [65,68]. The electrochemical methods often involve lengthy deposition steps. However, CFA methods can also be used to study the speciation of metals [69]. In addition, the analyses can be performed quite rapidly.

Zamzow *et al.* [70] have demonstrated that copper binding in seawater can be determined using the chemiluminescence reaction of 1,10–phenanthroline [71,72] with uncomplexed Cu(II) in an FIA system (FIA-CL). Early work with this system required a column to separate the copper from the salt background in seawater [73]. Further modifications to the hardware and reaction manifold have eliminated the interferences, which allows copper to be determined by directly injecting seawater into the analytical system. The chemiluminescence produced is proportional to the amount of labile copper in the sample. Copper bound to organic ligands does not dissociate fast enough to produce chemiluminescence. The labile copper (Cu′) in most seawater samples is too low to be determined directly. However, the amount of ligand in a sample and the equilibrium constant for the reaction can be determined by adding increasing amounts of copper to a series of samples and measuring the labile copper in solution after allowing sufficient time for the added copper to equilibrate with any uncomplexed ligand (Figure 12). If the total copper concentration in the original sample is known, then this titration can be modeled mathematically to determine ligand concentration, ligand equilibrium constant and the free copper concentration in the original sample [70]. Acidification of samples to pH 3 dissociates the metal from the ligand and allows total concentrations to be determined.

4.2.2 Redox Speciation

Redox speciation of trace elements can also be performed with FIA methods. For example, the determination of iron in an FIA-CL system using the reagent brilliant sulfoflavine is selective for Fe(II) [74]. This system was used to determine the speciation of iron in the +II and +III oxidation states in samples from the vicinity of a deep sea hydrothermal system. All of the iron in the hydrothermal source solutions is present in the +II oxidation state. The half-life for transformation of Fe(II) to Fe(III) is approximately 32 h at the low temperatures and pH values found in this environment [75]. The ratio of Fe(II)

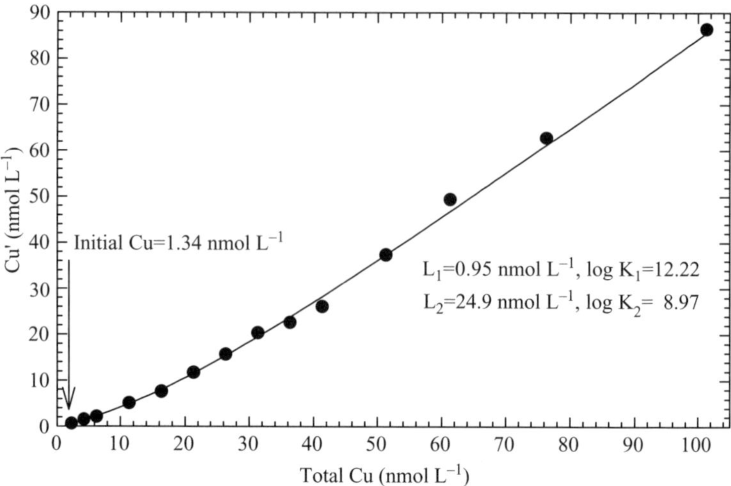

Figure 12. Labile copper (Cu′) detected by FIA with chemiluminescence detection is plotted versus total copper in a copper titration of a seawater sample [70]. Varying amounts of copper were added to subsamples of surface seawater sample from Monterey Bay and allowed to equilibrate for several hours before determination of Cu′ in each subsample. The solid line shows the results of a model fit to this titration data that incorporates two copper binding ligand classes with concentrations L_1 and L_2. The equilibrium constant is a conditional stability constant ($K = [CuL]/[Cu'][L_T{-}CuL]$)

to Fe(III) can be combined with the half-life to estimate the time that has elapsed since the hydrothermal solution entered the ocean [76].

The kinetics of metal oxidation at low nanomolar concentrations have been studied using FIA-CL systems. The ability to measure reaction rates at very low concentrations in natural seawater samples is an important goal because it ensures that small concentrations of organic ligands do not alter metal reaction rates. King *et al.* [77] measured the oxidation rate of Fe(II) in seawater samples with less than 100 nmol L^{-1} total iron concentrations. Von Langen *et al.* [78] measured the oxidation rate of Mn(II) in seawater at concentrations less than 20 nmol L^{-1}. In both cases, they found good agreement with measurements performed in solutions that contained micro molar concentrations of the metal. Organic ligands that strongly complex iron exist in seawater at concentrations near 1 nmol L^{-1} [79]. The oxidation and chemical scavenging rate of Fe(II) may be substantially reduced when iron drops below that level [1].

4.3 *IN SITU* APPLICATIONS

One of the greatest challenges in measurements of trace element concentrations in environmental samples is the elimination of contamination during sampling

[7,13]. *In situ* measurements of metal concentrations greatly reduce the risk of contamination and allow much higher sampling resolution. The earliest applications of CFA to *in situ* analysis of metals in seawater were based on spectrophotometric measurements produced by forming metal–ligand charge transfer complexes that absorb light [76,80]. These studies focused on systems which had elevated metal concentrations, therefore. The analytical manifolds used for this work (Figure 13) are similar to those described above for nitrate measurements [81]. Samples and reagents are propelled with a peristaltic pump through a manifold of 0.5 mm i.d. Teflon tubing and passed through a photometric detector. *In situ* systems may not filter samples to discern between dissolved and particulate chemicals. However, the reaction conditions are relatively mild, and most of the reactive chemical that is detected is in the dissolved form. Measurements of dissolvable (dissolved + easily leached at pH near 3) trace metal concentrations made *in situ* are in good agreement with measurements of dissolved metal concentrations made by atomic absorption spectrophotometry in shore-based laboratories [76].

In situ measurements of dissolvable metals using unsegmented CFA can yield much higher resolution measurements than are possible by any other technique. The response time for analytical systems is on the order of 20 s. At typical instrument lowering rates (20–60 m min^{-1}), this gives a spatial resolution of 6–20 m. A profile of dissolvable Mn through a hydrothermal vent plume demonstrates the resolution that is attainable (Figure 14).

Subsequent work has resulted in the development of CFA systems with higher sensitivity. For example, Klinkhammer [82] has developed a CFA system for the analysis of Mn that uses fluorescence detection. This very simple analytical system uses a novel system to add the reagents KIO_4 and N,

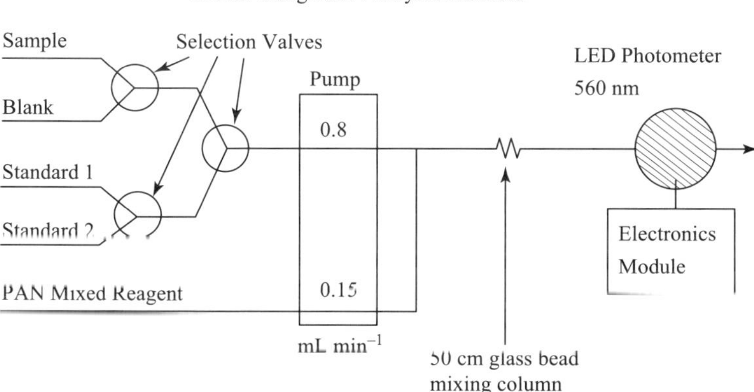

Figure 13. Reaction manifold used for the determination of dissolvable manganese *in situ* by complex formation with 1-(2-pyridylazo)-2-naphthol (PAN) [81].

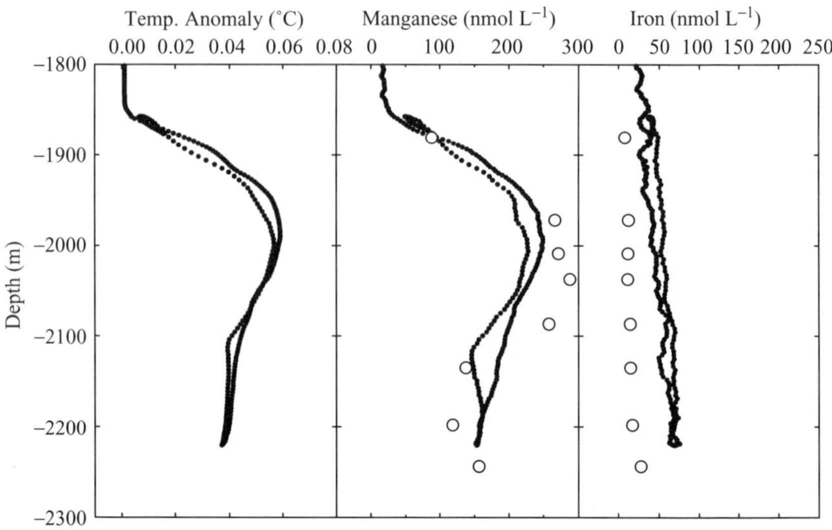

Figure 14. Vertical profile of temperature anomaly, dissolvable manganese and iron measured *in situ* over hydrothermal vent systems on the Juan de Fuca Ridge [76] with a CFA system (closed circles). Measurements of dissolved manganese and dissolved iron in samples collected with Niskin bottles and analyzed by electrothermal atomization atomic absorption spectrophotometry in a shore laboratory are shown for comparison (open circles)

N′-diethylaniline (DEA). Acrylic beads impregnated with KIO_4 and a solid rod of fibrous fluorocarbon with the DEA reagent adsorbed to it slowly leach the reagents into a flowing stream of seawater. Mn(II) reacts with the reagents to suppress fluorescence of the DEA. This system has a detection limit of $< 0.1\,\text{nmol}\,L^{-1}$ and is capable of detecting manganese at ambient levels ($< 1\,\text{nmol}\,L^{-1}$) in seawater. It has been deployed to measure manganese in hydrothermal systems [83] and coastal environments [84].

5 OTHER TRACE COMPOUNDS

Flow injection methods have been applied to the on site determination of other trace compounds. These methods are particularly important because the highly reactive nature of many trace compounds precludes storage of samples for later analysis in the laboratory. For example, the concentration of hydrogen peroxide in samples stored in the dark can decay by half in 1 d [31]. Measurements of hydrogen peroxide have been made on ships at sea using spectrophotometric- [31], chemiluminescence- [32] and fluorescence- [85] based detection methods in CFA systems. Studies of the photochemical processes that control this compound would not be possible without methods that can be deployed in the field.

FIA is also becoming used more frequently for the determination of a suite of organic compounds. Early applications focused on non-specific analyses, such as the determination of total organic carbon [86]. More recently, FIA methods have become widely applied to the determination of specific groups of organic compounds. Polycyclic aromatic hydrocarbons have been determined in FIA systems using chemiluminescence detection [87]. Fluorescence-based detection has been used in FIA systems for the determination of nucleic acid content in fish larvae [88]. Biosensors for organic compounds are often incorporated into flow injection systems to automate sample delivery to the sensor. For example, a sensor for amines that can be applied to assessment of freshness in fish has been used in a flow injection system [89]. Immunoassays have been developed for a variety of organic compounds and incorporated into FIA systems, as well. These systems incorporate antibodies, which have been expressed to compounds such as pesticide residues, to detect specific organic molecules or classes of molecules [90]. These methods have not yet been widely used for either on site or *in situ* analyses. However, there are no inherent limitations that would restrict their use to laboratory environments.

6 CONCLUSIONS

CFA methods have proven to be an excellent tool for on site analyses of nutrients and metals in studies of aquatic systems. They are a promising tool for *in situ* analyses, as well. The apparatus of a CFA or FIA system can be adapted to a wide range of reactions and detectors. It is feasible to carry out all of the main components of a chemical analysis with these systems, including sample treatment (filtration, buffering, preconcentration, matrix separation) and calibration with blanks and standard solutions. Although this chapter has focused on optical methods of detection using molecular spectroscopy, chemiluminescence or fluorescence, CFA systems can and have been adapted to operate with virtually all of the detection methods discussed in the chapters of this book. Given the flexibility of these systems, it may not be proper to consider these systems as sensors, but rather as miniature laboratories designed to automate the various steps of a chemical analysis. They represent an intermediate step towards the development of micro-electromechanical systems such as those disscused in Chapter 12.

The advantages and limitations of using CFA and FIA for on site and *in situ* analysis relate primarily to the flexibility of their architecture. They are mechanical systems that comprise pumps, hydraulic systems, valves and detectors. These systems allow the analyst to incorporate many more steps in an on site analysis, relative to simple sensors. This can improve sensitivity or eliminate interferences. Stability of the detector systems is not as critical in these instruments because they can be recalibrated as frequently as required, even when deployed at full ocean depth for *in situ* chemical determinations.

The limitations of using CFA are also related to this mechanical complexity. CFA systems are inherently more complex than a simple sensor. They have many potential failure modes that affect their reliability for on site or *in situ* deployments over long time periods. Robust hydraulic components, primarily reliable, low power pumps and, where necessary, simple valve systems, are required for these systems to become widely accepted as routine monitoring and research tools. Long term deployments of CFA systems will also require stable chemical reagent systems. It is common practice for many laboratory analyses to have reagent systems that must be prepared daily. This is not as easy to accomplish when systems are deployed in remote areas. Overcoming this problem will require development of reagent solutions with long term stability, or the development of automated systems, perhaps based on electrochemistry, to generate highly unstable reagents from stable reactants. Finally, continued effort must be made to reduce the volumes of reagents required in CFA systems for long term deployments.

GLOSSARY

Chemical analyzer A device in which mass transport carries the chemical species to be determined to the detector.

Chemical sensor A device in which passive diffusion transports the chemical species to be determined to the detector.

Conditional stability constant An equilibrium constant expressed as a concentration product and valid only for a medium similar to that in which it was measured.

Continuous flow analysis (CFA) An analytical process in which the sample is introduced by aspiration and the concentration of analyte is measured uninterruptedly in the stream of liquid. The sample stream may be segmented by gas bubbles or unsegmented.

Flow injection analysis (FIA) An analytical process in which the sample is periodically injected into a flowing carrier liquid and the concentration of analyte is measured uninterruptedly in the stream of liquid. The carrier stream is unsegmented.

Solid-phase reactor Solid phase, such as immobilized chelate or redox reactive metal, that reacts with the analyte in the flowing sample stream.

Submersible chemical analyzer (Scanner) A continuous flow analyzer configured to operate underwater.

REFERENCES

1. Johnson, K. S., Gordon, R. M. and Coale, K. H. (1997) What controls dissolved iron in the world ocean?, *Mar. Chem.*, **57**, 137–161.
2. Dickey, T., Marra, J., Granata, T., Langdon, C., Hamilton, M., Wiggert, J., Siegel, D. and Bratkovich, A. (1991). Concurrent high resolution bio-optical and physical time series observations in the Sargasso Sea during the spring of 1987, *J. Geophys. Res.*, **96**, 8643–8663.
3. Zuehlke, R. W. and Kester, D. R. (1985). Development of shipboard copper analyses by atomic absorption spectroscopy. In *Mapping Strategies in Chemical Oceanography*, ed. Zirino, A., Advances in Chemistry Series, No. 209, American Chemical Society, Washington, DC, pp. 117–137.
4. Johnson, K. S. and Jannasch, H. W. (1994). Analytical chemistry under the sea surface: monitoring ocean chemistry *in situ*, *Naval Res. Rev.*, **XLVI** (3), 4–12.
5. DeGrandpre, M. D. and Bellerby, R. G. J. (1995). Chemical sensors in marine science, *Oceanus*, Spring/Summer, 30–32.
6. Johnson, K. S., Coale, K. H. and Jannasch, H. W. (1992). Analytical chemistry in oceanography, *Anal. Chem.*, **64**, 1065A–1075A.
7. Bruland, K. W., (1983) Trace elements in seawater. In *Chemical Oceanography*, Vol. 8, ed. Riley, J. P. and Chester, R., Academic Press, London, pp. 157–220.
8. Brendel, P. J. and Luther, G. W., (1995). Development of a gold amalgam voltammetric microelectrode for the determination of dissolved Fe, Mn, O sub(2), and S(−II) in porewaters of marine and freshwater sediments, *Environ. Sci.Technol.*, **29**, 751–761.
9. Walt, D. R. (1992). Designing new sensors with old chemistry, *ChemTech*, November, 658–663.
10. Windom, H. L., Byrd, J. T., Smith, Jr., R. G. and Huan., F. (1991). Inadequacy of NASQAN data for assessing metal trends in the nation's rivers, *Environ. Sci. Technol.*, **25**, 1137–1142.
11. Coale, K. H. and Flegal, A. R. (1989). Copper, Zinc, cadmium and lead in surface waters of Lakes Erie and Ontario, *Sci. Total Environ.*, **87–88**, 297–304.
12. Borg, H. (1995). Trace elements in lakes. In *Trace Elements in Natural Waters*, ed. Salbu, B. and Steinnes, E., CRC Press, pp. 177–201.
13. Benoit, G., Hunter, K. S. and Rozan, T. R. (1997). Sources of trace metal contamination artifacts during collection, handling and analysis of freshwaters, *Anal. Chem.*, **69**, 1006–1011.
14. Luque de Castro, M. D. and Tena, M. T. (1993). Solid interfaces as analytical problem solvers in flow injection analysis, *Talanta*, **40**, 21–36.
15. Brewer, P. G. and Riley, J. P. (1966). The automatic determination of nitrate in seawater, *Deep-Sea Res.*, **12**, 765–772.
16. Mee, L. D. (1986). Continuous flow analysis in chemical oceanography: principles, applications and perspectives, *Sci. Tot. Environ.*, **49**, 27–87.
17. Ruzicka, J. and Hansen, E. H. (1975). *Anal. Chim. Acta*, **78**, 145.
18. Ruzicka, J. and Hansen, E. H., *Flow Injection Analysis*, 2nd edn., Wiley, New York, p. 498.
19. Valcarcel, M., and Luque de Castro, M. D. (1987). *Flow-Injection Analysis*, Horwood, pp. 354–377.
20. Newman, A. (1996). In the swim of flow injection analysis, *Anal. Chem.*, 68, 203A–206A.
21. Kirkwood, D. S. (1992). Stability of solutions of nutrient salts during storage, *Mar. Chem.*, **38**, 151–164.

22. Piro, A. and Rossi, G. (1969). Determinazione dei nutrienti mediante autoanalyzer, *Pubbl. Staz. Zool. Napoli*, **37**, suppl., 290.
23. Anderson, L. (1979). Simultaneous spectrophotometric determination of nitrite and nitrate by flow injection analysis, *Anal. Chim. Acta*, **110**, 123.
24. Johnson, K. S. and Petty R. L. (1982). Determination of phosphate in seawater by flow injection analysis with injection of reagent, *Anal. Chem.*, **54**, 1185–1187.
25. Daniel, A, Birot, D., Blain, S, Treguer, P, Leielde, B. and Menut, E. (1995). A submersible flow-injection analyser for the in-situ determination of nitrite and nitrate in coastal waters, *Mar. Chem.*, **51**, 67–77.
26. Dasgupta, P. K., Bellamy, H. S., Liu, H., Lopez, J. L., Loree, E. L., Morris, K., Petersen, K. and Mir, K. A. (1993). Light emitting diode based flow-through optical absorption detectors, *Talanta*, **40**, 53–74.
27. McKelvie, I. D., Peat, D. M. W., Matthews, G. P. and Worsfold, P. J. (1997). Elimination of the Schlieren effect in the determination of reactive phosphorous in estuarine waters by flow-injection analysis, *Anal. Chim. Acta*, **351**, 265–271.
28. Atienza, J., Herrero, M. A., Maquieira, A. and Puchades, R. (1991). Flow injection analysis of seawater: anionic and organic species, *Crit. Rev. Anal. Chem.*, **22**, 331–344.
29. Atienza, J., Herrero, M. A., Maquieira, A. and Puchades, R. (1991). Flow injection analysis of seawater. Part II. Cationic species, *Crit. Rev. Anal. Chem.*, **23**, 1–14.
30. Johnson, K. S., Petty, R. L. and Thomsen, J. (1985). Flow injection analysis for seawater micronutrients. In *Mapping Strategies in Chemical Oceanography*, ed. Zirino, A., Advances in Chemistry Series, No. 209, American Chemical Society, Washington, DC, pp. 7–30.
31. Johnson, K. S., Willason, S. W., Wiesenburg, D. A., Lohrenz, S. E. and Arnone, R. A. (1989). Hydrogen peroxide in the Western Mediterranean Sea: a tracer for vertical advection, *Deep-Sea Res.*, **36**, 241–254.
32. Price, D., Mantoura, R. F. C. and Worsfold, P. J. (1988). Shipboard determination of hydrogen peroxide in the western Mediterranean sea using flow injection with chemiluminescence detection, *Anal. Chim. Acta*, **371**, 205–215.
33. Johnson, K. S., and Petty, R. L. (1983). Determination of nitrate and nitrite in seawater by flow injection analysis, *Limnol. Oceanogr*. **28**, 1260–1266.
34. Gisin, M. and Thommen, C. (1989). Contemporary wet-chemical flow analyzers for process control, *Trends Anal. Chem.*, **8**, 62–66.
35. Blundell, N. J., Worsfold, P. J., Casey, H. and Smith, S. (1995). The design and performance of a portable, automated flow injection monitor for the in-situ analysis of nutrients in natural waters, *Environ. Int.*, **21**, 205–209.
36. Benson, R. L., Truong, Y. B., McKelvie, I. D., Hart, B. T., Bryant, G. W. and Hilkman, W. P. (1996). Monitoring of dissolved reactive phosphorous in waste-waters by flow injection analysis. Part 2. On-line monitoring system, *Wat. Res.*, **30**, 1965–1971.
37. Hall, P. O. J. and Aller, R. C. (1992). Rapid, small-volume, flow injection analysis for CO_2 and NH_4^+ in marine and freshwaters, *Limnol. Oceanogr.*, **37**, 1113–1119.
38. Gay, Jr., S. M. and Folb, R. (1981). Physics and practical aspects of towing, with emphasis on underway water sampling. In *Water Sampling While Underway*, ed. Tulin, M. P., National Academy Press, pp. 85–127.
39. Johnson, K. S., Beehler, C. L. and Sakamoto-Arnold, C. M. (1986). A submersible flow analysis system, *Anal. Chim. Acta*, **179**, 245–257.
40. Johnson, K. S., Sakamoto-Arnold, C. M. and Beehler, C. L. (1990). Continuous determination of nitrate concentrations *in situ*, *Deep-Sea Res.*, **36**, 1407–1413.

41. Taylor, C. D., Howes, B. L. and Doherty, K. W. (1993). Automated instrumentation for time-series measurement of primary production and nutrient status in production platform-accessible environments, *Mar. Tech. Soc. J.*, **27**, 32–44.
42. David, A. R. J., McCormack, T., Morris, A. W. and Worsfold, P. J. (1998). A submersible flow injection-based sensor for the determination of total oxidised nitrogen in coastal waters, *Anal. Chim. Acta* **361**, 63–72.
43. Johnson, K. S., Beehler, C. L., Sakamoto-Arnold, C. M. and Childress, J. J. (1986). *In situ* measurements of chemical distributions in a deep-sea hydrothermal vent field, *Science*, **231**, 1139–1141.
44. Childress, J. J. and Fisher, C. R. (1992). The biology of hydrothermal vent animals: physiology, biochemistry, and autotrophic symbioses, *Oceanogr. Mar. Biol. Annu. Rev.*, **30**, 337–441.
45. Gamo, T., Sakai, H., Nakayama, E., Ishida, K. and Kimoto, H. (1994). A submersible flow-through analyzer for *in situ* colorimetric measurement down to 2000 m depth in the ocean, *Anal. Sci.*, **10**, 843–848.
46. Massoth, G. J., Baker, E. T., Feely, R. A., Butterfield, D. A., Embley, R. E., Lupton, J. E., Thomson, R. E. and Cannon, G. A. (1995). Observations of manganese and iron at the CoAxial seafloor eruption site, Juan de Fuca Ridge, *Geophys. Res. Lett.*, **22**, 151–154.
47. Jannasch, H. W., Johnson, K. S. and Sakamoto, C. M. (1994). A submersible, osmotically pumped analyzer for continuous determination of nitrate, *Anal. Chem.*, **66**, 3352–3361.
48. McGillicuddy, Jr., D. J., Robinson, A. R., Siegel, D. A., Jannasch, H. W., Johnson, R., Dickey, T. D., McNeil, J., Michaels, A. F. and Knap, A. H. (1998). Influence of mesoscale eddies on new production in the Sargasso Sea, *Nature*, **394**, 263–266.
49. Weeks, D. A. and Johnson, K. S. (1996). Solenoid pumps for flow injection analysis, *Anal. Chem.*, **68**, 2717–2719.
50. Aminot, A. and Kerouel, R. (1996). Stability and preservation of primary calibration solutions of nutrients, *Mar. Chem.*, **52**, 173–181.
51. Coale, K. H., Johnson, K. S., Fitzwater, S. E., Gordon, R. M., Tanner, S., Chavez, F. P., Ferioli, L., Sakamoto, C., Rogers, P., Millero, F., Steinberg, P., Nightingale, P., Cooper, D., Cochlan, W. P., Landry, M. R., Constantinou, J., Rollwagen, G., Trasvina, A. and Kudela, R. (1996). The IronEx-II mesoscale experiment produces massive phytoplankton blooms in the equatorial Pacific, *Nature*, **383**, 495–501.
52. Watson, A. J., Law, C. S., Van Scoy, K. A., Millero, F. J., Yao, W., Friederich, G. E., Liddicoat, M. I., Wanninkhof, R. H., Barber, R. T. and Coale, K. H. (1994). Minimal effect of iron fertilization on sea-surface carbon dioxide concentrations, *Nature*, **371**, 143–145.
53. Moffett, J. W., Brand, L. E., Croot, P. L. and Barbeau, K. A. (1997). Cu speciation and cyanobacterial distribution in harbors subject to anthropogenic Cu inputs, *Limnol. Oceanog.*, **42**, 789–799.
54. Shiller, A. M. (1997). Manganese in surface waters of the Atlantic Ocean, *Geophys. Res. Lett.*, **24**, 1495–1498.
55. Yeats, P. A. (1998). An isopycnal analysis of cadmium distributions in the Atlantic Ocean, *Mar. Chem.*, **61**, 15–24.
56. Gill, G. A. and Bruland, K. W. (1990). Mercury speciation in surface freshwater systems in California and other areas, *Environ. Sci. Technol.*, **24**, 1392–1400.
57. Measures, C. I. and Edmond, J. M. (1989). Shipboard determination of aluminum in seawater at the nanomolar level by electron capture detection gas chromatography, *Anal. Chem.*, **61**, 544–547.

58. Bruland, K. W., Coale, K. H. and Mart, L. (1985). Analysis of sea water for dissolved cadmium, copper and lead: an intercomparison of voltammetric and atomic absorption methods, *Mar. Chem.*, **17**, 285–300.
59. Blain, S. and Treguer, P. (1995). Iron(II) and iron(III) determination in sea water at the nanomolar level with selective on-line preconcentration and spectrophotometric determination, *Anal. Chim. Acta*, **308**, 425–432.
60. Mottola, H. A. (1988). *Kinetic Aspects of Analytical Chemistry*, Wiley, New York, 285 p.
61. Townshend, A. (1990). Solution chemiluminescence—some recent analytical developments, *Talanta*, **115**, 495–500.
62. Bright, F. V. (1988). Bioanalytical applications of fluorescence spectroscopy, *Anal. Chem.*, **60**, 1031A–1039A.
63. McIntire, G. L. (1990). Micelles in analytical chemistry, *CRC Rev. Anal. Chem.*, **21**, 257–278.
64. Sakamoto-Arnold, C. M. and Johnson, K. S. (1987). Determination of picomolar levels of cobalt in seawater by flow injection analysis with chemiluminescence detection, *Anal. Chem.* **59**, 1789–1794.
65. Donat, J. R. and Bruland, K. W., (1995). Trace elements in the oceans. In *Trace Metals in Natural Waters*, ed. Steinnes, E. and Salbu, B., CRC Press, Boca Raton, FL, pp. 247–281.
66. Obata, H., Karatani, H., Matsui, M. and Nakayama, E. (1997). Fundamental studies for chemical speciation of iron in seawater with an improved analytical method, *Mar. Chem.*, **56**, 97–106.
67. Coale, K. H. and Bruland, K. W. (1988). Copper complexation in the northeast Pacific, *Limnol. Oceanogr.*, **33**, 1084–1101.
68. Van Den Berg, C. M. G., (1988). Electroanalytical chemistry of seawater. In *Chemical Oceanography*, Vol. 9, ed. Riley, J. P., Academic Press, London, pp. 197–245.
69. Luque de Castro, M. D. (1986). Speciation studies by flow-injection analysis, *Talanta*, **33**, 45–50.
70. Zamzow, H., Coale, K. H., Johnson K. S. and Sakamoto, C. M. (1998). Determination of copper complexation in seawater using flow injection analysis with chemiluminescence detection, *Anal. Chim. Acta*, **377**, 133–144.
71. Yamada, M. and Suzuki, S. (1984). Micellar enhanced chemiluminescence of 1,10-phenanthroline for the determination of ultratraces of copper(II) by flow injection method, *Anal. Lett.*, **17**, 251–263.
72. Sunda, W. G. and Huntsman, S. A. (1991). The use of chemiluminescence and ligand competition with EDTA to measure copper concentration and speciation in seawater, *Mar. Chem.*, **36**, 137–163.
73. Coale, K. H., Stout, P. M., Johnson, K. S. and Sakamoto, C. M. (1992). Determination of copper in seawater by the flow injection method with chemiluminescence detection, *Anal. Chim. Acta*, **266**, 345–351.
74. Elrod, V., Johnson, K. S. and Coale, K. H. (1991). Determination of subnanomolar levels of iron(II) and total dissolved iron in seawater by flow injection analysis with chemiluminescence detection, *Anal. Chem.*, **63**, 893–898.
75. Millero, F. J., Sotolongo, S. and Izaguirre, M., (1987). The oxidation kinetics of Fe(II) in seawater, *Geochim. Cosmochim. Acta*, **51**, 793–801.
76. Chin, C. S., Coale, K. H., Elrod, V., Johnson, K., Massoth, G. and Baker, E., (1994). *In situ* observations of dissolved iron and manganese in hydrothermal vent plumes, Juan de Fuca Ridge, *J. Geophys. Res.*, **99**, 4969–4984.

77. King, D. W., Lounsbury, H. A. and Millero, F. J. (1995). Rates and mechanism of Fe(II) oxidation at nanomolar total iron concentrations, *Environ. Sci. Technol.*, **29**, 818–824.
78. von Langen, P. J. , Johnson, K. S., Coale, K. H. and Elrod, V. A. (1997). Oxidation kinetics of manganese(II) in seawater at nanomolar concentrations, *Geochim. Cosmochim. Acta*, **61**, 4945–4954.
79. Rue, E. L. and Bruland, K. W. (1995). Complexation of iron(III) by natural organic ligands in the Central North Pacific as determined by a new competitive ligand equilibration/adsorptive cathodic stripping voltammetric method, *Mar. Chem.*, **50**, 117–138.
80. Coale, K. H., Chin, C. S., Massoth, G. J., Johnson, K. S. and Baker, E. T. (1991). *In situ* chemical mapping of dissolved iron and manganese in hydrothermal plumes, *Nature*, **352**, 325–328.
81. Chin, C. S., Johnson, K. S. and Coale, K. H. (1992). Spectrophotometric determination of dissolved manganese in natural waters with 1-(2-pyridylazo)-2-naphthol: application to analysis *in situ* in hydrothermal plumes. *Mar. Chem.*, **37**, 65–82.
82. Klinkhammer, G. P. (1994). Fiber optic spectrometers for in-situ measurements in the oceans: the ZAPS probe, *Mar. Chem.*, **47**, 13–20.
83. Klinkhammer, G. P., Chin, C. S., Wilson, C. and German, C. R. (1995). Venting from the Mid-Atlantic Ridge at 37° 17′N: the Lucky Strike hydrothermal site. In *Hydrothermal Vents and Processes*, ed. Parson, L. M., Walker, C. L. and Dixon, D. R., Geological Society Special Publication No. 87, pp. 87–96.
84. Klinkhammer, G. P., Chin, C. S., Wilson, C., Rudnicki, M. D. and German, C. R. (1997). Distributions of dissolved manganese and fluorescent dissolved organic matter in the Columbia River estuary and plume as determined by *in situ* measurement, *Mar. Chem.*, **56**, 1–14.
85. Amouroux, D. and Donard, O. F. X. (1995). Hydrogen peroxide determination in estuarine and marine waters by flow injection with fluorescence detection, *Oceanol. Acta*, **18**, 353–361.
86. Edwards, R. T., McKelvie, I. D., Ferrett, P. C., Hart, B. T., Bapat, J. B. and Koshy, K. (1992). Sensitive flow-injection technique for the determination of dissolved organic carbon in natural and waste waters, *Anal. Chim. Acta*, **261**, 287–294.
87. Andrew, K. N., Sanders, M. G., Forbes, S. and Worsfold, P. J. (1997). Flow methods for the determination of polycyclic aromatic hydrocarbons using low power photomultiplier tube and charge coupled device chemiluminescence detection, *Anal. Chim. Acta*, **346**, 113–120.
88. Canino, M. F. and Caldarone, E. M. (1995). Modification and comparison of two fluorometric techniques for determining nucleic acid contents of fish larvae, *Fish. Bull.*, **93**, 158–165.
89. Chemnitius, G. C., Suzuki, M. and Isobe, K. (1992). Thin-film biosensor: substrate specificity and application to fish freshness determination, *Anal. Chim. Acta*, **263**, 93–100.
90. Kraemer, P. M. (1997). Automation of immunochemical analysis—a prototype instrument to determine diuron or atrazine in water, *Lab. Robot. Autom.*, **9**, 81–89.
91. Powell, R. T., King, D. W. and Landing, W. M. (1995). Iron distributions in surface waters of the south Atlantic, *Mar. Chem.*, **50**, 13–20.
92. Obata, H., Karatani, H. and Nakayama, E. (1993). Automated determination of iron in seawater by chelating resin concentration and chemiluminescence detection, *Anal. Chem.*, **65**, 1524–1528.

93. Bowie, A. R., Achterberg, E. P., Mantoura, R. F. C. and Worsfold, P. J. (1998). Determination of sub-nanomolar levels of iron in seawater using flow injection with chemiluminescence detection, *Anal. Chim. Acta*, **361**, 189–200.
94. Measures, C. I., Yuan, J. and Resing, J. A. (1995). Determination of iron in seawater by flow injection analysis using in-line preconcentration and spectrophotometric detection *Mar. Chem.*, **50**, 3–12.
95. Nakayama, E., Isshiki, K., Sohrin, Y. and Karatani, H. (1989). Automated determination of manganese in seawater by electrolytic concentration and chemiluminescence detection, *Anal. Chem.*, **61**, 1392–1396.
96. Chapin, T. P., Johnson, K. S. and Coale, K. H. (1991). Rapid determination of manganese in seawater by flow injection analysis with chemiluminescence detection. *Anal. Chim. Acta*, **249**, 469–478.
97. Kolotyrkina, I. Ya., Shpigun, L. K., Zolotov, Y. A. and Tsyin, G. I. (1991). Shipboard flow injection method for the determination of manganese in seawater using in-valve preconcentration and catalytic spectrophotometric detection, *Analyst*, **116**, 707–710.
98. Resing, J. A. and Mottl, M. J. (1992). Determination of manganese in seawater by flow injection analysis using online preconcentration and spectrophotometric detection, *Anal. Chem.*, **64**, 2682–2687.
99. Mallini, L. J. and Shiller, A. M. (1993). Determination of dissolved manganese in seawater by flow injection analysis, *Limnol. Oceanogr.*, **38**, 1290–1295.
100. Nowicki, J. L., Johnson, K. S., Coale, K. H., Elrod, V. A. and Lieberman, S. H. (1994). Determination of zinc in seawater using flow injection analysis with fluorometric detection, *Anal. Chem.*, **66**, 2732–2738.
101. Resing, J. A. and Measures, C. I. (1994). Fluorometric determination of Al in seawater by flow injection analysis with in-line preconcentration, *Anal. Chem.*, **66**, 4105–4111.
102. Martin, J. H., Knauer, G. A. and Broenkow, W. W. (1985). VERTEX, the lateral transport of manganese in the northeast Pacific, *Deep-sea Res.*, **32**, 1405–1427.
103. Bruland, K. W. (1980). Oceanographic distributions of cadmium, zinc, nickel and copper in the North Pacific, *Earth Planet. Sci. Lett.*, **47**, 176–198.

8 Dynamic Aspects of *In Situ* Speciation Processes and Techniques

H. P. VAN LEEUWEN
Wageningen University, The Netherlands

1 INTRODUCTION

Within a chemical context, speciation is the notion to indicate the distribution of an element, say a metal, over different chemical species. The distinction is of great practical importance since speciation may greatly affect chemical and biological reactivities (see e.g. ref. [1] on metal speciation and bioavailability). In many situations, the speciation of an element is taken as an equilibrium characteristic, reflecting the species distribution in an unperturbed system. However, in order to understand the reactive properties of complex systems, it is mandatory to have knowledge on the kinetics of interconversion of the species present. Only with such knowledge will it be possible to understand fully the dynamic behavior of such systems and to apply non-equilibrium analytical techniques to them. Involvement of kinetic features in speciation analysis may, at first sight, seem to be an unwanted complicating factor. However, we should realize that in many instances the kinetic properties are of importance.

In Situ Monitoring of Aquatic Systems: Chemical Analysis and Speciation Edited by J. Buffle and G. Horvai.

If, for instance, we measure the distribution of a metal over different complex species with the intention of explaining metal uptake characteristics of organisms, then unavoidably we must consider speciation together with its dynamics. In such a situation the use of a dynamic speciation analysis technique is sheer necessity.

This chapter is intended to outline the background of the dynamic nature of existing methods for *in situ* speciation analysis, and to establish the relation between the operational time-scale of a given technique and the dynamic response of the complex system in terms of the rate parameters of the pertaining association/dissociation reactions. In doing so, it will appear helpful to define ranges of values of the relevant thermodynamic and kinetic speciation parameters where limiting types of behavior are expected. Along these lines we shall pay due attention to the definition of *lability*, a notion which indicates the contribution from complex species to the supply of uncomplexed species involved in a surface reaction. Important examples are of course found in the complete field of techniques for non-equilibrium speciation analysis: various types of voltammetry [2], flux analysis in permeation liquid membrane (PLM) systems [3], different types of gel probe techniques [4], etc. Special attention will be paid to techniques for *in situ* speciation analysis. Thus the cases of microelectrodes and thin film techniques will be treated in detail. The chapter is formulated in terms of complex metal ion species, but has in fact a more general validity, in that it applies to any type of association reaction involving the reactive species.

2 DYNAMIC NATURE OF THE VOLUME REACTION

Let us consider the simplest scheme of a metal ion M which can associate with a ligand L to form the complex species ML

$$\mathrm{M} + \mathrm{L} \underset{k_d}{\overset{k_a}{\rightleftharpoons}} \mathrm{ML} \tag{1}$$

where charges are omitted for the sake of clarity.

The rate constant for the formation reaction is denoted as k_a and that for the dissociation reaction by k_d. The ratio k_a/k_d equals the thermodynamic stability K. For ligand concentrations in sufficient excess over the metal concentration, the association reaction becomes quasi-monomolecular with a rate constant k'_a, equal to $k_a c_L$.

The metal species in reaction (1) have characteristic lifetimes that derive from the reaction rate constants. Thus for M it is $1/k'_a$ and for ML $1/k_d$. On a time-scale t much larger than the lifetimes of M and ML, each individual metal ion frequently changes from M to ML and vice versa. Such a situation is denoted as '*dynamic*' and obeys the condition

$$k_{a}'t, k_{d}t \gg 1 \tag{2}$$

The other limiting case is where the lifetimes of M and ML are much larger than the operational time-scale t. Perturbations of species concentrations are then not followed by significant (re)equilibration. This limit is defined by

$$k_{a}'t, k_{d}t \ll 1 \tag{3}$$

which is referred to as 'static', and the corresponding complex species ML are often denoted as 'inert'.

It is important to emphasize that the conditions (2) and (3) are concerned with the rate characteristics of a *volume reaction*. They not only tell us whether the system is capable of attaining equilibrium within a certain time, but also tell us what is detected by consumptive bulk analytical methods such as exchange or titration techniques. Even if one is only interested in equilibrium speciation, and applies non-consumptive methods of analysis, it is useful to consider the above conditions since they are helpful in confirming whether or not equilibrium exists at all. It has been shown by Eigen and Tamm [5] that in aqueous systems the rate of the complex formation reaction (1) is determined by the rate of the removal of the first water molecule from the inner hydration shell. This is true for a remarkably wide range of ligands and allows us to make *a priori* estimations of the dynamic character of complex systems from tabulated values of k_a and the stabilities K (from which k_d values derive for given k_a). Some examples will be given at the end of the next section.

3 DYNAMIC NATURE OF INTERFACIAL REACTIONS

Now we shall consider what happens with the simple complex system given by reaction (1) if the free metal species M is taken up at an interface which is in contact with a homogeneous solution containing both M and ML. The interface may be the surface of some analytical collector or sensor such as an electrode or a species-selective membrane.

First we shall analyze the limiting case where the interface acts as a perfect sink for M. This implies that transfer processes at the very interface are much faster than transport processes and/or kinetics of the volume reaction in the solution and hence that the concentration of M at the interface is zero. The scheme is

$$\begin{array}{c} \mathrm{M} + \mathrm{L} \rightleftarrows \mathrm{ML} \\ \downarrow \\ \text{fast interfacial transfer} \end{array} \tag{4}$$

and is depicted in Figure 1. The flux of M towards a planar interface can be found by solving the conservation equations for the different species [6]:

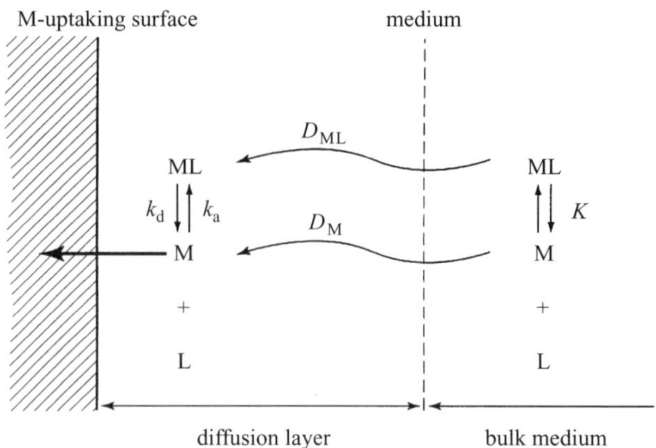

Figure 1. Schematic representation of the different steps involved in the uptake of M at a surface in contact with a medium containing free M and complex ML

$$\begin{aligned}\partial c_i/\partial t &= \partial J_i/\partial x \\ &= D_i\partial^2 c_i/\partial x^2 + \partial(v c_i)/\partial x \pm k_d(c_{ML} - K c_M c_L)\end{aligned} \tag{5}$$

(i = M, L or ML; the sign of the final term is + for M and L and − for ML)

where J_i is the flux of species i, c_i its concentration, D_i its diffusion coefficient and v the velocity of the liquid (of importance under convective conditions). The three terms on the right-hand side of equation (5) represent the diffusional term, the convective contribution and the chemical reaction source term, respectively.

We shall analyze two cases here, i.e. the case of diffusion coupled to the chemical reaction (where the convective term vanishes) and the case of steady-state transport (where $\partial c/\partial t$ is zero).

3.1 THE DYNAMIC CASE OF DIFFUSION AND HOMOGENEOUS REACTION KINETICS

For semi-infinite diffusion as the sole mechanism of transport towards the planar surface equation (5) reduces to:

$$\partial c_i/\partial t = D_i\partial^2 c_i/\partial x^2 \pm k_d(c_{ML} - K c_M c_L) \tag{6}$$

with the boundary conditions

$$t = 0, x \geq 0 \qquad c_{ML}/c_M c_L = K \tag{7}$$

$$t = 0, x \to \infty \qquad c_M = c_M^*,\ c_L = c_L^*,\ c_{ML} = c_{ML}^* \tag{8}$$

where c^* denotes the bulk concentration. In the so-called limiting flux regime, where the flux has its maximum value, further boundary conditions for the system under consideration are

$$t > 0,\ x = 0 \begin{cases} c_{\mathrm{M}} = 0 & (9) \\ \partial c_{\mathrm{ML}}/\partial x,\ \partial c_{\mathrm{L}}/\partial x = 0 & (10) \end{cases}$$

reflecting that the interface is a perfect sink for M and impermeable to ML and L.

Here, we will restrict ourselves to the case of an *excess of ligand* L, i.e. $c_{\mathrm{L}}^* \gg (c_{\mathrm{M}}^* + c_{\mathrm{ML}}^*)$, so that the complex association reaction becomes quasi-monomolecular. Then one is allowed to replace the constant K by K', given by

$$K' = Kc_{\mathrm{L}}^* = c_{\mathrm{ML}}^*/c_{\mathrm{M}}^* \tag{11}$$

and the transport of L, equation (5) for i = L, can be ignored since the change of c_{L} is negligible.

For a dynamic system, that obeys condition (2), the resulting flux J is derived as [7–9]

$$J(t) - \frac{k_{\mathrm{d}}^{1/2} D_{\mathrm{M}}^{1/2} c_{\mathrm{T}}^* (1 + \varepsilon K')^{3/2}}{\varepsilon^{3/2} K' (1 + K')} \exp(\Lambda^2 t)\ \mathrm{erfc}\left(\Lambda t^{1/2}\right) \tag{12}$$

where $c_{\mathrm{T}}^* = c_{\mathrm{M}}^* + c_{\mathrm{ML}}^*$, ε is $D_{\mathrm{ML}}/D_{\mathrm{M}}$ and

$$\Lambda = \frac{k_{\mathrm{d}}^{1/2}(1 + \varepsilon K')}{\varepsilon^{3/2} K' (1 + K')^{1/2}} \tag{13}$$

On the basis of the magnitude of the argument $\Lambda t^{1/2}$ in the experfc of equation (12), we can find some useful limiting cases. The most important one is where the condition

$$\Lambda t^{1/2} \gg 1 \tag{14}$$

is obeyed. Since $\exp(x^2)\mathrm{erfc}(x)$ approaches $\pi^{-1/2}x^{-1}$for$x \gg 1$, the flux J then reduces to

$$J(t) = \frac{D_{\mathrm{M}}^{1/2} c_{\mathrm{T}}^* (1 + \varepsilon K')^{1/2}}{(\pi t)^{1/2} (1 + K')^{1/2}} \tag{15}$$

which can be rewritten as

$$J(t) = \frac{\bar{D}^{1/2} c_T^*}{(\pi t)^{1/2}} \tag{16}$$

where D is the mean diffusion coefficient of M and ML:

$$\bar{D} = \frac{D_M c_M^* + D_{ML} c_{ML}^*}{c_T^*} = D_M \frac{1 + \varepsilon K'}{1 + K'} \tag{17}$$

The flux is now characterized by the purely diffusion-controlled coupled transport of M and ML. There are no limitations due to finite values of the chemical reaction rate parameters k_a and k_d. Such a system is denoted as *labile* and the pertaining condition, expressed by equation (14), is usually denoted as the lability criterium. A short-hand notation for the complete lability criterium parameter is L, in this case equal to $(\Lambda t^{1/2})$.

The other kinetically extreme case, still for a dynamic system, is found for small values of the argument in equation (12):

$$\Lambda t^{1/2} \ll 1 \tag{18}$$

Then equation (12) reduces to

$$J(t) = \frac{k_d^{1/2} D_M^{1/2} c_T^* (1 + \varepsilon K')^{3/2}}{\varepsilon^{3/2} K' (1 + K')} \tag{19}$$

which represents a kinetically controlled flux. For very large K' ($\varepsilon K' \gg 1$) equation (19) further reduces to

$$J(t) = k_d c_T^* \left(D_M / k_a' \right)^{1/2} \tag{20}$$

This flux expression will be easily recognized by electrochemists as similar to the 'kinetic current', determined by k_d and an effective reaction layer thickness μ (equal to $(D_M/k_a')^{1/2}$) [10]. The concept of a reaction layer is based on the finite lifetime of a free M after its formation by dissociation of ML. This lifetime is given by $1/k_a'$, i.e. by the rate of re-association of M with L. According to Einstein's fundamental equation for diffusional mean displacement, a free M can travel over a distance $(D_M \times 1/k_a')^{1/2}$. The layer of solution, adjacent to the receiving surface, with this thickness, is called the reaction layer. Systems obeying condition (18) are denoted as *non-labile*. It is to be emphasized that both labile and non-labile systems obey condition (2) and are therefore dynamic (a qualification referring to the volume reaction). The distinction between labile and non-labile derives from a second condition which concerns the nature of the flux towards an interface: if it is purely diffusion-controlled (because the kinetics are fast enough to create this) then we have a labile system; if it is controlled by kinetics (because they are slow compared with the diffusional transport) then the system is non-labile. Note that there is a fundamental difference between non-labile systems and inert systems (Table 2). The former category obeys condition (2), whereas the latter category satisfies the opposite condition (3).

The dependence of the flux J on time provides us with a further illustration of the two most extreme situations, i.e. the labile and static limits. Figure 2 shows

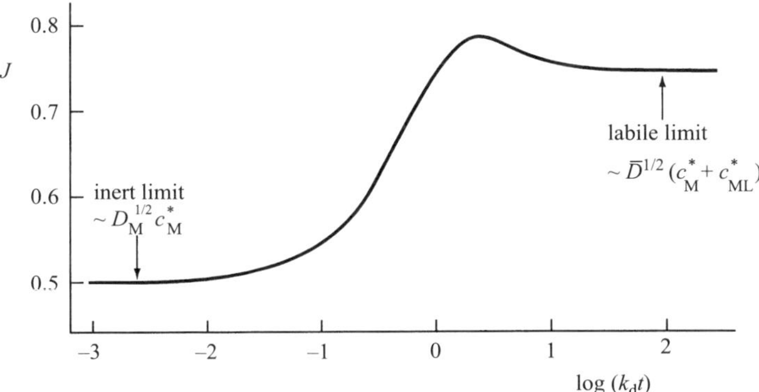

Figure 2. The flux J, as given by equation (12) and normalized with respect to $D_M^{1/2} c_T^*$, as a function of the normalized time $k_d t$. $D_{ML}/D_M = 0.1$; $K' = c_{ML}/c_M = 1$ (adapted from ref. [7])

that for $k_d t \ll 1$ the flux is purely controlled by the free metal species M, i.e. we are in the static or inert limit. The system behaves as if the complex ML were not present at all. At the other extreme, for $k_d t \gg 1$ (which for the chosen example with $K' = 1$ implies that also $k'_a t \gg 1$), the flux is deteremined by the simultaneous coupled diffusion of M and ML, i.e. we are in the labile limit. We note that in the case of more involved systems with different types of ligand L_i, the labile limit is still given by $\bar{D}^{1/2}(c_M^* + c_{ML}^*)$ where $\bar{D}$ is the weighted average over the contributions of the different labile species:

$$\bar{D} = \frac{D_M c_M^* + \sum_i D_{ML_i} c_{ML_i}^*}{c_M^* + \sum_i c_{ML_i}^*} \tag{21}$$

A detailed discussion of $\bar{D}$ in practical systems can be found in ref. [2]. We further note that for $D_{ML} = D_M$ the mean diffusion coefficient is identical to D_M and consequently the flux J does not contain any information on the species distribution over M and ML. For more detailed explanations of the dependence of J on $k_d t$, the reader is referred to the original literature [8,9].

As mentioned before, we can compute values for the association rate constant k_a on the basis of the dehydration rate constants of the inner hydration shell (k_{-w}). Eigen and Tamm [5] reported, e.g., a value of $O(10^{8.5})s^{-1}$ for k_{-w} of Cd(II)†. If we combine this with the value of $O(1–10)\ M^{-1}$ for the equilibrium constant of the fast preceding step in the complexation mechanism, i.e. the formation of the outer-sphere complex [11], an effective association rate constant k_a of $O(10^9)M^{-1}s^{-1}$ results. For various combinations of values of the parameters ε, K, c_L and t different types of behavior result. Table 1 gives some

† The symbol O is used to denote order of magnitude.

Table 1. Dynamics and labilities of Cd(II) complexes of various strengths on various time-scales, computed using equation (13). $k_a = 10^9 mol^{-1} Ls^{-1}; \varepsilon = 0.1$

K (L.mol^{-1})	c_L^* (mol L^{-1})	t (s)	k_a'(s^{-1})	k_d(s^{-1})	$\Lambda t^{1/2}$	Status
10^7	10^{-3}	10^{-1}	10^6	10^2	$10^{-1/2}$	dynamic: non-labile
10^7	10^{-3}	10	10^6	10^2	$10^{1/2}$	dynamic; labile
10^9	10^{-3}	10^{-1}	10^6	1	$(10^{-2.5})$*	static
10^6	10^{-4}	10^{-1}	10^5	10^3	10	dynamic; labile

* The parameter is actually only meaningful for dynamic systems.

examples. Note the confirmation of the existence of systems which are dynamic but non-labile.

3.2 THE STEADY-STATE CASE

In the most rigorous approach, the treatment of the case of a steady-state flux towards the receiving interface would require a return to equation (5) which would have to be solved for $\partial c_i/\partial t = 0$ and some specified velocity function v. This is an involved exercise that so far has not been realized on the same level as for the dynamic case described above (see e.g. the classical monograph by Levich [12]). However, there is an escape via the exploitation of the dynamic results. To a reasonable approximation, we may transform the variable t in the preceding equations to its equivalent under steady-state conditions, i.e. the diffusion layer thickness δ, via their basic Einstein-type relationship for linear diffusion:

$$t = \frac{\delta^2}{\pi D} \tag{22}$$

Thus we may transform the lability condition (14) into its steady-state analog by substituting equation (22) into condition (14), using equation (13) and taking $\bar{D}$ as the effective D for the labile system (equation (17)):

$$\Lambda_{ss} \gg 1 \tag{23}$$

where Λ_{ss} is given by

$$\Lambda_{ss} = \frac{k_d^{1/2}(1 + \varepsilon K')^{1/2}}{D_{ML}^{1/2} \varepsilon K'} \tag{24}$$

(The coefficient $\pi^{1/2}$ is ignored because we are dealing with inequalities here.) For sufficiently strong complexes, with $\varepsilon K' \gg 1$, lability condition (23) reduces to

$$\left(\frac{k_d}{D_{ML} \varepsilon K'}\right)^{1/2} \delta \gg 1 \tag{25}$$

which for complex systems with not too different diffusion coefficients of the different metal species (ε of order unity) comes to

$$\frac{k_d \delta}{k_a'^{1/2} D_M^{1/2}} \gg 1 \quad \left(or \ \frac{k_d^{1/2} \delta}{D_M^{1/2} K'^{1/2}} \gg 1 \right) \tag{26}$$

This condition is identical to the well-known Davison criterion [13]. It should be noted that for macromolecular environmental metal complexes ε may be as low as 10^{-2} or 10^{-3} [14], so that the condition $\varepsilon K' \gg 1$ may be quite restrictive. It will therefore be of practical value to consider condition (23) as the origin of condition (26).

The expressions for the steady-state fluxes $J(\delta)$ are easily formulated using equation (22) again. For a labile complex system we get

$$J(\delta) = \frac{\bar{D} c_T^*}{\delta} \tag{27}$$

which is in fact a trivial result.

The overall scheme of kinetic classification of metal complex systems is given in Table 2. The lability criterion parameter L is indicated by the ratio $J_{kin}/J_{dif'}$, i.e. the ratio between the purely kinetically-controlled flux and the purely diffusion-controlled flux. This ratio carries in fact the physical meaning of the lability parameters $\Lambda t^{1/2}$ and $\Lambda_{ss}\delta$. Derivations less rigorous than the above treatment leading to condition (14) usually contain formulations of J_{kin} and J_{dif} as the essential elements. All intermediate cases in Table 2 are of a kinetic nature, involving bulk reaction kinetics, or kinetics related to interfacial consumption, or both simultaneously.

Table 2. Schematic representation of the characteristic kinetic categories of metal complex systems. L equals J_{kin}/J_{dif}, where J_{kin} and J_{dif} denote the purely kinetic flux and the purely diffusion-controlled flux respectively. Further explanation in the text.

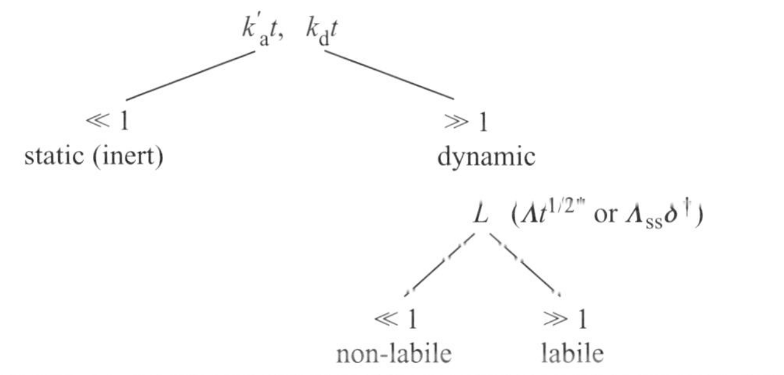

* The transient case, with t as the variable (equation (13)).

† The steady state case, with the diffusion layer thickness δ instead of t as the variable (equation (24)). Note that δ is a function of D_{ML} and that it depends on the hydrodynamic conditions.

4 LABILITIES OF CHEMICALLY HETEROGENEOUS SYSTEMS

Considering its importance in environmental analysis where complexes with, e.g., fulvics and humics are important, it seems useful to extrapolate the above treatment to the case of a chemically heterogeneous metal complex system. For such a system the apparent stability constant K is not a singular parameter but defined by some distribution which characterizes stability as a function of the degree of occupation of the complexing sites under the given conditions. A detailed treatment of heterogeneous metal complexes with natural organic matter can be found in [15–17].

For ligands of type i the association and dissociation rate constants $k_{a,i}$ and $k_{d,i}$ for association/dissociation of ML_i are related to the individual stability K_i

$$K_i = \frac{k_{a,i}}{k_{d,i}} = \frac{c_{ML_i}}{c_M c_{L_i}} \tag{28}$$

The overall stability can be expressed through the average stability parameter $\bar{K}$, which is defined by

$$\bar{K} = \frac{\sum_i c_{ML_i}}{c_M \sum_i c_{L_i}} \quad \left(= \frac{\sum_i K_i c_{L_i}}{\sum_i c_{L_i}} \right) \tag{29}$$

For detailed treatments on the modeling of metal ion binding by heterogeneous ligands, the reader is referred to the literature [16–21].

The lability issue has been considered for the case of a Freundlich type of distribution function for which

$$\bar{\theta} = A c_M^{\Gamma} \tag{30}$$

where

$$\bar{\theta} = \frac{\sum_i c_{ML_i}}{\sum_i (c_{ML_i} + c_{L_i})} \tag{31}$$

and A and Γ are constants with $0 < \Gamma < 1$. The Freundlich relation (30) holds for low $\bar{\theta}$; the relation between $\bar{K}$ and the distribution parameters is [21,22]

$$\bar{K} = A^{1/\Gamma} \left(\frac{\sum_i c_{ML_i}}{\sum_i c_{L_i}} \right)^{(\Gamma-1)/\Gamma} \tag{32}$$

For natural macromolecular ligands, the power Γ usually has a value in the range 0.3–0.7 [19,21]. Note that for $\Gamma = 1$ the homogeneous case is retrieved.

As outlined above, the association rate constant k_a is largely independent of the nature of the ligand and therefore it is reasonable to assume that $k_{a,i} = k_a$. Thus the distributed nature of K will be fully reflected by that of the dissociation rate constant k_d. For the distribution given by equation (30) it follows that the overall dissociation rate $(\sum_i k_{d,i} c_{ML_i})$ is given by

$$\sum_{i} k_{d,i} c_{ML_i} = k_a c_M^* \sum_{i} c_{L_i}^* \left(\frac{\sum_i c_{ML_i}}{\sum_i c_{ML_i}^*} \right)^{1/\Gamma} \tag{33}$$

where it has been used that in the bulk solution there is equilibrium. Formulation of equation (33) is equivalent to writing the average dissociation constant $\bar{k}_d$ in terms of Γ:

$$\bar{k}_d = \bar{k}_d^* \left(\frac{\sum_i c_{ML_i}}{\sum_i c_{ML_i}^*} \right)^{(1-\Gamma)/\Gamma} \tag{34}$$

where $\bar{k}_d^*$ represents the value of $\bar{k}_d$ in the bulk. Equation (34) shows how in the diffusion layer the average k_d is modified by a change in the concentration of complexed metal by consumption at the interface. Using equation (33) and the reaction layer principle, the kinetically controlled flux J_{kin} is found to be [22]:

$$J_{kin} = D_M^{1/2} k'^{1/2}_a c_M^* \left(\frac{\sum_i c_{ML_i}^{\phi}}{\sum_i c_{ML_i}^*} \right)^{1/\Gamma} \tag{35}$$

where c^{ϕ} denotes the mean concentration in the reaction layer which is defined here as $(D_M / k_a \sum_i c_{L_i}^*)^{1/2}$. The diffusion-controlled flux for the heterogeneous system would be

$$J_{dif} = (\pi t)^{-1/2} \bar{D}^{*1/2} \left(\sum_i c_{ML_i}^* + c_M^* \right) \tag{36}$$

which is a maximum estimate since for a heterogeneous system $\bar{D}$ is not constant in the diffusion layer, but decreasing with decreasing distance from the electrode surface [23]. For $\varepsilon < 1$, it decreases with decreasing $\sum_i c_{ML_i} / \sum_i c_{ML_i}^*$.

The above expressions hold for $\varepsilon \bar{K}' * \gg -1$. For arbitrary $\varepsilon \bar{K}'^*$ values, the equations become quite involved [24]. By comparing J_{kin} and J_{dif} as given by equations (35) and (36), we obtain for the lability criterion

$$\Lambda^* \chi^{1/\Gamma} t^{1/2} \gg 1 \tag{37}$$

where the bulk lability parameter Λ^* and the dimensionless complex concentration χ are short-hand notations, defined by [22]

$$\Lambda^* = \frac{(\pi D_M / \bar{D}^*)^{1/2} k_a'^{1/2}}{K^* \sum_i c_{L_i}^*} \tag{38}$$

$$\chi = \frac{\sum_i c_{ML_i}^{\phi}}{\sum_i c_{ML_i}^*} \tag{39}$$

The steady-state equivalent of condition (37) is

$$\Lambda^* \chi^{(1-\Gamma)/\Gamma} \bar{\delta} \gg 1 \tag{40}$$

with

$$\Lambda^* = \frac{\bar{k}_{\mathrm{d}}^*}{k_{\mathrm{a}}'^{1/2} \varepsilon^{1/2} D_{\mathrm{ML}}^{1/2}} \tag{41}$$

It is useful to verify that for $\Gamma = 1$ all expressions reduce to those given for homogeneous systems. For criterion (37) this is not immediately obvious since it is based upon the formulation of dissociation rates in terms of *local* complex concentrations in the reaction layer. The net effect of heterogeneity is a factor of $\chi^{(1-\Gamma)/\Gamma}$ which vanishes for $\Gamma \to 1$ (see for details refs. [22,24]).

Practical application of the lability criteria given requires estimation of the value of χ, the effective complex concentration in the reaction layer. This is crucial in defining the kinetic flux J_{kin}. Chemical heterogeneity of the metal complex system gives rise to an increasing effective stability $\bar{K}$ with decreasing distance from the uptaking surface. Under limiting flux conditions, that is, with condition (9) fulfilled, the ratio $\sum_{\mathrm{i}} c_{\mathrm{ML_i}}/c_{\mathrm{M}}$ also increases and approaches $c_{\mathrm{T}}/c_{\mathrm{M}}$ while $c_{\mathrm{T}}/c_{\mathrm{L,t}}$ approaches zero. Thus the average value of c_{T}^{ϕ} within the reaction layer can be utilized as a reasonable approximation for the value of $\sum_{\mathrm{i}} c_{\mathrm{ML_i}}^{\phi}$ in χ, certainly for use in criteria based on inequalities. The value of c_{T}^{ϕ} can be deduced from the concentration profile as developed on the effective time scale (δ^2/D) of the technique, calculated on the basis of $\bar{D}^*$ [24]. The difference between the lability criteria for homogeneous and heterogeneous systems is essentially governed by a term $\chi^{(1-\Gamma)/\Gamma}$, which vanishes for $\Gamma \to 1$. Under limiting flux conditions the average value of $\sum_{\mathrm{i}} c_{\mathrm{ML_i}}/\sum_{\mathrm{i}} c_{\mathrm{ML_i}}^*$ in the reaction layer near the uptaking surface is generally small and χ can have values as low as 0.01. If the heterogeneity of the sample is modest, with a Γ of e.g. 0.7, $\chi^{(1-\Gamma)/\Gamma}$ comes to $(0.01)^{0.43} = 0.14$. For very heterogeneous systems, with a Γ as low as 0.3, we get $(0.01)^{2.33} = 2.2 \times 10^{-5}$, showing that an increasing degree of heterogeneity lowers the term $\chi^{(1-\Gamma)/\Gamma}$, thereby decreasing the lability (see ref. [22] for more detail). The results clearly show that heterogeneity may count heavily in the lability features of heterogeneous systems. This general picture emerges under conditions where the transport in the medium is flux determining. If some other step in an overall uptake process is rate limiting, then the effects of heterogeneity may become much less severe.

5 LABILITIES OF COMPLEX SYSTEMS IN CONTACT WITH MICROSURFACES

If the dimensions of the uptaking surface are not large compared with diffusion layer and reaction layer thicknesses, δ and μ respectively, the scheme of a linear transport model is no longer valid. This is for instance the case with microelectrodes in monitoring units. We shall therefore now consider the case of non-linear diffusional transport and non-linear kinetic fluxes. Three limiting situations are distinguished and these are outlined in Figure 3. In the macroscopic case, the electrode dimension R is much larger than both μ and δ, and the corresponding lability parameter L was given above by equation (23). We continue here on the level of approximation given by equation (25), i.e. for sufficiently strong complexes. With decreasing electrode dimension, the diffusion layer and the reaction layer become non-linear. Since μ is smaller than δ we have to distinguish between two limiting situations, i.e. the case of a hemispherical diffusion layer and a linear reaction layer (Figure 3B) and the case where

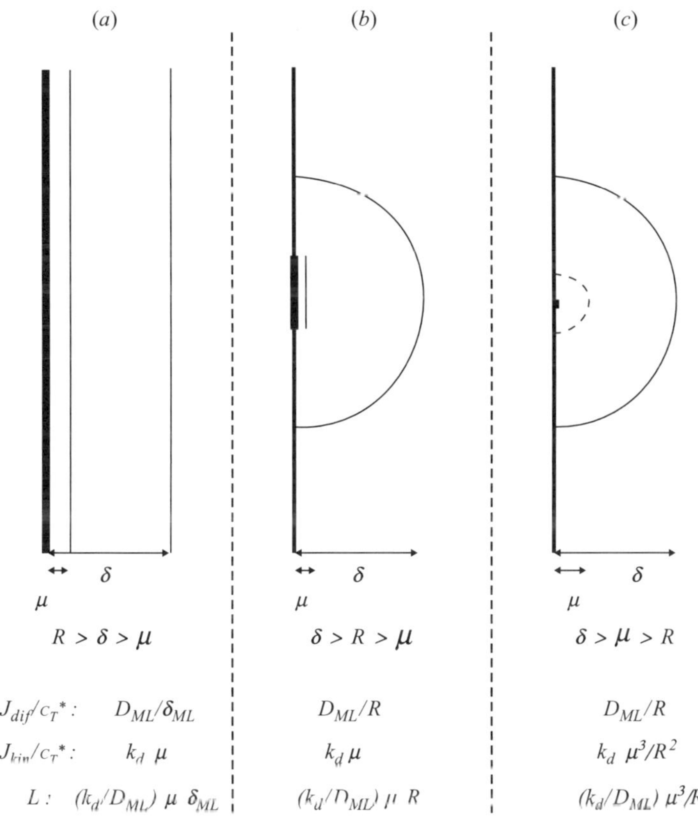

Figure 3. Limiting kinetic situations. Planar electrode, with planar diffusion (δ) and reaction (μ) layers (a). Circular micro-electrode with radius $\delta < R < \mu$ embedded in a large plane, i.e. with planar reaction layer and spherical diffusion layer (b), and with $\delta < \mu < R$, i.e. spherical reaction and diffusion layers (c). The kinetic flux J_{kin}, the diffusional flux J_{dif} and the lability criterion parameter L, are given for each case (Reprinted from [26], with permission by Elsevier)

both the reaction layer and the diffusion layer are hemispherical (Figure 3C). Since the formulation of L is based on comparing J_{kin} with J_{dif} we need expressions for these fluxes for the case of a hemispherical geometry. The diffusional limiting flux towards a microdisc with radius R is well known [25]

$$J_{\text{dif}} = \frac{2D_{\text{ML}}c^*_{\text{ML}}}{R} \tag{42}$$

For the kinetic flux J_{kin} we follow the reaction layer concept which leads to a hemispherical reaction volume with a characteristic *radius* μ, equal to $(D_{\text{M}}/k'_{\text{a}})^{1/2}$. The kinetic production of free M from ML is basically given by the dissociation rate $k_{\text{d}}c^*_{\text{ML}}$. Hence for a hemispherical reaction volume of $\frac{2}{3}\pi(D_{\text{M}}/k'_{\text{a}})^{3/2}$ the resulting flux to the receiving electrode with surface area πR^2 equals

$$J_{\text{kin}} = \frac{2\left(D_{\text{M}}/k'_{\text{a}}\right)^{3/2}k_{\text{d}}c^*_{\text{ML}}}{3R^2} \tag{43}$$

which is the hemispherical analogue of the linear J_{kin}, i.e. $(D_{\text{M}}/k'_{\text{a}})^{1/2}k_{\text{d}}c^*_{\text{ML}}$, as given by equation (20). In passing we note that kinetic fluxes as defined by equations (20) and (43) overestimate the real flux since they count every free metal ion formed within the reaction layer as reaching the surface. Because of the convergent character of spherical diffusion, the extent of this overestimation may be somewhat larger for a micro-electrode than for a macroscopic electrode. By comparing spherical and linear reaction layers with thickness $\frac{1}{2}\mu$, the difference may be expected not to exceed a factor of 4 (comparing $(\frac{1}{2}\mu))^3$ with $(\frac{1}{2}\mu))$. In the inequalities to be given below we shall anyway ignore such numerical coefficients including those appearing in equations (42) and (43).

With the above equations at hand it is straightforward to derive the lability criteria based on $L \gg 1$ [26]. For case b the result is:

$$\mu \ll R \ll \delta: \quad L = \frac{k_{\text{d}}R}{k_{\text{a}}^{\prime 1/2}\varepsilon^{1/2}D_{\text{ML}}^{1/2}} \gg 1 \tag{44}$$

and for case c:

$$R \ll \mu, \delta: \quad L = \frac{k_{\text{d}}D_{\text{M}}^{1/2}}{k_{\text{a}}^{\prime 3/2}\varepsilon R} \gg 1 \tag{45}$$

As expected, these conditions do not contain δ (or t), which is the consequence of the nature of the spherical diffusion flux, equation (42). Still the criteria (44) and (45) only apply if the complex system is dynamic, i.e. if it obeys the equivalent of condition (2). This corresponds to the condition $\mu < \delta$, and expresses the fact that the kinetic flux loses physical significance if the reaction layer is thicker than δ. We further note that the lability criteria (23), (44) and (45) are not fully rigorous, in the sense that they hold for $\varepsilon K' \gg 1$ [8], i.e. for sufficiently strong complexes.

Criteria (23), (44) and (45) show that the lability of a metal complex will generally change upon entering the micro-electrode regime. Figure 3 collects the different regimes with the relevant expressions for L written in terms of the reaction layer thickness μ. Starting point is the macroscopic case where the dimension of the electrode (R) is much larger than δ and μ, i.e. where the diffusion towards the electrode is linear. The lability criterion L is then proportional to $\mu\delta_{ML}$, and thus depends on the effective time scale. Upon reducing the size of the electrode, the first major difference occurs when R becomes smaller than δ and the diffusion towards the electrode becomes spherical. In the case where R is still much larger than μ and the reaction layer is planar, L becomes proportional to μR. Hence the lability of the complex as compared with the macroscopic case decreases since $R < \delta_{ML}$. This is the immediate consequence of the spherical diffusion which is more effective than the planar diffusion in the macroscopic case for the same system. The most extreme case occurs when R becomes smaller than μ so that the reaction layer also becomes spherical. The lability criterion L is then proportional to μ^3/R. Compared with the previous case the lability increases since the diffusional flux is still spherical, but now the reaction flux is also spherical and hence more effective. From these observations a picture emerges where the lability criterion L for a certain metal complex system has a constant value in the macroscopic case, and by decreasing the size of the electrode R decreases from the point where the diffusion becomes spherical ($R \approx \delta_{ML}$) until the point where the reaction layer becomes spherical ($R \approx \mu$) where it starts to increase again.

It is interesting to apply these equations to different systems to see where they are relevant. In choosing the systems to analyze, μ is the crucial parameter. It is primarily related to the k_a value which depends on the type of metal ion. As mentioned before, the rate of the (inner-sphere) complex formation in reaction (1) is determined by the rate of the removal of the first water molecule from the inner hydration shell. This allows us to make *a priori* estimations of the dynamic character of complex systems from tabulated values of k_a (see e.g. ref. [11]) and the stabilities K (from which k_d values derive for given k_a). For the present purpose we selected three types of metal ion M_1, M_2 and M_3 with significantly different kinetic characteristics. Table 3 collects them together with the ensuing reaction layer thicknesses taking the k_a values chosen on the basis

Table 3 Some selected values of the complex formation rate constant k_a and ensuing reaction layer thicknesses μ at ligand concentrations of 10^{-5} and 10^{-4}mol L^{-1}

	M_1	M_2	M_3
k_a(mol^{-1}L s^{-1})	10^{10}	10^7	10^5
μ(m) (for $c_L^* = 10^{-5}$ mol L^{-1})	1×10^{-7}	3×10^{-6}	3×10^{-5}
μ(m) (for $c_L^* = 10^{-4}$ mol L^{-1})	3×10^{-8}	1×10^{-6}	1×10^{-5}

of the data in refs [5,11]. M_1 has an extremely high k_a, M_2 an intermediate value and M_3 has a relatively low complex formation rate. Reaction layer thicknesses reflect these differences. Typical examples of the type M_1 are lead (II), copper (II) and zinc (II); for type M_2 iron (II) and manganese (II); and for type M_3 nickel (II) and vanadium (II). For the three types of metals mentioned, we computed the lability criterium parameter L for complex systems with $K = 10^7 \text{mol}^{-1}\text{L}, c_L^* = 10^{-5}\text{molL}^{-1}, D_M = 10^{-9}\text{m}^2\text{s}^{-1}, D_{ML} = 10^{-10}\text{m}^2\text{s}^{-1}$ and $\delta_{ML} = 10^{-4}$m. The results are given in Figures 4–6. Metal concentrations are

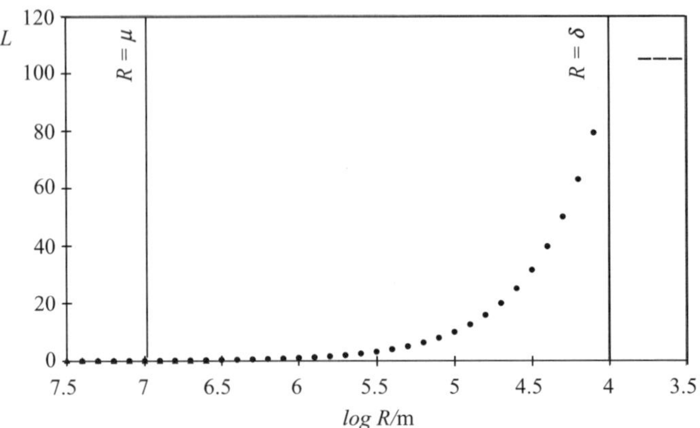

Figure 4. L versus $\log R/\text{m}$ for M_1 (Table 3). Parameter values: $k_a = 10^{10}\,\text{mol}^{-1}\text{L s}^{-1}$; $c_L^* = 10^{-5}\,\text{mol L}^{-1}$; $K = 10^7\,\text{mol}^{-1}\,\text{L}$; $k_d = 10^3\,\text{s}^{-1}$; $D_M = 1 \times 10^{-9}\,\text{m}^2\,\text{s}^{-1}$, $D_{ML} = 1 \times 10^{-10}\,\text{m}^2\,\text{s}^{-1}$, $\delta_{ML} = 1 \times 10^{-4}$ m; - - -, macroscopic limit (Reprinted from [26], with permission by Elsevier)

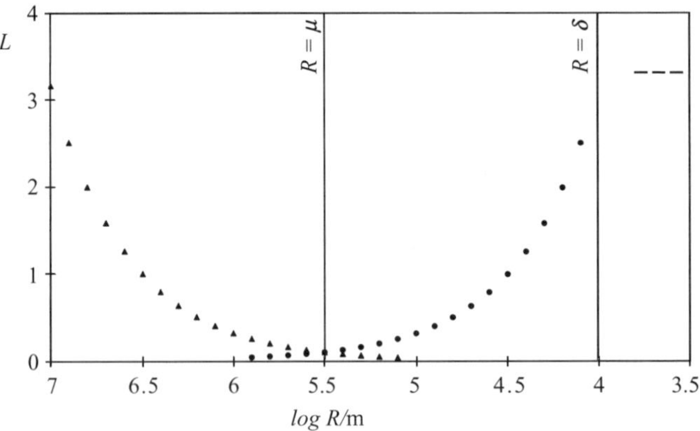

Figure 5. L versus $\log R/\text{m}$ for M_2 (Table 3). Parameter values: $k_a = 10^7\,\text{mol}^{-1}\text{L s}^{-1}$; $c_L^* = 10^{-5}\,\text{mol L}^{-1}$; $K = 10^7\,\text{mol}^{-1}\,\text{L}$; $k_d = 1\,\text{s}^{-1}$; $D_M = 1 \times 10^{-9}\,\text{m}^2\,\text{s}^{-1}$, $D_{ML} = 1 \times 10^{-10}\,\text{m}^2\,\text{s}^{-1}$, $\delta_{ML} = 1 \times 10^{-4}$ m; - - -, macroscopic limit (Reprinted from [26], with permission by Elsevier)

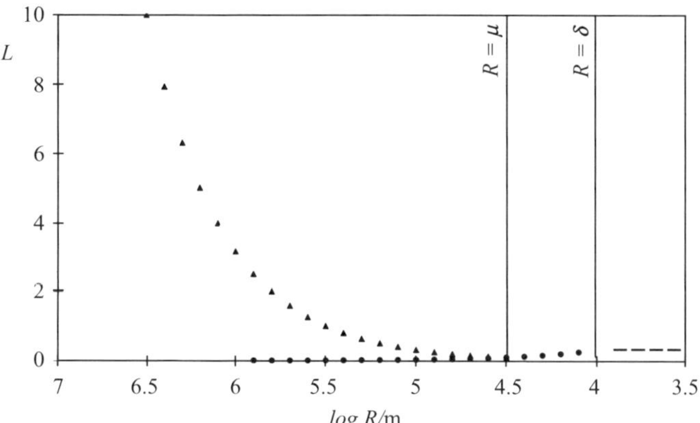

Figure 6. L versus $\log R/\mathrm{m}$ for M_3 (Table 3). Parameter values: $k_a = 10^5\,\mathrm{mol}^{-1}\,\mathrm{L\,s}^{-1}$; $c_L^* = 10^{-5}\,\mathrm{mol\,L}^{-1}$; $K = 10^7\,\mathrm{mol}^{-1}\,\mathrm{L}$; $k_d = 10^{-2}\,\mathrm{s}^{-1}$; $D_M = 1 \times 10^{-9}\,\mathrm{m}^2\,\mathrm{s}^{-1}$, $D_{ML} = 1 \times 10^{-10}\,\mathrm{m}^2\,\mathrm{s}^{-1}$, $\delta_{ML} = 1 \times 10^{-4}\,\mathrm{m}$; - - -, macroscopic limit (Reprinted from [26], with permission by Elsevier)

not specified, but should be well below $10^{-5}\mathrm{molL}^{-1}$, not to violate the excess ligand condition. Values for $\varepsilon K'$ are well above unity which is also a prerequisite for the applicability of the criteria (23), (44) and (45).

Figure 4 shows the lability of the M_1 complex. In the macroscopic limit it is labile which is a consequence of the very high k_a value. With the electrode size R decreasing below the diffusion layer thickness of 10^{-4} m, the lability parameter L approaches the value prescribed by equation (44) and decreases with decreasing R. L reaches unity for $R = 10^{-6}$ m and continues to decrease for lower R. Only for R below 10^{-7} m do we reach conditions where R is also less than μ so that condition (45) starts to apply and the lability increases again with decreasing electrode size. The current situation in the application of micro-electrodes, however, takes us down to dimensions of about 1 μm which means that for M_1 the practical significance of the increasing lability at very small electrode sizes is not important, yet.

In Figure 5 we move on to an appreciably lower k_a value for the M_2 complexes. With decreasing R, L decreases from above unity (largely labile) to $O(10^{-1})$ (non-labile) around an electrode size of a few μm, before increasing again with further decreasing R. This case clearly demonstrates the functionalities of the different flux generating mechanisms: first, lability decreases because of the diffusion layer changing from linear to (hemi-)spherical, and then lability increases again when the reaction layer also becomes spherical. This feature might open the way to control the lability of a certain complex by changing conditions, i.e. R, in such a way that a complex is deliberately rendered non-labile, or just inert.

A further case, with again much lower k_a, is represented by Figure 6 for M_3 complexes. This metal ion has such slow complexation kinetics that the reaction layer thickness is not much smaller than δ (i.e. condition (2) is hardly obeyed). In this case reduction of the electrode size will only reduce the lability a bit, before it increases again. The practical significance may be appreciable since here the lability may be increased to a substantially higher level as compared with the macroscopic limit.

Given a type of complex system there are not many means to influence its dynamic properties. Apart from changing the temperature and the stirring conditions (δ_{ML}), only the ligand concentration c_{L}^* is, sometimes, an externally controllable variable. Variation of c_{L}^* leads to derived changes of the reaction layer thickness μ, according to $(D_{\mathrm{M}}/k_{\mathrm{a}}c_{\mathrm{L}}^*)^{1/2}$. Figure 7 shows the effect of increasing c_{L}^* for the M_1 complex system. The lability shows an overall decrease due to the decreasing value of μ; for cases a and b (Figure 3) it goes down by a square root dependence and for case c, L decreases even by $(c_{\mathrm{L}}^*)^{3/2}$. This feature might be one of the key elements of forthcoming experimental studies.

Increasing the stability (K) leads to an overall decrease of lability. However, the nature of the dependence of L on R remains the same because, according to the Eigen mechanism principles, the reaction layer thickness μ does not change. Thus, increases in K lead to corresponding decreases of k_{d} which appear in the same way in all lability criteria (23), (44) and (45). Essentially the same can be said about the influence of the diffusion coefficient of the complex, D_{ML}. With decreasing D_{ML} we see an overall increase of the lability parameter L but no change in the nature of the L, R dependence since the value of μ is not affected by a change in D_{ML}.

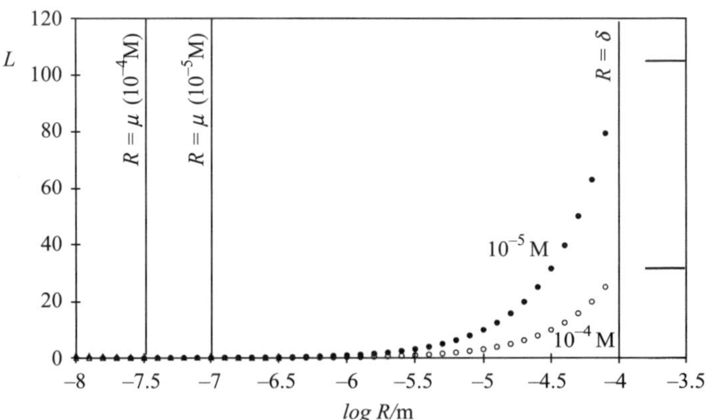

Figure 7. L versus log R/m for M_1 (Table 3) and different values of c_{L}^* (•, 10^{-5} mol L^{-1}; ○: 10^{-4} mol L^{-1}) Parameter values: $k_{\mathrm{a}} = 10^{10}$ mol^{-1} L s^{-1}; $K = 10^7$ mol^{-1} L; $k_{\mathrm{d}} = 10^3$ s^{-1}; $D_{\mathrm{M}} = 1 \times 10^{-9}$ m^2 s^{-1}, $D_{\mathrm{ML}} = 1 \times 10^{-10}$ m^2 s^{-1}, $\delta_{\mathrm{ML}} = 1 \times 10^{-4}$ m; —, macroscopic limit. (Reprinted from [26], with permission from Elsevier)

6 APPLICATION TO *IN SITU* SPECIATION SENSORS

In order to exploit the above theory, we shall now apply it to two types of sensors, discussed elsewhere in this volume. The two types are depicted in Figure 8. Both of them have a gel membrane layer which functions as a species selector and a well-defined diffusion medium. In one case, the gel layer is in contact with a micro-electrode [2,27], in the other with a planar layer of ion exchange resin particles [4,28,29].

6.1 THE MEMBRANE-COVERED MICRO-ELECTRODE SENSOR

The experimental conditions are outlined in Figure 8. In the operating mode proposed by Tercier and Buffle [2,27] the actual measurement with the micro-electrode is performed after equilibration of the gel layer with the sample solution. This basically means that the exchange between these two phases is governed by equilibrium properties only, i.e. by the different distribution properties of the different species. Lability characteristics of complexes in the sample solution (including the non-penetrating ones!) may only be of importance for the time necessary to reach the equilibrium with the gel layer. The real and remaining lability question is related to the actual measurement with the micro-sensor, which is in contact with a gel layer where the free metal ions and the penetrating complexes are uniformly distributed (equilibrium established). We first have to establish whether situation (a), with $R \ll \mu$, δ, or situation (b) with $\mu \ll R \ll \delta$, is applicable. The diffusion layer thickness δ depends on the type of experiment, or—more precisely—on the effective time-scale of the experiment (see e.g. ref. [2] for details). For most voltammetric experiments δ usually is O(10–100) μm. The radius of the micro-electrode area is usually O(1–10) μm.

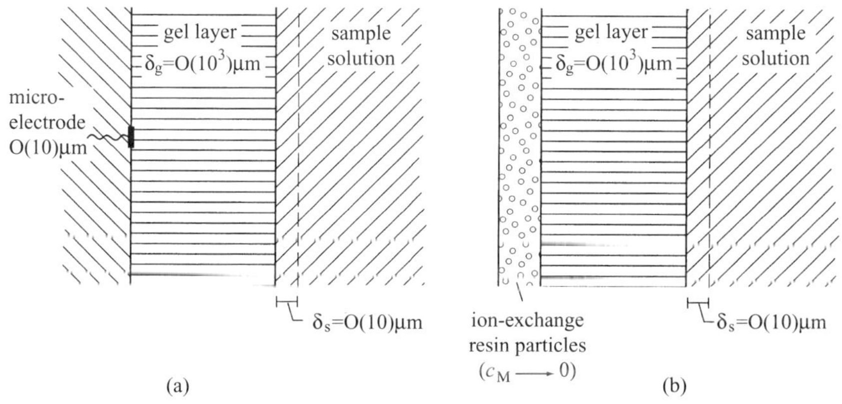

Figure 8. Two types of sensors employing a gel layer as diffusion medium: (a) the membrane-covered microelectrode; (b) the membrane-covered ion-exchange layer

For the reaction layer thickness μ, we have to distinguish between different types of ions. For heavy metal ions such as Pb^{2+} and Cd^{2+}, the complex association rate constant k_a is as high as 10^9 to 10^{10} M^{-1} s^{-1}. For a ligand concentration of 10^{-5} M and a D_M of 10^{-9} m^2 s^{-1}, this leads to a μ of $O(10^{-1})$ μm. Hence we have to apply equation (45) under the given conditions. Taking $\varepsilon = 10^{-1}$, we find that for $k_d > 10^3$ s^{-1} the lability criterion is obeyed. This means that complexes with $K < 10^{6.5}$ mol^{-1} L will exert labile behavior. Note that this thermodynamic window derives from a corresponding kinetic window thanks to the circumstance that k_a is fixed for a given type of metal ion. For monovalent ions such as Tl^+ the kinetics will be fast [5] and complexes will be weaker, so generally a high degree of lability may be expected. In contrast, higher-valency ions such as Fe^{3+} and Al^{3+} have much slower kinetics and higher complex stabilities, so they will be more likely to suffer from lability limitations.

6.2 THE MEMBRANE-COVERED ION-EXCHANGE LAYER

Davison and Zhang [28] proposed the set-up given in Figure 8B. It incorporates a gel layer of a few hundreds to one thousand μm thickness. The pores in the gel cover a range of sizes much larger than simple ions and small complexes which are therefore able to penetrate the gel phase. The mobilities of the small ions in the gel seem to be practically the same as those in water, as confirmed by independent data [30]. The technique depends on a steady state flux through the gel layer being set up. This situation is sketched in Figure 9. The profile is drawn on the basis of the simplifying assumptions that Donnan effects are negligible, that the diffusion coefficients of all penetrating species are the same as in the sample solution, that the metal ion binding by the resin is much stronger than for any of the penetrating ligands, and that all of the penetrating complexes behave labilely (so that the actual metal concentration at the resin–gel interface is zero). We will come back to the explanation of several of these assumptions below. The technique, known as diffusive gradients in thin-films (DGT), relies on deployment for at least several hours. The mean accumulated mass of metal on the resin is then measured and used to calculate a mean flux through the gel layer [4].

Non-penetrating complexes in the sample solution may contribute to the penetration process by dissociation into penetrating ones within the aqueous diffusion layer (with thickness δ_s). For the estimation of the relevant labilities, we can use equation (23) with δ_s for δ. The effect vanishes for $\delta_s \ll \delta_g$, where the concentrations of penetrating species at the gel–solution interface approach the concentrations in the bulk solution. However, this takes us to one of the practical dilemmas of the system: increasing δ_g enhances simplicity in several respects, but seriously increases the time necessary to reach steady state. The typical time τ necessary for realizing the steady-state profile in Figure 9 for a gel layer of 400–1000 μm and an overall D of 10^{-9} m^2 s^{-1} is of order δ_g^2/D, i.e.

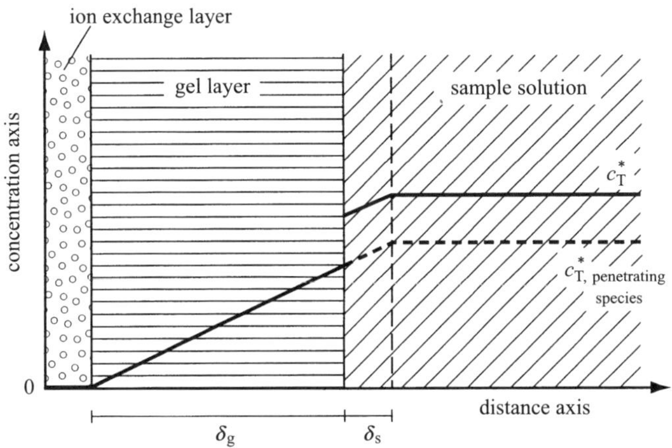

Figure 9. Steady state concentration profile in the membrane-covered ion-exchange layer (DGT mode). Explanation in the text

10^2–10^3 s. Increasing δ_g quadratically increases τ and thus readily leads to very long measurement times, the more so since the time required to reach steady state should be negligible compared with the overall accumulation time. For supposedly labile penetrating complexes, the flux eventually attained is given by

$$J = \frac{\bar{D}_{ps}c^*_{T,ps}}{\delta_g} \tag{46}$$

where the index ps refers to the penetrating species.

The amount of metal accumulated in the resin layer per unit of surface area (Γ_M) is given by

$$\Gamma_M = J[t - O(\tau)] \tag{47}$$

where a term of order τ accounts for the finite time necessary to fill up the gel layer to the profile of the steady state [6]. Obviously for times much greater than τ we come to the simplified accumulation equation

$$\Gamma_M = Jt \qquad (t \gg \tau) \tag{48}$$

We have seen that τ is 10^2–10^3 s, so the applicability of equation (48) usually requires time-scales in excess of an hour [31,32]. This is not a problem as long as DGT is used to provide time-averaged concentrations over days or weeks [33]. Another noteworthy aspect of this steady-state flux technique is that labilities of the penetrating complex species are derived from the gel layer thickness δ_g. So, for considering these labilities, we have to apply equation (23) with δ_g for δ. In fact, the overall speciation performance of the method is not simple: it exploits

(i) a size window somehow related to the gel structure, but not very well defined yet [27–29], coupled to
(ii) a lability window for the sample solution – gel layer interface (operational for δ_g not $\gg \delta_s$), and
(iii) a *different* lability window for the gel layer–resin interface.

Using simply a gel layer (without a backing resin) in contact with the sample solution, one could also wait for the distribution equilibrium to be established (this mode has been denoted as 'DET') [4,29,34]. Then, in contrast to DGT, δ_g is not relevant for the labilities of the complex species. In the DET mode we are dealing with an effective penetration (size) window, *not* coupled to any lability criterion (because we are waiting for equilibrium). The dynamic speciation features of the two modes DET and DGT are thus fundamentally different which, in certain cases, might make their deliberate combination helpful.

7 CONCLUSION AND OUTLOOK

The principles of labile behavior of relatively simple complex systems are now fairly well understood. It is useful to think about lability in terms of two simultaneous conditions: one on the volume reaction, which should be dynamic, and one on the interfacial reaction, which should be fully unlimited by the complex dissociation/association reactions in solution. Rigorous lability criteria have been derived for both the transient case, with t as the variable, and the steady state case, with δ as the variable. On a lower level of rigor, criteria for chemically heterogeneous systems, i.e. with distributed complex stabilities, have been given.

On the basis of existing theory, some approximate lability criteria for complex systems in contact with microsurfaces are proposed. Considering the growing importance of micro-sensors, it seems worthwhile to spend further efforts on this subject. For the retrieval of speciation information from analytical data obtained by *in situ* microsensors, a further development of the methodological background is mandatory. Thus we recommend that the practical development of *in situ* speciation sensors be accompanied by a simultaneous analysis of the underlying theoretical principles.

LIST OF SYMBOLS

A	Freundlich distribution constant, equation (30)
c_i	concentration of species i
c_i^*	bulk concentration of species i
c_T^*	$c_M^* + c_{ML}^*$

c_i^ϕ mean concentration of species i in reaction layer
D_i diffusion coefficient of species i
$\bar{D}$ mean diffusion coefficient, equations (17) and (21)
$\bar{D}^*$ $\bar{D}$ in bulk solution
J flux
J_{dif} diffusion-controlled flux
J_{kin} kinetically controlled flux
k_a association rate constant
k'_a $k_a c_L^*$
$k_{a,i}$ k_a for ligand i
k_d dissociation rate constant
$k_{d,i}$ k_d for ligand i
$\bar{k}_d$ average dissociation rate constant, equation (34)
$\bar{k}_d^*$ $\bar{k}_d$ in bulk
k_{-w} inner sphere dehydration rate constant
K complex stability constant, equation (7)
K' Kc_L^*
K_i individual K for ligand i
$\bar{K}$ mean stability, equation (29)
L lability criterion parameter J_{kin}/J_{dif}
R radius of circular microelectrode surface
t time
v liquid flow velocity
x axis perpendicular to the surface
Γ Freundlich distribution constant, equation (30)
δ steady-state diffusion layer thickness
δ_g gel layer thickness, Figure 9
δ_i δ for species i
δ_s thickness of diffusion layer in solution, Figure 9
ε D_{ML}/D_M
θ mean degree of occupation of ligands in heterogeneous system
Λ lability parameter, equation (13)
Λ_{ss} lability parameter, equation (24)
Λ^* bulk lability parameter, equation (38)
μ reaction layer thickness $(D_M/k'_a)^{1/2}$
τ time necessary to reach steady state, equation (47)
γ_i normalized c_i^ϕ, equation (39)

REFERENCES

1. Tessier, A. and Turner, D. R., ed. (1995). *Metal Speciation and Bioavailability in Aquatic Systems*, IUPAC Series on Analytical and Physical Chemistry of Environmental Systems, Vol. **3**, Wiley, Chichester.

2. Buffle, J. and Tercier, M. -L. (2000). *In situ* voltammetry: concepts and practice for trace analysis and speciation Chapter 9, this volume.
3. Buffle, J., Parthasarathy, N., Djane, N-K. and Mathiasson L. (2000). Permeation liquid membranes for field analysis and speciation of trace compounds in waters, Chapter 10, this volume.
4. Davison, W., Fones, G., Harper, M. Teasdale, P. and Zhang, H. (2000). Dialysis, DET and DGT: *in situ* diffusional techniques for studying water, sediments and soils, Chapter 11, this volume.
5. Eigen, M. and Tamm, K. (1963). Fast elementary steps in chemical reaction mechanisms, *Pure Appl. Chem.*, **6**, 97.
6. Crank, J. (1964). *The Mathematics of Diffusion*, Clarendon Press, Oxford.
7. Tamamushi, R. and Sato, G. P. (1972). Application of polarography and related electrochemical methods to the study of labile complexes in solution. In *Progress in Polarography*, ed. Zuman, P., Vol. 3, Interscience, New York, p. 1.
8. de Jong, H. G., van Leeuwen, H. P. and Holub, K. (1987). Voltammetry of metal complex systems with different diffusion coefficients of the species involved. I. Analytical approaches to the limiting current for the general case including association/dissociation kinetics, *J. Electroanal. Chem.*, **234**, 1.
9. de Jong, H. G. and van Leeuwen, H. P. (1987). Voltammetry of metal complex systems with different diffusion coefficients of the species involved. II. Behavior of the limiting current and its dependence on association/dissociation kinetics and lability, *J. Electroanal. Chem.*, **234**, 17.
10. Heyrovský, J. (1941). *Polarographie*, Springer, Wien; Heyrovský, J. and Kůta, J. (1965). *Principles of Polarography*, Nakladatelstvi Ceskoslovenské Akademie Ved, Praha.
11. Morel, F. M. M. and Hering, J. G. (1993). *Principles and Applications of Aquatic Chemistry*, Wiley, New York, p. 405 ff.
12. Levich, V. G. (1962). *Physicochemical Hydrodynamics*, Prentice Hall, Englewood Cliffs, NJ.
13. Davison, W. (1978). Defining the electroanalytically measured species in a natural water sample, *J. Electroanal. Chem.*, **87**, 395.
14. van Leeuwen, H. P., Cleven, R. F. M. J. and Buffle, J. (1989). Voltammetric techniques for complexation measurements in natural aquatic media. Role of the size of macromolecular ligands and dissociation kinetics of complexes, *Pure Appl. Chem.*, **61**, 255.
15. Buffle, J. (1988). *Complexation Reactions in Aquatic Systems*, Ellis Horwood, Chichester.
16. Buffle, J., Altmann, R. S., Filella, M. and Tessier, A. (1990). Complexation by natural heterogeneous compounds: site occupation distribution functions, a normalised description of metal complexation, *Geochim. Cosmochim. Acta*, **54**, 1535.
17. Buffle, J., Filella, M. and Altmann, R. S. (1994). Binding models concerning natural organic substances in performance assessment. In *OECD Documents*, NEA Workshop, Switzerland, p. 149.
18. Gamble, D. S. Underdown, A. W. and Langford, C. H. (1980). Copper(II) titration of fulvic acid ligand sites with theoretical, potentiometric, and spectrophotometric analysis, *Anal. Chem.*, **52**, 1901.
19. Buffle, J., Altmann, R. S. and Filella, M. (1990). Effect of physicochemical heterogeneity of natural complexants II. Buffering action and role of background sites, *Anal. Chim. Acta*, **232**, 225.

20. van Riemsdijk, W. H. and Koopal, L. K. (1992). In *Ion binding by Natural Heterogeneous Colloids*, *Environmental Particles*, ed. Buffle, J. and van Leeuwen, H. P. Vol. **1**, Lewis, Boca Raton, FL.
21. Filella, M., Buffle, J. and van Leeuwen, H. P. (1990). Effect of physicochemical heterogeneity of natural complexants I. Voltammetry of labile metal-fulvic complexes, *Anal. Chim. Acta*, **232**, 209.
22. van Leeuwen, H. P. and Buffle, J. (1990). Voltammetry of heterogeneous metal complex systems. Theoretical analysis of the effect of association/dissociation kinetics and the ensuing lability criteria, *J. Electroanal. Chem.*, **296**, 359.
23. Pinheiro, J. P., Mota, A. M. and Simões Gonçalves, M. L. S. (1996). Voltammetry of labile heterogeneous complexes with low diffusion coefficients, *J. Electroanal. Chem.*, **402**, 47.
24. Pinheiro, J. P., Mota, A. M. and van Leeuwen, H. P. (1999). On lability of chemically heterogeneous systems. Complexes between trace metals and humic matter, *Colloids Surf. A*, **151**, 181.
25. Bard, A. J. and Faulkner, L. R. (1980). *Electrochemical Methods; Fundamentals and Applications*, Wiley, New York.
26. van Leeuwen, H. P. and Pinheiro, J. P. (1999). Lability criteria for metal complexes in microelectrode voltammetry, *J. Electroanal. Chem.*, **471**, 55.
27. Tercier, M. -L. and Buffle, J. (1996). Antifouling membrane-covered voltammetric microsensor for in-situ measurements in natural waters, *Anal. Chem.*, **68**, 3670.
28. Davison, W. and Zhang, H. (1994). In Situ speciation measurements of trace components in natural waters using thin-film gels, *Nature*, **367**, 546.
29. Davison, W., Zhang, H. and Grime, G. W. (1994). Performance characteristics of gel probes used for measuring the chemistry of pore waters, *Environ. Sci. Technol.*, **28**, 1623.
30. Tong, J. and Anderson, J. L. (1996). Partitioning and diffusion of proteins and linear polymers in polyacrylamide gels, *Biophys. J.* , **70**, 1505.
31. Harper, M. P., Davison, W., Zhang, H. and Tych, W. (1998). Kinetics of metal exchange between solids and solutions in sediments and soils interpreted from DGT measured fluxes, *Geochim. Cosmochim. Acta*, **62**, 2757.
32. Zhang, H. and Davison, W. (1995). Performance characteristics of diffusion gradients in thin films for the in situ measurements of trace metals in aqueous solution, *Anal. Chem.*, **67**, 3391.
33. Zhang, H., Davison, W., Miller, S. and Tych, W. (1995). In situ high-resolution measurements of fluxes of Ni, Cu, Fe, and Mn and concentrations of Zn and Cd in porewaters by DGT, *Geochim. Cosmochim. Acta*, **59**, 4181.
34. Harper, M. P., Davison, W. and Tych, W. (1997). Temporal, spatial, and resoluting constraints for in-situ sampling devices using diffusional equilibration: dialysis and DET, *Environ. Sci. Technol.*, **31**, 3110.

9 *In Situ* Voltammetry: Concepts and Practice for Trace Analysis and Speciation

J. BUFFLE AND M.-L. TERCIER-WAEBER
University of Geneva, Switzerland

In Situ Monitoring of Aquatic Systems: Chemical Analysis and Speciation Edited by J. Buffle and G. Horvai.

1 INTRODUCTION (AND ADVICE TO THE READER)

Compared with the analysis of compounds or ions present in significant concentrations in waters (e.g. Ca^{2+}, O_2, NO_3^-), the analysis of trace compounds or elements implies specific requirements, in particular (i) using highly sensitive techniques (to determine concentrations in the range 10^{-12}–10^{-7} mol L^{-1}), (ii) minimizing sample – vessel interactions leading to contaminations or losses of analyte by adsorption, (iii) data interpretation in terms of evolved speciation concepts (as most organic or inorganic trace compounds are present in a large variety of different forms) and in terms of possible species transformation during sample handling and storage (including coagulation, microbial degradation etc.).

The development of direct *in situ* methods for trace compound analysis may solve most of the problems related to contaminations, losses on vessels, or transformations during sample storage and handling. The possible instrumental techniques usable for this purpose are not numerous, however, as they should combine the following criteria:

- high sensitivity;
- capability of detecting the test species with no or minimum sample manipulation;
- capacity for multicompound analysis, as environmental quality control usually requires the measurement of many compounds, and the development of one probe per compound would lead to bulky analytical systems difficult to handle;
- low cost instrument with low energy consumption;
- possibility of miniaturization.

Only very few techniques meet these requirements. Voltammetric techniques belong to them. They are based on the recording of the current–potential curves produced by the electrochemical oxido-reduction of the test compound. Because these techniques can be modulated in many ways, the same instrument can potentially provide a wealth of information on the various species of a large number of different compounds. The two major difficulties of these techniques are the following.

(1) Owing to their large flexibility, these techniques can be adapted to many different problems, but, as a counterpart, the theoretical background necessary for correct interpretation is larger than with other techniques.
(2) The solution–electrode interface is the key part of the sensing system. For routine environmental monitoring, long-term stability of the physical and chemical structure of this interface must be achieved. In particular, electrode surface fouling by adsorbing compounds, or oxidation by oxygen (even for noble metals), are major problems.

At present, few *in situ* voltammetric probes have been developed, and only trace metals, plus Fe(II), Mn(II), S(−II) and O_2 have been considered. This chapter therefore will focus on the following aspects:

- mainly *trace metals*, plus Fe(II), Mn(II), O_2 and S(−II) will be discussed, though potential applications to trace organics are numerous and will be mentioned;
- the *major theoretical principles,* in particular those required for correct speciation interpretation, will be summarized from the perspective of natural water analysis (sections 2.1 and 3)
- the characteristics of the major *components of voltammetric cell* (electrode types, flow-through cells) developed for laboratory applications will be compared with the constraints of *in situ* measurements, such as pressure, natural convection effects, or long-term stability problems (section 2.2);
- finally *on-site* (section 4) *and in situ* (section 5) *developments* will be described, to exemplify possible solutions to problems and remaining problems to solve for *in situ* applications.

ADVICE TO THE READER

Although each section is written in perspective of *in situ* measurements, the chapter is organized in such a way that experts in laboratory applications of voltammetry and speciation can pass directly to section 5, while persons experienced in classical voltammetric analysis but not in speciation aspects can pass directly to section 3.

2 GENERAL CONCEPTS AND VOLTAMMETRIC CELL COMPONENTS

2.1 PRINCIPLES OF VOLTAMMETRIC TECHNIQUES

Only the key concepts of voltammetric techniques, important for *in situ* applications in water, will be discussed here as they are described in detail in a number of books, e.g. ref. [1]: influence of chemical reactions on voltammetric signal; refs [2,3]: concepts and theory of the various voltammetric techniques; ref. [4]: interpretation of voltammetric results obtained in natural waters; ref. [5]: laboratory applications of voltammetric techniques; ref. [6]: on-going series in electroanalytical chemistry.

2.1.1 Faradaic and Capacitive (Background) Current

A voltammetric cell usually includes (Figure 1) a working electrode (WE), at which the analyte is measured, an auxiliary (AE) and a reference (RE) electrode. In most voltammetric techniques, the electronic circuit imposes a potential difference, *E*, between RE and WE and measures the flowing current, *i*. The circuit is constructed in such a way that the absolute potential of RE is constant, irrespective of the flowing current *i*, so that any change in *E* is applied to the working electrode. To reach that goal, the current passing through RE should be very low ($< 10^{-13}$ A); the auxilliary electrode (AE) is thus required to enable the flow of current in the whole circuit. The voltammetric curves are obtained by varying *E* and measuring *i* (see examples in Table 1). In addition, *E* may be modulated with time, *t*, (Table 1) so that a number of different voltammetric techniques, with different characteristics can be used (see section 2.1.2).

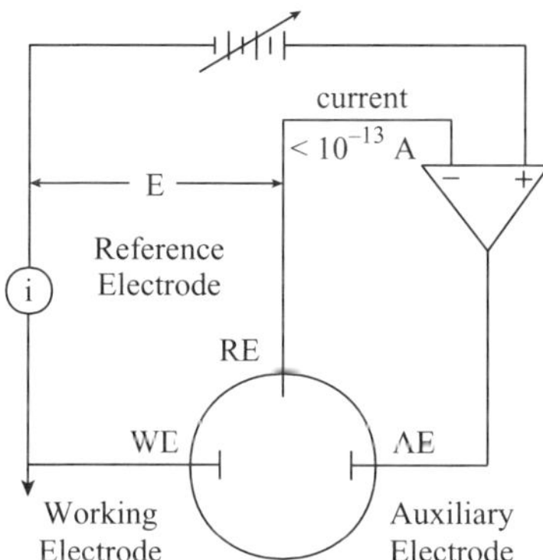

Figure 1. Schematic drawing of the electronic circuit of a potentiostat. E = imposed potential; i = measured current, WE, RE, AE = working electrode, reference electrode, auxiliary electrode respectively

In potentiometric stripping analysis (PSA), the current is imposed and potential recorded as a function of time. For simplicity, in the following, PSA will be included in the general term of 'voltammetric techniques' even though it strictly does not belong to this group. In all cases it is important to recognize that the current is a flux (of electrons) and thus depends on *dynamic* processes in solution; i is thus a function of E *and* time, even at constant E (see below):

$$i = f(E, t)$$

The solution contains the solvent (water), an electrolyte (to ensure the flow of current in solution by ion transport), and the test compound which will undergo the redox reaction at the electrode surface (the so-called depolarizer). In absence of depolarizer, the full curve in Figure 2A is obtained (its exact shape depends on the voltammetric technique used). In the presence of depolarizers, additional curves (dotted curves in Figure 2A), either waves or peaks, depending on the technique, are observed. The key characteristics of Figure 2 are the following:

Polarization range; limits of potential The full curve depicted in Figure 2A is valid for the Hg electrode. The potential window in which the oxido-reduction of a depolarizer is measurable is limited (i) on the negative potential side, by the reduction current of the cation of the electrolyte (Na^+) or of H_2O (into H_2), and (ii) on the positive potential side, by the oxidation current of Hg. The so-called polarization range of Hg (i.e. the domain in which the potential imposed to Hg can be changed) is thus limited at +0.2 and −1.7 V versus Ag/AgCl/sat. KCl reference electrode (section 2.2.2). The polarization ranges of the other major substrates used as working electrodes are shown in Figure 2B. The reason for the negative limit (i.e. reduction of H^+ or Na^+) is similar for all of them. The positive limit is due to the oxidation of either the metal (Pt, Ag) or water (into O_2). Note that since both reduction and oxidation potentials of water depend on pH, so do the corresponding polarization limits. Ag is usually not convenient for analysis, as its polarization range is small. Clearly Pt, Au, Ir and C are complementary to Hg: test compounds with oxido-reduction potentials < 0 will be better analysed on Hg, whereas the other substrates are necessary in the positive potential range (Figure 2B–D).

Depolarizers; environmental medium limitations A depolarizer is a compound which hinders the polarisation of the electrode towards either negative potential, by giving rise to a reduction current (e.g. $O_2 \longrightarrow H_2O_2$), or positive potential, by giving rise to an oxidation current (e.g. $S(-II) + Hg \longrightarrow HgS$). In electroanalysis, the test compounds are depolarizers. The corresponding characteristic potentials (i.e. half-wave potentials for waves (Figure 2A); peak potentials for peaks (Table 1)) are directly related to the normal potentials of the corresponding redox couples (section 3.1). The limiting current of the wave

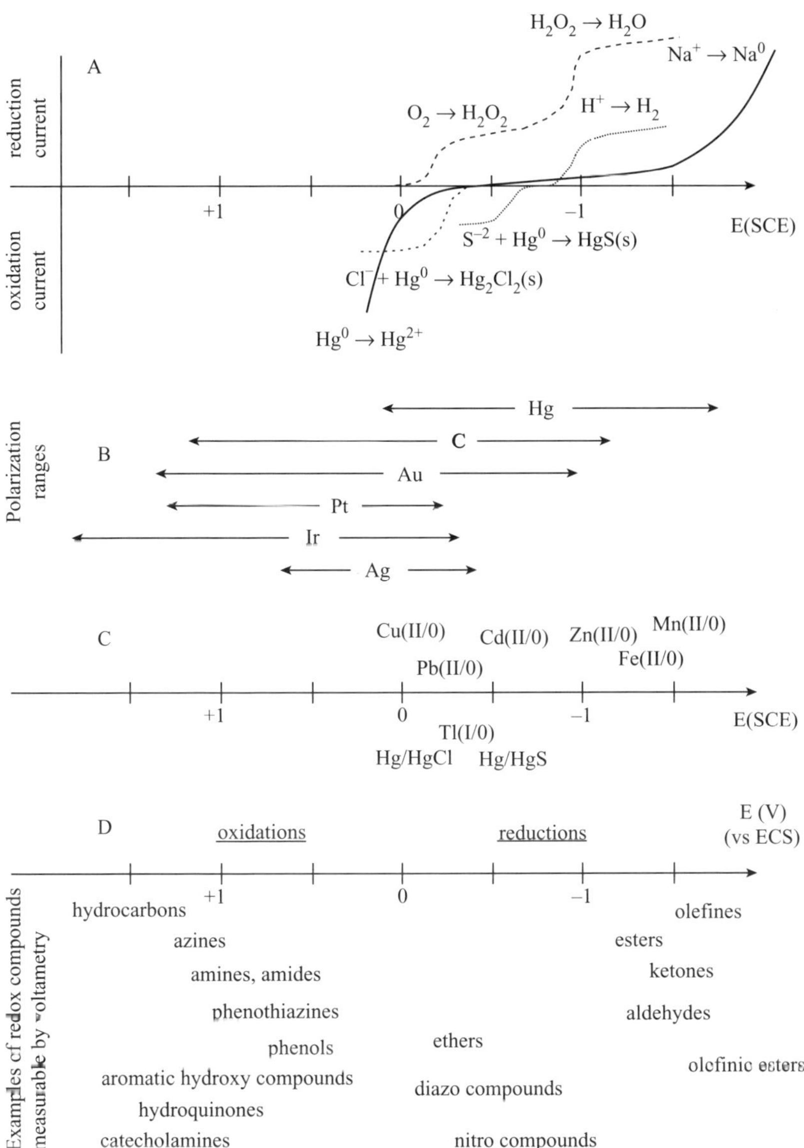

Figure 2. (A) Schematic depolarization curves of the solvent, O_2, S(−II), Cl^-, H^+, and Na^+ on Hg electrode. (B) Polarization ranges of Pt, Au, C , Ag and Ir electrodes. (C) and (D) Oxido-reduction potentials of organic and inorganic depolarizers

plateau (Figure 2A) or the peak current (Table 1) are proportional to the concentration of the depolarizer and are therefore the major parameters used for in analytical voltammetry.

It is of key importance to realize that *in situ* analysis in surface water is most often done in the presence of O_2, at a concentration (3×10^{-4} mol L^{-1} for saturated water) much higher than that of the trace metals or compounds of interest (often $< 10^{-7}$ mol L^{-1}). The reduction current of O_2 will then mask those of trace compounds whose redox potentials are < 0 V (Figure 2A), unless special precautions are taken (section 5.1.2.). Similarly in strongly anoxic waters, H_2S/HS^- may be present at high concentrations (10^{-6}–10^{-3} mol L^{-1}), in particular in sediments, and mask analyte curves at potentials < -0.6 V (Figure 2A), on Hg (see section 5.3.1.). The reduction wave of H^+ is usually not a problem, as it is very low for pH > 5.

Capacitive current; the sensitivity limit Even in the absence of depolarizer, the current in the polarization range is not zero (full line in Figure 2A). This is because, by imposing a potential difference (e.g. negative) between the working electrode and the solution, a capacitor is formed at the interface which includes a (negatively) charged plane on the electrode side of the interface and a layer of oppositely charged ions (positive), on the solution side. As for any capacitor, a change of potential difference, E, between the two planes will create a so-called capacitive current, i_c, to readjust the charges of the two planes. Contrary to the faradaic current, i_f, due to the redox process of the depolarizer, i_c does not involve any charge transfer through the interface. At a potential where a depolarizer is reduced or oxidized and gives rise to a faradaic current i_f, the net measured current is thus:

$$i_t = i_c + i_f \quad (1)$$

As i_c is independent of the analyte concentration, it will be a major cause (with oxygen reduction, see above) of the sensitivity limit of voltammetric techniques. i_c can be minimized (i) by using an appropriate $E(t)$ modulation during voltammetry, and (ii) by correct control of the electrode – solution interface. Much efforts have been devoted to develop special modulations (section 2.1.2) that maximize the ratio i_f/i_c, and hence improve the detection limit down to $\sim 10^{-11}$ mol L^{-1} (Table 1). As far as the interface itself is concerned, the Hg – water interface is much more reliable and uniformly controlled than any solid substrate – water interface. For that reason, Hg is still by far the best electrode substrate in voltammetry.

2.1.2 The Various Modulated Voltammetric Techniques

The most important modulation techniques for water analysis are summarized in Table 1. For other techniques and more details, the reader is referred to refs

[2–5]. The principle of SWV is discussed in ref. [7] and thorough study of capacitive current elimination by various pulse and mathematical transformation techniques is discussed in refs [3,6,8]. Table 1 assumes that the analyte is an oxidized species which is reduced at the electrode. This is often the case for inorganic elements, but not necessarily so. In particular many organic compounds are reduced species which are measured by electro-oxidation. In such cases potential scans would be in the opposite direction to those shown in Table 1. Contrary to all other techniques listed in Table 1, which can use static electrodes such as solid or hanging mercury drop electrodes, d.c. polarography [1] must be performed using dropping Hg or rotating disc electrodes and is thus not convenient for *in situ* analysis; it remains useful, however, for comparison purposes, as it is the simplest reference technique. Cyclic voltammetry is not sensitive enough for *in situ* analysis but is useful to study the electrode processes and thus to develop optimum analytical conditions (section 3). All other techniques of Table 1 are used for water analysis; their selection depends on the analyte (see section 4 for typical examples) and conditions. In particular:

(a) LSV, DPV and SWV are usable when analyte concentration is not too low (Table 1). Their advantage is that measurement time is very short (0.1 s to a few seconds).
(b) ASV or AdSV techniques are necessary for concentrations $< 10^{-8}$ mol L^{-1}. They require a longer time as they need a preconcentration step (also called deposition step), during a time t_d, before the recording of voltammetric curve (scanning step). To reach the highest sensitivity (10^{-11} mol L^{-1}), t_d may last up to 15 min. The preconcentration step can be done in various ways (see Table 1 and section 4, for examples), as follows.
    (i) Reduction of metal ions to the metallic state and deposition on (or in for Hg) the electrode. The metal is then reoxidized which gives rise to a peak whose i_p is proportional to the concentration of metal ion in solution. This technique, called *anodic stripping voltammetry (ASV)*, works very well with amalgamable metals on Hg (Tables 1 and 8; [9]), but is presently largely limited to this electrode for reliability reasons (see Tables 2 and 8 for exceptions).
    (ii) Adsorption of the analytes which are naturally adsorbed on the electrode, followed by a cathodic or an anodic scanning step depending on the redox properties of the analyte. This is applicable in particular to organic compounds.
    (iii) Adsorption of the analyte, A, at a potential where Hg is oxidized and forms an insoluble salt HgA at the electrode surface. HgA is then reduced into $Hg^0 + A$ during the scanning step, which gives rise to a peak proportional to the concentration of A. This is feasible with organic compounds and inorganic anions such as S(−II) and halides (Table 9).

Table 1. The major voltammetric techniques useful for *in situ* measurements in water, and their characteristic features. D.c. polarography is performed on a dropping Hg electrode (DME) or a rotating disc electrode; other techniques are performed on stationary electrodes. ΔE = pulse height (mV), f = frequency (Hz), t_p = pulse duration (s), t_d = deposition time (s), t_g = Hg drop time (s), v= scan rate (mV s^{-1}), i_p = peak current (A), τ = transition time for oxidation in PSA (s). Rows 1–4, techniques based on direct measurements. Rows 5–8, techniques with preconcentration steps. Column 4: in all cases i is proportional to analyte concentration; other factors are given in the column. DP = differential pulse, SW = square wave (see also glossary)

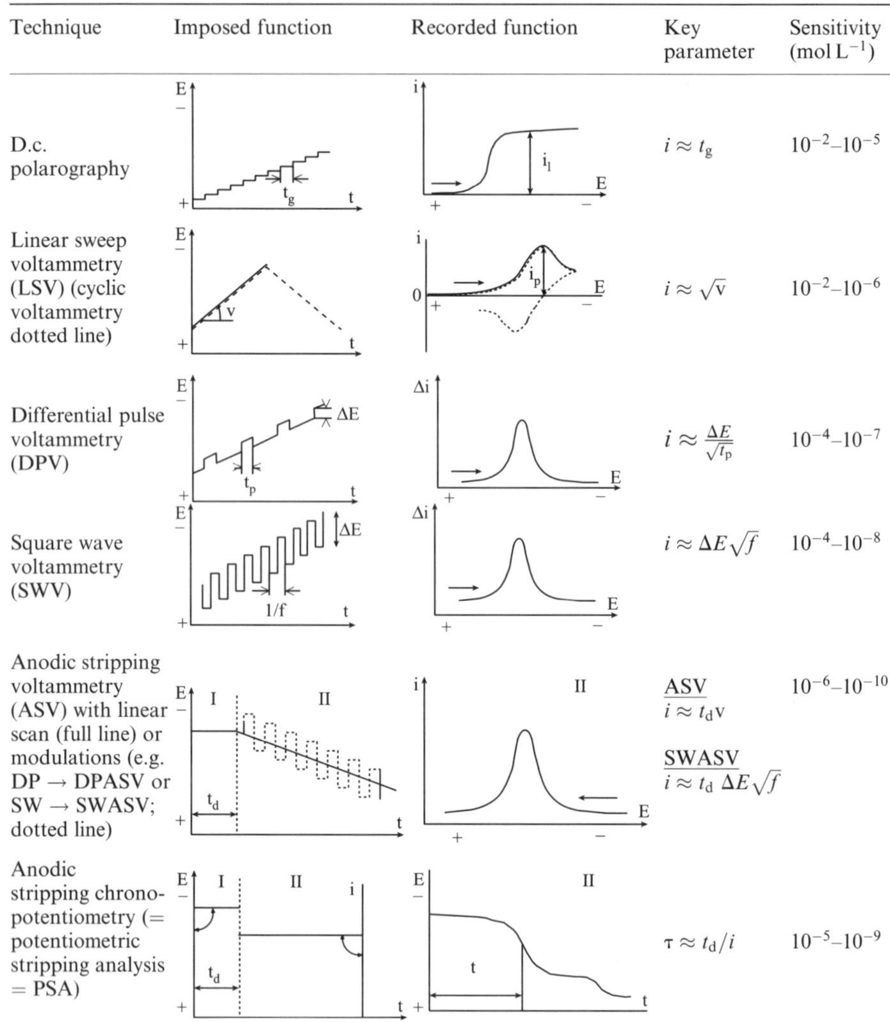

Technique	Imposed function	Recorded function	Key parameter	Sensitivity (mol L^{-1})
D.c. polarography			$i \approx t_g$	10^{-2}–10^{-5}
Linear sweep voltammetry (LSV) (cyclic voltammetry dotted line)			$i \approx \sqrt{v}$	10^{-2}–10^{-6}
Differential pulse voltammetry (DPV)			$i \approx \frac{\Delta E}{\sqrt{t_p}}$	10^{-4}–10^{-7}
Square wave voltammetry (SWV)			$i \approx \Delta E \sqrt{f}$	10^{-4}–10^{-8}
Anodic stripping voltammetry (ASV) with linear scan (full line) or modulations (e.g. DP → DPASV or SW → SWASV; dotted line)			ASV $i \approx t_d v$ SWASV $i \approx t_d \Delta E \sqrt{f}$	10^{-6}–10^{-10}
Anodic stripping chronopotentiometry (= potentiometric stripping analysis = PSA)			$\tau \approx t_d / i$	10^{-5}–10^{-9}

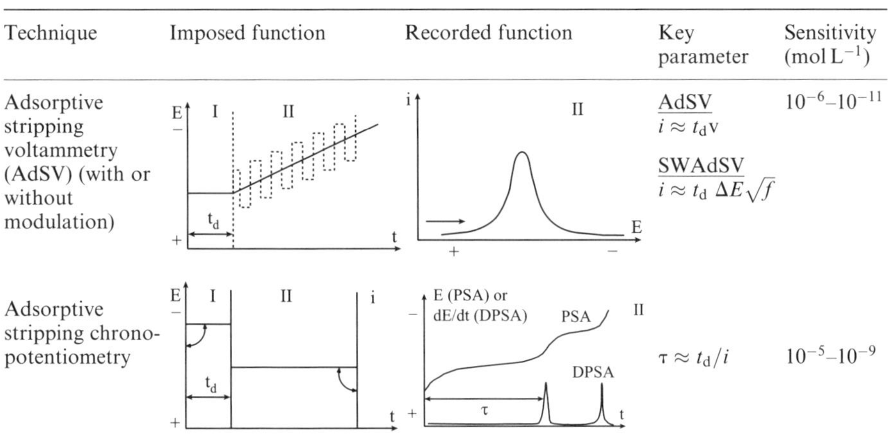

Technique	Imposed function	Recorded function	Key parameter	Sensitivity ($mol\ L^{-1}$)
Adsorptive stripping voltammetry (AdSV) (with or without modulation)			AdSV $i \approx t_d v$ SWAdSV $i \approx t_d\ \Delta E \sqrt{f}$	10^{-6}–10^{-11}
Adsorptive stripping chronopotentiometry			$\tau \approx t_d / i$	10^{-5}–10^{-9}

(iv) Addition in solution of a ligand L forming, with the analyte M, an adsorbable metal complex ML which is reduced during the scanning step. This technique can be applied to a number of trace metals (Tables 1 and 8).

All techniques based on preconcentration by adsorption (techniques (ii)–(iv) above) are called below *adsorptive stripping voltammetry (AdSV)* even though this term is sometimes restricted to techniques (iv), in the literature. The various compounds which can be analysed by AdSV have been reviewed in refs [10,11]. ASV techniques have been reviewed in ref. [9]. In both AdSV and ASV, the scanning step can be performed by means of various modulations, the differential pulse (DP) and square wave (SW) modulations being the most widely used for very sensitive detections. Typical parameters are $\Delta E = 20$–$50\,mV$, $t_p =$ 20–50 ms and $v = 2$–$5\,mV\,s^{-1}$ for DP modulation, and $\Delta E = 20$–$50\,mV$ and $f =$ 50–200 Hz, for SW modulation. The optimum conditions, however, depend on the test compound. The scanning step can also be performed at constant current, i, by recording $E = f(t)$ (PSA in Table 1). The analyte concentration is then proportional to the plateau length, τ (for theory of PSA see ref. [12]).

The techniques with a preconcentration step enable improvement of sensitivity by factors of 3–4 orders of magnitudes compared with direct measurements. This is a key characteristic for water analysis of many trace compounds and elements. The drawback for *in situ* analysis is that the electrode surface is modified by the preconcentration step and should be cleaned up before subsequent analysis. This is not difficult for ASV techniques applied to amalgamable metals deposited in Hg, as these metals are readily eliminated by reoxidation. It is more difficult for ASV of metals not or sparingly soluble in

Hg (Mn, Fe, Ni) and above all for AdSV techniques, as history effects may occur. In most cases, for laboratory applications of AdSV (Table 8), a Hg drop is used and renewed between analysis, which is not applicable *in situ*. Alternatively, the adsorbed compounds should be desorbed before subsequent analysis, by changing either the potential or the solution composition. In addition, waters often contain natural surface active compounds (section 3.4) which may interfere with the adsorption of the analyte. Thus rather drastic clean-up procedures requiring solution changes must often be allied to AdSV measurements. This makes AdSV more difficult to apply *in situ*. In particular:

- the voltammetric cell must be coupled to a flow injection system to change the solution before and after analysis;
- speciation is either not possible (only total concentrations are measured) or more difficult to interpret (see section 3.3.3), since reagent addition may perturb the sample.

2.1.3 Hydrodynamic Conditions and Electrode Geometry (Micro- and Macroelectrodes)

Because the analyte is consumed at the electrode, its transport mode from the bulk solution to the electrode surface influences the voltammetric signal and should be well controlled. Figure 3 shows that the dynamic factors influencing the reduction of a metal ion M are the charge transfer rate constant, k_0, the diffusion coefficients of M and ML species, D_M, D_{ML}, and the formation/dissociation rate constants of the metal complexes, k_f, k_d. In conditions of fast chemical reaction compared with diffusion, the diffusional transport of M+ML is determined by an average diffusion coefficient, $\bar{D}$, which depends on D_M and D_{ML} as discussed in section 3.1.1. In addition, for reductions at sufficiently negative potentials ($E - E_0 \ll 0$) the charge transfer is very fast compared with diffusion; the reduction flux, dN/dt, or the current, i, at constant E, on a spherical stationary electrode is then given by [1]:

$$i = nFA\frac{dN}{dt} = nFA\bar{D}c\left(\frac{1}{\delta} + \frac{1}{r}\right) \qquad (2)$$

where n = number of exchanged electrons per mole of M, dN/dt = number of moles of M reduced per time unit t, A = surface area, c = total metal concentration in the bulk solution, δ = thickness of the diffusion layer, and r = electrode radius. Equation (2) is strictly applicable only for a sphere in a quiescent solution [1]. It can be shown, however, [13] that for time > 10 s, equation (2) also applies to a disc-shaped electrode by replacing the term $1/r$ by $1.27/r$, even for the complicated SW modulation. Equation (2) is thus general enough to enable discussion of the role of electrode size and geometry for *in situ* applications (Figure 3).

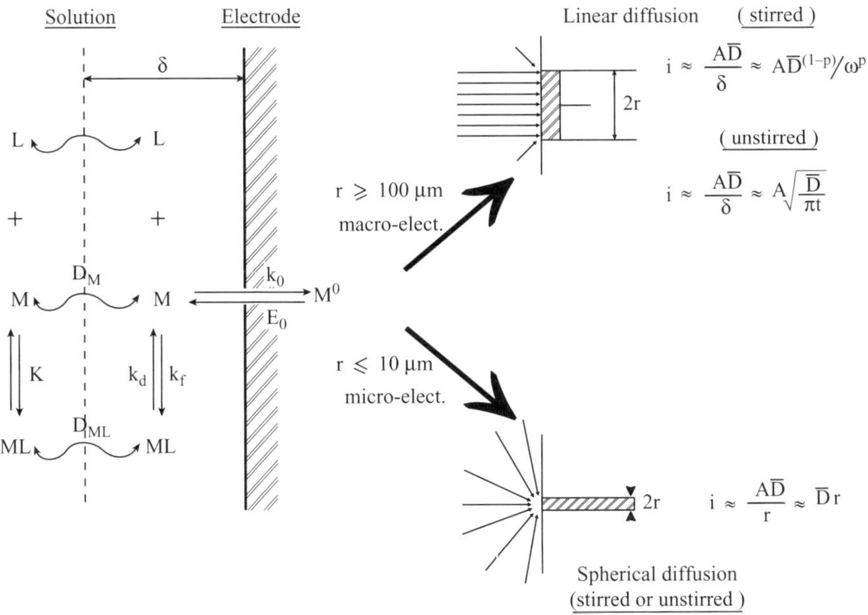

Figure 3. Major processes and parameters influencing the voltammetric current on macro- and microelectrodes. K = equilibrium complexation constant; k_0, k_d, k_f = charge transfer, dissociation and formation rate constants respectively; D_M, D_{ML}, = diffusion coefficients of M, ML respectively. $\bar{D}$ = average diffusion coefficient; E_0 = standard redox potential; r, A = radius and surface area of the electrode; t = time, δ = diffusion layer thickness, ω = rotation (or stirring) rate; p = constant ($0 < p < 1$)

For *large electrodes* (macroelectrodes), $1/r \ll 1/\delta$, so that i is primarily dependent on δ, which itself depends on hydrodynamic conditions. Two cases are worth discussing:

- In *quiescent solutions*, $\delta = \sqrt{\pi \bar{D} t}$, so that *i is time dependent and controlled by diffusion.* This is true even at constant E (no potential modulation) and in the absence of chemical control of the dynamic electrode process. An additional time dependence may result when a potential modulation is used.
- In *convective conditions*, e.g. at a rotating disc electrode or in solution stirred with a magnetic stirrer with a rotation speed ω, it can be shown ([14]; this book, Chapter 8) that $\delta \sim (\bar{D}/\omega)^p$ where $p \approx 0.3$–0.5. *i is then convection dependent,* and only varies with time when $E(t)$ modulation is used.

For *small electrodes* (microelectrode), $1/r \gg 1/\delta$ (equation (2)) and *i is independent of both time and stirring conditions.*

The general properties of microelectrodes are described in refs [15, 16] and a review of their analytical applications is given in ref. [17]. The size limit between macro- and microelectrodes depends on the time-scale of measurement when δ is time dependent. Typically, $r \leq 10\,\mu m$ corresponds to micro-behaviour and $r > 100\,\mu m$ corresponds to macro-behaviour. The major difference between micro- and macroelectrodes is that the curvature radius of macroelectrodes is larger than the diffusion layer thickness, so that linear diffusion can be considered to apply to all of them irrespective of their geometry, contrary to microelectrodes for which spherical (or semi-spherical) diffusion occurs (Figure 3) During the last 40 years, a large number of geometries and hydrodynamic conditions have been studied to get reliable, theoretically predictable, currents [2,5,6,18,19] using macroelectrodes (Figure 4A), in particular in HPLC measurements. However, it has proved difficult to get very high sensitivity with such electrodes. Two other limitations of macroelectrodes are important for *in situ* applications: (i) uncontrolled convection usually occurs in natural waters so that δ may vary, leading to unreliable results; (ii) to apply ASV and AdSV techniques, the flux of analyte towards the electrode should be time independent during the preconcentration step. With macroelectrodes, this is only possible in well-controlled stirred solution (Figure 4A). These techniques thus are not applicable in media where controlled convection cannot be imposed such as sediment. All these problems are solved with microelectrodes (Figure 4B,C), which are therefore much more convenient for *in situ* applications. Other advantages of microelectrodes for *in situ* measurements are the following:

- *Measurement at very low ionic strength (soft waters)* is possible without adding electrolyte. Indeed, because of their very small surface area, i is usually in the range pA–nA, i.e. the ohmic drop, $i.R$ of the test solution, (R = resistance of the solution) is very small even at low ionic strength [20].
- *Higher sensitivity* is obtained because of their higher signal-to-noise ratio.
- *High spacial resolution (100–300 μm) measurements* can be made in sediments or mats, with minimum perturbation (section 5.3), because of the very small dimension of the electrode and its electrode body (glass, plastic).
- *Discrimination between analyte bound to mobile ligands and colloids* is made easier, which is a key feature for speciation studies in waters (section 3.3.1). This is because i depends on well-defined molecular diffusion on microelectrodes instead of less reproducible convective transport on macroelectrodes.

2.2 THE MAJOR COMPONENTS OF VOLTAMMETRIC CELLS

2.2.1 Working Electrodes

The ideal working electrode should have a reproducible surface, a reproducible area, and a low residual current. Surveys of electrodes used for voltammetric analysis [21, 22] and for trace element speciation [23,24] have been published.

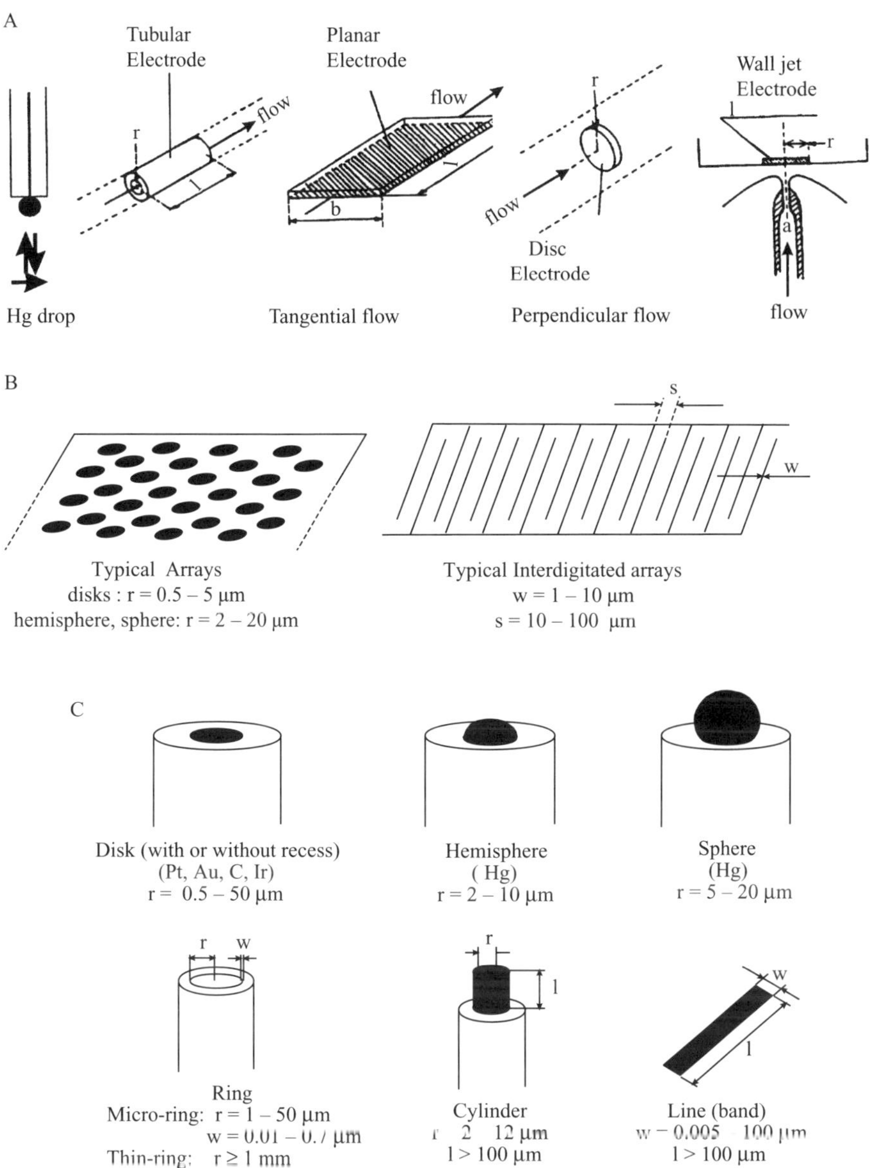

Figure 4. (A) Stationary macroelectrode geometries and hydrodynamic conditions for which theoretical expressions of current are available [2,3,5,18,19]. The arrows show the direction of flow; for the Hg drop, three directions of flow have been studied. (B) Commonly used geometries of microelectrode arrays (modified from ref. [17]. (C) Most commonly used geometries of microelectrodes (modified ref. [17])

Only the key aspects relevant for *in situ* analysis in waters are mentioned in this section, based on laboratory applications.

2.2.1.1 Macroelectrodes and electrode substrates The most common materials for analytical macroelectrodes are Hg, Au, Pt, and C [1,5,6,37]. The choice largely depends on the desired polarization range (Figure 2B). Hg electrodes are still the most widely used macroelectrodes in polarography and voltammetry. This is largely due to their much better reliability, as their surface is renewable for each analysis (for Hg drop electrodes), and to their very low capacitive current. In addition, many compounds can be analysed by reacting with Hg, which is not possible with other electrodes. This is the case for halides, sulfides (Table 9) and organic compounds such as cysteine or homocysteine [26], or organometallic compounds [27, 28]. In all cases they can accumulate at the Hg surface by forming an insoluble product which is subsequently oxidized or reduced (adsorptive voltammetry, Table 1). Hg electrodes include the dropping mercury electrode (DME), the static mercury drop electrode (SMDE), the hanging mercury drop electrode (HMDE) and the mercury film electrode (MFE) [1,3–6,9]. The first three are easily renewed, which avoids or minimizes memory effects, but they are more difficult to handle in non-laboratory conditions, in particular on site or *in situ*. They have been largely used for the analysis of compounds in waters [29, 210] (Tables 8 and 9), but presently the most widely used are the HMDE and MFE in either stirred or flowing solutions, and the rotating MFE [23, 24, 29]. The MFE has a lower detection limit than the HMDE for trace analysis and a better selectivity in ASV techniques, as peaks are sharper. However, it is more drastically affected by adsorption of fouling compounds. In addition, metal concentration in Hg is usually higher in MFE owing to the smaller volume of Hg, which may cause more problems with intermetallic compound formation or oversaturation of Hg with the test metal (pp. 515, 516 in ref. [5]; [30,31]). Amongst metals typically analysed, In, Cd, Tl, Zn, Sn, Pb and Bi have high solubility in Hg ($>$ 1 wt%), and Cu, Mn and Ni have intermediate solubility ($2–8 \times 10^{-3}$ wt%). Others such as Co and Fe have solubilities which are too low ($< 10^{-4}$ wt%) for analysis by ASV (p. 512 in ref. [5]; [32]). Formation of a number of intermetallic compounds has been reported in Hg (p. 516 in ref. [5]; [30, 31]). The most relevant for natural waters are Cu–Cd, Cu–Mn, Cu–Zn, Fe–Mn, Ni–Zn, Pt–Zn and Pt–Cd. The last two are an important limitation for the use of Pt as a substrate for MFE electrodes [33]. Because of the very low concentrations of trace metals in unpolluted waters, most of the other intermetallic compounds do not form significantly. Cu–Zn, however, may be an important exception, depending on conditions [29,34]. Formation of the Hg insoluble species CuZn and the Hg soluble complexes $CuZn_2$ and their stability constants have been reported [e.g. 35], allowing prediction of the importance of this effect in varying water conditions. It has been found experimentally [34] that Cu–Zn compounds do not form significantly on Hg microelectrodes for solution concentrations of Zn(II) or Cu(II) $< 20\,\mathrm{nmol\,L^{-1}}$. For higher concentrations, formation of these compounds can be prevented by adding Ga(III) to the solution [29,36], as Cu–Ga intermetallic compounds are more stable than Cu–Zn compounds. Zn(II) can thus be measured in the presence of Cu(II). The opposite is possible by depositing Cu(II) in unmodified samples at a potential where Zn(II) is not reduced.

The most common MFE substrates are Pt, Au, Ag, Ir or C (glassy carbon or graphite). Pt, Au and Ag are not very useful for reliable, long-term applications, as they dissolve in mercury, eventually converting the Hg film into a concentrated amalgam film as for Ag and Au [37,38]. They may also form intermetallic compounds with the test metals (e.g. Cd and Zn with Pt [33]). Au has been used as a substrate for MFE measurements in sediments [39]; stability of such electrodes, however, is less than 1 d. Glassy carbon substrates also have limitations as only Hg droplets are formed on this substrate instead of a true stable Hg film. Although they have been frequently used for laboratory analysis, such films are unusable for long term *in situ* operation. Graphite substrates have the same Hg film problem as glassy carbon, and in addition, they may oxidize when the redox potential of the test medium is too high, which decreases their lifetime [40]. Iridium is presently the only substrate which has a low solubility in Hg, on which a true film can be obtained [41,42], and which is very resistant to oxidation. On Ir macroelectrodes, however, Hg films are not very stable, but break into a number of droplets when the electrode is disconnected from the potentiostat, which also limits their application. As discussed below, a stable Hg film can be obtained on Ir microelectrodes.

Solid electrodes such as Pt, Au, Ag and various types of carbon substrates have also been used without Hg film, for measurements of either elements (Tables 2, 8 and 9) or organic compounds (Table 10) with redox potentials more positive than the oxidation of Hg (Figure 2D). Voltammetric analysis of metals with solid electrodes are usually less reproducible and sensitive than with Hg electrodes as the surface state of the electrode is not as well defined and capacitive current is higher. In addition, in stripping techniques, the crystalline form of the deposited metal (ASV) or compound (AdSV) may play an important role on the reoxidation peak shape. For metals, intermetallic compounds are more easily formed than in Hg. A major problem for long term use of solid electrodes is their surface renewal. In particular noble-metal electrodes adsorb hydrogen [37] and Pt and Ag may form oxide films. Several methods have been described for surface renewal, including mechanical polishing [43], laser activation [44], plasma treatment [45], chemical treatment [46], electrochemical cleaning [47,48], and vacuum heat treatment [49]. Apart from electrochemical and possibly chemical cleaning , they are not readily applicable *in situ*. In spite of these difficulties, solid electrodes might be useful for some *in situ* applications, although further developments are still needed. The present state of the art is summarized below.

Table 2. Main inorganic compounds which may be analysed directly on Au electrodes (laboratory conditions). In columns 4 and 5, M means mol L^{-1}. See glossary for acronyms

Compound	Electrode type	Technique	Media or natural sample	Sample treatment	References
As(III)	Au fibre microelectrode ($\phi = 25\ \mu m$)	CCPSA	Seawater	5 M HCl	50
As(III) + As(V)	Au fibre microelectrode ($\phi = 25\ \mu m$)	CCPSA	Seawater	5 M HCl, 0.01 M KI	50
As(V)	Au disc electrode ($\phi = 1.5$ mm) or Au film plated on Pt disc electrode ($\phi = 0.8$ mm)	CCPSA	2 M HCl		51
As(III) + As(V), Hg(II)	Au film plated on glassy carbon disc electrode ($\phi = 16$ mm)	ASV	Lake water	0.1 M HCl	52
Bi(III), Cu(II)	Au film plated on Au fibre microelectrode ($\phi = 10\ \mu m$)	CCPSA	Urine ref. sample	4 M HNO_3, 1 mM $KMnO_4$ (heating 10 min, 70 °C)	53
Bi(III), Cu(II), Pb(II), Sb(III)	Au film plated on glassy carbon disc electrode	PSA	0.1 M HCl		54
Cu(II)	Au film plated on glassy carbon disc electrode ($\phi = 3$ mm)	CCPSA	NBS ref. water sample	0.5 M HNO_3	55
Cu(II)	Au film plated on glassy carbon disc electrode $\phi = 3$ mm)	CCPSA	River and seawater	0.1 M HNO_3	55
Cu(II)	Au microcylinder electrode ($\phi = 100\ \mu m$, length = 5 mm)	CCPSA	Seawater / groundwater		56, 57
Cu(II)	Au fibre microelectrode ($\phi = 25\ \mu m$)	CCPSA	Blood	0.8 M HCl	58
Hg(II)	Au film plated on glassy carbon disc electrode ($\phi = 3$ mm.)	CCPSA	River and seawater	HNO_3	55

Hg(II)	Au fibre microelectrode	DPASV	Tap and river water	HCl or HNO_3, H_2O_2 UV irradiation	59
Hg(II)	Au film plated on Au fibre microelectrode ($\phi = 10\,\mu m$)	CCPSA	Urine ref. sample	4 M HNO_3, 1 mM $KMnO_4$ (heating 10 min at 70 °C)	53
Hg(II)	Au fibre microelectrode ($\phi = 10\,\mu m$)	CCPSA	Tap water	0.5 M HNO_3, 1 mM $KMnO_4$ (medium exchange for stripping)	60
Hg(II)	Au disc electrode ($\phi = 5$ mm)	DPASV	Seawater	UV irradiation	61
Hg(II)	Au disc electrode	DPASV	Seawater	HNO_3, UV irradiation	62
Hg(II)	Au cylinder electrode ($\phi = 100\,\mu m$, length = 5 mm)	CCPSA	Seawater/groundwater		56
Pb(II)	Au film plated on glassy carbon disc electrode ($\phi = 3$ mm)	CCPSA	NBS ref. water sample River water	HNO_3	55
Pb(II)	Au cylinder electrode ($\phi = 100\,\mu m$, length = 5 mm)	CCPSA	Seawater/groundwater		57
Sb(III)	Au fibre microelectrode ($\phi = 10\,\mu m$)	CCPSA	Seawater/groundwater	20 μM KI, 0.1 M HCl (medium exchange for stripping)	63
Sb(V)	Au fibre microelectrode ($\phi = 10\,\mu m$)	CCPSA	Seawater/groundwater	20 μM KI, 4 M HCl (medium exchange for stripping)	63
Se(IV)	Au film plated on glassy carbon disc electrode	CCPSA	NBS ref. water sample	0.5 M HNO_3	55
Se(IV)	Au fibre microelectrode ($\phi = 25\,\mu m$)	ASV	0.1 M $HClO_4$	0.5 M HNO_3	64

- *Gold* electrodes are the most widely used solid electrodes for inorganic compounds (Tables 2,8 and 9). A few heavy metals such as Hg, Cu, Bi, Pb and Cd have been measured on Au, as well as a number of elements more difficult to determine on mercury, such as As(III), Sb(V), Se(IV), and I_2. *Ag electrodes* have been used to determine halides and S(−II) [65] and it has been shown that I_2, Br_2, Cl_2, and H_2S may be analysed after separation into the gas phase from complex matrices and preconcentration on Ag electrodes [66,67]. The reliabilities of Au as well as Ag and Pt electrodes are limited by the fact that the electrochemical reactivity of the test compound depends on the crystal plane of the electrode exposed to the solution. For ideal reliability, a well-defined crystal plane should be used as electrode surface. In most cases, however, electrodes are made of polycrystalline material providing a statistical signal which may depend on electrode preparation. In particular, different results might be obtained using Au discs made of gold wires or gold films deposited on glassy carbon or other substrates.
- *Platinum and C* are mostly used for organic compounds (Table 10). A major drawback of Pt in aqueous solutions is its formation of surface oxides at positive potentials [68], which severely limits its applications. Carbon is much used for organic compounds [19,69–71], particularly in combination with HPLC, even though Hg electrodes are still predominantly used for both organic and inorganic compounds. C electrodes can be made of graphite, pyrolytic graphite, glassy carbon or reticulated carbon and have been extensively reviewed [5,72]. In particular, the polarization range of the various types of C electrodes are discussed in detail in ref. [5]. The reliability of such electrodes is limited by a number of electrochemical and oxidation reactions which may occur inside their pores or at their surface [73]. For that reason, both glassy carbon and C paste electrodes, which are a mixture of graphite and hydrophobic component such as wax, are the most widely used as their porosity is low.

Chemically modified electrodes (CMEs) form a vast category of non-Hg electrodes. They have attracted much attention during the past 15 years and numerous reviews have been published [19,71,74–82]. As CMEs may be very specific to the analysis of particular compounds, only the general principle is mentioned here. In all cases the electrode surface (often a carbon paste substrate) is modified, e.g. by direct chemical linkage of a substance to the surface, by physical adsorption, plasma reaction or any other means. The modification may result in a molecular change at the surface, or in the formation of a thick coating. Chemical modification of surfaces may have various functions such as enabling preferential preconcentration of the analyte at the surface, immobilizing in the coating a catalyst which may enhance the electrochemical response, or altering the physical properties of the electrode surface, e.g. to shift and separate overlapping signals. CME electrodes can be used, not only for analytical purposes, but also to study processes such as the interaction of natural organic compounds (e.g. polysaccharides) with metal ions [83]. Reviews of applications of CMEs to analysis of metals, anions and organic compounds, with detailed lists of analytes and conditions can be found in refs [75–82]. Although CMEs are potentially interesting, it must be emphasized that their applications are still limited to research areas, as further developments with respect to sensitivity, stability and/or regeneration of the surface reactive functional groups are required for routine applications.

2.2.1.2 Microelectrodes Since the late seventies, the development of microelectrodes (size $\leq 10\mu m$) has received considerable attention because of their striking advantages (section 2.1.3). In particular, owing to their reduced capacitive current and increased mass transport rate, microelectrodes exhibit excellent signal-to-noise characteristics, which, in combination with the recent developments in amplifiers, allow them to be used for the determination of analytes at ultra-low concentrations ($< 10^{-9}$ mol L^{-1}) [84–90].

The same substrates as for macroelectrodes (Pt, Au, Ag, Ir, C) are used for microelectrodes. Typical geometries are shown in Figure 4B, C. They can be produced by classical mechanical techniques or modern photolitographic, LIGA (see list of acronyms and Chapter 12) and screen-printing technologies [91,92]. Conventional mechanical techniques have been used mainly to produce single microelectrodes. The main steps involve sealing a C or Au fibre or an electroetched metal wire in an insulating material (often a glass capillary, but also epoxy resin or polyethylene) followed by polishing of the tip [88,91–93]. Electrode arrays (Figure 4B) can also be prepared by conventional mechanical means [94,95] such as filling nuclepore membrane cylindrical pores with metal [92,96–100] or sandwiching several metal foils between insulating layers [101–103]. Although reliable *single* microelectrodes can be produced with such classical techniques [93], microelectrode *arrays* produced in this way usually exhibit an important capacitive current because perfect sealing between the electrodes and the insulating material is very difficult to achieve for each electrode [96]. This problem has been largely eliminated [104] by using microtechnology-based methods, such as photolithographic, LIGA and screen-printing methods for fabrication (Chapter 12). In addition, these techniques offer other important advantages, in particular (i) automatic low cost production at industrial scale and (ii) wide flexibility in the choice of electrode geometry and material. Moreover, arrays of individually addressable microelectrodes with centre to centre spacing as low as a 10 μm can be produced with these techniques. Voltammetric, amperometric or potentiometric signals on each individually addressable microelectrode can be read successively [105–107] or by group of microelectrodes, using a multiplexing approach [108–109]. This type of sensor is of great interest for environmental purpose, in particular for *in situ* measurements of analyte gradients at the water–sediment interface (section 5.3; [110]).

Mercury-coated single microelectrodes and microelectrode arrays have been prepared using the same substrates as for macroelectrodes (Pt, Au, Ag, Ir, C) [87,92,93,111]. They show the same limitations. Ir is an exception, however, as the stability of the mercury coating is greatly improved on microdiscs compared with conventional size Ir electrodes. Indeed, the surface tension of Hg and the Hg–Ir cohesion forces are such that the stability of the hemispherical Hg cap formed on micrometer-sized Ir discs is optimum. Such microelectrodes are stable in open circuit and can be transferred from one solution to another, shaken, and rinsed without significantly affecting their performance [87,88,93,104]. It has also been shown that reliability and reproducibility close to 100% can be obtained for continuous measurements in low ionic strength (0.1–10 $mmol\,L^{-1}$) solutions for at least several days without renewal of the mercury layer [93].

2.2.2 Reference Electrodes

The reference electrode (RE) is the second key component of the voltammetric cell. The principle of REs for laboratory physico-chemical potentiometric studies is described in detail in various textbooks (e.g. refs [112,113]). For such applications, it is usually essential that the potential of the RE is well known and stable to within $\pm$ 0.1 mV. For *in situ* voltammetric trace analysis, potential stability within $\pm$ 1 mV may be sufficient (although speciation measurements may require better precision). Furthermore a number of additional requirements should be met.

(i) *The electrical resistance* of the electrode (in particular of its liquid junction) should be low and stable (typically $\leq 50\,k\,\Omega$) when fast modulation techniques such as SWASV are used.
(ii) *Pressure compensation devices* must be used for working at depth in natural waters with classical REs, including internal solutions (Table 3a). To avoid such mechanical complications, reference electrodes constructed of polymers and solid-state reference electrodes of various types have been proposed (Table 3b, c).
(iii) *Contamination* of the test sample by the RE is a major potential problem with all of them for the measurement of very low concentrations of analytes ($< 10^{-8}\,mol\,L^{-1}$):

- REs based on Cl^- (or other halides) may release Cl^- into the test solution, which gives rise to a voltammetric peak with Hg electrodes. This may interfere either when Cl^- is the test compound or when the Cl^- peak overlaps with the peak of the analyte;
- any RE includes chemical reagents at high concentration in the internal solution (e.g. AgCl, KCl for the Ag/AgCl/KCl// RE). They may include

many impurities, in low proportions, whose concentrations, however, may be high enough to contaminate the sample by diffusion through the liquid junction, for ultratrace metal analysis;
- compounds from the test sample may diffuse inside the internal solution or gel of the RE, and be released in a subsequent sample (history effect).

(iv) *Small* voltammetric cells are desirable for *in situ* measurements, in particular to minimize reagent addition, when voltammetric detection must be preceded by a chemical reaction. Small REs are then needed. There is, however, a lower size limit, which depends on the nature and construction of the electrode, below which the RE performance becomes unacceptable.

(v) *Fabrication procedure, mechanical fragility and long-term stability* are additional criteria. There is a lot of accumulated experience in the fabrication of reliable classical types of REs (Table 3a,b). However, these types of RE may not always fulfil the aforementioned criteria *(ii)–(iv)*. Newer solid-state mini- or micro-REs, can be fabricated based on microtechnologies (Chapter 12). Sometimes they are directly integrated with the working and auxiliary electrode in a complete mini-voltammetric cell (mainly type (d), (e) REs in Table 3). Such recent developments, however, need further systematic tests before being applied to routine analysis.

Table 3 summarizes various procedures which have been proposed to overcome problems *(i)–(v)*. In type (b) REs, the internal solution is retained in the pores of a strongly hydrophilic polymer. Mixing with the test solution and pressure problems are thus decreased. Contamination remains a major problem and only large electrodes with conventional geometry are produced commercially. Type (c) REs are completely solid-state. Their major drawback is their relatively limited lifetime owing to slow diffusion and dissolution of KCl through the hydrophobic solid in contact with the test solution. Their advantage is that millimetre-sized electrodes can be made (e.g. ref. [123]). Type (d) REs can be called 'pseudo' reference electrodes, as they include an Ag/AgCl redox couple, but no KCl to fix a precise potential value. As a result the potential is less stable and reliable than those of others. The Ag/AgCl wire is coated by a thin polymer layer which decreases the diffusion rate of Cl^- but may adsorb trace metals and possibly release them into subsequent samples. Their advantage is that mini-sized REs can be fabricated easily. Systematic stability tests have not been performed, however. Finally a number of metal/metal oxide couples can be used to build reference electrodes (type (e) REs). For instance, the couples mentioned in Table 3 exhibit Nernstian pH behaviour, i.e. the potential change of such electrodes is well known and equal to $60\,mV\,pH^{-1}$. By coating them with a polymer containing a pH buffer, a constant potential, independent of the test solution may be obtained. Such a minielectrode has been

Table 3. Various types of reference electrodes, in particular those (b–e), which have been proposed for miniaturization and avoiding the use of internal solution.

Electrode type	Advantages	Drawbacks	References
(a) Reference electrodes with internal liquid			
(1) Ag / AgCl / x mol L^{-1} KCl (2) Hg / Hg_2Cl_2/x mol L^{-1} KCl (Calomel)	• well controlled and stable E • commercially available	• large size • contamination of the media by Cl^- • liquid junction problem due to the pressure frequently encountered for measure in low ionic strength (even for P compensated electrode)	114,115
(b) Reference electrodes with polymers			
(3) Ag / AgCl / sat. KCl in solid gel (no ceramic junction) (4) Ag / AgCl / KCl in Xerolyt polymer (no ceramic junction)	• well controlled and stable E • withstand P up to: (3) 7×10^7 Pa (700 bar) and (4) 1.6×10^6 Pa (16 bar) without P compensation • commercially available	• large size • risk of contaminations due to diffusion through the gel	*, †
(5) Ag / AgCl / sat. KCl in gel (Metrohm or 1 % LGL agarose); zirconium oxide ceramic junction	• well controlled and stable E • small size • low diffusion through the ceramic junction • withstand P up to 3×10^6 Pa (30 bar) (max. P tested) without P compensation • readily prepared • lifetime > 1 year	• influence of T not tested	116,117

Electrode type	Advantages	Drawbacks	References
(c) Solid-state reference electrodes			
(6) Ag/AgCl/vinyl ester resin doped with KCl (Reflex low resistance material) (7) Hg / Hg_2Cl_2 / KCl / graphite paste in a PVC tube closed with a composite Teflon glass membrane.	• stable E • small size • readily prepared	• limited lifetime: (6) ~ 3 months, (7) ~ 2 weeks • influence of T and P not tested	118–122
(8) Ag / AgCl wire inserted in a pellet of KCl–alumina–Teflon pressed around it	• stable E • small size • systematic tests of the influence of T (5–25 °C) and P (10^5–3×10^6 Pa, i.e. 1–30 bar) on E	• limited lifetime: ~ 2 weeks	123–126
(d) Ag / AgCl covered with a thin film			
(9) Ag / AgCl / Nafion film (~ 20 μm) (10) Ag / AgCl / polyacrylamide hydrogel (30 μm) (11) Ag / AgCl / melted glass layer (few 100 μm)	• small size • readily prepared • no internal electrolyte	• $E = f$ (media, pH) • limited lifetime (9) ~ 2 weeks, (10) (11) ~ 1 month • (11) fragile	127–130
(e) Metal / metal oxide pseudo references			
(12) Sb / Sb_2O_3; (13) Ru / RuO_2; (14) Ti / IrO_2; (15) Ir / IrO_2; (16) Pd / PdO_2; (17) Si / MO_2 with M = Pt, Pd, Ru, Ir	• small size • no internal electrolyte	• test media must be pH buffered or the electrode covered with a pH-buffered gel • $E = f$ (pH) • oxide layer preparation requires care	131–137

* (3) Idronaut Srl (Via Monte Amiata 10, I-20047 Brugherio (Mi)).
† (4) Ingold Electrodes Inc. (261 Ballardvale St., Wilmington/MA, USA).

used for instance in ref. [138]. Their advantage is that minielectrodes integrated in mini-voltammetric cells can easily be produced.

2.2.3 Flow-through Electrochemical Cells

Flow-through cells (sometimes wrongly called flow-through electrodes) were initially developed for applications in liquid chromatography and flow-injection analysis. They are now widely used in routine analysis, as they enable measurements to be easily automated [19,139–141]. They are essential for *in situ* measurements when a chemical reaction must be performed before the voltammetric detection, and they offer a number of advantages even for the analysis of unperturbed samples (section 5.1.1.). The main advantages are the following:

- easy automation leading to a high frequency of sample analyses with improved precision [19, p.126];
- lower risk of contamination due to minimum sample handling;
- improved selectivity and sensitivity owing to medium exchange capabilities. This can be used, for example, in ASV techniques to perform the deposition and stripping steps in different solutions [142,143], e.g. to complex an interfering species or to shift the peaks of overlapping substances;
- improved selectivity and accuracy in PSA, when a chemical oxidant is used for the stripping step, owing to better control of the oxidant concentration;
- improved sensitivity and accuracy in voltammetries with MFE, solid electrodes or in AdSV, by enabling complete automation of the conditioning of the working electrode (e.g. mercury film deposition for MFE [29], desorption of previously adsorbed analyte or ligand in AdSV between analysis [144,145], reduction of metal oxides on Pt);
- broadening of voltammetric applications; indeed flow-through voltammetric cells are readily coupled to flow injection analysis systems [146], which enables chemical pretreatment of the sample before voltammetric measurement to transform electroinactive compounds to electroactive species, or to get more information on species distribution of a given analyte in the test sample;
- capability of analysing very small volumes (μL) of solution [147]. This is an important requirement for *in situ* analysis with chemical pretreatment, to minimize reagent consumption and the volume of waste solution to be collected.

Detailed discussions of advantages and drawbacks of various cell configurations can be found in refs [5 (Chapter 22), 19,140,148,149]. Below are summarized the key features which should be considered when designing a voltammetric flow-through cell for flow injection analysis, HPLC analysis and *in situ* measurements of trace compounds (the last criterion is specific to the last type of application):

- choice of the most appropriate reference electrode (section 2.2.2) and working electrode substrate (section 2.2.1), in particular with respect to polarization range, capacitive current, and surface state long-term stability;
- well-defined geometry (and hydrodynamics for macroelectrodes) enabling theoretical prediction of faradaic current (section 2.1.3, Figure 4);
- high sensitivity and therefore high mass transfer rate; for macroelectrodes the latter depends on hydrodynamic conditions (Figure 4); for microelectrodes, mass transfer rate depends only on the geometry of the electrode;
- small volume of the voltammetric cell (preferably < 10–$100\,\mu$L for the reason given above;
- robustness, low noise, achieved by minimum electrical resistance and optimal electrode positioning;
- simplicity of the cell design and easy maintenance;
- automated valves and pumps functioning at depth with low energy requirements.

A general discussion of features related to these criteria is given below. Aspects specifically related to *in situ* measurements are discussed in section 5.1.1.

The *theoretical equations* for the current at thin layer, tubular and wall jet flow through cells (Figure 4A), which are the most commonly used systems with macroelectrodes, are given in the references of Figure 4A and summarized in ref. [19]. The following general equation has been derived [19,144,150] for the limiting current, i_ℓ at constant potential. It is applicable to all macroelectrodes in flowing conditions of Figure 4A:

$$i_\ell = knFDcw\,\mathrm{Sc}^{\beta}\,\mathrm{Re}^{\alpha} \tag{3}$$

where k, α and β are constants which depend on the cell geometry, w is the characteristic electrode width, F is the Faraday constant, and D and c are the analyte diffusion coefficient and concentration respectively. $\mathrm{Sc} = \nu/D$ is the Schmidt number (ν = kinematic viscosity of the solution) and $\mathrm{Re} = U/\ell\nu$ = Reynolds number (U = characteristic velocity, ℓ = characteristic length of the cell). Equation (3) shows that i_ℓ (thus the sensitivity) increases with Re (i.e. U). There is a limit to this behaviour, however, when Re is too large and the flow breaks up into a turbulent regime. The specific equations for thin layer, tubular and wall jet systems (see refs in Figure 4A) have been compared with experiments. Differences often occur and can be attributed to (i) uncertainties in the values of diffusion coefficients and constants used in the theoretical equations, (ii) non-ideal behaviour of the electrode (e.g. adsorption, or passivation effects), and (iii) deviations from hydrodynamic laminar flow in the systems. The use of special spectroelectrochemical systems allowing visualization of the flow pattern in the vicinity of the working electrode, as a function of cell and working electrode geometry, is useful for developing the desired flow-through cell. Such systems have been described for studying flow-patterns at Hg drop electrodes, in stagnant and flowing solutions [151,152], as well as at stationary and rotating disc glassy carbon electrodes [153]. It is worth emphasizing that since the measured current on microelectrodes (Figure 4B, C) is independent of hydrodynamic conditions when the size of the electrode is small enough, the design and practical use of corresponding flow-through cells is more flexible.

Applications of the various flow-through cells with the corresponding references are listed in [19]. Flow-through cells with dropping mercury electrode (DME), hanging mercury drop electrode (HMDE) and sessile mercury drop electrode (SMDE), have been built under conditions of parallel, opposite and perpendicular flow of solution (Figure 5A–D). Although Hg drops allow easy renewal of the electrode surface, major difficulties with such cells are (i) the rather large dead volumes, and (ii) the incorporation of devices to enable reliable Hg drop formation/dislodgment. For that reason, most flow-through cells use either MFEs or solid electrodes. Nevertheless the SMDE has been used for *in situ* voltammetric measurements in water ([154] section 5.2) and an

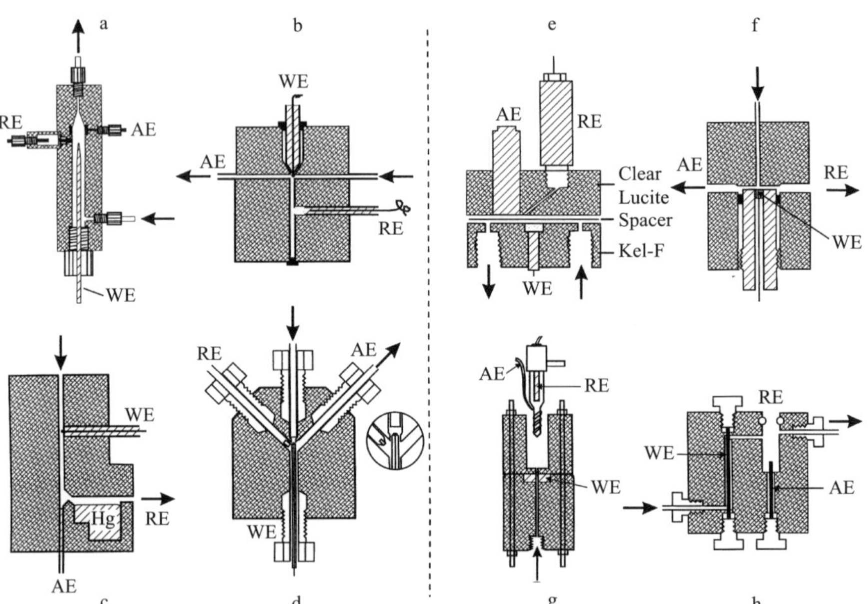

Figure 5. Most commonly used voltammetric flow-through cells configurations, in FIA or HPLC literature (modified from refs [116], a; [140], b,c,f; [19], d,e; [141], g,h). WE, AE, RE = working, auxiliary and reference electrodes, respectively. Arrows=solution flow. (a)–(d) cells based on Hg drop WE; (e)–(h) cells based on Hg film or solid WE

HMDE in a perpendicular flow has been used on board a ship for seawater analysis ([155]; section 4).

Wall jet systems are represented in Figures 5d, f. [19]. In these systems, a laminar jet of solution is expelled against the working electrode surface. For theoretical equations to be applicable, the diameter of the nozzle must be at least 10 times smaller than the electrode size, and the auxiliary and reference electrodes should be far enough away not to perturb the flow of solution close to the electrode. As the cell volume cannot be very small for such conditions, this configuration does not seem to be appropriate for *in situ* measurements. Smaller systems can be used (Figure 5d,f), but theoretical equations are no longer valid.

Figure 5e shows a typical *thin-layer detector* (TLD) [Chapter 22 in 5, 19], whose channel width is fixed by a Teflon spacer. Such systems are much used as they are easy to handle and clean up, and the dead volume is a few microlitres. For practical purposes, it is important to discriminate two types of working conditions of a TLD: the amperometric or voltammetric conditions (when $<5\%$ of the analyte of the solution reacts at the electrode) and the coulometric conditions, (when $> 95\%$ of the analyte reacts at the electrode). Amperometric conditions occur at comparatively large channel width and flow rate and low electrode surface area and the corresponding current depends on $U^{1/3}$, whereas coulometric conditions are achieved at small channel width and flow rate and the corresponding current depends on U. In practice, cells for amperometric conditions are easier to build but have lower detection limits. This problem can be overcome by using voltammetric detection.

Figure. 5g–h shows *tubular and cylindrical flow-through cells*. In cylindrical systems, solutions flows outside the rod-shaped electrode. Reproducible laminar flow seems difficult to achieve in this case as solution inlet coaxial to the electrode is not possible. A major drawback of these electrodes is the difficulty of cleaning the electrode surface, and the fact that some electrode materials (in particular Hg) are difficult to use. For these reasons, these two types of flow-through cells are not much used, although tubular electrodes were used in an early example of automatic voltammetry on board a ship for seawater analysis [29].

The *relative positions* of the working, reference and auxiliary electrodes in small flow-through cells, specially the TLD, should be selected carefully [5,19,156]. It may not be so crucial for measurements at constant potential in highly conducting solutions. However, it is a key aspects for scanning techniques with fast scan rates or high pulse frequency [157,158], in particular for square wave or pulse voltammetry in low ionic strength natural waters. In TLDs a number of different electrode positions have been tested for amperometric detection, two important ones being shown in Figure 6a and b. In most cell design, the auxiliary electrode is placed downstream of the working electrode to prevent electrolytic products from the former interfering with the detection. In such a case, however, a significant proportion of the iR drop takes

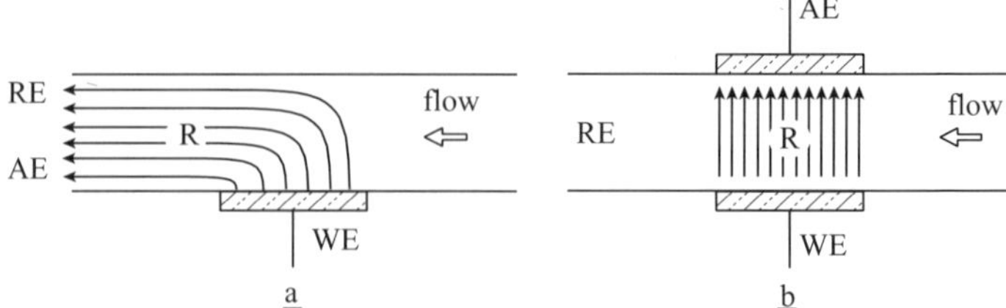

Figure 6. Electric field patterns for two electrode configurations in thin layer detector (TLD). R = resistance; WE, RE, AE = working, reference, auxiliary electrodes. Modified from ref. [5], p. 621

place along the electrode surface (Figure 6a) so that the potential at the downstream edge of the electrode is closer to the value controlled by the potentiostat than that on the upstream edge [140,156]. Different redox reactions may then occur on the various parts of the working electrode. Another possibility is to position the auxiliary electrode opposite the working electrode (Figure 6b). Sensitivity is usually increased, since the resistance is minimized, but the risk that the products of the auxiliary electrode interfere with the process occurring at the working electrode is greater (see above). However, when measurements can be performed in flowing conditions with linear velocity high enough compared with the diffusion time across the cell, such interferences are negligible unless gas bubbles are formed (Chapter 22 in ref. [5]). Another option is a cylindrical cell in which the working micro- or spherical electrode (WE) is positioned in the middle of the channel and the auxiliary electrode (AE) is a ring downstream of the WE as shown in Figure 5a. The latter configuration prevents interference from the AE's products and ensures that a uniform potential is applied to the WE surface. Its drawback, as for the wall jet cell, is that the cell volume is larger than a few hundreds of microlitres. In general, configurations of Figure 6a can be used for most trace analysis in highly conductive solutions. Configurations of Figures 6b or 5a, depending on the requirements on the minimum volume and hydrodynamic conditions, are preferred in cases of resistant solutions. Note that there is no unique optimal electrochemical flow-through cell, but that the most appropriate one will depend on the specific problem at hand.

The source of *noise* may be the electronics, the electrodes or the flow system. The various sources have been reviewed in ref. [159] and discussed in ref. [156]. It is worth mentioning that decreasing the TLD cell thickness may increase the noise [160] and that the noise intensity strongly depends on the polishing quality of the body cell material and the type of hydrodynamic flow (i.e. laminar or turbulent).

A major advantage of TLD is its capability to be combined with dual (or even multiple) electrode detection, as shown in Figure 7a and b. In such systems, two

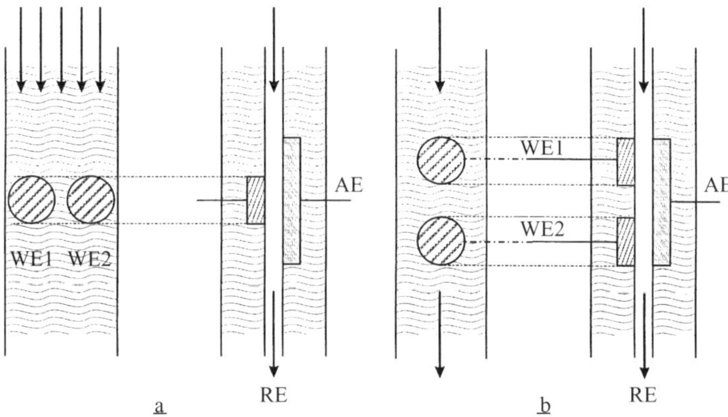

Figure 7. Two configurations (parallel in (a) and series in (b)) of dual working electrode flow-through cells. WE1, WE2 = first and second working electrodes. AE, RE = auxilliary and reference electrodes. Arrows indicate the flow

or more electrodes are positioned in parallel (Figure 7a) or in series (Figure 7b) in the stream, with a corresponding auxiliary electrode across the channel and a reference electrode downstream. The amperometric or voltammetric currents at each electrode can then be monitored in different electrochemical conditions (Chapter 16, 22 in ref. [5]; [19]). Parallel electrodes can be used e.g. (i) to detect two compounds which otherwise give rise to overlapping peaks, (ii) to detect simultaneously an irreversible and a reversible redox couple with different modulations, or (iii) to determine simultaneously ASV peaks and their background current to perform automatic background substraction and improve sensitivity [161]. In the series configuration, the products formed on the first electrode can be detected, sometimes more easily, on the second one. Such multiple electrode detection approaches open a wide range of capabilities to improve sensitivity, selectivity and multicomponent analysis in complex media such as natural waters. Multiple detection on an individually addressable electrode array (section 2.2.1.2) has been reported in ref. [162].

3 INTERPRETATION OF CURRENT–POTENTIAL–TIME CURVES; SPECIATION

Voltammetric techniques are powerful not only to measure ultratrace analyte concentrations, but also to determine the nature of physico-chemical processes and chemical reactions in solution, and the distribution of the various species of a given compound or element. The influence of such reactions on $i = f(E, t)$ curves may be diverse and complicated. It has been discussed in several books (in particular, refs [1–4] for general approaches; ref. [4] for applications to water; ref. [163] for applications to biomolecules; ref. [164] for application to organic chemistry). Only a summary of key aspects will be given below, to enable the non-specialist to enter the field with a critical mind and to use the key concepts for interpreting as correctly as possible the results obtained by *in situ* voltammetry. This aspect is particularly important as the purpose of *in situ* measurements is to perform analysis, with no or minimal sample perturbation. Any natural chemical reaction in solution will thus significantly affect the voltammetric data. For detailed discussion of chemical components and processes relevant to natural waters, the reader is referred to ref. [4]. The concepts of speciation are discussed below for voltammetries (i.e. techniques where potential is imposed and current is measured). The same concepts, however, are applicable to PSA (Table 1; imposed current and measured potential).

3.1 REVERSIBILITY, LABILITY AND MOBILITY

3.1.1 General Concepts

The interpretation of voltammetric curve provided by any analyte, with respect to its speciation in solution, should consider the key concepts of redox reversibility, chemical lability and physical mobility. This is schematized below (Figure 8) for the reduction of a metal ion M forming a complex ML in solution and reducible into M^0 at the electrode. (Note that conceptually, in the following discussion, M may be replaced by any anion or organic compound and L by any type of reagent.)

In the presence of two or several species of M, $i = f(E, t)$ curves are always the result of at least three processes in series, namely:

1) the diffusion in solution of M species, whose flux is characterized by the diffusion coefficients D_{ML}, D_M and $\bar{D}$ and the diffusion layer thickness, δ; $\bar{D}$ is the average diffusion coefficient (see below);

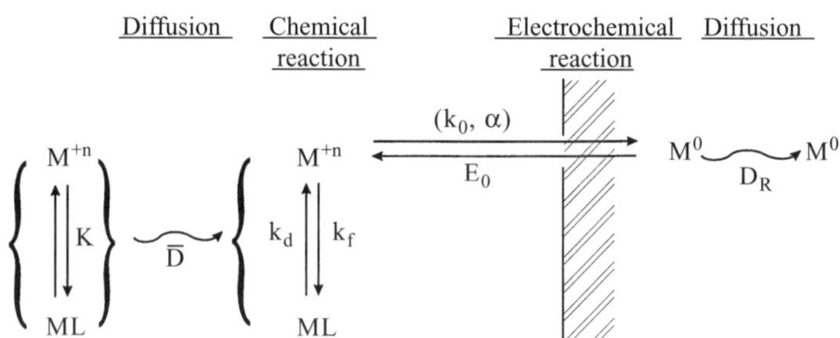

Figure 8. The successive processes controlling the voltammetric signal, resulting from the reduction of a metal ion M^{n+} complexed by a ligand L (here α = charge transfer coefficient)

2) the chemical reaction(s) of M, characterized by its equilibrium constant K, and its formation and dissociation rate constants, k_f and kd;
3) the electrochemical (electron transfer) reaction characterized by its equilibrium standard potential, E_0, and its kinetic parameters, k_0 (charge transfer rate constant) and α (charge transfer coefficient);
4) for amalgamable metals in Hg, the diffusion of M^0 in Hg, with diffusion coefficient, D_R, must also be considered.

Because diffusion in solution is always present, voltammetric techniques are dynamic and interpretation of results must be based on flux considerations. In particular, the relative importance of the first three kinetic processes above is essential. Data interpretation is considerably simplified when only one (at maximum two) step is rate limiting.

- *The redox reaction* is called *rapid* or *reversible* when the charge transfer rate is much higher than the diffusion rate, i.e. $k_0 \gg \bar{D}/\delta$ (in m s^{-1}), for macroelectrodes (linear diffusion; Figure 3) and $k_0 \gg \bar{D}/r$ for microelectrodes (spherical diffusion; Figure 3). In such a case $i = f(E, t)$ curves only depend on E_0, and not on k_0 and α. It is worth emphasizing that unambiguous interpretation of speciation in complicated media such as natural waters is usually restricted to reversible redox systems (see section 3.4) and that tests must be used to check the validity of this condition (section 3.1.2–3.1.3).
- With respect to *chemical process*, the complex ML is called *labile*, *non-labile* or *inert* respectively, when the formation/dissociation reactions of ML are very fast (labile complex) or slow (non-labile complex) compared with diffusion, or when dissociation does not occur at all (inert). It can be shown [165; Chapter 8 of this book] that for dynamic systems ($k_d t$, $k_a[L]_t \gg 1$) and linear diffusion (macroelectrodes), the lability criterion Λ is given by:

$$\Lambda = \frac{k_d^{1/2}(1 + \varepsilon K[\mathrm{L}])^{1/2}\,\delta}{D_{\mathrm{ML}}^{1/2}\,\varepsilon K[\mathrm{L}]} \quad (4)$$

where $\varepsilon = D_{ML}/D_M$. Complexes are labile or inert when $\Lambda \gg 1$ or $\ll 1$, respectively. Again, in complicated natural waters, unambiguous speciation interpretation is usually possible only for these two extreme conditions, in which the voltammetric curves do not depend on k_d or k_f. In the former case, the value of K can be determined from the peak potential shift (section 3.2), whereas, in the latter case, the voltammetric peak current specifically provides $[M^{n+}]$ and K can only be obtained indirectly. It is worth emphasizing that lability tests (section 3.1.4) are required to ensure correct interpretation of data.

- with respect to *physical transport* (diffusion), the *mobility* of the complex species ML depends on its diffusion coefficient, D_{ML}, compared with that of M^{n+}, D_M. The characteristic parameter is $\varepsilon = D_{ML}/D_M$ and ML is said to be immobile when $\varepsilon = 0$. For labile complexes, M^{n+} and ML are always in equilibrium with each other, even in the diffusion layer at the electrode surface, and it can be shown [1,4; Chapter 8; equation (2)] that the overall current, i, due to the reduction of the whole of M species, is given by:

$$i = nFA\bar{D}c_{\mathrm{M}}\left(\frac{1}{\alpha} + \frac{1}{r}\right) \tag{5}$$

where c_{M} is the total concentration of M, $\bar{D}$ the average diffusion coefficient of M species and r the radius of the electrode. When there is only M^{n+} and ML in solution, $\bar{D}$ is a weighted average of D_{M} and D_{ML}:

$$\bar{D} = D_{\mathrm{M}}\frac{[\mathrm{M}]}{c_{\mathrm{M}}} + D_{\mathrm{ML}}\frac{[\mathrm{ML}]}{c_{\mathrm{M}}} \tag{6}$$

$$\bar{D} = D_{\mathrm{M}}\frac{1}{1 + K[\mathrm{L}]} + D_{\mathrm{ML}}\frac{K[\mathrm{L}]}{1 + K[\mathrm{L}]} \tag{7}$$

In natural waters, which often contain colloidal complexants, the conditions are usually such that $\varepsilon = D_{\mathrm{ML}}/D_{\mathrm{M}} < 1$ and $\bar{D} < D_{\mathrm{M}}$. As a consequence, the reduction current of M will be lower in the presence rather than in the absence of L, owing to the low mobility of M complexes. Lower current is also observed for mobile ($\varepsilon = 1$) but chemically non-labile or inert complexes (section 3.1.4), albeit for completely different reasons (slow chemical reaction for non-labile complexes and slow physical transport for non- or slowly mobile complexes). Different information on speciation (ML dissociation rate constants in the former case and size of the complex in the second one) can be gained in the two cases. Therefore checking mobility by experimental tests (section 3.1.4) is also very important before trying to interpret the data.

3.1.2 Reversibility Tests Based on Peak Current and Potential

Because of the large number of different voltammetric techniques available, many experimental tests can be done to check reversibility, lability and mobility, as well as the possible occurrence of additional 'secondary phenomena' (section 3.4) [1–6]. The simplest ones applicable to water analysis are summarized in Table 4. Note that the criteria of Table 4 are usually also valid, albeit sometimes with adaptation, for analogous techniques (e.g. mentioned in parenthesis in the first column). The table assumes that oxidized forms of M species exist in solution and that only M^{n+} is reduced at the electrode. For oxidation processes, the currents would be of opposite signs, but potential shifts and current changes would be similar.

The third column of Table 4 shows the signals obtained with reversible systems in the absence of complexant L. In cyclic voltammetric curves (CV), the first step corresponds to the reduction of M^{n+} (cathodic peak; $i_{\mathrm{p}}^{\mathrm{c}}$, $E_{\mathrm{p}}^{\mathrm{c}}$) and the second step to the reoxidation of M^0 produced during the first step (anodic peak; i_p^a, E_p^a). Differential pulse (DP) voltammetry can be used in the direct mode (pulses applied towards negative potentials for a reduction as shown in Table 4) or in the reverse mode (pulses opposite to the scan direction). This gives rise respectively to the normal reduction (cathodic) peak (shown in Table 4) or the reoxidation (anodic) peak of the metal reduced just before the pulse. In square wave (SW) voltammetry, the reduced metal is reoxidized at each cycle and both the reduction and oxidation currents are recorded. Thus a reduction (cathodic) peak and a reoxidation (anodic) peak are obtained simultaneously in one scan (not shown in Table 4). Commercial apparatuses sometimes give only the algebraic sum of the two peaks (shown in Table 4), but as discussed below the individual reduction and oxidation scans are useful for reversibility and lability tests. Note that in both DPV and SWV, only the amplitude of the current change due to pulse application is recorded and not the total reduction (or oxidation) current. For all the techniques of Table 4, reversibility can be checked by comparing anodic and cathodic peak currents (i_p^a, i_p^c) or potential (E_p^a, E_p^c) as well as peak width ($W_{1/2}$) with theoretical relationship or values.

3.1.3 The Technique Measurement Time and the Corresponding Reversibility Tests on Macroelectrodes

The changes of peak currents and potentials with parameters related to measurement time of the techniques, t_m, are also good reversibility tests on macroelectrodes. t_m is the time during which the analyte M is reduced at the electrode, corresponding to the formation of a diffusion layer by depletion of M at the electrode surface. On macroelectrodes, the thickness δ of the diffusion layer increases with t_m as (Chapter 9 in ref. [4]): $\delta = \sqrt{\pi \bar{D} t_m}$. In d.c. polarography t_m is related to the frequency of drop renewal (i.e. drop time). In modulated techniques t_m is fixed by the rapidity of potential change of the technique. The corresponding relevant parameters are $\Delta E/\mathrm{v}$ for CV, pulse duration, t_p, for DPV, or 1/frequency $= 1/f$, for SWV and sinusoidal a.c. voltammetry.

The reversibility of a redox system is relative to the rate of diffusion, $\bar{D}/\delta$ for macroelectrodes and $\bar{D}/r$ for microelectrodes (section 2.1.1.). On macroelectrodes, a given redox system will thus be apparently more reversible with slow techniques (for which t_m and δ are larger) than with faster techniques. Since t_m can be varied by means of potential modulation, each technique is able to cover a given range of t_m (Figure 9) and about three orders of magnitudes can be covered by using several complementary techniques. Reversibility can thus readily be tested. On microelectrodes where spherical diffusion occurs, the diffusion flux at the electrode surface is proportional to $\bar{D}/r$ and does not depend on t_m and the potential modulation. The above reversibility tests are then not applicable in such conditions. But reversibility then depends on electrode size, r.

Theoretical equations valid for *reversible systems on macroelectrodes* (linear diffusion) show that i_p^a and i_p^c vary with $(t_m)^{-1/2}$, i.e. with v, t_p and f, as shown in

Table 4. Summary of the changes of peak parameters with metal complex formation ($M + L \rightleftharpoons ML$) for the voltammetric techniques relevant for *in situ* applications. The most useful tests of reversibility, lability and mobility are given at the bottom of the Table (see text for the various possible cases). Symbols are defined on the curves

Method	**E = f(t)**	**Without L Reversible systems**	**With L (I) . Labile complexes $D_M = D_{ML}$**
CV (CSV, ASV)	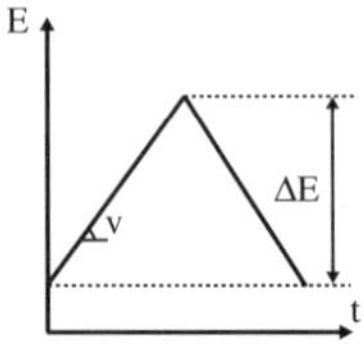 E; v; ΔE; t	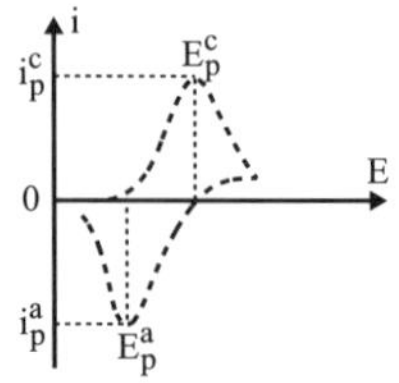 i; i_p^c; E_p^c; 0; E; i_p^a; E_p^a	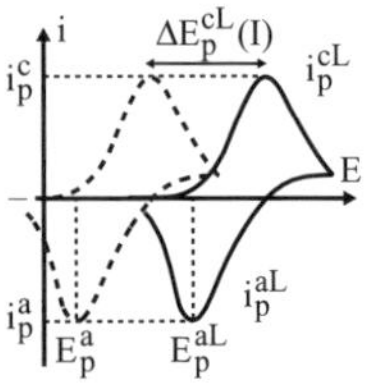 i; ΔE_p^{cL}(I); i_p^c; i_p^{cL}; E; i_p^a; E_p^a; E_p^{aL}; i_p^{aL}
DPV (DPASV)	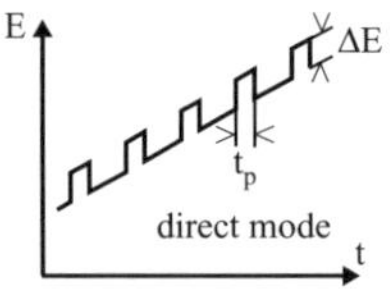 E; ΔE; t_p; direct mode; t	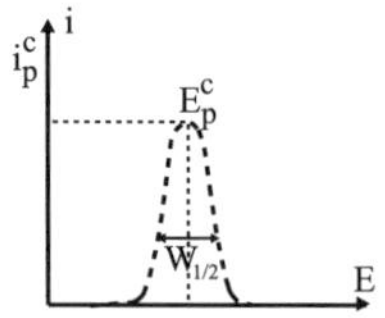 i; i_p^c; E_p^c; $W_{1/2}$; E	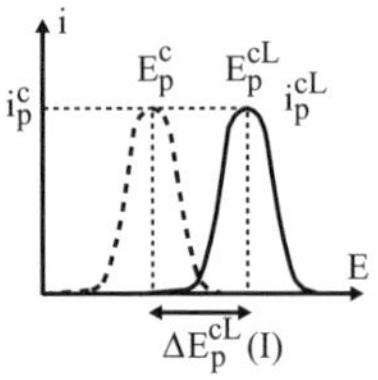 i; E_p^c; E_p^{cL}; i_p^c; i_p^{cL}; E; ΔE_p^{cL}(I)
SWV (SWASV)	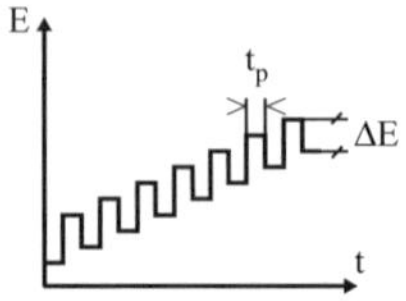 E; t_p; ΔE; t	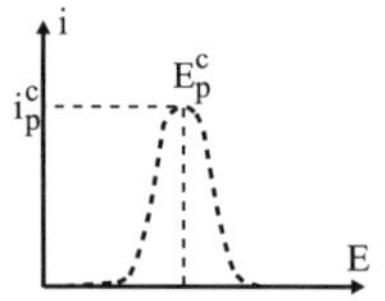 i; i_p^c; E_p^c; E	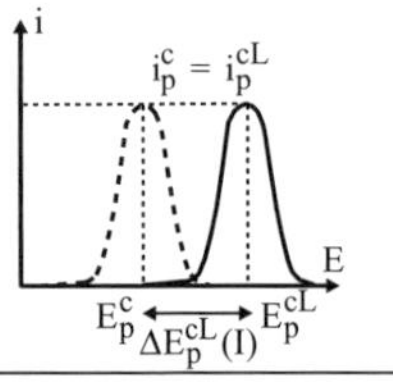 i; $i_p^c = i_p^{cL}$; E; E_p^c; ΔE_p^{cL}(I); E_p^{cL}
	Criteria based on:		
CV	i_p^c/i_p^a	$i_p^c/i_p^a = 1$	$i_p^{cL}/i_p^{aL} = 1$
CV; DPV; SWV	i_p^{cL}/i_p^c (*)	–	$i_p^{cL}/i_p^c = 1$
CV	$E_p^a - E_p^c = \Delta E_p^{ac}$	$\Delta E_p^{ac} = 59/n(\text{mV})$	$\Delta E_p^{ac} = 59/n(\text{mV})$
CV; DPV; SWV	$E_p^c - E_p^{cL} = \Delta E_p^{cL}$ (I–IV)	–	$\Delta E_p^{cL}(\text{I}) > 0$ (*)
CV	$E_p^a - E_p^{aL} = \Delta E_p^{aL}$ (I–IV)	–	$\Delta E_p^{aL}(\text{I}) = \Delta E_p^{cL}(\text{I}) > 0$
DPV	Peak width = $W_{1/2}$	$3.5RT/nF(\Delta E \rightarrow 0)$	$3.5RT/nF$
CV	t_m†	$i_p^c, i_p^a \approx \sqrt{v}$	$i_p^{cL}, i_p^{aL} \approx \sqrt{v}$
DPV	t_m†	$i_p^c \approx 1/\sqrt{t_p}$	$i_p^{cL} \approx 1/\sqrt{t_p}$
SWV	t_m†	$i_p^c \approx 1/\sqrt{f}$	$i_p^{cL} \approx 1/\sqrt{f}$
CV; DPV; SWV	t_m†	$E_p^c \approx t_m^0$	$E_p^{cL} \approx t_m^0$

Notes: * Valid for both anodic and cathodic peaks.
† Only valid for macroelectrodes.
In DPV and SWV both anodic and cathodic peaks can be recorded and used for reversibility tests, as for CV.

With L (II) . Labile complexes $D_{ML} < D_M$	**With L (III) . Non labile complexes $D_M = D_{ML}$**	**With L (IV) . Inert complexes**	**Method**
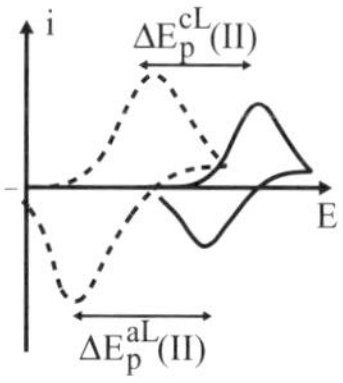	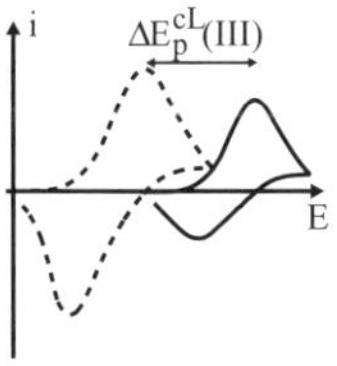	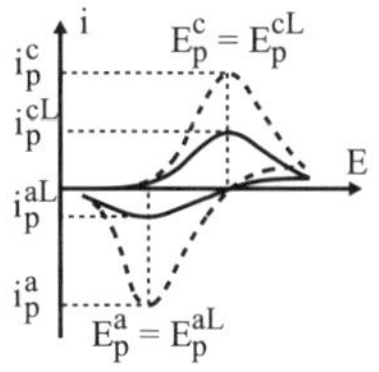	CV (CSV, ASV)
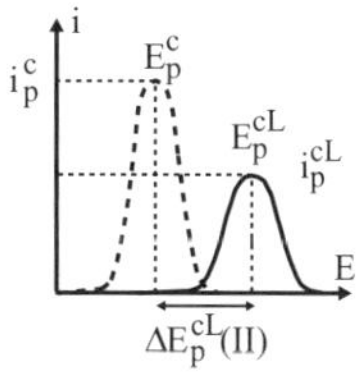	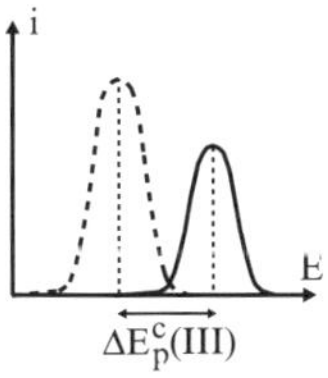	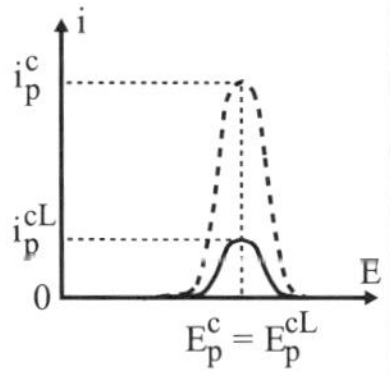	DPV (DPASV)
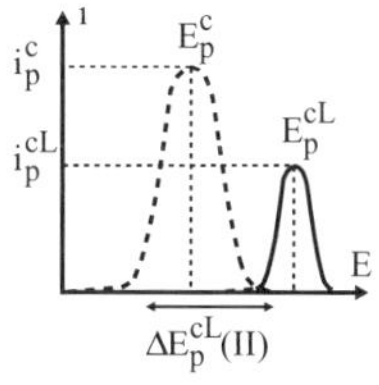	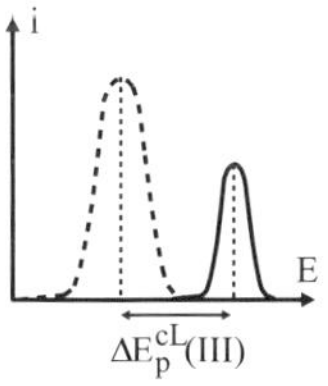	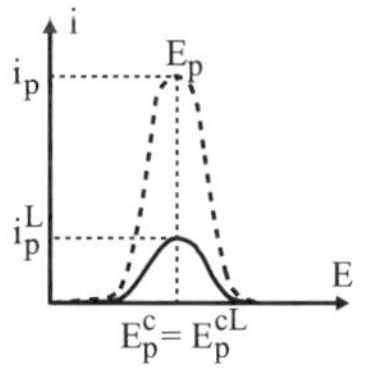	SWV (SWASV)
			Criteria based on:
$i_p^{cL}/i_p^{aL} = 1$	$i_p^{cL}/i_p^{aL} \neq 1$	$i_p^{cL}/i_p^{aL} = 1$	i_p^c/i_p^a
$i_p^{cL}/i_p^c < 1$	$i_p^{cL}/i_p^c < 1$	$i_p^{cL}/i_p^c < 1$	i_p^{cL}/i_p^c
$\Delta E_p^{ac} = 59/n$(mV)	$\Delta E_p^{ac} > 59/n$(mV)	$\Delta E_p^{ac} = 59/n$(mV)	$E_p^a - E_p^c = \Delta E_p^{ac}$
$0 < \Delta E_p^{cL}(\text{II}) < \Delta E_p^{cL}(\text{I})$	$0 < \Delta E_p^{cL}(\text{III}) < \Delta E_p^{cL}(\text{I})$	$\Delta E_p^{cL}(\text{IV}) = 0$	$E_p^c - E_p^{cL} = \Delta E_p^{cL}$ (I–IV)
$0 < \Delta E_p^{aL}(\text{II}) = \Delta E_p^{cL}(\text{II}) < \Delta E_p(\text{I})$		$\Delta E_p^{aL}(\text{IV}) = 0$	$E_p^a - E_p^{aL} = \Delta E_p^{aL}$ (I–IV)
$3.5\ RT/nF$	$> 3.5RT/nF$	$3.5RT/nF$	Peak width$= W_{1/2}$
$i_p^{cL}, i_p^{aL} \approx \sqrt{v}$		$i_p^{cL}, i_p^{aL} \approx \sqrt{v}$	t_m†
$i_p^{cL} \approx 1/\sqrt{t_p}$	complicated changes	$i_p^{cL} \approx 1/\sqrt{t_p}$	
$i_p^{cL} \approx 1/\sqrt{f}$	with v, t_p, f	$i_p^{cL} \approx 1/\sqrt{f}$	
$E_p^{cL} \approx t_m^0$		$E_p^{cL} \approx t_m^0$	

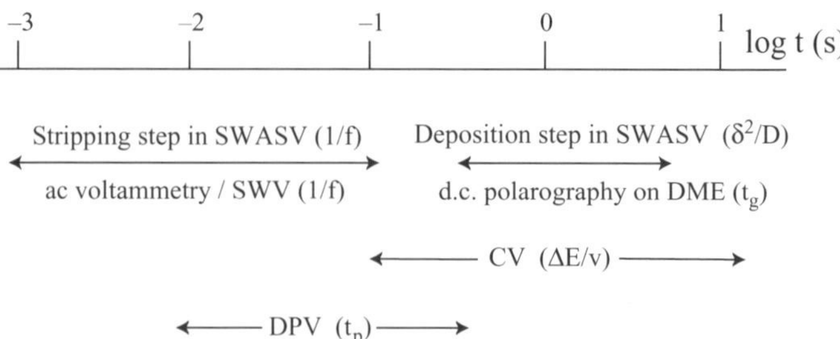

Figure 9. The measurement time ranges (t_m) of various voltammetric techniques and the corresponding time controlling parameters in parenthesis (see text for explanations)

table 4, and that E_p^a and E_p^c are t_m independent (Chapter 9 in ref. [4]). Three additional considerations should be emphasized:

(1) As already mentioned, the criteria of Table 4, based on t_m, are valid only for macroelectrodes, i.e. when $1/\delta \gg 1/r$ in equation (2). This consideration also shows that the limit between micro- and macroelectrodes is relative and depends itself on the rapidity of the technique used. A given electrode may behave as a macroelectrode ($1/\delta \gg 1/r$) with slow techniques and as a microelectrode ($1/\delta \ll 1/r$) with fast techniques.

(2) The fulfillment of one single criterion of Table 4 is rarely enough to ensure that a system is reversible. This is particularly the case when tests are done at low concentrations of analyte (as it is often imposed by environmental conditions) where significant interferences by capacitive current or large errors in E_p may occur.

(3) t_m should not be confused with other voltammetric time periods which are also important, albeit for other reasons. They are depicted in Figure 10 for SWASV. Apart from $t_m (= t_p$ in SWASV or DPASV, one to tens of ms), the other major time periods are the following:

- the sampling time, t_s, i.e. the time during which the current is recorded at the end of $t_m (t_s \ll t_m)$;
- the deposition time, t_d (∼ seconds to tens of minutes) in all ASV or AdSV techniques. This is the time during which the reduced metal is accumulated in the electrode before reoxidation (ASV) or the complex is absorbed at the electrode surface before its reduction (AdSV);
- waiting times t_w (seconds to hours). Waiting times may correspond to various steps, e.g. (i) the 'pause' duration for which the electrode has been in contact with the solution, at rest, before use or between analyses (e.g. t_{w2} in Figure 10), or (ii) the duration of a possible 'cleaning step' (t_{w1} in Figure 10) for which the electrode is left at a potential enabling a reliable surface to be obtained before analysis.

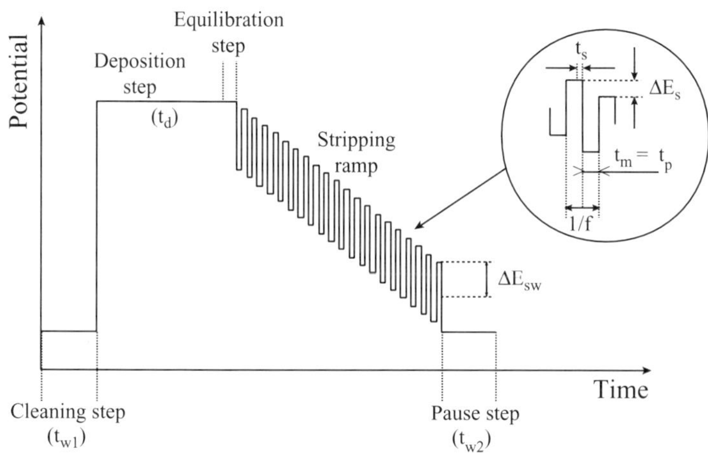

Figure 10. The important time scales in voltammetry (exemplified by means of SWASV: t_m = measurement time (= t_p = pulse duration in square wave or pulse techniques), t_d = deposition time in ASV and AdSV techniques, t_w (t_{w1}, t_{w2}) = waiting times (see text); t_s = sampling time; ΔE_s = step amplitude, ΔE_{SW} = pulse amplitude

3.1.4 Lability and Mobility Tests

Columns 4–7 of Table 4 show the differences in $i = f(E, t)$ curves, in the presence (full curves) and absence (broken curves) of complexant L. Cases I–IV correspond to: (I) fully labile and mobile ($\varepsilon = 1$) ML complex, (II) labile, but slowly mobile complex, (III) fully mobile but non-labile complex, and (IV) inert complex (either mobile or non-mobile). Similar criteria to those for reversibility can be used to test these various cases (Table 4). The following general comments can be made.

(a) Additional criteria based on the comparison of i_p and E_p values in the presence and absence of L enable the lability and mobility criteria to be tested. Variations of i_p and E_p with [L] may provide precise information on lability, as well as complexation constants and complex stoichiometry (section 3,2; [1,2,4]).
(b) Tests using parameter t_m must be interpreted carefully, since an overall 'non-reversible' behaviour may indicate either slow charge transfer at the solution electrode interface (as in the absence of ligand), or slow chemical kinetics. Detailed study of the influence of t_m can be used in principle to discriminate between these two cases; in complicated media, however, unambiguous interpretation of such results is not simple.
(c) A decrease in peak current in a complexing medium ([L] > 0) compared to non-complexing one ([L] = 0) does not necessarily imply the existence of

chemically inert or non-labile complexes. Slowly mobile complexes may have the same effect and are likely to be present in most aquatic systems because of the ubiquity of colloidal or polymeric complexants. Discrimination between non-mobile (case II) and non-labile or inert (cases III and IV) complexes is rather simple, as case II is characterized by a reversible behaviour, with decrease in current and potential shift compared with a non-complexing medium. Formation of fully inert complexes (case IV) results in a reversible wave or peak, and a decrease (sometimes full suppression) of current, but without potential shift. Non-labile complexes (case III) exhibit an apparent non-reversible behaviour which in reality is due to slow chemical kinetics.

3.2 RELATION BETWEEN THE VOLTAMMETRIC SIGNAL AND LABILE AND MOBILE COMPLEXES IN UNPERTURBED COMPLICATED MEDIA

It is most important to perform the aforementioned tests to get correct speciation interpretation. Because of various difficulties arising in natural samples (low concentration levels, interferences etc.), data interpretation published in the literature is too often based on untested oversimplified assumptions which sometimes lead to incorrect conclusions (e.g. ref. [166] and discussion in ref. [167]). Reversibility, lability and mobility tests, as well as those for checking the presence or absence of 'secondary' phenomena, are discussed in section 3.4. Some of those may profondly influence $i = f(E, t)$ curves. In this section we shall assume that there is no such secondary effect, the redox system is reversible, the complex ML is labile and the ligand is in excess. In such conditions, it can be shown [1–4] that the potential shift, $\Delta E_{\mathrm{p}}^{\mathrm{L}} = E_{\mathrm{p}} - E_{\mathrm{p}}^{\mathrm{L}}$, and the ratio $i_{\mathrm{p}}^{\mathrm{L}}/i_{\mathrm{p}}$ in any voltammetric technique are equal to those, $\Delta E_{1/2}^{\mathrm{L}} = E_{1/2} - E_{1/2}^{\mathrm{L}}$, and $i_{\ell}^{\mathrm{L}}/i_{\ell}$ in d.c. polarography. They are given by the theory of De Ford and Hume which has been very clearly discribed by Crow [168] for d.c. polarography. For peak shape voltammograms, the key expressions are the following *(only valid for reversible and labile systems in absence of secondary reactions)* [1–4]:

$$\Delta E_{\mathrm{p}} = \frac{RT}{nF} \ln \frac{i_{\mathrm{p}}^{\mathrm{L}}}{i_{\mathrm{p}}} + \frac{RT}{nF} \ln \alpha \tag{8}$$

For macroelectrodes (linear diffusion):

$$\frac{i_{\mathrm{p}}^{\mathrm{L}}}{i_{\mathrm{p}}} = \sqrt{\frac{\bar{D}}{D_{\mathrm{M}}}} \tag{9}$$

i.e., when only M^{n+} (hereafter denoted by M) and ML are present (Figure 8; eqs 6, 7) [1,4]:

$$\frac{i_p^L}{i_p} = \sqrt{\frac{1}{\alpha} + \frac{D_{ML}}{D_M}\frac{\alpha - 1}{\alpha}} \tag{10}$$

where $\alpha = c_M/[M]$, D_R is the diffusion coefficient of M^0 in Hg and the other symbols have their usual meaning. For spherical or disc microelectrodes (spherical diffusion), equation (9) becomes

$$i_p^L/i_p = \bar{D}/D_M \tag{11}$$

As D_M and D_R are usually known in synthethic systems, equations (6–8) and (10) allow determination of D_{ML} and α, and then $K = [ML]/[M][L]$, provided [L] is in excess and also known. Note that, when $D_{ML} < D_M$, $\bar{D} < D_M$ and the first term of equation (8) is important for computing correct values of α from ΔE_p values. This term is most important when colloidal or macromolecular ligands are present.

In complicated aquatic media, the total concentration of M, c_M, is the sum of free M and a large number of species with various labilities and mobilities. Similarly $\bar{D}$ will include a large number of corresponding terms:

$$\bar{D} = D_M f_M + \sum_m D_m f_m \tag{12}$$

where f_m is the fraction ($f_m = [M]_m/c_M$) of the mth metal complex, mobile enough to contribute significantly to the overall current. There is presently not enough information in the literature on the diffusion and lability properties of natural complexes to provide detailed f values of specific M species. Nevertheless, rough estimates of the proportions (i) of free M, $f_M = 1/\alpha$, (ii) of labile and mobile M species, $f_m = \sum[M]_m/c_M$, and (iii) of that of inert M species, $f_i = \sum[M]_i/c_M$, can be obtained by assuming that most immobile M species are also inert. Although this assumption is presently somewhat arbitrary and would need detailed study, it is reasonable as explained in section 3.5.2. Under such conditions, c_M and $\bar{D}$ are given by:

$$c_M = [M] + \sum_m [M]_m + \sum_i [M]_i \tag{13}$$

Combining equations (11) and (12) leads to equation (14), valid for microelectrodes:

$$\frac{i_p^L}{i_p} = \frac{[M]}{c_M} + \frac{\sum_m D_m[M]_m}{D_M c_M} \tag{14}$$

In many aquatic systems, most mobile and labile species are complexes with small inorganic ligands (carbonate, sulfate, chloride etc.), for which $D_m \approx D_M$. In addition, with microelectrodes, the contribution of species with sizes > 3–4 nm is very small even when they represent a large fraction of the labile complexes (section 3.5.2) Then as a first approximation, equation (14) becomes

$$\frac{i_p^L}{i_p} = \frac{\alpha_m}{\alpha} = \frac{[M] + \sum[M]_m}{c_M} \tag{15}$$

Combination of equations (8) and (15) gives:

$$\Delta E_p \approx \frac{RT}{nF} \ln \alpha_m \tag{16}$$

Thus by measuring (i) E_p^L and i_p^L and in the test medium, (ii) the corresponding total concentration c_M by a separate technique (e.g. AAS or ICP-MS) and (iii) the values of E_p and i_p of a standard solution at concentration c_M, equations (15) and (16) provide

$$\alpha_m = \frac{[M] + \sum_m [M]_m}{[M]} \tag{17}$$

and $\alpha = c_M/[M]$. Under such conditions, the separate values of [M], $\sum_m[M]_m$ and $[M]_i$ can be computed. Equations (15) and (16) are valid for voltammetry on microelectrodes. For macroelectrodes, equations are slightly different: in principle the same parameters can be estimated, but in practice, experimental errors may be much larger (section 3.3.1.(c)).

3.3 DISCUSSION OF VARIOUS SPECIATION APPROACHES USING VOLTAMMETRIC TECHNIQUES

3.3.1 Measurement of Free M, Mobile M and Immobile or Inert M Concentrations

The above discussion shows that presently only estimations of these metal fractions are possible, as detailed information on the physicochemical properties of aquatic ligands and complexes (e.g. dissociation rate constants, size distribution) [4,169,170] is insufficient to state unambiguously the limitations of the assumptions used. The latter, however, are supported by the few data available on aquatic ligand and complex properties [4], and *in situ* voltammetric measurements (e.g. section 5.2). There are few studies on aquatic systems where interpretation is based on systematic rigorous tests of the electrode process. They have been performed in laboratory conditions. Discussing detailed examples would be too long. The reader is referred to Chapter 9 in ref. [4] for such examples. A selected list of systematic studies is given in Table 5. They may serve as reference studies and are recommended to the beginner in the field. They also illustrate the nature of the major problems and ligand properties specifically related to environmental systems, and emphasize that voltammetric signals depend not only on the reversibility, lability and mobility factors discussed above, but also on 'secondary phenomena', whose importance and possible minimization will be briefly discussed in section 3.4 (see ref. [4] for detailed discussion).

The following conditions are required or desirable to obtain accurate information on free M, mobile M and immobile/inert M fractions.

(a) *Precise and accurate measurements of both* i_p^L/i_p *and* $E_p - E_p^L$ (with errors $\leq 5\%$ and 1 mV respectively) are required, which implies (i) using a stable and reliable reference electrode, (ii) recording the voltammograms with extended potential scale and (iii) performing careful and accurate calibrations of i_p and E_p in non-complexing medium, under otherwise similar conditions (in particular ionic strength and temperature) as the test sample. Calibration is a key aspect of *in situ* measurements and will be discussed in section 5.3.3 and 6.
(b) *The total metal concentration*, c_M, must be determined by other means (AAS, ICP-MS) or by voltammetry after decomplexation (e.g. by acidification). Though c_M is not required to compute α_m and α from ΔE_p, and i_p^L/i_p it is necessary to compute the absolute concentrations of free, mobile and inert M.
(c) *The selection of either macro- or microelectrode* has an important consequence on the accuracy of the measured mobile fraction. This is more easily understood with the binary system (M + ML), for which equations (10) and (11) can be rewritten as:

$$\frac{i_p^L}{i_p} = \sqrt{f_M + \frac{r_M}{r_{ML}}(1 - f_M)} \qquad (10')$$

$$\frac{i_p^L}{i_p} = f_M + \frac{r_M}{r_{ML}}(1 - f_M) \qquad (11')$$

for macro- and microelectrodes, respectively. In these equations, $f_M = [M]/c_M = 1/\alpha$, and $r_M/r_{ML} = D_{ML}/D_M$ (based on the Stokes–Einstein equation) is the ratio of radii of M and ML. Obviously, the contribution of ML to i_p^L depends on both its proportion and size. The relation between i_p^L/i_p and r_{ML}/r_M at a microelectrode is shown on Figure 11 for $f_M = 1$ (no ML), $f_M = 0$ (hypothetical case where $[M] = 0$), and the intermediate situation, $f_M = 0.5$ (50% of M and ML). For comparison, if ML did not contribute at all to the current, irrespective of its size, horizontal lines at $i_p^L/i_p = 1, 0$, and 0.5 would be obtained respectively. Let us consider the case of $f_M = 0$ ($[M] = 0$). The current provided by ML is equal to that due to the same concentration of M when $r_{ML} = r_M$, but decreases quickly for $r_{ML}/r_M > 1$ and becomes negligibly small for $r_{ML}/r_M > 8$–10, i.e. $r_{ML} >$ 3–4 nm for $r_M = 4$ Å. In mixtures of M and ML, equation (11′) enables us to compute that the contribution of ML to the total current is for example: (i) 55% for $[M]/c_M = 0.5$ and $r_{ML} = 4nm$, (ii) 20% for $[M]/c_M = 0.1$ and $r_{ML} = 4$ nm and (iii) 10% for $[M]/c_M = 0.1$ and $r_{ML} = 40$ nm. Clearly for a mixture of a large number of components, the ratio i_p^L/i_p on a

Table 5. Selected list of speciation studies using voltammetry or polarography, performed on (or relevant for) aquatic systems, and based on detailed electrode process examinations. In (G) the electrode processes have not been studied, but intercomparison of different methods have been used. (R)NPP= (reversed) normal pulse polarography, DPP= differential pulse polarography; for the nature of these techniques as well as pseudopolarography, see ref. [4] or the cited references

Studied element —— Sample nature	Voltammetric technique —— Speciation procedure	Primary characteristics of the redox system and complexes —— *Secondary phenomena studied in detail*	References
(A) Pb(II) —— Synthetic seawater	ASV, Pseudopolarography —— Direct: shift of E_p	reversible; labile, mobile, complexes —— *no secondary phenomena*	171, 172
(B) Cd(II), Zn(II), Pb(II) —— NTA or EDTA in synthetic seawater	DPP —— Direct: shift of E_p and decrease of i_p	reversible; inert + labile; mobile complexes —— *no secondary phenomena*	173–176
(C) Pb(II), Cd(II), Zn(II) —— Fulvic and humic solutions	NPP, RNPP —— Direct: decrease of current	reversible; labile, slowly mobile complexes —— *adsorption of fulvic/humic on electrode*	170, 177
(D) Pb(II), Cd(II) —— Fulvic solution	CV, DPP, NPP, ASV, DPASV —— Direct: shift of potential and decrease of current	reversible, labile, slowly mobile complexes —— • *adsorption of fulvics and induced complexes at electrode surface* • *change of pH at electrode surface* • *non-excess of ligand at electrode surface* • *chemical heterogeneity of ligands*	178–183

(E) Fe(II), S(−II) —— anoxic lake water	CV, DPP, NPP, d.c. polarography —— Direct: changes in potential and current	non-reversible, non-labile, mobile complexes —— *adsorption of FeS complex at electrode*	184, 185
(F) Cd(II), Cu(II) —— Ethylendiamine in synthetic seawater	ASV, pseudopolarography —— Direct: potential shift Indirect: ligand exchange	reversible, labile, mobile complexes —— *no secondary effect*	186, 187
(G) Cu(II), Zn(II) —— Complete sea or lake water sample	DPAdSV, DPASV —— Indirect: ligand exchange	no study of primary processes at electrode —— *no study of secondary phenomena at electrode* *intercomparison of different methods*	188–190

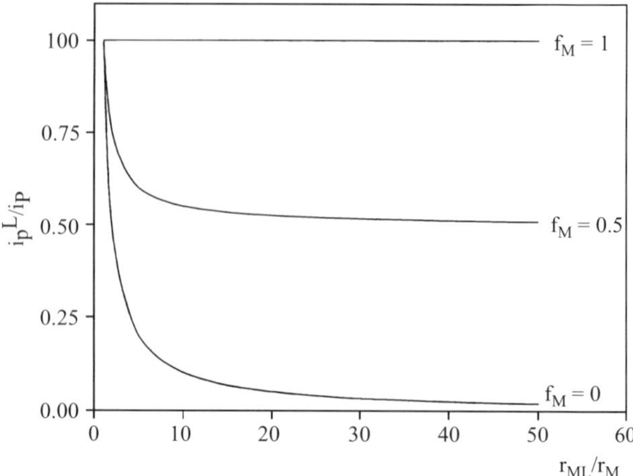

Figure 11. Relative importance of the contribution of the slowly diffusing species (ML) and the fast diffusing one (free M), on the peak current ratio i_p^L/i_p, on microelectrode (equations. (11) and (11′)) for various proportions of M and ML at $C_M = [M] + [ML] =$ const. The solution contains only M, ML and an excess of L; ML is labile $f_M = [M]/C_M$.

microelectrode will be a weighted average of all species but with a large weight for the species with $r_{ML} \leq \sim 4$ nm. The latter include in particular all the small inorganic complexes (with ligands such as carbonate, chloride, etc.) for which $r_{ML} \sim r_M$. Microelectrodes thus approximate quite well a discrimination between small dissolved molecules and colloidal species whose size limit is ~ 1 nm [191]. Such a discrimination is much more difficult with macroelectrodes, as usually ill-controlled hydrodynamic conditions are used resulting in less reproducible values of i_p^L/i_p. In addition, the corresponding errors on f_M or r_M/r_{ML} are amplified by the fact that they are proportional to $(i_p^L/i_p)^2$ (equation (10′)). Consequently interpretation of i_p^L/i_p for size discrimination is rather ambiguous on macroelectrodes.

(d) *Integrating a size separation device with the voltammetric sensor* can greatly help the interpretation. For instance, microelectrodes integrated in gels have been developed for *in situ* measurements (section 5.1.4). It has been shown that much of the colloidal and particulate material is excluded from the gel [192] and thus does not contribute to the voltammetric signal. As TEM and AFM pictures [169,193] suggest that small size colloids (< 10–20 nm) are often bound to larger aggregates, it can be expected that the voltammetric signal measured after equilibration of the gel with the test sample will be mostly due to free ions plus the small (often inorganic) complexes. In such case, the assumptions of section 3.2 are largely fulfilled.

3.3.2 Sample Titration by Metal; Complexation Capacity Measurements

Titration of the sample by the test metal ion cannot be performed *in situ*. However, as it is frequently performed in the laboratory, its meaning is briefly discussed in the light of the previous concepts. Average complexation capacity, C_c, and equilibrium constants, K, of the whole of the ligands of the sample (symbolized by L) are determined by titrating the solution with the test metal M and measuring its voltammetric peak height:

$$L + M \rightleftharpoons ML \tag{18}$$

A basic assumption of this procedure is that ML is fully inert and non-electroactive, so that i_p is proportional to free M in solution. Ideally, for infinitely stable complexes, $i_p = 0$ as long as $c_M < c_L$, and $i_p =$ const. $\times$ [M] for $c_M > c_L (c_M, c_L =$ total concentrations of M and L respectively). In practice a curvature is obtained around the equivalent point ($c_M \approx c_L$) which is usually interpreted as being due to the non-infinite stability of the ML complex. Various mathematical procedures have been developed to extract the corresponding equilibrium constant K from such curves, assuming the formation of either one or several complexes [e.g. 188,189,194].

The application of such measurements to water samples with a large number of complexants is limited by the fact that the basic assumption (fully inert electroinactive ML) is often fulfilled either only in part of the titration (usually at the beginning) or not at all. Major problems to consider in the interpretation of such curves are the following [4, 170].

(1) Unfractionated aquatic samples (as well as isolated polyfunctional complexants such as fulvic or humic compounds) include a very large number of complexing site types, the strongest of which are saturated first, with occupation of progressively weaker sites occurring successively. Rigorous interpretation of titration data should then consider continuous distributions of complexing site, and the meaning of average complexation constants attributed to one or a few arbitrary sites is rather limited [4,195–197].
(2) The wide distribution of equilibrium constants of natural complexing sites is paralleled by a corresponding wide distribution of dissociation rate constants [4]. As a consequence, the complexes become progressively more labile as the saturated sites become weaker from the beginning to the end of the titration, and therefore the basic assumption (inert complexes) is less and less fulfilled [170];
(3) Aquatic samples include complexants with a wide size distribution which is not related to the stability of their complexes. As a consequence, mobile and non-mobile complexes are formed in an unpredictable manner along the titration curve, which complicates even more its interpretation.

(4) When ASV techniques are used, a break point in the titration curve is sometimes observed, which corresponds to saturation of the ligands inside the diffusion layer and not in the bulk solution [4,183,196]. The relation between this break point and the true C_c value in the bulk solution is complicated and depends on the mobility of complexes, but in any case it usually occurs at total metal concentrations 1–2 orders of magnitude lower than C_c [4,196].

This combination of factors makes unambiguous interpretation of titration curves in terms of complexation parameters very difficult. The basic assumption (inert, electroinactive complex formation) is valid only when the following tests are fulfilled all along the titration range [4,170].

- The E_p value should be constant and, in principle, equal to that obtained in non-complexing medium.
- The titration curve shape should not change by using voltammetric techniques with different t_m (e.g. ASV, DPASV, SWASV).
- In any ASV mode, the titration curve shape must not change by changing the deposition time t_d. Such a change suggests that the ligand is saturated in the diffusion layer, at the electrode surface, and not in the bulk solution.

3.3.3 Speciation by Ligand Exchange and Voltammetry

Ligand exchange is another approach to determine metal complexation in complicated media. A ligand Y, forming with M a complex of known stoichiometry (e.g. MY) and stability constant (K_{MY}), is added to the sample, in such a concentration that $[Y] \gg [MY]$, while equilibria with natural ligands L are only slightly perturbed:

$$\mathrm{ML} + \mathrm{Y} \rightleftharpoons \mathrm{L} + \mathrm{MY} \tag{19}$$

For solutions containing one single ligand L, the determination of [MY], of the total metal concentration, c_M, along with knowledge of K_{MY} enables the computation of [ML], [M] and K_{ML}.Three cases may occur for the voltammetric determination of MY, as follows.

(a) *Y (e.g. ethylendiamine) forms labile complexes with M (e.g. Cu, Cd* [186,187]). In such a case, MY is measured by an ASV technique. Reversibility, lability and mobility tests should be done to ensure that the complex formed in the sample has the same properties as that in synthetic solution, in particular that no mixed, YML, complex is formed. Checking that i_p^L (and i_p) varies linearly with t_d is also useful to ensure that Y is in sufficiently large excess, even in the diffusion layer, and is not saturated at the electrode surface during the stripping step (section 3.4).

(b) *Y (e.g. EDTA) forms inert, electroinactive complexes with M (e.g. Zn* [188, 189]). In such a case, the information is obtained by using an ASV technique to measure the signal provided by the natural mobile/labile complexes, ML and its change by addition of EDTA. Tests on reversibility, lability and mobility are now required in the presence and absence of Y, to ensure that (i) simple interpretation is possible with natural ligands only, and (ii) the voltammetric properties of ML complexes are not perturbed by the excess of Y. As in case (a), linear variation of i_p^L with t_d is a good indication that peak shape and current are not affected by the possible saturation of natural ligands, by M produced inside the diffusion layer, during the stripping step [182,183];

(c) *MY (e.g. Cu–catechol) adsorbs on the electrode and is measured by adsorptive stipping voltammetry (AdSV)* [188,190,198,199]. In this case, lability and mobility tests are not important. On the other hand, the isotherm and kinetics of adsorption of MY on the electrode surface must be known in synthetic solution, and compared with the kinetic curve in the test solution to ensure that the same complexes are formed and that competitive adsorption with natural ligands is not operative. This last problem is a major one in AdSV and its role may be estimated by varying the contact time, t_w, between the solution and the electrode (e.g. with a HMDE) at a potential where MY does not adsorb while natural complexants do.

Apart from the above tests which are specific for each type of voltammetric detection, the following major possible problems must be considered and tested in all cases.

- Natural ligands, L, are chemically heterogeneous, and correct interpretation of data must consider continuous distribution of K_{ML} values [4,195,197] rather than a few discrete binding sites. It is then expected [195,196] that the measured K_{ML} values will increase with [Y] as found in ref. [190]. The natural complexation conditions are then found at relatively low [Y] while keeping a stoichiometric excess compared with [ML] at the electrode surface. Low [Y] is also preferable to minimize mixed complex formation with L. The occurrence of effects due to heterogeneous ligands and mixed complex formation can be assessed by measuring complexation as a function of [Y], but discriminating between these effects is not an easy task.
- K_{ML} values obtained by ligand competition methods are often very high (e.g. $\sim 10^{10}\,\mathrm{L\,mol^{-1}}$ for Zn(II), 10^{12}–$10^{15}\,\mathrm{L\,mol^{-1}}$ for Cu(II)) [188,189]. Dissociation of such very stable ML complexes should be very slow (hours to months) and attainment of a true equilibrium might be a major problem. Such kinetic effects are easily tested by measuring the voltammetric signal as a function of time over the expected dissociation time scale.

- Finally, adsorption of natural aquatic components on the electrode surface may either compete with Y in case of AdSV or interfere with the charge transfer process in ASV measurements. Such effects can be tested by varying the rest time, t_w, before the deposition time, t_d, (Figure 10) while imposing a potential where no redox reaction occurs. Change in i_p or E_p would suggest an adsorption effect.

In general, as for metal titration methods (section 3.3.2.) interpretation of ligand exchange methods must be done with great care as many artefacts are possible. Checking the results by as many tests as possible and different techniques, ref. e.g. [188], is highly recommended.

3.4 SECONDARY PHENOMENA INFLUENCING VOLTAMMETRIC MEASUREMENTS

A number of 'secondary' factors (i.e. additional to reversibility, lability and mobility) can influence the voltammetric curves. They have been described in various books (e.g. ref. [4] for applications to waters) and it is beyond the scope of this chapter to describe them in detail. This section only summarizes the most important ones, by giving possible tests to check their occurrence and procedures to minimize them (Table 6).

(1) For ligands forming labile complexes the free natural ligand concentration, [L], may be higher at the electrode surface than in the bulk, during the reduction step of ASV techniques (because of ML dissociation and M reduction), and lower during the reoxidation step because of the release of large quantity of M accumulated in the electrode during the deposition [200,201]. In both cases, this leads to broadening of peaks, and values of i_p^L/i_p smaller than the expected value. This may be a major problem when titrating the ligand by metal (section 3.3.2). The effect may be particularly acute during the stripping step in ASV, as the surface concentration of reoxidized M may be up to 1000 times higher than in the bulk, especially for MFE and Hg-coated microelectrodes [200,201]. In such cases, the condition $c_L/c_M > 10^4$ (with c_L = total ligand concentration) should be fulfilled for equations (8), (9) and (11) to be valid in ASV mode. This is rarely the case for strong ligands in aquatic systems. The occurrence of a sufficient excess of ligand forming labile complexes is demonstrated by a linear increase of i_p with t_p, and the independence of E_p on t_d [200,201]. If [L] is not high and cannot be increased, the smallest possible t_d should be used or solution should be exchanged between deposition and stripping, to reoxidize the metal in a well-controlled solution (section 2.2.3).
(2) Natural organic ligands may adsorb on the electrode surface and induce the formation of adsorbed ML complexes. This effect may drastically affect i_p^L

Table 6. Major secondary phenomena which may influence the voltammetric curves, and possible tests and remedies. L = natural ligands

Secondary phenomena	Possible test with ASV techniques	Possible remedy
(1) L concentration at electrode surface, is different from the bulk solution e.g. due to ML reduction or M^0 reoxydation	$i_p^L = f(t_d)$ $E_p^L = f(t_d)$	• increase L concentration to $[L] \gg [ML]$ • work at small t_d • exchange solution between deposition and stripping step in ASV
(2) adsorption of L on the electrode surface and formation of adsorbed ML	$i_p^L = f(t_d)$ $E_p^L = f(t_d)$	• use gel-integrated or other protected microelectrode
(3) adsorption of non-complexing, non-electroactive natural compounds modifying the redox process	$i_p = f(t_w)$ $E_p = f(t_w)$	• use gel-integrated or other protected microelectrodes
(4) pH increase at electrode surface due to O_2 reduction	$i_p, E_p = f$ (pH, buffer capacity)	• use *in situ* deoxygenation • increase pH buffer
(5) oversaturation of Hg by M^0 in ASV techniques, leading to peak deformation	$i_p, E_p = f(t_d)$	• work at small t_d
(6) chemically heterogeneous natural ligand leading to peak broadening	$i_p, E_p = f([M])$ at constant $[L] \gg [M]$	• interpret the data based on complexation affinity distribution

and ΔE_p^L values as surface complexation may be much stronger than bulk complexation, owing to higher concentration of complexing site [183]. i_p^L and E_p^L are then controlled by surface complexation and not bulk complexation. This effect is detected by non linear $i_p^L = f(t_d)$ curve, and strong decrease of ΔE_p^L by increasing t_d. This problem can be overcome by means of protected electrodes such as the gel-integrated microelectrodes (section 5.1.4).

(3) The blocking of the redox reaction by adsorption of non-complexing compounds may be tested by following i_p and E_p as function of the contact time, t_w, between the electrode and the solution, at controlled potential [202,203]. A HMDE or a cell enabling solution exchange (section 2.2.3) should preferably be used for this purpose. Integration of the microelectrode in protective gel may also help to solve this problem (section 5.2.6.1).

(4) Trace metals may be measured in non-deoxygenated solution, by SWASV (section 5.2.6.2 and 5.2.6.3), because SW modulation suppresses the oxygen signal. Nevertheless O_2 is also reduced and the pH may drastically increase at the electrode surface, possibly leading to large local changes in speciation,

in particular with the formation of metal hydroxydes. Measurements should be made in the presence and absence of O_2 in laboratory conditions, and the influence of pH and buffer capacity on i_p and E_p should be tested.

(5) Metals with low solubility in Hg (e.g. Mn, Cu, Ni) may be accumulated, during the deposition step of ASV techniques, in excess of their solubility limit. Asymetrical or even more deformed reoxidation peaks are then obtained. Such an effect may be recognized by non-linear changes of i_p, positive shifts of E_p and changes in peak shape, on increasing t_d. The remedy is to use only small t_d values. Cu–Zn intermetallic compounds may also be formed in Hg for too large t_d values. Criteria to minimize this effect have been discussed in ref. [34].

(6) With chemically heterogeneous complexants, the complexation strength depends on the metal/ligand ratio, as stronger sites are occupied at a low ratio and weaker sites ar higher ratios [195–197]. When M is reduced or oxidized at the electrode, a concentration profile of M, and thus a profile of metal/ligand ratio is formed within the diffusion layer. As a result, the degree of complexation $\alpha = c_M/[M]$, is not constant, but varies within the diffusion layer and with E, which results in broadening of waves or peaks [204]. This effect is characterized by a change of E_p with c_M, while maintaining $[L] = \text{constant} \gg c_M$. (For a simple ligand in excess, E_p is independent of c_M.) This effect reflects a basic property of the test complexant and must be interpreted using appropriate theory, e.g. complexation affinity spectra [4,195,197].

3.5 CLASSIFICATION OF METALS AND NATURAL LIGANDS WITH RESPECT TO SPECIATION MEASUREMENTS

3.5.1 Voltammetric Properties of Metals

As discussed in section 3.1 and 3.2, and exemplified in Table 5, direct voltammetric speciation based on i_p and E_p measurements is usually possible only with reversible redox systems, since otherwise, interpretation is often too complicated to give unambiguous results. For a redox system to be reversible, conditions (19) or (20) must be fulfilled (Figure 3):

$$\text{For macroelectrodes:} \quad k_0 \gg \bar{D}/\delta \tag{19}$$

$$\text{For microelectrodes:} \quad k_0 \gg \bar{D}/r \tag{20}$$

where k_0 = charge transfer rate constant at the solution/electrode interface ($m\,s^{-1}$), $\bar{D}$ is the average diffusion coefficient of the test Ox or Red species in solution ($m^2\,s^{-1}$), δ is the diffusion layer thickness (m), and r the radius of the electrode (m). In typical conditions on macroelectrodes, $\bar{D}/\delta = 10^{-6}$–$10^{-4}\,m\,s^{-1}$, so that for reversible systems: $k_0 \geq 10^{-3}\,m\,s^{-1}$. For microelectrodes typically $k_0 \gg 10^{-2}\,m\,s^{-1}$ is required. Values of k_0 of relevant redox couples at water/Hg interface can be found in ref. [1] and extensive critical lists of redox properties of inorganic elements are given in ref. [205]. Three further points should be considered:

- conditions (19) and (20) and the above estimated values of k_0 and $\bar{D}/\delta$ or $\bar{D}/r$ show that a given redox system may change from reversible on a macroelectrode to quasi-reversible or irreversible on a microelectrode.
- for a given redox couple and electrode, k_0 may also depend on the composition of the double layer at the electrode surface. The character of redox systems at the borderline of reversibility may thus pass from reversible to quasi-reversible or irreversible, depending on ionic strength. For example, for Zn^{+2}/Zn^0, k_0 decreases from 5×10^{-4} to 2.8×10^{-5} m s^{-1} when electrolyte concentration increases from 0.1 to 1.0 M $NaNO_3$, and it even decreases to 4×10^{-6} m s^{-1} in 0.25 M Cs_2SO_4 [1].
- $\bar{D}/\delta$ depends on the technique used and the size of the natural ligands and of the corresponding complexes. For macroelectrodes, a quasi-reversible system may be changed to reversible or irreversible by using a slower (larger δ) or faster technique (smaller δ), respectively [2–4]. Similarly, the reversibility limit may be slightly different in solution containing colloidal ligands compared with calibration solutions with synthetic ligands, for both micro- and macroelectrodes.

In most practical cases, the following rule of thumb can be used for the voltammetric classification of redox systems at the Hg electrode:

- *Reversible systems*, for which *direct speciation approaches* (section 3.2 and 3.3.1; Table 5) are applicable. They include: Ag(I)/Ag^0, Cd(II)/Cd^0, In(I)/In^0, Hg(II)/Hg^0, Tl(I)/Tl^0 and Pb(II)/Pb^0;
- *Borderline systems* for which *direct speciation approaches* (section 3.2) may be applicable but by carefully checking redox reversibility since it depends on conditions. They include: Mn(II)/Mn^0, Cu(II)/Cu^0 and Zn(II)/Zn^0;
- *Irreversible systems*, for which concepts of sections 3.2 and 3.3.1 are not applicable, and for which only *indirect speciation* methods such as ligand exchange methods (section 3.3.3) are applicable. They include:

$$\mathrm{Cr(III)/Cr^0, Fe(II)/Fe^0, Co(II)/Co^0, Ni(II)/Ni^0, As(III)/As^0,}$$
$$\mathrm{Sb(III)/Sb^0, Se(IV)/Se^0.}$$

Note that indirect methods can also be applied to reversible or borderline systems [188,189], but direct speciation approaches are preferable when they are applicable. Fe is a special case. The Fe(II)/Fe^0 couple is strongly irreversible, but some information on soluble or colloidal FeS complex formation in aquatic media can be gained by careful study [184]. The charge transfer of Fe^{+3}/Fe^{+2} is usually reversible. Fe(III), however, is strongly hydrolysed which shifts the redox potential from $\sim +0.77\,\text{V}$ in non-complexing media to a potential in the range from 0 to $-0.6\,\text{V}$ [206,207] in neutral complexing solutions, with a well-defined dependence of potential peak on pH [207]. Detailed interpretation of Fe(III) speciation, however, is very complicated as polynuclear hydroxo complexes and mixed ligand–hydroxo complexes are formed at circumneutral pH [206, 207].

3.5.2 Voltammetric Properties of Aquatic Ligands

Although an infinite number of different ligands may be found in aquatic systems, they may be separated into a few groups of homologous complexants with similar properties. These properties are still ill known, but some general trends relevant to interpretation of voltammetric speciation data is available in Table 7 [4,169]. The following general comments can be made.

- Metal complexes formed with the major inorganic anions except OH^- and S^{-2} (group 1 in Table 7) are typical examples of fully mobile and fully labile complexes, for which, in addition, the condition of excess of ligand is usually fulfilled in waters.
- Information on specific, small but strong ligands released by organisms (group 8) is still rare and it is not even very clear whether or not they must be discriminated from the fulvic fraction which also includes small and strong ligands.
- Most colloidal complexants (groups 3–6) are either large or aggregated into large entities, so that they can often be considered as immobile with respect to voltammetry. In addition, the release of metal ions included inside colloids larger than a few nm may be slow, both because of slow chemical dissociation, and slow diffusion of M inside the particule. The combination of these two processes is referred to as 'effective' lability in Table 7, as the voltammetric signal is affected by its net result. For groups 3–5, this effective lability can be considered as low or negligible [34].
- Fulvic compounds (group 7) are special cases, as the stability of their complexes strongly depends on metal/ligand ratio and pH [4,196]. Their lability varies accordingly, ranging from fully labile to inert. Their mobility is also variable. The diffusion coefficient of single fulvic molecules is 0.5–3 $\times 10^{-10}\,\text{m}^2\,\text{s}^{-1}$ [4,208], i.e. lower than D_M but non-negligible. However, they may aggregate with other colloids or with themselves at pH$<$ 6 and high concentrations, (typically $> 50\,\text{mg}\,.\,\text{L}^{-1}$), and be effectively immobile [4,169].
- OH^- and S^{-2} (group 2) may form with metals, mononuclear (presumably mobile and labile) complexes, polynuclear (partially mobile and labile) complexes and colloidal solids (low or negligible mobility and lability), but information is scarce for these compounds.

Table 7. Voltammetrically relevant properties of the major types of natural complexants and their complexes with trace metals in natural systems [4,169]; r = particle radius; ℓ = length of macromolecule; M_w = molecular weight

Complexant type	Information related to mobility of complexes	'Effective' complex lability	Thermodynamic stability of complexes	Adsorption of complexants and complexes on electrodes
(1) Inorganic anions (CO_3^{2-}, Cl^-, SO_4^{2-}, F^- etc.) except OH^- and S^{-2}	D_{ML} close to D_M	high	weak	no
(2) OH^-, S(−II), polysulfides				
• soluble M complexes	D_{ML} close to D_M	sometimes labile	strong	sometimes
• colloidal species	$1 < r < 500$ nm	low or nil	strong	often
(3) Clay colloids	$r > 10$ nm often aggregated	low	intermediate	weak or nil
(4) Fe(III) oxyhydroxide	$1 \leq r < 500$ nm often aggregated	low	intermediate to strong	intermediate to strong
(5) Mn(IV) oxides		low	strong	
(6) Fibrillar polysaccharides	ℓ = 10–1000 nm often aggregated	high	low	possible
(7) Soil-derived fulvics (FA) and humics (HA)	$0.5 < r < 5$ nm partly aggregated. Typically for FA: $D_{ML} \sim (0.5\text{–}3) \times 10^{-10} m^2/s$	From fully labile (high M/L; pH < 7) to fully inert (low M/L; pH ≥ 7)	From low (high M/L; pH < 7) to very high (low M/L; pH ≥ 7)	usually strong
(8) Small specific organic complexants possibly released by organisms	$M_w < 1000$		weak to strong	

- Adsorption of complexants and complexes on electrode surface occurs with all colloidal complexants (groups 2–7). Most of these compounds, however, are retained by the gel of integrated microelectrode (section 5.1.4). This secondary process is thus largely minimized with this type of electrode, as well as the corresponding complications in data interpretation (sections 3.2 and 3.3.1), owing to low mobility and lability of these complexes. Small surface active compounds found in surficial sediments or in water treatment plants, might, however, pass through the gel and interfere at the electrode surface.

4 LABORATORY AND FIELD VOLTAMMETRIC ANALYSIS OF WATER

As already mentioned, the range of substances that can be effectively or potentially monitored by voltammetric techniques in aquatic systems is broad and includes inorganic, organometallic and organic substances. Detection limits below 10^{-10} mol L^{-1} are available, and the dynamic range of these techniques embraces several orders of magnitudes, so that analyte concentrations at the millimolar level are also measurable. In addition to measuring analytes with redox properties, it is also possible to measure surface active compounds, using tensammetric techniques [145,209]. Because of these features, voltammetric techniques are currently used in laboratory conditions in nearly all fields of analytical chemistry, e.g. biochemical and pharmaceutical analysis, food, agricultural, ore and environmental analysis.

4.1 LABORATORY OR IN FIELD ANALYSIS OF METAL IONS

Analysis of metal ions (Tables 2 and 8) is the major type of applications of voltammetry to natural waters, probably because direct analysis is possible with no or minimum change of the sample. Direct reduction methods such as DPP, DPV or SWV, have been used both in the laboratory and on field to measure Mn(II) and Fe(II) in anoxic freshwater [223–225]. In most cases, however, direct reduction methods are not sensitive enough for trace metals in water, so that either ASV or AdSV techniques with various modulations (usually DP or SW; Table 1) have been the most widely used [71,257,258]. PSA (Table 1) is also a method of choice although less frequently used because of its lower sensitivity. Table 8 summarizes the metals, samples and techniques reported in the literature for water analysis since 1984. Similar tables for reports before 1984 can be found in [Chapter 19 in ref. [5] and 145,210,211]. Some elements not mentioned in Table 8 may also be measured by AdSV [258,259], but improvements in the procedures, in particular their detection limits, are still needed for their application to natural waters. Applications of AdSV to water analysis will broaden the application range of voltammetry and recent publications have emphasized its promising potential [258,260,261]; see refs [11,145,262] for reviews. However more experience is still needed to evaluate its reliability and usefulness for routine analysis. In contrast, ASV techniques have been used for more than 30 years for laboratory trace metal measurements in waters [260,263–267] and routine instruments have been developed. Although it is in principle technically simpler, PSA [268,269] has been less developed than ASV or AdSV techniques [270]. This may be because, until recently, chemical reoxidation of the previously deposited metal was required, making automation less easy. More recently commercial instruments have been developed to perform the reoxidation at constant current; oxygen, however, may be an important interferant. PSA is believed somewhat controversially to be less prone than ASV to interferences due to adsorption on the electrode [271]. The list of procedures of Table 8 is useful to get an overview of capabilities of voltammetric techniques; it is worth mentioning that their application to *in situ* measurements is often not straightforward, because of lengthy preparation

Table 8. Voltammetric conditions which have been used to determine metal ion concentrations in water samples, in laboratory and on board ship, since 1984. For older reviews, see refs [5,21,145,210,211]. Acronyms of techniques and electrodes: see list of acronyms

Compound	Electrode type	Technique	Natural water sample	Added ligand and sample treatment	References
As(III)/As(V)/Hg(II)	Au plated on glassy carbon electrode	ASV	lake water	acidified pH 1	52
Ag(I)	Carbon paste CME	CSV	waste water	evaporation followed by redissol. in HNO_3	212
Ag(I)	Glassy C-based thin Hg film electrode	ASV	river water	Hg(II), acetate buffer pH 4.5	213
Cu(II)/Pb(II)	Rotating glassy C-based thin Hg film electrode	SWASV	melted snow	Acetate, pH of nat. samples	214
Cu(II)/Pb(II)	Rotating glassy C-based thin Hg film electrode	DPASV	Antarctic snow	HNO_3, $Mg(NO_3)_2$, $HgNO_3$	215
Cu(II)/Pb(II)	Hg-plated Ir-based single or array microelectrode	SWASV	river water	–	93,104
Cu(II)/Pb(II)	Gel-integrated single or array microsensor	SWASV	river water	–	192,216
Cu(II)/Pb(II)/Cd(II)	Hg-plated Ir-based microelectrode arrays	SWASV	spring water	–	217
Cu(II)/Pb(II)/Cd(II)	HMDE	ASV	wastes water, sediments	chemical sequential extraction	218
Cu(II)/Pb(II)/Cd(II)/Zn(II)	MFE coated with Nafion or cellulose acetate/Nafion membrane	DPASV	seawater, polluted freshwater	–	219
Pb(II)	Hg-plated Ir-based microelectrode arrays	SWASV	tap water	–	217
Pb(II)/Cd(II)	Glassy C-based thin Hg film electrode	SWASV	tap water	acidified pH 2	220
Pb(II)/Cd(II)	Rotating glassy C-based thin Hg film electrode with and without Nafion coating	SWASV	river water	acetate pH 7.7	221
Cu(II)/Pb(II)/Cd(II)/Zn(II)	HMDE, MFE, Au electrode	SWASV	sea and freshwaters	–	36
Fe(II)	Carbon paste electrode doped with 1-10-phenanthroline and Nafion	DPP	natural waters	acidified pH 2	222
Fe(II)/Mn(II)	DME, SMDE	DPP, NPP	Lake water (on-field meas.)	–	223–226
Hg(II)	Carbon paste electrode	DPASV	river and sea waters	Na_2SO_4, Na-tetraphenylborate	227
Mn(II)	Rotating glassy carbon electrode	DPCSV	seawater	HEPPES buffer pH 7–8	228
Mn(II)	graphite electrode	DPCSV	mineral water	UV irradiation, acidification	229
Cu(II)/Pb(II)/Cd(II)	Rotating glassy carbon electrode	PSA	seawater	$HgNO_3$ as chemical oxidant acidified pH 3	392

Table 8 (*cont.*)

Compound	Electrode type	Technique	Natural water sample	Added ligand and sample treatment	References
Cu(II)/Pb(II)/ Cd(II)/Zn(II)	Glassy C-based thin Hg film electrode	PSA	creek and tap waters	Hg(II) as chemical oxidant acidified pH 2	393
Cu(II)/Pb(II)/ Cd(II)/Zn(II)	Rotating glassy C-plated thin film Hg electrode	PSA	waste and rain waters	UV irradiation, acidified pH 1	394
Cu(II)/Pb(II)/ Hg(II)/Se(IV)	Au microelectrode	PSA	sea and groundwaters	–	55
Al(III)	HMDE	AdSV	freshwater	lumogallion	395
Al(III)	Graphite carbon electrode	AdSV	NBS ref. water samples	8-hydroxyquinoline	396
As(III)	HMDE	AdSV	river water	Hydrazine sulfate, 2M HCl	397
Cu(II)	HMDE	AdSV	seawater	Catechol or 1, 2 - dihydroxyanthra quinone-3-sulfonic acid	398,399
Cu(II)	SMDE	AdSV	freshwater	1-(2-thiazolylazo)-2-naphthol Acetate buffer pH 3.7	400
Cu(II)	Carbon paste electrode doped with benzoinoxime	AdSV	lake, river, tap waters	–	401
Cu(II)/Pb(II)/ Cd(II)	HMDE	AdSV	seawater (ship-board meas.)	8-quinolinol HEPES buffer pH 7.9	402
Cu(II)/Pb(II)/ Cd(II)/Ni(II)/ Co(II)	HMDE	AdSV	sea, river, groundwaters	1-phenylpropane-1-1pentylsulfonyl hydrazone-2-oxime, ammonium buffer pH 8–9	403
Cu(II)/Ni(II)/ Fe(III/V(V)	HMDE	AdSV	sea water	dimethylglyoxime (Ni), catechol(Cu, V), 1-nitroso-2-naphthol (Fe) with/without UV irradiation, pH 2.5	404
Co(II)	HMDE in flow-through cell	AdSV	sea water (ship-board)	nioxime UV digestion, ammonium buffer pH 9	155
Co(II)	HMDE	AdSV	natural waters	1,10-phenanthroline	405
Co(II)	HMDE	AdSV	estuarine water	dimethylglyoxime triethanolamine/ NH_2Cl	406
Co(II)/Ni(II)	HMDE	AdSV	lake water	dimethylglyoxime	230
Cr(VI)	HMDE	AdSV	seawater	diethylenetriamino pentaacetic acid acetate buffer pH 5.2	231
Cr(VI)/U(VI)	HMDE	AdSV	groundwater	Pr gallate + diethylenetria-minepenta-acetic acid, acetate pH 5.8	232
Cr(VI)/U(VI)/ Fe (III)	HMDE	AdSV	groundwater	cupferron, phosphate buffer pH 6	233

Compound	Electrode type	Technique	Natural water sample	Added ligand and sample treatment	References
Fe(III)	HMDE	AdSV	groundwater	4-(2-thiazolylazo)-2-naphthol EDTA, acetate buffer	234
Fe(II)/Fe(III)	HMDE	AdSV	lake water	1-nitroso-2-naphthol	235
Mn(II)/ Mn(total)	Carbon paste electrode doped with 1-(2-pyridylazo)-2-naphthol	AdSV	seawater	$H_2SO_4 + H_2O_2$ for Mn tot	236
Mo(VI)	HMDE	AdSV	rain water	dichlorooxine, NaCl + HCl pH 2	238
Mo(VI)	HMDE	AdSV	seawater	2-(2′-thiazolylazo)-p-cresol acetate pH 3^-	238
Mo(VI)	HMDE	AdSV	drinking water	3-methoxy-4-hydroxy mandelic acid $KBrO_3 + K_2SO_4 + H_2SO_4$	239
Ni(II)	Sol–gel C-composite electrode doped with dimethylglyoxime	AdSV	river water	acetate, pH 7.7	240
Pb(II)	SMDE	AdSV	tap water	calcein blue acidified pH 2	241
Ru(III)/Pt(II)	HMDE	AdSV	tap water	4-(2-pyridylazo)-2-naphthol BR-buffer pH 9.3	242
Sb(III)	HMDE	AdSV	sediments	digestion, 1-phenyl-3-methyl-4-benzoyl-pyrazolone-5 acetate buffer	243
Sb(III)/Sb(V)	HMDE	AdSV	river water	chloranilic acid, HCl pH 1–3	244
Se(IV)	Rotating silver disc electrode	AdSV	synthetic solution	HCl, HNO_3	245
Ti(IV)	HMDE	AdSV	sea, river, rain waters	solochrome violet RS acetate buffer pH 5.1	246
Ti(IV)	HMDE	AdSV	sea and freshwaters	mandelic acid, HCl UV irradiation	247
Tl(III)	CME	AdSV	freshwater	thioridazine, acidified pH 2	248
U(VI)	HMDE	AdSV	seawater	2-thenoyltri-fluoroacetone EDTA, acetate buffer pH 4.5	249
U(VI)	HMDE	AdSV	river, drainage waters	chloranilic acid, NaOH/HCl pH 2.5	250
U(VI)	HMDE	AdSV	seawater	oxime, PIPES buffer pH 6.8	251
V(V)	HMDE	AdSV	tap water	5-bromo-(2-pyridylazo)-2,7-dihydroxyna-phthalene, acetate buffer pH 4.5	252
V(V)	MFE (in flow-through cell)	AdSV	tap water	Cupferron-BrO_3, acetate buffer pH 4.2	253
Zr(IV)	HMDE	AdSV	seawater	solochrome violet RS	254

Table 8 (*cont.*)

Compound	Electrode type	Technique	Natural water sample	Added ligand and sample treatment	References
Co(II)/Ni(II)/ Zn(II)	HMDE	ASV/ AdSV	estuary waters	2-quinolinethiol (for AdSV)	255
Cu(II)/Pb(II)/ Cd(II)/Zn(II)/ Cr(VI)	HMDE	ASV, PSA AdSV	soils, sediments	microwave digestion	256

steps, insufficient reliability for long term routine measurements or too complicated procedures. The reader is referred to the cited literature, to make his own opinion.

4.2 ANALYSIS OF INORGANIC ANIONS AND NEUTRAL COMPOUNDS

Laboratory measurements of anions and neutral inorganic species (such as S^0) of environmental relevance have also been performed by polarographic and voltammetric techniques (Table 9). In this case, chemical change of the sample is often required which makes *in situ* measurements more difficult. Important exceptions are Cl^-, Br^-, I^-, SCN^- and S(−II) which can be measured in unmodified samples based on their reaction with Hg^0 to form an adsorbed compound. In particular S(−II) has been successfully determined on site, in anoxic lake water samples, by DPP [223–225] or DPAdSV (207). Until now, these measurements have been performed mainly by using mercury drop electrodes, whose surface is easily renewed. Their application *in situ* requires systematic tests with Hg film or Hg amalgam electrodes, ultimately to enable reliable determination of these compounds.

4.3 ANALYSIS OF ORGANIC COMPOUNDS

The number of organic compounds which can be analysed by voltammetric techniques is almost infinite. This results from the fact that a large number of different functional groups can be reduced or oxidized [295] (Figure 2d, Table 10). Detailed reviews and lists of compounds have been published for all organic compounds [11,19,71,145,297,315,316] as well as for specific fields, in particular pharmaceuticals [69,70,295,296,316], biological molecules [163,316], or environmental pollutants [11,297,317,318]. An exhaustive list of compounds measurable by voltammetry is beyond the scope of this chapter. Table 10, however, lists the main groups of measurable compounds, in particular those relevant for environmental purposes. It can be seen that voltammetry is applicable to most pollutant groups. Specific values of $E_{1/2}$ and other relevant electrochemical

Table 9. Inorganic anions and neutral compounds analysed by voltammetry in water samples, in laboratory. Acronyms: see list of acronyms

(a) Analysis after sample pretreatment

Anion	Electrode / Technique	Media	Detection limit (μmol L^{-1})	References
NO_2^-	DME / DPP	Diphenylamine, KSCN, $HClO_4$, pH ~ 1.5	0.1	272
NO_2^-	HMDE / AdSV	Sulphanilamide, 1-naphthylamine, HCl pH 2.5	3×10^{-4}	273
NO_2^-	Carbon paste CME/CSV	10^{-2} mol.L^{-1} NaOH	2×10^{-2}	274
NO_3^-	DME / a.c. or d.c. polarography	Phenol, H_2SO_4pH ~ 0.5	80	275
NO_3^-	SMDE / CV	0.04 M MCl_3 (M = Ce(III), La(III) or Gd(III))	1	276
NO_3^-	DME / DPP	$[UO_2(CH_3COO)_2]$, K_2SO_4, HCl pH1	1	277
NO_3^-	Au-disc Au-ring electrode / CSV	$Cd(ClO_4)_2/HClO_4$ pH 3	0.5	278
CN^-/S(−II)	DME / d.c. polarography	KOH, borate buffer pH 9.75	40	279
PO_4^{3-}	DME / oscillographic polarography	HCl–Sb(III)–ammonium, molybdate–acetone–butanone	0.1	280
Cl^-	Ag rod electrode / AdSV	KNO_3, HNO_3 pH1	10	281
Br^-	DME / d.c. polarography	NaOCl	?	282
$Cl^-/Br^-/I^-$	DME, HMDE / DPP, AdSV	0.1 mol.L^{-1} HNO_3/1.8 mol.L^{-1} H_2SO_4 / acetate buffer	10	283–287
I^-	Ag disc electrode / AdSV	Acetate buffer pH 4.7	4×10^{-2}	288
I^-	HMDE / AdSV	Acetate buffer pH 4.7	5	283
ClO_3^-	DME / DPP	NaCl, Fe_2SO_4, pH5	10	283
IO_3^-	DME/DPP	$CaCl_2$	0.2	283
F^-	DME / d.c. or a.c. polarography	Zr-alizarin S, HCl	?	289
SO_3^{-2}	DME / DPP	Acetate buffer	1	283
$S_2O_8^{2-}$	DME / DPP	H_3BO_3, KOH, KCl pH 9.75	100	283
CN^-	DME / DPP	KOH / Borate pH = 9.8	0.3	283

(b) Analysis without sample pretreatment

Anion	Electrode / Technique	Electrode process	Detection limit (μmol L^{-1})	References
S^0	DME, SMDE, HMDE / DPP, CSV, AdSV, SWAdSV	$S^0 + Hg \rightarrow HgS_{(ads)}$ $HgS_{(ads)} + H^+ + 2e^- \longrightarrow Hg + HS^-$	3×10^{-3}	290–293
S^{2-}	DME, SMDE, HMDE / DPP, AdSV, SWAdSV	$S^{2-} + Hg \rightleftharpoons HgS + 2e^-$	5×10^{-2}	226,290–293
S^{2-}	Hg-Au amalgam electrode / SWAdSV	$S^{2-} + Hg \rightleftharpoons HgS + 2e^-$	0.1	39
S_n^{2-}	DME, SMDE, HMDE / DPP, CSV	$S_n^{2-} + Hg \rightleftharpoons HgS + (n-1)S + 2e^-$	0.1	291,293
SO_3^{2-}	DME, SMDE, HMDE / DPP, CSV	$2SO_3^{2-} + Hg \rightleftharpoons Hg(SO_3)_2^{2-} + 2e^-$	0.5	291,293
$S_2O_3^{2-}$	DME, SMDE, HMDE / DPP, CSV	$2S_2O_3^{2-} + Hg \rightleftharpoons Hg(S_2O_3)_2^{2-} + 2e^-$	1	291,293
$Cl^-/Br^-/I^-$	HMDE, SMDE / AdSV	$2X^- + Hg \rightleftharpoons Hg_2(X)_2 + 2e^-$ ($X^- = Cl^-, Br^-$ or I^-)	Cl^-, Br^-, I^- ·5	285, 294

Table 10. List of groups of organic compounds which may be analysed by voltammetric techniques in biological, pharmaceutical and environmental media. The list is not exhaustive. See the references mentioned for details

Biological molecules [71, 163]
Proteins
Purines, Pyrimidines, Nucleotides, Nucleic acids
Flavins, Flavin nucleotides, Riboflavin
Pyrrol, porphyrin, pyridin nucleotides
Pharmaceuticals [69–71, 145, 296]
Alcaloids, anethethics
Cardiac drugs, Anticancer drugs
Antibiotics, antidepressants, sulfonamides
Sex hormones, catecholamines
Environmentally relevant compounds
Pesticides [70, 71, 145, 297–300]
DDT
Carbamate-based pesticides
Triazine-based pesticides
Thiourea-based pesticides
Pesticides with nitro groups [298]
Surfactants, detergents [302–304]
Toxic or carcinogenic compounds
Oil products [145, 305, 306]
Aromatic amines [297, 307, 311]
Phenol derivatives [297, 308, 311]
Nitrosamines [309, 310]
Aflatoxins [311]
Polycyclic aromatics [11, 301, 311–313]
Polychlorobiphenyls [11, 145, 311, 301, 312]
Organometallics [159, 314]

parameters can be found in compilations, e.g. in refs [319, 320] (exhaustive list of compounds) and [70] (pharmaceuticals), and in the specific papers cited in references of Table 10. Detection of organic compounds may be performed directly or after chemical modification to introduce an electroactive group into the molecule when necessary. With or without chemical modification, four major modes of voltammetric techniques are used for trace levels:

- oxidation without preconcentration, by DPP, DPV, or SWV, usually on glassy carbon electrodes [19,70,295];
- reduction without preconcentration by DPP, DPV, or SWV, usually on DME, HMDE or MFE;
- reduction or oxidation after adsorptive preconcentration (AdSV), usually on HMDE, but also on MFE or solid electrodes (11,71,145];

- use of tensammetric techniques after adsorptive preconcentration (AdST = adsorptive stripping tensammetry), usually on HMDE [71,145]. This technique is applicable to compounds which do not undergo an electron charge transfer, but which adsorb/desorb at a well-defined potential. This process gives rise to a significant change in capacitive current (section 2.1.1) which can be recorded as a peak whose height is directly related to the concentration of adsorbed compound. For the principle of this technique, see ref. [321]. It is particularly applicable to surfactants, oil products and PCB which are not oxidizable or reducible and which cannot be easily chemically transformed into a redox depolarizer.

Flow-through cells, often coupled to HPLC, have been largely used for voltammetric detection of organic compounds, either with glassy carbon electrode or MFE [19]. The sensitivity of voltammetric techniques for organic compounds is typically 10^{-7}–$10^{-6}\,mol\,L^{-1}$ for direct reduction or oxidation methods, and $10^{-9}\,mol\,L^{-1}$ or even lower for AdSV techniques. Maximum sensitivities of $2 \times 10^{-11}\,mol\,L^{-1}$ for riboflavin [322] and $5 \times 10^{-10}\,mol\,L^{-1}$ for the pesticide DNOK (2-methyl-4,6-dinitrophenol) [298] have been reported [69,145]. This brief survey mainly shows that voltammetric techniques have a great potential for the analysis of trace organics in waters. The major limitation is that a preliminary step is needed to separate the test compounds and sometimes to modify it chemically before analysis. The coupling of flow injection analysis types of techniques with voltammetry is then required in most cases. Developments in this area, specifically for *in situ* applications, are highly needed.

4.4 INSTRUMENTATION FOR ON SITE MEASUREMENTS

Compared with the very large number of analytical procedures developed for laboratory analysis of inorganic and organic compounds of environmental relevance by voltammetry, little has been done to develop voltammetric equipments for on site or *in situ* measurements in waters. This is somewhat surprising as there is a clear need of instruments for automatic, continuous, real-time monitoring of pollutants in order in particular to overcome the problems related to sample handling (e.g. contaminations and losses by adsorption [323] or sample degradations [324]). Voltammetry enables one to build low cost, miniaturized equipment with low energy requirement and is thus well suited for this purpose. It requires good cooperation between electrochemists, electronics and mechanical technicians, but major limitations are probably (i) the insufficient training of chemists and technicians in voltammetric techniques and (ii) the insufficient long-term reliability of most electrodes (see section 5.1). A few systems for on site measurements have been reported, however, based on procedures developed for laboratory analysis (Tables 8–10). They can be classified as (a) field-portable voltammetric instrument [325–327] and (b) non-portable on-line analysers for in field or on shore continuous automatic measurements of metals by ASV and AdSV [29,155,328].

The portable instruments (category (a)) are based on commercial, laboratory-type computer-controlled potentiostats [329,330], the major improvements being the adaptation of the polarograph to work on batteries and its compact construction, to enable its transport, with reagent and conventional electrodes, in a portable box. A conventional plastic disposable cell with a standard three-electrode configuration (usually an Hg-plated glassy carbon working electrode, an Ag/AgCl/ sat. KCl reference electrode and a platinum rod auxiliary electrode) is generally used [325,326]. A plexiglas cell of similar geometry but with the working electrode fixed at the bottom has also been proposed [327]. The mercury film is plated before a set of analyses (typically daily) and the measurements performed immediately after sampling.

In on-line analysers for in field measurements (category (b)), the various analytical steps, e.g. sampling, UV irradiation, mixing of controlled volume of reagents or standards, degassing, stirring, electrode conditioning and voltammetric measurements are computer controlled. This enables one to reduce the analysis time, to improve the reproducibility and to minimize contamination, in particular when ship-board measurements are performed [29,155,328]. In most reported systems, a standard three-electrode cell configuration has been used with a mercury drop working electrode and the conventional reference and auxiliary electrodes. The cell was only modified to incorporate input and output for the solution as in refs [329,330] and in the Tacussel APASA system. In two cases special flow-through cells were built. Ref. [155] reports a mini plexiglass cell enabling one to adapt an HMDE in a configuration similar to that in Figure 6B. In the system reported by Clarell and Zirino [328], a tubular Hg-plated graphite working electrode was connected to a tubular pseudo Ag/AgCl reference electrodes with the auxiliary electrode placed downstream. Both working and reference electrodes were a part of the flow-through cell body. This system was the very first automated system enabling real-time measurement on ship, for several days. The main drawback of all the above systems is that they are based on standard commercially available laboratory equipment (i.e. electrodes, potentiostats, pumps and valves) and they are not optimized for in field conditions.

5 *IN SITU* VOLTAMMETRIC PROBES

5.1 DESIGNING AN *IN SITU* VOLTAMMETRIC CELL

The main components of the whole submersible instrument (called the probe) are the hardware and software (not discussed here) and the voltammetric cell which itself includes the three electrodes, in particular the working electrode or voltammetric sensor. This section discusses the development of cells and sensors for reliable, continuous *in situ* (at depth) voltammetric measurements of mobile and total concentrations of metals and their speciation. Contrary to the measurement of mobile species, that of total concentration requires some sample transformation (e.g. heating, addition of acid or synthetic ligand). Similarly analysis of anions or organic compounds also requires sample transformation (sections 4.2 and 4.3). Overall, the following important conditions must be fulfilled to develop a versatile voltammetric probe:

for the whole voltammetric cell and probe

- minimum sample transformation during analysis, to minimize both the possible artefacts and instrument complexity, as well as reagent and energy consumption;
- optimization and simplification of analytical steps, to minimize the number of valves, pumps and their energy requirements as well as their electronic control;
- flexibility and easy calibration;
- ease of deployment and maintenance in field conditions;

for the sensor

- choice of the most reliable working and reference electrodes and cell geometry including electrode positioning (section 2.2.3);
- choice of electrochemical conditions to maintain the sensor surface unaltered or recondition it *in situ* when necessary;
- overall analysis time as short as possible;
- elimination of fouling problems;
- elimination of interferences by O_2 (in oxic waters) and S(−II) (in anoxic waters).

These aspects are discussed below in more detail, based on the information of Section 2.

5.1.1 Choice of the Working and Reference Electrodes and Cell Geometry

5.1.1.1 Working electrode The working electrode is the heart of the voltammetric systems as the quality of the measurements depends primarily on the reliability of its surface, in particular for long term *in situ* applications. This requirement is not readily fulfilled (see section 5.1.3). Working electrodes can be divided in two main groups (section 2.2.1), the mercury and the solid electrodes (including the chemically modified electrodes), and two geometries, macro- (> 100 μm) and microelectrodes (< 10 μm; typical values, see §2.2.3). Microelectrodes are of particular interest for *in situ* measurements as (i) they can be used even in freshwaters without addition of electrolyte and (ii) stirring is not required during the preconcentration steps of stripping techniques which greatly simplifies the instrumentation and cell design. In addition, under optimum fabrication conditions, their excellent signal-to-noise ratio enables the determination of ultralow concentrations (10^{-11}–10^{-9} mol L^{-1}) with shorter preconcentration times than with macroelectrodes. Their main drawback is that some of the fabrication steps are more tricky than for macroelectrodes and may lead to high background current under bad conditions. The most important characteristics for a good signal/noise ratio are (i) an excellent sealing between the metal wire and the surrounding electrode body at the tip of capillary-based microelectrodes (Figure 13a), (ii) the absence of organic layer deposition on metal surface after photolithography, for microelectrode arrays and a good adhesion of the insulating layer (Si_3N_4 in Figure 13b), and (iii) for both type of electrodes the absence of electrical resistance between the working electrode and the potentiostat (good electric contacts and sealing) particularly when fast techniques, such as SWV or SWASV, are used. The advantages and limitations of the Hg and solid working electrodes for *in situ* applications are summarized below.

- *Mercury drop electrodes* are not suitable because reliable devices to control the drop size are technically complicated to adapt *in situ*. In addition, the relatively large volume of Hg required for long term measurements would represent an important risk of pollution in case of system failure. Mercury film electrodes (MFEs) are presently more promising. They are produced by plating Hg^0 on a substrate (C, Ir, Pt) by electrolysing a Hg(II) solution. To simplify the intrumentation and to avoid possible pollution problems, the mercury film has to be plated before deployment and must be stable and reproducible over an extended period of time. Until now, stable and reproducible Hg film have been reported for 1 day on C-based [154] and Au-based [39] macroelectrodes and for ≥ 10 days for Ir-based microelectrodes [93, 345]. Improvement can be expected in the latter case and in general microelectrodes are also recommended for the other reasons given above. The possible formation of intermetallic compounds in the Hg film during the preconcentration steps of stripping techniques may be a drawback of MFEs for both macro- and microelectrodes for the analysis of samples with high concentration. It is usually not significant when trace metal concentrations are below 10^{-8} mol L^{-1} [34], as is often the case in most unpolluted waters.
- *Solid electrodes* are mainly used for the analysis of elements or organic compounds with redox potentials more positive that the oxidation of Hg (section 2.2.1.1). They have also been proposed to replace Hg. They present, however, a number of drawbacks, in particular related to the non-reproducibility of their surface: they oxidize easily and it is difficult to renew their surface under controlled conditions. In addition, they often present high capacitive current, and consequently their detection limit is above nanomo-

lar. Finally intermetallic compounds are readily formed which limits their reproducibility. Further development and systematic studies are thus required before applying these electrodes to long term *in situ* measurements. Amongst solid electrodes, the most promising is Au (Tables 2 and 8). Ir is interesting as it is difficult to oxidize, but it has never been tested systematically for trace analysis. *Chemically modified electrodes (CMEs)* are potentially interesting as solid eletrodes with well-controlled surface. However, their development is still at the stage of infancy (section 2.2.1.1.) and they are not yet reproducible enough to be usable either for routine or for *in situ* applications.

5.1.1.2 Reference electrodes The various types of reference electrodes are listed in Table 3 and discussed in section 2.2.2. Their advantages and limitations for *in situ* applications can be summarized as follows.

- The commercially available reference electrodes with conventional liquid junctions (type a in Table 3) are not suitable because they are pressure dependent and not easy to handle because of the internal solution.
- Type d (polymer-coated) electrodes presently do not seem to be reliable enough for routine analysis.
- Metal/metal oxide electrodes (type e) are promising provided they are included in pH buffered gel, but they need more systematic study before being used *in situ*;
- Solid-state reference electrodes (type c) are pressure independent, and stable, but their lifetime is limited to 2–3 weeks.
- Ag/AgCl/sat. KCl in gel (type b) are stable, reliable and have a long lifetime. They can be used up to 60 000 kPa (= 600 bar). Their drawback is their large size except for type b5, and the possible problems linked to contamination or history effects (section 2.2.2). The use of a slightly porous ZrO_2 ceramic junction (type b5) helps to solve the contamination problem.

Type b5 and c8 electrodes have been successfully used for *in situ* voltammetric measurements, for periods of several days and seem to be at present the most reliable electrodes.

5.1.1.3 Choice of the measurement voltammetric cell Two types of electrode assembly and handling can be considered:

- direct immersion in the water body of the three electrodes, either isolated from each other (a), or combined at the tip of an integrated sensor (b), or
- assembly of the three electrodes in a flow-through cell (c) such as those discussed in section 2.2.3, and pumping of the test water through the cell.

System (a) is used for the measurement of high resolution concentration profiles at the sediment–water interface [39,110]. For such purpose, the working electrode body must be as small as possible (< 1–2 mm in thickness or diameter), to minimize perturbation of the sediment structure. The working electrode is fixed on a micromanipulator and is introduced in the sediment while the reference and auxiliary electrodes usually remain in the supernatant solution; distances of a few centimeters between the electrodes do not seem to be problematic.

System (b) has been used in surface water [56] and ground water [331]. It is a simpler, potentially more compact system than flow-through cells (c), but it is much more limited in terms of measurable analytes and long term operation, as it does not enable transformation of the sample before voltammetric detection (see introduction of section 5) or *in situ* reconditioning of electrodes with chemical reagents.

Flow-through cells (c) have many advantages for measurements in the water column, in particular for automatic recording of parameters for long periods of time. Fouling by large particles, plankton cells etc. is easily avoided by using a filter with large pores at the entry of the cell. For ultratrace measurements, contamination by the sensor body is easily avoided (as well as the losses by adsorption on cell walls) by extensive flushing of the flow-through cell. The optimum analytical conditions in this respect can be checked by recording the analyte concentration as a function of flushing time; a constant value must be obtained. An additional major advantage of a flow-through cell is the possibility to adapt a flow injection system [146] to perform, e.g., (i) *in situ* chemical transformation before voltammetric measurement, (ii) ASV with medium exchange, or (iii) AdSV. Such facilities may widely extend the number of analytes measurable by voltammetry (Tables 8–10) including anions and organic compounds. As mentioned in section 2.2.3, there is no a unique optimal design for electrochemical flow-through cells, as it must be adapted to the specific problem at hand. Some important criteria must be taken into account, however. In particular, minimization of the cell volume is important for *in situ* measurements, to minimize possible reagent consumption; but the corresponding increase of electrical resistance must also be carefully considered (section 2.2.3). The key criteria discussed in section 2.2.3 apply to flow-through submersible cells. A few additional ones specific to *in situ* measurements are the following:

- robustness, easy maintenance and/or replacement of the working and reference electrodes, in field conditions ;
- pressure compensation when necessary;
- automated valves and pumps functioning at depth, with low energy requirements and easy maintenance.

To meet all the requirements is not straightforward. The submersible cell of the VIP system (section 5.2.5; Figure 16 [116]) has been developped by considering them in details. It is discussed below as an example. A three-electrode configuration consisting of a working gel-integrated microelectrode (section 5.1.4 for details), an Ag/AgCl/sat. KCl gel reference electrode (Table 3, type b5)) and a built-in platinum ring auxiliary electrode has been used. A microelectrode has been chosen because of the advantages mentioned in sections 2.2.1.2 and 5.1.1.1, in particular the fact that it enables ASV measurements to be performed in low ionic strengh freshwaters without stiring. The later characteristic enables both the reproducibility of measurements to be improved and energy consumption to be minimized, which is important for autonomous long term operation. The positioning chosen for the three electrodes was that of Figure 5a. This configuration (i) prevents the products formed at the auxiliary electrode from interfering at the working electrode, (ii) minimizes the attachement, on the working electrode tip, of gas microbubbles formed by temperature changes and (iii) minimizes the *iR* drop because of the short distance between the working and auxiliary electrodes. The major drawback of this configuration is that the internal compartment (A) volume (Figure 16b) is larger than a few hundreds of microlitres, which may be a limitation when reagent addition is required. The external compartment B (Figure 16B) is filled up with a gel of agarose in 1 mol L^{-1} $NaNO_3$ in contact with the test solution and the reference electrode at the bottom of the cell by means of ceramic porous junctions. This external compartment plays several roles: (i) it is a double bridge for the reference electrodes and thus avoids Cl^- leakage, (ii) it shields both the working and the auxiliary electrodes, and (iii) it acts as a pressure equalizer by means of the rubber pressure compensator. The cell is screwed, with o-ring seals, to the cover of a pressure case base which incorporates the preamplifier for the microsensor and on which the submersible peristaltic pump is fixed (Figure 16b). The pressure case base is mechanically connected to the electronic housing via two titanium rod connectors, through which the electrical connections of the electrodes, the preamplifier and the pump are performed. The advantages of this configuration are the following: (i) the microsensor and the internal compartment are protected against shocks, (ii) maintenance and replacement of the working and reference electrodes are easy since they are simply screwed with o-ring seals, (iii) watertightness of all the electrical connections is made possible using normal connectors and (iv) the preamplifier is isolated from the main electronic part which minimizes an important source of noise.

5.1.2 Dissolved Oxygen

O_2 is a major electroactive compound in oxic water. Its concentration in O_2-saturated water ($\sim 3 \times 10^{-4}$ mol L^{-1}) is usually much higher than those of the trace compounds of interest. In voltammetry, O_2 can be either the analyte or an interfering compound when determining other species. As shown in Figure 2a, oxygen gives two reduction waves at Hg electrodes, corresponding to the reduction of O_2 to H_2O_2 (~ -0.1 V on Hg, versus Ag/AgCl/sat. KCl; equation (21)) and of H_2O_2 to H_2O (~ -0.9 V; equation (22)) successively.

$$2O_2 + 4H^+ + 4e^- \longrightarrow 2H_2O_2 \tag{21}$$

and

$$H_2O_2 + 2H^+ + 2e^- \longrightarrow 2H_2O \tag{22}$$

Similar reduction processes occur on non-Hg electrodes. These two waves (or peaks depending on the technique) can be used to analyse O_2 and/or H_2O_2 concentrations in water [226] or sediment [39]. The oxygen electrode (Chapter 2), because of its protective hydrophobic membrane, is much more specific to O_2 than voltammetry, and therefore is recommended in most cases. Voltammetric analysis of O_2, however, is useful (i) for the determination of waterborne H_2O_2 and (ii) for the simultaneous determination of concentration profiles of O_2 and other chemicals in sediments. In such cases it may be important to determine the profiles of all the test compounds (including O_2) on the same electrode, since different profiles can be obtained at different locations owing to the horizontal heterogeneity of sediment. When analysing O_2 by voltammetry, it must be noted that the peak (or half-wave) potentials of O_2 and H_2O_2 reductions depend on pH (reactions (21) and (22)).

In most cases, however, O_2 plays the role of interferant during the voltammetric analysis of trace compounds. Important cases where this interference does not occur are the analysis of compounds in anoxic water at the bottom of eutrophic lakes or seas, or in sediments, where O_2 is depleted by natural processes. In such cases, e.g. for the analysis of Fe(II) and S(−II), *in situ* voltammetry [39,226] has a major advantage over all other techniques as these strongly reduced species are readily oxidized by air contamination in classical sampling and sample handling procedures. *In situ* voltammetry completely obviates such problems.

In oxygenated waters, the reduction of O_2 produces two types of interferences for trace metal analysis:

(a) production of a current, 4–6 orders of magnitude higher than that produced by the trace metals of interest;
(b) pH increase at the electrode surface owing to the consumption of H^+ during the reduction of O_2 (equations (21) and (22)). For air-saturated unbuffered water, pH values may then rise to ~ 11 at the electrode surface.

Consequently, in the laboratory, dissolved oxygen is generally removed by purging the sample with N_2, He or Ar gas. Since such degassing also removes other gases, including CO_2, it may change the pH and metal speciation. To avoid this problem, N_2/CO_2 mixtures in well-controlled proportions can be used in the laboratory [332], but such a degassing procedure is not applicable *in situ*.

Type (a) interference has been overcome in laboratory applications of ASV techniques by using various modes of subtractive voltammetry [161,333–335] and/or SWASV. Their combination is required to determine trace metals at the 10^{-11}–10^{-9} mol.L^{-1} level in oxygenated solution, and they are briefly discussed below. Substractive voltammetry is performed by using two working electrodes to which different potential sequences and/or conditions are applied (e.g. Figure 12a and b). The correction is done by substracting the currents recorded at the two electrodes. Many of these studies make use of a bipotentiostat and a four-electrode system but stripping and background currents may also be recorded successively on the same electrode and the two current–voltage curves subtracted numerically. Different procedures have been proposed. In general, it has been found [154] that the substraction technique on a single macroelectrode can only be used if the electrode is subject to conditions as close as possible during normal and background current measurements; in that respect, the substractive mode (b) of Figure 12 was found to be the most appropriate. For mercury film microelectrodes [336], mode (a) (Figure 12), without stirring can be used. It has also been shown [337] that the interfering current due to the reduction of O_2 is very much minimized in SWASV, because this technique is little sensitive to irreversible processes particularly at high frequencies. SWASV has been successfully used in the laboratory for the determination of trace metals in oxygenated saturated seawater or acidified samples, by using classical macro-SMDE [338] and MFE [154,339]. It has also been used successfully for *in situ* measurements of trace metals in oxygen-saturated seawater, below the nanomolar range, using mercury film on either macro- or microelectrodes [116,154]

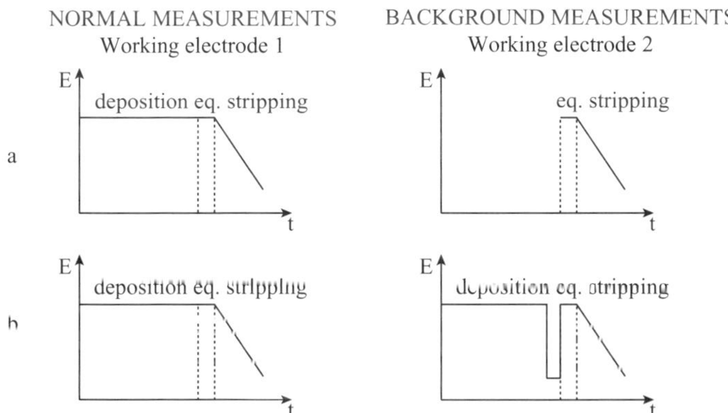

Figure 12. Two modes of substractive voltammetry. Measurement cycles applied to working electrodes 1 and 2. eq. = equilibration time. Note that stirring is only necessary for macroelectrodes

(section 5.2). Metal concentration measurements below nanomolar level requires the combination of both substractive mode and SWASV.

The above techniques substract the background current and minimize the alternating component of the O_2 current produced by SW modulation, but they do not eliminate the reduction process of O_2 and therefore the increase of pH at the electrode surface. They are thus applicable in well-buffered aquatic media, such as seawater (section 5.2.6.3), but not in freshwater with low buffer capacity ($\leq 10^{-3}$ mol L^{-1}). In such cases a large pH increase at the electrode surface may lead to the formation of hydroxide, carbonate or sulfide precipitates during the stripping step [200]. Voltammetric peaks (in particular those of Pb and Cu) may then be drastically deformed or even disappear. In such cases, either (i) a pH buffer must be added to the solution, which is not recommended for ultratrace analysis, because of possible contaminations, or (ii) O_2 should be eliminated from the sample. A number of systems (Table 11) have been proposed for laboratory on-line oxygen removal from solutions, based on chemical, electrolytic, photochemical or physical methods, in particular in combination with luminescence spectroscopic detection or liquid chromatography coupled to UV or amperometric detectors. Chemical methods (Table 11) are based on the reduction of oxygen by a reducing agent added to the solution (e.g. sulfite in alkaline solutions), immobilized in a reactor (e.g. Zn–oxygen scrubber column), or both (e.g. glucose addition to a solution flowing through a reactor containing co-immobilized glucose oxidase and catalase). These methods, as well as those based on the reaction of oxygen with organics, such as citric, formic and oxalic acids under UV irradiation (Table 11), are not appropriate for trace metal analysis since contamination may be introduced and the addition of reagent may change metal speciation. On-line electrolytic reduction of oxygen (Table 11) is not appropriate either since the test metal ions will also be removed at negative potentials. In contrast, on-line oxygen removal systems based on diffusion through membranes (Table 11) have been succesfully used for laboratory applications and is promising for in field and *in situ* adaptation. In those methods, the sample is pumped along a semi-permeable membrane [346–348] or tubing [349–351] surrounded by chemical reducing agents or inert gases at reduced pressure. Such systems have been used on site [155] to analyse total metal concentration after acidification and UV irridiation. For *in situ* analysis and speciation in unperturbed samples, optimal oxygen permeable material and efficient chemical scavenging solutions must be found, to maximize the rate of O_2 removal while minimizing the diffusion of CO_2 and thus pH variation in the test sample. Such a development has been performed [345], by using glucose/glucose oxidase to reduce O_2 outside the diffusion tubing. To eliminate O_2 more selectively than by pure diffusion, carrier-aided transport through a hollow fibre-supported liquid membrane (Chapter 10) could be used (Table 11) [345,352,353]. This technique, however, needs further development before being applied routinely.

Table 11. Major modes of elimination of dissolved oxygen or minimization of its interference in voltammetry

Technique	Comments (advantages and *drawbacks*)	References
Off-line removal		
(1) Purging with inert gaz (N_2, He, Ar)	• most often used in lab condition • *not applicable in situ* • *pH change due to concommitant elimination of CO_2*	5 5
(2) Purging with $N_2 - CO_2$ mixture	• as (1) but pH = const. owing to CO_2/HCO_3^- buffer	332
On-line O_2 removal by reaction in the sample		
(3) Chemical consumption of O_2 by added reductive agent (e.g. SO_3^{2-}, Zn°, glucose)	• efficient • *possible interference of reagent with voltammetric signal* • *risks of contamination due to large reagent additions* • *speciation not possible; total concentration measurements only* • *often difficult to apply in situ*	340–342
(4) Photochemical reduction by organic compounds (e.g. UV with citric or oxalic acid)	• as (3) • simultaneous elimination of organic matter	343
(5) Electrochemical reduction of O_2	• efficient; none of the drawbacks of (3) and (4) but: • *simultaneous removal of some of the test trace metals* • *applicable only to well pH buffered media*	344
On-line O_2 removal without reaction in the sample		
(6) Diffusion through hydrophobic membranes	• applicable to both speciation and total concentration measurements • *possibly less efficient as methods (3)–(5)* • *possible pH increase due to concommitant CO_2 removal*	154,155, 345–351
(7) Carrier aided diffusion through supported liquid membranes	• as (6) but • more efficient than (6) • potentially selective to O_2 transport (no pH change) • *needs further improvements before application*	352,353
Voltammetric oxygen signal minimization without O_2 removal		
(8) Substraction of background current (including O_2 signal)	• applicable *in situ* • measurement of both speciation and total concentration possible • no sample perturbation • *only applicable to anodic stripping techniques* • *only applicable in well-pH-buffered waters*	154,161, 333–335
(9) Square wave ASV with high frequency ($\geq$ 50 Hz)	• as (8); in addition • *only fast redox system can be analysed*	154,338, 339

5.1.3 Elimination of Interferences Due to Adsorption of Organic and Inorganic Compounds

A major problem of direct voltammetric measurements in complicated matrices such as natural waters is the fouling of the working electrode surface by organic biopolymers and inorganic colloids (section 3.4). In particular, it may (i) increase significantly the capacitive current [203], (ii) impede the electron transfer of redox reactions and thus decrease or suppress the voltammetric signals, and (iii) induce surface complexation of metals and thus modify peak currents and potentials [4,183]. Several studies of the effects of organic compounds on voltammetric curves have been published for the DME [354], HMDE [183,203,355,356] and the MFE [357] (see ref. [4] for a review). Several methods have been proposed to eliminate the adverse effect of organic substances on voltammograms. They are briefly outlined below.

The oxidative elimination of organic compounds by ultraviolet irradiation [358] is commonly used in the laboratory. This procedure is too difficult to apply *in situ*, but Brainina and co-workers have proposed an alternative electrochemical method for the oxidation of organic matter that can be used *in situ* in seawater [40,359]. Oxidants are produced at a graphite anode by the following reactions :

$$2Cl^- \rightarrow Cl_2 + 2e^- \qquad (23)$$

$$Cl_2 + H_2O \rightarrow HCl + HClO \qquad (24)$$

$$2H_2O \rightarrow O_2 + 4H^+ + 4e^- \qquad (25)$$

Oxidation of organic matter is facilitated owing to the acidification of the solution by reactions (24) and (25). Trace metals can then be measured voltammetrically in the organic-free solution on an MFE, immediately after the oxidation step. This mode of elimination of organic matter is useful for measuring total metal concentration in seawater. It is not suitable, however, for speciation measurements or freshwater analysis. In addition it does not protect the electrode against fouling by inorganic compounds.

Thin semipermeable protective membranes (typical thickness = 5–20 μm), covering Hg film electrodes, have been proposed to prevent the diffusion of interfering compounds toward the electrode surface, by size exclusion and/or electric charge repulsion effects. The most commonly used thin protective membranes are cellulose acetate [82,360,361] and Nafion [362–364]. However other thin membranes such as composite Nafion–cellulose acetate [219], polyaniline [365], poly(ester sulfonic acid) [366] and poly(ethyl 3–thiopheneacetate) [367] have been also proposed. Although some interesting results have been obtained with cellulose acetate and Nafion, both in laboratory tests with synthetic solutions containing differents surfactants [354,368], and in direct voltammetric determination of trace metals in body fluids [363], these membranes are not effective enough for complete elimination of fouling in direct measurements in natural water samples containing small molecular weight humic and fulvic acids [368,369]. Another problem is that most of the membranes proposed are not fully inert towards the analyte and may cause a memory effect [360,368]. This is the case for instance with Nafion coatings on which trace metals may be adsorbed by electrostatic attraction. In addition, the preparation modes of these membranes is not well controlled, which limits the reliability of any protective effect and therefore of the measured signal.

It has been shown that *'thick' membranes* have special advantages in the protection of microelectrodes (section 2.2.1; [192,216]). An Hg-plated microelectrode (radius of a few micrometres) has been covered by an agarose gel layer that is relatively thick (200–400 μm) compared with the electrode size and the diffusion layer thickness formed during the voltammetric measurement (section 5.1.4. for details). It has been shown that pure agarose gel is fully inert towards mono-and divalent cations as well as anions (i.e. they diffuse freely through it). On the other hand, systematic tests on organism culture media and river water samples, as well as *in situ* tests in a eutrophic lake (section 5.2.6.1), have shown that the gel is efficient for protecting the electrode against polysaccharides [216] released by organisms, river clay particles [192, 216], lake-borne iron oxyhydroxides [117] and some fulvic compounds [192], although the effect might depend on the nature of fulvics in the last case. The properties of this gel have been used to develop a novel gel-integrated microsensor, as explained below.

5.1.4 Gel-integrated Microsensors

In fact the thick gel mentioned above plays an even more important role than just as a protective layer against fouling. The combination of the gel with a voltammetric microelectrode (Figure 13c) should rather be seen as an integrated microanalytical system that performs the following two separate operations.

- The gel ensures (i) a separation of the electroactive species (mobile and labile) from most of the other species, by dialysis equilibration between the gel and the test solution, and (ii) well-controlled physical and chemical conditions at the electrode surface (see below). Both actions are important in facilitating the interpretation of voltammetric data.
- The microelectrode inside the gel detects the separated species. It is worth noting that macroelectrodes cannot be used in such a system for two reasons. First the diffusion layer thickness at the electrode surface during voltammetric detection should be much smaller than the gel thickness (Figure 13C); very thick gel layers (> millimetres) should then be used with macrosized electrodes, which would imply very long equilibration times with the test solution. Even more important is the fact that solution stirring is required to maintain a constant reduction flux in the deposition step of ASV on macroelectrodes but is not possible inside the gel. With microelectrodes a constant reduction flux is obtained in quiescent solutions as a result of spherical diffusion (sections 2.1.3 and 2.2.1; [371]), and stirring is not required.

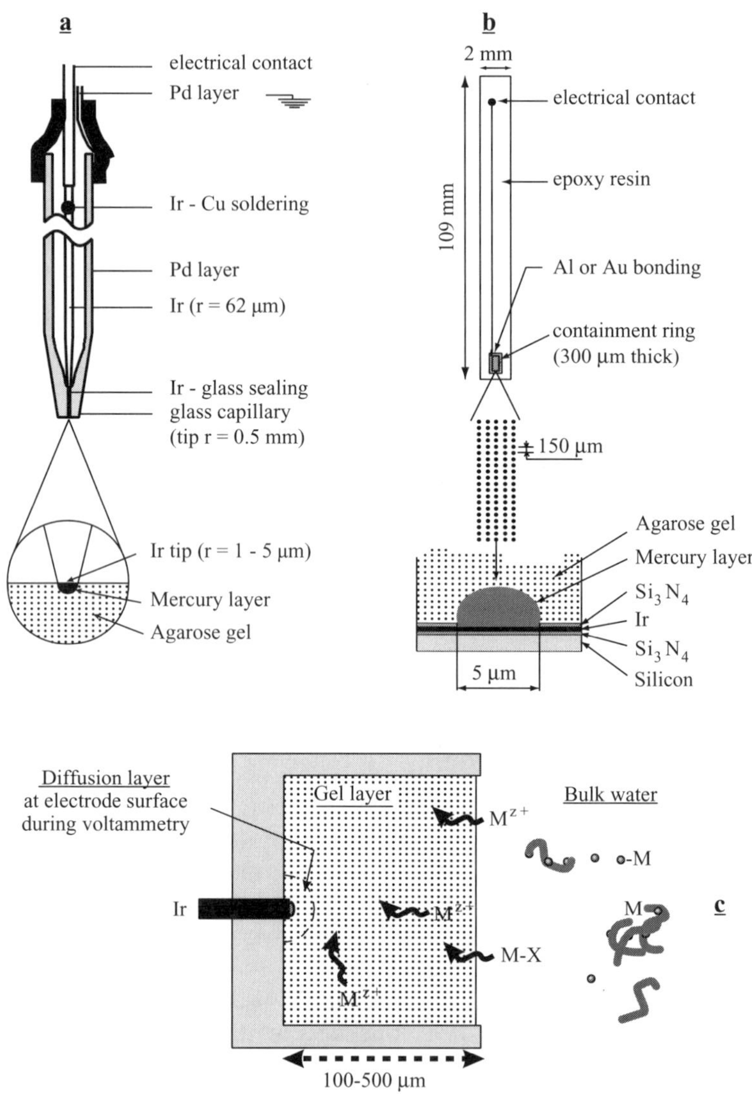

Figure 13. Schematic description (c) [370] of the gel-integrated microelectrodes based on electroetched Ir wire sealed in glass capillary (a) [192], or microelectrode array produced by microtechnology (b) (modified from ref. [216]).Only small species penetrate the gel. After equilibration they are measured by voltammetry inside the gel (no gradient at the gel–water interface)

Two types of gel-integrated microelectrodes (GIMEs) have been developed (Figures 13a and b; [192,216]): the glass capillary-based *a*garose gel *m*embrane *m*ercury-plated *I*r-based micro*e*lectrode (μ-AMMIE; Figure 13A) and the *a*garose gel *m*embrane *m*ercury-plated *I*r-based microelectrode *a*rray (μ-AMMIA; Figure 13B) fabricated by microtechnology. The details of fabrication steps are given in refs [93,104,192,216]. Briefly, the capillary-based microelectrode is built by sealing an electroetched Ir wire with a final tip diameter of few micrometres in a shielded glass capillary followed by mechanical polishing [93]. The microelectrode arrays are produced by means of thin film technology on chips and photolitographic techniques (Chapter 12; [104,216,217]). Their geometry may be changed. 5×20 interconnected iridium microdisc electrodes with a diameter of 5 μm and a centre to centre spacing of 150 μm have been used in ref. [104], with the array surrounded by a 300 μm thick Epon SU8–8 ring, to contain the gel. Both array and single microelectrodes were covered with a 1.5 % LGL type of agarose gel membrane. The use of high purity gel (< 0.03 % sulfur and ash) is essential. Mercury hemispheres are plated onto Ir substrate by electrochemical reduction of Hg^{2+} or reoxidized electrochemically in KSCN solution for their renewal. Both processes can be performed through the gel layer; in both cases the currents are recorded and the diameters of the mercury hemispheres are determined from the corresponding electrical charge. Electrochemical yields for Hg hemisphere formation are close to 100 % which ensures high reproducibility of drop size and surface area (variability ≤ 5 % [93,104,192]). A given agarose membrane can be used over an extended period of more than 1 month and it has been shown that diffusion through it is independent of pressure up to 6×10^4 kPa (600 bar) (Figure 14D; [216]). Hg hemispheres are stabilized by the gel and can be used for at least up to 10 days without renewal ([345], Figure 14C).

Measurements are performed in two successive steps: (a) equilibration of the agarose gel with the test solution (typically 5 min for a membrane thickness of 300 μm) and (b) voltammetric analysis inside the gel. Examples of sensitivity and reproducibility are given in Figures 14 and 17 and Table 12. The major advantages of GIMEs for *in situ* measurements are the following [117,192, 216,370]:

(a) natural biopolymers and inorganic colloids are efficiently excluded from the gel and do not interfere with voltammetric measurements ([192,216], sections 5.1.3 and 5.2.6.1);
(b) well-controlled molecular diffusion of mobile species occurs in the gel, i.e. ill-controlled hydrodynamic conditons of the test water body do not influence the voltammetric signal [192];
(c) well-controlled chemical conditions (such as pH buffering) can be maintained in principle inside the gel and thus at the electrode surface;

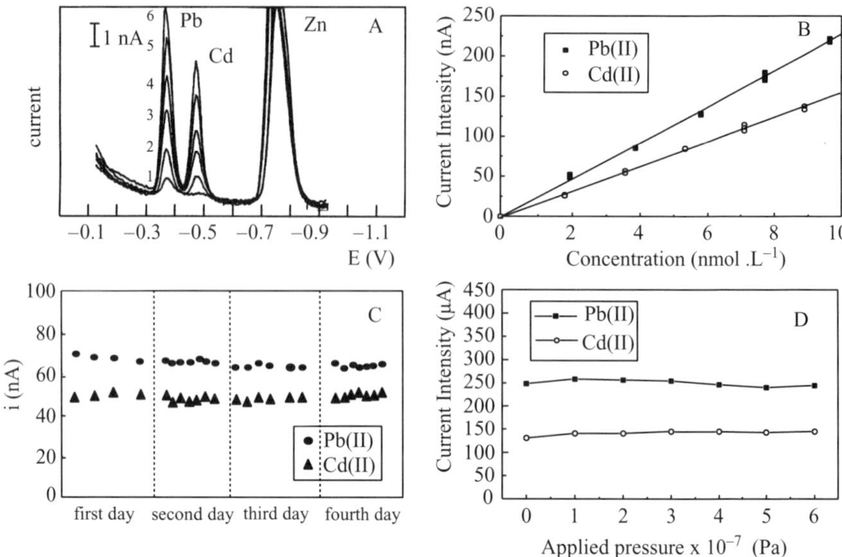

Figure 14. Examples of data for Pb(II) and Cd(II) measurement at the nanomolar level, showing the reliability of the sensor. (A) Example of SWASV curves for direct measurements (curve 1) of Pb(II) and Cd(II) in Arve river (Switzerland), filtered on 0.2 μm membrane ([Pb]=0.8 nmol L^{-1}; [Cd] = 0.35 nmol L^{-1}) and the corresponding standard additions (curves 2–6); μ-AMMIE. (B) Triplicate calibration curves in freshwaters; μ-AMMIA. (C) Stability of the SWASV signal over 4 days, using the same Hg microelectrode (μ-AMMIE. (D) Influence of pressure on the SWASV measured with a μAMMIA) (1 bar = 10^5 Pa) [93,104,216]

(d) because of advantages (a)–(c), interpretation of voltammetric data in terms of speciation is significantly simplified compared with voltammetry in the absence of gel; in particular, discrimination between mobile and non-mobile species is easier (sections 3.2 and 3.3.1);

(e) the external medium is not modified by the voltammetric measurement. This is particularly relevant for *in situ* measurements in sediments, where the fluxes at the electrode might influence the sediment concentration profiles in absence of gel, as may be the case in DGT (Chapter 11). In a GIME this effect does not occur since the voltammetric detection is done after equilibration of the gel with the test medium and the voltammetric diffusion layer is small compared with the gel thickness. This simplifies interpretation of data in terms of environmental processes (see also section 5.3.2.).

To enable sound interpretation of voltammetric curves of trace metals in terms of speciation, the diffusion properties of free metal ion, and environmentally relevant mobile species inside the gel should be known. There is presently little available information for the mobile complexes [372], but systematic studies have been done with free metal ions and various gels [192,373–376] (see also Chapter 11). Diffusion measurements in agarose gels with Tl^{+}, Cd^{2+}, Pb^{2+}, NO_3^-, and SO_4^{2-} have shown that for these ions (i) diffusion is not restrained by complexation inside the gel, provided it is pure enough, but (ii) slowing down of diffusion may occur [192,216,373] as a result of physical effects, which depends on agarose concentration and type. Diffusion coefficients in 1.5 % LGL agarose gel may be typically 50 % lower than in water. Lower gel concentrations are not desirable because of gel structure irreproducibility and insufficient mechanical resistance. Fick's law is obeyed for diffusion in the gel [192,216,375,376], which enables accurate prediction of the response time of the system for equilibration of the gel with an homogeneous test sample such as water. Typically, 5 min is necessary for a gel thickness of 300 μm. It is worth mentioning that in non-homogeneous systems such as sediments, equilibration between the GIME and porewater may be much longer than 5 min owing to the slow desorption of the analyte from the solid phase of the sediment itself (sometimes $\geq$ days; chapter 11). The advantage of GIME is that it enables this slow release to be recorded, which is an interesting characteristic of the test medium. Since the gel thickness should be much larger than the voltammetric diffusion layer at the microelectrode surface, it might be difficult to decrease the intrinsic response time of the GIME to values much lower than a few minutes. Note that the full voltammetric measurement response time includes this equilibration time ($\sim$5 min) plus the ASV deposition time (typically 30 s to 15 min), which depends on the analyte concentration.

5.2 *IN SITU* VOLTAMMETRIC MEASUREMENTS, IN THE WATER COLUMN, IN THE ABSENCE OF SULFIDE

There have been very few attempts to develop submersible voltammetric probes. Most have been prototypes, usable for 1 day in surface water (i.e. depth < 15 m) and based on the immersion of a flow-through voltammetric cell or sensor connected through a long cable to a standard commercial potentiostat located at the surface. The first prototypes were based on flow-through cells and commercial mercury macroelectrodes and were reported in 1990 by Tercier *et al.* [154] and Brainina *et al.* [377]. The various reported systems are discussed below.

5.2.1 Flow-through Voltammetry of Seawater Using Macroelectrodes, with Sample Pretreatment

Brainina *et al.* [377] developed their system for total trace metal concentration measurement in seawater. Although some description of the system is given, it has not been published in full detail. It consists essentially of two units for (i) oxidation of organic matter by electrochemically produced Cl_2 and simultaneous sample acidification (section 5.1.3; [40,359]), and (ii) voltammetric measurement of trace metals. Both units are integrated into a flow-through submersible device. The system also includes the possible addition of an internal standard. A thin-layer voltammetric cell is used as a measuring unit and an impregnated graphite electrode with predeposited mercury coating is used as working electrode.

5.2.2 Flow-through Voltammetry Using Macro- and Microelectrodes, Without Sample Treatment

Such a probe was reported by Tercier *et al.* [154] and was composed of three sub-units: a flow-through plexiglas voltammetric cell, a plexiglas submersible housing, and a control box with its communication cord. Various working electrodes were employed: a sessile mercury drop electrode (SMDE), a mercury-plated glassy carbon-based macroelectrode (Hg droplets rather than Hg film are formed on glassy carbon) and an Hg-plated Ir-based microelectrode. To allow proper operation of the electromagnetic valves, pump and step by step motor down to 20 m, a nitrogen counterpressure (0.5 to 0.7 atm in excess of the pressure at the measurement depth) was imposed inside the submersible housing, from the surface through the communication cord. This probe has been successfully used to measure Cu, Pb, Cd and Zn *in situ* in seawater without sample pretreatment [154]. This prototype has shown the advantages and prospects of *in situ* determinations as well as the developments required for a routine probe. Trace metal concentrations as low as 3×10^{-11} mol L^{-1} could be determined in oxygen-saturated seawater by square wave anodic stripping voltammetry using the Hg-plated glassy carbon-based macroelectrode. Contaminations were found to be much less of a problem than in classical sample collection and handling for laboratory analysis, even by minimizing sample storage (1 h) and handling, and using the same voltammetric instrument for laboratory as for *in situ* measurements. Continuous monitoring for a period of 1 month from a coastal pier (San Diego,USA) revealed significant daily changes in metal concentrations depending on physical and biological parameters (in particular, temperature, tide, turbidity, and chlorophyll). These would have been very difficult to measure with classical methods, as their orders of magnitude were similar to contamination levels observed in laboratory measurements. The results in lake water pointed out that measurements in freshwater are more difficult than in seawater, because of the low ionic strength, the lower pH buffer capacity and the larger content of colloidal fouling material. They highlighted the necessity of developing a protected microsensor and a deoxygenation system coupled on-line with the voltammetric cell.

5.2.3 Potentiometric Stripping Analysis Using a Combined Sensor, Without Sample Pretreatment

In situ potentiometric stripping analysis (PSA) at constant current by direct immersion of a combination sensor including the reference, auxiliary and working electrodes, has been reported by Wang *et al.* [56]. The combined sensor was housed in a PVC tube, and connected to the end of a 15 m long shielded communication cable (Figure 15a). A gold cylinder working macroelectrode (100 μm diameter, 5 mm long), previously proved [379–381] to be convenient for the PSA measurement of Hg, Cu and Pb, has been used. Laboratory tests were done by performing standard addition calibration curves in uncontaminated seawater samples. Linear curves were observed for the three metals in the concentration range 50–150 nmol L^{-1}. Detection limits of 5, 10, 25 nmol L^{-1} (corresponding to a signal-to-noise ratio of 3) were estimated for 5 min preconcentration time for copper, lead and mercury respectively, with a standard deviation of 2.8 % for Cu^{2+} (based on 20 replicate measurements in 200 nmol L^{-1} Cu(II) solution). *In situ* analysis of Cu(II) was performed in surface waters of San Diego Bay (ca 0.6 m depth), at different locations along the bay, without pretreatment, by immersing the electrode from the side of a small boat. Copper concentrations in the range 20–60 nmol L^{-1} could be measured. Although these values are quite high compared with other coastal areas, they are in good agreement with previous measurements in San Diego Bay. Note that, like the other voltammetric techniques applied to

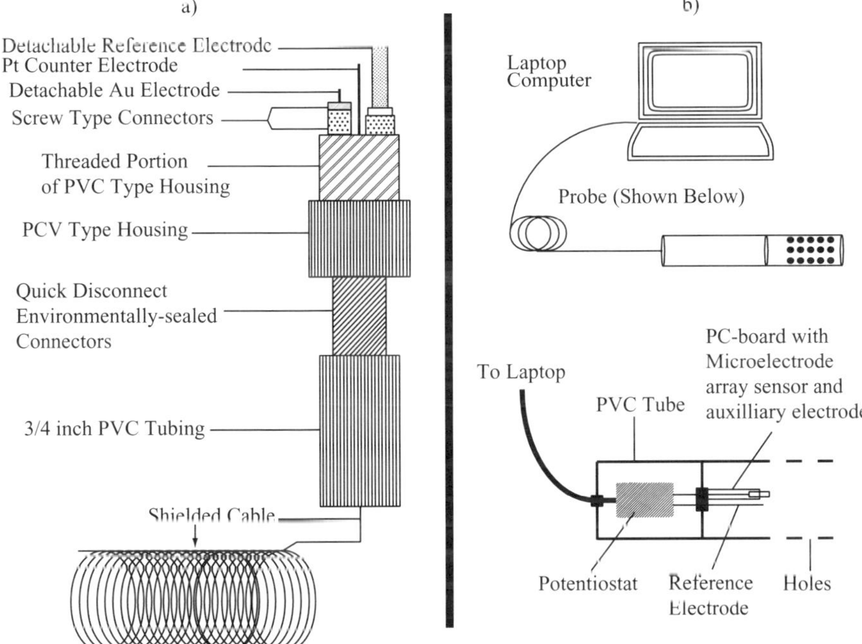

Figure 15. Schematic representation of the electrode assembly reported by Wang *et al.* [56] for *in situ* PSA analysis (a), and of the probe with integrated micropotentiostat reported by Herdan *et al.* [331] (b)

unmodified samples, PSA measures selectively the most mobile and labile species (sections 3.2 and 3.3.1). Interestingly, fouling of the electrode has been reported with a similar gold electrode and PSA, in raw river water, but not in seawater [380].

5.2.4 ASV on Directly Immersed Microsensor, Without Sample Treatment

A probe based on SWASV, with direct immersion of the sensor has been proposed by Herdan *et al.* [331], for fast *in situ* screening of heavy metals in groundwater. The working electrode consisted of a lithographically fabricated microelectrode array of 20 interconnected iridium discs of 10 μm diameter, electrochemically coated with mercury hemispheres [217] similar to that of Figure 13B. An on-chip iridium line auxiliary electrode and an external solid-state Ag/AgCl reference electrode were used (Figure 15b). As this probe has been developed for measurements in boreholes, the electrode assembly was physically protected against shocks by a PVC tube bearing holes close to the sensor (Figure 15b). The whole probe is formed by the submersible body of the electrode assembly on which on-chip miniaturized potentiosat and microcontroller are also fixed. The probe is linked to a laptop computer by a 33 m shielded electrical cable. Power to the potentiostat is supplied by four 9 V batteries. The utility of this probe for rapid on site screening of heavy metals was tested by conducting a proof-of-concept field demonstration at a metal-contaminated landfill site, located at a military area (Bedford, MA, USA). The system was used to measure, by SWASV, the mobile and labile fraction of Cu(II) and Pb(II) *in situ* and *on site* and their total concentration after acidification to pH 2 on site and in laboratory. Cu(II) and Pb(II), in the range 0.06–0.2 $\mu mol\ L^{-1}$ and 0.15–1.5 $\mu mol\ L^{-1}$, respectively, have been measured *in situ* in three different wells at 6–9 m depths. Comparison of on site and *in situ* measurements with laboratory electrochemical and ICP analysis showed that the tested probe can give order of magnitude estimations of metal concentrations and their evolution trends. Interferences probably due to adsorption of suspended particles were observed in some cases. The great novelty of this probe is the integration of a miniaturized potentiostat on the sensor body, which greatly reduces the dimensions of the whole probe ($\oslash = 5$ cm). Its limitation is that the noise of such a potentiostat is much higher than that of classical potentiostats, which limits the sensitivity. Lower reproducibility and accuracy were also observed in comparison with classical devices, which might be linked to the poor renewal of solution with the present protective cover, as well as to the fact that the electrodes are not protected against fouling. The authors claim that, by adding an antifouling membrane on the sensor, this probe may be useful for rapid, low-cost, screening of sites contaminated with heavy metals. Although further improvements are needed, the development of micropotentiostats is most interesting for *in situ* measurements.

5.2.5 Flow-through Voltammetry Using a Gel-integrated Microelectrode, in Unmodified Samples; the Voltammetric *In Situ* Profiling System (VIP)

The systems described in sections 5.2.1–5.2.4 have shown the feasibility and usefulness of *in situ* voltammetric probes, but, in most cases, they have insufficient sensitivity for trace metal determinations in unpolluted waters, and none of them can be used for continuous monitoring over extended periods of time, or for large depth profiling. The VIP system tries to overcome these problems by means of an improved sensor and advanced microprocessor and telemetry technology. It is built to enable measurements up to 500 m depth [116, 117]. The heart of the submersible probe is the agarose gel-integrated Hg-plated Ir-based microelectrode or microelectrode array described in section 5.1.4 (Figure 13). The whole profiling system includes several units (Figure 16): the voltammetric submersible probe, a complementary multiparameter submersible probe (for depth, temperature, conductivity, pH, oxygen), a calibration deck unit, and a surface deck unit connected to an IBM compatible PC. The voltammetric probe itself includes the following modules: an electronic probe housing, a pressure-compensated flow-through plexiglas voltammetric cell (internal volume = 1.5 ml) with a platinum ring auxiliary electrode and a Ag/AgCl/sat. KCl gel reference electrode, a pressure case base incorporating the preamplifier for the voltammetric microsensor and a sampling submersible peristaltic pump. The electronic housing contains all the hardware and firmware necessary to manage the voltammetric measurements, the data transfer by telemetry and the interfacing of the multiparameter probe, the calibration deck unit and the submersible peristaltic pump. A management software allows the user to control and configure the operating conditions for the voltammetric technique, the data acquisition, the calibration and automatic maintenance operations, in order to make maximum use of voltammetric capabilities (sections 2.1 and 3), albeit in a user friendly manner. Data files are stored in a non-volatile memory having its own battery which ensures high data retention and protection. The multiparameter probe allows control of the exact depth of the voltammetric probe and simultaneous measurement of the complementary basic major parameters (temperature, conductivity, pH, O_2) which are necessary for both speciation and environmental interpretation. The calibration deck unit enables the following procedures to be performed in the laboratory or on site: (i) the renewal of the microsensor Hg layer, (ii) the calibration of the probe, and (iii) the analysis of standards or collected samples (possibly after chemical modification). The surface deck unit powers and interfaces the measuring system, by telemetry, with a PC. The telemetric signal is superimposed to the system power supply and flows along the probe holding cable. This unit allows an autonomy of about 35 h and can be recharged either in continuous mode using solar power or after use. A detailed technical description of the VIP system is given in ref. [116].

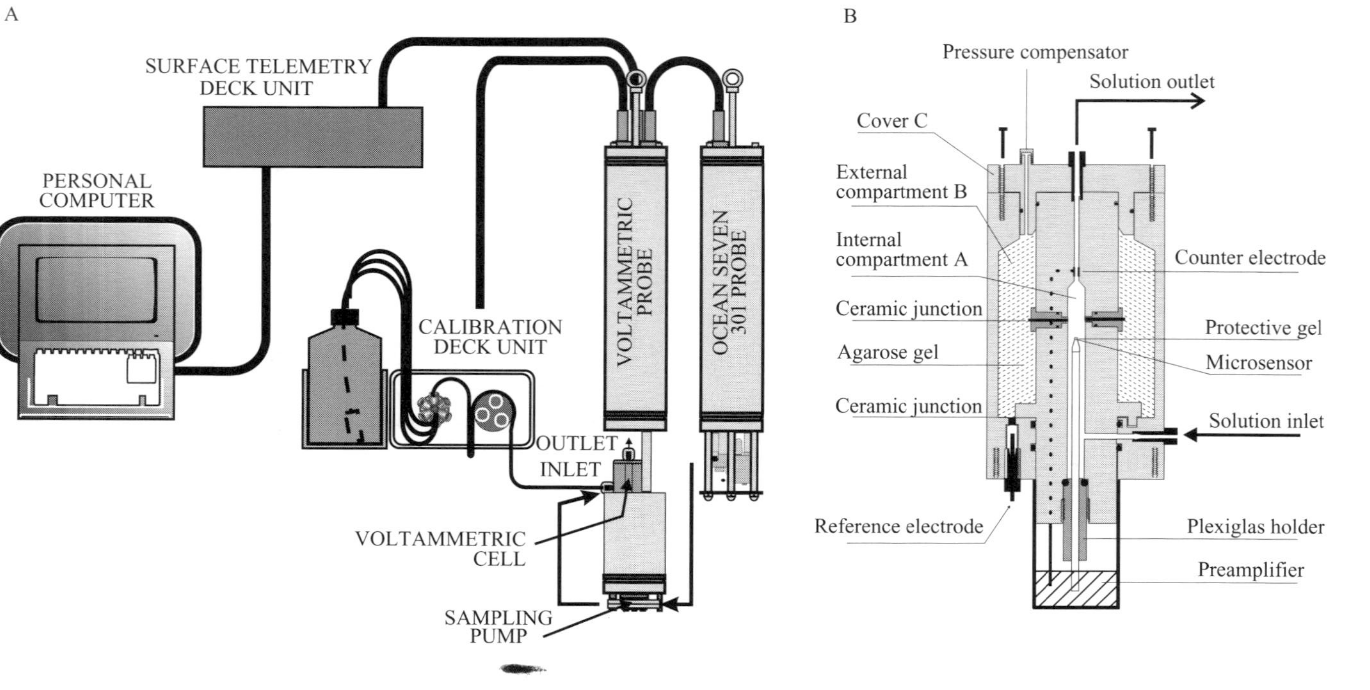

Figure 16. Schematic representation of the voltametric *in situ* profiling system: VIP (A), and the details of its flow-through cell (B) modified from refs [116,117]

5.2.6 Examples of Applications of the VIP System

Until now, detailed systematic tests of *in situ* voltammetric measurements have only been reported with VIP. Hence evaluation of the performance of *in situ* voltammetry will be based on these results. Additional specific information obtained with other systems can be found in the corresponding sections.

5.2.6.1 Stability and reliability of voltammetric measurements: Mn(II) in anoxic water Mn(II) concentration profiles were measured within the anoxic hypolimnion of the shallow eutrophic Lake Bret (Switzerland), with the objective of testing the stability, accuracy and reliability of the measurements performed with the VIP system [117]. For this purpose, *in situ* VIP measurements with microelectrode array were compared with voltammetric measurements performed on site (on a platform at the lake surface) with microsensor arrays. In all cases measurements with and without protective gel were performed; the samples for surface measurements were pumped from exactly the same depth as the VIP position and transferred into an air-tight cell, thermostated at 20 °C. Comparison was also done with ICP-AES analysis performed in the laboratory, on the same acidified samples (section 5.2.6.2).

In situ measurements were done without changing the Hg layer during the 3 days of operation, including calibrations of the probe the day before and the day after each deployment. Calibrations were performed by standard additions in 0.2 μm filtered lake water in otherwise the same conditions as for on site measurements (see above). Calibration mode is an important problem which is discussed in more detail in sections 5.3.2.4 and 6. The major results of these studies were the following.

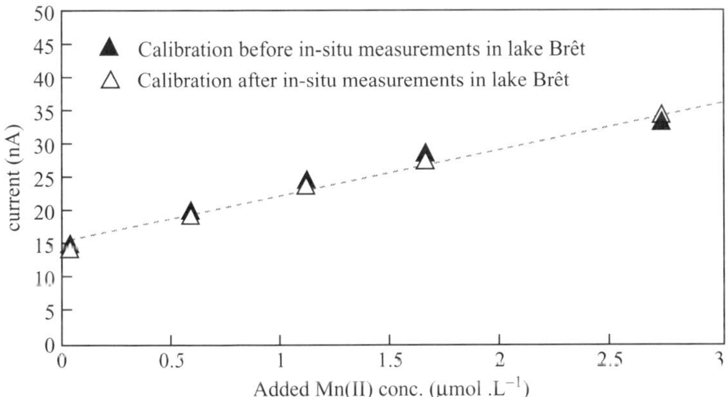

Figure 17. Calibration curves of Mn(II) obtained by standard addition in lake water, before and after deployment of the probe [117]. All measurements in triplicate

- Calibration curves were reproducible before and after deployment (Figure 17), and from one deployment to another. Standard deviations of the slopes of calibration curves were in the range 5.2–6.9 % (95 % probability) for seven calibrations of each of the following operation modes in various deployments: SWASV on μ-AMMIE, SWCSV on μ-AMMIE and SWASV on a μ-AMMIA. This demonstrates the good reproducibility of the Hg layer formation, its stability and the absence of memory effects during deployment.
- Comparison between voltammetric on site measurements with and without gel (Figure 18a) clearly shows the importance of the protective role of the agarose gel. Concentrations that are biased to systematically too low values were obtained for the unprotected microsensor. A detailed study [117] showed that the peak current attenuations observed with unprotected sensors are due to adsorption on Hg, of lake-borne iron(III) hydroxide colloids [382]. It must be emphasized that, apart from the peak current lowering, no other perturbation in voltammetric peaks (peak potential or shape) was

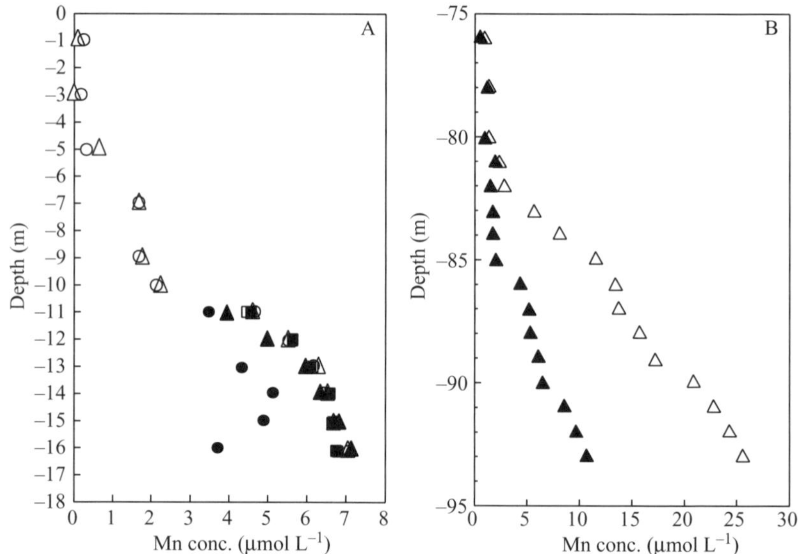

Figure 18. Concentration profiles of Mn(II) measured *in situ* with the VIP, in Lake Bret, Switzerland (August 20, 1997) (A), and Lake Lugano, Switzerland (November 14, 1996) (B). Explanations: see text (modified from ref. [117]). (A) ■, *in situ* SWASV measurements (VIP–μ-AMMIA); ▲, on-field SWASV measurements (μ-AMMIA); ●, on-field SWASV measurement (microelectrode array without gel); □, ICP-AES measurements on ultracentrifuged samples (30 000 rpm for 15 h); Δ, ICP-AES measurements on samples filtered on 0.2 μm filters; ○, ICP-AES on raw water samples. (B) ▲, *In situ* SWCSV measurements (VIP–μ-AMMIE); Δ, colorimetric measurements on samples filtered on 0.45 μm pore size membranes

observed. Thus measurements in the absence of a protective gel would have led to significant underestimations of Mn(II) concentrations, in variable proportions depending on Fe(III) hydroxide concentration in the water, i.e. to wrong values *and shapes* of the concentration–depth profiles in the lake.

- Voltammetric current varies with temperature (3–8 % $^{\circ}C^{-1}$, depending on the technique, analyte and conditions used), which is of significant concern for concentration profiling in water columns where temperature may vary from 4 to 25 °C (sometimes more), depending on system and season. Temperature calibration can be performed in the laboratory with synthetic solutions [407]. The results of Figure 18a show that such calibrations can be used for correcting *in situ* Mn(II) peaks, since on site thermostated measurements and corrected data obtained from *in situ* measurements at variable temperatures (between 4 and 14 °C) give very similar concentrations.
- The good correspondence (Figure 18A) between the concentration values obtained from *in situ* measurements after temperature correction, and both on site measurements and ICP-AES analysis (see section 5.2.6.2), also suggests that valid information can be obtained by *in situ* measurements, and in particular that pressure has no effect on the results, as laboratory tests had also shown (Figure 14).

5.2.6.2 Intercomparison of methods and speciation: (a) Mn(II) in an anoxic lakes; (b) Cu(II), Pb(II), Cd(II) and Zn(II) in deoxygenated freshwater
Validation of *in situ* voltammetric measurements in a complicated medium by intercomparison with other methods is far from easy, as differences may arise both from (i) wrong measurements with the new (*in situ*) method or (ii) different sensitivities of the various methods to the different species of the test analyte. In particular, for metal ion analysis, voltammetry is mostly sensitive to the mobile metal species (< 4 nm; section 3.2), while the other detection techniques which can be used for comparison such as AAS or ICP-MS have little or no sensitivity to metal speciation. Comparisons must therefore include size fractionation, e.g. by ultrafiltration on low porosity membranes or by ultracentrifugation. Even when artefacts of both methods are minimal, none of these techniques has size limits which correspond exactly to that of voltammetry. Consequently, it is impossible to get completely unambiguous results from such intercomparisons. Careful studies may, however, give useful information. Two examples are briefly explained below.

Intercomparisons of methods were performed for the analysis of Mn(II) in anoxic water of Lake Bret, Switzerland (depth 11–18 m; Figure 18a) and Lake Lugano, Switzerland (80–95 m; Figure 18b) [117]. Laboratory measurements were done by AAS, ICP-AES or colorimetry, on acidified (pH=2) samples of raw water, filtered water (0.2 or 0.45 μm pore size) and ultracentrifuged water (124 000 g for 15 h, corresponding to a size limit of ~ 6 nm [383]. Figure 18A shows that in Lake Bret all the methods give the same results, which suggests that in this case (a) the whole Mn(II) is voltammetrically mobile, (probably most of it is in the Mn^{2+} form), and (b) *in situ* voltammetry provides valid results compared with classical laboratory techniques. By contrast, Figure 18b shows that colorimetry on samples filtered through 0.45 μm membranes give higher values than *in situ* voltammetry. Other available information [117] indicates that the differences were due to the resuspension from the sediments of Lake Lugano of colloidal Mn (1 nm < Mn < 450 nm) which is measured by colorimetry but not by voltammetry (section 3.2). These two cases exemplify the additional information which can be gained from *in situ* voltammetry: by combining it with spectrometry on both filtered and raw water, mobile, colloidal and particulate metal species can thus be determined.

Comparison of voltammetric data with methods not (or slightly) affected by speciation can only be done in conditions where total metal concentrations are measured. For metals such as Zn, Cd, Pb and Cu, this may be obtained by acidifying the samples to pH $\leq$ 2. In such cases, the above spectrometric techniques and voltammetry should give the same results. In fact even more information can be gained by voltammetry (Figure 19) by acidifying the sample gradually and recording the change in peak current with pH [34]. At sufficiently

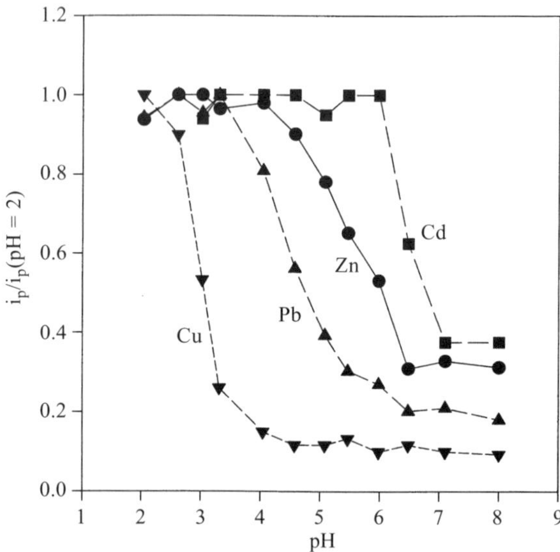

Figure 19. Change with pH of Cu(II), Pb(II), Cd(II) and Zn(II) SWASV peak currents on μ-AMMIA. Measurements performed directly in Arve river water, filtered on 0.2 μm filter, without other sample treatment except for acidification. $i_p(pH{=}2)$ = peak current at pH = 2. i_p = peak current at any pH. Total concentrations: Zn(II) = 28.8 nmol L^{-1}; Cd(II) = 0.27 nmol L^{-1}; Pb(II) = 3.9 nmol L^{-1}; Cu(II) = 16.1 nmol L^{-1} [34]

acidic pH values a plateau is obtained which corresponds to total metal concentration also measured by the other methods.

At higher pH, the proportion of metal bound to non-mobile colloids increases with pH, and the current decreases. The pH corresponding to the inflection point of the jump ($pH_{1/2}$) gives useful information on the adsorption energy of the metal on the natural particles or colloids [34,384]. Figure 19, for instance, shows that in the Arve river (Switzerland), surface complexation energy by colloids increases in the order Cd(II)<Zn(II)<Pb(II)<Cu(II) since the corresponding $pH_{1/2}$ decreases in this order. This is fully in line with the accepted complexation properties of these metals. The interest of voltammetry, is that (i) mobile species can be measured specifically in presence of colloidal and particulate species by *in situ* measurements and (ii) detailed speciation study can be performed with minimum sample handling (e.g. acidification).

5.2.6.3 Sensitivity limit in oxygenated water: Cu(II), Pb(II) and Cd(II) in seawater As mentioned in section 5.1.2 oxygen is an interferent for the measurement of trace metals because of (i) its high reduction current, and (ii) the pH increase at the electrode surface due to its reduction. The latter problem is overcome in well-pH-buffered systems, such as seawater. The efficiency of substractive mode SWASV in eliminating the first problem (section 5.1.2.) is consequently best tested in seawater where, in addition, trace metal concentrations can be very low.

The sensitivity of the VIP system was tested at two seawater sites: in the Venice Lagoon (Italy) and the Gullmar Fjord (Sweden) [116]. *In situ* measurements in Venice Lagoon were performed in three different stations: (1) Breda (industrial area), (2) Fusina (wide artificial navigation channel) and (3) Bocca di Porto (main entrance of the Lagoon). The *in situ* VIP measurements of Cd(II), Pb(II) and Cu(II) (average of three replicates), and the total metal concentrations measured by voltammetry in UV-irradiated, acidified samples in laboratory are given in Table 12. The mobile and labile fraction of trace metals measured by VIP only represents 10–20% of the total metal concentration. Similar proportions were observed in the Arve river (Figure 19, [34,93]). The colloidal and particulate fractions are thus predominant in both cases. Table 12 also shows that concentrations (as well as concentration changes) at the ppt level (10^{-11}–10^{-10} mol L^{-1}) can be measured *in situ* with a good reproducibility using the VIP system. Based on a signal-to-noise ratio of 2, the detection limits for Cu(II), Pb(II) and Cd(II) in seawater are 0.1, 0.05 and 0.05 nmol L^{-1}, respectively, for a preconcentration time of 15 min.

In Gullmar Fjord, two types of tests have been carried out: (i) *in situ* measurements at different depths and stations, and (ii) unattended, autonomous *in situ* measurements. Similar results and sensitivities were obtained as in Venice Lagoon. In unattended autonomous monitoring, measurements were performed every hour for 52 h. Similar sensitivity to that in manual operation

Table 12 Concentrations of the mobile fraction (c_m) of trace metals, measured *in situ* by voltammetry (VIP) in three locations of Venice Lagoon. Total concentration of these metals (c_t), measured by voltammetry after UV irradiation of acidified samples [116]

Station	Cd (II) (nmol L^{-1})		Pb(II) (nmol L^{-1})		Cu(II) (nmol L^{-1})	
	c_m (VIP)	c_t (lab.)	c_m (VIP)	c_t (lab.)	c_m (VIP)	c_t (lab.)
Breda	0.18 ± 0.04	1.33 ± 0.03	1.64 ± 0.11	3.03 ± 0.17	3.93 ± 0.86	10.57 ± 1.08
Fusina	0.25 ± 0.04	1.20 ± 0.01	0.42 ± 0.05	3.10 ± 0.10	1.56 ± 0.39	23.33 ± 2.60
Bocca di Porto	0.02 ± 0.005	0.23 ± 0.03	0.06 ± 0.01	1.0 ± 0.10	–	5.97 ± 0.81

was obtained and the tests showed the capabilities of VIP to record relatively fast changes [116].

5.3 *IN SITU* VOLTAMMETRIC MEASUREMENTS OF Fe(II), Mn(II), S(−II) AND TRACE COMPOUNDS IN ANOXIC/SULFIDIC WATERS AND SEDIMENTS

5.3.1 Properties of the Major Voltammetrically Active Components of Anoxic Waters

Measurements in the presence of sulfide are significantly more difficult than in its absence, for a number of reasons briefly discussed below. Difficulties result both from complicated processes at the electrode surface, and from the large number of chemical reactions that S(−II) may undergo in solution. In particular, it is readily oxidized by O_2 and other oxidants and produces a large number of sulfur species with intermediate redox states. It may also form complexes or solids of intermediate to very high stability, including polynuclear and colloidal species, with Fe(II) and many trace metals. Data interpretation must consider equally carefully the processes occurring in the bulk solution and at the electrode surface, and be based on detailed corresponding tests, to provide correct conclusions. It is not the purpose and largely out of the scope of this chapter to discuss the geochemistry of Fe, Mn and S elements in water and sediment. The reader is referred to the excellent, thorough review by Davison [225] and to ref. [384] for a general textbook. Field and *in situ* voltammetric measurements of concentration profiles of S(−II), Fe(II) and Mn(II) in lake waters have been reported in refs [117,184,185,207,224,226]. These elements have also been measured at millimetric resolution in salt marsh sediment cores on an amalgamated Au electrode [39,385] and *in situ* in lake sediment, on GIME (Section 5.1.4; [A. Tessier, unpublished results]). The following discussion is based on the whole of these data.

O_2, Mn(II), Fe(II) and S(−II) occur in anoxic waters, at intermediate concentrations (10^{-7}–10^{-3} mol L^{-1}). They are usually found in water or sediment layers of increasing depth in the afore-mentioned order. O_2 and S(−II) are usually not found together, but in some cases either O_2 and Mn(II) or Mn(II), Fe(II) and S(−II) may be present simultaneously at the same water layer. The predominance of these species is highly dependent on biological productivity, and the depths of the aforementioned layers follow significant seasonal and sometimes daily changes (Figure 13 in ref. [386]). Consequently, interpretation of voltammetric signals in anoxic/sulfidic waters should be based on a detailed knowledge of the redox processes involving each of the above species. Figure 20 gives an overview of the most important voltammetric peaks which can be obtained in anoxic lake and sediment waters with different voltammetric techniques and electrodes. Peak shapes (and sometimes potentials) may depend on techniques and this may be used to improve the desired selectivity for the test compound. In addition, as already mentioned, it is significantly more difficult to get reproducible results on solid or Hg film electrodes than on renewable electrodes (HMDE or DME), and great care must be used when interpreting results obtained with the former (section 5.3.2.). Nevertheless, Figure 20 points out some important general features regarding the use of voltammetric peaks for the determination of the concentrations of the corresponding compound (see below), the choice of appropriate analytical conditions (section 5.3.2.), and the interpretation of speciation (section 5.3.3.).

Analysis of O_2 and H_2O_2 They have been measured voltammetrically in water column [226] and sediment pore waters [39,385]. H_2O_2 is only measurable in the absence of Fe(II) because of peak overlap and because catalytic waves [1], which are difficult to control in natural medium, can be formed when both H_2O_2 and Fe(II) are present. In principle specifically designed microelectrodes (Chapter 2) are better for the measurement of O_2. Voltammetric determination, however, is useful in obtaining the concentrations profiles of O_2, Fe(II), Mn(II) and S(−II) simultaneously at exactly the same location. The maximum sensitivity for O_2 (a few μmol L^{-1}) will be limited by other major redox reactions, in particular the oxidation of Hg, which may occur at the electrode, in the same potential range (i.e. $\sim +0.2$ to -0.2 V; Figure 2A) as the peak of O_2 (~ -0.05 to -0.2 V; Figure 20A).

Analysis of Mn(II) Field measurements have been perfomed (Figure 20B–D) on lake water, by DPP on DME [207,223,224,226] and more recently by ASV on SMDE and SWASV on a GIME [117,154]; section 5.2.6. SWV on an amalgam Au electrode has also been used in salty marsh sediments [39,385]. Good analytical reproducibility has been shown for techniques based on

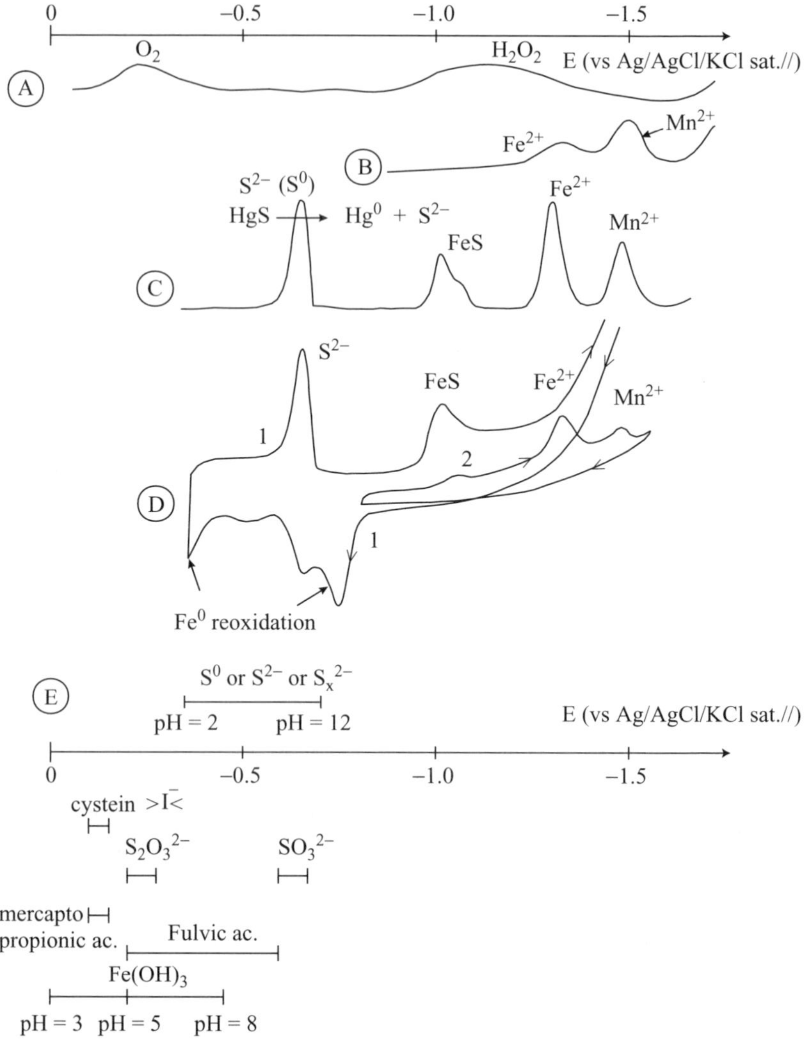

Figure 20. Examples of voltammetric curves in anoxic waters determined in different conditions. (A) O_2 and H_2O_2 SWV reduction peaks in a sediment core (amalgamated Au electrode: [39]); (B) Fe(II) and Mn(II) SWV reduction peaks in absence of S(−II) *in situ* in Lake Croche sediment, Québec (μ-AMMIE) [A. Tessier, unpublished results]. (C) DPP peaks of Mn(II), Fe(II), S(− II) and the 'FeS' species (see text) in hypolimnion of Lake Bret, Switzerland (on-line analysis on DME of water pumped to the lake surface; [224]); (D) curves 1 and 2, cyclic voltammetry on HMDE of the same water as in curve (C), with starting potentials at −0.4 V (curve 1) and −0.8 V (curve 2) [De Vitre R., Buffle J., unpublished results]; (E) ranges of peak potentials of important electroactive components of anoxic waters others than Fe(II) and Mn(II) ; for S(−II), S^0, S_x^{2-} and $Fe(OH)_3$, potentials increase negatively with pH (from [207,226,292,385,387])

DME, HMDE and GIME in lake waters. The sensitivity is slightly increased with SWASV (limit at 0.1 instead of 1.0 μmol L^{-1}), but high frequency or long deposition time cannot be used, as Mn(II)/Mn^0 is not a reversible system, and the solubility of Mn^0 in Hg is low. As Mn(II) is reduced at a rather negative potential (~ −1.5 V), the sensitivity is strongly influenced by the baseline which in turn is very dependent on the adsorption of compounds at the electrode surface (compare Figure 20D, curves 1 and 2; discussion below). This baseline problem may be most important in sediments, unless electrodes protected from adsorbing compounds are used.

Analysis of S(−II) It is based on the reaction of S(−II) with Hg:

$$S(-II) + Hg^0 \Leftrightarrow HgS_{ads} + 2e^-$$

When the potential of an Hg electrode is scanned from negative to positive values (so called direct scan), a peak corresponding to the above reaction occurs at ~ −0.6 V at circumneutral pH values (Figure 20C–E). The redox reaction is reversible, but the product (HgS) adsorbs at the electrode surface and modifies it. To increase sensitivity, S(−II) may be accumulated as HgS at the electrode surface, for a fixed period of time (seconds to minutes) at a potential less negative than −0.6 V (E > − 0.6V). After that period, the potential is scanned negatively ('reverse scan') and a peak is produced corresponding to the reduction of HgS_{ads} to Hg^0 and S(−II). The difficulty of this analysis is that HgS films that are too thick (thicker than a monolayer) give rise to double and triple peaks that are unusable analytically. In addition, thick films are not completely reduced during the scan so that the electrode surface may change from one analysis to the other. This problem must be checked carefully. It can be overcome by using fast direct scans at high concentrations of sulphide (>1–10 μmol L^{-1}) and reverse scans (albeit with accumulation time as short as possible) for lower concentrations. Although S(−II) has been analysed directly in sediment on a Hg/Au amalgam electrode [39,385], detailed studies of analytical reproducibility have only been reported for measurements in the water column by DPP and d.c. polarography [207,224,226], and a.c. polarography [293,388], using the renewable DME and voltammetric cells which enable immediate measurement after pumping. As S(−II) is very sensitive to oxidation by air after sampling, the adaptation of *in situ* voltammetry for reliable measurement of S(−II) is highly desirable. Voltammetry provides information which is different from and complementary to those given by either the sulphide ion-sensitive electrode or the H_2S electrode (Chapter 4; section 5.3.3).

Analysis of Fe(II) In the ***absence of S(−II)***, a peak occurs at −1.35 V for the reduction of Fe^{2+} and its labile complexes into Fe^0, and can be used for analytical purposes. The conditions of reproducibility have been well studied for field application in lake waters with d.c. polarography, DPP on DME and CV on HMDE (Figure 20C and D; [207,223,224,226]. The occurrence of a SWV peak on an amalgamated gold electrode in porewaters of sediment cores [39,385] and on GIME, *in situ*, in both sediment and lake water (A. Tessier, unpublished results) has also been demonstrated, but detailed analytical conditions have not been studied. Important aspects to consider are that the Fe^{2+}/Fe^0 couple is highly irreversible and Fe^0 is completely insoluble in Hg [389,390]. Consequently, slow voltammetric techniques (e.g. low frequency SWV, or preferably CSV) must be used, and ASV techniques are not applicable. In addition, since Fe^0 deposits at the electrode surface and is not easily reoxidized (section 5.3.2) care must be used to reduce as little Fe^{2+} as possible during the analysis on non-renewable electrodes in order not to change their surface.

In the ***presence of S(−II)***, a small size, voltammetrically inert 'FeS' complex of unknown stoichiometry, not dissociable in the time frame of voltammetry, can form in solution and reduce at the Hg electrode at ~ -1.1 V (Figure 20C and D). The chemical properties of this species and its electrode behaviour have been studied in detail [184,224]. Although they are not yet fully understood, a few characteristics are clear.

(i) Not only can this complex form in solution, but it can also be produced at higher concentrations at the electrode surface, when S(−II) is accumulated as HgS_{ads} by imposing an electrode potential $E > -0.6$ V (Figure 20D curve 1). During the subsequent negative scan, a high concentration of S(−II) is produced at the electrode surface by reduction of HgS_{ads}, and the local formation of 'FeS' is thus substantially enhanced (compare curves 1 and 2 of Figure 20D; for the latter no HgS_{ads} is formed). In such conditions, the 'FeS' peak is much larger than it would be if it only reflected the concentration in solution, and the Fe^{2+} peak may disappear. In the presence of S(−II), preconditioning the electrode at potentials < -0.7 V is therefore essential for correct measurements of both Fe^{2+} and 'FeS'.

(ii) Even when it is measured using the correct conditions, the current of the 'FeS' peak is not always simply proportional to the concentration of the 'FeS' complex, because the electrode process involves an adsorption step and possibly slow chemical kinetics [184,390].

(iii) The reoxidation of Fe° is easier in the presence of 'FeS' absorbed on the electrode (peak at ~ -0.7 V on Figure 20D curve 1; [390]), than in its absence (reoxidation peak at $E > -0.4$ V, [389]. This aspect is important for the conditioning of the electrode (section 5.3.2.).

Analysis of other species Other species which may be detected voltammetrically are the following (Figure 20E): $S_2O_3^{2-}$ [226,292,387], SO_3^{2-} [226], mercapto compounds (e.g. cystein and mercaptopropionic acid, [292]), elemental sulfur and polysulfides [292,293,388], and I^- [385]. Although these peaks can be used for reliable analytical measurements in the field or in the laboratory, detailed analytical studies of their use *in situ* have not yet been performed. If sulfide is present in solution, the analysis of trace metals discussed in sections 3 and 4 of this chapter will be affected. Many metals will form either metal sulfide colloids or particles not measurable voltammetrically, or strong complexes with concommittant significant negative potential shifts (see section 5.3.3.).

5.3.2 Optimizing Analytical Conditions and Data Interpretation

The above discussion clearly shows that direct *in situ* measurements of the above species is feasible but not straightforward. This is particularly so in sediments, where a high spatial resolution is needed, calibration cannot be performed *in situ*, and chemical transformation of the analyte before voltammetry is not possible. The demonstration of the occurrence of voltammetric peaks, as in refs [39,385] is thus potentially useful, but it does not imply that they can be used in a simple way for reliable measurement of concentrations. Before fully quantitative interpretation can be achieved, it is necessary to understand and interpret in detail the electrode process and subsequently to optimize the analytical conditions. Such a substantiated study is likely to be difficult. Unfortunately the feasibility of analytical measurements in sediments has been claimed in the literature [39,385] before performing any careful study of this type. This may lead to disappointment of users who are non-specialists of voltammetry. A few key aspects to consider for optimization of measurement conditions are given below.

5.3.2.1 Micro-versus macroelectrodes in sediments General advantages and limitations of micro- and macroelectrodes have been discussed in sections 2.2.1, 3.2, 3.3 and 5.1–5.2. It has been shown that microelectrodes have a number of advantages, in particular for analysis and speciation in the water column. Application to sediment deserves a brief discussion. In this case, the terms micro- and macroelectrodes are often used by comparison with the spatial resolution needed to interpret concentration profiles in terms of the microstructure of the sediment. In that respect, electrodes with *tips* $\leq 100\,\mu m$ can be referred to as microelectrodes. This definition, however, is different and less demanding than that based on voltammetric principles. In this latter case, micro- and macroelectrodes differ in the fact that spherical and linear diffusion respectively occur on the *sensing element* (Figure 3). As discussed in section 2.2.1, according to this definition (which is used throughout this chapter), the sizes of micro- and macroelectrodes are such that $r/\delta \ll 1$ (typically $r \lesssim 10\,\mu m$) and $r/\delta \gg 1$ (typically $r \gtrsim 100\,\mu m$), respectively. Thus, the amalgamated Au electrodes reported in refs [39,385], the size of whose sensing element is $\sim 100\,\mu m$ are macroelectrodes (with linear diffusion behaviour), contrary to what is claimed in the papers. Clear discrimination between voltammetrically micro- and macroelectrodes, with the sense used in this chapter, is important for several reasons, in particular:

- only microelectrodes can be used reliably in freshwater with low ionic strength, owing to their low ohmic drop;
- only microelectrodes are independent of hydrodynamic convection;
- only microelectrodes enable the preconcentration step of ASV, PSA or AdSV (Table 1) to be done reliably without stirring. This is essential for the analysis of low concentrations of S(−II), of mercapto compounds, thiosulfate, I^-, or trace metals in sediment, which cannot be stirred.

With microelectrodes it is possible to use gel coatings (section 5.1.4.). This greatly helps to minimize the fouling problem which may be drastic when measurements are performed *in situ*, especially in biologically active sediments.

5.3.2.2 Interpretation of voltammetric peaks Figure 20 shows that in an unperturbed anoxic water or sediment, a significant number of voltammetric peaks will be recorded simultaneously.This is a useful capability of the technique, but special care is required for interpretation. The following comments can be useful in that respect.

- O_2 and S(−II) may be in much higher concentrations than the other species and consequently may mask some of their peaks. As discussed in section 5.1.2. the signal of O_2 may be largely minimized in SWASV, but it is not the case in direct SWV. Measurement of trace metals by SWASV in non-sulfidic water or sediment is therefore possible, but measurement of mercapto compounds or thiosulfate in the presence of oxygen will probably be difficult. Similarly, in the presence of S(−II), analysis of S^0, Pb(II) and Cd(II) is likely to be difficult or impossible. (Note, however, that there is no specific information on these analyses.)
- Even in the absence of O_2 and S(−II), a large number of species may produce peaks in the range from −0.2 to −0.9 V. Attribution of a peak to a particular species must then be done very carefully. Intercomparisons with other methods are most useful.
- The baseline can significantly deteriorate in the potential range from −0.2 to −0.7 V, (Figure 20E) owing to the capacitive current resulting from the adsorption of fulvic or other types of organic matter (see refs [202,203] for detailed study of adsorption). In some cases a broad hump overlapping with other peaks is formed, in the presence of Fe(III) hydroxide, in the potential range from 0 to −0.45 V. It has been studied in detail in synthetic solution [206]. It results from the reduction of polynuclear Fe(III) hydroxo complexes, albeit with a complicated electrode process including an adsorption step. The peak is shifted by $\sim -100\,\mathrm{mV\,pH^{-1}}$ between pH=3 ($E_p \sim 0$ V) and

pH=8 ($E_p \sim -0.5$ V) (Figure 20E; [207]). i_p and E_p values of this peak can in principle be related to the stability and concentration of the corresponding Fe(III) species, but because it depends on adsorption and other complicated reduction steps this dependence is rather involved [206]. In most cases, the peaks or humps due to organic compounds and Fe(III) hydroxides interfere with other measurements and must be minimized, e.g. by using GIMEs (section 5.1.3).

5.3.2.3 Conditioning of the electrode As mentioned in sections 5.3.1 and 5.3.2.2, a number of electrode processes occurring in anoxic and sulfidic waters lead to the deposition of compounds at the electrode surface, in particular Fe° due to the reduction of Fe(II), adsorbed HgS or Hg_2I_2 due to oxidation of Hg in the presence of S(– II) or I^-, absorbed S^0 or S_x^{2-}, adsorbed Hg(II) mercapto or thiosulfato complexes, adsorbed organics or $Fe(OH)_3$. Such layers change the electrode surface. For electrodes with a renewable surface, such as DME or HMDE, such films will not be a major problem, but it is a key aspect to consider if reliable results are to be achieved with solid or mercury film electrodes. In some cases, the adsorbed compounds may drastically change the electrode process, and consequently the peak potential and current. The reader is referred to ref. [1] for an excellent detailed discussion of these effects, and to ref. [4] for a briefer description in the context of water analysis. An example of this effect is shown in Figure 20D, curve 1 and 2. Because of the adsorption of FeS in curve 1, the baseline current is drastically increased for $E < -1.5$ V, owing to a facilitated reduction of the electrolyte, and consequently the reduction peak of Mn^{2+} is masked and not measurable. Similarly in the presence of adsorbed 'FeS' (curve 1), Fe^0 is deposited at the electrode surface in a form which is reoxidized more readily (peak at –0.7 V) than in absence of absorption [390,391]. In the latter case, in particular at $E < -1.5$ V, Fe(II) is reduced to Fe^0 crystals which penetrate Hg without being wetted by it, so that reoxidation is incomplete and occurs at $E > -0.4$ V [389]. As a consequence, renewing the electrode surface after Fe(II) analysis is easier in sulfidic waters than in the absence of sulfide. Nevertheless, there is presently little information on this reoxidation process which should be studied in detail to enable optimization of electrode conditioning.

All the above processes must be considered simultaneously to select the optimun conditions of electrode conditioning. Such a detailed study remains to be done and optimum conditions will somewhat depend on the nature of the electrode. General rules, however, are given below.

- The electrode potential should not be maintained at too positive values between analyses (E should be < -0.7 V in the presence of S(–II)) to avoid oxidation of Hg and formation of adsorbed Hg salts; see above).

- Similarly, the electrode potential should not be too negative between analyses ($E \geq -0.85$ V) to avoid accumulation of Fe^0 (when Fe(II) is present) and Zn^0 ($E_p \approx -0.9$ V) on or in the electrode.
- A rest potential of −0.8 V is thus usually the best compromise. It must be realized, however, that some reduction (in particular of Cu(II), Pb(II) and Cd(II)) and adsorption of organic matter may occur at this potential. Performing a set of fast cyclic scans may help to ensure that the electrode surface is reproducible before analysis.
- Although each specific case must be studied for itself, it is not recommended to leave the electrode in open circuit between operations. In such a case, the potential will take a value for which the sum of reduction and oxidation currents is nil. The exact nature of processes occurring at the electrode are then ill controlled and they may change with time, depending on the biogeochemical processes occurring in the sediment.

5.3.2.4 Calibration Even with well-conditioned electrodes, the effects of a number of factors which influence the peak current and potential should be studied in detail, in order to ensure ultimately a correct transformation of peak currents into concentrations. These operations are collectively referred to here as calibration and are discussed in more general terms for *in situ* voltammetric measurements in section 6. The present section considers redox processes which are irreversible, quasi-reversible or influenced by chemical or adsorption reactions, as is the case for most test compounds in anoxic/sulfidic waters. The major factors which influence such processes are outlined below.

- *pH* influences significantly the potential (and often the current) of all the redox reactions discussed in this section (except the reduction of Mn^{2+} and Fe^{2+}). Variations from 0 to $-60\,mV\,pH^{-1}$ (sometimes more) may occurr, depending on the redox reaction. A peak potential may thus vary along the depth versus concentration profiles recorded in the same lake or sediment. Detailed knowledge of this pH dependance is thus essential for data interpretation.
- *Temperature* may have a very strong influence on irreversible, or quasi-reversible redox reactions [1], or redox reactions coupled to chemical reactions. Changes of 5–10 % $^{\circ}C^{-1}$ and sometimes more, can then be expected. Most of the processes discussed in this section fall in this category. Again, temperature may vary with depth in oceans, lakes and sediments. So a detailed knowledge of the temperature dependence of the peak of interest is a key requirement for rigorous analytical measurements.
- Influence of *ionic strength* must be studied in details for the general reasons given in section 6. In addition, it may play an important role, in particular at low ionic strength, on peak currents and potentials of quasi-reversible or irreversible processes, and on processes coupled to adsorption reactions. Both the concentration and the charge of the ions of the electrolyte may

influence the effective rate constant of charge transfer [1,4]. As electrolyte concentration and nature may change from one environmental system to another (e.g. on both sides of the water–sediment interface) the influence of this effect should be known.

- *Voltammetric parameters*, in particular those related to measurement time (scan rate, pulse duration, square wave frequency; section 3.1.3.) and, for ASV or AdSV techniques, to deposition time, are key factors to study. Proportionality between current and concentration is only valid for a limited range of deposition times in AdSV. Parameters related to measurement time may strongly influence the peak current, potential and width of non reversible processes or processes including chemical reations [4].

5.3.3 Present Speciation Capabilities

As discussed in section 3.5, it is difficult to perform speciation studies with the compounds relevant for this section, because they are involved in non-reversible redox couples or they are linked to complicated chemical reactions. Nevertheless, a few comments are worth mentioning.

- The peak height of the reaction $HgS_{ads} + 2e^- \rightleftharpoons Hg^0 + S(-II)$, (or that of the reverse reaction), obtained at deposition times short enough to avoid saturation of the surface, results from the diffusion and reaction with Hg, of all mobile species ($\sum[S(-II)]$) in labile equilibrium with S^{2-}. In most cases they include $[S^{2-}]$, $[HS^-]$ and $[H_2S]$, since labile metal sulfide complexes are negligible. As the above three species have the same diffusion coefficient, the peak height is proportional to $\sum[S(-II)]$ irrespective of their proportions, i.e. irrespective of pH. The De Ford and Hume equation (equation (8)) is applicable to this system of labile species, with $i_p^L = i_p$ (because diffusion coefficients are the same) and $\alpha = 1 + \beta_1 [H^+] + \beta_2 [H^+]^2$. In principle the species distribution can thus be determined from the peak potential, provided the solution is buffered enough so that the pH at the electrode surface is the same as in solution, even during the potential scan. Usually, however, the accuracy is not very good. Even so the voltammetric measurement provides information ($\sum[S(-II)]$) complementary to the S(−II) ion-selective electrode ($[S^{2-}]$) and H_2S sensor ($[H_2S]$).
- A speciation approach based on the discrimination between mobile, colloidal and particulate species (section 3.2) is applicable to the Mn(II), Fe(II) and S(−II) systems. This has been demonstrated by field DPP measurements on DME of mobile ($<\sim 4$ nm) species of Fe(II) and S(−II) in the anoxic layer of Lake Bret (Switzerland) [207,224]. Differences between voltammetric and colorimetric measurements with and without filtration on 0.45 μm provided the colloidal and particulate Fe(II) and S(−II) fractions. Results [185,207,224] showed (Figure 21) dramatic seasonal changes of these

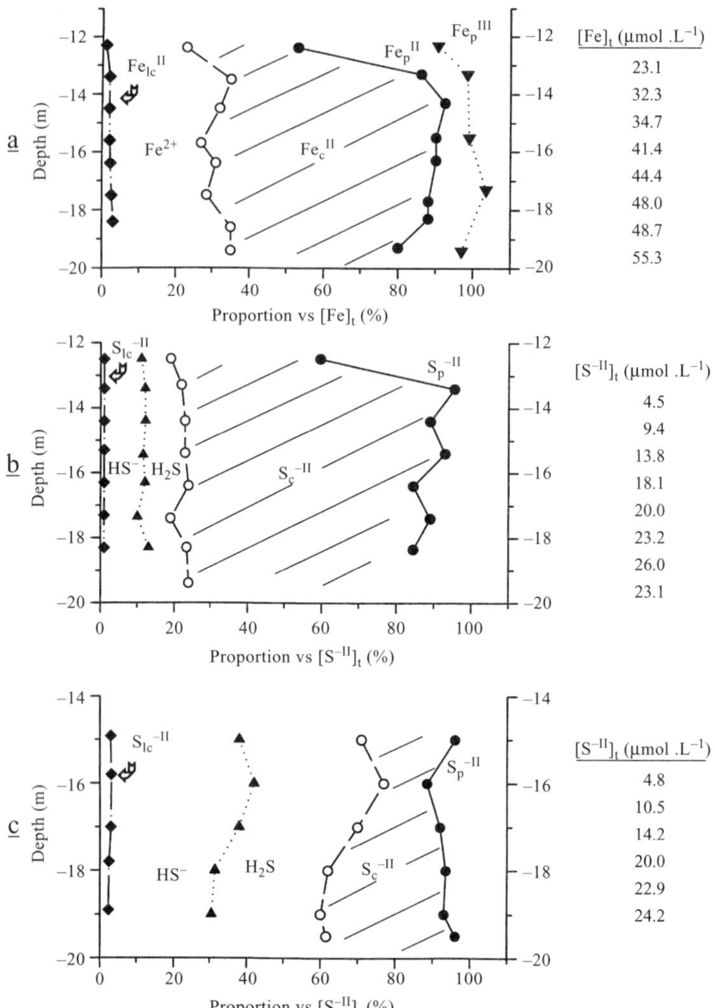

Figure 21. Cumulative graph showing the proportions (%) of mobile, colloidal (subscript c) and particulate (subscript p) species of Fe(II) and S(−II) in hypolimnion of Lake Bret, Switzerland. All measurements were performed on site. The mobile species are measured by DPP. Colloidal and particulate species have been measured by difference between DPP and colorimetric measurements with and without filtration on 0.45 μm membrane respectively. The distribution of mobile species of S(−II) into S^{2-}, HS^-, and H_2S , and the distribution of Fe(II) into Fe^{2+} and the sum of its labile inorganic complexes (Fe_{lc}^{II}) have been computed by computer code. (a, b) Distributions of Fe and S species respectively, on the 21st September 1981; (c) distribution of S species on the 6th August 1981 illustrating a drastic seasonal change (modified from refs [185,224]). The total concentration values for Fe, $[Fe]_t$, and S(−II), $[S(-II)]_t$, given on the right of the graphs, correspond to the same depths and in the same order, as the data points on the graphs

fractions, which suggests that colloidal FeS might play an important role in sedimentation of solid FeS. As for Mn(II) in the water column (section 5.2.6.2), these results illustrate that voltammetric data are not equivalent to spectrometric measurements of total concentrations, but rather give complementary information.

- As discussed above, in some conditions, an 'FeS' species is formed and gives a specific signal at the electrode (Figure 20D). Although it is chemically inert in the voltammetric time frame, detailed investigations have shown [184] that it is in slow equilibrium (minutes to hours) with S(−II) in solution, and that its molecular weight is low, since its diffusion coefficient is similar to that of aquo Fe(II). It might be a small polynuclear complex of Fe(II) and S(−II), possibly a precursor of amorphous solid FeS. At any rate it seems likely to be environmentally relevant and thus worthy of further studies, for which voltammetry might be particularly helpful.

6 RECOMMENDATIONS AND PROSPECTS

Voltammetric techniques are potentially capable of analysing a large number of inorganic and organic compounds, with a sensitivity which competes with the most sensitive spectrometric techniques. Because of their basic principles, they can give information on speciation without combination with additional separation technique. Miniaturized instruments with low energy requirement can be easily built. Because of these unique characteristics, voltammetric techniques are methods of choice for *in situ* automatic, real-time monitoring of chemical compounds in water. The major limitations and developments required for their routine use are discussed below.

6.1 EDUCATION IN ELECTROCHEMISTRY AND EXPERTISE IN VOLTAMMETRY

Use of voltammetric methods and interpretation of the results is based on a number of evolved physical chemistry concepts: the structure of the solid–water interface, electron transfer kinetics at interfaces, adsorption of molecules at interfaces, kinetics of chemical reactions in solution and diffusion processes. Several of these concepts form only minor parts of the classical curriculum of chemists, and their more complicated coupling is very rarely taught. Furthermore when voltammetry is taught in analytical chemistry courses, it is usually done as a brief description of technical aspects without detailed explanation of the underlying physico-chemical processes which may affect the signal. This generally poor appreciation of the key principles is a major difficulty for the development of these techniques, in particular for their application to direct measurements *in situ*, where a number of complicated processes have to be considered simultaneously. Without this background it is difficult to have sufficient understanding of the factors affecting the signal to be able to make the correct choice of the appropriate conditions for analysis and calibration.

6.2 CALIBRATION OF VOLTAMMETRIC PROCEDURES

Because many factors may affect the signal, calibration must be done with great care, and with due regard for the physico-chemical processes involved. Several major factors must be studied in detail.

- *pH, temperature* and *ionic strength* have already been discussed specifically for measurements in sediments (section 5.3.2.4) in relation to redox systems with irreversible behaviour or coupled to chemical or adsorption processes. Their influence is discussed below with respect to factors others than the electrode process. *pH* strongly influences many redox reaction either directly (section 5.3) or indirectly, e.g. in the case of metal complex measurements, and pH buffering capacity is important in maintaining a well-defined surface pH at the electrode, e.g. in cases of reduction of O_2 (section 5.1.2) or accumulation of a compound with acid–base properties on the electrode (e.g. S(− II)). The *temperature dependence* of the recorded peak current may enable one to discriminate between diffusion-controlled systems (small current variation), non-reversible system or systems controlled by chemical kinetics (large current variation), which is required to understand the processes underlying the analytical procedure and choose the appropriate calibration conditions. Not only may the *ionic strength* and nature of electrolyte significantly influence the current of non-reversible and chemically controlled systems, particularly at low ionic strengths ($\leq 10^{-2}$ M; [1, 4]), but even with reversible systems an undesired migration current [1] and significant resistance (ohmic drop) may develop at too low ionic strength.
- *Hydrodynamic convection* must be systematically controlled to get reliable results. Convective effects may be a major problem with electrode size $\geq 100\,\mu m$ and even with a diameter of $10\,\mu m$ they are not negligible [93]. For *in situ* use, only gel-integrated microsensors ensure that the current is purely diffusion controlled with an electrode size of $\sim 10\,\mu m$.
- Parameters related to the *measurement time of the technique* (scan rate, pulse duration, frequency) may dramatically influence the peak current and potential. These factors must be studied in detail both for calibration purposes and because any deviation from the dependence expected for the diffusion-controlled case may give useful information on the nature of the electrode process [2].
- It is worth emphasizing that obtaining *reproducible straight lines* for classical calibration curves (peak current versus concentration of test compound) in pure electrolyte solution is usually a necessary but by no means a sufficient

condition to claim that these curves can be used to relate the peak current measured in environmental media to concentration. An understanding of the electrode mechanism obtained by systematic variation of the parameters described above is first required. Calibration must be performed in a synthetic solution which mimics closely the test solution (e.g. in a solution with the same major salts). A calibration by standard additions in the real test medium must then be performed in the same concentration range, and the slopes of the two curves compared. It is useful to compare at the same time the curves corresponding to peak potential versus test compound concentration. When the curves of peak current and potential versus concentration are similar in both synthetic and natural solutions, the test medium can be considered as non-complexing, and the concentration can be computed from the calibration curve in synthetic or natural medium. If either peak current or peak potential dependences are not the same in synthetic and natural solutions, the difference is most likely due to the presence in the test medium either of complexing agents of the test compound or of compounds which may adsorb on the electrode surface and modify the electrode process. Further detailed studies are then necessary (sections 3.1–3.4) before developing an analytical procedure.

6.3 DEVELOPMENT OF RUGGED AND RELIABLE SENSORS

The development of sensors negligibly affected by adsorption phenomena is a key requirement for *in situ* measurements. HMDE may solve this problem in many cases but it is difficult (although not impossible; [154]) to use *in situ* in waters, and impossible to use in sediment. Presently three types of electrodes have been used: the amalgamated Au electrode, whose major limitation is that it lasts for a maximum of 12 h [39], the Au electrode whose reliability is not clear in particular in anoxic or sulfidic media and the gel-integrated Hg-plated Ir-based microelectrodes (GIME; [93], [192], [216]) which has proved to work reliably for at least 3–4 days even *in situ* in sediments. In all cases, more work is needed to advance understanding of the advantages and limitations of these electrodes. Further work is also needed to increase understanding of the molecular processes at interfaces and to develop new types of sensors. Chemically modified electrodes based on nanotechnologies offer promising advances, but there is still much work to do before the resulting sensors can be used routinely.

6.4 DEVELOPMENTS FOR THE ANALYSIS OF FURTHER COMPOUNDS

This chapter has focused on those compounds which can be analysed by direct measurements, namely O_2, H_2O_2, some trace metals and anions. As mentioned in section 4, in principle a large number of other inorganic and organic compounds might be analysed by coupling voltammetric detection to a preliminary step that performs a separation, a preconcentration and/or a chemical reaction. In order to undertake these operations *in situ*, the whole system must be miniaturized. Integration of various analytical steps, including the detection, in a single small analytical system can be realized by means of microtechnologies (Chapter 12). The resulting integrated unit is called a micro total analytical system = μTAS. There is little doubt that such developments will be the challenge of the next 10–15 years.

ACKNOWLEDGMENTS

The authors warmly thanks W. Davison, for his detailed constructive comments, and R. Menghetti for the preparation of the figures.

LIST OF ACRONYMS

Electrodes

AE	auxiliary electrode
μ-AMMIE	agarose gel membrane mercury-plated Ir-based microelectrode
μ-AMMIA	agarose gel membrane mercury-plated Ir-based microelectrode array
CME	chemically modified electrode
DME	dropping mercury electrode
GIME	gel-integrated microelectrode
HMDE	hanging mercury drop electrode
MFE	mercury film electrode
RE	reference electrode
SMDE	sessile mercury drop electrode
WE	working electrode

Voltammetric techniques

AdSV	adsorptive stripping voltammetry (see section 2.1.2 for the definition used here)
ASV	anodic stripping voltammetry (see section 2.1.2)
CCPSA	constant current potentiometric stripping analysis
CSV	cathodic sweep voltammetry
CV	cyclic voltammetry
DPAdSV	differential pulse adsorptive stripping voltammetry
DPASV	differential pulse anodic stripping voltammetry
DPCSV	differential pulse cathodic sweep voltammetry
DPP	differential pulse polarography

DPV	differential pulse voltammetry
NPP	normal pulse polarography
PSA	potentiometric stripping analysis
RNPP	reverse pulse polarography
SWASV	square wave anodic stripping voltammetry
SWAdSV	square wave adsorptive stripping voltammetry
SWV	square wave voltammetry

Other techniques

AAS	atomic absorption spectrometry
ICP-AES	inductively coupled plasma atomic emission spectrometry
ICP-MS	inductively coupled plasma mass spectrometry
LIGA	'Lithographie, Galvanoformung und Abformung' (=lithography, electroforming and moulding)

MATHEMATICAL SYMBOLS

A	surface area
c_X	total concentration of X
$\bar{D}$	average diffusion coefficient
D_X	diffusion coefficient of compound X
E	potential
E_0	standard redox potential
F	Faraday constant
f	frequency or fraction of metal complex
i	current
K	equilibrium constant
k_d, k_f	dissociation, formation rate constant
k_0	charge transfer rate constant in redox reactions
l	length
M_w	molecular weight
n	number of exchanged electron
P	pressure
r	radius
T	absolute temperature
t	time
t_d	deposition time
t_g	(Hg) drop time
t_m	measurement time
t_p	pulse duration
t_s	sampling time
t_w	waiting time

v	potential scan rate
[Y]	concentration of the specific species Y
α	charge transfer coefficient in redox reaction or degree of complexation (= c_M/[M]) in complexation reaction
δ	thickness of the diffusion layer

Superscripts

a	anodic
c	cathodic
L	in presence of ligand L

Subscripts

ℓ	labile
m	measurement or mobile
p	peak or particulate
t	total

REFERENCES

1. Heyrowsky, J. and Kuta, J. (1966). *Principles of Polarography*, Academic Press, London.
2. Galus, Z. (1994). *Fundamentals of Electrochemical Analysis,* 2nd edn, Ellis Horwood, Chichester.
3. Bard, A. J. and Faulkner, L. R. (1982). *Electrochemical Methods : Fundamentals and Applications*, Masson, Paris.
4. Buffle, J. (1988). Studies of complexation properties by voltammetric methods. In *Complexation Reactions in Aquatic Systems. An Analytical Approach*, Ellis Horwood, Chichester, Chapter 9.
5. Kissinger, P. T. and Heineman, W. R., ed. (1984). *Laboratory Techniques in Electroanalytical Chemistry*, Marcel Dekker, New York.
6. Bard, A. J., ed. (1966–1998). *Electroanalytical Chemistry*; Series of 20 volumes, Marcel Dekker, New York.
7. Osteryoung, J. G. and Osteryoung, R. A. (1985). Square-wave voltammetry, *Anal. Chem.*, **57**, 101A–110A.
8. Champagne, G. Y. and Chevalet, J. (1997). High sensivity multiple waveform voltammetric method and instrument. PCT Int. Appl. Patent No. Au 97–40057, US 97–798016, 101 pp.
9. Vydra, F., Stulik, K. and Julakova, E. (1976). *Electrochemical Stripping Analysis*, Ellis Horwood, Chichester.
10. Kalvoda, R. A. and Kopanica, M. (1989). Adsorptive stripping analysis in trace analysis, *Pure Appl. Chem.*, **61,** 97–112.
11. Kalvoda, R. (1988). Ecoelectrochemistry: Prospects for selective electrochemical sensing and removal/destruction of pollutants, *Selective Electrode Rev.*, **10**, 127–183.
12. Abul, H. and Coetzee, J. F. (1985). Potentiometric stripping analysis: theory, experimental verification and generation of stripping polarograms, *Anal.Chem.*, **57**, 581–585.

13. Aoki, K., Tokuda, K., Matsuda, H. and Osteryoung, J. (1986). Reversible square-wave voltammogramms. Independance of electrode geometry, *J. Electroanal. Chem.*, **207**, 25–39.
14. Levitch, V. G. (1962). *Physicochemical Hydrodynamics*, Prentice Hall, Englewood Cliffs, NJ.
15. Pletcher, D. (1991). Why microelectrodes? In *Microelectrodes. Theory and Applications*, ed. Montenegro, M. I., Queiros, M. A., Daschbach, J. L., NATO AST Series, The Netherlands, Chapter 1.
16. Pons, S. and Fleischmann (1987). The behavior of microelectrodes, *Anal. Chem.*, **59**, 1391A–1399A.
17. Bond, A. M. (1994). Past, present and future contributions of microelectrodes to analytical studies employing voltammetric detection. A review, *Analyst*, **119**, R1–R21.
18. Stulik, K. and Pacakova, V. (1984). Electrochemical detection in high-performance liquid chromatography, *Crit. Rev. Anal. Chem.*, **14(4)**, 297–351.
19. Gunasingham, H. and Fleet, B. (1989). Hydrodynamic voltammetry in continuous-flow analysis. In *Electroanalytical Chemistry*, Vol. 16, ed. Bard, A. J., Marcel Dekker, New York, p. 89.
20. Howell, J. O. and Wightman, R. M. (1984). Ultrafast voltammetry and voltammetry in highly resistive solutions with microvoltammetric electrodes, *Anal. Chem.*, **56**, 524–529.
21. Bond, A. M., Scholz, F. (1990). A survey of electrodes used for voltammetric analysis, *Z. Chem.*, **30**, 117–129.
22. Goyal, R. N. and Mittal, A. (1993). Application of electroanalytical techniques in monitoring metal ions pollution in water. A review, *J. Sci. Ind. Res.*, **52**, 607–623.
23. Florence, T. M. (1986). Electrochemical approaches to trace element speciation in waters. A review, *Analyst,* **111**, 489–505.
24. Smart, R. B. (1987). Electrode systems for measurement of environmental pollutants, *Hazard Assess. Chem.*, **5**, 1–27.
25. Meites, L., ed. (1965). *Polarographic Techniques, 2nd edn*, Wiley, New York.
26. Bond, A. M., Thomson, S. B., Tucker, D. J. and Briggs, M. H. (1984). Electrochemical studies of homocysteine and homocystine at mercury electrodes, *Anal. Chim. Acta*, **156**, 33–42.
27. Bond, A. M. and MacLachlan, N. M. (1985). Oxidation processes at mercury electrodes for tetraphenyllead and related compounds in dichloromethane, *J. Electroanal. Chem.,* **182**, 367–382.
28. Bond, A. M. and MacLachlan, N. M. (1987). Reversible oxidation processes in dichloromethane for triethyl-, diethyl-, trimethyl-, and dimethyllead compounds at mercury electrodes, *J. Electroanal. Chem.,* **218**, 197–211.
29. Zirino, A. (1981). Voltammetry of natural sea water. In *Marine Electrochemistry,* ed. Whitfield, M. and Jagner, D., Wiley, New York, Chapter 10.
30. Barendrecht, E. (1967). Stripping voltammetry, *Electroanal. Chem.*, **2**, 53–109.
31. Koslovsky, M. and Zebreva, A. (1972). In *Progress in Polarography*, Vol. 3, ed. Zuman, P. and Meites, L., Wiley-Interscience, New York, p.166.
32. Galus, Z. (1984). Diffusion coefficients of metals in mercury, *Pure Appl. Chem.*, **56,** 635–644.
33. Cox, J. A. (1967). *Analytical utility of mercury film electrode*, Ph.D. Thesis, University of Illinois, Urbana, IL.
34. Pei, J., Tercier-Waeber, M.-L. and Buffle, J. (2000). Simultaneous determination and speciation of zinc, cadmium, lead and copper in natural waters with minimum

handling and artefacts by voltammetry on gel-integrated microelectrode array, *Anal. Chem.*, **72**, 161–171.
35. Piccardi, G. and Udisti, R. (1987). Intermetallic compounds and the determination of copper and zinc by anodic stripping voltammetry, *Anal. Chim Acta*, **202**, 151–157.
36. Keis, H. and Piibar, U. (1994). Determination of heavy metals in sea and freshwaters by anodic stripping voltammetry, *Tartu. Ulik. Toim.*, **975**, 119–127.
37. Yoshida, Z. (1981). Structure of mercury layer deposited on platinum and hydrogen-evolution reaction at the mercury-coated platinum electrode, *Bull. Chem. Soc. Jpn.*, **54**, 556–561.
38. Stojek, Z. and Bublik, Z. (1977). Silver based mercury film electrode. Part III : comparison of theoretical and experimental anodic stripping results obtained for lead and copper, *J. Electroanal. Chem.*, **77**, 205–224.
39. Brendel, P. J. and Luther III, G. W. (1995). Development of a gold amalgam voltammetric microelectrode for the determination of dissolved Fe, Mn, O_2 and S(−II) in porewater sediments, *Environ. Sci. Technol.*, **29**, 751–761.
40. Brainina, K. Z., Vilchinskaya, E. A. and Khanina, R. M. (1990). Influence of the redox potential of the medium on stripping voltammetric measurement results, *Analyst*, **115**, 1301–1304.
41. Kounaves, S. P. and Buffle, J. (1986). Deposition and stripping properties of mercury on iridium electrodes, *J. Electrochem. Soc.*, **133**, 2495–2498.
42. Kounaves, S. P. and Buffle, J. (1987). An iridium-based mercury film electrode. Part I : selection of substrate and preparation, *J. Electroanal. Chem.*, **216**, 53–69.
43. Berge, H. and Strübing, B. (1969). Verwendung von Kohleelektroden bei der voltammetrischen Anordnung mit kontinuierlich aktivierter Oberfläche, *Fresenius' Z. Anal. Chem.*, **247**, 12–15.
44. Poon, M. and McCreery, R. L. (1987). Repetitive in situ renewal and activation of carbon and platinum electrodes : applications to pulse voltammetry, *Anal. Chem.*, **59**, 1615–1620.
45. Evans, J. and Kuwana, Th. (1979). Introduction of functional groups onto carbon electrodes via treatment with radio-frequency plasmas, *Anal. Chem.*, **51**, 358–365.
46. Rice, M. E., Galus, Z. and Adams, R. N. (1983). Graphite paste electrodes. Effects of paste composition and surface states on electron-transfer rates, *J. Electroanal. Chem.*, **143**, 89–102.
47. Koile, R. C. and Johnson, D. C. (1979). Electrochemical removal of phenolic films from a platinum anode, *Anal. Chem.*, **51**, 741–744.
48. Engstrom, R. C. and Strasser, V. A. (1984). Characterization of electrochemically pretreated glassy carbon electrodes, *Anal. Chem.*, **56**, 136–141.
49. Fagan, D. T., Hu, I. F. and Kuwana, Th. (1985). Vacuum heat treatment for activation of glassy carbon electrodes, *Anal. Chem.*, **57**, 2759–2763.
50. Hua, C., Jagner, D. and Renman, L. (1987). Automated determination of total arsenic in sea water by flow constant current stripping analysis with gold, *Anal. Chim. Acta*, **201**, 263–268.
51. Jagner, D., Renman, L. and Stefansdottir, S. H. (1994). Determination of arsenic by stripping potentiometry on regression calibration, *Electroanalysis*, **6**, 201–208.
52. Viltchinskaia, E. A., Zeigman, L. L., Garcia, D. M. and Santos, P. F. (1997). Simutaneous determination of mercury and arsenic by anodic stripping voltammetry, *Electroanalysis* **9**, 633–640.
53. Huang, H., Jagner, D. and Renman, L. (1987). Simultaneous determination of mercury(II), and bismuth(III) in urine by flow constant-current stripping analysis with a gold fiber electrode, *Anal. Chim. Acta*, **202**, 117–22.

54. Wang, E., Sun, W. and Yang, Y. (1984). Potentiometric stripping analysis with a thin-film gold electrode for determination of copper, bismuth, antimony and lead, *Anal. Chem.*, **56** (11), 1903–1906.
55. Pinilla, G. E. and Ostapczuk, P. (1994). Potentiometric stripping determination of mercury(II), selenium(IV), copper(II) and lead(II) at a gold film electrode in water samples, *Anal. Chim. Acta*, **293**, 55–65.
56. Wang, J., Foster, N., Armalis, S., Larson, D., Zirino, A. and Olsen, K. (1995). Remote stripping electrode for in situ monitoring of labile copper in the marine environment, *Anal. Chim. Acta*, **310**, 223–231.
57. Wang, J., Larson, D., Foster, N., Armalis, S., Lu, J., Rongrong, X., Olsen, K. and Zirino, A. (1995). Remote electrochemical sensor for trace metal contaminants, *Anal. Chem.*, **67**, 1481–1485.
58. Wang, J., Sucman, E. and Tian, B. (1994). Stripping potentiometric measurements of copper in blood using gold microelectrodes, *Anal. Chim. Acta*, **286**, 189–195.
59. Rievaj, M., Mesaros, S. and Bustin, D. (1993). Application of gold fiber voltammetric microelectrode in trace analysis for mercury in waters and fertilizers, *An. Quim.*, **89** (3), 347–350.
60. Huang, H., Jagner, D. and Renman, L. (1987). Flow potentiometric and constant-current stripping analysis for mercury(II) with gold, platinum and carbon fiber working electrode. Application to the analysis of tap water, *Anal. Chim. Acta*, **201**, 1–9.
61. Gustavsson, I., (1986). Determination of mercury in sea water by stripping voltammetry, *J. Electroanal. Chem.*, **214**, 31–36.
62. Goto, M., Ikenoya, K. and Ishii, D. (1980). Anodic stripping semidifferential electroanalysis of mercury(II) at gold disk electrode and its application to environmental analysis, *Bull. Chem. Soc. Jpn.*, **53**, 3567–3572.
63. Huang, H., Jagner, D. and Renman, L. (1987). Flow constant-current stripping analysis for antimony(III) and antimony(V) with gold fiber working electrodes. Application to natural waters, *Anal. Chim. Acta*, **202**, 123–129.
64. McLaughlin, K., Boyd, D., Hua, C. and Smyth, M. R. (1992). Anodic stripping voltammetry of selenium(IV) at a gold fiber working electrode, *Electroanalysis*, **4**, 689–693.
65. Opekar, F. and Bruckenstein, S. (1984). Determination of gaseous hydrogen sulfide by cathodic stripping voltammetry after preconcentration on a silver metalized porous membrane electrode, *Anal. Chem.*, **56**, 1206–1209.
66. Scholz, F., Nitschke, L and Hensrion, G. (1987). Sorption from gazeous phase in stripping voltammetry of halogens, *J. Electroanal. Chem.*, **224**, 303–306.
67. Scholz, F., Nitschke, L. and Hensrion, G. (1988). Sorption from the gazeous phase as a new method for preconcentration in stripping voltammetry, *Zh. Anal. Khim.* **43**, 1166–1169.
68. Adams, R. N. (1969). *Electrochemistry at Solid Electrodes*, Marcel Dekker, New York.
69. Bersier, P. M. and Bersier, J. (1985). Applied polarography and voltammetry of organic compounds in practical day to day analysis. Part I. Applied polarographic and voltammetric techniques, *Crit. Rev. in Anal. Chem.*, **16**, 15–79.
70. Patriarche, G. J. and Zjang, H. (1990). Electroanalytical techniques for drug analysis, *Electroanalysis* **2**, 573–579.
71. Wang, J. (1989). Voltammetry following nonelectrolytic preconcentration, in *Electroanalytical Chemistry*, Vól. 16, ed. Bard, A. J., Marcel Dekker, New York, pp. 1–88.

72. Kinoshita, K. (1988). *Carbon: Electrochemical and Physico-chemical Properties*, Wiley, New York.
73. Besenhard, J. O. and Fritz, H. P. (1983). The electrochemistry of black carbons, *Angew. Chem. Int. Ed. Engl.*, **22**, 950–975.
74. Taylor, R. F. and Schultz, J. S., ed. (1996). *Handbook of Chemical and Biological Sensors*, IOP Publishing, Bristol.
75. Murray, R. W. (1984). Chemically modified electrodes. In *Electroanalytical Chemistry*, Vol. 13, ed. Bard, A. J., Marcel Dekker, New York, pp. 191–368.
76. Abruña, H. D. (1988). Coordination chemistry in two dimensions : chemically modified electrodes, *Coord. Chem. Rev.*, **86**, 135–189.
77. Schreurs, J. and Barendrecht, E. (1984). Surface-modified electrodes (SME), *Recueil Trav. Chim., Pays Bas*, **103**, 205–219.
78. Albery, W. J. and Hillman, A. R. (1981). Modified electrodes. In *Annual Reports on the Progress of Chemistry*, Vol. 78, The Royal Society, London, pp. 377–437.
79. Patriarche, G. J. (1986). New trends on modified electrodes: applications to drug analysis, *J. Pharm. Biomed. Analysis* **4**, 789–797.
80. Dong, S. and Wang, Y. (1989). The application of chemically modified electrodes in analytical chemistry. A review, *Electroanalysis*, **1**, 99–106.
81. Kalcher, K. (1990). Chemically modified carbon paste electrodes in voltammetric analysis. A review, *Electroanalysis*, **2**, 419–433.
82. Baldwin, R. P. and Thomsen, K. N. (1991). Chemically modified electrodes in liquid chromatography detection. A review, *Talanta*, **38**, 1–16.
83. Wang, J., Taha, Z. and Naser, N. (1991). Electroanalysis at modified carbon-paste electrodes containing natural ionic polysaccharides, *Talanta*, **38**, 81–88.
84. Daniele, S., Baldo, M. A., Ugo, P. and Mazzocchin, G. A. (1989). Determination of heavy metals in real samples by anodic stripping voltammetry with mercury microelectrodes. Part 2 : Application to rain and sea waters, *Anal. Chim. Acta*, **219**, 19–26.
85. Wang, J., Tuzhi, P. and Zadeii, J. M. (1987). Evaluation of differential-pulse anodic stripping voltammetry at mercury-coated carbon fiber electrodes. Comparison to analogous measurements at rotating disk electrodes, *Anal. Chem.*, **59**, 2119–2122.
86. Sottery, J. P., and Anderson, C. W. (1987). Short-pulse rapid-scan stripping voltammetry at a thin mercury film carbon fiber electrode, *Anal. Chem.*, **59**, 140–144.
87. Kounaves, S. P. and Deng, W. (1991). An iridium-based mercury ultramicroelectrode. Fabrication and characterization, *J. Electroanal. Chem.*, **301**, 77–85.
88. De Vitre, R. R., Tercier, M.-L., Tsacopoulos, M. and Buffle, J. (1991). Preparation and properties of a mercury-plated iridium-based microelectrode, *Anal. Chim. Acta*, **249**, 419–425.
89. Tercier, M.-L. and Buffle, J. (1993). In situ voltammetric measurements in natural waters: Future prospects and challenges. Review, *Electroanalysis*, **51**, 187–200.
90. Harmann, A. R. and Baranski, A. S. (1990). Fast cathodic stripping analysis with ultramicroelectrodes, *Anal. Chim. Acta*, **239**, 35–44.
91. Wightmann, R. M. and Wipf, D. O. (1989). Voltammetry at ultramicroelectrodes. In *Electroanalytical Chemistry*, Vol. 15, ed. Bard, A. J., Marcel Dekker, New York, pp. 267–353.
92. Fleischmann, M., Pons, S., Rolison, D. R. and Schmidt, P. P. (1987). *Ultramicroelectrodes, Datatech Systems*, Morganton, NC.
93. Tercier, M.-L., Parthasarathy, N. and Buffle, J. (1995). Reproducible, reliable and rugged Hg-plated Ir-based microelectrode for in situ measurements in natural waters, *Electroanalysis*, **7**, 55–63.
94. Magee, Jr., L. J. and Osteryoung, J. (1989). Fabrication and characterization of glassy carbon linear array electrodes, *Anal. Chem.*, **61**, 2124–2126.

95. Strobhen, W. E., Smith, D. K. and Evans, D. H. (1990). Characterization of arrays of microelectrodes for fast voltammetry, *Anal. Chem.*, **62**, 1709–1712.
96. Cheng, I. F., Schimpf, J. M. and Martin, C. R. (1990). Ultramicroelectrode ensembles. Part V : sealing defects between the ensemble elements and the host membrane with octadecyltrichlorosilane, *J. Electroanal. Chem.*, **284**, 499–505.
97. Penner, R. M. and Martin, C. R. (1987). Preparation and electrochemical characterization of ultramicroelectrode ensembles, *Anal. Chem.*, **59**, 2625–2630.
98. Cheng, I. F. and Martin, C. R. (1988). Ultramicrodisk electrode ensembles prepared by incorporation of carbon paste into a microporous host membrane, *Anal. Chem.*, **60**, 2163–2165.
99. Cheng, I. F., Wighteley, L. D. and Martin, C. R. (1989). Ultramicroelectrode ensembles. Comparison of experimental and theoretical responses and evaluation of electroanalytical detection limits, *Anal. Chem.*, **61**, 762–766.
100. Bramlick, C. J., Martin, C. R. and Tokuda, K. (1992). Microhole array electrodes based on microporous alumina membranes, *Anal. Chem.*, **64**, 1201–1203.
101. Odell, D. M. and Bowyer, W. J. (1990). Fabrication of band microelectrode arrays from metal foil and heat-sealing fluoropolymer film, *Anal. Chem.*, **62**, 1619–1623.
102. Bartelt, J. E., Deakin, M. R., Amatore, C. and Wightman, R. M. (1988). Construction and use of paired and triple band microelectrodes in solutions of low ionic strength, *Anal. Chem.*, **60**, 2167–2169.
103. Wu, H. P. (1993). Fabrication and characterization of a new class of microelectrode arrays exhibiting steady-state current behavior, *Anal. Chem*, **65**, 1643–1646.
104. Belmont-Hébert, C., Tercier, M.-L., Buffle, J., Fiaccabrino, G. C. and Koudelka-Hep, M. (1996). Mercury-plated iridium-based microelectrode arrays for trace metals detection by voltammetry: optimum conditions and reliability, *Anal.Chim. Acta*, **329**, 203–214.
105. Meyer, H., Drewer, H., Gruendig, B., Cammann, K., Kakerow, R., Manoli, Y., Mokwa, W. and Rospert, M. (1995). Two-dimensional imaging of O_2, H_2O_2 and glucose distributions by an array of 400 individually addressable microelectrodes, *Anal. Chem.*, **67**, 1164–1170.
106. Frisbie, C. D., Fritsch-Faules, I., Wollman, E. W. and Wrighton, M. S. (1992). Preparation and characterization of redox active molecular assemblies on microelectrode arrays, *Thin Solid Films*, **210–211**, 341–347.
107. Glass, R. S., Perone, S. P. and Ciarlo, D. R. (1990). Application of information theory to electroanalytical measurements using a multielement microelectrode array, *Anal. Chem.*, **62,** 1914–1918.
108. Hermes, T., Buehner, M., Buecher, S., Sundermeier, C., Dumschat, C., Borchardt, M., Cammann, K. and Knoll, M. (1994). An amperometric microsensor array with 1024 individually addressable elements for two-dimentional concentration mapping, *Sens. Actuators B*, **21**, 33–37.
109. Fiaccabrino, G. C., Koudelka-Hep, M., Jeanneret, S., van den Berg, A. and de Rooij, N. F. (1994). Array of individually addressable microelectrodes, *Sens. Actuators B*, **19**, 675–677.
110. Tercier-Waeber, M.-L., Pei, J., Buffle, J., Fiaccabrino, G. C., Koudelka-Hep, M., Riccardi, G., Confalonieri, F., Sina, A. and Graziottin, F. (2000). A novel voltammetric probe with individually addressable gel-integrated microsensor arrays for real-time high spatial resolution concentration profile measurements, *Electroanalysis*, **12**, 27–34.
111. Wechter, C. and Osteryoung, J. (1990). Voltammetric characterization of small platinum–iridium-based mercury film electrodes, *Anal. Chim. Acta*, **234**, 275–284.
112. Quintin, M. (1970). *Electrochimie*, Presses Universitaires de France, Paris

113. Covington, A. K. (1969). Reference electrodes. In *Ion-selective Electrodes*, ed. Durst, R. A., NBS Special Publication 314, US Governmental Printing Office, Washington, DC, Chapter 4.
114. Caton, Jr., R. D. (1973). Topics in chemical instrumentation. LXXIII . Reference electrodes, *J. Chem. Educ.*, **50**, A571–A578.
115. Caton, Jr., R. D. (1974). Topics in chemical instrumentation. LXXIII . Reference electrodes (concluded), *J. Chem. Educ.*, **51**, A7–A8, A14–A19.
116. Tercier, M.-L., Buffle, J. and Graziottin, F. (1998). A novel voltammetric in situ profiling system for continuous, real-time monitoring of trace elements in natural waters, *Electroanalysis*, **10**, 355–363.
117. Tercier-Waeber, M.-L., Belmont-Hébert, C. and Buffle, J. (1998). Real-time continuous Mn(II) monitoring in lakes using a novel voltammetric in situ profiling system, *Environm. Sci. Technol.*, **32**, 1515–1521.
118. Ruzicka, J., Lamm, C. G. and Tjell, J. C. (1972). Selectrode^{TM}—the universal ion-selective electrode. Part 3 : Concept, constructions and materials, *Anal. Chim. Acta*, **62**, 15–28.
119. Lamm, C. G., Hansen, E. H. and Ruzicka, J. (1972). In situ use of the ion selective electrode, the selectrode^{TM}, in studies of soil–plant relationships, *Anal. Lett.*, **5**, 451–459.
120. Ruzicka, J., Hansen, E. H. and Tjell, J. C. (1973). Selectrode^{TM}—the universal ion-selective electrode. Part VI: the calcium (II) selectrode employing a new ion exchanger in a nonporous membrane and a solid-state reference system, *Anal. Chim. Acta*, **67**, 155–178.
121. Rehm, D., McEnroe, E. and Diamond, D. (1995). An all solid-state reference electrode based on a potassium chloride doped vinyl ester resin, *Anal. Proc. Ind. Anal. Commun.*, **32**, 319–322.
122. Diamond, D., McEnroe, E., McCorrick, M. and Lewenstam, A. (1994). Evaluation of a new solid-state reference electrode junction material for ion-selective electrodes, *Electroanalysis*, **6**, 962–971.
123. Jermann, R., Tercier, M.-L. and Buffle, J. (1992). Pressure insensitive solid state reference electrode for in situ voltammetric measurements in lake water, *Anal. Chim. Acta*, **269**, 49–58.
124. Bound, G. P. and Fleet, B. (1977). The development of a solid state reference electrode for use in soil measurements, *J. Sci. Food. Agric.*, **28**, 431–437.
125. Zhao, S., Hu, N., Zhao, J. and Wang, J. (1988). Study on solid-state reference electrode, *Zhonggao Kexue Jisha Daxue Xuebao*, **18**, 312–318.
126. Müller, B. and Hauser, P. C. (1996). Effect of pressure on the potentiometric response of ion-selective electrodes and reference electrodes, *Anal. Chim. Acta*, **320**, 69–75.
127. Moussy, F., Jakeway, S., Harrison, D. J. and Rajotte, R. W. (1994). In vitro and in vivo performance and lifetime of perfluorinated ionomer-coated glucose sensors after high-temperature curing, *Anal. Chem.*, **66**, 3882–3888.
128. Moussy, F. and Harrison, D. J. (1994). Prevention of the rapid degradation of subcutaneously implanted Ag/AgCl reference electrode using polymer coatings, *Anal. Chem.*, **66**, 674–679.
129. Brehier, D. C. and Belford, R. E. (1995). Thick-film reference electrodes for solid-state pH measurement, *Anal. Proc. Ind. Anal. Comm.*, **32**, 323–326.
130. Arquint, P., Koudelka-Hep, M., de Rooij, N.-F., Bühler, H. and Morf, W. E. (1994). Organic membranes for miniaturized electrochemical sensors: fabrication of a combined pO_2, P_{CO2} and pH sensor, *J. Electroanal. Chem.*, **378**, 177–183.

131. Kreider, K. G., Tarlov, M. J. and Cline, J. P. (1995). Sputtered thin-film pH electrodes of platinum, palladium, ruthenium and iridium oxides, *Sens. Actuators B*, **28,** 167–172.
132. Katsube, T., Lauks, I. and Zemel, J. N. (1982). pH-sensitive sputtered iridium oxide films, *Sens. Actuators* , **2,** 399–410.
133. Papeschi, G., Bordi, S., Beni, C. and Ventura, L. (1976). Use of an iridium electrode for direct measurement of pI of proteins after isoelectric focusing in polyacrylamide gel, *Biochim. Biophys. Acta*, **453**, 192–199.
134. Bordi, S., Cara, M. and Papeschi, G. (1984). Iridium/iridium oxide electrode for potentiometric determination of proton activity in hydroorganic solutions at sub-zero temperatures, *Anal. Chem.*, **56**, 317–319.
135. Ardizzone, S., Carugati, A. and Trasatti, S. (1981). Properties of thermally prepared iridium dioxide electrodes, *J. Electroanal. Chem.*, **126**, 287–292.
136. Galizzoli, D., Tantardini, F. and Trasatti, S. (1974). Ruthenium dioxide: a new electrode material. Part I: behaviour in acid solutions of inert electrolytes, *J. Appl. Electrochem.*, **4**, 57–67.
137. Glab, S., Hulanicki, A., Edwall, G. and Ingman, F. (1989). Metal/metal oxide and metal oxide electrodes as pH sensors, *Crit. Rev. Anal. Chem.*, **21**, 29–47.
138. Keller, O. C. and Buffle, J. (2000). Voltammetric and reference microelectrodes with integrated microchannels for flow-through microvoltammetry: Part I; The microcell, *Anal. Chem.*, **72**, 936–942.
139. Heineman, W. R. and Kissinger, P. T. (1980). Analytical electrochemistry. methodology and applications of dynamic techniques, *Anal. Chem.*, **52**, 138R–151R.
140. Stülik, K. and Pacáková, V. (1984). Electrochemical detection in high performance liquid chromatography. Review, *Crit. Rev. Anal. Chem.*, **14**, 297–351.
141. Trojanowicz, M. (1996). Electrochemical detectors in automated analytical systems. In *Modern Techniques in Electroanalysis*, ed. Vanysek, P., Wiley, New York, Chapter 5.
142. Buchanan, Jr., E. B. and Soleta, D. D. (1982). Automated square-wave anodic stripping voltammetry with a flow-through cell and matrix exchange, *Talanta*, **29**, 207–211.
143. Wang, J. and Freiha, B. A. (1983). Selective voltammetric detection based on adsorptive preconcentration for flow injection analysis, *Anal. Chem.*, **55**, 1285–1288.
144. Gunasingham, H., Ang, K. P. and Ngo, C. C. (1986). The d.c. anodic stripping voltammetry at the mercury-film wall-jet electrode, *J. Electroanal. Chem.*, **215** , 123–137.
145. Kalvoda, R. (1987). Polarographic determination of adsorbable molecules, *Pure Appl. Chem.*, **59**, 715–722.
146. Ruzicka, J. and Hansen, E. H., ed. (1988). *Flow Injection Analysis,* 2nd edn, Wiley, New York.
147. Wang, J., Dewald, H. D. and Green, B. (1983) Anodic stripping voltammetry of heavy metals with a flow injection system, *Anal. Chim. Acta*, **146**, 45–50.
148. Stulik, K. and Pacàkovà, V. (1992). Selectivity in flow electroanalysis, *Selective Electrode Rev.*, **14**, 87–142.
149. Stulik, K. and Pacàkovà, V. (1987). *Electrochemical Measurements in Flowing Liquids*, Ellis Horwood, Chichester.
150. Elbicki, J. M., Morgan, D. M. and Weber, S. G. (1984). Theoretical and practical limitations on the optimization of amperometric detectors, *Anal. Chem.*, **56**, 978–985.

151. Tercier, M.-L., Bernard, C., Bujard, F., Rodak, S. and Buffle, J. (1990). Photomicroscopie measurement of the diffusion layer around mercury drop electrodes. Part I: description of the electrochemical system, *Electroanalysis*, **2**, 89–97.
152. Tercier, M.-L. and Buffle, J. (1990). Photomicroscopie measurement of the diffusion layer around mercury drop electrodes. Part II: choice of the optimal conditions, *Electroanalysis* **2**, 99–105.
153. Brains, C. H. P., Doornbos, D. A. and Brunt, K. (1982). The hydrodynamics of the amperometric detector flow cell with a rotating disk electrode, *Anal. Chim. Acta*, **140**, 39–49.
154. Tercier, M.-L., Buffle, J., Zirino, A. and De Vitre, R. R. (1990). In situ voltammetric measurement of trace elements in lakes and oceans, *Anal. Chim. Acta*, **237**, 429–437.
155. Colombo, C., van den Berg, C. M. G. and Daniel, A. (1997). A flow cell for on-line monitoring of metals in natural waters by voltammetry with a mercury drop electrode. *Anal. Chim. Acta*, **346**, 101–111.
156. Kissinger, P. T., Bruntlett, C. S., Bratin, K. and Rice, J. R. (1979). Trace organic analysis: A new frontier in analytical chemistry, *NBS Spec. Publ.*, **519**, 705.
157. Swartzfager, D. G. (1976). Amperometric and differential pulse voltammetric detection in high performance liquid chromatography, *Anal.Chem.*, **48**, 2189–2192.
158. Mc Crehan, W. A. and Durst, R. A. (1978). Measurement of organomercury species in biological samples by liquid chromatography with differential pulse electrochemical detection, *Anal. Chem.*, **50**, 2108–2112.
159. Johnson, D. C., Weber, S. G., Bond, A. M., Wightman, R. M., Shoup, R. E. and Krull, I.S. (1986). Electroanalytical voltammetry in flowing solutions, *Anal. Chim. Acta,* **180**, 187–250.
160. Stulik, K. and Pacáková, V. (1981). Comparison of several voltammetric detectors for high-performance liquid chromatography, *J. Chromatogr.*, **208**, 269–278.
161. Wang, J. and Ariel, M. (1977). Anodic stripping voltammetry in a flow-through cell with fixed mercury film glassy carbon disk electrodes. Part (II): the differential mode (DPASV), *J. Electroanal. Chem.*, **85**, 289–297.
162. Matson, W. R., Langlais, P., Volicer, L., Gamache, P. H., Bird, E. and Mark, K. A. (1984). N-electrode three-dimensional liquid chromatography with electrochemical detection for determination of neurotransmitters, *Clin. Chem.*, **30**, 1477–1488.
163. Dryhurst, G. (1977). *Electrochemistry of Biological Molecules*, Academic Press, New York.
164. Mairanovskii, S. G. (1968). *Catalytic and Kinetic Waves in Polarography*, Plenum, New York.
165. De Jong, H. G. and van Leeuwen, H. P. (1987). Voltammetry of metal complex systems with different diffusion coefficients of the species involved. Part II: Behaviour of the limiting current and its dependence on association/dissociation kinetics and lability, *J. Electroanal. Chem.*, **234**, 17–29
166. Luther III, G. W., Rickard, D. T., Theberge, S. M. and Oldroyd, A. (1996). Determination of metal (bi)sulfide stability constants of Mn^{++}, Fe^{++}, Co^{++}, Ni^{++}, Cu^{++}, and Zn^{++} by voltammetric methods, *Environ. Sci. Technol.,* **30**, 671–679.
167. Davison, W. (1996). Comment on 'Determination of metal (bi)sulphide stability constants of Mn^{2+}, Fe^{2+}, Co^{2+}, Ni^{2+}, Cu^{2+} and Zn^{2+} by voltammetric methods', *Environ. Sci. Technol.*, **30**, 3638–3639.
168. Crow, D. R. (1969). *Polarography of Metal Complexes*, Academic Press, London.
169. Buffle, J., Wilkinson, K., Stoll, S., Filella, M. and Zhang, J. (1998). A generalized description of aquatic colloidal interactions: the 3–colloidal component approach, *Environ. Sci. Technol.,* **32**, 2887–2899.

170. van Leeuwen, H. P., Cleven, R. and Buffle, J. (1989). Voltammetric techniques for complexation measurements in natural aquatic media. Role of the size of macromolecular ligands and dissociation kinetics of complexes, *Pure Appl. Chem.,* **61,** 255– 274.
171. Bilinski, H., Huston, R. and Stumm, W. (1976). Determination of the stability constants of some hydroxo and carbonato complexes of Pb(II), Cu(II), Cd(II) and Zn(II) in dilute solutions by anodic stripping voltammetry and differential pulse polarography, *Anal. Chim. Acta*, **84**, 157–164.
172. Nurnberg, H. W., Valenta, R., Mart, L., Raspor, B. and Sipos, L. (1976). Applications of polarography and voltammetry to marine and aquatic chemistry. II. The polarographic approach to the determination and speciation of toxic trace metals in the marine environment, *Fresenius' Z. Anal. Chem.*, **282**, 357–367.
173. Raspor, B., Nurnberg, H. W., Valenta, P. and Branica, M. (1980). The chelation of lead by organic ligands in sea water. In *Lead in the Marine Environment*, ed. Branica, M. and Konrad, Z., Pergamon, Oxford, pp. 181–195.
174. Raspor, B., Valenta, P., Nurnberg, H. W. and Branica, M. (1978). The chelation of cadmium with NTA in sea water as a model for the typical behaviour of trace heavy metal chelates in natural waters, *Science Total Environ.*, **9**, 87–109.
175. Raspor, B., Nurnberg, H. W., Valenta, P. and Branica, M. (1981). Voltammetric studies of the stability of Zn(II) chelates with NTA and EDTA and the kinetics of their formation in Lake Ontario water, *Limnol. Oceanogr.*, **26**, 54–66.
176. Raspor, B., Valenta, P., Nurnberg, H. W. and Branica, M. (1977). Application of polarography and voltammetry to speciation of trace metals in natural waters. II. Polarographic studies on the kinetics and mechanism of Cd(II) chelate formation with EDTA in sea water, *Thalassia Jugoslavica*, **13**, 79–91.
177. Cleven, R. F. M. J. (1984). *Heavy metal polyacid interactions*, Ph.D. thesis, Agricultural University of Wageningen, The Netherlands.
178. Buffle, J. and Greter, F.-L. (1979). Voltammetric study of humic and fulvic substances. Part II; Mechanism of reactions of the Pb-fulvic complexes on the mercury electrode, *J. Electroanal. Chem.,* **101,** 231.
179. Buffle, J., Greter, F.-L., Nembrini, G., Paul, J. and Haerdi, W. (1976). Capabilities of voltammetric techniques for water quality control problems, *Z. Anal. Chem.*, **282**, 339.
180. Greter, F.-L., Buffle, J. and Haerdi, W. (1979). Voltammetric study of humic and fulvic substances. Part I; Study of the factors influencing the measurement of their complexing properties with lead, *J. Electroanal. Chem.*, **101**, 211.
181. Pinheiro, J. P., Mota, A. M. and Gonçalves, M. L. S. (1994). Complexation study of humic acids with Cd(II) and Pb(II), *Anal. Chim. Acta,* **284**, 525–537.
182. Bugarin, M. G., Mota, A. M., Pinheiro, J. P. and Goncalves, M. L. S. (1994). Influence of metal concentration at the electrode surface in differential pulse anodic stripping voltammetry in the presence of humic matter, *Anal. Chim. Acta*, **294**, 271–281.
183. Buffle, J., Vuilleumier, J. J., Tercier, M.-L. and Parthasarathy, N. (1987). Voltammetric study of humic and fulvic substances. Part V: interpretation of metal ion complexation measured by anodic stripping voltammetric methods, *Sci. Total Environ.*, **60**, 75–96.
184. Davison, W., Buffle, J. and De Vitre, R. (1998). Voltammetric characterization of a dissolved iron sulphide species by laboratory and field studies, *Anal. Chim. Acta*, **377**, 193–203.
185. Buffle, J., De Vitre, R. R., Perret, D. and Leppard, G. G. (1988). Combining field measurements for speciation in non perturbable water samples. In *Metal*

Speciation: Theory, Analysis and Applications, ed. Kramer, J.R. and Allen, H.E., Lewis, Michigan, Chapter 5.

186. Scarano, G., Morelli, E., Seritti, A. and Zirino, A. (1990). Determination of copper in sea water by anodic stripping voltammetry using ethylenediamine, *Anal. Chem.*, **62**, 943–948.
187. Kounaves, S. P. and Zirino, A. (1979). Studies of cadmium–ethylenediamine complex formation in sea water by computer assisted stripping polarography, *Anal. Chim. Acta,* **109**, 327–339.
188. Donat, J. R. and Bruland, K. W. (1990). A comparison of two voltammetric techniques for determining Zn speciation in Northeast Pacific ocean waters. *Mar. Chem.,* **28**, 301–323.
189. Xue, H. B. and Sigg, L. (1994). Zinc speciation in lake waters and its determination by ligand exchange with EDTA and differential pulse anodic stripping voltammetry, *Anal. Chim. Acta,* **284,** 505–515.
190. van den Berg, C. M. G., Nimmo, M., Daly, P. and Turner, D. R. (1990). Effects of the detection window on the determination of organic copper speciation in estuarine waters, *Anal. Chim. Acta,* **232,** 149–159.
191. Buffle, J. and van Leeuwen, H. P., ed. (1992). *Environmental Particles*, Vol. 1, IUPAC Series in Analytical and Physical Environmental Chemistry, Lewis, Boca Raton, FL.
192. Tercier, M.-L. and Buffle, J. (1996). Antifouling membrane-covered voltammetric microsensor for in situ measurements in natural waters, *Anal. Chem.*, **68**, 3670–3678.
193. Wilkinson, K. J. Balnois, E., Leppard, G. G. and Buffle, J. (1999). Characteristic features of the major components of freshwater colloidal organic matter revealed by transmission electron and atomic force microscopy, *Collids Surf. A*, **155**, 287–310.
194. Scarponi, G., Capodaglio, G., Barbante, C. and Cescon, P. (1996). The anodic stripping voltammetric titration procedure for study of trace metal complexation in sea water. In *Element Speciation in Bioinorganic Chemistry,* ed. Caroli, S., Wiley-Interscience, New York, Chapter 11.
195. Buffle, J., Filella, M. and Altmann, R. S. (1994). Polyfunctional description of metal complexation by natural organic matter. In *OECD Document; Proceeding of an NEA Workshop*, Bad Zurzach, Switzerland.
196. Buffle, J., Tessier, A. and Haerdi, W. (1984). Interpretation of trace metal complexation by aquatic organic matter. In *Complexation of Trace Metals in Natural Waters*, ed. Kramer, C. J. M. and Duinker J. C., Martinus Nijhoff/Dr. W. Junk, p. 301.
197. Buffle, J. and Altmann, R. S. (1987). Interpretation of metal complexation by heterogeneous complexants. In *Aquatic Surface Chemistry*, ed. Stumm W., Wiley, New York, Chapter 13.
198. Xue, H. B. and Sigg, L. (1993). Free cupric ion concentration and Cu(II) speciation in a eutrophic lake, *Limnol. Oceanogr.*, **38**, 1200–1213.
199. Xue, H. and Sunda, W. (1997). Comparison of Cu^{2+} measurements in lake water determined by ligand exchange and cathodic stripping voltammetry and by ion selective electrode. *Environ. Sci. Technol.,* **31**, 1902–1909.
200. Buffle, J. (1981). Calculation of the surface concentration of the oxidized metal during the stripping step in the anodic stripping techniques and its influence on speciation measurements in natural waters, *J. Electroanal. Chem.*, **125,** 273.
201. Almeida Mota, A. M., Buffle, J., Kounaves, S. P. and Simoes Gonçalves, M. L. (1985). The importance of concentration effects at the electrode surface, in anodic

stripping voltammetric measurements of complexation of metal ions in natural water concentrations, *Anal. Chim. Acta*, **172**, 13–30.
202. Mota, A. M., Pinheiro, J. P. and Gonçalves, M. L. (1994). Adsorption of humic acids on a mercury aqueous solution interface, *Wat. Res.*, **28,** 1285–1296.
203. Buffle, J. and Cominoli, A. (1981). Voltammetric study of humic and fulvic substances. Part IV : behaviour of fulvic substances at the mercury–water interface. *J. Electroanal. Chem.*, **121**, 273–299.
204. Filella, M., Buffle, J. and van Leeuwen, H. P. (1990). Effect on physico-chemical heterogeneity of natural complexants. Part I. Voltammetry of labile metal–fulvic complexes, *Anal. Chim. Acta,* **232,** 209–223.
205. Bard, A. J., Parsons, R. and Jordan, J., ed. (1985). *Standard Potentials in Aqueous Solutions*, Marcel Dekker, New York.
206. Buffle, J. and Nembrini, G. (1977). Study of the mechanism of the electrochemical reduction of hydrolysed iron(III) species, in connection with their colloidal properties, *J. Electroanal. Chem.*, **76**, 101.
207. Zali, O. (1983). *Cycles chimiques dans un lac eutrophe*, Ph.D. thesis No. 2090, University of Geneva.
208. Lead, J., Wilkinson, K. J., Startchev, K., Canorica, S. and Buffle, J. (2000). Diffusion coefficients of humic substances as determined by fluorescence correlation spectroscopy: Role of solution conditions, *Environ. Sci. Technol.*, **34**, 1365.
209. Bersier, P. M. and Bersier, J. (1988). Polarographic adsorption analysis and tensammetry: toys or tools for day-to-day routine analysis?, *Analyst*, **113**, 3–14.
210. Davison, W. and Whitfield, M. (1977). Modulated polarographic and voltammetric techniques in the study of natural water chemistry, *J. Electroanal. Chem.*, **75**, 763–789.
211. Whitfield, M. (1975). The electroanalytical chemistry of sea water. In *Chemical Oceanography*, Vol. 4, ed. Riley, J. P. and Skirrow, G., Academic Press, London, Chapter 20.
212. Li, P., Gao, Z., Xu, Y., Wang and G., Zhao (1990). Determination of trace amounts of silver with chemically modified carbon paste electrode, *Anal. Chim. Acta*, **229**, 213–219.
213. Wang, J., Li, R. and Huiliang, H. (1989). Improved anodic stripping voltammetric measurements of silver by codeposition with mercury, *Electroanalysis*, **1**, 417–421.
214. Chakrabarti, C. L., Cheng, J., Lee, W. F., Back, M. H. and Schroeder, W. H. (1996). Rotating disk electrode voltammetry for studying kinetics of metal complex dissociation in model solutions and snow samples, *Environ. Sci. Technol.*, **30,** 1245–1252.
215. Volkening, J. and Heumann, K. G. (1988). Determination of heavy metals at the pg/g level in Antartic snow with DPASV and IDMS, *Fresenius' J. Anal. Chem.*, **331**, 174–181.
216. Belmont-Hébert, C., Tercier, M.-L., Buffle, J., Fiaccabrino, G. C. and Koudelka-Hep, M. (1998). Gel-integrated microelectrode arrays for direct voltammetric measurements of heavy metals in natural waters and other complex media, *Anal. Chem.*, **70**, 2949–2956.
217. Kounaves, S. P., Deng, W., Hallok, P. M., Kovacs, G. T. A. and Storment, C. W (1994). Iridium-based ultramicroelectrode array fabricated by microlithography, *Anal. Chem.*, **66**, 418–423.
218. Aualiitia, T. U. and Pickering, W. F. (1988). Sediment analysis—lability of selectively extracted fractions, *Talanta*, **35**, 559–566.
219. Morrison, G. M. P. and Florence, T. M. (1989). Electrochemical speciation analysis of metals at membrane-coated electrodes, *Electroanalysis*, **1**, 485–491.

220. Wojciechowski, M. and Balcerak, J. (1990). Square-wave anodic stripping voltammetry at glassy carbon-based thin mercury film electrodes in solutions containing dissolved oxygen, *Anal. Chem.*, **62**, 1325–1331.
221. Lam, T. M., Chakrabarti, C. L., Cheng, J. and Pavski, V. (1997). Rotating disk electrode voltammetry / anodic stripping voltammetry for chemical speciation of lead and cadmium in freshwaters containing dissolved organic matter, *Electroanalysis*, **9**, 1018–1029.
222. Gao, Z., Li, P., Wang, G. and Zhao, Z. (1990). Preconcentration and differential-pulse voltammetric determination of iron(II) with Nafion-1,10–phenanthroline-modified carbon paste electrodes, *Anal. Chim. Acta*, **241**, 137–146.
223. Davison, W. (1976). Comparison of differential pulse and D.C. sampled polarography for the determination of ferrous and manganous ions in lake water, *J. Electroanal. Chem.*, **72**, 229–237.
224. De Vitre, R. R., Buffle, J., Perret, D. and Baudat, R. (1988). A study of iron and manganese transformations at the $O_2/S(-II)$ transition layer in a eutrophic lake (Bret, Switzerland): a multimethod approach, *Geochim. Cosmochim. Acta*, **52**, 1601–1613.
225. Davison, W. (1993). Iron and manganese in lakes. A review, *Earth-Sci. Rev.*, **34**, 119–163.
226. Davison, W., Buffle, J. and De Vitre, R. R. (1988). Interpretation of speciation measurements: a case study. Direct polarographic determination of O_2, Fe(II), Mn(II), S(−II) and related species in anoxic waters, *Pure Appl. Chem.*, **60**, 1536–1548.
227. Svancara, I., Vytras, K., Hua, C. and Smith, M. R. (1992). Voltammetric determination of mercury(II) at a carbon paste electrode in aqueous solutions containing tetraphenylborate ion, *Talanta*, **39**, 391–396.
228. Roitz, J. S. and Bruland, K. W. (1997). Determination of dissolved Mn(II) in coastal and estuarine waters by differential pulse cathodic stripping voltammetry, *Anal. Chim. Acta*, **344**, 175–180.
229. Labida, J., Vanickova, M. and Beinrohr, E. (1989). Determination of dissolved manganese in natural waters by differential pulse cathodic stripping voltammetry, *Mikrochim. Acta*, **1**, 113–120.
230. Gassama, N., Sarazin, G. and Evrard, M. (1994). The distribution of Ni and Co in a eutrophic lake: an application of square-wave voltammetry method, *Chem. Geol.*, **118**, 221–233.
231. Boussenant, M., van den Berg, C. M. G. and Ghaddaf, M. (1992). The determination of the chromium speciation in sea water using catalytic cathodic stripping voltammetry, *Anal. Chim. Acta*, **262**, 103–115.
232. Wang, J., Lu, J., Wang, G., Luo, D. and Tian, B. (1997). Simultaneous measurements of trace chromium and uranium using mixed ligand adsorptive stripping analysis, *Anal. Chim. Acta*, **354**, 275–281.
233. Wang, J., Lu, J., Luo, D., Wang, G. and Tian, B. (1997). Simultaneous adsorptive stripping voltammetric measurements of trace chromium, uranium and iron in the presence of cupferron, *Electroanalysis*, **9,** 1247–1251.
234. Wang, G., Ma, C., Zhang, X. and Wang, J. (1994). Adsorptive voltammetric measurement of trace level of iron(III) with 4–(2–pyridylazo) resorcin, *Anal. Lett.*, **27**, 1165–1173.
235. Abollino, O., Mentasti, E., Sarzanini, C., Porta, V. and van den Berg, C. M. G. (1991). Speciation of iron in Antarctic lake water by adsorptive cathodic stripping voltammetry, *Anal. Proc. (London)*, **28**, 72–73.

236. Khoo, S. B., Soh, M. K., Cai, Q., Kham, M. R. and Guo, S. X. (1997). Differential pulse cathodic stripping voltammetric determination of Mn(II) and Mn(VII) at the 1-(2-pyridylazo)-2-naphthol-modified carbon paste electrode, *Electroanalysis*, **9**, 45–51.
237. Quentel, F., Riso, R. and Madec, C. (1988). Determination of trace concentrations of molybdenum(VI) in aqueous media by adsorptive differential pulse polarography, *Analusis*, **16**, 507–513.
238. Farrias, P. A. M., Ohara, A. K., Nobrega, A. W. and Gold, J. S. (1994). Ultratrace determination of molybdenum in the presence of 2–(2'-thiazolylazo)-p-cresol by catalytic–adsorptive stripping voltammetry, *Electroanalysis*, **6**, 333–339.
239. Wang, J., Lu, J. and Taha, Z. (1992). Catalytic–adsorptive stripping voltammetric determination of ultratrace levels of molybdenum in the presence of organic hydroxy acids, *Analyst*, **117**, 35–37.
240. Wang, J., Pamidi, P. V. A., Nascimento, V. B. and Angnes, L. (1997). Dimethylglyoxime-doped sol–gel carbon composite voltammetric sensor for trace nickel, *Electroanalysis*, **9**, 689–692.
241. Yokoi, K., Yamaguchi, A., Mizumachi, T. and Koide, T. (1995). Direct determination of trace concentrations of lead in freshwater samples by adsorptive cathodic stripping voltammetry of a lead–calcein blue complex, *Anal. Chim. Acta*, **316**, 363–369.
242. El-Shahawi, M. S., Abu Zuhri, A. Z. and Kamal, M. M. (1994). Adsorptive stripping voltammetric measurements of trace amounts of platinum(II) and ruthenium(III) in the presence of 1–(2–pyridylazo)-2–naphthol, *Fresenius' J. Anal. Chem.*, **348**, 730–735.
243. Wang, J., Wu, Y. and Liao, Z. (1995). Studies on adsorptive wave of antimony(III)-PMBP complex, *Xua. Yan. Yu Ying.*, **7**, 64–67.
244. Wagner, W., Sander, S. and Henze, G. (1996). Trace analysis of antimony(III) and antimony(V) by adsorptive stripping voltammetry, *Fresenius' J. Anal. Chem.*, **354**, 11–15.
245. Takashi, I. and Tatsuhiko, T. (1996). Cathodic stripping voltammetry of selenium(IV) at a silver disk electrode, *Anal. Chem.*, **68**, 3789–3792.
246. Wang, J. and Mahmoud, J. S. (1986). Stripping voltammetry with adsorptive accumulation for trace measurements of titanium, *J. Electroanal. Chem.*, **208**, 383–394.
247. Li, H. and van den Berg, C. M. G. (1989). Determination of titanium in sea water using adsorptive cathodic stripping voltammetry, *Anal. Chim. Acta*, **221**, 269–277.
248. Cai, Q. and Khoo, S. B. (1995). Determination of trace thallium after accumulation of thallium(III) at a 8–hydroxyquinoline-modified carbon paste electrode, *Analyst*, **120**, 1047–1053.
249. Naganuma, T. and Jin, J. (1990). Determination of uranyl ion in seawater by adsorptive voltammetry, *Aichi Kyoiku Daigata Kenkyu Hokoku Shizen Kagako*, **39**, 35–43.
250. Sander, S., Wagner, W. and Henze, G. (1995). Direct determination of uranium traces by adsorptive stripping voltammetry, *Anal. Chim. Acta*, **305**, 154–158.
251. van den Berg, C. M. G. and Nimmo, M. (1987). Direct determination of uranium in water by cathodic stripping voltammetry, *Anal. Chem.*, **59**, 924–928.
252. Jin, W., Shi, S. and Wang, J. (1990). Determination of ultra-trace vanadium by 1.5^{th} and 2.5^{th} order derivative adsorption voltammetry, *J. Electroanal. Chem.*, **292**, 41–47.
253. Greenway, G. M. and Wolfbauer, G. (1995). On-line determination of vanadium by adsorptive stripping voltammetry, *Anal. Chim. Acta*, **312**, 15–25.

254. Wang, J., Peng, T. and Varughese, K. (1987). Adsorptive preconcentration for voltammetric measurements of trace levels of zirconium, *Talanta*, **34**, 561–566.
255. Panneli, M. G. and Voulparopoulos, A. N. (1994). Simultaneous voltammetric determination of cobalt, nickel and labile zinc using 2–quinolinethiol in the presence of surfactants without prior digestion, *Fresenius' J. Anal. Chem.*, **348**, 837–839.
256. Olsen, K. B., Wang, J., Setiadji, R. and Lu, J. (1994). Field screening of chromium, cadmium, zinc, copper and lead in sediments by stripping analysis, *Environ. Sci. Technol.*, **28**, 2074–2079.
257. Wang, J. (1985). *Stripping Analysis: Principles, Instrumentation and Applications*, VCH, Deerfield Beach, FL.
258. van den Berg, C. M. G. (1991). Potentials and potentialities of cathodic stripping voltammetry of trace elements in natural waters, *Anal. Chim. Acta*, **250**, 265–276.
259. Wang, J. (1988). Adsorptive stripping voltammetry. A new electroanalytical avenue for trace analysis, *J. Res. Natl. Bur. Stand. U.S.*, **93**, 489–490.
260. van den Berg, C. M. G. (1989). Electroanalytical chemistry of sea water. In *Chemical Oceanography*, **Vol. 9**, ed. Riley, J.P., Academic Press, London, Chapter 51.
261. van den Berg, C. M. G., Khan, S. H., Daly, P. J., Riley, J. P. and Turner, D. R. (1991). An electrochemical study of Ni, Sb, Se, U and V in the estuary of the Tamar, *Estuarine, Coast. Shelf Sci.*, **33**, 309–322.
262. Abuzuhri, A. Z. and Voelter, W. (1998). Applications of adsorptive stripping voltammetry for trace analysis of metals, pharmaceuticals and biomolecules. A review, *Fresenius' J. Anal. Chem.*, **360**, 1–9.
263. Nürnberg, H. W. (1984). The voltammetric approach in trace metal chemistry of natural waters and atmospheric precipitation. Review, *Anal. Chim. Acta*, **164**, 1–21.
264. Valenta, P. (1988). Potentialities and applications of voltammetry in the determination of ecotoxic trace metals in natural waters body fluids and foods, *GIT Fachz. Lab.*, **32**, 312–320.
265. Nürnberg, H. W. (1985). Applications and potentialities of voltammetry in environmental chemistry of ecotoxic metals. In *Electrochemistry in Research and Development*, ed. Kalvoda, R. and Parsons, R., Plenum, New York, p. 121.
266. Whitfield, M. and Jagner, D. (1981). *Marine Electrochemistry. A Practical Introduction*, Wiley- Interscience, New York.
267. Nürnberg, H. W. (1983). Investigations on heavy metal speciation in natural waters by voltammetric procedures, *Fresenius' Z. Anal. Chem.*, **316**, 557–565.
268. Jagner, D. and Granéli, A. (1976). Potentiometric stripping analysis, *Anal. Chim. Acta*, **83**, 19–26.
269. Jagner, D. (1982). Potentiometric stripping analysis. A review, *Analyst*, **107**, 593–599.
270. von Wandruszka, R. (1994). Trace element determination by electrochemical methods. In *Determination of Trace Elements*, ed. Alfassi, Z. B., VCH, New York, Chapter 10.
271. Town, R. M. (1997). Potentiometric stripping analysis and anodic stripping voltammetry for measurement of Cu(II) and Pb(II) complexation by fulvic acid: a comparative study, *Electroanalysis*, **9**, 407–415.
272. Chang, S.-K., Kozeniauskas, R. and Harrington, G. W. (1977). Determination of nitrite ion using differential pulse polarography, *Anal. Chem.*, **49**, 2272–2275.
273. van den Berg, C. M. G. and Li, H. (1988). The determination of nanomolar levels of nitrite in fresh and sea water by cathodic stripping voltammetry, *Anal. Chim. Acta*, **212**, 31–41.
274. Kalcher, K. (1986). A new method for the voltammetric determination of nitrite, *Talanta*, **33**, 489–494.

275. Bartik, M. and Kupka, J. (1960). Polarographische Nitratbestimmung im biologischen Material, *Collect. Czechoskov. Chem. Commun.*, **25**, 3356–3362.
276. Markusova, K. (1989). Determination of nitrate in natural waters by voltammetry at a stationary mercury drop electrode, *Anal. Chim. Acta*, **221**, 131–138.
277. Hemmi, H., Hasebe, K., Ohzeki, K. and Kambara, T. (1984). Differential pulse polarographic determination of nitrate in environmental materials, *Talanta*, **31**, 319–323.
278. Xing, X. and Scherson, D. A. (1987). Rotating ring-disk electrode method for the quantitative determination of nitrate ions in aqueous solutions in the submicromolar range, *Anal. Chem.,* **59**, 962–964.
279. Canterford, D. R. (1975). Simultaneous determination of cyanide and sulfide with rapid direct current polarography, *Anal. Chem.*, **47**, 88–92.
280. Chen, L. (1992). Oscillographic adsorptive wave determination of microamounts of orthophosphate, *Talanta*, **39**, 765–768.
281. Laitinen, H. A. and Lin, Z.-F. (1963). Coulometric efficiency of anodic deposition and cathodic stripping of chloride at silver electrodes, *Anal. Chem.*, **35**, 1405–1412.
282. Berge, H. and Brügmann, L. (1971). Polarographische Methoden zur Bestimmung von Bromidionen im Meerwasser, *Beitr. Meereskd.*, **28**, 19–32.
283. Grether, C., Bruttel, P. and Hadinger, O. (1977). Determination polarographique des anions inorganiques. Application Bulleting Metrohm, No. 111 f.
284. Herring, J. R. and Liss, P. S. (1974). A new method for the determination of iodine species in seawater, *Deep-sea Res.*, **21**, 777–783.
285. Manandhar, K. and Pletcher, D. (1977). Determination of halide ions by cathodic stripping analysis, *Talanta*, **24**, 387–390.
286. Colovos, G., Wilson, G. S. and Moyers, J.-L. (1974). Simultaneous determination of bromide and chloride by cathodic stripping voltammetry, *Anal. Chem.*, **46**, 1051–1054.
287. Butler, E. C. and Smith, J. D. (1980). Iodine speciation in seawaters—the analytical use of ultra-violet photo-oxidation and differential pulse polarography, *Deep-sea Res.*, **27A**, 489–493.
288. Shain, I. and Perone, S. P. (1961). Application of stripping analysis to the determination of iodide with silver microelectrodes, *Anal. Chem.*, **33**, 325–329.
289. Berge, H. and Brügmann, L. (1972). Indirekte Bestimmung von Fluoridionen im Meerwasser durch Wechselstrompolarographie, *Beitr. Meereskd.*, **29**, 115–127.
290. Ciglenecki, I. and Cosovic, B. (1997). Electrochemical determination of thiosulfate in seawater in the presence of elemental sulfur and sulfide, *Electroanalysis*, **9**, 775–780.
291. Luther III, G. W., Church, T. M., Giblin, A. E. and Howarth, R. W. (1986). In *Organic Marine Geochemistry*, ed. Sohn M., ACS Symp. Ser. **305**, American Chemical Society, Washington, DC, Chapter 20.
292. Wang, F., Tessier, A. and Buffle, J. (1998). Voltammetric determination of elemental sulphur in porewater, *Limnol. Oceanogr.*, **43**, 1353–1361.
293. Ciglenecki, I. and Cosovic, B. (1996). Electrochemical study of sulfur species in seawater and marine phytoplankton, *Mar. Chem.*, **52**, 101–110.
294. Ortiz, R., De Marquez, O. and Marquez, J. (1988). Cathodic stripping determination of halides in mixtures, *Anal. Chim. Acta*, **215**, 307–310.
295. Bersier, P. M. and Bersier, J. (1985). Applied polarography and voltammetry of organic compounds in practical day-to-day analysis. Part II. *Rev. Anal. Chem.*, **16**, 81–128.
296. Bersier, P. M. and Bersier. J. (1994). Polarographic, voltammetric and HPLC-EC of pharmaceutically relevant cyclic compounds, *Electroanalysis*, **6**, 171–191

297. Smyth, W. F., ed. (1980). *Electroanalysis in Hygiene, Environmental, Clinical and Pharmaceutical Chemistry*, Elsevier, Amsterdam.
298. Benadikova, H. and Kalvoda, R. (1984). Adsorptive stripping voltammetry of some triazine- and nitro group-containing pesticides, *Anal. Lett.*, **17**, 1519–1531.
299. Koen, J. G. and Huber, J. F. K. (1970). A rapid method for residue analysis by column liquid chromatography with polarographic detection. Application to the determination of parathion and methylparathion on crops, *Anal. Chim. Acta,* **51**, 303–307.
300. Volke, J. and Slamnik, M. (1981). Polarography and related methods. In *Pesticide Analysis,* ed. Dask, G., Marcel Dekker, New York.
301. Lam, N. K. and Kopanica, M. (1984). Determination of trichlorobiphenyl by adsorption stripping voltammetry, *Anal. Chim. Acta,* **161**, 315–324.
302. Lankelma, J. L. and Poppe, H. (1970). Design of a tensammetric detector for high speed high efficiency liquid chromatography in columns and its evaluation for the analysis of some surfactants, *J. Chromatogr. Sci.*, **14**, 310–315.
303. Zutic, V., Cosovic, B. and Kozarac, Z. (1977). Electrochemical determination of surface active substances in natural waters. On the adsorption of petroleum fractions at mercury electrode / seawater interface, *J. Electroanal. Chem.*, **78**, 113–121.
304. Bednarkiewicz, E., Donten, M. and Kublik, Z. (1981). Determination of trace surfactants in distilled, potable and untreated waters and in supporting electrolytes by tensammetry with accumulation on the HMDE, *J. Electroanal. Chem.*, **127**, 241–253.
305. Kalvoda, R. and Novotny, L. (1986). Polarographic behaviour of petroleum components in aqueous solutions: a study using DPP and convective adsorption accumulation, *Collect. Czech. Chem. Commun.*, **51**, 1587–1594.
306. Kalvoda, R. and Novotny, L. (1984). Monitoring petroleum and petroleum products in waters on the basis of their activity at the solution-mercury interface, *Vodni hospodarstvi*, **34**, 291–296.
307. Concialini, V., Chlavari, G. and Vitali, P. (1983). Electrochemical detection in high-performance liquid chromatographic analysis of aromatic amines, *J. Chromatogr.*, **258**, 244–251.
308. Lores, E. M., Edgerton, T. R. and Moseman, R. F. (1981). Method for the confirmation of chlorophenols in human urine by LC with an electrochemical detector, *J. Chromatogr. Sci.*, **19**, 466–469.
309. Samuelsson, R. and Osteryoung, R. A. (1981). Determination of *N*-nitrosamines by high-performance liquid chromatographic separation with voltammetric detection, *Anal. Chim. Acta,* **123**, 97–105.
310. Samuelsson, R. and Rydstrom, T. (1980). Pulse polarographic studies of *N*-nitrosamines, *Anal. Chem. Symp. Ser.*, **2** (Electroanal. Hyg., Environ., Clin. Pharm. Chem.), 435–444.
311. Chey, W. E., Adams, R. N. and Yllo, M. S. (1977). Anodic differential pulse voltammetry of aromatic amines and phenols at trace levels, *J. Electroanal. Chem.*, **75**, 731–738.
312. Gunasingham, H., Tay, B. T., Ang, K. P. and Koh, L. L. (1984). Electrochemical detection of polynuclear aromatic hydrocarbons following reversed-phase gradient high-performance liquid chromatography using a large-volume wall-jet detector, *J. Chromatogr.,* **285**, 103–114.
313. Smyth, M. R. and Osteryoung, J. (1980). Electroanalysis of environmental carcinogens, *Anal. Chem. Symp. Ser.*, **2** (Electroanal. Hyg., Environ., Clin. Pharm. Chem.), 423–434.

314. Bond, A. M. and McLachlan, N. M. (1986). Direct determination of tetraethyllead and tetramethyllead in gasoline by high-performance liquid chromatography with electrochemical detection at mercury electrodes, *Anal. Chem.*, **58**, 756–758.
315. Pietrzyk, D. J. (1974). Organic polarography, *Anal. Chem.*, **46**, 52R–73R.
316. Brezina, M. and Zuman, P., (1958). *Polarography in Medicine, Biochemistry and Pharmacy*, Interscience, New York.
317. Krizan, V. (1981). *Analysis of the Atmosphere,* Alpha-SNTL, Bratislava, Prague.
318. Kalvoda, R., ed. (1987). *Electroanalytical Methods in Chemical and Environmental Analysis*, Plenum, New York.
319. Siegerman, H. (1975). Oxydation and reduction half-wave potentials of organic compounds. Appendix. In *Techniques of Chemistry*, Vol. V, Part II: *Techniques of Electroorganic Synthesis*, Wiley, New York, pp. 667–1052.
320. Meites, L., Zuman, P. and Rupp, E. B., ed. (1976–1982). *Handbook Series in Organic Electrochemistry*, Vols. 1–5, CRC Press, Boca Raton, FL.
321. Breyer, B. and Bauer, H. H. (1963). *Alternating Current Polarography and Tensammetry*, Interscience, New York.
322. Wang, J., Luo, D. B., Farias, P. A. M and Mahmoud, J. S. (1985). Adsorptive stripping voltammetry of riboflavin and other flavin analogues at the static mercury drop electrode, *Anal. Chem.,* **57**, 158–162.
323. Mart, L. (1982). Minimization of accuracy risks in voltammetric ultratrace determination of heavy metals in natural waters, *Talanta*, **29**, 1035–1040.
324. Chen, Y. and Buffle, J. (1996). Physicochemical and microbial preservation of colloid characteristics of natural water samples. Part II: physicochemical and microbial evolution, *Wat. Res.*, **69**, 2185–2192.
325. Mann, A. W. and Lintern, M. J. (1984). Field analysis of heavy metals by portable digital voltammeter, *J. Geochem. Explor.*, **22**, 333–348.
326. Mann, A. W. (1987). Portable digital voltammetry and its application to ultratrace analysis in exploration, *Symp. Ser.—Australas Inst. Min. Metall*, **54**, 119–122.
327. Behnert, J. and Raezke, K. P. (1990). Schwermetallanalytik mittels Voltammetrie, *LaborPraxis*, **14**, 508–511.
328. Clavell, C. and Zirino, A. (1985). On-line shipboard determination of trace metals in seawater with a computer-controlled voltammetric instrument. In *Mapping Strategies in Chemical Oceanography*, ed. Zirino, A., American Chemical Society, Washington, DC, Chapter 8.
329. Valenta, P., Sipos, L., Kramer, I., Krumpen, P. and Rützel, H. (1982). An automatic voltammetric analyzer for the simultaneous determination of toxic trace metals in water, *Fresenius' Z. Anal. Chem.*, **312**, 101–108.
330. Bond, A. M. and Svestka, M. (1993). Developments, trends and commercial availability of instrumentation (hardware and software) in microcomputer based voltammetry. Review, *Collect. Czech. Chem. Commun.*, **58**, 2769–2812.
331. Herdan, J., Feeney, R., Kounaves, S. P., Flannery, A. F., Storment, C. W. and Kovacs, G. T. A. (1998). Field evaluation of an electrochemical probe for in situ screening of heavy metals in groundwater, *Environ. Sci. Technol*, **32**, 131–136.
332. Lecomte, J., Mericam P., Astruc, A. and Astruc, M. (1981). Oxygen elimination in the direct polarographic determination of trace metals in natural waters, *Anal. Chem.*, **53**, 2371–2374.
333. Kemula, W. (1967). Polarographic methods of analysis, *Pure Appl. Chem.*, **15**, 283–296.
334. Wang, J. and Ariel M. (1981). Subtractive anodic stripping voltammetry with twin identical mercury film electrodes differing in their convection transport during deposition, *Anal. Chim. Acta*, **128**, 147–153.

335. Sipos, L., Kozar, S., Kontusic, F. and Branica, M. (1978). Subtractive anodic stripping voltammetry with rotating mercury coated glassy carbon electrode, *J. Electroanal. Chem.*, **87**, 347–357.
336. Baranski, A. S. (1987). Rapid anodic stripping analysis with ultramicroelectrodes, *Anal. Chem.*, **59**, 662–666.
337. Osteryoung, J. and O'Dea, J. (1986). Square wave voltammetry. In *Electroanalytical Chemistry*, Vol. 14, ed. Bard, A. J., Marcel Dekker, New York, pp. 209–308.
338. Wojciechoski, M., Go, W. and Osteryoung, J. (1985). Square-wave anodic stripping analysis in the presence of dissolved oxygen, *Anal. Chem.*, **57**, 155–158.
339. Wojciechoski, M. and Balcerzak, J. (1990). Square-wave anodic stripping voltammetry at glassy-carbon-based thin mercury film electrodes in solutions containing dissolved oxygen, *Anal. Chem.*, **62**, 1325–1331.
340. Persson, B. and Rosen, L. (1981). Flow injection determination of isosorbide nitrate with polarographic detection, *Anal. Chim. Acta*, **123**, 115–123.
341. Maccrehan, W. A. and May, W. E. (1984). Oxygen removal in liquid chromatography with a zinc oxygen-scrubber column, *Anal. Chem.*, **56**, 625–628.
342. Olsson, B., Ögren, L. and Johansson, G. (1983). An enzymatic flow injection method for the determination of oxygen, *Anal. Chim. Acta*, **145**, 101–108.
343. Barisci, J. N. and Wallace, G. G. (1992). Removal of oxygen in flowing solutions using a photochemical process, *Electroanalysis*, **4**, 323–326.
344. Hanekamp, H. B., Woogt, W. H., Bos, P. and Frei, R. W. (1980). An electrochemical scrubber for the elimination of eluent background effects in polarographic flow-through detection, *Anal. Chim. Acta*, **118**, 81–86.
345. Tercier-Waeber, M.-L. and Buffle, J. (2000). Submersible on-line oxygen removal system coupled to an in situ voltammetric probe for trace element monitoring in freshwater, *Environ. Sci. Technol.*, **34,** 4018–4024.
346. Trojanek, A. and Holub, K. (1980). The continuous removal of oxygen from flowing solutions, *Anal. Chim. Acta,* 1 **21,** 23–28.
347. Bessarabov, D. G., Jacobs, E. P., Sanderson, R. D. and Beckman, I. N. (1996). Use of nonporous polymeric flat-sheet gas-separation membranes in a membrane-liquid contactor: experimental studies, *J. Memb. Sci*, **113**, 275–284.
348. Moskvin, L. N., Rodinkov, O. V., Katruzov, A. N., Grigorev, G. L. and Kromovborisov, S. N. (1995). Dissolved oxygen removal from aqueous media by the chromatomembrane method, *Talanta*, **42**, 1707–1710.
349. Pedrotti, J. J., Angnes, L and Gatz, G. R. (1994). A fast, highly efficient, continuous degassing device and its application to oxygen removal in flow-injection analysis with amperometric detection, *Anal. Chim. Acta*, **298**, 393–399.
350. Rollic, M. E., Ho, C.-N. and Warner, I. M. (1983). Sample deoxygenation for fluorescence spectrometry by chemical scavenging, *Anal. Chem.*, **55**, 2445–2448.
351. Chai, X. S. and Danielsson, L. G. (1996). Approaches to in-line removal of dissolved oxygen in flow systems for process analysis, *Anal. Chim. Acta*, **332**, 31–38.
352. Johnson, B. M., Baker, R. W., Matson, S. L., Smith, K. L., Roman, I. C., Tuttle, M. E. and Lonsdale, H. K. (1987). Liquid membrane for the production of oxygen enriched air. II facilitated-transport membranes, *J. Memb. Sci.*, **31**, 31–67.
353. Niederhoffer, E. C., Timmons, J. H. and Martell, A. E. (1984). Thermodynamics of oxygen binding in natural and synthetic dioxygen complexes, *Chem. Rev.*, **84**, 137–203.
354. Smart, R. B. and Stewart, E. E. (1985). Differential pulse anodic stripping voltammetry of cadmium (II) at a membrane-covered electrode: measurements in the presence of model organic compounds, *Environ. Sci. Technol.*, **19**, 137–140.

355. Brezonik, P. L., Brauner, P. A. and Stumm, W. (1976). Trace metal analysis by anodic stripping voltammetry: effect of sorption by natural and model organic compounds, *Wat. Res.*, **10**, 605.
356. Sagberg, P. and Lund, W. (1982). Trace metal analysis by anodic stripping voltammetry. Effect of surface-active substances, *Talanta*, **29**, 457–460.
357. Wang, J. and Lao, D. B. (1984). Effect of surface-active compounds on voltammetric stripping analysis at the mercury film electrode, *Talanta*, **31**, 703–707.
358. Laxen, D. H. and Harrison, R. M. (1981). A scheme for physicochemical speciation of trace metals in freshwater samples, *Sci. Total Environ.*, **19**, 59–82.
359. Brainina, Kh. Z., Stozhko, R. M., Yu, N. and Chernysheva, A. V. (1984). Electrochemical wet ashing and separation of interfering elements in stripping voltammetry of natural waters, *Zh. Anal. Khim.*, **39**, 1660–1663 (English translation).
360. Aldstadt, J. H. and Dewald, H. D. (1993). Effect of model organic compounds on potentiometric stripping analysis using a cellulose acetate membrane-covered electrode, *Anal. Chem.*, **65**, 922–926.
361. Wang, J., Bonakdar, M. and Pack, M. M. (1987). Glassy carbon electrodes coated with cellulose acetate for adsorptive stripping voltammetry, *Anal. Chim. Acta*, **192**, 215–223.
362. Vidal, J. C., Viñao, R. B. and Castillo, J. R. (1992). Binding capacity of casein to lead and voltammetric speciation of lead in milk with a Nafion coated electrode, *Electroanalysis*, **4**, 653–659.
363. Hoyer, B. and Florence, T. M. (1987). Application of polymer-coated glassy carbon electrodes to the direct determination of trace metals in body fluids by anodic stripping voltammetry, *Anal. Chem.*, **59**, 2839–2842.
364. Hoyer, B. and Florence, T. M. (1987). Application of polymer-coated glassy carbon electrodes in anodic stripping voltammetry, *Anal. Chem.*, **59**, 1608–1614.
365. Dam, M. E. R., Thomsen, K. N., Pickup, P. G. and Schroder, K. H. (1995). Comparative study of polymer-coated mercury film electrodes for voltammetric analysis of lead and cadmium in the presence of surfactants, *Electroanalysis*, **7**, 70–78.
366. Wang, J. and Taha, Z. (1990). Poly(ester-sulfonic-acid)-coated mercury film electrodes for anodic stripping voltammetry, *Electroanalysis*, **2**, 383–387.
367. Wang, Z., Galal, A., Zimmer, H. and Mark, H. B. (1992). Anodic stripping voltammetry at mercury films deposited on conducting poly(3–methylthiophene) electrodes, *Electroanalysis*, **4**, 77–85.
368. Dalangin, R. R. and Gunasingham, H. (1994). Mercury(II) acetate-Nafion modified electrode for anodic stripping voltammetry of lead and copper with flow-injection analysis, *Anal. Chim. Acta*, **291,** 81–87.
369. Zen, J. M., Hsu, F. S., Chi, N. Y., Huang, S. Y. and Chung, M. J. (1995). Effect of model organic compounds on square-wave voltammetric stripping analysis at the Nafion/chelating agent mercury film electrodes, *Anal. Chim. Acta*, **310,** 407–417.
370. Buffle, J., Wilkinson, K. J., Tercier, M.-L. and Parthasarathy, N. (1997). In situ moniotring and speciation of trace metals in natural waters, *Ann. Chim.*, **87**, 67–82.
371. Zoski, C. G. J. (1990). A survey of steady-state microelectrodes and experimental approaches to a voltammetric steady-state, *J. Electroanal. Chem.*, **296**, 317–333.
372. Zhang, H. and Davison, W. (1999). Diffusional characteristics of hydrogels used in DGT and DET techniques. *Anal. Chim. Acta* **398**, 329–340.
373. Kapur, V. (1995). *Transport in polymer/gel modified micropores*, Ph.D. thesis, Carnegie Mellon University, Pittsburgh, PA.

374. Cizkowska, M. and Osteryoung, J. G. (1996). Voltammetric studies of transport of thallium (I) counterions in solutions of polyelectrolytes in mixed sovents, *J. Phys. Chem.*, **100**, 4630–4636.
375. Davison, W., Zhang, H. and Grime, G. W. (1994). Performance characteristics of gel probes used for measuring the chemistry of pore waters, *Environ. Sci. Technol.,* **28,** 1623–1632.
376. Zhang, H., Davison, W., Miller, S. and Tych, W. (1995). In situ high resolution measurements of fluxes of Ni, Cu, Fe and Mn and concentrations of Zn and Cd in porewaters by DGT, *Geochim. Cosmochim. Acta*, **59**, 4181–4192.
377. Brainina, Kh. Z., Khanina, R. M., Forhstadt, V. M., Vilchinskaya, E. A. and Gaponenko, G. L. (1990). In *Proc. of J. Heyrovsky Centenial Congress on Polarography. 41st Meeting of International Society of Electrochemistry*, Prague, p.17.
378. Wang, J. and Tian, B. (1993). Gold ultramicroelectrodes for on-site monitoring of trace lead, *Electroanalysis,* **5**, 809–814.
379. Wang, J., Sacman, E. and Tian, B. (1994). Stripping potentiometric measurements of copper in blood using gold microelectrode, *Anal. Chim. Acta*, **286**, 189–195.
380. Gil, E. P. and Ostapczuk, P. (1994). Potentiometric stripping determination of mercury (II), selenium (IV), copper (II) and lead (II) at a gold film electrode in water samples, *Anal. Chim. Acta*, **293**, 55–65.
381. Jagner, D., Josefson, M. and Aren, K. (1982). Flow potentiometric stripping analysis for mercury (II), *Anal. Chim. Acta*, **141**, 147–156.
382. Buffle, J., De Vitre, R. R., Perret, D. and Leppard, G. G. (1989). Physico-chemical characteristics of a coloidal iron phosphate species formed at the oxic–anoxic interface of a eutrophic lake, *Geochim. Cosmochim. Acta*, **53**, 399–408.
383. Perret, D., Newman, M., Nègre, J. C., Chen, Y. and Buffle, J. (1994). Submicron particles in the Rhine river. I Physico-chemical characterisation, *Wat. Res.*, **28**, 91–106.
384. Stumm, W. and Morgan, J. J. (1995). *Aquatic Chemistry*, Wiley, New York.
385. Luther, G. W., Brendel, P. J., Lewis, B. L., Sunby, B., LeFrançois, L., Silverberg, N. and Nuzzio, D. B. (1998). Simultaneous measurements of O_2, Mn, Fe, I^- and S(−II) in marine pore waters with a solid-state voltammetric microelectrode, *Limnol.Oceanogr.*, **43**, 325–333.
386. Buffle, J. and Stumm, W. (1994). General chemistry of aquatic systems. In *Chemical and Biological Regulation of Aquatic Systems*, Lewis, Boca Raton, FL, Chapter 1.
387. Ciglenecka, I. and Cosovic, B. (1997). Electrochemical determination of thiosulfate in sea water in the presence of elemental sulfur and sulfide, *Electroanalysis,* **9**, 775–780.
388. Batina, N., Ciglenecki, I. and Cosovic, B. (1992). Determination of elemental sulfur, sulfide and their mixtures in electrolyte solutions by ac voltammetry, *Anal. Chim. Acta*, **267**, 157–164.
389. Haerdi, W., Buffle, J. and Monnier, D. (1969). Contribution à l'étude du dosage de submicrotraces de fer par polarographie inverse sur goutte de mercure pendante. Parts I et II, *J. Electroanal.Chem.,* **23,** 81–98.
390. Winkler, K., Krogulec, T. and Galus, Z. (1985). Formation of FeS and its effect on the electrode reactions of the Fe(II)/Fe^0 system in thiocyanate solutions at mercury electrode, *Electrochim. Acta*, **30**, 1055–1062.
391. Stojek, Z. and Kublik, Z. (1976). Investigation of iron(II)/iron(mercury) system in thiocyanate media by cyclic and stripping voltammetry with hanging mercury drop and mercury film electrodes, *J. Electroanal. Chem.*, **70**, 317–331.

392. Riso, R. D., Le Coore, P. and Chaumery, C. J. (1997). Rapid and simultaneous analysis of trace metals (Cu, Pb and Cd) in seawater by potentiometric stripping analysis, *Anal. Chim. Acta*, **351**, 83–89.
393. Adeloju, S. B., Sahara, E. and Jagner, D. (1996). Anodic stripping potentiometric determination of Cu, Pb, Cd and Zn in natural waters on a novel combined electrode system, *Anal. Lett.*, **29**, 283–302.
394. Ostapczuk, P. (1993). Present potentials and limitations in the determination of trace elements by potentiometric stripping analysis, *Anal. Chim. Acta*, **273**, 35–40.
395. Quentel, F., Elleouet, C. and Madec, C. L. (1997). Determination of traces of aluminium(III) in natural freshwater by cathodic redissolution flollowing adsorption of aluminium– lumogallion complex, *Analusis*, **25**, 222–225.
396. Cai, Q. and Khoo, S. B. (1993). Determination of trace aluminium by differential pulse adsorptive stripping voltammetry of aluminium(III)–8–hydroxyquinoline complex. *Anal. Chim. Acta*, **276**, 99–108.
397. Li, H. and Smart, R. B. (1996). Determination of sub-nanomolar concentration of arsenic(III) in natural waters by square wave cathodic stripping voltammetry, *Anal. Chim. Acta*, **325**, 25–32.
398. van den Berg, C. M. G. (1984). Determination of copper in seawater by cathodic stripping voltammetry of complexes with catechol, *Anal. Chim. Acta*, **164**, 195–207.
399. Quentel, F., Elleouet, C. and Madec C. (1994). Determination of copper in seawater by adsorptive voltammetry with 1,2–dihydroxyanthraquinone-3–sulfonic acid, *Electroanalysis*, **6**, 683–688.
400. Farias, A. M. P., Ferreira, S. L. C., Ohara, A. K., Bastos, B. M. and Goulart, S. M. (1992). Adsorptive stripping voltammetric behavior of copper complexes of some heterocyclic azo compounds, *Talanta*, **39**, 1245–1253.
401. Zhang, G. and Fu, C. (1991). Adsorptive voltammetric determination of copper with a benzoin oxime paste electrode, *Talanta*, **38**, 1481–1485.
402. Collado-Sanchez, C., Perez-Pena, J., Gelado-Caballero, M. D., Herrera-Melian, J. A. and Hernandez-Brito, J. J. (1996). Rapid determination of copper, lead, cadmium in unpurged seawater by adsorptive stripping voltammetry, *Anal. Chim. Acta*, **320**, 19–30.
403. Iliadou, E. N., Girousi, S. T., Dietze, U., Otto, M., Voulgaropoulos, A. N. and Papadopoulos, C. G. (1997). Simultaneous determination of nickel, cobalt, cadmium, lead and copper by adsorptive voltammetry using 1-phenylpropane-1-pentylsulfonylhydrazone-2-oxime as a chelating agent, *Analyst*, **122**, 597–600.
404. Nimmo, M., van den Berg, C. M. G. and Brown, J. (1989). The chemical speciation of dissolved nickel, copper, vanadium and iron in Liverpool Bay, Irish sea, *Estuar., Coast. Shelf Sci.*, **29**, 57–74.
405. Bedeki, V. I. and Voulgaropoulos, A. N. (1992). Investigation of the determination of ultratrace amounts of cobalt(II) complexed with 1,10–phenanthroline by adsorptive voltammetry, *Fresenius' J. Anal. Chem.*, **342**, 352–356.
406. Zhang, H., Vire, J. C., Patriarche, G. J. and Wollast, R. (1988). Determination of cobalt ions in natural waters using differential pulse adsorptive stripping voltammetry, *Anal. Lett.*, **21**, 1409–1424.
407. Tercier-Waeber, M. L., Buffle, J., Confalonieri, F., Riccardi, G., Sina, A., Graziottin, F., Fiaccabrino, G. C., Koudelka-Hep, M. (1999). Submersible voltammetric probes for in situ real-time trace element measurements in surface water, ground-water and sediment-water interface. *Meas. Sci. Technol.* (Special issue: Subsea measurement and intrumentation), **10**, 1202–1213.

10 Permeation Liquid Membranes for Field Analysis and Speciation of Trace Compounds in Waters

J. BUFFLE AND N. PARTHASARATHY
University of Geneva, Switzerland
N.-K. DJANE
ABB Corporate Research, Sweden
L. MATTHIASSON
University of Lund, Sweden

In Situ Monitoring of Aquatic Systems: Chemical Analysis and Speciation Edited by J. Buffle and G. Horvai.

1 INTRODUCTION

In environmental systems trace compounds, either organic or inorganic, are found in various chemical forms, such as free hydrated ions, complexes with organic natural biopolymers, or associated with colloidal particles. These species play various roles in geochemical and biological cycling of elements including transport, bioaccumulation and toxicity [1]. In natural water, the concentrations of many chemicals of environmental relevance are low ($< 100\,\text{nM}$) and that of their free forms is still lower ($< 100\,\text{pM}$). Their determination remains a challenge to analytical chemists, since very few techniques combine both speciation capability and high sensitivity [2].

Permeation liquid membrane (PLM) techniques are newly emerging procedures for separation and preconcentration of target elements or species [3–13]. They are often called in the literature supported liquid membrane (SLM) techniques, based on the type of devices which is the most widely used for liquid membrane formation (Figure 4). SLM should be used when referring specifically to this technical device. The more general word PLM is recommended and will be used below when the general concept is of interest, as it does not depend on the device but clearly points out the fact that the flux (permeation) through the membrane is the key factor of interest.

Although PLM analytical techniques have only been used sparsely in the field and not yet at depth in waters, rugged system have been developed and they have proven to be very promising techniques for *in situ* applications. PLMs basically preconcentrate the analyte from the test solution into a so-called strip or receiver solution. The chemical composition of the system is such that facilitated transport occurs through the membrane, in other words the preconcentration occurs automatically as soon as the permeation membrane is in contact with the test sample. The PLM approach is thus a preconcentration technique that is easy to handle *in situ* and, in this respect, it is very similar to the technique of diffusive gradient in thin films (DGT; [2]) described in Chapter 11. In addition, as the analyte is preconcentrated in a solution, the coupling of the PLM with various detectors is possible. Thus a PLM can also be combined with *in situ* probes. Although the application of PLMs to in field or *in situ* analysis of waters is still in its infancy, the present results show many potential advantages: in particular a high selectivity, minimum sample handling, easy automation, real time analysis and application to wide ranges of metal ions, inorganic anions and organic compounds. A last important feature is that it does not provide total analyte concentration in water but it is selective for the free and/or lipophilic forms of the test compounds, i.e. it provides particularly useful information for bioavailability and toxicity studies of chemicals. By coupling a PLM to sensitive detectors, determination of very low concentrations (down to 10^{-13} M or below) is possible.

In this chapter the fundamental principles of PLM, the mechanism of transport, and applications of this technique to trace compound analysis will be presented. A particular emphasis is given to the role of speciation of the test metals in the sample, on their transport flux through the PLM and the corresponding influence on preconcentration. Although this part is treated for trace metals, the same concepts are applicable to organic compounds.

2 DESCRIPTION OF PERMEATION LIQUID MEMBRANE SYSTEMS

2.1 PRINCIPLE

Liquid membrane (LM) systems consist (Figure 1) of a water immiscible organic liquid, usually containing a carrier, C, selective for the species of interest, interposed between two aqueous phases called respectively the source and strip (or receiver) solutions. The transport is based on liquid–liquid extraction coupled with diffusion. The analyte, M, (Figures 1 and 2; M is supposed to be any analyte in Figure 1 and a metal ion in Figure 2 and later in the chapter) diffuses to the source–membrane interface, and either reacts with the carrier C (if present) to form a lipophilic complex, or just dissolves in the solvent. The carrier mainly helps to extract hydrophilic compounds. The test compound or its complex diffuses across the membrane and is released at the membrane–strip interface. The driving force for this facilitated transport is a gradient of chemical potential of the test compound. This potential should be lower in the strip than in the source solution. This can be achieved in various ways. For trace metal transport (Figure 2), the stripping agent S is a ligand (stronger than C and

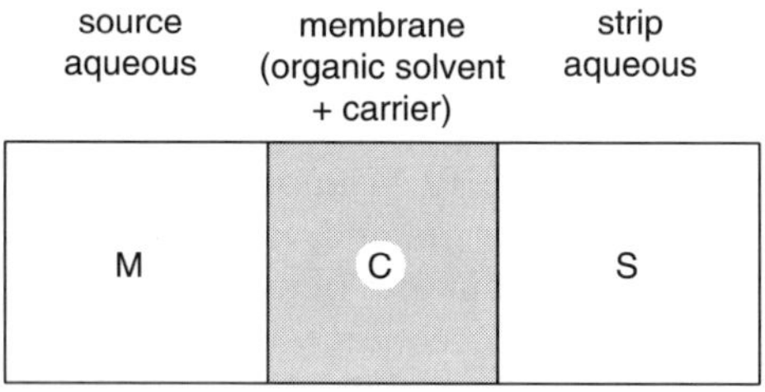

source aqueous | membrane (organic solvent + carrier) | strip aqueous

M | C | S

Figure 1. Schematic diagram of a liquid membrane system. M = test analyte (M may be a metal ion but also an anion or an organic compound (see Section 3); S = stripping agent; C = carrier

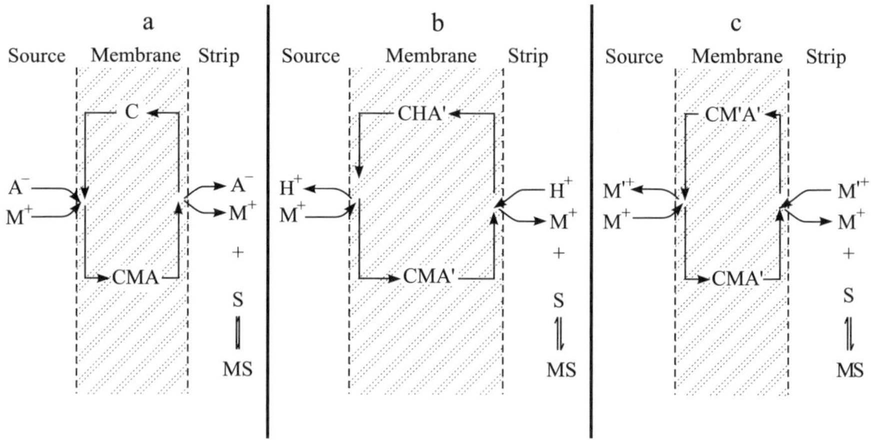

Figure 2. Mechanisms for maintaining electroneutrality during the transport of a cation M^+ through liquid membranes: (a) cotransport of anions, A^-; (b) counter-transport of H^+; (c) counter-transport of another (alkaline) cation, M'^+. A' = lipophilic anion on the membrane.Other symbols as in Figure 1 (modified from ref. [18])

any ligand of the source phase), dissolved in the strip solution (Section 5.2.1; Figure 18). For the transport of organic compounds, acid–base reactions in the source and strip solutions are often used to decrease the chemical potential in the strip phase. In addition, during the transport of charged species such as metal ions, electroneutrality must be fulfilled in the whole system. For transport of cations, this can be achieved by (a) the co-transport of a counter-anion (Figure 2a), (b) the counter-flux transport of H^+ (Figure 2b), or (c) the counter-flux transport of any other cation (Figure 2c). Conditions (a), (b) or (c) should be carefully selected when developing the analytical procedure, in particular the choice of the strip phase composition. Usually scheme (a) is not recommended, as the flux of M will depend on the test (= source) composition. Scheme (b) must also be used with caution as the pH (and thus speciation) in the test solution may vary if conditions are not appropriate.

Depending on the device used (section 2.2.), an accumulation time of a few minutes to a few hours (Figure 7 a and b) is needed for the system to reach a full equilibrium. Flux through the system is then nil. Note that the technique can be used in several ways, to get different types of information, by measuring the total concentration of the analyte in the strip solution, c_{st}:

- at equilibrium, the corresponding concentration, c_{st}^{e}, is proportional to either the total concentration or to that of the free (uncomplexed) analyte species in the source solution, depending on the nature of the stripping phase and the volumes of strip and source phases (section 5.2.1);
- the flux through the membrane (i.e. dc_{st}/dt) over a short accumulation time may provide the concentration of free analyte species, of the whole of the labile complexes or of lipophilic complexes depending on the characteristics of the device used (sections 5.2.2 and 5.2.3).

These various approaches are discussed in detail in section 5.

2.2 DEVICES

Three basic PLM systems can be used: bulk (BLM), emulsion (EML) and supported (SLM) liquid membranes (Figures 3 and 4). A *bulk liquid membrane* (BLM) (Figure 3a) consists of a stirred organic phase that separates source and strip solutions. It is useful for physicochemical studies, but, because of its low flux, it is unsuitable for analytical applications [7–9]. *Emulsion liquid membranes* (ELMs) were first developed by Li and Cussler [14,15]. The solvent is dispersed in the source solution containing a surfactant which stabilizes an emulsion made of aqueous strip droplets coated by a thin film of solvent (Figure 3b). After extraction, the emulsion is collected and broken. Since transport occurs through an extremely thin membrane, it is very fast. This system is well suited for industrial application [16,17], but not for either on site or *in situ*

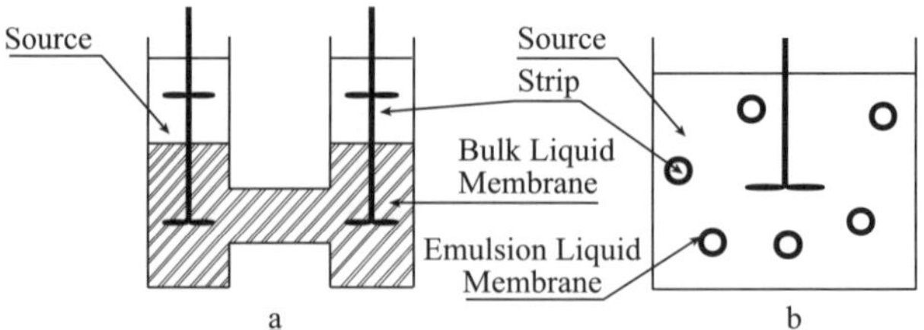

Figure 3. Schematic description of (a) a bulk liquid membrane (BLM) and (b) an emulsion liquid membrane (ELM)

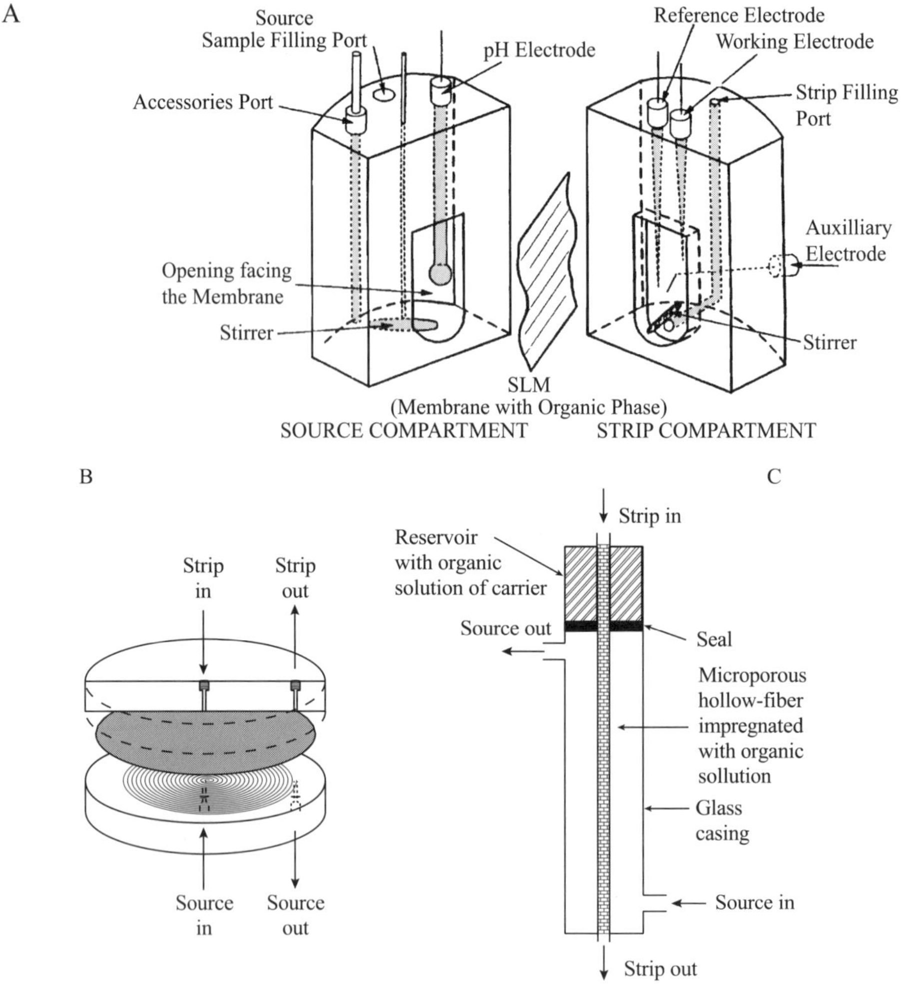

Figure 4

D

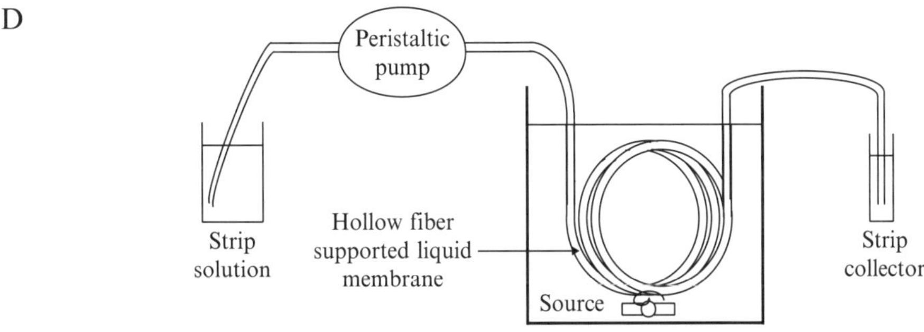

E

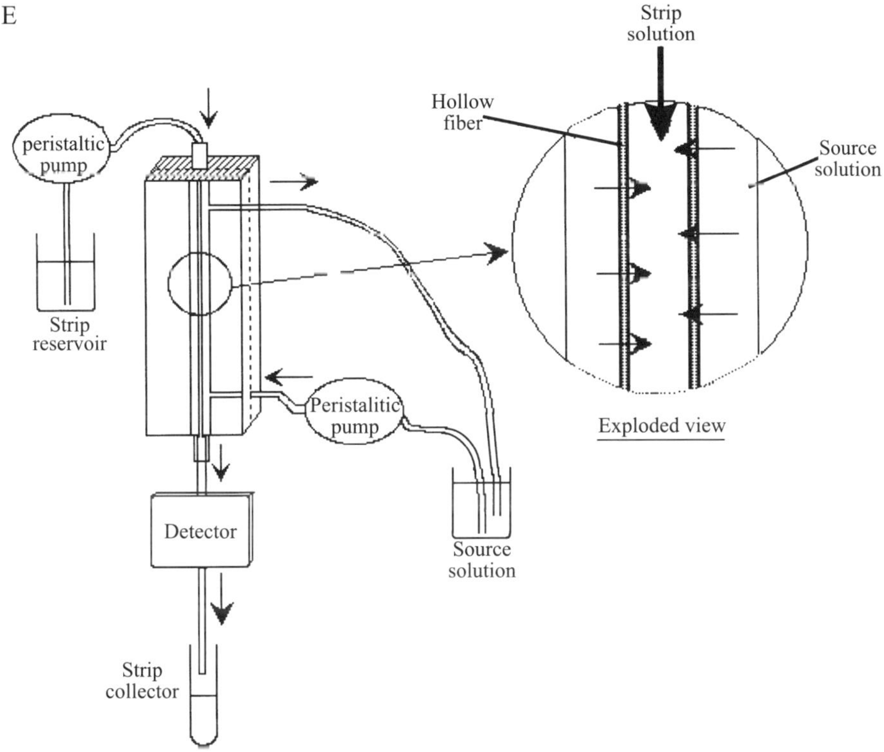

Figure 4. Various types of cells used with flat sheet and hollow fibre permeation SLMs. (A) Flat sheet with stirred source and strip solutions [18]; (B) flat sheet with a flow-through cell made of two spiral channels machine grooved in PTFE blocks [5]; (C–E) single hollow fibre with strip solution inside the lumen. In (C) and (E) [21] the source solution is circulated around the fiber. System C includes a reservoir for continuous regeneration of the liquid membrane in the support (modified from refs [5,18,21])

environmental measurements. In the *supported liquid membrane* (SLM) system (Figure 4A–E), the source and strip solutions (Figure 2) are separated by a chemically inert microporous support impregnated with the hydrophobic solvent and carrier. The organic liquid is held in the pores of the membrane by capillary forces. This type of LM configuration is best suited for separation, preconcentration, and extraction of test compounds or elements from real world samples. The attractive feature is that the fluxes are high owing to large surface area and, since small quantities of carrier are required, expensive carriers can be used. Long term stability of the membrane is the major limitation. This point will be discussed in detail in section 3.4.

Different types of support geometries [3,4,7,12] can be used for an SLM. The simplest system, in which a flat sheet support separates two stirred compartments (Figure 4A) is useful for physicochemical but not analytical studies, as the preconcentration factor, $F = c_{st}/c_s^0$ (c_{st} = strip concentration at time, t, and c_s^0 = source concentration at time $t = 0$) is low with such systems (section 5.2.1; [18]). Hollow fibre (Figure 4C–E; [3,7]) and spiral wound geometries (Figure 4B; [19]) are more amenable to analytical applications at trace levels. The surface area to volume ratios of these systems are high (up to $10^{+4}\,m^2 m^{-3}$; [20]), and hence high fluxes and short analysis time can be achieved. The strip solution is introduced in the lumen of the fibre (or in one channel of the spiral system) and left stagnant, while the source solution is stirred (Figure 4D) or circulated (Figure 4C and E) on the outside of the fibre, or in the second channel of the spiral system (Figure 4B). Preconcentration factors of the order of 100–1000 [21] can be obtained with such systems. In the rest of the chapter, the discussion on PLM principles will refer often to SLM devices even though they are applicable to other LM types.

2.3 THE COMPONENTS OF PLM SYSTEMS

The major components of PLM are: the inert membrane support, the aqueous solution and the organic phase (carrier and solvent) constituents.

2.3.1 Organic Solvent

The role of the solvent is to form a hydrophobic barrier between the two aqueous solutions. Some of the commonly used organic solvents in liquid–liquid extractions such as ethers, ketones, alkanes, alcohols, fatty acids, aromatic hydrocarbons and haloalkanes can be used for PLM. However, to design stable membranes, certain physicochemical criteria must be met in addition to those needed for the extraction step. The choice of solvent in fact is critical, as it influences membrane stability, rate of mass transport and extraction efficiency.

The solvent must be very insoluble in water since only a thin film of solvent with a large area is in contact with large volumes of aqueous solutions. Slight solubility would cause solvent losses resulting in risks of leakage between the two aqueous phases. An intermediate viscosity is required. Indeed, the diffusion coefficients of the analytes in the liquid membrane, and in turn the mass transfer rates across it, are inversely proportional to the viscosity of the solvent. Thus a solvent with low viscosity is preferable to obtain high fluxes. However, when the membrane is in contact with a turbulent flow of source solution, it may deteriorate as a result of emulsion formation. This effect may be minimised using high viscosity solvents that also exhibit low surface tensions with water. Finally volatile solvents should also be avoided.

As far as chemical properties of the solvent are concerned, in addition to the requirement of low solubility of solvent in water, two factors have to be considered: the rate of mass transfer and the selectivity. A high partition coefficient for the analyte between the source solution and the organic solvent is favorable, as this gives rise to high mass transfer rates (K_p' in equation (5)). However, the partition coefficient of potential interfering species should be low to enhance the selectivity of the system. In addition, for analytes with too high partition coefficients in the membrane phase, the transfer rate from the membrane to the strip solution may be slow. It may then be necessary to wait for some time, after the preconcentration experiment, to ensure that the membrane is empty of the test compound and no memory effect will occur. Extensive work has been done on the effect and polarity of the solvents on the overall performance of the PLM, especially on membrane stability [22–25]. Examples of some commonly used organic solvents are given in Table 1.

Dozol *et al.* [23] studied the transport of strontium by DC18C6 across an SLM impregnated with different solvents and investigated their influence on SLM stability (Table 2). They found that isotridecanol in aromatic solvents with aliphatic chains having at least six carbon atoms (e.g. *n*-hexylbenzene)

Table 1. Physicochemical properties of a few commonly used solvents ($T = 25\,°C$). [210]

Solvent	Surface tension, $\gamma \times 10^3 (N\,m^{-1})$	Viscosity, $\eta \times 10^3$ $(kg\,m^{-1}\,s^{-1})$	solvent–water interfacial tension, $\gamma \times 10^3 (N\,m^{-1})$	Density, $\rho \times 10^3 (kg\,m^{-3})$
n-Heptane	19.6	0.382	50.8	0.684
iso-Heptane	–	0.430	32.7	0.705
Methylcyclohexane	23.3	0.659	41.1	0.771
Toluene	27.9	0.539	35.7	0.866
Kerosene	25.3	1.24	41.8	0.790
Nitrobenzene	43.2	1.87	25.7	1.20
Ethyl chloroacetate	31.2	1.13	15.9*	1.159
l-Octanol	27.1	7.47	8.4†	0.827

* At $T = 27°C$.
† At $T = 20°C$.

Table 2. Influence of the nature of the solvent on the lifetime of a flat-sheet SLM in the transport of radionuclides from nuclear waste waters [23]. Carrier 0.5mol L^{-1} DC18C6 (dicyclohexane 18 crown 6). Source solution = synthetic nuclear waste water containing Sr nuclide. Membrane = Celgard 2500; Cell system = type of Figure 4A. Stirring rate = 500 rpm. Strip solution = demineralised water. In column 2, M means mol L^{-1}

Solvent			Lifetime (h)
p-Diisopropylbenzene	+	0.6 M 4-nonylphenol	> 200
p-Diisopropylbenzene	+	0.9 M isotridecanol	> 200
n-Butylbenzene	+	0.4 M 4-nonylphenol	70–140
n-Hexylbenzene	+	0.5 M 4-nonylphenol	> 200
n-Hexylbenzene	+	0.7 M 1-decanol	> 200
n-Hexylbenzene	+	0.7 M 1-isotridecanol	> 200
n-Hexylbenzene	+	0.8 M 1-isotridecanol	> 200
n-Octylbenzene	+	0.9 M 1-isotridecanol	> 200
n-Decylbenzene	+	1.0 M 1-isotridecanol	> 200
n-Dodecylbenzene	+	1.1 M 1-isotridecanol	> 200
n-Tridecylbenzene	+	1.3 M 1-isotridecanol	> 200
2-Ethyl-1-hexanol			30-90
Alcohol phenetyl			6
1-Decanol			48-72
Isotridecanol			> 200
2-Nitropropane			1
1-Nitrohexane			7-24
Nitrocyclohexane			5-6
4-Nitro-*m*-xylene			53-70
2-Nitro-*p*-cumene			168-192*
1,6-Dibromohexane			130
1,2,4-Trichlorobenzene			40

* Organic phase without extractant.

provided the longest lifetimes (> 200 h). Audunsson [22] investigated the influence of five different solvents on the mass transfer rates of benzylamine as a model substance. It was found that the rate of transfer of benzylamine decreases with increased viscosity among the alkanes. This observation is in agreement with other studies [26,27]. Barnes [25] evaluated the influence of different organic solvents containing 40% di-2–ethylhexylphosphoric acid (DEHPA) on the general performance of the SLM, for the transport of Cu(II). Some of the results are shown in Figure 5 [25] and point out a few general trends in agreement with the results of Izatt *et al.* [28]. The expected theoretical curve is an initial linear increase in flux followed by a curvature reaching a plateau which corresponds to the equilibrium state (Figure 6). The salient features of the results obtained by Barnes (Figure 5) are the following: (i) some of the solvents indeed show the expected behaviour (e.g. diethylbenzene); (ii) others show a faster initial increase (higher flux) due to the lower viscosity of the solvent (e.g. pentane) while (iii) a third class of solvents shows an initial increase

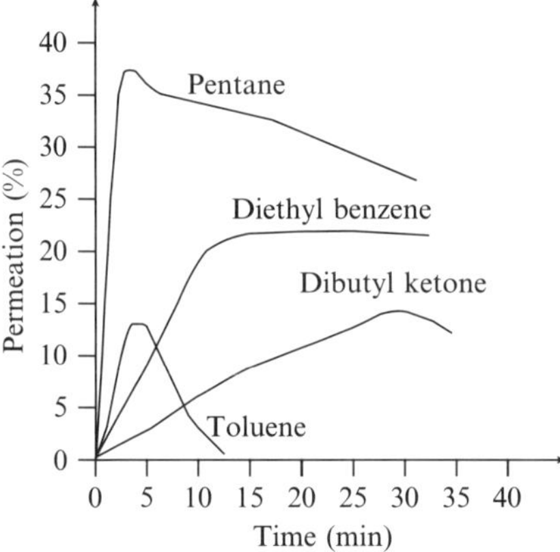

Figure 5. Permeation versus time curves for the transport of Cu(II) by diethylhexylphosphoric acid (40 %) in various solvents (modified from ref. [25]). System used: see Figure 4B type. Flow rates = $1.5\,\text{ml}\,\text{min}^{-1}$ (source solution) and $0.75\,\text{ml}\,\text{min}^{-1}$ (strip solution). Source solution: $1.57 \times 10^{-3}\,\text{mol}\,\text{L}^{-1}$ Cu(II) in $0.25\,\text{mol}\,\text{L}^{-1}$ acetate buffer. Strip solution: $0.25\,\text{mol}\,\text{L}^{-1}\,HNO_3$

followed by a decrease (e.g. toluene, pyridine) which is due to loss of the solvent either by dissolution in water or volatisation. When these last two effects become very important (e.g. acetone, dichloromethane, pyridine), no initial increase in permeation rate is observed.

2.3.2 Carriers

A large number of potential organic carriers exist for PLM. Neutral, cation-exchange and anion-exchange extractants are commonly used. The reagents used for analytical metal determination include in particular the macrocyclic extractants. Most studies on extraction by macrocyclic compounds are devoted to oxygen-containing macrocycles [6,8,12]. However, a variety of macrocyclic compounds with more than one ring or with donor atoms such as nitrogen, oxygen–nitrogen combinations, sulfur or phosphorus in place of oxygen, have been synthesized [6,8,12,13]. The most important property of these compounds is their unique selectivity for metal ions [6,8]. These compounds, especially the diaza crown ethers, are known to complex well with transition metals as well as with alkali and alkaline earth metals and can be used as metal transport carriers [28–32]. However, the unsubstituted diaza compounds are very hydrophilic and hence are unsuitable for PLM applications. The addition of alkyl side chains on the ring such as in 1,10–didecyl-1,10–diaza-crown-6 is required for PLM purposes [33]. The most important carriers used in PLM are listed in Tables 5, 6 and A1–A3. The thermodynamic and kinetic properties of their complexes may be found in refs [34,35].

A theoretical relationship has been found [36,37] between flux and thermodynamic stability constants (log *K*) of metal complexation with macrocyclic ligands. The optimal log *K* values (in methanol) for effective transport of monovalent ions and divalent ions were found to be in the ranges 5.5–6.0 and 6.5–7.0, respectively. This was confirmed by the work of Kirch and Lehn [38], theoretically as well as experimentally. For instance, for Cu(II) transport with 22DD (section 2.4), the log *K* value for the Cu(II) complex is 7.8 in aqueous medium, i.e. close to the optimum value. Stability constants which are not too high ($\log K < 10$) or too low ($\log K > 4$) would be a good choice. For high selectivity, tailor-made macrocyclic compounds must be synthesised with a cavity which best fits with the target metal. Unfortunately, log *K* values of many lipophilic macrocycles in low dielectric solvents are still unknown.

2.3.3 Membrane Support

The physical and morphological properties of the inert membrane material play a significant role in the stability of SLMs, and may have some effect on solute permeation. The inert membrane support must have a critical surface tension both greater than that of the organic solution and lower than that of the aqueous solutions [23]. Furthermore, the pore diameter must be as small as possible (≤ 0.1 μm; [24]). Audunsson [22] investigated the influence of different support matrices (with varying pore size) for the extraction of amines. It was observed that membranes with pore diameters ≥ 3.0 μm showed instantaneous leakage between the source and strip solutions. Table 3 lists the flat sheet (3a) and hollow fibre (3b), materials which are the most widely used as supports in SLM.

2.4 MODELS OF TRANSPORT PROCESSES THROUGH PLMs

Metal transport rates in PLM can be limited by diffusion or by chemical kinetics. Various theoretical models for predicting fluxes in flat sheet and hollow fibre PLMs have been developed (Table 4). Only a model based on the most widely used assumptions (Models I of Table 4) will be discussed in some detail below; for more information, the interested reader should consult the references given in Table 4 and in section 5.2.

Experimental transport results show that, amongst the various possible processes occurring in PLM transport (Figure 17), diffusion of the metal complex across the membrane is often the rate limiting factor. For practical applications, two experimental systems are worth comparing: (i) the system of Figure 4A, where a flat sheet membrane is separating two well-stirred compartments of similar volumes, and (ii) systems where the source compartment is homogeneous but infinitely large compared with the strip compartment. The latter situation corresponds for instance to the configuration of Figures 4B,D and E where the volume of the strip channel is very small (1–100 μL) whereas an unlimited volume of source solution can be circulated in the source channel (Figure 4B) or around the hollow fibre (Figure 4D and E). In both cases the analyte, M, (a metal ion, an inorganic anion or an organic compound) will be

Table 3. Characteristics of microporous flat sheets (a) and hollow fibre (b) used as supporting membrane in an SLM (specifications from manufacturer)

(a)

Membrane	Material	Pore diameter d_p(μm)	Thickness ℓ(μm)	Porosity (%)
Fluropore	polytetrafluoroethylene			
FP-200		2.00	100	83
FP-045		0.45	80	75
FP-010		0.10	60	55
FG		0.2	60	70
FH		0.5	60	85
FA		1.0	60	85
FS		3.0	125	85
Duragard 2500	polypropylene	0.04	25	45
Nuclepore	polycarbonate	0.40	10	125
Durapore GV (white)	poly(vinylidene fluoride)	0.22	125	75
Miltex	polytetrafluoroethylene			
LS		5.0	125	60
LC		10.0	125	68
Hoechst	polypropylene			
Celgard 2500		0.04	25	45
Celgard 2400		0.02	25	45

(b)

Hollow fibre	Material	Pore diameter d_p (μm)	Thickness ℓ (μm)	Lumen diameter (mm)	Porosity (%)
Gore-Tex TA001	PTFE	2	400	1.00	50
Mitsubishi Rayon					
KPF-190M	Polypropylene	0.16	22	0.20	45
EHF-207T	Polypropylene	0.27	55	0.27	70
Akzo-Nobel (Accurel)					
plasmaphan	Polypropylene	0.2	170	0.33	–
Q3/2	Polypropylene	0.2	100	0.60	–
S6/2	Polypropylene	0.2			
Hoechst					
Celgard X-20	Polypropylene	0.03	25	0.40	40
Celgard X-10	Polypropylene	0.03	25	0.20	20

transported through the membrane, as long as its chemical potential in the strip solution is lower than that in the source solution. Assuming the activity coefficients to be equal to unity and the chemical potentials of M at standard state in source and strip solutions to be equal to each other, the transport will stop when the bulk concentrations of free M in the two solutions are equal, i.e. $[M]_s^b = [M]_{st}^b$. If M is in the free form in the source solution and mostly combined to S, as MS, in the strip solution, then M concentrations will vary with time as follows (Figure 6), in systems (i) and (ii).

(i) When $V_s \sim V_{st}$, M is depleted from the source solution, $[M]_{st}^b$ increases with time, but $[MS]_{st}^b$ remains ~ 0 because of the strong complexation of M by S (Figure 6A). Consequently the gradient in the membrane decreases to 0 and equilibrium is reached when $[M]_s^b = [M]_{st}^b \sim 0$. Figure 6B clearly shows that the maximum total concentration of M in the strip solution ($= [MS]_{st}^b(t = \infty)$) is equal to the initial concentration of M in the source ($[M]_s^b(t = 0)$), if V_s and V_{st} are equal. The initial slopes of the concentration versus time curves (Figure 6B) are proportional to the corresponding gradients in the membrane (see below), and therefore also to ($[M]_s^b(t = 0)$).
(ii) When $V_s = \infty$ (Figure 6C), M never depletes in the source, even when a large number of moles are consumed by the strip solution. Thus $[M]_s^b$ remains independent of time (Figure 6D), whereas M continues to accumulate in the strip, until $[M]_{st}^b = [M]_s^b$. This may correspond to extremely high values of total M concentration in the strip ($= [MS]_{st}^b$) if S is a strong reagent. $[M]_{st}^b$ at equilibrium and the initial slope of $[M]_{st}^b$ versus time curve are both proportional to $[M]_s^b$.

The mathematical derivation of the simplest model for case (i) is given below as example. Similar mathematical expressions for case (ii) can be derived by analogy.

Danesi ([3,39]; Models I in Table 4) has developed a simple model for case (i), for both counter-(Figure 2b) and cotransport (Figure 2a) systems by considering a series of rate limiting steps, i.e. (i) steady-state diffusion of metal species in the aqueous boundary Nernst layer, at the source–membrane interface (not shown in Figure 6; see Figure 19); (ii) steady-state diffusion of metal complex in the membrane and (iii) chemical kinetics for the formation/dissociation of MC at the source–membrane interface. The concentration gradients and symbols are those of Figure 19, but the model does not take into account the diffusion of M in the strip solution and assumes $[M]_{st}^j \to 0$. The model also assumes that M is not complexed by ligands L in the source solution. This latter condition is specifically discussed in section 5.2 (Figure 19).

The overall flux, J, of M through the membrane is equal to the number of moles of M leaving the source compartment, per unit of time, per unit area:

$$J = -\frac{V_s}{A}\frac{d[M]_s^b}{dt} \tag{1}$$

where A is the surface area of the membrane.

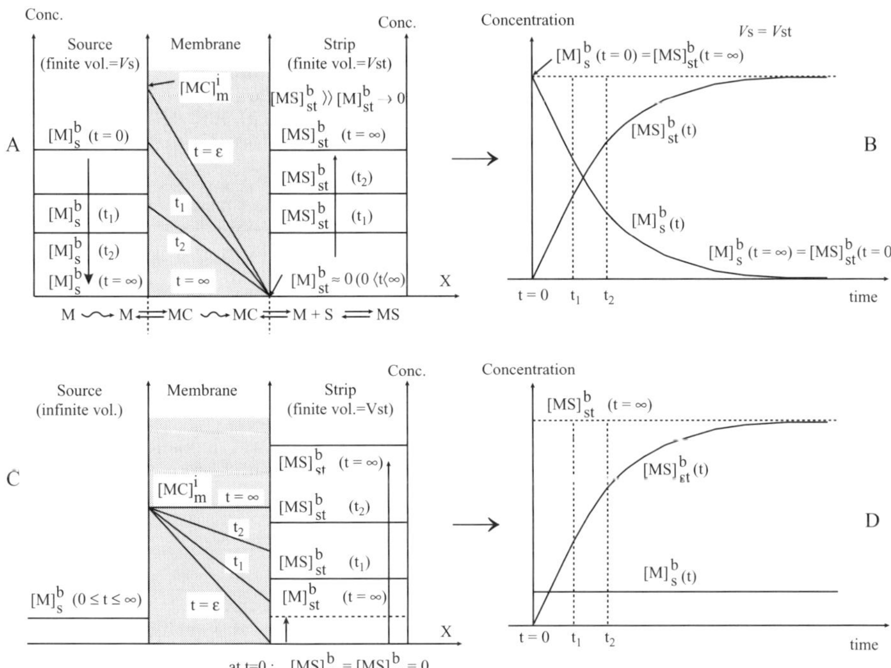

Figure 6. Schematic representations of the evolutions of concentrations and concentration gradients (A, C) and the resulting concentration versus time curves (B,D) in strip and source solutions of PLM, for two different conditions:

- (A, B) equal volumes of source and strip solutions ($V_{st} = V_s$) and excess of stripping agent S in the strip solution (e.g. configuration of Figure 4A);
- (C, D) infinite volume of source solution and finite (small) volume of the strip solution with an excess of stripping agent (e.g. configuration of a hollow fibre; Figure 4C–E) with an infinite reservoir of source, such as lake or ocean water).

Non-linear gradients (not shown) are formed at short times ($0 < \text{time} < \varepsilon$). Linear gradients in the membrane are simplified representations and diffusion layers in source and strip solutions are neglected for clarity (see Figures 17 and 19 for more details). Source and strip compartments are supposed to be stirred (homogeneous). There is no complexing agent in the source solution. $[MC]_m \gg [M]_m$ in the membrane and $[MS]_{st} \gg [M]_{st}$ in the strip solution. Subscripts and superscripts: s = source; st = strip; b = bulk; i = interface; m = membrane.

Table 4. Characteristics of the major theoretical models developed for predicting metal ion flux by carrier-aided transport through the PLM (M = metal; C = carrier; MC = metal–carrier complex; $[M]_s$, $[M]_{st}$ = concentrations of M in source and strip phases respectively)

Kinetic assumptions	Equilibrium conditions	Other conditions and comments	References
Models I. Rate limiting step = diffusion in the membrane (and in solution) at steady state			
• Diffusion in the membrane and source and strip solutions • Steady-state linear concentration gradient in the membrane	• Fast equilibrium of M partitioning at interface • Negligible concentration of free M in the membrane	• Flat sheet membrane or hollow fibre • Analytical equations • Tests with Eu^{3+}/HDEP and other systems	3,39, 40,41 42,43
• Diffusion in the membrane only • Steady-state linear concentration gradient	• Fast equilibrium of MC in the membrane • Formation of ion pairs with M in source solution • Electroneutrality considered in the whole system	• Flat sheet membrane or hollow fibre • Loss of carrier is considered • Analytical equations	36,43, 212
Models II. Rate limiting step = diffusion in the membrane and solutions; non-steady state			
• Diffusion in the membrane and source and strip solution • Non-steady state in the membrane • Steady- state in aqueous solutions	• Fast partitioning at interface • $[M]_s$ = const.; $[M]_{st} = 0$	• Flat sheet • Numerical simulation • Tests with Cu^{II}/oxime	44
• Diffusion in the membrane and source and strip solutions • Non-steady state in the membrane and strip solution • Linear gradient in source solution	• Fast partitioning at interfaces • $[M]_s$ = constant • $[M]_{st}$ = increases with time	• Hollow fibre • Numerical simulation • Tests with Cu^{II}/22DD	45
Models III: Rate limiting steps = diffusion in the membrane and/or interfacial reactions			
• Slow diffusion in membrane only • Steady-state conditions • Slow rate of M release at membrane–strip interface	• Fast partitioning at source membrane interface • Electroneutrality maintained in the membrane • Carrier saturated by M	• Flat sheet • Analytical equations • Tests with K, Na/ Calix 4 arene/NPOE (and other solvents in ref. [36])	42,46, 47
• Slow adsorption of carrier and desorption of metal–carrier complex at source/ membrane interface • Diffusion *not* limiting	• Fast MC complexation/ decomplexation reactions • Source–membrane interface saturated with carrier	• Bulk liquid membrane (BLM) • Analytical equations • Tests with K/carboxylic acid-crown ether	48

J is governed by the following processes:

(i) the diffusion of M through the source phase diffusion boundary layer (Figures 6 and 19 for symbols)

$$J_s = \frac{D_s^M}{\delta_s}\left([M]_s^b - [M]_s^i\right) \tag{2}$$

where J_s is the corresponding flux in the source phase (in mol m^{-2} s^{-1});

(ii) chemical reaction kinetics at the source–membrane interface

$$J_i = k_f^C[M]_s^i - k_d^C[MC]_m^i \tag{3}$$

where J_i is the interfacial flux and k_f^C, and k_d^C are (pseudo) first order reaction rate constants;

(iii) diffusion of MC complex in the membrane

$$J_m = \frac{D_m^{MC}}{\ell}\left([MC]_m^i - 0\right) \tag{4}$$

At steady state, $J_s = J_i = J_m = J$. By combining equations (2)–(4), the interfacial concentrations can be eliminated to give the overall flux J:

$$J = \left[\frac{1}{(\delta_s/D_s^M) + \left(\ell/K_p' D_m^{MC}\right) + \left(k_d^C\right)^{-1}}\right][M]_s^b \tag{5}$$

where the term in square brackets is the overall permeability coefficient, $P = J/[M]_s^b$, expressed in m s^{-1} and $K_p' = K_p^M[C]_m = [MC]_m^i/[M]_s^i$ (Figure 18). The three limiting cases of equation (5) are the following:

- diffusion in source solution is rate limiting when

$$\frac{\delta_s}{D_s^M} \gg \frac{\ell}{D_m^{MC}K_p'} + \frac{1}{k_d^C} \tag{6}$$

- diffusion in the membrane is rate limiting when

$$\frac{\ell}{D_m^{MC}K_p'} \gg \frac{\delta_{rs}}{D_s^M} + \frac{1}{k_d^C} \tag{7}$$

- partitioning at source–membrane interface is limiting when

$$\frac{1}{k_d^C} \gg \frac{\delta_s}{D_s^M} + \frac{\ell}{D_m^{MC}K_p'} \tag{8}$$

In all cases, by combining equations (1) and (5) and integrating the number of moles of M passing through the membrane over time t, the following relationship is obtained for the decrease of $[M]_s^b$ with time in the source solution:

$$[\mathrm{M}]_{\mathrm{s}}^{\mathrm{b}} = [\mathrm{M}]_{\mathrm{s}}^{\mathrm{b}}(t=0) \cdot \exp\left(-\frac{A}{V_{\mathrm{s}}}Pt\right) \quad (9)$$

Equation (9) describes the theoretical curve of Figure 6B and the experimental curve of Figure 7a1. The validity of the above equations and those derived from similar models (Models I in Table 4) were experimentally verified by systematic transport studies of several PLM systems using various metal ions and membrane carriers.

The above model provides analytical equations for the flux, but is restricted mostly to diffusion-limited transport under steady-state conditions, i.e. conditions in which linear concentration gradients are established in the aqueous diffusion layers or in the membrane (Figures 6 and 19). It is also limited to homogeneous source solutions and in particular it is not applicable to source solutions flowing through channels, for which the analyte concentration varies along the channel. Other models have been proposed in the literature (Table 4) based on different assumptions:

- diffusion-limited transport but under non-steady-state conditions (Models II in Table 4);
- slow physico-chemical processes at the membrane–strip interface (Models III, [42,47]). In particular, the role of the solvent in the rate limiting step is discussed based on this model [47];
- slow adsorption/desorption reaction of the carrier at the source–membrane interface (Model III, [48]). This is particularly relevant, since most carriers have strong surface active properties. Note that in this model the formation/dissociation of MC is assumed to be fast.

Although, as mentioned above, most papers report diffusion-limited controlled systems, it must be stressed that a careful check of the rate limiting step has not been done very often, and those systems which demonstrate diffusion-limited transport are usually preferred because of simplicity of interpretation.

3 APPLICATIONS OF PLMs AND STABILITY OF SLMs

3.1 NON-ANALYTICAL APPLICATIONS OF PLMs

The PLM principle has been much more widely used for industrial applications (e.g. food, pharmaceutical, clinical, and environmental fields [3,5,7,9,11,39, 49–51,211] than for analytical chemistry. Tables A1, A2 and A3 summarise the applications relevant to water treatments to provide an overview of the type of test compounds and conditions used for transport and to promote analytical applications, e.g. by simply adapting, where possible, the industrial application procedures. Tables A1–A3 and 5–7 also provide an overview of the major carriers, solvents and driving forces used for transport of organic and inorganic compounds.

There is an extensive literature on the use of SLMs for recovery of metals from low grade ores and their removal from groundwater, nuclear wastes and waste water treatment [3 and refs cited therein; 12] (Table A1). In all cases, the transport is carrier aided and in most cases it is coupled to counter-transport of proton (pH gradient in the membrane; (Figure 2b, Table A1). Numerous Cu(II) transport studies have been reported in the literature ([10,18,39,52–54]; Table A1), because of its industrial importance and the availability of adequate carriers.

In contrast to applications for cations, very few studies deal with anion transport (see ref [9] for a review). The transport is coupled to proton cotransport or to coanion counter-transport (Table 2a). It is usually carrier aided, sometimes with additional ion pair formation. Transport through a PLM is often more selective towards the more lipophilic anions [55], but anion size is more important for quaternary ammonium salts and macrocyclic carrier-aided transport [9,56–58]. The carriers employed for anion separation include tetra alkyl ammonium salts or phosphonium salts [55,59–61], metalloporphyrins (or phthalocyanines; [62]) and vitamin B12 derivatives [63]. The most systematic investigations of anion transport through PLMs has been made for nitrate ions using quaternary ammonium chloride, phenate or sulphonamidate as carrier [55,59], a major application being in decontamination of nitrate polluted industrial, ground, and surface waters.

Applications of PLMs to the preconcentration/separation of organic compounds have also been investigated (Table A3), although much less than for metals. In particular, enantiomeric separation of amino acids, and sugar and pesticide separations are important in pharmaceutical and food industries and in environmental applications [9,5,64,65]. Enantiomeric separation of amino acids was first performed using trialkylphosphate carrier [66] but it is now achieved much more selectively by using chiral crown ethers [9,67,68]. Smith [65] has given an overview of the potential application of boronic acid carriers for sugar separation. Other important organics of analytical importance are discussed in section 3.2.

3.2 ANALYTICAL APPLICATIONS OF PLMs

In contrast to the large number of industrial applications of PLMs, only a few analytical applications are reported. Nevertheless it has been shown that the PLM technique has powerful capabilities for analytical separation and preconcentration of trace metal ions, inorganic anions, and trace organics in biological and environmental samples (Tables 5–7) prior to their determination by sensitive instrumental analytical methods such as atomic absorption spectrometry, voltammetry, inductively coupled plasma–mass spectromery, fluorimetry, or chromatographic techniques. Analysis of trace elements or compounds can be done off-line, or on-line by coupling the PLM device to instrumental detection systems. Fully automated systems for such analyses has recently been described (sections 4.3 and 4.4).

3.2.1 Metal Ions

Only few reports deal with the use of PLMs for trace metal ion analysis (Table 5). Compared with industrial applications, a 100 % recovery (or at least a well-controlled and reproducible recovery) is required in analytical applications, which imposes significant constraints on the choice of carrier and solvent. Constraints are even more drastic for *in situ* applications where long term stability is essential (section 4.1). In addition, a close coupling of PLM with the detector may be difficult, depending on the nature of the detector, owing to compatibility problems with the hydrophobic components of the PLM. A final limitation is related to the lack of appropriate commercially available carriers or the difficulty in their synthesis. Although these problems can be overcome, they explain why trace metal analyses with PLM techniques are still in their infancy, even though they are a promising analytical tool.

The utility of PLMs for analytical preconcentration of metals was first reported in 1986 by Cox *et al.* [10] for Cu (II) preconcentration using oxime and kerosene as carrier and solvent. They found preconcentration factors, *F*, of 10 after 60 min and showed that *F* was independent of Cu(II) concentrations, ionic strength and pH over the ranges tested. The latter observation is of utmost importance for applications to real water samples with varying and unknown compositions. Uto *et al.* [69], in the same year, reported the preconcentration of Cu(II), Cd(II), and UO_2^+, using respectively bathocuproine, trioctylmethyl-ammonium chloride, and tributylphosphate as carriers. Ohki *et al.* [70] reported an interesting system, in which the uphill transport of Cu(II) was aided by redox reactions. Hydroxylamine is added to the sample, to reduce Cu(II) to Cu(I) which then binds to the carrier. In the strip phase, dissolved oxygen oxidises Cu(I) to Cu(II), making the transport irreversible. After 3 h, however, *F* was found to be only 3.6. Uto *et al.* [69] also reported the first example of on-line coupling of SLM to a voltammetric detector, for Cd analysis. They noted adsorption of organic solvent on the electrode surface during voltammetric measurements.

Papantoni *et al.* [26] have used a flow-through SLM system for the preconcentration of Cu, Cd, Co, Ni, and Zn and the analysis of total metal concentrations in natural waters and biological samples, by using a spiral wound (Figure 4B) and a PTFE flat sheet SLM membrane system comprising di-hexylether as solvent, methyl trioctyl ammonium chloride as carrier and 8–hydroxyquinoline or SCN^- as strip solution. Preconcentration factors over 500 were obtained after 30 min. Moreover, systematic tests indicated that memory effects and

Table 5. Analytical applications of SLMs for metal ion preconcentration from natural waters, waste waters, and ground waters. F = preconcentration factor; V_s/V_{st} = source/strip volume ratio. In column 4, M means mol L^{-1}. HDEHP = hexyl diethyl phosphonic acid

Metal	Carrier (solvent used in brackets)	Support	Strip solution	Analytical conditions	References
Cu	Oxime (kerosene)	Flat sheet (Celgard 2500)	0.1 M HCl	$F = 10$ after 1 h	10
Cu	Bathocuproine (Chloroform)	Flat sheet (PTFE)	H_2O	$F = 3.6$ after 3 h (not practical for *in situ* application)	69,70
Cd	Trioctyl methylammonium chloride (chloroform)	Flat sheet (PTFE)	0.1 M NH_4NO_3/NH_3 buffer pH = 9.7	Synthetic solution $F = 20.6$ after 4 h $V_s/V_{st} = 50$	69
Cu, Cd, Co, Ni, Zn	Methyl trioctylammonium chloride HDEHP (kerosene)	Flat sheet (PTFE)	0.1 M HNO_3	$F > 500$ after 24 h $V_s/V_{st} = 1500$; stable 4 d; adaptable for *in situ* trace analysis	26,71
Cu, Pb, Cd, Zn	Didecyl 1,10-diaza 18-crown-6 ether + lauric acid (phenylhexane/toluene)	Flat sheet (Celgard 2500)	5×10^{-4} M CDTA or 10^{-3} M $Na_4P_2O_7$	$F = 14$ after 3 h $V_s/V_{st} = 16$ for Cu and Pb	18,21
Cu, Pb, Cd, Zn	Didecyl 1,10-diaza 18-crown-6 ether + lauric acid (phenylhexane/toluene)	Hollow fibre (Accurel Q3/2) length = 21 cm i.d. = 600 μm; o.d. = 800 μm	5×10^{-4} M CDTA	$F = 80–4000$ after 2 h for Cu and Pb	18,21,72
Cyano complexes of Au, Fe, Ni, Co, Pd, Pt, Cr, As, Cs	Methyl trioctylammonium chloride (*t*-butylether)	Flat sheet (PTFE)	10^{-2} M $NaClO_4$	$F = 50–600$ after 2 h $V_s/V_{st} = 5000$	73

adsorption of humics on the membrane as well as membrane fouling are negligible, these being key requirements for natural water applications. Djane and co-workers [74,75] have also reported the preconcentration of Pb(II) with di-2–ethylhexyl phosphoric acid in kerosene for total metal analysis of river water, blood and urine samples, by coupling SLM with either AAS [74] or potentiometric stripping analysis [75]. In general, combinations of SLM with instrumental techniques (section 4) such as flow injection analysis and atomic absorption spectrometry [76], capillary electrophoresis [73], graphite furnace atomic absorption [26,71], potentiometric stripping analysis [75] or anodic stripping voltammetric techniques [21] enable one to analyse a large number of metals (Table 5), and F values of 50–5000 can be reached within 5–60 min, which suggests that detection limits well below 10^{-10} mol. L^{-1} may be achieved.

In most of the examples cited above, PLM separation requires a modification of the sample (pH adjustment, adding a complexing agent etc.), which largely limits its utility for metal speciation studies. Recently a PLM system was specifically designed ([18,21,72]; Table 5) to enable direct simultaneous speciation of trace metals (Cu, Cd, Pb, Zn) in natural waters without any sample perturbation, in particular for the selective determination of the free metal ions. The carrier solution used in this system consists of 1,10–didecyl diaza crown ether (22DD) with lauric acid, in a 1/1 mixture of phenylhexane and toluene as solvent. Laurate acts as a counter-anion in the membrane, and the transport of test metal is coupled to the counter-transport of Na^+ co-cations. 22DD is a macrocycle containing oxygen and nitrogen donor atoms, which can complex both alkali and transition metal ions. Preconcentration factors of 100–3000, i.e. high sensitivities, are achievable.

3.2.2 Anions

Very few applications of PLMs for anion analysis have been reported. Amongst the non-analytical applications (Table A2), separation of nitrate and cyanide are environmentally relevant. Analytical applications of SLMs have been developed for two environmentally important anions, i.e. phosphate and nitrate (Table 6). Oxychloro-molybdenum(V) tetraphenyl porphyrin [77], non-cyclic neutral metallocleft compounds [56], macrocyclic polyamines [57] and macrocyclic compounds with a metal bipyridyl pendant group [58] have been used as phosphate carriers. A Co(III) complex is used for nitrite. Systematic tests are still needed, however, to evaluate the usefulness of this approach for routine anion analysis.

3.2.3 Organic Compounds

Analysis of trace organics often requires isolation and preconcentration of target compounds. Audunsson in 1986 was the first [78] to employ an SLM

Table 6. Analytical applications of SLMs for anion preconcentration from waters. In column 4, M means mol L^{-1}. NPOE = nitro phenyloctyl ether

Anion	Carrier (solvent used in brackets)	Transport mechanism or driving force	Strip solution	Application	Reference
$H_2PO_4^-$	Oxomolybdenum (V)-tetraphenyl pophyrin complex (trichlorobenzene)	Chloride anion gradient	0.5 M NaCl	Phosphate determination in river or ground water	77
$H_2PO_4^-$	Uranyl salen metalocleft compound (NPOE)	Cation cotransport	H_2O	Phosphate determination in natural water	56
NO_2^-	Dicyano cobalt(III) heptapropylcobyrinate (isopropylbenzene)	SCN^- counter-anion gradient	10^{-2}–10^{-3} M NaSCN	Nitrite determination in anoxic waters	63

for preconcentration and separation of aliphatic and aromatic amines in urine prior to analysis by gas chromatography. As for all organics (Table 7), the transport was driven by acid–base chemistry and not carrier aided. In this type of transport, the weakly acid or basic test compounds are first converted to neutral form by adjusting the pH of the sample. The neutral compound is then transported by diffusion through the membrane and back extracted with an acidic or basic strip solution. Using a flat sheet PTFE support (Millipore) impregnated with undecane carrier, triethyl amine at low ppb levels was detected in urine [5,78]. Based on this principle, Mathiasson and Jonsson have done pioneering work on the application of SLMs to analytical determinations of organic compounds in environmental and biological samples ([5]; Table 7). These include the determination of: sulphonyl urea in herbicides [5], acidic herbicides, e.g. phenoxy acids, in the presence of humic substances [5], chlorinated phenols [80], and aniline derivatives [81] in natural waters, for water quality control. Using typical enrichment times of 10–20 min, detection limits ranging from few tens of ppt to ppb level of amines, depending on the specific compounds, have been obtained, which are below the EEC stipulated permissible levels for drinking and surface water. Other applications of amine analysis are discussed in ref. [5]. These references provide a wealth of information related to selectivity, extraction efficiency, flow rates, and the number of analyses using the same membrane without regeneration or memory effects.

Analysis of several triazine herbicides has been performed by SLM–HPLC coupling in river and lake water with detection limits of 0.03 ppb [79]. No memory effects were observed. SLMs coupled to capillary zone electrophoresis have been applied to the determination of basic drugs such as Brombutcrol, in

Table 7. Analytical applications of SLMs to preconcentration of organic compounds. In all cases, transport is not carrier aided but pH driven. In column 4, M means mol L^{-1}

Target compounds	Solvent	Support	Strip solution	Application	Reference
Herbicides, phenoxy acids	Undecane	Flat sheet (PTFE)	Phospate buffer (pH=9)	Field sampling of acidic herbicides in river water	5
Herbicides, sulphonyl urea	Undecane	Flat sheet	Phospate buffer (pH=8.5)	River water analysis	5
Herbicides, alkylthio s-triazine	*n*-Hexylether	Flat sheet (PTFE)	0.1 M H_2SO_4	River water analysis	79
Chlorinated phenol	Undecane	Flat sheet (PTFE)	Phosphate buffer (pH=13)	Quality control of industrial waste water	80
Aniline derivatives	Undecane	Flat sheet (PTFE)	0.3 M H_2SO_4	Monitoring water for quality control	81
Triethyl amine	Undecane	Flat sheet (PTFE)	0.05 M H_2SO_4	Analysis of urine, plasma and animal manure	5
Basic drugs; Racemic brombuterol	Undecane	Hollow fiber (Polypropylene)	5×10^{-4} M H_3PO_3	Blood plasma analysis	82
Herbicides (simazine, propazine phenylurea)	Dihexyl ether or undecane	Flat sheet (PTFE)	0.5 M H_2SO_4	Surface waters in particular lake water	83

plasma [82]. A miniaturised SLM device has been developed to reduce the volume of strip to 1.2 μL, in order to lower the limits of detection (~ 4 nmol.L^{-1} for 10 min preconcentration), and to get maximum preconcentration factors with only small sample volumes (~ 0.5 mL).

3.3 OPERATIONAL PROCEDURES

The appropriate carrier, selective for the target species, is dissolved in a water-immiscible solvent. An example of a carrier selective for Cu(II) is the neutral 1,10-diaza crown ether (22DD) [18,21,72,84]. The liposoluble anion laurate (LA) was also added to the carrier solution phase to avoid the need for a simple counter-ion to cross the aqueous/organic interface. A solution containing 0.1 mol L^{-1} 22DD and 0.1 mol L^{-1} LA in a 1/1 mixture of phenylhexane/toluene was used as carrier solution for metal transport studies across the SLM.

3.3.1 Preconcentration with Flat Sheet SLM Devices

Celgard 2500 flat sheet polypropylene was used in refs [18,21,84] as the support. It was cut into rectangular strips (5.5 cm × 4.9 cm) to place it in the diffusion cell (see Figure 4A). A small amount of carrier solution (ca 100–200 μL) was pipetted onto a clean watch glass. The membrane pores are immediately filled and the membrane becomes transparent. The excess solvent coating the surface can be removed by simply immersing the membrane in a small volume of water (5 mL) or by gentle wiping with a tissue paper.

The impregnated membrane is fixed in the diffusion type of cell with the membrane separating the two aqueous solutions. The cell (Figure 4A) consists of two half-cylindrical Plexiglas chambers each one with a magnetic stirring rod at the bottom, and a vertical rectangular PTFE window which serves two purposes. It protects the Plexiglass from chemical attack by non-polar solvents used for making carrier solutions. Secondly it prevents any leakage of aqueous solution from one compartment to the other. The impregnated flat sheet membrane is inserted between the two half cells by placing the membrane on one of the openings and tightening the two halves together. For proper fixation to avoid leakage problems, the two half cells should be well aligned. It is better to use membrane sizes slightly larger than the openings to ensure no leakage.

After assembling the cell with the SLM, the source and strip solutions are filled through the sampling ports. The metal transport characteristics can be determined by using equal volumes of strip and source solutions. 80 mL strip and source solutions were used for instance for Cu(II) transport studies. The sample and strip solutions are stirred with a magnetic stirrer. The stirring speed are preferably kept constant at 480 rpm in both source and strip solutions. Metal transport was found to be independent of stirring at stirring rates > 400 rpm. During each experiment, aliquots of strip and source solutions are periodically withdrawn and analysed to determine the Cu(II) concentration, e.g. by flame or flameless atomic absorption spectrometry (Figure 7a). From the Cu concentration versus time data, the flux and permeability coefficient can be evaluated as described in section 2.4 and Figure 6.

In order to perform preconcentration in the strip phase, a smaller volume Plexiglas half cell (8 mL) must be used for the strip solution. As before, source and strip solutions are filled through sampling ports and stirred. In all cases great care must be taken to maintain the source and strip solutions at the same level to avoid leakage due to hydrostatic pressure differences.

3.3.2 Preconcentration with Hollow Fibre SLM (HFSLM) Devices

For obtaining higher preconcentration factors at shorter times, hollow fibre SLMs (HFSLMs) are more suitable, since μL volumes of strip and large volumes of source solutions can be used. Accurel pp q5/2 (Akzo) hydrophobic hollow fibre (HF) was used as the support in refs [21, 72]. A single fibre module was used in transport experiments. Either of the SLM systems shown in Figure 4D or E can be used. In both cases the inside of the fibre acts as the strip solution compartment. The strip solution is filled in the lumen by means of a

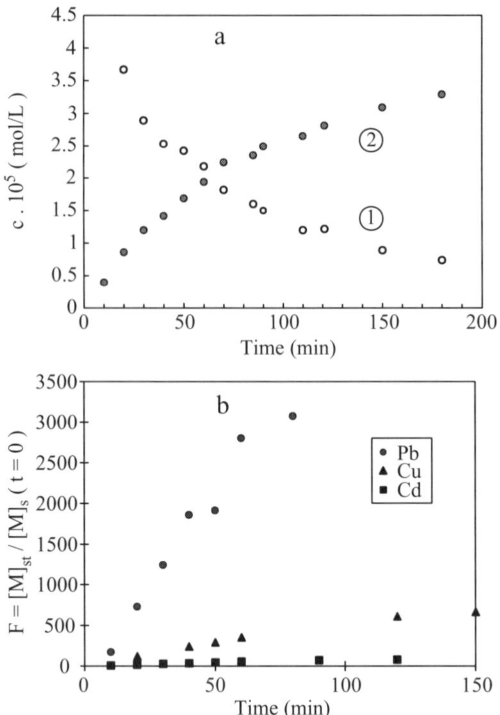

Figure 7. Transport of metal ions through a SLM composed of the carrier 22DD (1,10-didecyl-1,10-diazacrown ether), lauric acid, in 1/1 mixture of toluene/phenylhexane. Conditions used: *source*, 10^{-2} mol.L^{-1} MES buffer (pH=6); *strip*, 5×10^{-4} mol L^{-1} CDTA (pH=6.56). Volumes of source and strip solutions = both 80 mL in (a) and 250 and 0.068 mL, respectively, in (b). (a) Transport of Cu^{2+} with system of Figure 4A. Polypropylene flat sheet membrane, Celgard 2500; characteristics in Table 3a. Open and filled circles = concentration in the source and the strip solution respectively. Initial concentration in the source = 5×10^{-5} mol L^{-1} Cu^{2+}. (b) Measurement with system of Figure 4D. Polypropylene Akzo q3/2 Accurel hollow fibre membrane (length = 25.4 cm; other characteristics see Table 3b). Concentrations in the source: 10^{-6} mol L^{-1} Pb^{2+}, 5×10^{-6} mol L^{-1} Cu^{2+}, 5×10^{-5} mol L^{-1} Cd^{2+} (modified from ref. [21])

peristaltic pump. For this purpose, the inlet and outlet ends of the fibre are joined to short lengths of PTFE connection tubings (0.8 mm i.d.) via a heat shrinkable polypropylene tubing which provides an air- and water-tight seal. Connection is done in such a way that the ends of the hollow fibre and PTFE tubings are aligned without any gap between the two tubes. The tube is shrunk with a hot air blower to obtain a tight sealing. Good air-tight sealing between the Teflon and the hollow fibre is difficult to achieve because of its flexibility and low heat resistance and the high temperatures needed for the PTFE to fit air-tight into the heat shrinkable tubing. In order to obtain good sealing, the corresponding parts of the hollow fibre had to be made rigid by coating and impregnating them with the resin Nanoplast and hardening it as described by the resin manufacturer (Rolf Bachhuber, Chemikalien, Zubehör für Electronenmikroskopie, Germany), prior to inserting them into the heat shrinkable tubing for sealing. Air-tight sealing between the hollow fibre ends and Teflon tubings is a prerequisite for proper functioning of an HFSLM. Retraction of solution in the tubing during measurement, for instance, is indicative of a poor seal.

Hollow fibres with lengths varying from 14 to 25 cm have internal volumes of 50–60 μL. The fibre is cleaned twice with toluene and is left to air dry. It is then impregnated with the carrier solution from the shell side by slowly flowing the carrier using a pasteur pipette. Well-impregnated membranes are transparent (unwetted membrane is opaque) when the organic solvent penetrates the membrane.The excess solvent in the fibre is removed by first rinsing on the outside with Milli-Q water and dipping in a beaker (250 mL). This step is repeated twice. The water in the beaker is renewed, the HFSLM is immersed in it and Milli-Q water is then passed through the lumen side slowly at a flow rate of 0.1 mL min^{-1} using a peristaltic pump to remove excess in the lumen side (this step requires ca 3 mL). The impregnated membrane can be stored in water.

In system shown in Figure 4D, the impregnated membrane is immersed in distilled water. The lumen side is drained and filled with the strip solution at 0.1 mL min^{-1} flow rate. At time $t = 0$, the distilled water is replaced by the test sample stirred at 480 rpm. An aliquot of the source is removed from the source to determine the initial concentration at $t = 0$. The strip solution is left to stand for a prescribed time, t, and it is then collected in the collecting tube by running the peristaltic pump for 4 min at a flow rate of 0.1 mL min^{-1}, which pushes the solution out of the fibre and simultaneously fills the lumen with a new solution. Some accumulation of test compound continues during the time the preconcentrated sample is being collected. Its contribution to the preconcentration factor is not important at times > 20 min and can be calibrated for shorter times. Aliquots of the source solution are withdrawn at various times for analysis to determine the concentration of the target metal. Typical preconcentration factor versus time plots (e.g. Figure 7b) show that high preconcentration factors can be obtained even for short times. The preconcentration factor was found to be independent of the initial concentration of metal in the sourcre solution in the studied range 10^{-7}–5×10^{-6} mol L^{-1}. This is a key feature for determining the concentration of an analyte in an unknown solution.

In the HF system of Figure 4E, the fibre is held in the centre of the groove with PTFE connection tubing and polypropylene nuts for tightening at both ends. The impregnated membrane is placed in the cell. The strip solution is filled in the lumen as described above. The strip solution is kept stagnant and the source solution is circulated on the outside of the fibre by means of a peristaltic pump at various flow rates. Preconcentration and analysis are done as described before.

3.3.3 Operational Conditions

3.3.3.1 Cailibration of HFSLMs At short times, preconcentration factors are proportional to preconcentration time (Figure 6B and D), and the concentration of analyte in the source remains practically constant. The initial slope, s, of the F versus time curve, using a standard solution of metal for a given SLM system is independent of the initial concentration of analyte in the source solution and can be used for the determination of an unknown concentration, c, from the expression

$$c = c_{st}/st$$

where t is the preconcentration time, and c_{st} the concentration in the strip solution at time t. Note that when $V_s \gg V_{st}$, the above equation may be valid over fairly long periods of time.

3.3.3.2 Effect of flow rate of the source solution on preconcentration; reproducibility The metal fluxes across the HFSLM in the cell shown in Figure 4E were determined from the slopes of the linear portion of F versus t plots for various flow rates (slope is directly proportional to flux). The plot of slope versus flow rate (Figure 8) shows that the flux passes through a maximum at ca 100–150 mL min^{-1} corresponding to a linear velocity of 0.055–0.09 m s^{-1} [21]. The increase in flux with flow rate is due to the decrease in the aqueous diffusion layer thickness, δ_s (equation (5)). The subsequent decrease in flux is indicative of the instability of the HFSLM at source flow rates > 400 mL min^{-1}. Therefore

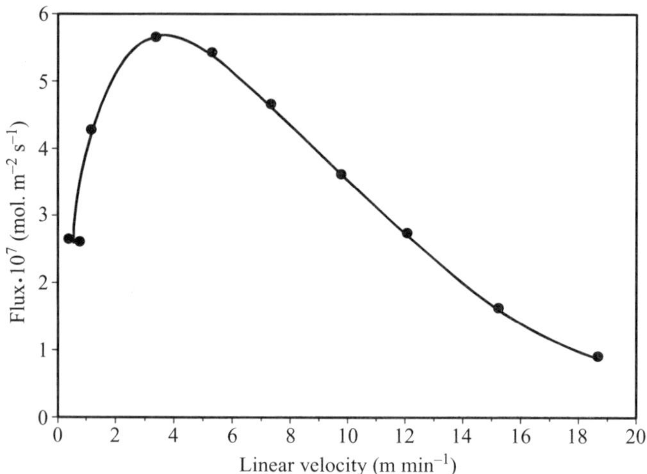

Figure 8. Effect of linear velocity (or flow rate) of the source solution on the transport flux of Cu(II) with HFSLM system of Figure 4E. Conditions as in Figure 7B. In the system used, flow rate (mL min^{-1}) = 0.036 linear velocity (m min^{-1}) (modified from [21])

the optimum flow rates should be determined carefully, for the flow type HFSLM system used.

Replicate measurements of flux at the optimal flow rate show that fluxes are reproducible to within 5% ($n = 5$; 95% probability; Figure 9). This makes the HFSLM preconcentration device suitable for reliable trace analysis.

3.3.3.3 Stability of HFSLMs For continuous monitoring of trace metals, long term stability of the HFSLM is a key parameter. Cu(II) transport and preconcentration measurements were made by operating the HFSLM system of Figure 4E for ca 12 h every day for 1 week without reimpregnating the membrane. While not in use, the membrane was stored in Milli-Q water. The results in Figure 9 show that the device can be used for at least 4 d without significant loss in sensitivity towards target ions [21]. Djane *et al.* [27] also report similar stability with their systems working with real samples. Stability over 30 d of continuous operation has been reported by Danesi *et al.*[85].

At optimal stirring or flow rate conditions (480 rpm or 0.09 m s^{-1}), preconcentration factors versus time obtained with the HFSLM systems depicted in Figure 4D and E show that both these systems are reliable for separation and preconcentration. Both can be coupled on-line to sensitive detectors.

3.4 STABILITY OF THE SLMs

PLM long-term stability is a key issue for continuous real-time monitoring in water. Stability of SLMs depends on both the physical and chemical properties

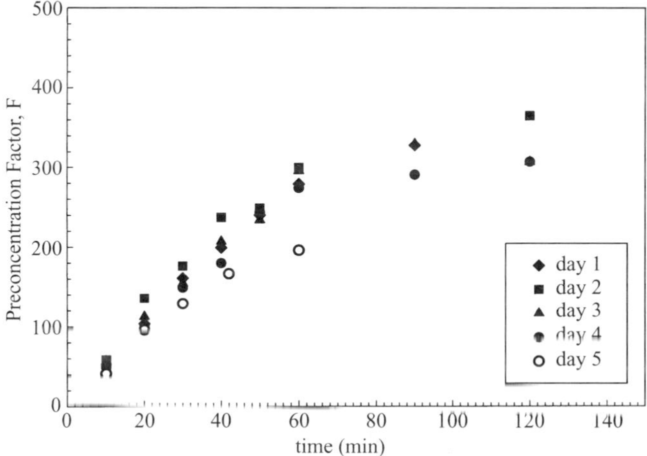

Figure 9. Reproducibility and lifetime of HFSLM system of Figure 4E, for the transport of Cu(II) in the conditions of Figure 7b. Flow rate = 144 mL min^{-1}. The various curves were repeated every day without reimpregnation of the hollow fibre [21]

of the constituents. Causes of instability may vary from one system to another, but loss of the impregnating liquid (carrier and solvent) from the membrane into the source and strip solutions is the major one [7,86,87]. This phenomenon has been the topic of several papers and reviews [7,23,56,85–96]. An excellent critical, up-to-date review, together with several ways to improve stability of SLMs, have been presented by Kemperman *et al.* [91]. Caution should be used when comparing literature data, as the criteria used for assessing membrane instability are not clear cut [91 and refs cited therein]. For instance, decline in the flux of the target species without membrane regeneration has been used by most authors as a diagnosis for membrane instability [86,90,93], but several others consider their membrane stable when constant flux is obtained by continuously regenerating the impregnating liquid phase [85,88,89,97]. Progressive wetting of the membrane [90,92], or decrease in selectivity [43,91,93,98], are other factors used to determine membrane stability.

3.4.1 Mechanisms of Membrane Degradation

The loss of impregnating liquid is chiefly due to the following factors [7,86,87].

Hydrostatic pressure difference across the membrane It may result from the pumping of source and strip solutions [3,99]. This pressure difference is more important in hollow fibre devices than in flat sheet systems. The impregnating liquid is expelled from the pores when the pressure difference exceeds a critical value, ΔP_c, fixed by the capillary forces inside the pores [90]:

$$\Delta P_c = 2\gamma \cos \Theta / r \tag{10}$$

where γ is the water / solvent interfacial surface tension, Θ is the contact angle between the support pore wall and the liquid membrane phase, and r is the pore radius (correction for irregular pore sizes can also be included [93]). Equation (10) shows that the smaller the pore size, the higher will be ΔP_c. Typical values for flat sheet membranes with pore sizes in the range 0.04–0.4 μm are in the range 50–250 kPa (0.5–2.5 atm), depending on the SLM system used. This is significantly higher than most hydrostatic pressure differences normally used in analytical systems. Thus pressure difference is usually not the sole, or even the major, cause for carrier and solvent loss [21,24,86,100].

Osmotic pressure gradient across the membrane Under an osmotic pressure gradient, water can flow from dilute to concentrated aqueous solutions. According to Danesi *et al.* [90] and Fabiani *et al.* [92] when the salt concentrations of source and strip solutions are sufficiently different, water may then push the organic solvent out of the membrane pores. Independent investigations [23,101–103] have indeed shown that, if the ionic strengths on both sides of the membrane are unequal, membrane degradation occurs. Zha *et al.* [103] also observed that the organic phase loss was severe on the source side as a result of emulsion formation in a system where the ionic strength of the source and strip phases were 0.3 and 4.5 mol L^{-1}, respectively. The addition of salt to balance the ionic strength in the system drastically reduced membrane degradation. Osmotic pressure by itself, however, is usually too small to explain SLM instability [86,91]. The effect of ionic strength is more likely linked to emulsion stability as observed by Zha *et al.* [103] (see below).

Progressive wetting of the support membrane pore walls The interfacial tension between aqueous and organic phases and between the aqueous phase and support membrane might decrease with time, in particular due to contamination of the aqueous–organic interface by surface active compounds [90,92]. The aqueous phase might then enter the membrane pores and replace the organic phase [91,94], or emulsions can be formed (see below).

Blockage of pores The support pores can be blocked in two ways. Local precipitation of carrier with target ions has been observed for trilauryl amine/nitrate [104], and calcium phosphonic acid systems [105]. It has also been suggested [91,95,106] that water can enter the pores and replace organic phase by forming micelles at the surface of the membrane, thus decreasing the flux. This mechanism has, however, been ruled out by Kemperman *et al.* [91].

Solubility of carrier and solvent in the strip and source solutions The higher the solubility of solvent and carrier in water, the shorter the lifetime of the SLM [91 and refs cited therein]. For example, in the SLM removal of nitrate, techneciate and chromate from groundwater by primary, secondary and tertiary aliphatic amines as carriers, the order of SLM stability was tertiary > secondary > primary amines. This order is opposite to that of their solubility in water. Similar examples have been observed for carrier [104] as well as solvent [23,36,91,107] solubility in water. Although there is no doubt that increases in the water solubility of SLM components decrease membrane stability, it is not clear whether the loss of solvent or carrier is due to real solubilisation or to the fact that an increased solubility also corresponds to a facilitated formation of emulsion droplets. At any rate, the choice of very slightly water soluble solvents and carriers is required.

Emulsion formation of the liquid membrane phase in water, induced by shear forces Neplenbroek and co-workers [86,108] proposed that formation of emulsion droplets composed of surface-active carriers and organic solvent could be a major cause for loss of carrier and solvent. The emulsion is formed as a result of the presence of a surface tension gradient [108] combined with lateral shear forces, resulting in the local deformation of the interfacial meniscus in the pore of the support and consequently the formation of droplets. According to Kemperman *et al.* [91], Kelvin–Helmoltz instabilities arising from the movements of two phases at different velocities parallel to their interface [93,109,110], and membrane vibrations due to small pulsations from aqueous solutions [91 and refs cited therein] are the most likely causes for deformation of the liquid membrane meniscus. Instabilities due to turbulent flow and density differences might also play a role [91,108–110]. The stability of the emulsions resulting from these deformations is strongly dependent on the concentration and composition of the salts present in the aqueous phase.

Therefore, to maximise SLM lifetimes, organic liquid phases and carriers showing low water solubility and high organic–water interfacial tensions should be used. It must be noted that the latter criterion is not easily fulfilled since good carriers are usually surface-active and thus tend to decrease the surface tension at water–solvent interface. In addition, supports with low water wettabilty, source and strip solutions with similar ionic strengths and minimal differences in hydrostatic pressure are preferable. Several methods to improve the lifetimes of SLMs have been reported [16,85,90,91 and refs cited therein; 92,111] and are briefly discussed below.

3.4.2 Methods for SLM Lifetime Improvements

Reimpregnation of support Continuous reimpregnation of the membrane is possible for hollow fibres by incorporating a reservoir in the hollow fibre module (Figure 4C). In this way SLMs have been stable for more than 30 d. Various systems of continuous reimpregnation have been described [91 and refs cited therein].

Sandwich SLMs An SLM can be placed between two thin hydrophilic membranes such as lens paper or ultrafiltration membrane (15 μm thick) [112]. No significant decrease in flux was observed compared with the unprotected SLM. The stability was better but loss of carrier and solvent was not completely eliminated. Ion exchange membranes have also been used in the same manner [113]. Cu transport, studied using such a system, exhibited fluxes similar to those for uncoated SLMs but with greater stability.

Gelation A gel is an elastic rubbery solid, forming a cross-linked polymer network in a swollen organic liquid [91,114]. The liquid prevents the polymer from collapsing into a compact mass while the polymer network prevents the liquid from oozing out of the pores. SLM-based separations have been performed with gel membranes without membrane support, by using PVC [115] or cellulose triacetate [116] as gels. However, poor mechanical strength was observed with thin membranes and thicker membranes yielded low fluxes [117].

Gel formation in the pores of the supporting membrane has been used by Babock *et al.* [106], Neplenbroeck *et al.* [100] and others. For instance PVC (a gel forming polymer) is dissolved in tetrahydrofuran with carrier. The membrane support is impregnated with this solution and the solvent is evaporated. Such a gelled membrane was stable for 28 d [106] but the flux was low [106,117,118]. Membranes coated with other polymers, such as polyurethane [119], polymethacrylate [86 and refs cited therein], polystyrene [120] and silicon rubber [120], exhibit poorer stability [100,119] than PVC-coated membranes.

Another way of preparing a gel-stabilised SLM is to apply a thin PVC layer on the source side of the membrane and to cross-link it afterwards [100]. This was done by dissolving a carboxylated PVC and the carrier in THF, adding an activator and cross-linking agent, and spreading this mixture, with a tissue paper, on the source side of the flat sheet support preimpregnated with the carrier solution. The THF evaporates and a gel network with smaller mesh size is formed. Only the source side has been coated in refs [91,100], because emulsion droplets were observed on this side only. Comparisons with uncoated membranes showed that such gel-coated membranes prevent emulsion droplet formation without significant decreases in fluxes. Membrane lifetimes greater than 1 year have been obtained. The drawback is the poor reproducibility of the gel coating and the fact that the coating procedure is difficult to adapt to hollow fibres. In addition, this gel coating does not seem to prevent loss of carrier and solvent by dissolution [52].

Interfacial Polymerisation Interfacial polymerisation reactions occur between two very reactive monomers at the interface of two immiscible solvents. For example, when aqueous piperazine solution is added to water-immiscible trimesoylchloride (TMCl), polymers are formed at the interface. To prepare thin composite SLMs, an Accurel flat sheet membrane was impregnated with TMCl solution. It was then dipped in a stirred aqueous piperazine solution for a set time and the resulting composite membrane with an interfacial gel layer was removed and rinsed with water. It was impregnated with the carrier and transport measurements were made in the usual way [121]. Fluxes were found to be constant over a longer period of time than for the uncoated membrane and, most important, the dissolution of carrier and solvent were considerably reduced. Hollow fibres and flat sheet membranes can be coated in this way. The shortcoming is the difficulty in obtaining reproducible coatings.

In conclusion, promising ways of improving stability for *in situ* applications seem to be (i) the sandwich membrane , (ii) gelation at the membrane surface, using cross-linked polymers, if emulsion formation is the cause of instability and (iii) interfacial polymerisation in the case of carrier dissolution. These and other technologies enabling one to coat the SLM with hydrophilic polymer layers, or to attach the carrier covalently inside the membrane, are likely to produce long lifetime SLMs in the future.

4 CAPABILITIES OF PLMs FOR IN FIELD AND *IN SITU* ANALYSIS

Sampling and laboratory analysis is presently the most commonly used approach for measuring traces of organic and inorganic pollutants in waters. It is usually performed by collecting aqueous samples in carefully cleaned flasks, at fixed time intervals, followed by storage and laboratory handling and analysis. For trace compounds large sample volumes must be collected and preconcentration and separation techniques must be performed in the laboratory, before instrumental determination. Furthermore, to estimate the time evolution and spatial distribution of pollutants, a large number of samples must be collected and analysed, which is both time consuming and expensive. In addition, contamination, losses or sample changes during storage and sample handling in laboratory may cause significant analytical errors. A procedure that minimises such errors is therefore highly desirable.

Like DGT (diffusive gradient in thin film) and DET (diffusive equilibration in thin films) (Chapter 11), the SLM technique makes it possible to perform simultaneously *in situ* preconcentration and separation of the test compound from the sample matrix, which may interfere in the detection step. Most of the aforementioned problems are thus avoided. In addition, the preconcentration step is relatively short (minutes to hours). In principle, PLMs can be used *in situ* in two ways:

- as independent *in situ* preconcentration/separation devices, enabling subsequent concentration measurements in the laboratory [122]. The devices shown in Figure 4B,D and E can be used for *in situ* or in field preconcentration, and the strip solution is brought back to the laboratory and analysed by instrumental detection (AAS, ICP-MS, voltammetry, GC, LC, CE). This is presently the only way this technique has been used in the field (see section 4.1 for examples).
- as a preconcentrating/separating module of an integrated *in situ* probe, including also a detector for *in situ* real-time analysis of the test compound in the strip solution. This is technically more difficult and no *in situ* analysis has yet been done in this way. Laboratory integrated systems have been developed, however, and a brief discussion is given in section 4.5.

At any rate, miniaturisation of the PLM device and its easy coupling to analytical detectors is a requirement for both types of operations. These aspects are discussed in sections 4.3–4.5. At present, only SLM systems have been used for in field analysis. This is discussed below.

4.1 FIELD AND *IN SITU* PRECONCENTRATION/SEPARATION BY MEANS OF PLMs

Few field applications of PLMs, all based on SLMs, have been performed until now, but the following examples demonstrate its capabilities in this field. A field, time integrating method for preconcentrating herbicides from waters prior to laboratory analysis has been reported in refs [5,122,123]. With this system, detection limits $\leq 10^{-8}$ mol L^{-1} can be achieved for some herbicides. The system is shown in Figure 10. The water was pumped at an optimum flow rate of 0.8 mL min^{-1} and diluted with 0.5 mol L^{-1} H_2SO_4 before passing over the membrane. The enriched strip solutions were then collected in vials and stored in an ice box prior to analysis in the laboratory. Recently an automated on-line coupling of SLM with liquid chromatography has been reported [81,124] for the determination of seven aniline derivatives in waters, with detection limits ranging from 0.5 to 1.5 nmol L^{-1}. Aniline derivatives originate from many industrial activities and are priority pollutants in the EEC list; their maximum permissible limit in drinking waters is 1 nmol L^{-1} [124].

In field preconcentration and separation of Cu(II), Pb(II) and Cd(II) have also been performed by using both flat sheets and hollow fibre systems (Figure 4A,D and E). The carrier solution was the 1,10-didecyl diaza crown ether (22DD) and lauric acid, dissolved in a 1/1 mixture of toluene and phenylhexane [18,21] (Table 8). Typical accumulation–time curves of Cu(II) in the strip phase are shown in Figure 7a and b for devices of Figure 4A (equal volumes of source and strip solution) and Figure 4D (HFSLM), respectively. No preconcentration occurs in the former system, but very high preconcentrations (up to a few

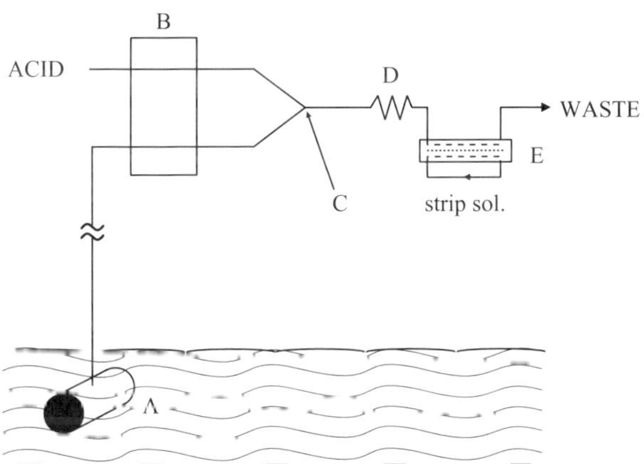

Figure 10. Field sampling set-up for enrichment by an SLM of acidic herbicides in natural waters: (A) sampling point; (B) peristaltic pump; (c) confluence point of sample stream (0.8 mL min^{-1}) and a stream of 0.4 mol L^{-1} H_2SO_4(0.15 mL min^{-1}); (D) mixing coil; (E) SLM devices (Figure 4B) with stopped flow in the strip channel [122,123]

Table 8. Trace metal concentrations measurements in various unperturbed waters after preconcentration/separation by SLM using 22DD as carrier in 1/1 toluene/phenylhexane combined to conventional methods of detection in the strip

Sample	Test element	Total dissolved metal concentration after acidification (nmol L^{-1})	Direct measurements without acidification* (nmol L^{-1})		Type of membrane
			SLM-based measurement	SWASV	
River water (Arve, GE, Switzerland)	Pb	3.5 ± 0.08†	0.5 ± 0.05‡	0.58nmol.L^{-1}	Flat sheet
	Cu	40.0 ± 0.5§	8.3 ± 0.1	7.16 ± 0.1	Flat sheet
Drinking water (Tap water; Geneva, Switzerland)	Pb	0.8 ± 0.2†	0.3 ± 0.1	0.4 ± 0.2	HFSLM
	Cu	128 ± 0.5§	16.8 ± 0.5	9.24 ± 0.1	HFSLM
Lake Bret** (VD, Switzerland)	Pb (13.5 m)††	0.8 ± 0.2†	0.02 ± 0.01		HFSLM
	Pb (14 m)	n.d.‡‡	0.5 ± 0.3		HFSLM
	Pb (13 m)	1.5 ± 0.1†	0.2 ± 0.05		HFSLM
	Cu (13.5 m)††	21.4 ± 1.1†	3.8 ± 0.2		HFSLM
	Cu (14 m)	30.3 ± 1†	8.0 ± 0.1§§		HFSLM
	Cu (13 m)	43 ± 1	5.9 ± 0.1		HFSLM
Atlantic Ocean (Brittany shore, France)	Pb	0.4 ± 0.1†	0.2 ± 0.1		HFSLM
	Cu	1000 ± 50†	2.4 ± 0.1		HFSLM

* Direct measurements in non-deoxygenated sample.
† Measurements by ICP-MS.
‡ Comparative measurements were also performed directly in the sample by ultrafiltration on PM10 membrane followed by square wave voltammetry (0.3 nmol L^{-1}).
§ Measurements performed by FAAS.
** On site SLM measurements of samples collected at different dates and depths (as indicated in parentheses).
†† Colloidal iron hydroxide in the water column observed during this period.
‡‡ n.d. = not determined.
§§ Comparative measurements by ultrafiltration on PM10 membrane and FAAS gave 9 nmol L^{-1}

thousands) could be achieved in the second. Field preconcentration with a hollow fibre device was performed in a home-made vessel completely filled up by pumping the test water, in which the hollow fibre was immersed. The test solution was stirred during preconcentration with a magnetic stirrer After field preconcentration, the strip solutions were collected and analysed in the laboratory by flameless AAS and ICP-MS. This system has been applied to river, lake, sea and tap waters (Table 8). No membrane fouling or memory effects were observed.

It is worth noting that PLM-based results of Table 8 provide much lower metal concentrations than those obtained by FAAS or ICP-MS after acidification. This suggests that a high proportion of metal is bound to colloidal or particulate matter (section 5.2.2). On the other hand, they are closer or equal to metal concentrations obtained by voltammetry (chapter 9), or those obtained in the filtrate of 2 nm pore size ultrafiltration membranes [125]. This is because, under the conditions used (preconcentration of unperturbed source sample; negligible depletion of metal ion in the source solution; well-stirred source solution), only the free metal ions and possibly the small size labile or lipophilic metal complexes are measured (section 5.2), with the latter being likely to be absent in the waters tested. For reasons discussed in Chapter 9, voltammetry also provides a signal proportional to the free metal ion plus the mobile (small size) and labile (quickly dissociated) complexes, and, in addition, in the freshwaters tested, this fraction was also dominated by the free metal ion. Thus the fact that SLM and voltammetric results are similar is in agreement with expectations based on the speciation capabilities of the techniques and the composition of the test water (see section 5.2. for discussion).

Rigorous speciation interpretation of data of Table 8 would need more systematic results. It is important, however, to point out that PLMs are very much species sensitive, and in particular highly selective to free ions and lipophilic complexes (section 5), when appropriate hydrodynamic conditions are used. Because of this characteristic, the SLM technique is complementary to (i) voltammetry or DGT techniques (which provide mainly the concentration of the whole of the mobile and labile species; Chapters 9 and 11) and (ii) AAS or ICP-MS (which give total metal concentrations). Systematic tests with well-characterized ligands confirm these results [18,72].

For reliable metal speciation measurements in complex environmental systems, however, a stringent requirement is that the preconcentration factor F must be independent of sample composition, since sample modification is not allowed. Hence, the metal ion transport in PLMs should be coupled to the counter-transport of a chemically inert co-cation (Figure 2c) such as Na^+ as in the system discussed above [18,21]. Systems using counter-transport of protons (Figure 2b) should be employed with caution, as sample pH and speciation modifications could occur during analysis. Cotransport of sample anions (Figure 2a) must be avoided as F is then dependent on sample composition.

4.2 THE NEED FOR MICRO- OR MINIANALYTICAL SYSTEMS FOR *IN SITU* APPLICATIONS OF PLMs

For *in situ* applications of PLM in waters, the development of micro- (or at least mini-) analytical systems is important for several reasons. An obvious one is that small devices are more readily handled in site or *in situ*. There are, however, more fundamental reasons.

(1) When speciation is of interest (section 5), the test sample should not be modified, and in particular decreases in analyte concentration in the source solution due to membrane extraction should be negligible. The maximum depletion of the test species in the source solution $[(c_s^0 - c_s)/c_s]_{max}$ will occur at

infinite time (equilibrium conditions), and it can be shown (section 5; equation (19)) that

$$[(c_s^0 - c_s)/c_s]_{max} = V_{st}\alpha_{st}/V_s\alpha_s \quad (11)$$

where the c's denote concentrations of the analyte, in the source solution before (c_s^0) and after preconcentration(c_s), and α its degree of complexation (= total /free concentration) and the other symbols have their usual meaning. Because of the principle of the method, $\alpha_{st} \gg \alpha_s$; Equation (11) then implies that the condition $V_{st} \ll V_s$ must be fulfilled to maintain $c_s \sim c_s^0$. V_{st} should be $\sim$ 100 times smaller than V_s for $\alpha_{st}/\alpha_s \sim 1$, and even smaller for higher α values.

(2) The maximum value of the preconcentration factor, F, (at time $t = \infty$) is inversely proportional to V_{st} (equation 18)):

$$1/F = c_s^0/c_{st} = V_{st}/V_s + \alpha_s/\alpha_{st} \quad (12)$$

To get high sensitivity (F values of hundreds to thousands), V_{st} values should be in the microlitre range to work with reasonable source solution volumes (millilitre range). Note that the ratio α_s/α_{st} should also be very low to obtain high F values.

(3) In the absence of complicated processes at the interfaces (Section 5.2), the response time of the system is controlled by the molecular diffusion of the analyte (Section 2.4; Figure 19), (i) on the solution side of the membrane, (ii) in the membrane, and (iii) in the strip solution, since stirring of the strip solution is not possible when strip volumes are very small. For planar systems (linear diffusion) the characteristic transport time, τ, is given by

$$\tau = d^2/D$$

where d is the thickness of the diffusion layer and D is the diffusion coefficient of the analyte inside this layer. Using typical values of $D_s = D_{st} = 8 \times 10^{-10}\,m^2\,s^{-1}$ in the aqueous source and strip solutions, $D_m = 5 \times 10^{-10}\,m^2\,s^{-1}$ inside the membrane [84] and $\delta_s = 20\,\mu m$, $\ell = 100\,\mu m$ and $d_{st} = 200\,\mu m$ (= thickness of the strip solution), one obtains

$$\tau_s = 0.5\,s; \; \tau_m = 20\,s, \text{ and } \tau_{st} = 50\,s$$

These values are significantly shorter than the true response time of the system, because the transport of one molecule through the membrane or the strip solution is much shorter than the accumulation time of a large number of molecules in the strip. Nevertheless they show that diffusion in the unstirred strip solution is often the slow step when its thickness is in the range from 0.1 to a few mm. Theoretical calculations can be performed to optimise the system dimensions [45]. For example, both theoretical and experimental results show that for hollow fibres having 200 μm internal radius and a membrane thickness of 100 μm, $F = F_{max}/2$ after about 30 min. Since the response time depends on d_{st}^2, strip solutions with thicknesses in the range 50–100 μm are desirable to obtain accumulation times of the order of minutes.

(4) In PLM techniques, the strip solution must be renewed for each analysis. When the PLM is incorporated in an autonomous *in situ* probe for continuous analysis, a stock of strip solution must be stored in a reservoir of capacity not exceeding 1 L for practical reasons. This implies the use of a microlitre scale volume strip compartment (i.e. in the range 1–100 μL), to allow autonomy for a large number of analyses.

(5) Integration of a microsensor directly inside the strip solution will be required to monitor the full F versus time curve in real time (Figures 6B,D and 7). Indeed, although the analyte concentration in the source phase can be obtained from the single value of F_{max} (at $t = \infty$), it can be obtained more rapidly from the slope, dF/dt, of the initial linear portion of the F versus time curve (Figures 6 and 7), at shorter times. However, with systems where the PLM device and detector are positioned successively on-line (e.g. Figure 4E), a new preconcentration experiment is needed for each point of Figure 7, which is time consuming. By integrating the detector in the strip solution, the F versus time curve can be recorded in one run. This, however, requires that species transport by diffusion in the strip solution not be the limiting step and therefore that the volume of strip solution be small.

4.3 COUPLING OF SLMs TO DETECTORS FOR THE DETERMINATION OF ORGANIC COMPOUNDS

When a PLM is used as an independent preconcentration device, two procedures can be used after the preconcentration step on site or *in situ* for quantitative analysis: (i) the strip solution is collected on site in vials and transferred to the laboratory for classic automatic (or manual) analysis, or (ii) the PLM device can be connected to a sensitive detector for on-line analysis, in the field or in laboratory. Option (ii) is preferable when using very small strip solution volumes (microlitres).

On-line connection of PLMs to specific detectors has been achieved only with SLMs, but for many classes of compounds, and is described below. The requirements of such on line coupling are reproducibility, robustness, computer control and user-friendly operation. Detailed technical aspects of the coupling may play an important role in meeting these conditions, but their description is beyond the scope of this chapter. The reader should refer to the original papers.

For inorganic elements, very selective detection techniques exist (AAS, ICP MS, voltammetry), so that direct injection of the strip solution into such detectors is possible. In the case of organic molecules, since a large number of compounds belonging to the same class can be simultaneously preconcentrated by the SLM device, a further separation technique with higher resolution must usually be used for the analysis of the strip solution. In such cases, SLM has been coupled on-line to gas chromatography (GC), liquid chromatography (LC, HPLC), capillary electrophoresis (CE) and ion chromatography (IC). Coupling of an SLM to these techniques is reviewed below.

4.3.1 Coupling of SLMs to Liquid Chromatography

Coupling of an SLM to LC is the most widely used approach. For analysis of organics, for instance, this combination has been applied for the determination of acidic herbicides in natural waters [123], aromatic anionic surfactants [126], aliphatic amines in blood [127], basic drugs and drug metabolites in model solutions [128] and in blood plasma [129]. An example of such an on-line SLM–LC system is shown in Figure 11. Before the sample enters the source channel of the SLM, the analyte is converted into its neutral form (Section 3.2.3) by on-line mixing with an acidic or alkaline solution. The conversion is necessary to allow diffusion of analyte species through the organic solvent and into the strip side where it is irreversibly trapped by acid–base reaction. After a prescribed time, the enriched analytes in the strip solution are transported to the chromatographic system for separation and detection.

This system has been used for the analysis of amines in air [131,132] and rainwater [132]. Norberg *et al.* [133] have also used a fully automated SLM–LC system for the determination of aniline derivatives in environmental water samples. Lindegård *et al.* [134] developed a miniaturised SLM unit incorporated into an ASTED (Gilson) apparatus and coupled to an LC system for the

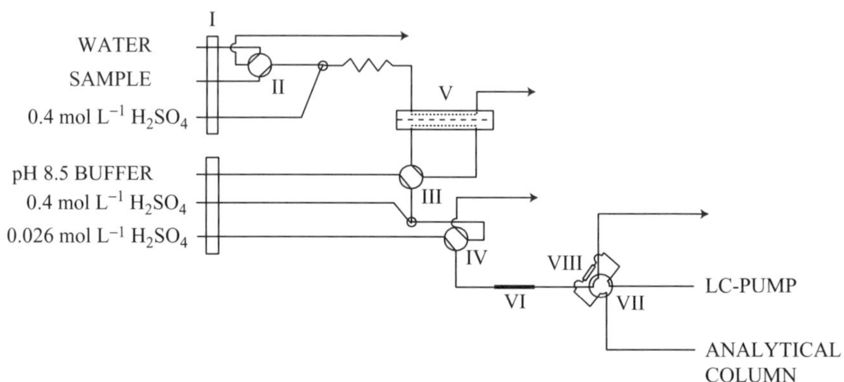

Figure 11. Typical on-line coupling of an SLM to LC for the preconcentration and analysis of herbides [130]. The roman numerals indicate the various components of the analytical procedure; the positioning of valves corresponds to the sample preconcentration step. I (peristaltic pump), II–IV and VII (switching valves), V (SLM separator unit; Figure 4B), VI (mixing column) and VIII (pre-column) [37]

evaluation of amperozide and its metabolites in model solutions and blood plasma. The set-up was arranged in such a way that, after the enrichment, the analyte was transfered into a loop to simplify and enhance the direct injection into the LC system. A lower extraction efficiency was obtained in plasma samples than in synthetic solutions, this result being attributed to protein binding of amperozide. The degree of protein binding was estimated with the SLM. A similar system has been used [135] for the determination of urinary *trans*-muconic acid. A miniaturised system coupling an SLM, LC and a selective phenol oxidase-based biosensor has been reported by Norberg *et al.* [136] for the determination of phenol, *p*-cresol and 4–chlorophenol, with a detection limit of 0.5 μmol L^{-1}. Nilvé *et al.* [130] developed an on-line system for the determination of herbicides in natural water by incorporating a pre-column loaded with C_{18} sorbent in the strip circuit (Figure 11). The analytes (sulfonylureas) were extracted from acidic aqueous sample solutions into an alkaline receiver solution. Prior to the loading of the enriched analytes onto the C_{18} sorbent, the solution pH was adjusted to convert the analytes into their protonated form.

4.3.2 Coupling of SLMs to Other Chromatographic Systems

On-line coupling of an SLM–LC–GC system has been reported by Audunsson [78] and used for the determination of urinary amines. A similar system was used by Grönberg *et al.* [137] for the determination of free amines in aqueous salt solutions and by Lindergård *et al.* [127] for the determination of aliphatic amines in blood plasma. The combination of SLM and ion chromatographic (IC) systems has also been reported for the determination of carboxylic acids in soils [138–140] and its combination with solid-phase extraction (Figure 12) and gas chromatography has been reported by Shen *et al.* [141] for the determination of local anaesthetics in plasma samples.

4.3.3 Coupling of SLMs to Capillary Zone Electrophoresis and Electrochemical Detectors

Capillary zone electrophoresis is another attractive technique which has been interfaced with SLMs for the determination of drugs such as bambuterol in plasma samples [142]. A hollow fibre membrane rather than the usual flat sheet membrane was used in this work. The hollow fibre device was more convenient as the sample volume could be reduced by at least a factor of 10 (from 20 to 2 μL), thus yielding an enrichment factor of about 40 000. Knutsson *et al.* [80] analysed low levels of chlorinated phenols with an electrochemical detector. In this investigation, a 30 min extraction of five chlorinated phenols (1–5 chlorine atoms) from natural water samples gave detection limits below 10^{-9} mol L^{-1}. Table 9 summarises the various types of SLM couplings developed for organic compound analysis.

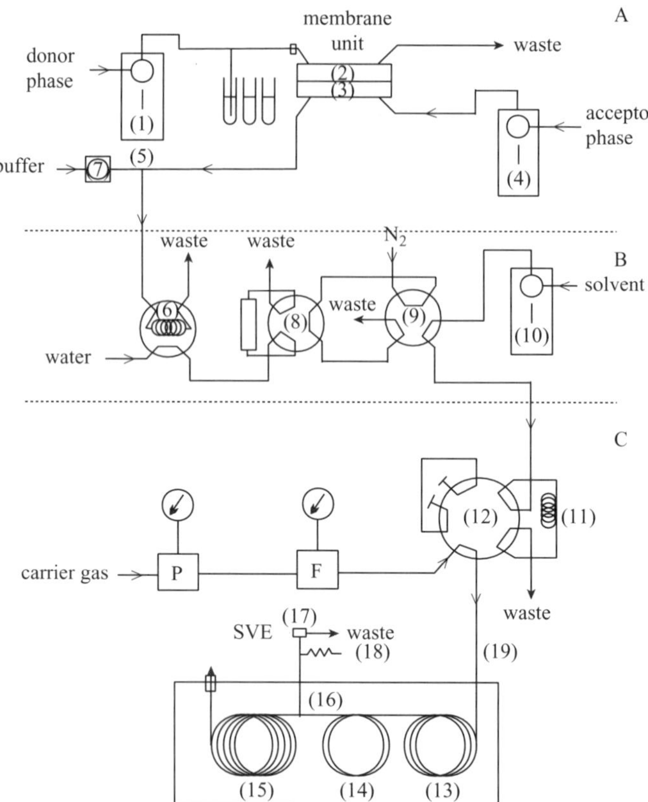

Figure 12. On-line coupling of an SLM (A) to solid phase extraction (B) and gas chromatography (C) [141]. In unit (A) isolation of target analytes takes place. In unit (B) the aqueous solution is exchanged with hexane as the target analytes are eluted with it. In the final analysis stage (C), the total volume of hexane is transferred into the loop of the chromatographic system. (1) autosampler syringe; (2) donor channel; (3) acceptor phase; (4) second syringe; (5) mixing tee loop; (6) sample processor; (7) peristaltic pump; (8), (9) switching valve; (10) dilutor; (11) loop; (12) pneumatically controlled ten-port valve; (13) column (retention gap); (14) retaining precolumn; (15) analytical column; (16) fused silica capillary; (17) silicon rubber tubing; (18) fused silica restrictor; (19) capillary; P = pressure; F = flow; SVE = solvent vapour exit

4.4 COUPLING OF SLMs TO DETECTORS FOR THE DETERMINATION OF INORGANIC ELEMENTS

In contrast to the large number of examples of coupling of SLMs with chromatographic techniques, there is only one report on the coupling of SLMs with spectrometric determinations, especially graphite furnace atomic absorption spectrometry (GFAAS). Malcus *et al.* [71] used this combination with four

Table 9. On-line coupling of SLM separation and preconcentration of organic compounds with their chromatographic determination. HPLC = high pressure liquid chromatography; GC = gas chromatography; IC = ionic chromatography

Compound	Sample matrix	Pretreatment system	References
Aniline derivatives	Environmental waters	SLM–LC	133
Phenols	Blood plasma/Environmental waters	SLM–LC–biosensor	81,136
Carboxylic acids	Air/water/soil	SLM–IC	138,139
Organic acids	Soil	SLM–IC	140
Drugs	Blood plasma	SLM–GC	129
Drugs	Blood plasma	SLM–CZE	142
Herbicides	Environmental waters	SLM–SPE–LC	123,130
Anionic surfactants	Reagent water	SLM–LC	126
Amines	Urine	SLM–GC	5,78
Amines (aliphatic)	Blood	SLM–GC	127

parallel SLM devices for the determination of aluminium, copper and cadmium in synthetic solutions at concentrations of nmol L^{-1} level. Figure 13 shows the automated, on-line system developed in this work. In this system, the automated SLM processing of the sample is done parallel with the ETAAS analysis of the previous sample including the standards, to enhance the sample throughput. Djane and co-workers [27,143] used the same set-up for the determination of total copper, cadmium and lead in natural water samples after acidification. Detection limits of 1.9, 0.24 and 0.9 nmol L^{-1} were obtained for the three metals respectively.

The coupling of HFSLMs to voltammetric methods is rarely reported. The first microsystem with close coupling of an HFSLM to voltammetry for trace metal determination was described in refs [144,145]. It is described in section 4.5. In the two different types of HFSLM system described in Figure 4D and E, the lumen of the hollow fibre can also be coupled on-line to a custom-built flow-through voltammetric detector. This has been used for the simultaneous analysis of dissolved Pb(II) and Cu(II) in river water [146]. Detection limits of ~ 0.1 nmol L^{-1} have been obtained with such a system. As discussed in section 4.5, the protection of the working electrode against the adsorption of traces of solvent and carrier leached out from the SLM is a major aspect when building such systems. This and other aspects linked to the coupling or integration of SLMs with voltammetric detection are discussed in section 4.5.

4.5 INTEGRATION OF SLMs AND DETECTORS IN MICROANALYTICAL SYSTEMS

Few developments have been attempted in this area, up to now, although important progress may be expected, based on modern microtechnology

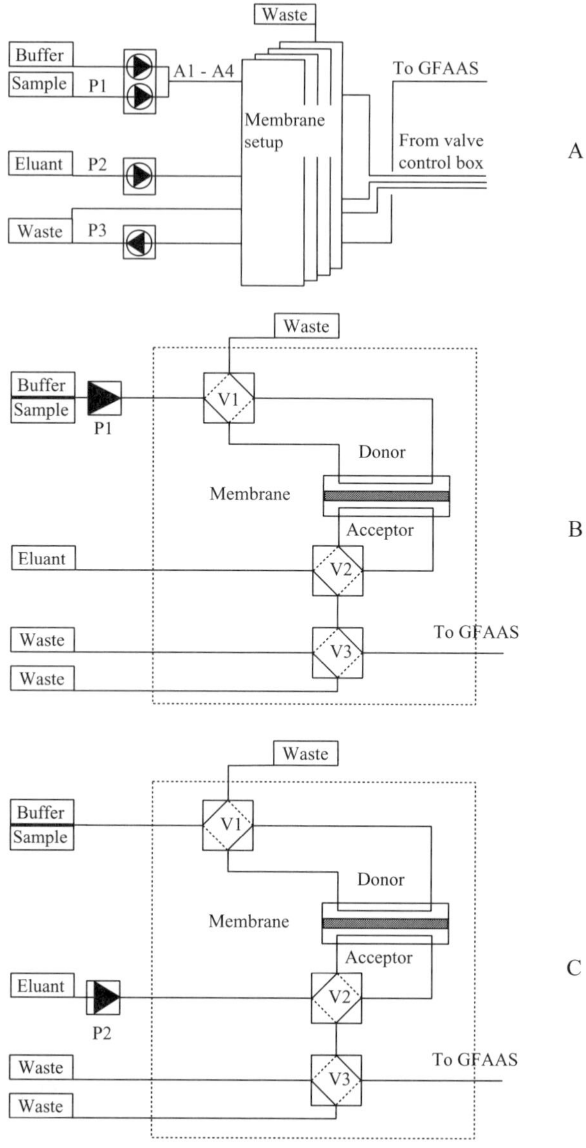

Figure 13. Set-up for automated trace enrichment by an SLM and on-line determination of metals by graphite furnace AAS. (A) Schematic drawing for four SLM systems used to preconcentrate four samples in parallel [71]. Sample solutions are pumped through lines (A1–A4), where each line leads to one of the four membrane devices. After sample enrichment, the target analytes are pumped into vials in the autosampler of the GFAAS prior to analysis. (B) Detailed scheme for the enrichment step. (C) Detailed scheme for the elution step. A third (washing) step is not shown. P1–P3 = pumps; V1–V3 = valves

(Chapter 12). The reason of developing micro- or mini-integrated systems are discussed in Section 4.2. Uto *et al.* [69] were the first to report a sensor integrating an SLM and voltammetry, but dimensions were not small and their system has apparently not been improved. A microsystem is described below. It has not been employed *in situ*, but it is a useful basis for discussing the major problems inherent in such developments.

Figure 14 shows this integrated microsystem based on the use of an HFSLM with an internal diameter of 400 μm, and a membrane thickness of 25 μm, combined with microelectrodes for voltammetric detection [144,145]. This system is based on the classic technology of microsensors: two capillary-based sensing devices are positioned at each end of the hollow fibre. The first device includes a channel, a Pt auxiliary electrode and a reference Ti/IrO_2 electrode in a pH buffered agarose gel. The pH-buffered gel maintains the potential of the Ti/IrO_2 electrode constant, irrespective of the pH of the strip solution [144,147] (An Ag/AgCl reference electrode is not used to avoid the interference of Cl^- ions with the Hg voltammetric electrode.). The second capillary device includes a channel and an Ir microelectrode, with a diameter of a few microns,

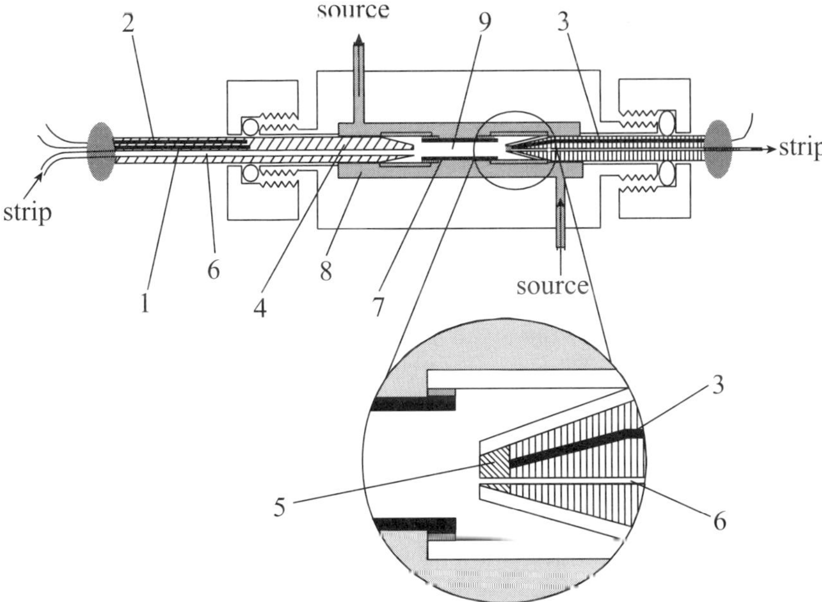

Figure 14. Diagram of a microanalytical system [144,145] combining a hollow fibre SLM (7), and voltammetric microelectrodes: (1) Ti/IrO_2 reference electrode in pH buffered agarose gel (4); (2) Pt auxiliary electrode in the agarose gel; (3) Hg-coated Ir-based microelectrode covered with a C_{18}-containing agarose gel (5); (6) circulation channel for the strip solution; (7) hollow fibre SLM membrane, (8) source solution; (9) strip solution

ultimately coated with Hg. It is integrated in a 400–500 μm thick agarose gel layer including C_{18}-modified silica particles, which hinders the adsorption on the electrode of traces of hydrophobic solvent and carrier leached out from the SLM (see below). It is coated with an Hg semi-drop formed by electrodeposition of Hg(II) through the agarose gel [144]. The Hg semi-drop is usually stable for a few days and can be renewed through the gel [144]. The channels in the two capillaries enable the renewal of the strip solution between each analysis. The source solution is circulated in the compartment outside the HFSLM, at a prescribed flow-rate. Operational procedure is described in refs [144,145].

Microanalytical systems of the type shown in Figure 14 allow one to minimise the strip solution dimensions (volume of 1–10 μL and channel radius down to 200 μm in Figure 14). On the other hand, by using such classic technology, it is difficult to produce capillaries including built-in channels and one or two reliable microelectrodes, with an overall tip diameter of less than 500 μm. In particular, smaller capillaries break easily when mounting them with the HFSLM. For that reason, the two capillaries in the system shown in Figure 14 are not inserted into the HFSLM, but are positioned very close to its ends. In such a configuration, the preconcentrated strip solution must be pushed by pumping towards the voltammetric electrode for analysis. Therefore the F versus time curve cannot be recorded in one run, because of this set-up, and also because of the time required for equilibration between the protective gel of the electrode and the preconcentrated strip solution, prior to detection.

These difficulties can be overcome by means of systems based on flat-sheet membranes and modern microtechnology, using in particular the Hg-coated Ir-based microelectrode arrays described in Chapter 9. Other systems can be imagined, in particular based on fibre optics for absorbance or fluorescence detection. The technical fabrication of optodes for organic and inorganic compounds, based on fibre optics technology, is well described [148,149].

The aforementioned developments of SLM coupling to unattended analytical methods have highlighted the major problems that must be solved to develop robust SLM-based microanalytical systems for long term monitoring. They are briefly summarized below.

- *Long-term membrane stability* has been discussed in section 3.4. It is linked to the loss of performance of the membrane, often due to solvent or carrier dissolution in the water phases, or to the formation of microemulsions at the membrane–water interface. It is a general problem for SLM-based analytical systems, but it is more important for microsystems such as those discussed in this section where a very small membrane area may be in contact with a large volume of source solution. For instance, it has been observed that an SLM composed of a 1/1 mixture of toluene/phenylhexane with 0.1 mol L^{-1} lauric acid +0.1 mol L^{-1} 22DD leads to stable measurements for 4 d, for 10–50 cm long HFSLM (Section 3.3, Figure 4C and E) or for 1–10 cm^2 flat-sheet SLM systems [18,21], whereas membrane transport decreases significantly after 1–2 h with the microsystem shown in Figure 14 [145]. In this case, the replacement of the solvent (toluene/ phenylhexane) by another one (phenyldecane) less miscible with water and less volatile, and of lauric acid by palmitic acid (less soluble in water) solved the problem, at the expense of a lower flux of analyte due to the higher viscosity of phenyldecane [145].

- *The choice of materials* for building the microsystems may be difficult, as materials resistant to the non-polar solvent used in PLM (silica, glass) are not always inert towards the test trace compounds (in particular with respect to adsorption of trace metals), and many plastics are attacked by solvents (Section 2.3.). Plastics such as PEEK, Kel-F and polyvinylidene fluoride, can be used for SLM microsystem fabrication. When microtechnologies are used, the choice of materials may be even more restricted (Chapter 12).
- *Protection of the voltammetric sensor.* Leaching of SLM solvent and/or carrier may be low enough not to cause problems with long term stability of the membrane, but sufficient to interfere with the detector. This is particularly severe with metallic (hydrophobic) electrodes used for voltammetric detection, on which the hydrophobic solvent and the surface active carrier (e.g. 22DD and lauric acid, [144,145]) tend to adsorb strongly. Since a monolayer may be enough to block the redox process (Chapter 9; Figure 15a), very small amounts of these compounds in the strip solution may create interferences. As discussed above, one way to solve this problem is by using C_{18}-containing protective layers [144] (Figure 15b). For systems where an HFSLM is coupled on-line with a flow-through voltammetric cell, a plug of reticulated carbon powder can be inserted between the HFSLM and the detector [146]. Other solutions are to use carriers that are even less soluble in water, or to attach the carrier covalently inside the membrane [150].
- *Operating conditions of the SLM microsystem.* The choice of operating conditions is more restricted in an integrated microsystem than in on-line SLM-detector systems. In particular, the choice of the complexing agent of the analyte in the strip solution should meet several criteria: (i) it should be strong enough to create the driving force for analyte transport (Section 2.1.), (ii) it should be hydrophilic enough (usually with a sufficiently high electric charge) not to be soluble in the SLM, and (iii) it should not interfere with metal detection (e.g. it should not form electroinactive metal complexes in the case of voltammetric detection, or create spectral interferences in spectrometry). For the microsystems shown in Figure 14, pyrophosphate and NTA at pH = 6 have been found to be good stripping agents for Pb and Cd. CDTA may also be used for Cu. Similarly reagents introduced in the strip compartment, for preconditionning the sensor, must be compatible with the SLM and their impact on the flux of analyte must be carefully checked. This is the case, for example, for the formation of the Hg semi-drop by electro-

deposition of Hg(II), in the system shown in Figure 14. Systematic studies showed [143,144] that the usual procedure of deposition of Hg(II) from acidic solution is not possible but that in $1.5\,\mathrm{mol\,L^{-1}}$ acetate at pH = 4.5 a good Hg deposition can be obtained without perturbing the SLM.

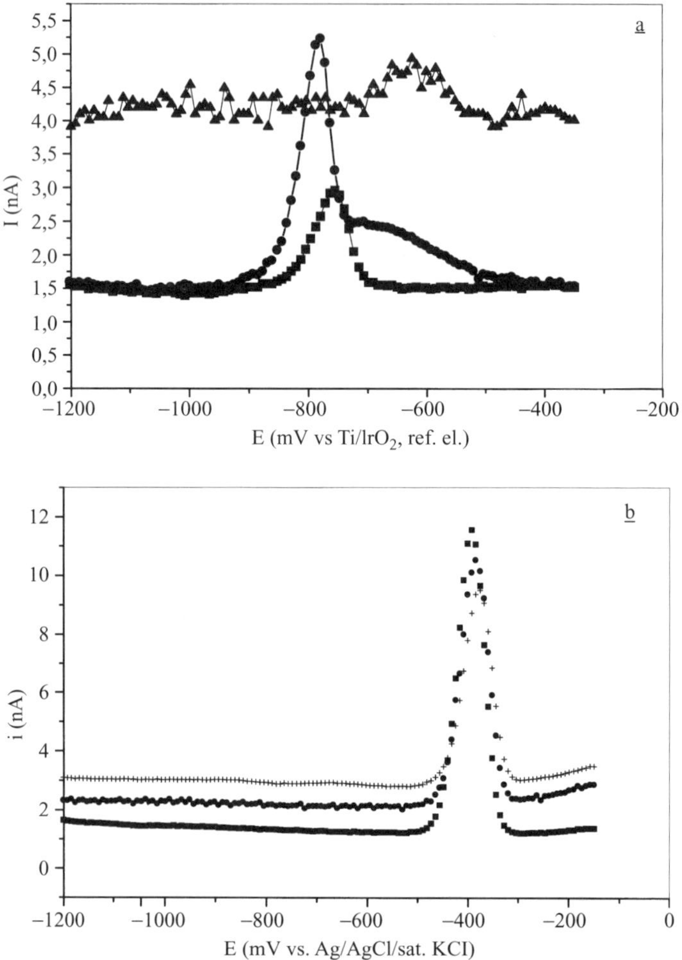

Figure 15. Efficiency of C_{18} containing agarose gel as protective membrane of the Hg-coated Ir-based microelectrode against fouling by the SLM solvent and carrier. (a) SWASV peaks of Pb(II) in the lumen of HFSLM of the microsystem of Figure 14, without protective gel; (●) immediately after Hg Film formation, (■) 120 min after Hg film formation, (▲) 150 min after Hg film formation. (Reference electrode = Ti/IrO_2 in 4.5% agarose +$\mathrm{mol\,L^{-1}}$ MES, pH = 7). (b) SWASV peak of an electrode protected by C_{18} containing agarose gel (Figure 14) before (●) and after one (■) and two (+) dipping in the concentrated SLM organic phase.

5 INTERPRETATION OF PLM-BASED MEASUREMENTS IN CHEMICAL SPECIATION AND BIOLOGICAL UPTAKE STUDIES

5.1 THE CHEMICAL SPECIES OF WATER AND SEDIMENTS IN RELATION TO TRANSPORT THROUGH PLMs

Figures 16 and 17 (see also Chapter 9, section 3.5) summarise the behaviour of the major classes of chemical species of a given element or compound denoted by M. Note that M may represent any chemical such as a metal ion, an inorganic anion (e.g. phosphate), or an organic compound (e.g. pesticide, PAH, etc.) (Section 3.2). In all cases, it is useful to discriminate between different forms of M:

- *the free form*, i.e. not associated with any ion or compound other than the solvent (e.g. hydrated metal ion, unprotonated inorganic anions or organic compounds);
- *molecular sized complexes*, ML or MA, where A or L may be any small size species such as the proton, metal ions, inorganic anions or organic compounds;
- *colloidal*, MQ, or *particulate species*, MP, where M is adsorbed on natural colloids ($1\ \text{nm} < Q < 1\ \mu\text{m}$) or particles ($P > 1\ \mu\text{m}$).

Note that this classification is somewhat arbitrary, in particular the specific values of 1 nm and 1 μm mentioned above, since, rigorously, continuous distributions in size or degree of hydrophobicity must be considered. Experience shows, however, that (i) such semi-quantitative discrimination greatly helps developing sound analytical techniques and (ii) the orders of magnitude of a few nanometres and a few microns, for size limit values between 'molecular' and colloidal size species on the one hand, and between colloidal and particulate

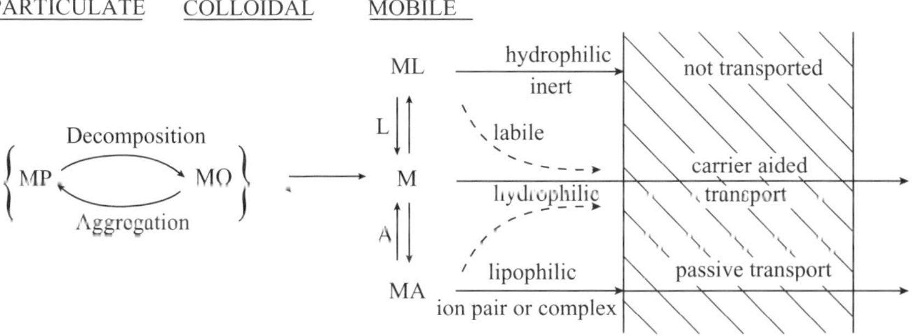

Figure 16. Schematic representation of the behaviour of free metal ion (M) and the major types of M complexes in aquatic systems: complexes with particles (MP) and colloids (MQ); mobile hydrophilic (ML) and lipophilic (MA) complexes

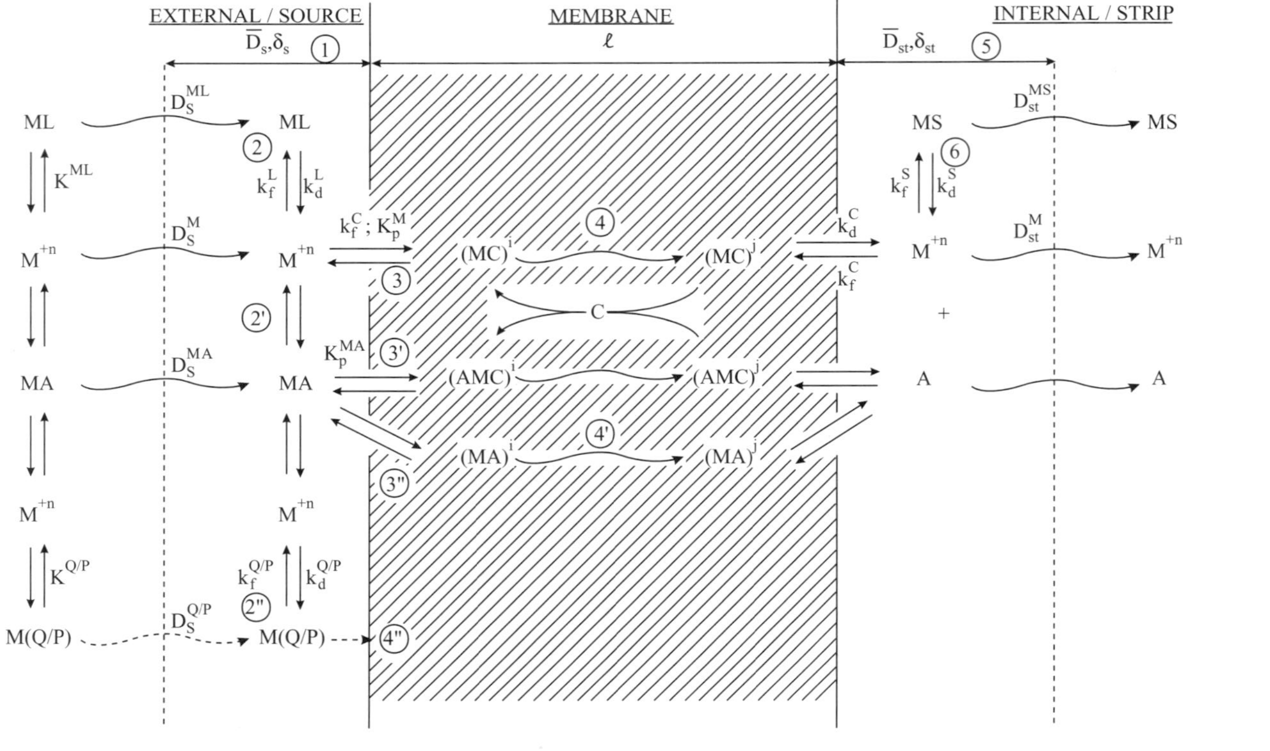

Figure 17. Major processes and physico-chemical parameters controlling the flux of M from aqueous samples through the PLM membrane (numbers refer to Table 11). The vertical dotted lines schematically represent the diffusion layers in the source and strip solutions at the membrane surface. Symbols for complexes and ligands: see Figure 16

Table 10. Typical examples of particulate, colloidal and mobile (hydrophilic and lipophilic) metal complexes in waters. (The transport of the mentioned lipophilic complexes through biological membranes is discussed in the corresponding references.)

Particulate species (> 1 μm) [151,152]	Colloiddal species (1 nm–1 μm) [153,154]	Mobile species (< 4 nm)			
		Hydrophilic labile complexes [151]	Hydrophilic inert complexes	Lipophilic complexes	Ion pairs and mixed complexes
Transition and heavy metal ions adsorbed on:] Ca,CO_3 Aluminiosilicates (clays) SiO_2 FeOOH Organism detris	Transition and heavy metal ions adsorbed on: clays SiO_2 FeOOH MnO_x Fulvic/humics proteins polysaccharides	Charged inorganic complexes $Cu(CO_3)_2^{-2}$, $CuCl^+$, Charged complexes with hydrophilic organic ligands: Cu (citrate) Cu $(oxalate)_2^{2-}$ Cu $(salicylate)_2^{-2}$ Cu $(fulvic)^{n-}$ (at high Cu/fulvic ratio)	Cu–EDTA [155] Cu–fulvics at very low Cu/fulvic ratio [154]	Uncharged inorganic complexes: $CuCO_3^0$ [155] $HgCl_2^0$ [156,158,159] $CdCl_2^0$ [158] $Zn(OH)_2^0$ [158] Weakly charged or uncharged complexes with liophilic organic ligands: $Cu(oxine)_2$ [155,163] $Cu(DDC)_2$† [155] $Cd(DDC)_2$ [155,164] $Pb(DDC)_2$ [155] R_2 = Hg* [165]	Ion pairs: NaCl [156,157] $NaCO_3^-$[156,157] RSnX* [160] RPbX* [160] RHgX* [161] Mixed complexes: F–Al–memb. site [162]

* R = alkyl chain. X = OH^- or Cl^-
† DDC = diethyldithiocarbamate.

size species on the other, correspond to domain boundaries between which different physicochemical processes occur (Chapter 9, section 3.5).

Because of their large size, MQ and MP (also denoted together as M(Q/P) in Figure 17) diffuse very slowly and cannot pass through the PLM. Molecular sized species on the other hand have comparatively high diffusion coefficients. Molecular and small colloidal species ($\leq\sim$ 4 nm) whose mobilities are large or non-negligible are called the mobile species (Chapter 9, section 3). In addition to the free form, the mobile species may include several types of complexes whose nature depends on M. When M is a metal ion, the important types of species relevant for both PLM analysis and bioaccumulation (Figure 16) are the following (See Table 10 for examples):

- the hydrated ('free') metal ion, hydrophilic, which can only pass through the membrane by means of an hydrophobic carrier, C;
- the hydrophilic labile complexes, ML, whose dissociation rate is significant with respect to the transport rate of M through the membrane. Under certain conditions (Section 5.2.2) M may pass through the membrane, after dissociating from ML and binding to C, but the complex ML itself, which is hydrophilic and/or charged, is not soluble in the membrane;
- the lipophilic, usually neutral ion pairs or complexes, MA, formed between M^{n+} and either an anion A^{n-} or a lipophilic ligand. MA may pass the membrane either by passive diffusion or by forming a mixed complex, AMC with the carrier C.

It is useful to compare the transport processes through PLMs and biological membranes, in order to evaluate in which respect PLM-based measurements can provide information on metal bioavailability. Refs [156,166–171] describe biological membrane structures and transport mechanisms. Figure 17 and Table 11 give an overview of the major processes and reactions occurring outside or inside the membrane or at the water–membrane interface, which may limit the overall flux from the external to the internal solution. Processes occurring *inside* the membrane are not shown in Figure 17, but are briefly listed in Table 11, which compares the various limiting transport steps reported for PLM and biological membranes. The following remarks can be made, regarding the analogies and differences between the two types of membranes.

- Transport processes *in the test solution* are almost identical. This is a key feature of PLMs, since under well-chosen conditions (Section 5.2) it provides precise information on the chemical speciation factors of the test medium which directly influences bioavailability (as well as other environmental processes). Interestingly, only one mathematical model has dealt in detail with chemical speciation coupled to diffusive fluxes under environmental conditions [178], and in this model, slowly diffusing colloids and macromolecules

Table 11. Comparison of the various limiting transport steps which have been reported or discussed in the literature for PLMs and biological plasma membranes. Reaction or process numbers refer to Figure 17. General references: [154,156,166,167,169,170,171]

Biological membrane	Supported liquid membrane
External (or source) solution	*External (or source) solution*
(a) Diffusion of M species in solution (1) may be limiting for both major ions [156,167] and trace metals [172].	(a) Diffusion in the aqueous boundary layer is significant in many cases [39,84], but systematic studies of metal complexation effects are limited [18,21].
(b) Consideration of coupled chemical kinetics and diffusion (steps 1, 2-2″,3) has been shown theoretically to possibly play a significant role [172].	
External solution/membrane interface	*External solution/membrane interface*
(c) Thermodynamic role of a mixed complex formation (3′) (≡AMC) between solution metal and anion, and membrane site [162].	(c) Thermodynamic stability of mixed complex formation (3′), between the solution anion, A, the test metal, M, and the carrier, C, can influence the transport rate through the membrane [3,9,37,42,173].
	(d) Kinetics of adsorption/desorption of the MC complex in relation with carrier protonation, on the membrane side of the interface can be a limiting factor (detailed discussion in ref [174]).
(e) Slow chemical binding of M (3) on membrane surface sites, C_i, of carrier protein or ion channel imbedded in the membrane and/or saturation of binding sites [172]. (A similar kinetic limitation is speculated in [172] at the internal solution/membrane interface).	(e) Kinetic of MC complex formation at the interface may be slow [42,44]. This, however, seems to be more an exception than a rule.
	(e′) Carrier concentration in the membrane may limit the flux of major ion transport (K^+, NO_3^- through the membrane, but is unlikely to play a role for trace compounds [3,8,28]).
Bulk of the membrane (see also Figure 19)	*Bulk of the membrane*
(f) Passive diffusion of lipophilic complexes (4′) due to concentration gradient through the membrane [156,167,169,175].	(f) Passive diffusion of the lipiphilic complexe (4′) due to concentration gradient [146]
(g) Diffusion of a metal–carrier protein complex (4) due to electrochemical gradient through the membrane [156,167,169].	(g) Diffusion of the MC complex (4) due to electrochemical gradient of M in the whole system (section 2.2; [39,44,45]).
(h) Diffusion of partially hydrated metal ion inside an internally hydrophilic proteic ion channel [156].	(h) In membranes with complexants covalently attached to the backbone polymer [150], M may pass from site to site ('Fixed site hoping') when the site density is large enough [176].

Table 11. (*cont.*)

Biological membrane	Supported liquid membrane
(i) Other carrier mechanisms (rotating proteins inside the membrane, sliding surface active complexant [167], and multi-steps mechanisms [169]).	(i) Retarded diffusion has been observed [84], where the MC complex adsorbs on the polymeric membrane, which slows down diffusion but increases the partition of metal between the solution and the membrane. The resulting flux is not affected.
(j) Pinocytosis, i.e. invagination of the membrane fraction coated by a complex of M with a hydrophilic macromolecule and internalisation of the resulting vacuole (e.g. Fe–transferrin; [156]).	
(k) Phagocytosis, i.e. engulfing of metal bearing colloids or particles by the membrane, and internalisation of the resulting vacuole (4″) [156, 167].	
Internal (or cell) solution	*Internal solution*
(m) Diffusion in the finite, small volume of the biological cell, implying accumulation of M and non steady-state conditions [177].	(m) Diffusion in the strip compartment becomes a slow step typically when its dimensions are larger than the membrane thickness, but too small to enable efficient convection by stirring (typically 0.1 to a few mm) [44,45].
(n) Influence of the internal degree of complexation of M on the accumulation flux inside the cell [155].	(n) Slow release of M at the membrane–strip interface has been considered [46,47].

were not taken into account. The only difference between PLMs and biological membranes with respect to transport in external solution is the size or geometry of the systems. In microorganisms with radius less than a few microns, spherical diffusion has to be considered, for which the rate parameter is D/r [$\mathrm{m\,s^{-1}}$](D = diffusion coefficient [$\mathrm{m^2\,s^{-1}}$]; r = microorganism radius [m]; [172,177]; Chapter 8; Chapter 9, section 2.1.3). In most cases, the present PLM systems either are flat or have size $> 100\,\mu\mathrm{m}$, so that linear diffusion is operative, for which the rate parameter is D/δ [$\mathrm{m\,s^{-1}}$] (Chapter 8; Chapter 9, section 2.1.3) (δ = aqueous diffusion layer thickness [m]). δ depends on convection and varies with D^p(p = constant ~0.3–0.7, often close to 0.5). Correct quantitative comparison of fluxes at PLM and *microorganism* surfaces would require either the construction of PLM systems with spherical diffusion (e.g. based on ELM, Figure 2), or the mathematical corrections of experimentally determined fluxes, based on our knowledge of r and δ.

- Some of the PLM and biological transport steps *at the water–membrane interface or inside the membrane* are conceptually similar (in particular the adsorption on carrier site, diffusion of metal–carrier complex, or passive diffusion of lipophilic complexes and ion pairs), and the corresponding mathematical models may be similar. Although this analogy is conceptually useful, it must be remembered that (i) the nature of carrier binding sites (therefore the stability constants and metal fluxes) may be very different, (ii) facilitated metal transport in biological membranes is usually a multistep process and is thus much more complicated than in a PLM [156,166], and (iii) passive diffusion fluxes depends on $D_m K_p^M/\ell$ (see Figure 17 for symbols). D_m values in a PLM and a plasma membrane may be drastically different owing to differences in viscosities [167]. K_p^M will also be different because of the different natures of the complexing sites in the two types of membranes, and ℓ is usually much smaller for plasma membranes ($\sim$ 5–10 nm) than for PLMs (typically 20–100 μm).

Even though there is little hope of developing PLM-based sensors in the near future that will match processes occurring in a biological membrane, PLMs are presently the only technique, along with voltammetry (Chapter 9), capable of relating the chemical speciation of a compound in the test medium to its flux at interfaces. In particular, it will be shown (Section 5.2) that the fractions of free M, its labile hydrophilic complexes, and its lipophilic complexes can be determined from their contribution to the PLM flux, by carefully engineering the PLM conditions, by varying the carrier concentration and thus the transport rate through the membrane. Using this information to estimate the flux of bioavailable metal into a particular type of organism is much more difficult, because of differences in membrane transport. Order of magnitude estimations, however, seem feasible by calibration, and are expected to give results at least as good as, and probably better than, estimations of permeabilities based on purely thermodynamic oil–water partition coefficients [155,156,179]. Better understanding of some of the factors influencing transfer through biological membranes can also be expected by manipulating both PLMs and biological membranes. In particular, it will be shown in section 5.2 that studies with PLMs may improve our understanding of the limits of the so-called ‘free ion activity model’ often used to explain metal–microorganism interactions [175], according to which the bio-uptake flux is proportional to the free metal ion activity in solution.

5.2 A SPECIATION-BASED MODEL PREDICTING THE METAL FLUX THROUGH PLMs

5.2.1 General Considerations

Before trying to relate the metal flux through PLMs quantitatively to metal speciation in solution, it is useful to bear in mind the following concepts. Note that in the subsequent sections, activities will be considered as equal to concentrations for simplicity.

5.2.1.1 Equilibrium metal concentrations in the strip and source solutions Figure 18 shows definitions of symbols which will be used later, as well as the relation between metal concentrations in complexing, source and strip solutions (L and S are the complexing agents in the source and strip solutions, and C is the carrier in the membrane). The important concept is that, when equilibrium is reached (superscipt e) in the complete system, the *free* M concentrations (or more precisely its activities) are equal in the source and strip solutions (equation (14)), provided the compositions of the strip and source solutions are close enough, so that the standard chemical potentials of M can be taken as equal. This equilibrium condition is independent of the dissociation and diffusion rates of the complexes; the latter will only influence the time needed to reach the equilibrium. At equilibrium, the ratio of total concentrations of M in the source and strip solutions is then equal to the ratio of the degrees of complexation, α_s^e/α_{st}^e, in each solution (equation (15)). This ratio is thus the major driving force for metal accumulation in the strip phase. Before equilibrium, M will

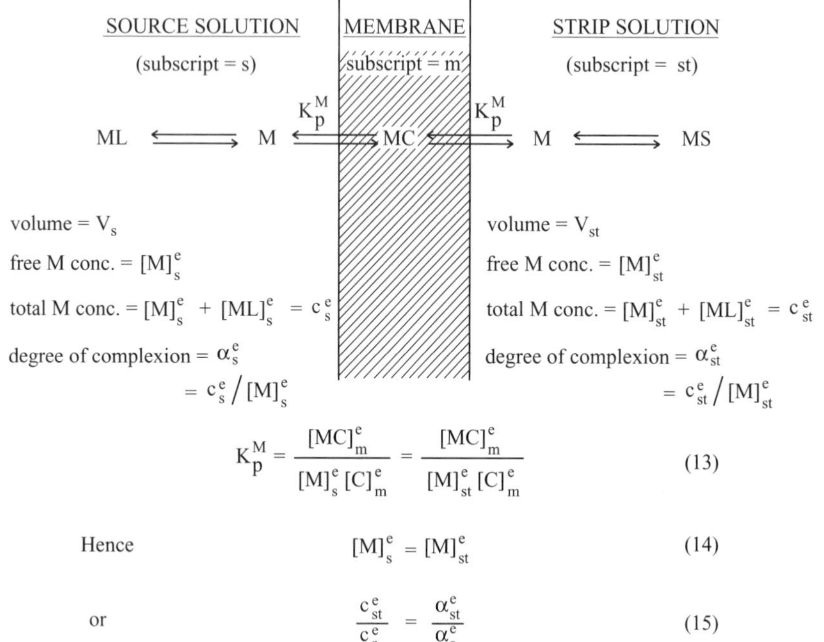

Figure 18. State of the PLM system at equilibrium (superscript e) and definitions of concentrations, [X], volumes, *V*, and degrees of complexation, α, in the system

diffuse through the membrane, as long as $c_{st} < c_s\alpha_{st}/\alpha_s$. This principle imposes a few important limiting conditions for successful application of PLM to speciation measurements.

(1) During the accumulation process, c_s decreases and c_{st} increases, and consequently α_s and α_{st} increases and decreases, respectively, owing to progressive release of free L and saturation of S (Note that such an effect has also been noticed in microorganisms [155]). If the total concentration of S, $[S]_t$, in the strip solution is too low, accumulation will stop owing to saturation of S; c_{st} then will only reflect the value of $[S]_t$ in the strip solution, but it will not give any information on metal speciation in the source solution. Thus for speciation applications, $[S]_t$ must be high enough, so that α_{st} = constant during the accumulation step, irrespective of c_{st}. This condition is assumed to be fulfilled hereafter.

(2) The important speciation parameters one has to determine in the source phase are the total and free metal ion concentrations, c_s^0, $[M]_s^0$, or their ratio α_s^0, at time $t = 0$ (superscript 0), i.e. for the conditions where the test sample has not been perturbed by the depletion of M due to its accumulation in the strip solution. This perturbing effect can be minimised in two ways (Section 2.4):

- measurement of c_s and $[M]_s$ from the *initial flux* through the membrane, i.e. the initial slopes in Figures 6B and D or 7 (see also section 5.2.2);
- measurement of c_{st} at equilibrium (i.e. $t = \infty$), but for $V_s = \infty$. This is the situation for a PLM integrated sensor (very small V_{st}), dipped in lakes or oceans (infinite Vs). Then no depletion occurs ($c_s^0 = c_s^e = c_s$) and equation (15) (Figure 18) becomes

$$c^e_{st} = \alpha_{st} c_s / \alpha_s = \alpha_{st} [M]_s \tag{16}$$

The minimum value of V_s necessary to satisfy equation (16) can be obtained from equation (17), which states that the number of moles of M in the strip solution is equal to that released from the source (assuming that the number of moles of M in the membrane is negligible). At equilibrium

$$c_s^0 V_s - c_s^e V_s = c^e_{st} V_{st} \tag{17}$$

Combining equations (15) and (17) yields the following expressions for the maximum values (at equilibrum) of the preconcentration factor in the strip solution, F^e, and the depletion factor in the source solution, d^e, for any value of V_s and V_{st}:

$$F^e = \frac{c^e_{st}}{c_s^0} = \frac{V_s \alpha_{st}}{V_{st}\alpha_{st} + V_s \alpha_s} \tag{18}$$

$$d^{e} = \frac{c_s^0 - c_s^e}{c_s^e} = \frac{V_{st}\alpha_{st}}{V_s\alpha_s} \quad (19)$$

In particular these equations show that depletion is negligible and equation (16) is fulfilled when

$$V_{st}\alpha_{st} \ll V_s\alpha_s; \text{ then } F^e = \alpha_{st}/\alpha_s \text{ and } d^e \to 0$$

Equation (19) also enables one to choose the maximum volume of the strip solution. For instance, if $\alpha_{st}/\alpha_s = 10$ (a minimum value under normal conditions), one should get $V_{st} \leq V_s/1000$ for 1% depletion in the source, i.e. $V_{st} \leq 100\,\mu L$ when $V_s = 100\,mL$. Such calculations show that developing microSLM integrated systems is useful not only to decrease the response time (Section 4.2), but also to facilitate interpretation of results at equilibrium.

(3) For the opposite condition ($V_{st}\alpha_{st} \gg V_s\alpha_s$,), F^e no longer depends on metal speciation in the source, but only on the ratio V_s/V_{st}, since equation (18) becomes

$$c_s^0 = c_{st}^e V_{st}/V_s \quad (20)$$

This condition can be used when total concentration in the source (c_s^0) is of interest, irrespective of its speciation (note, however, that fully inert complexes will not be measured by a PLM unless they are liposoluble; see sections 5.2.2. and 5.2.3).

5.2.1.2 Parameters related to metal speciation and flux in the test water The parameters controlling the dynamic speciation of M in the source solution are schematically represented in Figure 17. They are (i) the equilibrium constants (K) and (ii) kinetic rate constants (k_f and k_d) of all M complexes, and (iii) their diffusion coefficients (D). In Figure 17 and sections 5.2.2–5.2.3, only one ligand L is considered for simplicity. It must be borne in mind, however, that complexants in natural waters are chemically and physically heterogeneous, including a very large number of different compounds and complexing sites. This means that:

- Thermodynamic complexation properties are represented by a broad distribution of equilibrium constants [154]. As a result, the degree of complexation, α_s depends not only on total ligand concentration, but also on the metal/ligand total concentrations ratio (e.g. [154,180]). When the transport rate across the membrane is sufficient to set up a zone of local depletion at the outer face of the membrane, a gradient of metal/ligand ratio and therefore a gradient of α_s is created inside the diffusion layer, which will affect the flux, as discussed in detail for voltammetric curves [181].

- Similarly, realistic flux calculations should consider a broad distribution of complex dissociation rate constants, whose relative influences will also depend on metal/ligand ratio (as discussed for voltammetry [182]).
- Because of physical heterogeneity, broad distributions of size (and thus diffusion coefficients) should also be considered when computing fluxes. This is briefly discussed in section 5.2.2.

5.2.1.3 Parameters controlling the flux inside the membrane The flux through the membrane is governed by its permeability, $P_m (m\,s^{-1})$, which relates the flux $J(mol\,m^{-2}\,s^{-1})$, to the concentration difference of the diffusive species on both sides of the membrane. When transport through the membrane is controlled by diffusion of MC, i.e. when the formation and dissociation rates of MC at source–membrane and membrane–strip interfaces (Figure 17) are both very fast, P_m is given by (Section 5.2.2)

$$P_m^{MC} = D_m^{MC} K_p' / \ell \tag{21}$$

K_p' is the partition coefficient of M between the source and membrane solutions for a given value of $[C]_m (K_p' = K_p^M [C]_m = [MC]_m/[M]_s$; Figure 18). It is worth noting that equation (21) may also account for any reversible (i.e. not rate controlling) reactions of MC inside the membrane, such as adsorption of MC on the solid matrix of the membrane. It has been shown [84] that such a reaction results in an effective distribution coefficient, $K_D (= [MC]_m'/[M]_s$ where $[MC]_m'$ is the total concentration of MC, adsorbed and dissolved, in the membrane) higher than the thermodynamic partition coefficient, $K_p' (= [MC]_m/[M]_s$, where $[MC]_m$ is only the dissolved concentration of MC in the membrane phase). However, such an adsorption reaction also slows down the diffusion in the membrane and thus results in an effective diffusion coefficient, $\bar{D}_m$, lower than the molecular diffusion coefficient of MC in the solvent (D_m^{MC}). Provided the adsorption energy is strong enough, it can be shown [84] that $D_m^{MC} K_p' = \bar{D}_m K_D$, so that equation (21) is still applicable despite the existence of this reversible adsorption reaction. Interestingly, even though the condition $D_m^{MC} K_p' = \bar{D}_m K_D$ implies that P_m and the flux of M in the membrane are not affected by such intramembrane reversible reactions, they may play a useful role in decreasing the leaching of carrier from the membrane, which is a major cause of instability (section 3.4).

5.2.2 Solution Containing Only Labile Hydrophilic Complexes; Effect on the Carrier-aided Transport of M Through the PLM

A simplified picture depicting the transport of M from a solution containing only labile and semi-labile hydrophilic complexes (all represented by ML), as well as the concentration gradients of the various species, is shown in Figure 19.

Note that ML now includes MQ and MP of Figure 17 and the lipophilic complexes MA are assumed to be absent. Other important assumptions for the equations given below are the following.

- Equations correspond to *initial* fluxes, i.e. there is no accumulated M in the strip phase:

$$[\mathrm{MS}]^{\mathrm{b}}_{\mathrm{st}} = [\mathrm{M}]^{\mathrm{b}}_{\mathrm{st}} = 0$$

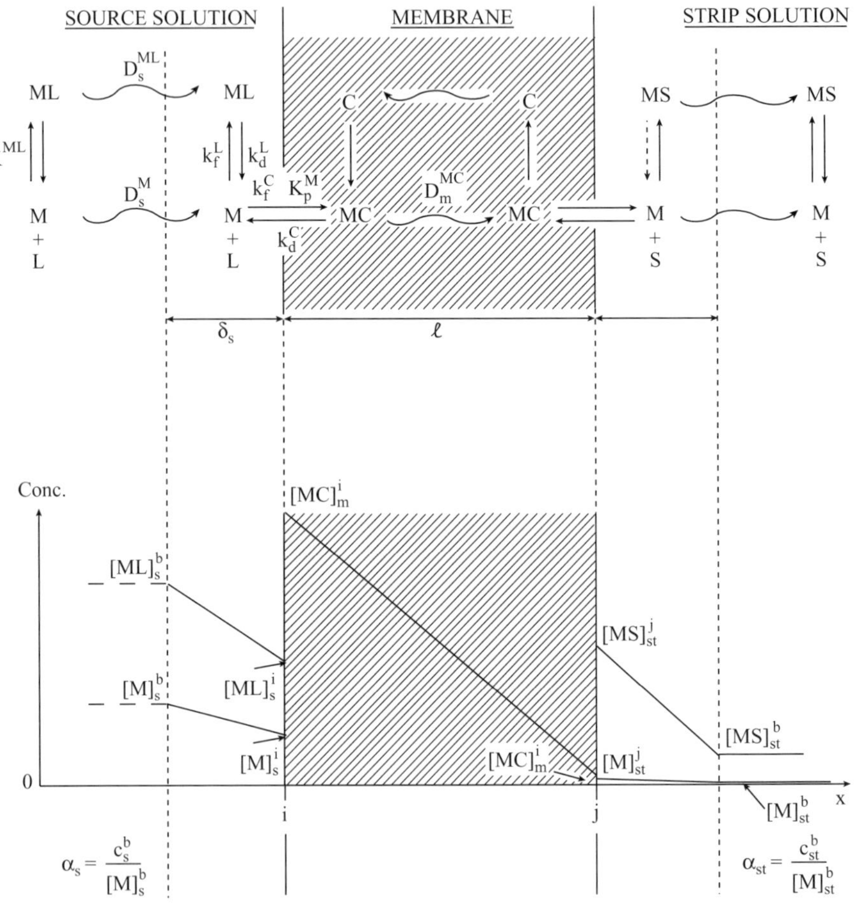

Figure 19. Processes controlling the flux of M through a PLM, from a solution containing only free M and mobile, labile complexes ML. Concentration gradients in the source solution, the membrane and the strip solution, are drawn for time close to zero (no accumulation of M in the strip solution) albeit just after the attainment of the steady state (linear gradients). Vertical dotted lines indicate the diffusion layers in the strip and source solutions at the membrane surfaces

- Linear diffusion is considered in the system, which implies that diffusion in the source and strip solutions is influenced by convection through the diffusion layer thicknesses, δ_s and δ_{st}, in both compartments.
- The partition reactions at both aqueous/membrane interfaces are not rate limiting.
- All complexes ML are labile, i.e. they can form and dissociate many times during their time of transport through the diffusion layer of the source solution (Figure 19). This implies that the first-order rate constant of dissociation of ML, (k_d^L;s^{-1}), and the pseudo first order rate constant of formation of ML ($k_f^L[L]_s = k_d^L K^{ML}[L]_s$;$s^{-1}$) are much higher than their effective diffusion rate constant (D/δ^2;s^{-1}). When the diffusion coefficients of M, and ML, D_s^M and D_s^{ML} are different (e.g. when L is a colloid or macromolecule), this condition becomes precisely (Chapters 8 and 9)

$$\frac{k_d^{1/2}(1 + \epsilon K^{ML}[L]_s)^{1/2}\delta_s}{(D_s^{ML})^{1/2}\epsilon K^{ML}[L]_s} \gg 1 \tag{22}$$

where $[L]_s$ = concentration of L (assumed to be in excess) in the test solution, K^{ML} is the complexation constant of ML (Figures 17 and 19), and $\epsilon = D_s^{ML}/D_s^M$. The metal flux is difficult to compute for non-fully labile complexes (i.e. when equation (22) is not fulfilled). A simple approximate equation has been derived in such conditions for fluxes into organisms [183], and numerical solutions have been given for fluxes at voltammetric electrodes [184], but rigorous flux equations cannot be obtained analytically. For this reason, only fully labile complexes, for which equation (22) is fulfilled, are considered below.

Under these conditions, by using a mathematical approach similar to that used in deriving equation (5) (Section 2.4), the following expression for the overall flux is obtained [185]:

$$J = \frac{c_s^b}{\alpha_s}\left[\frac{\delta_s}{\bar{D}_s\alpha_s} + \frac{\ell}{D_m^{MC}K_p^M[C]_m} + \frac{\delta_{st}}{\bar{D}_{st}\alpha_{st}}\right]^{-1} \tag{23}$$

where $[C]_m$ is the carrier concentration in the membrane, K_p^M the equilibrium constant for the reaction $[M]_s^i + [C]_m \rightleftharpoons [MC]_m^i$, and the other symbols have their usual meaning. For the scheme in Figure 19, $c_s^b = [M]_s^b + [ML]_s^b$, but in natural waters c_s^b includes the sum of all labile complexes of M. Similarly, for the scheme in Figure 19, $\alpha_s = 1 + K^{ML}[L]_s^b$, but in natural waters α_s would include the sum of the corresponding terms for all labile complexes of M.

$\bar{D}_s$ and $\bar{D}_{st}$ are weighted averages of the diffusion coefficients of free M and all labile M complexes, as defined in ref. [154] (see also Chapter 9, section 3). In particular, in the case of Figure 19:

$$\bar{D}_s = D_s^{M} f_{M} + D_s^{ML} f_{ML} = D_s^{M} \frac{1}{1 + K^{ML}[L]_s^{b}} + D_s^{ML} \frac{K^{ML}[L]_s^{b}}{1 + K^{ML}[L]_s^{b}} \tag{24}$$

For natural waters, it can be written as (Chapter 9):

$$\bar{D}_s = D_s^{M} f_{M} + \sum_{m} D_s^{m} f_m + \sum_{c} D_s^{c} f_c + \sum_{p} D_s^{p} f_p \tag{24'}$$

where the sub- and superscripts m, c and p refer to mobile, colloidal and particulate labile complexes in solution, and the fs are the corresponding fractions of metal bound to these complexants.

Since usually $\delta_s \approx \delta_{st}$, $\bar{D}_{st} \geq \bar{D}_s$, and $\alpha_{st} \gg \alpha_s$, the third term in equation (23) becomes negligible, and this equation can be rewritten as equation (25) or (25′) which describes the relationship between the total permeability (P_t), the permeability of the diffusion layer in the source solution, (P_s) and that of the membrane (P_m).

$$\frac{c_s^{b}}{J} = \frac{\delta_s}{\bar{D}_s} + \frac{\ell}{D_m^{MC}} \frac{\alpha_s}{K_p^{M}[C]_m} \tag{25}$$

$$\frac{1}{P_t} = \frac{1}{P_s} + \frac{1}{P_m} \tag{25'}$$

Clearly two limiting cases, summarised in Table 12, can exist depending on the value of the permeability criterion, π:

$$\pi = \frac{K_p^{M}[C]_m D_m^{MC} \delta_s}{\alpha_s \bar{D}_s \ell} \tag{26}$$

- *For* $\pi \ll 1$, ($P_m \ll P_s$; $P_t = P_m$), J is directly proportional to free metal ion concentration, [M], at constant membrane composition (equation (28); Table 12), irrespective of chemical and physical heterogeneity of the sample, since there is no gradient in the diffusion layer. Under this condition, PLM works as an ion-selective electrode, even though the flux through the membrane and not its potential difference is measured.
- *For* $\pi \gg 1$, ($P_s \ll P_m$;$P_t = P_s$), diffusion followed by dissociation of labile complexes in the source solution will control the flux and J provides the total metal concentration, c_s^{b}, albeit of labile complexes only, (equation (29); Table 12) irrespective of the degree of complexation. Chemical and physical heterogeneities, however, will play significant roles, and in particular the flux value will depend on the average size of complexants (influence on $\bar{D}_s$). For the ith metal complex, having diffusion coefficient D_i and radius r_i, the Stokes–Einstein law gives:

$$D_i / D_s^{M} = r_M / r_i \tag{27}$$

Table 12. Characteristic features for the two limiting cases of the permeability criterion, π, for the transport of mobile and labile ML complexes ($\pi = K_p^M[C]_m D_M^{MC}\delta_s/\alpha_s\bar{D}_s\ell$)

π value	Limiting flux	Flux value		Conditions in the membrane	M + ML gradients in the diffusion layer	Role of chemical and physical heterogeneity on J
$\pi \ll 1$	membrane	$J = D_m^{MC} K_p^M \frac{[C]_m}{\ell}[M]_s^b$	(28)	ℓ/δ_s high K_p low $[C]_m$ low	~ 0	$\sim$ nil
$\pi \gg 1$	source diffusion layer	$J = \frac{\bar{D}_s}{\delta_s} c_s^b$	(29)	ℓ/δ_s low K_p high $[C]_m$ high	maximum	large

where r_M is the radius of free hydrated M ion. Equation (24′) can then be rewritten as

$$\frac{\bar{D}_s}{D_s^M} = f_M + \sum_m \frac{r_M}{r_m} f_m + \sum_c \frac{r_M}{r_c} f_c + \sum_p \frac{r_M}{r_p} f_p \tag{30}$$

By using the conditions $r_M = 0.4\,\text{nm}$, $0.4\,\text{nm} < r_m < 1.0\,\text{nm}$, $1.0\,\text{nm} < r_c < 1.0\,\mu\text{m}$, and $1.0\,\mu\text{m} < r_p$, equation (30) shows that the contributions of colloidal and particulate metal species are negligible unless their proportion is very high, owing to their low diffusion coefficients. For example, even by using small colloid sizes, ($r_c = 10\,\text{nm}$), the 'colloid' term in equation (30) only represents 10% of the right-hand side, for $f_c = 1 - f_M = 0.75$ (assuming that mobile and particulate species are absent). The contribution of particulate species is even much smaller: with $r_p = 10\,\mu\text{m}$ and $f_p = 1 - f_M = 0.99$, the 'particulate' term represents 0.4% of the right-hand side of equation (30).

An interesting feature of PLM is that π *may be manipulated experimentally* by varying ℓ, δ_s (by convection), K_p^M (by changing the nature of the carrier) or the carrier concentration, $[C]_m$. The latter approach is the easiest one and presents the advantage that several orders of magnitude of π can be scanned, which may allow one to pass from one limiting case to the other; thus both $[M]_s^b$ and c_s^b can in principle be measured. Comparison of J, obtained by PLM for various $[C]_m$ values, with biological uptake may give information on the contribution of labile complexes to bioavailability. It must be emphasised that a particular PLM system (with given K_p^M, $[C]_m$, ℓ, and δ_s values) may exhibit either low permeability ($\pi \ll 1; J = \text{const.}\ [M]_s^b$) or high permeability ($\pi \gg 1$; $J = \text{const.}'\ c_M^b$) behaviour, depending on the value of α_s in the test solution, since π depends on α_s (equation (26)). Studying the influence of $[C]_m$ on the flux is the easiest way to test under which conditions the PLM is operating. Interestingly, 20 years ago Whitfield and Turner [178] proposed the use of ion-selective electrodes and voltammetry to determine $[M]_s^b$ and c_s^b in the two limiting cases discussed here. Because of technical limitations this could not be developed. A PLM is a more flexible tool not only for measuring these two parameters but also for monitoring the role of the dynamic response of metal complexes on metal flux, as function of P_m.

The linear dependence of $1/J$ on δ_s and ℓ (equation (25)) has been confirmed experimentally in a non-complexing source solution ($\alpha_s = 1$) using Cu(II) transported by 22DD as the carrier in a liquid membrane of 1/1 mixture of toluene/phenylhexane [84]. The proportionality between J and $[M]_s^b$ (equation (28)) has also been confirmed with the same transporting system and source solutions containing various hydrophilic labile complexants (Cu–sulfosalycilate, Cu–tiron, Cu–oxalate, Cu–fulvic acids, Pb–fulvic acids; [18,72], Figure 20). By substituting the values of the parameters corresponding to the above transport system in equation (26) and using the condition $\pi \ll 1$, the minimum value of α_s required for fulfillment of equation (28) was computed to be $\alpha_s \geq 7$. For Cu(II) and Pb(II), this condition is fulfilled in most natural waters. Speciation results obtained for Cu(II), Pb(II) and Cd(II) in the Arve river (Switzerland) (Table 8; [28]) using an SLM, voltammetry and ultrafiltration in the range of $0.5\,\text{nmol L}^{-1} < c_s^b < 5\,\text{nmol L}^{-1}$ suggest that free metal ion concentrations are indeed measurable in natural water samples (section 4.1). On the other hand,

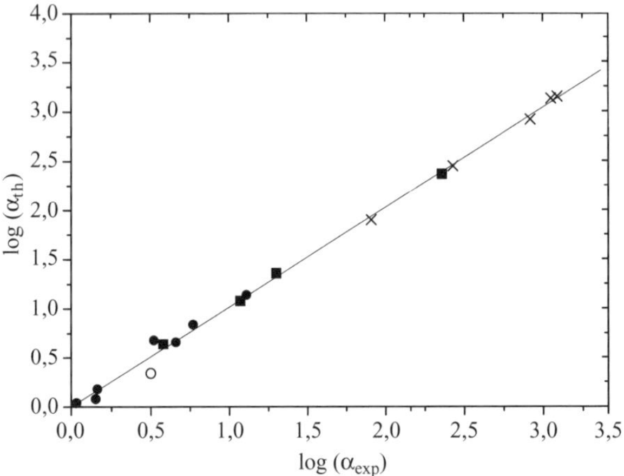

Figure 20. Transport of Cu(II) through the same PLM system as that of Figure 4 A,D, but the source solution contains successively various ligands (sulfosalycilate, tiron, oxalate, fulvic acids) in varying concentrations [N. Parthasarathy and J. Buffle, unpublished results]. Experimental α values ($\alpha_{exp} = c_s^{Cu}/[Cu^{2+}]$) were obtained from the ratio of fluxes in presence of ligands to those in absence of ligand (equation (28)). Theoretical α values, α_{th}, were obtained from the literature values for the stability constants of the corresponding Cu(II) complexes and acid–base equilibria. Both flat-sheet SLMs and HFSLMs have been used. In all cases these results show that diffusion in the membrane was the limiting step ($\pi \ll 1$; Table 12)

by using a complexant with low α_s and low D_s value (polyacrylate with $M_w = 4 \times 10^4$), the condition $\pi \gg 1$ could be reached where J was indeed observed to be proportional to c_s^b [186].

There is a strong conceptual analogy between equation (28) and the so-called 'Free ion activity model (=FIAM)' often used to interpret metal bioavailability measurements in environmental systems [175]. According to FIAM, the flux of metal uptake by organisms is proportional to (i) free metal ion activity in the test solution ($\equiv [M]_s^b$), (ii) free binding sites concentration, $[B]_m$, at the membrane surface ($\sim [C]_m$), (iii) the corresponding adsorption constant, $K_{ads}(\sim K_p^M)$ and an empirical 'internalisation' rate constant, k_{int}, ($\sim D_m/\ell$). Clearly the limiting condition $\pi \ll 1$ also applies to FIAM, where (for the labile complexes considered in this section) π is now given by

$$\pi = \frac{K_{ads}[B]_m k_{int} \delta_s}{\alpha_s \bar{D}_s} \tag{31}$$

Some literature results [172,175,178] suggest that this criterion is not always fulfilled for microorganisms and that transport in solution might be flux limiting in some cases. Under those conditions, equation (29) would be applicable to microorganisms as well. At any rate, comparison of PLM and bio-uptake fluxes under variable π conditions may provide useful insight into the factors affecting bioaccumulation processes.

5.2.3 Solutions Containing both Lipophilic and Hydrophilic Complexes

As natural waters contain a large number of complexes with varying properties, the existence of some lipophilic complexes, MA, is also likely (Table 10), even though they are expected to be present in low proportions compared with the hydrophilic complexes, ML. Their contribution to the overall flux of M should, however, be evaluated. Two limiting cases can be considered (Figure 21), depending on whether the MA complexes are labile (e.g. ion pairs) or inert (e.g. alkyl Hg or alkyl Sn), within the time scale of measurement. These two cases are briefly delt with below, on the basis of Figure 21.

Case of labile MA and ML complexes The flux J_s in the source solution is due to the diffusion of free M and all complexes, including MA, i.e.

$$J_s = \frac{\bar{D}_s}{\delta_s}(c_s^b - c_s^i) \tag{32}$$

where $\bar{D}_s$ is given by equation (24′) and includes the terms for M, ML and MA, and c_s^i is the total concentration of M at the source side of the interface.

Assuming that, inside the membrane, MA is not in equilibrium with MC and that steady-state concentration gradients are established, the total flux of M

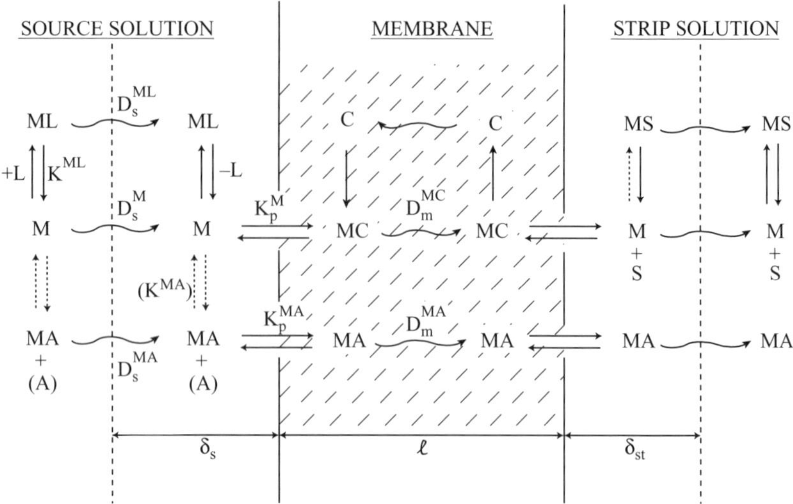

Figure 21. Processes controlling the flux of M through a PLM, from a solution containing free M, mobile and labile complexes ML, and mobile lipophilic complexes MA, either labile or inert. In this diagram, M does not form mixed complexes with A and C in the membrane

inside the membrane, J_m, is the sum of two independent diffusion terms for MC and MA:

$$J_m = \frac{D_m^{MC}}{\ell}\left([MC]_m^i - [MC]_m^j\right) + \frac{D_m^{MA}}{\ell}\left([MA]_m^i - [MA]_m^j\right) \tag{33}$$

where superscripts i and j indicates the source–membrane and membrane–strip solutions interfaces respectively (Figure 19). For time $t \sim 0$ (i.e. just after steady-state concentration gradients are established), $[MA]_m^j = [MC]_m^j = 0$. In addition, the various species in solution and in the membrane are related by equations (34)–(37):

$$K_p' = K_p^M [C]_m = \frac{[MC]_m^i}{[M]_s^i} \tag{34}$$

$$\alpha_s = \alpha_s^i = \frac{c_s^i}{[M]_s^i} \tag{35}$$

$$K_p^{MA} = \frac{[MA]_m^i}{[MA]_s^i} \tag{36}$$

$$K^{MA} = \frac{[MA]_s}{[M]_s [A]_s} \tag{37}$$

By combining these equations with the condition:

$$J = J_s = J_m \tag{38}$$

expressions can be obtained for c_s^i and J. The latter can be written in a form analogous to equation (25);

$$\frac{c_s^b}{J} = \frac{\delta_s}{\bar{D}_s} + \frac{\ell \alpha_s}{D_m^{MC} K_p^M [C]_m + D_m^{MA} K_p^{MA} K^{MA} [A]_s} \tag{39}$$

Comparison of equations (25) and (39) shows that the permeation of MA, in addition to carrier-aided transport of M, mostly results in a proportional increase of the overall permeability of the membrane.

Experimentally equation (39) shows that $[MA]_s$ can be determined by PLM in the absence of carrier ($[C]_m = 0$). In addition, diffusion in the membrane should be slow compared with diffusion in solution, otherwise ML would dissociate and transform into MA in the diffusion layer, because of the depletion of MA at the membrane surface. Combining $[C]_m = 0$ with the condition $\delta_s/\bar{D}_s$ much smaller than the second term in the right-hand side of equation (39) gives the theoretical condition under which labile MA complexes are measurable by PLM in presence of labile ML complexes:

$$\frac{D_m^{MA} K_p^{MA} K^{MA} [A]_s \delta_s}{\bar{D}_s \alpha_s \ell} \ll 1$$

Experimentally, one can find conditions under which $\delta_s/\bar{D}_s$ is negligible by recording $1/J$ as a function of ℓ. By using $K^{MA}[A]_s = [MA]_s/[M]_s$, equation (39) then becomes:

$$J = \frac{D_m^{MA} K_p^{MA}}{\ell} [MA]_s \tag{40}$$

Thus $[MA]_s$ can be determined, provided calibration can be done to measure the corresponding permeability. The contribution of lipophilic complexes to J being known, the overall flux measured for $[C]_m > 0$ can be corrected, using equation (39), to find the free metal ion concentration, $[M]_s^b$, as indicated in section 5.2.2. Note that in principle c_s^b can also be measured in the presence of labile lipophilic complexes, from equation (29) (Table 12), under conditions where the membrane permeability (including the contribution of MA) is high compared with $\bar{D}_s/\delta_s$. This can be tested by determining the roles of ℓ or δ_s on J.

Case of inert lipophilic complex, MA, in presence of labile hydrophilic complex, ML Assuming that MA is not in equilibrium with other M species, the degree of complexation of M, α_s, includes only the *labile* species, but not the fraction, f_{MA}, of inert complexes, $f_{MA} = [MA]_s/c_s^b$.

$$\alpha_s = (1 - f_{MA}) c_s^b / [M]_s^b \tag{41}$$

Since M is not at equilibrium with MA, the overall flux of M in solution comprises two independent terms, and equation (32) can be rewritten as:

$$J_s = \frac{\bar{D}_s}{\delta_s}(1 - f_{MA})(c_s^b - c_s^i) + \frac{D_s^{MA}}{\delta_s^{MA}}\left([MA]_s^b - [MA]_s^i\right) = J_s^M + J_s^{MA} \tag{42}$$

$\bar{D}_s$ (equation (24′)) includes only the terms for the labile ML species, by omitting the term for MA. δ_s and δ_s^{MA} may be different, since δ usually depends on the diffusion coefficient ([154,185]; Chapter 9). The other equations derived for the case of labile MA are still valid, except equation (37) which is ignored. On the other hand, the fluxes of MA in solution and in the membrane should be equal, irrespective of the fluxes of labile M species and vice versa:

$$J_s^M = J_m^M \tag{43}$$

$$J_s^{MA} = J_m^{MA} \tag{44}$$

Combining these equations gives the following expression for the overall flux :

$$J = c_s^b \left[\frac{1 - f_{MA}}{\frac{\delta_s}{\bar{D}_s} + \frac{\ell}{D_m^{MC}} \frac{\alpha_s}{K_p^M [C]_m}} + \frac{f_{MA}}{\frac{\delta_s^{MA}}{D_s^{MA}} + \frac{\ell}{D_m^{MA}} \frac{1}{K_p^{MA}}} \right] \tag{45}$$

As expected, although the equations for the overall flux are slightly different for labile and inert lipophilic complexes (equations (39) and (45)), the flux in the absence of carrier ($[C]_m = 0$) and for conditions where diffusion in solution is not rate limiting ($\delta_s/\bar{D}_s$ negligible), is the same for both and is given by equation (40). $[MA]_s$ can thus be found as before, but no information can be obtained on the lability or inertness of MA. Knowing the contribution of lipophilic complexes, equation (45) can be used in principle to correct the flux at $[C]_m > 0$ to determine $[M]_s^b$ and the sum of all labile complexes, $c_s^b(1 - f_{MA})$, as above.

It must be emphasised that the determination of $[M]_s^b$, in the presence of lipophilic complexes, may be difficult when α_s is very high, i.e. when the contribution of the carrier-aided transport of M to the overall membrane flux is small compared with the contribution of lipophilic complexes. Under the conditions where diffusion in solution is not rate limiting (δ/D very small), both equations (39) and (45) show that the membrane flux contribution of lipophilic complexes is predominant when

$$\frac{[MA]_s}{[M]_s} > \frac{D_m^{MC}}{D_m^{MA}} \frac{K_p^M [C]_m}{K_p^{MA}} \tag{46}$$

(Note that for labile complexes $[MA]_s/[M]_s$ can be computed by $[MA]_s/[M]_s = K^{MA}[A]_s$.) Therefore the limiting concentration above which MA interferes with the measurement of $[M]_s$ is the same for both inert and labile lipophilic complexes. Since often $D_m^{MC} \approx D_m^{MA}$, very high values of K_p^M and $[C]_m$ should be used to obtain a good selectivity of M over MA. There is a limit, however, to increasing these values, as membrane permeability increases and diffusion in solution may become limiting. It may also be noted that, in natural waters, $[M]_s^b$ may be very low, typically 10^{-9}–10^{-15} mol L^{-1} for transition and soft metals. Therefore, the presence of even a small fraction of MA may be sufficient to create an interference of MA over M when $f_M \ll f_{MA} \ll 1$. Although this may be considered as a problem from a purely analytical point of view, when $[M]_s$ is the parameter of interest, it might not be the case for environmental applications. Indeed, if lipophilic complexes produce a larger flux than M through PLM, it is likely that they will do so in biological membranes. Then $[MA]_s$ becomes the environmentally important parameter to measure.

5.3 COMPARISON OF PLM WITH OTHER SPECIATION TECHNIQUES

Presently the vast majority of metal (or other chemical) speciation techniques (ion-selective electrodes, ultrafiltration, chromatography, extraction, ligand competition; [154,187–190]), aim at measuring the concentrations of a specific species, such as the free metal ion, in equilibrium with all labile complexes, or of groups of complexes, such as size fractions or hydrophilic versus hydrophobic complexes, etc. Apart from voltammetry (Chapter 9), very few attempts (see ultrafiltration and ligand exchange chromatography in ref. [154]) have been made to determine the dynamic characteristics of metal complexes. In addition, most speciation techniques are either not sensitive enough or too unwieldy for direct *in situ* or even on site applications.

Ligand exchange techniques appear to be amongst the most promising approaches for measuring the free metal ion concentration (and hence the overall degree of complexation of the metal) at the very low concentration levels usually encountered in waters (10^{-11}–10^{-8} mol L^{-1}). In such techniques, classically, a known 'reference' ligand, R, forming a detectable complex MR with known stability constant, is added to the test sample, in substoichiometric concentration compared with the total metal concentration. The reaction below occurs:

$$\sum ML_i + R \rightleftharpoons MR + \sum L_i.$$

where ΣML_i represents only the metal complexes in the sample which can dissociate in response to addition of R. This represents only a small proportion of the whole family of labile metal complexes, owing to the substoichiometric conditions. MR can be measured by various techniques, either directly in solution, by cathodic stripping voltammetry (e.g. [191,192]), anodic stripping voltammetry [193] or fluorescence [194], or after separation by liquid-liquid extraction [195] or chromatography [196]. From the known equilibrium constant of MR formation, the free M concentration can then be computed. Values of free M down to 10^{-15} mol L^{-1} have been measured in this way (e.g. [197]). There are, however, the few following major potential drawbacks to adding the reference ligand R to the solution.

- Mixed complexes involving R and natural ligands may be formed, which are difficult to check.
- R may partly adsorb on colloids or associate with natural macromolecules. The concentration of free R is then lower than assumed, which may result in too low an estimate of the free M concentration.
- For very stable complexes (corresponding to free metal ion concentrations of 10^{-13}–10^{-15} mol L^{-1}), the dissociation kinetics of the complexes may be very slow (days to weeks), so that equilibrium between R and the sample is not necessarily reached within the experimental time.
- Natural components (in particular natural organic matter) may interfere with the determination of MR, e.g. by adsorption on the voltammetric electrode, or by producing fluorescence quenching.

These problems can be avoided by incorporating the competing (reference) ligand in a hydrophobic phase, non-miscible in water. Lipophilic reference ligands are then used and the competition reaction occurs at the water–solvent interface:

$$\sum ML_{i(\text{water})} + R_{(\text{solvent})} \rightleftharpoons MR_{(\text{solvent})} + \sum L_{i(\text{water})}$$

Three types of analytical devices (PLMs being one of these) have been developed based on this reaction. Although they are promising for *in situ* measurements in water, they are still in the development state and very few applications have been done in this field:

- *Ion-selective electrodes* based on ligands dissolved in hydrophobic (e.g. PVC) membranes. The free metal ion activity is determined based on the surface potential of the electrode, at equilibrium. Classical ISEs have a detection limit close to 10^{-6} mol L^{-1} total metal ion concentration, owing to contamination of the sample by the ISE membrane itself. This renders ISE far too insensitive, for water analysis. However, it has been shown recently that this limit can be extended to 10^{-11}–10^{-12} mol L^{-1} [198,199] provided a small metal flux is created inside the membrane towards the internal solution, to avoid the contamination problem.

- *Ion-selective optodes* based on the coating of an optical fibre by a hydrophobic film that contains R and forms the membrane. By equilibration with the test solution, MR is formed in the film and measured by absorbance or fluorescence, which provides detection limits of 10^{-8}–10^{-6} mol L^{-1}. By coupling a proton reaction in the membrane, free metal ion activities down to 10^{-11} mol L^{-1} can, however, be measured in pH-buffered test solutions. The response time, however, may be hours, below 10^{-8} mol L^{-1} [198].
- *Flux measurement through a PLM* (this chapter). A PLM is also based on competition reaction at an interface (R = carrier C), but in contrast to the above two techniques a flux is maintained through the membrane by means of a chemical gradient. This flux is speciation dependent and is thus the measured signal.

These three approaches are all complementary and promising for developing *in situ* sensors for speciation measurements.

6 CONCLUSION AND PROSPECTS

The literature results suggest that PLMs are a very promising speciation sensitive technique for *in situ* preconcentration and separation of trace compounds in waters. The major advantages of this approach are as follows.

- Its versatility: it can be used for analysis of inorganic cations and anions, as well as organic compounds, and it can be coupled to electrochemical, spectrometric and chromatographic techniques.
- Its experimental simplicity: it requires minimum sample handling and energy consumption and the corresponding devices can be readily miniaturised.
- Its capability of performing in relatively short times, preconcentrations with very high preconcentration factors: this makes it a tool of choice for studying the role of chemical compounds at ultratrace levels.
- Its sound physico-chemical principles: this facilitates the development of well-controlled analytical devices, and the determination of well-defined environmental parameters.
- Its sensitivity to speciation: in particular it allows determination of the three most important types of species for environmental interpretation, i.e. the free (or uncomplexed) ion or compound, the total concentration of all labile complexes, and the lipophilic complexes. These are key parameters for investigations of bioavailability, ecotoxicity and biogeochemical cycles.
- Its sensitivity to fluxes of compounds instead of bulk concentration: this feature, in combination with its physico-chemical principles, makes it a

useful analogue (even though it is a simple one) of biological membranes. Systematic comparative studies of SLMs and biological membranes might ultimately lead to the development of PLMs as alternative for biotests.

There are presently two major difficulties that hinder further development of SLMs for environmental *in situ* applications. Both are technical.

- The long-term stability of the membrane must be increased. A number of solutions have already been studied (Section 3.4), but it is likely that a real solution to this problem will come from the synthesis of multilayered composite membranes, and/or membranes with attached ligands. This implies the development of specific sophisticated synthesis procedures. It must be realised that the present development of analytical applications of PLM is very much limited by the fact that membranes are produced for industrial purposes, for which the requirements differ from those for analytical applications. In addition, analytical applications are presently not considered as a potential market by membrane producers, so that the quality and quantity of membrane production is a constant question mark. The production of high quality membranes specifically developed for analytical applications would be highly desirable.
- As explained in section 4, the development of microanalytical systems with sizes $< 100\,\mu m$ (at least for the source and strip channels) is required to obtain both fast response times and high preconcentration factors. Devices based on microtechnology are urgently needed in this respect, both for on-line coupling of PLMs with detectors and for fully integrated analytical systems.

ACKNOWLEDGMENT

The authors are grateful to P. G. C Campbell and H. P. v. Leeuwen for their detailed comments and to R. Menghetti, for drawing the figures.

LIST OF ACRONYMS

AAS	Atomic absorption spectrometry
BLM	Bulk liquid membrane
ELM	Emulsion liquid membrane
HPLC	High pressure liquid chromatography
ICP–MS	Inductively coupled plasma–mass spectrometry
GC	Gas chromatography
PLM	Permeation liquid membrane (PLM includes SLM, BLM and ELM)

SLM Supported liquid membrane
22DD 1,10-didecyl-1,10-diaza crown ether

LIST OF SYMBOLS

Chemical symbols

A lipophilic ligand forming complexes MA with M
L hydrophilic ligand forming a labile ML complex with M
M metal ion

Mathematical symbols

A surface area
c total concentration of a metal ion or of any organic compound, irrespective of its speciation
c_s^0 concentration of the analyte in the source solution before preconcentration
D^X diffusion coefficient of X
$\bar{D}$ average diffusion coefficient of several species in equilibrium
F preconcentration factor ($= c_{st}/c_s$)
J flux
k_f^c, k_d^c formation, dissociation rate constants of complex MC
K_p partition coefficient
ℓ membrane thickness
P permeation
t time
V volume
δ diffusion layer thickness
π permeation criterion
[X] concentration of the specific species X

subscripts

s source solution
st strip solution
m membrane

superscripts

b bulk solution
i interface (source/membrane)
j interface (membrane/strip)

APPENDIX

Table A1. Non-analytical applications of SLMs for metal removal or recovery, from natural waters, waste waters, ground waters

Metal	Carrier (solvent used in parenthesis)	Transport mechanism or driving force	Application	References
Cu	Oxime (kerosene)	pH gradient	Hydrometallugy; recovery of Cu	200
Cu	Oxime (toluene)	pH gradient	Hydrometallugy; recovery of Cu	39
Cu	Oxime (octanol)	pH gradient	Hydrometallugy; recovery of Cu	201
Cu	Oxime (kerosene, NPOE)*	pH gradient	Waste streams (*in situ* analytical adaptation possible)	52
Zn	Hexyldiethylphosphonic acid (dodecane)	pH gradient	Removal from industrial waste waters	39
Co	Di-2,4,4-trimethyl pentylphosphinic acid (decalin + diethylpropylbenzene)	pH gradient	Separation of Co from Co–Ni mixtures in hydrometallurgy(analytical application possible)	3,39,85
Cd	Trilaurylamine (triethylbenzene)	Co-anion transport and anion gradient	Waste streams	39,87
Zn	Trioctylamine (decane + 2-ethylhexylalcohol)	Co-ion transport and anion gradient	Waste streams	88
Cr(VI)	Trilaurylamine (dodecane)	pH gradient	Groundwater	104,203
U(VI)	Bis(2,4,4-trimethylpentyl) phosphinic acid (dodecane)	Complexing agent in strip solution	Groundwater	89,203
U(VI)	7-Dodecenyl-8-quinolinol (kerosene)	pH gradient	Seawater	204
Actinides	CMPO† (diethyl benzene or duodecane + TBP‡)	Co-anion transport	Nuclear waste water	3,12,39
Lanthanides, e.g. Eu^{3+}	HDEHP§ (duodecane)	Co-anion transport	Waste water	3,50
Cs	Calix 4-arene (NPOE)*	Co-ion transport	Nuclear waste water	205
Pt	Tofioctyl amine (xylene)	pH gradient	Recovery of metals from industrial waste waters	206
Au	Didecyl-1,10-diaza crown ether (decanol)	Cation gradient	Industrial recovery of precious metal	207
Cd	bis-1-Hydroxyheptyl cyclohexane-18-crown-6 (phenyl hexane)	Co-anion transport	Laboratory study	43
Li	14-Crown nitrophenol derivative (NPOE)*	pH gradient	Industrial recovery of Li	208

* NPOE = Nitrophenyloctylether.
† CMPO = *n* octylphenyl (*N,N*-diisobutyl carbomoyl) phosphine oxide.
§ HDEHP = hexyl diethyl phosphonic acid.
‡ TBP = tributylphosphate.

Table A2. SLMs for anions: selected examples of non-analytical applications to water treatment

Anion	Carrier (solvent used in parentheses)	Transport mechanism or driving force	Application	References
NO_3^-	Quarternary ammonium phenate, or sulphonamidate (trioctylphosphate)	pH gradient	Removal of nitrate from drinking water	59,60
NO_3^-	Tetraoctyl ammonium chloride (NPOE)	Cl^- counter-anion gradient	Removal of nitrate from polluted waters	55
CN^-	Tetraphenyl porphinato manganese(III) complex (NPOE)	Hydroxide counter-anion gradient	Removal from waste streams	62
Cl^-	Tridecylbutyl ammonium (1-octanol)	Hydroxide counter-anion gradient	Removal from brackish water	61

Table A3. Examples of non-analytical applications of SLMs for separation of organic compounds (for a review, see ref. [9]). NPPE = nitrophenylphenylether; NPOE; see Table A1

Target compound	Carrier (solvent used in brackets)	Transport mechanism or driving force	Application	Reference
Amino acids	Trialkyl phosphate (NPOE or NPPE)	Co-anion transport (PF_6^-)	Enantiomer separation in food and pharmaceutical industry	66
Amino acids	Chiral crown ethers (NPPE)	Co-anion transport (PF_6^-)	Enantiomer separation in food and pharmaceutical industry (in principle applicable for total aminoacid in waters analysis)	67,68
Amino acids	Quarternary ammonium salt (kerosene)	Counter-anion	Same as above	64
Sugars	Phenylboronic acid trioctyl ammonium chloride admixture (NPOE)	Cation gradient	Food industry (in principle adaptable to sugar analysis in natural waters)	65,209

REFERENCES

1. Batley, G. E., ed. (1989). *Trace element speciation: Analytical Methods and Problems*, CRC Press, Boca Raton, FL.
2. Davison, W. and Zhang, H. (1994). In situ speciation measurements of trace components in natural water systems, *Nature*, **367**, 546.
3. Danesi, P. R. (1984–1985). Separation of metal species by supported liquid membranes, *Sep. Sci. Technol.*, **19**, 857.
4. Izatt, R .M., Lind, G. C., Breuning, H. R. L., Bradshaw, J. S., Lamb, J. D. and Christensen, J. J. (1986). Design of cation selectivity into liquid membrane systems using macroçyclic carriers, *Pure Appl. Chem.*, **58**, 1453.
5. Jönsson, J.-Å. and Mathiasson, L. (1992). Supported liquid membrane techniques for sample preparation and enrichment in environmental chemistry, *Trends Anal. Chem.*, **11**, 106.
6. Nijenhuis, W. F., de Jong, F. and Reinhoudt, D. (1993). Macrocyclic carriers in supported liquid membranes, *Recl. Trav. Chim. Pays-Bas*, 1**12**, 317.
7. Noble, R. D. and Way, J. D., ed. (1987). *Liquid Membrane Theory and Applications*, ACS Symp. Ser. 347, American Chemical Society, Washington, DC.
8. McBride Jr., D. W., Izatt, R. M., Lamb, J. D. and Christensen, J. J. (1984). Cation transport in liquid membranes mediated by macrocyclic crown ether and cryptand compounds. In *Inclusion Compounds*, Vol. 3, ed. Atwood, J. L., Davis, J. E. D. and MacNicol, D. D., Academic Press, New York, Chapter 16, p. 571.
9. Araki, T. and Tsukube, H. ed. (1990). *Liquid membranes. Chemical Applications*, CRC Press, Boca Raton, FL.
10. Cox, J. A., Bhatnagar, A. and Francis, R. W. (1986). Evaluation of coupled transport across a liquid membrane as an analytical preconcentration technique, *Talanta*, **33**, 713.
11. Schulz, A., (1988). Separation techniques with supported liquid membranes, *Desalination*, **68**, 191.
12. Bartsch, R. A. and Way, J. D., ed. (1996). *Chemical Separation with Liquid Membranes*, ACS Symp. Ser. 642, American Chemical Society, Washington, DC.
13. Moyer, B.A., (1994). Complexation and transport. In *Comprehensive Supramolecular Chemistry*, Vol. 1, ed. Davies, J. E. D., MacNicol, D. D., Vogtle, I. F. Lehn, J. M. and Atwood, J.L., Pergamon, Oxford, Chapter 10, p. 377.
14. Li, N. N. (1968). *Separating hydrocarbons with liquid membrane*, US Patent, 3 410 794.
15. Hochhauser, A. and Cussler, E. L. (1975). Concentrating chromium with surfactant liquid membrane, *AIChE Symp. Ser.*, **71**, 136.
16. Marr, R. J. and Draxler, J. (1992). Applications. In *Membrane Handbook*, ed. Ho, W. S. and Sirkar, K. K., Chapman and Hall, New York, Chapter 39, p. 701.
17. Zhang, X. J., Liu, J. H., Fan, Q. J., Lin, Q. T., Zhang X. T. and Lu, T. S. (1988). Industrial application of liquid membrane separation for phenolic waste water treatment. In *Separation Technology*, ed., Li, N. N. and Strathman, H., United Engineering Trustees, New York, p.190.
18. Parthasarathy, N. and Buffle, J. (1994). Capabilities of supported liquid membranes for metal speciation in natural waters: application to copper speciation, *Anal. Chim. Acta*, **284**, 649.
19. Teramoto, M., Matsuyama, A., Takaya, H. and Asano, S. (1987). Development of spiral type supported liquid membrane module for separation and concentration of metal ions, *Sep. Sci. Technol.*, **22**, 2175.
20. Kesting, R. E. (1985). *Synthetic Polymeric membranes: A Structural Perspective*, Wiley-Interscience. New York.

21. Parthasarathy, N., Pelletier, M. and Buffle, J. (1997). Hollow fibre based supported liquid membrane: a novel analytical system for trace metal analysis, *Anal. Chim. Acta*, **350**, 183.
22. Audunsson, G. (1986). Aqueous/aqueous extraction by means of a liquid membrane for sample cleanup and preconcentration of amines in a flow system, *Anal. Chem.*, **58**, 2714.
23. Dozol, J. F., Casa, J. and Sastres, A. (1993). Stability of flat sheet supported liquid membrane in the transport of radionuclides from reprocessing concentrate solutions, *J. Memb. Sci.*, **82**, 237.
24. Takeuchi H., Takahashi, K. and Goto, W. (1987). Some observations on the stability of supported liquid membranes, *J. Memb. Sci.*, **34**, 19.
25. Barnes, D. E., (1992). Flow-injection analysis in the evaluation of supported liquid membranes, *Anal. Chim. Acta*, **261**, 441.
26. Papantoni, M., Djane, N.-K., Ndungu, K., Jönsson, J.-Å. and Mathiasson, L. (1995). Trace enrichment of metals using a supported liquid membrane technique, *Analyst*, **120**, 1471.
27. Djane, N.-K., Ndungú, K., Malcus, F., Johansson, G. and Mathiasson, L. (1997). Supported liquid membrane enrichment using an organophosphorus extractant for analytical trace metal determinations in river waters, *Fresenius' J. Anal. Chem.*, **358**, 822.
28. Izatt, R. M., Clark, G. A., Bradshaw, J. S., Lamb, J. D. and Cristenson, J. J. (1986). Macrocycle facilitated transport of ions in liquid membrane systems, *Sep. Purif. Methods*, **15**, 21.
29. Parthasarathy, N. and Buffle, J. (1991). Supported liquid membrane for analytical separation of transition metal ions. Part I. complexation properties of 1,10–didecyl-1,10–diaza-18–crown-6, *Anal. Chim. Acta*, **254**, 1.
30. Lamb, J. D., Breuning, R. L., Izatt, R. M., Hirashima, Y., Tse, P. K. and Christensen, J. J. (1988). Characterisation of a supported liquid membrane, *J. Memb. Sci.*, **37**, 13.
31. Fyles, T. M. (1985). On the rate limiting steps in the membrane transport of cations across liquid membranes, *J. Memb. Sci.*, **24**, 229.
32. Lindoy, L. F. and Baldins, D. S. (1989). Ligand design for selective metal-ion transport through liquid membranes, *Pure Appl. Chem.*, **61**, 909.
33. Parthasarathy, N. and Buffle, J. (1991). Supported liquid membrane for analytical separation of transition metal ions. Part II. appraisal of lipophilic 1,10–didecyl-1,10–diaza-18–crown-6 as metal ion carriers in the membrane, *Anal. Chim. Acta*, **254**, 9.
34. Izatt, R. M., Bradshaw, J. S., Nielson, S. A., Lamb, J. D., Christensen, J. J. and Sen, D. (1985). Thermodynamic and kinetic data for cation–macrocycle interaction, *Chem. Rev.*, **85**, 271.
35. Izatt, R. M., Pawlak, K., Bradshaw, J. S. and Bruening, R. L. (1991). Thermodynamic and kinetic data for macrocyclic interaction with cations and anions, *Chem. Rev.*, **91**, 1721.
36. Peterson, R. T. and Lamb, J. D. (1996). *Rational design of liquid membrane separation systems*. In *Chemical Separation with Liquid Membranes*, ed. Bartsch, R. and Way, J. D., ACS Symp. Ser. 642 American Chemical Society, Washington, DC, Chapter 4, p. 57.
37. Lamb, J. D., Christensen, J. L., Oscarson, J. L., Nielson, B. L., Asay, B. W. and Izatt, R. M. (1980). The relationship between complex stability constants and transport through liquid membrane, *J. Am. Chem. Soc.*, **102**, 6820.
38. Kirch, M. and Lehn, J. M. (1975). Selective transport of alkali metal cations through a liquid membrane, *Angew. Chem. Ind. Eng.*, **14**, 555.

39. Danesi, P. R. (1987). Science and application of SLM. In *Separation Science Technology, Proceedings 58th Engineering Foundation Conference*, Germany, ed. Li, N. and Strathamann, H., p. 261.
40. Danesi, P. R. and Yinger, L. R. (1986). Origin and significance of the deviations from a pseudo first order rate law in the coupled transport of metal species through supported liquid membranes, *J. Memb. Sci.*, **29**, 195.
41. Danesi, P. R., (1984). A simplified model for coupled transport of metal ions through hollow fibre supported liquid membranes, *J. Memb. Sci.*, **20**, 231.
42. Christoffel, J. J., de Jong, F. and Reinhoudt, D. N. (1996). Mechanistic studies of carrier mediated transport through supported liquid membranes. In *Chemical Separation with Liquid Membranes*, ed. Bartsch, R. A. and Way, J. D., ACS Symp. Ser. 642, American Chemical Society, Washington, DC, Chapter 3, p. 19.
43. Izatt, R. M., Breuning, R. L., Breuning, M. L., LindH, G. H. and Christensen, J. J. (1989). Modelling diffusion limited, neutral macrocycle mediated cation transport in supported liquid membrane, *Anal. Chem.*, **61**, 1140.
44. Yi, J. (1995). Effects of the boundary layer and interfacial reaction on the time lag, *Kor. J. Chem. Eng.*, **12**, 391.
45. Keller, O., Poitry, S. and Buffle, J. (1994). A hollow fibre supported liquid membrane system for metal speciation and presoncentration: calculation of the response time, *J. Electroanal. Chem.*, **378**, 165.
46. Reichwien-Buintenhuis, E. G., Visser, H. C., De Jong, F. and Reinhoudt, D. N. (1995). Kinetic versus diffusion control in carrier mediated co-transport of alkali cations throgh supported liquid membranes, *J. Am. Chem. Soc.*, **117**, 3913.
47. Visser, H. C., de Jong F. and Reinhoudt, D. N. (1995). Kinetics of carrier mediated alkali cation transport through supported liquid membranes: Effect of membrane solvent, co-transported anion and support, *J. Memb. Sci.*, **107**, 267.
48. Fyles, T. M. (1987). Extraction and transport of alkali metal salts by crown ethers and cryptands: estimation of extraction constants and their relationship to transport flux, *Can. J. Chem.*, **65**, 854.
49. Niederhoffer, E. C., Timmons, J. H. and Martell, A. E. (1984). Thermodynamics of oxygen binding in natural synthetic dioxygen complexes, *Chem. Rev.*, **2**, 137.
50. Misra, B. M. and Gill, J. S. (1996). SLM Supported liquid membranes in metal separation. In *Chemical Separations with Liquid Membrane*, ed. D. Way, and R. Bartsch, ACS Symp. Ser. 642, American Chemical Society, Washington, DC, Chapter 25, p. 361.
51. Johnson, B. M., Baker, R. W., Matson, S. L., Smith, K. L., Roman, I. C., Tuttle, M. E. and Lonsdale, H. K. (1987). Liquid membrane for production of oxygen enriched air, *J. Memb. Sci.*, **31**, 1.
52. Wijers, M. C., Bargeman, D., Van Den Burgard, Th., Leese, T. A. and Tasker, P. A. (1994). Extraction of copper from dilute solutions using Acorga P50% oxime in hollow fibre and flat sheet supported liquid membranes. In *Impurity Control Disposal, Hydrometallurgy Process*, 24th *Annual Hydrometallurgy Meeting* p. 137.
53. Yun, C. H., Prasad, R., Guha, A. K. and Sirkar, K. K. (1993). Hollow fibre solvent extraction removal of toxic heavy metals from aqueous waste stream, *Ind. Eng. Chem.*, **32**, 1186.
54. Bautista, R. (1993). Liquid membrane separation of metals in aqueous solution. *Emerging Separation Technologies for Metals and Fuels*, ed. Lakshmanan, V., Bautista, R. and Somasundaran, P., The Minerals, Metals and Materials Society, p. 257.
55. Neplenbroek, A. M., Bargeman, D. and Smolder, C. A. (1992). Nitrate removal using supported liquid membranes: transport mechanism, *J. Memb. Sci.*, **67**, 107.

56. Visser, H. C., Rudkevich, D. M., Verboon, W., de Jong, F. and Reinhoudt, D. N. (1994). Anion carrier mediated transport of phosphate selectivity of H2PO4– over Cl-, *J. Am. Chem. Soc.*, **116**, 11554.
57. Dietrich, B. (1993). Designs of anion receptors. Applications, *Pure Appl. Chem.*, **65**, 1457.
58. Beer, P. D., Drew, M. G. B., and.Graydon, A. (1996). Chloride and dihydrogeno phosphate selective anion recognition by new acyclic mono, bis and tris cobaltocenium receptors, *J. Chem. Soc., Dalton Trans.*, **4**, 129.
59. Kreevoy, M. M., Kotchevar, A. J. and Aften, C. W. (1987). Decontamination of nitrate polluted water, *Sep. Sci. Technol.*, **22**, 361.
60. Kreevoy, M. M. and Nitsche, C. I. (1982). Progress toward a solution to nitrate problem, *Environ Sci. Technol.*, **16**, 635.
61. Molnar, W. J., Wang, C. P., Evans, D. F. and Cussler, E. L. (1978). Liquid membranes for concentrating anions using hydroxide flux, *J. Memb. Sci.*, **4**, 129.
62. Sumi, K., Kimura, M., Kokufuta, E. and Nakamura, I. (1991). Selective uphill transport of cyanide ion mediated by tetraphenylporphinatomanganese (III) complex through bulk and supported liquid membranes, *J. Memb. Sci.,* **8**, 155.
63. Palet, C., Munoz, M., Daunert, S., Bachas, L. G. and Vallente, M. (1993). Vitamin B12 derivatives as anion carriers in transport through supported liquid membranes, *Anal. Chem.*, **5**, 533.
64. Molinari, R., De Bartolo, L. and Drioli, E. (1992). Coupled transport of amino acids through supported liquid membrane. I. Experimental optimization, *J. Memb. Sci.*, **73**, 205.
65. Smith, B. D. (1996). In *Chemical Separation with Liquid Membrane*, ed. Bartsch, R. and Way, D., ACS Symp. Ser. 642, American Chemical Society, Washington, DC, Chapter 14, p. 194.
66. Sugiura, M. and Yamaguchi, T. (1983). Coupled transport of amino acids through liquid membranes supported by microporous films, *Nippon Kagaku Kaisshi,* **6**, 854.
67. Yamaguchi, T., Nishiimura, K., Shimbo, T. and Sugiura, M. (1985). Enantiomer resolution of amino acids by a polymer supported liquid membrane containing a chiral crown ether, *Chem. Lett.*, **10**, 1549.
68. Shinbo, T., Yamaguchi, T., Yanagishita, H., Sabaki, K., Kitamato, D. and Sugiura, M. (1993). Supported liquid membranes for enantio selective transport of amino acids by chiral crown ethers: Effect of membrane solvent on transport rate and membrane stability, *J. Memb. Sci.*, **84**, 241.
69. Uto, M., Yoshida, H., Sugawara, M. and Umezawas, Y. (1986). Uphill transport membrane electrodes, *Anal. Chem.*, **58**, 1798.
70. Ohki, A., Takeda, T., Tagaki, M. and Ueno, K. (1983). Batho-cuproine mediated copper transport through liquid membrane driven by redox potential, *J. Memb. Sci.*, **15**, 231.
71. Malcus, F., Djane, N.-K., Johansson, G. and Mathiasson, L. (1996). Automated trace enrichment and determination of metals using a combination of supported liquid membrane for sample pretreatment and graphite furnace atomic absorption spectroscopy for the determination, *Anal. Chim. Acta*, **327**, 295.
72. Parthasarathy, N., Buffle, J., Gassama, N. and Cuenod, F. (1999). Speciation of trace metals in waters: Direct selective separation and preconcentration of free metal ion by supported liquid membrane, *Chem. Anal. (Warsaw*), **44**, 455.
73. Kuban, P., Buchberger, W. and Haddad, P. R. (1997). Determination of metallo cyanides by capillary electrophoresis after concentration on supported liquid membrane, *J. Chromatogr. A.*, **770**, 329

74. Djane, N.-K., Ndungu, K., Malcus, F., Johansson, G. and Mathiasson, L. (1997). Supported liquid membrane enrichment using an organophosphorus extractant for analytical trace metal determinations in river waters, *Fresenius J. Anal. Chem.*, **358**, 822.
75. Djane, N.-K., Armalis, S., Ndung'u, K., Johansson, G. and Mathiasson, L. (1998). Supported liquid membrane coupled on-line to potentiometric stripping analysis at a mercury-coated reticulated vitreous carbon electrode for trace metal determination in urine, *Analyst*, **123**, 393.
76. Barnes, D. E. (1995). Rapid optimisation of chemical parameters affecting supported liquid membranes, *Sep. Sci. Technol.*, **30**, 751.
77. Kokufuta, E. and Nobusawa, M. (1990). Uphill transport of phosphate ion through oxo-molybdenum (V)–tetraphenylporphyrin complex containing bulk liquid membrane: Effect of halogen ions on phosphate ion flux, *J. Memb. Sci.*, **48**, 141.
78. Audunsson, G. (1988). *Trace analysis of amines by GLC with special emphasis on sample pretreatment, particularly supported liquid membranes*, PhD Dissertation, Lund University, Lund.
79. Megersa, N. and Jonsson, J.-Å. (1998). Trace enrichment and sample preparation of alylthio-s-triazine herbicides in environmental waters using a supported liquid membrane technique in combination with high performance liquid chromatography, *Analyst*, **12**, 225.
80. Knutsson, M., Mathiasson, L. and Jönsson, J. A. (1996). Supported liquid membranes for sampling and sample preparation of pesticides in water, *Chromatographia*, **42**, 165.
81. Norberg, J., Zander, Å. and Jonsson, J. Å. (1997). Fully automated on line supported liquid membrane liquid chromatographic determination of aniline derivatives in environmental waters, *Chromatographia*, **46**, 483.
82. Palmarsdottir, S., Thordarson, E., Edholm, L. E., Jonsson, J.-Å. and Mathiasson, L. (1997). Determination of basic drug bambuterol in human plasma by capillary electrophoresis using double stacking for large volume injection and supported liquid membranes for sample pretreatment, *Anal. Chem.*, **69**, 1732.
83. Trocewicz, J. (1996). Determination of herbicides in surface waters by means of a supported liquid membrane technique and high performance liquid chromatography, *J. Chromatogr. A*, **725**, 121.
84. Guyon, F., Parthasarathy N. and Buffle J. (1999). Mechanism and kinetics of Cu(II) transport through diazo-crownether-fatty acid supported liquid membrane, *Anal. Chem.*, **77**, 819.
85. Danesi, P. R. and Rickert, P. G. (1986). Some observations of the performance of hollow fiber supported liquid membranes for Co–Ni separations, *Solv. Ion. Exch.*, **4**, 149.
86. Neplenbroek, A. M., Bargeman, D. and Smolders, C. A. (1992). Supported liquid membranes; instability effects, *J. Memb. Sci.*, **67**, 121.
87. Molinary, R., Drioli, E. and Pantano, G. (1989). Stability and effect of diluents in supported liquid membranes for Cr(III), Cr(VI), and Cd(II) recovery, *Sep. Sci. Technol.*, **24**, 1015.
88. Tanigaki, M., Uedo, M. and Eguchi, E. W. (1988). Facilitated transport of zinc chloride through hollow fiber supported liquid membrane, *Sep. Sci. Technol.*, **23**, 1161.
89. Chiarizia, R., Horowitz, E. P., Rickert, P. and Hodgson, K. M. (1990). Application of supported liquid membrane for removal of uranium from ground water, *Sep. Sci. Technol.*, **125**, 1571.
90. Danesi, P. R., Reichley-Yinger, L. and Rickerts, P. G. (1987). Lifetime of supported liquid membranes: The influence of interfacial properties, chemical composition and

water transport on the long-term stability of the membranes, *J. Memb. Sci.*, **31**, 117.

91. Kemperman, A. J. B., Bargeman, D., Van den Boomgard, Th. and Sratham, H. (1996). Stability of supported liquid membranes; state of the art, *Sep. Sci. Technol.*, **31**, 2733.
92. Fabiani, C., Merigiola, M., Scibona, S. and Castagnola, A. M. (1987). Degradation of supported liquid membrane under an osmotic pressure, *J. Memb. Sci.*, **30**, 97.
93. Zha, F. F., Fane, A. G., Fell, C. J. D. and Schofield, R. W. (1992). Critical displacement of a supported liquid membrane, *J. Memb. Sci.*, **75**, 69.
94. Takeuchi, H. and Nakano, M. (1989). Progressive wetting of supported liquid membranes by aqueous solutions, *J. Memb. Sci.*, **42**, 183.
95. Baker, R. and Blume, I. (1990). Coupled transport membranes. In *Handbook of Industrial Membrane Technology*, Noyes, NJ, Chapter 9, p. 511.
96. Noble, R. D., Koval, C. and Pellergino, J. (1989). Facilitated transport membrane systems, *Chem. Eng. Progr.*, **85**, 58.
97. Takeuchi, H., Takeuchi, K. and Nakano, M. (1990). Separation of heavy metals from aqueous solution by hollow fibre type supported liquid membrane, *Wat. Treat.*, **5**, 222.
98. Deblay, P., Delpine, S., Minier, M. and Renon, H. (1990). Selection of organic phases for optimal stability and efficiency of flat sheet supported liquid membranes, *Sep. Sci. Technol.*, **26**, 97.
99. Nakano, M., Takeuchi, K. and Takeuchi, H. (1987). A method for continuous operation of supported liquid membranes, *J. Chem. Eng.*, **20**, 26.
100. Neplenbroek, A. M., Bargeman, D. and Smolders, C. A. (1992). Supported liquid membranes, stabilization by gelation, *J. Memb. Sci.*, **67**, 149.
101. Neplenbroek, A. M., (1989). *Supported liquid membranes*; PhD. thesis of University Twente.
102. Zha, F. F., Fane, A. G., Fell, C. J. D. and Schofields, R. W. (1992). Critical displacement pressure of a supported liquid membrane, *J. Memb. Sci.*, **75**, 69.
103. Zha, F. F., Fane, A. G. and Fell, C. J. D. (1995). Effects of surface tension gradients on stability of supported liquid membranes, *J. Memb. Sci.*, **107**, 75.
104. Chiarizia, R. (1991). Application of supported liquid membrane for removal of nitrate, technetium (VII) and Chromium (VI) from ground water, *J. Memb. Sci.*, **55**, 39.
105. Belfer, S., Binman, S., Lati, Y. and Zoltov, S. (1990). Immobilized extractants: selective transports of Mg and Ca from chloride solution via hollow fibre, *J. Appl. Poly. Sci.*, **40**, 2073.
106. Babock, W. C., Baker, R. W., Kelly, D. J., Kleiber, W. and Lonsdale, H. K. (1979). Coupled transport membranes for metal separations, NTIS Rep. **621**, 59.
107. Izatt, R. M., Roper, D. K., Breuning, R, L. and Lamb, J. D. (1989). Macrocycle mediated cation transport using hollow fiber supported liquid membranes, *J. Memb. Sci.*, **45**, 73.
108. Neplenbroek, A. M., Bargeman, D. and Smolders, C. A. (1992). Mechanism of supported liquid membrane; degradation and emulsion formation, *J. Memb. Sci.*, **57**, 133.
109. Belcher, P. (1983). *Encyclopedia of Emulsion Technology*, Vol. 1, Basic Theory, Marcel Dekker, New York.
110. Gopal, E. (1968). Principles of emulsion formation. In *Emulsion Science*, ed. Shermmon, P., Academic Press, New York, p. 23.
111. Teramoto, M. and Teramoto, H. (1983). Mechanism of copper permeation through hollow fiber liquid membranes, *Sep. Sci. Technol.*, **18**, 871.

112. Saito, T. (1992). Deterioration of liquid membrane and its importance in permeation transport of zinc (II) ion through a supported liquid membrane containing batho-cuproine, *Sep. Sci. Technol.*, **27**, 1.
113. Kadem, O. and Bromberg, L. (1994). Ion exchange membranes in extraction, *J. Memb. Sci.,* **2**, *1255.*
114. Tanaka, T., ed. (1987). *Gels Encyclopedia of Polymer Science and Engineering*, Wiley, New York.
115. Bloch, R., Finkelstein, A., Kadem, O. and Vofsi, D. (1967). Metal ion separation dialysed through solvent membranes, *Ind. Eng. Chem. Process Des. Dev.*, **6**, 231.
116. Suguira, M., Kikawa, M. K. and Urita, S. (1987). Effect of plasticizer on carrier mediated transport of zinc ions through cellulose triacetate membranes, *Sep. Sci. Technol.*, **22**, 2263.
117. Bromberg, L., Levin, G. and Kadem, O. (1992). Transport of metals through gelled supported liquid membranes containing carriers, *J. Memb Sci.*, **71**, 41.
118. Levin, G. and Bromberg, L. (1993). Membrane composed of dioctyldithiocarbamate substituted on polyvinyl chloride and di(2ethyl hexyl dithio phosphoric acid), *J. Appl. Poly. Sci.*, **48**, 335.
119. Nielson, H. J. and Hansen, E. H. (1976). New nitrate ion selective electrodes based on quarternary ammonium compounds in non porous membranes, *Anal. Chim. Acta.*, **85**, 1.
120. Fiedler, U. and Ruzicka, J. (1973). The universal ion selective electrode, *Anal. Chim. Acta*, **67**, 179.
121. Kemperman, A. J. B., Rolevink, H. H. M., Van den Boomgard, Th. and Strathmann, H. (1997). Hollow fiber supported liquid membranes with improved stability for nitrate removal, *Sep. Purif. Technol.*, **12**, 119.
122. Thordardson, E., Sveinbjörg, P., Mathiasson, L. and Jönssons, J.-Å. (1996). Sample preparation using a miniaturized supported liquid membrane device connected on-line to packed capillary liquid, *Anal. Chem.*, **68**, 2559.
123. Knutsson, M., Nilve, G., Mathiasson, L. and Jönssons, J. Å. (1992). Supported liquid membrane technique for time-integrating field sampling of acidic herbicides at sub parts per billion level in natural waters, *J. Agric. Food Chem.*, **40**, 2413.
124. Hennion, M. C., Subra, P., Coquart, V. and Rosset, R. (1991). Determination of polar aniline derivatives in aqueous environmental samples using on-line liquid chromatographic preconcentration, *Fresenius' J. Anal. Chem.*, **339**, 488.
125. Tercier, M. L., Parthasarathy, N. and Buffle, J. (1995). Reproducible, reliable, and rugged Hg-plated Ir-based microelectrode for in-situ measurements in natural waters, *Euroanalysis*, **7**, 55.
126. Miliotis, T., Knutsson, M., Jönsson, J.-Å. and Mathiasson, L. (1996). Ion-pair extraction of aromatic anionic surfactants using the supported liquid membrane technique, *Int. J. Environ. Anal. Chem.*, **64**, 35.
127. Lindegård, B., Jönsson, J.-Å. and Mathiasson, L. (1992). Liquid membrane work-up of blood plasma samples applied to gas chromatographic determination of aliphatic amines, *J. Chromatogr.*, **573**, 191.
128. Jönsson, J.-Å., Mathiasson, L., Lindegård, B., Trocewicz, J. and Olssons, A.-M. (1994). Automated system for the trace analysis of organic compounds with supported liquid membranes for sample enrichment, *J. Chromatogr.*, **665**, 259.
129. Lindergård, B., Björk, H., Jönsson, J.-Å., Mathiasson, L. and Olssons, A.-M. (1994). Automated column liquid chromatographic determination of a basic drug in blood plasma using the supported liquid membrane technique for sample pretreatment, *Anal. Chem.*, **66**, 4490.

130. Nilvé, G., Knutsson, M. and Jönssons, J.Å. (1994). Liquid chromatographic determination of sulfonylurea herbicides in natural waters after automated sample pretreatment using supported liquid membranes, *J. Chromatogr. A*, **688**, 75.
131. Grönberg, L., Lövkvist, P. and Jonssons, J.-Å. (1992). Determination of aliphatic amines in air by membrane enrichment directly coupled to a gas chromatograph, *Chromatographia*, **33**, 77.
132. Grönberg, L., Lövkvist, P. and Jönssons, J.-Å. (1992). Measurement of alphatic amines in ambient air and rainwater, *Chemosphere*, **24**, 1533.
133. Norberg, J., Zander, Å. and Jönssons, J.-Å. (1997). Fully automated on-line supported liquid membrane–liquid chromatography determination of aniline derivatives in environmental waters, *J. Chromatogr.*, **46**, 483.
134. Lindegård, B. (1994). *Determination of amines and amine N-oxides in biological samples particularly with supported liquid membranes for sample pretreatment*, Ph.D. Dissertation, Lund University, Lund, p. 128.
135. Tiruye, D. (1997). *Application of supported liquid membrane technique to clean-up and enrich urinary trans, trans-muconic acid*, M.Sc. Dissertation, Lund University, Lund.
136. Norberg, J., Emnéus, J., Jönsson, J.-Å., Mathiasson, L., Burestedt, E., Knutsson, M. and Marko-Vargas, G. (1997). On-line supported liquid membrane–liquid chromatography with a phenol oxidase-based biosensor as a selective detection unit for the determination of phenols in blood plasma, *J. Chromatogr. B*, **701**, 39.
137. Grönberg, L., Lövkvist, P. and Jönssons, J.-Å. (1993). An interface between a liquid flow system and a packed column gas chromatography, *J. Chromatogr.*, **31**, 384.
138. Grönberg, L. (1993). *Membrane enrichment methods for the determination of aliphatic amines and carboxylic acids*, Ph.D. Dissertation, Lund University, Lund.
139. Shen, Y., Obuseng, V., Grönberg, L. and Jönssons, J.-Å. (1996). Liquid membrane enrichment for the ion chromatographic determination of carboxylic acids in soil samples, *J. Chromatogr.*, **725**, 189.
140. Shen, Y., Ström, L., Jönsson, J.-Å. and Tylers, G. (1996). Low-molecular weight organic acids in the rhizophere soil solution of beech forest (*Fagus Sylvatica* L.) Cambisols determination by ion chromatography using supported liquid membrane enrichment technique, *Soil Biol. Biochem.*, **28**, 1163.
141. Shen, Y. (1997). *On-line sample pretreatment based on membrane enrichment, mainly using supported liquid membranes, for chromatographic analysis*, Ph.D. Dissertation, Lund University, Lund.
142. Pálmarsdóttir, S., Lindegård, B., Deininger, P., Edholm, L.-E., Mathiasson, L. and Jönssons, J.-Å. (1995). Supported liquid membrane technique for selective sample workup of basic drugs in plasma prior to capillary zone electrophoresis, *J. Cap. Elec.*, **4**, 185.
143. Djane, N.-K., Bergdahl, I., Ndungú, K., Schültz, A., Johansson, G. and Mathiasson, L. (1997). Supported liquid membrane enrichment combined with atomic absorption spectrometry for the determination of lead in urine, *Analyst*, **122**, 1073.
144. Keller, O. C. and Buffle, J. (2000). Voltammetric and reference microelectrodes with integrated microchannels for flow through microvoltammetry. Part I: The microcell, *Anal. Chem.*, **72**, 936–942.
145. Keller, O. C. and Buffle, J. (2000). Voltammetric and reference microelectrodes with integrated microchannels for flow through microvoltammetry. Part II: Coupling the microcell to a SLM preconcentration technique, *Anal. Chem.*, **72**, 943–948.
146. Parthasarathy, N., Pelletier, M., Buffle, J. (2000). On line coupling of hollow fiber supported liquid membrane with a mini flow-through cell for square wave voltammetric detection, *Anal. Chim. Acta*, submitted

147. Glab, S., Hulanicki, A., Edwall, G. and Ingman, F. (1989). Metal/metal oxide and metal oxide electrodes as pH sensors, *Crit. Rev. Anal. Chem.*, **21**, 29.
148. Turner, A. P. F., Karube, I. and Wilson, G. ed. (1987). *Biosensors, Fundamentals and Applications*, Oxford University Press, Oxford.
149. Lambeck, P. (1992). Integrated optochemical sensors, *Sens. and Actuators B*, **8**, 103.
150. Wienk, M. J., Stolwijk, T. B., Sudholter, E. J. R. and Reinhoudt, D. N., (1990). Stabilization of crown ether containing supported liquid membranes, *J. Am. Chem. Soc.*, **112**, 797.
151. Stumm, W. and Morgan, J. J. (1996). *Aquatic Chemistry*, Wiley, New York.
152. Stumm, W. (1992). *Chemistry of the Solid–Water Interface*, Wiley, New York.
153. Pham, M. K. and Garnier, J.-M. (1998). Distribution of trace elements associated with dissolved compounds ($< 0.45\ \mu m$–1 nm) in freshwater using coupled (frontal cascade) ultrafiltration and chromatographic separations, *Environ. Sci. Technol.*, **32**, 440.
154. Buffle, J. (1988). *Complexation Reactions in Aquatic Systems*, Ellis Horwood, Chichester.
155. Phinney, J. T., Bruland, K. W. (1994). Uptake of lipophilic organic Cu, Cd and Pb complexes in the coastal diatom Thalassiosira weissflogii, *Environ Sci. Technol.*, **28**, 1781.
156. Simkiss, K., Taylor, M. G. (1995). Transport of metals across membranes. In *Metal Speciation and Bioavailability in Aquatic Systems*, ed. Tessier, A. and Turner, D.R., IUPAC Series on Analytical and Physical Chemistry of Environmental Systems, Vol. 3, Wiley, Chichester, Chapter 1.
157. Burton, R. F. (1983). Inorganic ion pairs in physiology: significance and quantitation, *Comput. Biochem. Physiol.*, **74A**, 781.
158. Simkiss, K. (1983). Lipid solubility of heavy metals in saline solutions, *J. Mar. Biol. Assoc., UK*, **63**, 1.
159. Gutnecht, J. (1981). Inorganic mercury transport through lipid bilayer membranes, *J. Memb. Biol.*, 61**, 61.**
160. Wieth, J. O. and Tosteson, M. T. (1979). Organotin mediated exchange diffusion of anions in human red cells, *J. Gen. Physiol.*, **75**, 765.
161. Saouter, E., Hare, L., Campbell, P. G. C., Boudon, A. and Ribeyre, F. (1993). Mercury accumulation in the burrowing mayfly Hexagenia rigida (ephemeroptera) exposed to Ch_3HgCl or Hg_2Cl_2 in water and sediment, *Wat. Res.*, **27**, 1041.
162. Wilkinson, K. J., Bertsch, P. M., Jagoe, C. H. and Campbell, P. G. C. (1993). Surface complexation of aluminium on isolated fish gill cells, *Environ Sci. Technol.*, **27**, 1132.
163. Florence, T. M., Lumsedn, B. G., Fardy, J. J. (1993). Evaluation of some physicochemical techniques for the determination of the fractionation of dissolved copper toxic to the marine diatom Nitzschia closterium, *Anal. Chim. Acta*, **151**, 281.
164. Poldoski, J. E. (1979). Cadmium bioaccumulation assays. Their relationship to various ionic equilibria in Lake Superior water, *Environ Sci. Technol.*, **13**, 701.
165. Hintelmann, H., Ebinghaus, R. and Wilken, R. D. (1993). Accumulation of Hg(II) and methylmercury by microbial biofilms, *Wat. Res.*, **27**, 237.
166. Ballatori, N. (1991). Mechanisms of metal transport across liver cell plasma membrane, *Drug. Met. Rev.*, **23**, 83.
167. Stein, W. D. (1986). *Transport and Diffusion across Cell Membrane*, Academic Press, New York.
168. Davies, A. G. (1978). Pollution studies with marine plankton 2 Heavy metals, *Adv. Mar. Biol.*, **15**, 381.
169. Hall, J. L. and Baker, D. A. (1977). *Cell Membranes and Ion Transport*, Longman, New York.

170. Seeling, J. and Seeling, A. (1980). Liquid conformation in model membranes and biological membranes, *Q. Rev. Biophys.*, **13**, 19.
171. Bloom, M., Evans, E., Mouritsen, O.G. (1991). Physical properties of the fluid liquid-bilayer components of cell membranes: a perspective, *Q. Rev. Biophys.*, **24**, 293.
172. Hudson, R. J. M. and Morel, F. M. M. (1993). Trace metal transport by marine microorganisms: implications of metal coordination kinetics, *Deep-sea Res.*, **40**, 129.
173. Lamb, J. D., Christensen, J. J., Izatt, S. R., Bedke, K., Astin, M. S. and Izatt, R. M. (1980). Effects of salt concentration and anion on the rate of carrier facilitated transport of metal cations through bulk liquid membranes containing crown ethers, *J. Am. Chem. Soc.*, **102**, 3399.
174. Fyles, T. M., Malik-Diemer, V. A., Mc Gavin, C. A. and Whitfield, D. M. (1982). Membrane transport systems. III A mechanistic study of cation–proton coupled counter-transport, *Can. J. Chem.*, **60**, 2259.
175. Campbell, P. G. C. (1995). Interactions between trace metals and aquatic organisms: a critique of the free ion activity model. In *Metal Speciation and Bioavailability in Aquatic Systems*, ed. Tessier, A. and Turner, D. R., IUPAC Series on Analytical and Physical Chemistry of Environmental Systems, Vol. 3, Wiley, Chichester, Chapter 2.
176. Munro, T. A. and Smith, B. D. (1997). Facilitated transportof amino-acids by fixed-site jumping, *J. Chem. Soc., Chem. Commun.*, **22**, 2167.
177. Jackson, G. A. and Morgan, J. J. (1978). Trace metal chelator interactions and phytoplancton growth in sea water media: theoretical analysis and comparisons with reported observations, *Limnol. Oceanogr.*, **23**, 268.
178. Whitfield, M. and Turner, D. R. (1979). Critical assessment of the relationship between biological thermodynamic and electrochemical availability. In *Chemical Modeling in Aqueous Ssystems*, ed. Jenne, E. A., ACS Symp. Ser. 93, American Chemical Society, Washington, DC, Chapter 29.
179. Neubert, R. (1992). Transport across artificial lipid membranes. Part 24. Influence on substance properties on diffusion accross membranes, *Pharmazie*, **46**, 442.
180. Buffle, J. and Altmann, S. (1987). Interpretation of metal complexation by heterogeneous complexants. In *Aquatic Surface Chemistry*, ed. Stumm, W., Wiley, New York, Chapter 13.
181. Filella, M., Buffle, J. and van Leeuwen, H. P. (1990). Effect of physico-chemical heterogeneity of natural complexants. Part I. Voltammetry of labile metal–fulvic complexes, *Anal. Chim. Acta*, **232**, 209.
182. Van Leeuwen, H. P. and Buffle, J. (1990). Voltammetry of heterogeneous metal complex systems. Theoretical analysis of the effects of association/dissociation kinetics and the ensuing lability criteria, *J. Electroanal. Chem.*, **296**, 359.
183. Tessier, A., Buffle, J. and Campbell, P. G. C. (1994). Uptake of trace metals by aquatic organisms. In *Chemical and Biological Regulation of Aquatic Systems*, ed. Buffle, J. and De Vitre E., Lewis, Boca Raton, FL Chapter 6.
184. De Jong, H., van Leeuwen, H. P. and Holub, K. (1987). Voltammetry of metal complex systems with different diffusion coefficients of the species involved. Parts I–III, *J. Electroanal. Chem.*, **234**, 1–29; **235**, 1–10
185. Agraz, R., van Leeuwen, H. P. and Buffle, J., unpublished results.
186. Pinheiro, J. P., Parthasarathy, N. and van Leeuwen, H. P., unpublished results.
187. Filella, M., Town, R. and Buffle, J. (1995) Speciation in freshwaters. In *Chemical Speciation in the Environment*, ed. Ure, A. M. and Davidson, C. M., Blackie, London, Chapter 7.

188. Caroli, S., ed. (1996). *Element Speciation in Bioinorganic Chemistry*, Wiley, New York.
189. Nijenhuis, W. F., van Doorn, A. R., Reichwein, A. M, de Jong, F, Reinhoudt and D. N. (1993). Selective transport of urea by diffusion by macrocycle carriers through supported liquid membrane, *J. Org. Chem*, **58**, 2265.
190. Apte, S. and Bately, G. E. (1995). Trace metal speciation of labile chemical species in natural waters and sediments: non electrochemical approaches. In *Metal Speciation and Bioavailability in Aquatic Systems*, ed. Tessier, A. and Turner, R. D., IUPAC Series on Analytical and Physical Chemistry of Environmental Systems, Wiley, Chichester, Chapter 6, p. 259.
191. Xue, H. and Sigg, L. (1997). Comparison of [Cu++] measurements in lake water determined by ligand exchange and cathodic stripping voltammetry and by ion-selective electrode, *Environ Sci. Technol.*, **31**, 1902.
192. Campos, M. L. A. M. and van den Berg, C. M. G. (1994). Determination of copper complexationin sea water by cathodic stripping voltammetry and ligand competition with salicylaldoxime, *Anal. Chim. Acta*, **284**, 481.
193. Xue, H. B. and Sigg, L. (1994). Zinc species in lake water and its determination by ligand exchange with EDTA and differential pulse anodic stripping voltammetry, *Anal. Chim. Acta*, **284**, 505.
194. Kim, J. I., Rhee, D. S., Wimmer, H., Buckau, G. and Klenze R. (1993). Complexation of trivalent actinide ions (Am^{3+}, Cm^{3+}) with humic acid. A comparison of different experimental methods, *Radiochim. Acta*, **62**, 35.
195. Miller, L. A. and Bruland, K. (1994). Determination of copper speciation in marine waters by competitive ligand equilibration liquid–liquid extraction. An evaluation of the technique, *Anal. Chim. Acta*, **284**, 573.
196. Hirose, K., Dokiya, Y. and Sugimura, Y. (1982). Determination of conditional stability constants of organic copper and zinc complexes dissolved in sea water, using ligand exchange method with EDTA, *Mar. Chem.*, **11**, 343.
197. Xue, H. B. and Sigg, L. (1994). Free cupric concentration and Cu(II) speciation in a eutrophic lake, *Limnol. Oceanogr.*, **38**, 1200.
198. Bakker, E., Buhlmann, P. and Pretsch, E. (1997). Carrier based ion-selective electrodes and bulk optodes. 1. General characteristics, *Chem. Rev.*, **97**, 3083.
199. Sokalski, T., Ceresa, A., Zwickl, T. and Pretsch, E. (1997). Large improvement of the lower detection limit of ion-selective polymer membrane electrodes, *J. Am. Chem. Soc.*, **119**, 11347–11348.
200. Baker, R. W., Tuttle, M. E., Kelly, D. J. and Lonsdale, H. K. (1977). Coupled transport membranes: copper separation, *J. Memb. Sci.*, **2**, 213.
201. Lee, K., Evans, D. F. and Cussler, E. L. (1978). Selective copper recovery with two type of liquid membranes, *AICHE J.*, **24**, 860.
202. Way, J. D., Noble, R. D. and Bateman, B. R. (1985). Selection of supports for immobilized liquid membrane. In ACS Symp. Ser. 269, ed. Lloyd, D. R., American Chemical Society, Washington, DC, Chapter 6, p. 121.
203. Chiarizia, R., Horowitz, E. P. and Hodgson, K. M. (1992). Removal of inorganic contaminants from ground water; Use of supported liquid membrane. In ACS Symp. Ser. **509**, American Chemical Society, Washington, DC, p. 22.
204. Akiba, K. and Hiroyoki, H. (1985). Extraction of uranium by a supported liquid membrane, *Talanta*, **32**, 824.
205. Asfari, Z., Bresset, C., Vicens, J., Hill, C., Dozol, J. F., Rouquette, H., Eyamard, S., Lamare, V. and Tournois, B. (1996). Cesium removal from nuclear waste water by supported liquid membranes containing calix-bis-crown compounds. In *Chemical Separation with Liquid Membranes*, ACS Symp. Ser. 642, ed. Way, J.D.

and Bartsch, R. A., American Chemical Society, Washington, DC, Chapter 26, p. 376.

206. Nishiki, T. and Bautista, R. G. (1985). Platinum IV extraction with supported liquid membrane containing trioctylamine carrier, *AICHE J.*, **31**, 2093.
207. Tromp, M., Burgard, M. and Le Roy, M. J. F. (1988). Extraction of gold and silver cyanide complexes supported liquid membranes containing macrocyclic extractants, *J. Memb. Sci.*, **38**, 295.
208. Sakamato, H., Kimura, K. and Shono, T. (1987). Lithium separation and enrichment by proton driven cation transport through liquid membrane lipophilic crown nitrophenols, *Anal. Chem.*, **59**, 1513.
209. Paugam, M. F., Morin, G. T. and Smith, B. D. (1993). Active transport of uridine through liquid organic membrane mediated phenyl boranic acid and driven by fluoride gradient, *Tetrahedon Lett.*, **34**, 3723.
210. Chemical Society of Japan (1984). *Kagaku Benran (Handbook of Chemistry): Fundamental*, 3rd edn, Maruzen, Tokyo, p. II–82; 2nd edn, 1966, pp. 505,535.
211. Mikulai, V., Rajec, P., Svec, A. and Mackovás, J. (1993). Transport of strontium cation through hollow fiber supported liquid membranes, *J. Radioanal. Nucl. Chem. Lett.*, **175**, 287.
212. Stolwijk, T. B., Sudholter, E. J .R. and Reinhoudt, D. N. (1987). Crown ether mediated transport: A kinetic study of potassium perchlorate transport through a supported liquid membrane containing dibenzo-18–crown-6, *J. Am. Chem. Soc.*, **109**, 7042.

11 Dialysis, DET and DGT: *In Situ* Diffusional Techniques for Studying Water, Sediments and Soils

W. DAVISON, G. FONES, M. HARPER, P. TEASDALE AND H. ZHANG
Lancaster University, UK

In Situ Monitoring of Aquatic Systems: Chemical Analysis and Speciation Edited by J. Buffle and G. Horvai.

1 INTRODUCTION

The common view of an *in situ* sensor is of a device such as an electrode [1–3] or optrode [1,4] which continuously responds to the concentration of defined chemical species in solution. This type of continuous *in situ* measurement represents one of three basic categories of *in situ* measurement classified according to the frequency and location of the analysis [5]. Rather than making measurements continuously, some procedures perform a series of discrete measurements *in situ*, either directly or after collection of discrete samples. Autonomous analysers based on flow injection analysis [6] fall into this second category, as do *in situ* voltammetric techniques where potential scans and current measurements are performed periodically [7]. In the third category, fractionation of chemical species occurs *in situ*, but analysis is delayed until the fractionated sample is returned to the laboratory. The three *in situ* techniques considered in this chapter, dialysis, diffusive equilibration in thin films (DET) and diffusive gradients in thin films (DGT), fall into the third category. In each case the *in situ* fractionation is accomplished by diffusion through a semi-permeable membrane.

The most common *in situ* application of dialysis is its deployment in sediments as a series of vertically spaced cells, commonly known as peepers [8]. Such devices are referred to as pore water samplers because at equilibrium the solute composition of the medium within the dialysis cell is very similar to that of the pore water contacting the membrane. However, peepers are not simply sampling devices. Only those chemical species that are sufficiently small to pass through the membrane are collected. This fractionation of chemical species occurs *in situ*. The collected solutes are analysed in the laboratory after retrieval of the dialysis assembly.

The recently developed technique of diffusive equilibrium in thin films, known as DET [9], is similar to dialysis, but the sampling medium is a hydrogel rather than a solution retained by a dialysis membrane. As for dialysis, solutes equilibrate with the solution held in the sampler (the hydrogel is typically 95 % water) and are analysed when the assembly is returned to the laboratory. Fractionation of chemical species again occurs *in situ*, but this time the collected species depends on the pore size of both the protective outer membrane and of the hydrogel itself. A key feature of both dialysis and DET is that they enable measurement of equilibrium concentrations of solutes if left in place for sufficient time.

The technique of diffusive gradients in thin films (DGT) appears superficially similar to DET, but it separates species kinetically and does not rely on establishment of equilibrium [10]. Solutes freely diffuse through a layer of hydrogel and are then immobilised in an underlying layer of binding agent. The device is deployed for a fixed period of time and then the mass of accumulated solute in the binding layer is measured. Because the diffusion layer thickness during the *in situ* deployment is well defined by the known thickness of the hydrogel, the mean concentration in solution can be calculated using Fick's laws. As with DET and dialysis the chemical separation is made *in situ*, but analysis is performed under controlled laboratory conditions. However, with DGT separation of chemical species occurs continuously. It is based on molecular size, as for DET and dialysis, and also on the kinetic availability (lability) of species. Although DGT can provide only a measure of the time-averaged effect of these *in situ* separation processes, in most cases a near steady-state situation is attained during deployment, allowing direct interpretation of the measurement as a mean flux or as a concentration.

In situ dialysis has been widely used for many years in waters [11], sediments [8] and groundwater [12]. Whereas DET and DGT are more recent arrivals, they are finding wide application in (a) determining solutes in waters [13], sediments [14,15] and soils [16], (b) directly measuring remobilisation fluxes [17,18] and (c) making measurements at high spatial resolution [9,19]. This chapter critically appraises the principles and applications of these three related *in situ* procedures within a unifying framework that considers their theoretical basis and practical usage. The accounts of dialysis and DET are integrated as they can be considered as variants of the same basic procedure. With the more established technique of dialysis a broad review of the topic is provided with more focused attention on recent developments, while principles and methodology receive more attention for the newly emerging DET and DGT procedures. The greater complexity and scope of DGT necessitates its deployment in waters being considered separately from its use in sediments and soils.

Organic compounds have been measured by semi-permeable membrane devices [20] relying on similar principles to DGT, but their detailed treatment is beyond the scope of this review which focuses on inorganic solutes. The related diffusion-based speciation technique that uses supported liquid membranes is dealt with in a separate chapter [21].

2 DIALYSIS AND DET: DIFFUSIONAL EQUILIBRATION SAMPLERS

Samplers that function by diffusional equilibration provide one of the simplest ways of collecting natural waters for analysis. Usually they comprise only an aqueous sampling medium (SM†) and a diffusional interface with the external water body. Dissolved constituents diffuse across the interface until the SM has the same composition as the external waters (i.e. equilibrium is reached). Ideally the composition will not have been changed by the process. Samplers are commonly deployed *in situ* and sufficient time is allowed for equilibrium to be approached for the species of interest. Only species able to diffuse across the interface (which is usually a membrane filter) or within the SM are collected (e.g. dissolved gases, free ions, most ionic complexes and molecules, and small colloids). Fractionation is therefore primarily according to size. In the continuum of particle sizes that exist in natural waters there may be species that diffuse only slowly through the filter or gel. Within realistic times those partially mobile phases may not have reached equilibrium and so the size cut-off will not be sharp unless very long deployment times are used.

Samplers based on diffusional equilibration can be divided into two broad categories: those using dialysis (diffusional equilibration across a membrane into water) and those using diffusional equilibration in a thin film (DET), the thin film usually being a hydrogel containing water.

Dialysis water samplers were the first diffusional samplers developed for natural waters. Although they have been used for direct fractionation of species in water columns [11] with simple dialysis bags, their most popular use has been to measure sediment porewater profiles. In this application a filter membrane is often used rather than a semi-permeable membrane. Strictly speaking the use of

† The sampling medium of the diffusional sampler will be referred to as SM to avoid confusion with references to the sampled waters, which may be described with the same term.

the term dialysis is questionable as the process is more akin to passive filtration. For convenience in this work we have chosen to use the term dialysis while recognising its limitations. In most studies, the SM was purified water, which was removed after retrieval to be preserved or analysed immediately. The earliest porewater samplers used Tygon tubing for dissolved gases [22], dialysis bags for dissolved ions [23] and a plastic block with multiple chambers covered with a dialysis membrane for both dissolved gases and ions [8]. This last design, commonly known as a peeper, has proved the most popular model. Dialysis samplers were also developed for soil groundwater sampling [12], where they have come to be known as multilayer samplers (MLS). It would appear from the nomenclature and citation patterns used in the literature that applications in sediments, groundwaters and water columns have developed almost independently of one another. It is hoped that this review will promote a greater cross-fertilisation of ideas between such related fields.

The popularity of *in situ* dialysis samplers is partly due to alternative methods for sampling both sediment and soil systems being prone to systematic errors. For sediments, changes in temperature, pressure and redox condition before or during separation of the pore waters can occur with *ex situ* techniques [24–28]. Suction samplers, traditionally favoured for soil studies, tend to disrupt the colloidal species, changing the water composition, and are unable to sample from different depths with adequate precision [29,30]. Furthermore adsorption losses, to vessels and membranes alike, can be a major problem with any technique requiring filtration [11].

DET is a more recently developed technique [9] that uses as the SM a thin polyacrylamide gel (typically less then 1 mm thick), which is about 95 % water. The gel is usually covered by a membrane to protect and minimise fouling of the gel surface and the two layers are placed in a plastic holder for deployment. Immediately after retrieval the solute must be chemically fixed or the gel sliced at the desired spatial resolution to prevent diffusional relaxation from obscuring pore water solute structure. More recently constrained DET samplers, in which the gel is in adjacent isolated compartments to limit diffusional exchange, have been used to overcome this difficulty [31]. Individual gel slices are then back-equilibrated into dilute acid before conventional analysis. Alternatively beam techniques can be used for measuring and mapping concentrations in the gel after drying.

The following sections describe the attributes and performance of various diffusional equilibration samplers, supported by both experimental observations and modelled simulations. Practical considerations and applications of dialysis and then DET samplers are discussed separately. Important performance features and practical details are summarised in Table 1.

Table 1. Comparison of performance for techniques

Technique*	**Response time†**	**Resolution**	**Speciation principle**	**Relaxation**	**Sample volume/dilution**
Dialysis	2 d (solution)‡ 2–3 weeks (sediment)‡	5–50 mm	Size fractionation (filter pore size) / mobility	none	Several ml—allows wide variety of analysis. No sample dilution
DET (wet, unconstrained)	30 min (solution) 1 d (sediment)	0.5–2.5 mm	Size fractionation (gel and filter pore size) / mobility	some (fixed)§ extensive (not fixed)§	$\sim 40\ \mu L$ (for 1mm resolution). Sample dilution (elution), typically 100 ×
DET (wet, constrained)	30 min (solution) 1 d (sediment)	1–2.5 mm	Size fractionation (gel and filter pore size) / mobility	none	as wet, unconstrained. Higher resolution ⇒ less sample volume
DET (solid, unconstrained)	30 min (solution) 1 d (sediment)	0.5–2.5 mm	Size fractionation (gel and filter pore size) / mobility	Some (only used for fixable elements)§	n.a.¶—analysis in solid phase
DET (solid, constrained)	30 min (solution) 1 d (sediment)	0.2–1 mm	Size fractionation (gel and filter pore size) / mobility	none	n.a.—analysis in solid phase
DGT (wet)	1 h (solution) 1 d (sediment)	0.5–5 mm	Size fractionation / mobility + lability	Minimal (resin immobilises species in situ)	$\sim 40\ \mu L$ Sample dilution (elution) typically 100× *in situ* preconcentration
DGT (solid)	1 h (solution) 1 d (sediment)	0.1–1 mm	Size fractionation / mobility + lability	as wet, but any relaxation more of a problem at high resolution	n.a.—analysis in solid phase

* Wet and solid refer to type of analysis: wet is sliced and back eluted prior to analysis, dry is measured in solid phase by PIXE, laser ablation, or radiation counting.

† Refers to minimum deployment times needed for typical sampler designs (see Figure 2.i). In sediments the time is dependent on the extent of mixing or resupply in the sediment; times given are conventional deployment times.

‡ Possibly less in saline waters.

§ *fixed* refers to species which are chemically immobilised, *not fixed* refers to species free to relax until the gel is sliced.

¶ Not applicable.

2.1 DESCRIPTION AND MODELLING OF DIFFUSIONAL EQUILIBRATION SAMPLERS

Figure 1 compares the design and features of some commonly used diffusional equilibration samplers[‡]: (i) dialysis bags; (ii) dialysis peepers; (iii) dialysis multilayer samplers; (iv) unconstrained DET; (v) constrained DET. The key feature determining equilibrational response time for DET is the thickness of the SM, comprising the gel and overlain filter. Thinner diffusional layers ensure a more rapid equilibration for any given species. For dialysis samplers the geometry may be more complicated. The ratio of the volume of the SM, V (cm^3), to the area of the diffusional interface, A (cm^2), called the design factor, F (cm), has been recognised as a more general indicator of relative equilibration rates [27,32]. Lower F values ensure more rapid equilibration. For any given F, however, equilibration time is inversely proportional to the diffusion coefficient, D (cm^2/s), of analyte species.

The volume of the SM also determines the number and type of analyses that can be performed with dialysis samplers and should be the starting point when considering a sampler design.

The sampling interval, I, is the distance between discrete data points obtained after measurement (the depth resolution) and is a consideration when vertical concentration profiles are required. Small sampling intervals (higher resolution) are required to ensure that the observed profile of solute in a sampler is an accurate representation of porewater concentrations. The sampling interval for dialysis samplers is determined by the separation of the adjacent sampling chambers and has a minimum practical value of about 0.5 cm. For unconstrained DET, it is determined by the vertical distance between the gel slices or the spatial resolution of the surface analysis technique used. The thickness of the gel also determines the profile fidelity (see section 2.1.2) and consequently influences the maximum spatial resolution attainable [33]. For constrained DET, resolution is determined by the separation of adjacent gel compartments. Both types of DET samplers are capable of much higher resolution (sub-mm) measurements than can be achieved with dialysis samplers. However, the optimal sampling interval depends upon the nature of the study. If vertical profiles spanning metres need to be measured (as is often the case with groundwater studies) sub-mm resolution is inappropriate. Groundwater studies with profiles measured at cm-scale resolution have reasonably been described as micro-scale [34].

The design of dialysis and DET samplers has largely evolved from practical considerations and the performance has been evaluated experimentally. However, the simple diffusion of ions through sediments and water is amenable to modelling. Consequently, it is possible to optimise or evaluate the performance

‡ Specific information on construction materials used are given in sections 2.3.1 (dialysis) and 2.4.1 (DET), along with other practical considerations.

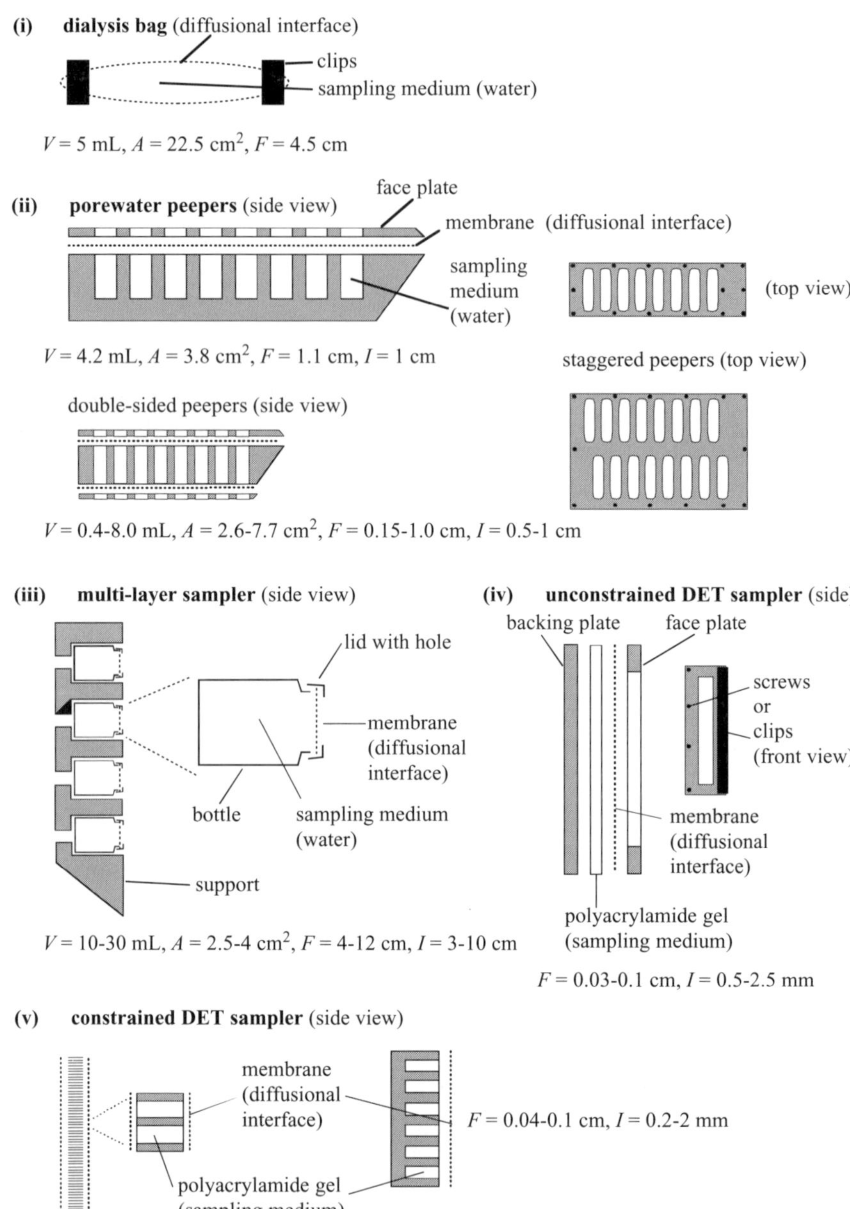

Figure 1. Designs and features of common diffusional equilibration samplers: (i) dialysis bags; (ii) dialysis peepers; (iii) dialysis multilayer sampler; (iv) unconstrained DET; (v) constrained DET. Typical ranges of the sampling medium volume (V, cm^3), diffusional interface surface area (A, cm^2), design factor ($F = V/A$, cm) and sampling interval (I, cm or mm) are given

of diffusional equilibration samplers. Relationships derived from Fick's law have been used to predict the concentration, C_t, of the analyte species in a well-mixed SM after deployment times, t (s), for samplers immersed in well-mixed solutions [27] (equation (1)):

$$t = F(1/k^{M}) \ln[C_0/(C_0 - C_t)] \quad (1)$$

F is the design factor, k^M (cm/s) is the membrane permeation coefficient for the analyte species (experimentally determined and is effectively the ratio of the diffusion coefficient of the species within the membrane to the thickness of the membrane), and C_0 is the initial analyte concentration in the sampled water body. The assumption that the SM in the dialysis assembly is well mixed limits the use of equation (1). It may apply for dialysis in special circumstances (see section 2.1.1). However, it will not occur under any conditions for DET as such mixing can not arise in gels.

Davison *et al.* [35] modelled the equilibration of DET samplers using Fick's Law by assuming that the external waters were well mixed, as in rivers, oceans and many lakes, with transport in the gel being diffusional. Experimental time-dependent measurements of Mn equilibration in gels immersed in well-stirred solutions agreed well with the modelled data [35]. Where the sampler geometry is simple, so that F is equivalent to the depth (or diffusional path length, L) of the sampler, Harper *et al.* [33] have calculated that the mean solute concentration in the SM reaches 95 % of that in the external solution within a time, t_{95}, given by equation (2):

$$t_{95} = 1.15L^2/D \quad (2)$$

99 % equilibration is achieved in 1.57 t_{95}.

More comprehensive models are required to describe the approach to equilibrium of solutes in diffusional samplers within sediments or other systems where mixing of the external waters cannot be assumed. Furthermore the effects of local depletion of solutes in porewaters and resupply of solutes by equilibration with ions sorbed to the particulate sediment phase must be considered for some species [33].

2.1.1 Equilibration Times in Sediment

Harper *et al.* [33] modelled (two-dimensionally) the equilibration of solutes in dialysis and DET samplers deployed in sediment. As solutes diffuse from the porewaters into the dialysis sampler, their concentrations in the pore waters adjacent to the membrane may become depleted, depending upon the resupply behaviour for each solute. Three alternative regimes of resupply to porewater solute concentrations were investigated: (a) the sustained case where solute concentrations were buffered by desorption or dissolution from the particulate phase (this is equivalent to immersion in a well-mixed solution, as for equation (2)); (b) the diffusive case where there was no resupply other than diffusion; and (c) the partially sustained case where there was partial resupply by a first order reaction from the solid phase. For typical dialysis and DET designs ((ii) and (iv) from Figure 1), times for 95% equilibration ranged from 12 min and 23 h respectively for the sustained case, to 15 h and 56 d respectively for the diffusive case. Partial resupply produced intermediate equilibration times, which were generally closer to the sustained case, consistent with most experimental observations [33]. In practice the extent of resupply will be unknown and given that equilibration times vary enormously depending upon which case prevails, they should be experimentally determined where possible.

Figure 2 demonstrates how the design factor and analyte resupply regimes affect the equilibration times (to 90%) for (i) DET and (ii) dialysis samplers. For all resupply regimes the equilibration time increases with *F*. Note the change in scale between the two axes for the DET and dialysis plots. The case of sustained resupply takes an order of magnitude less time to equilibrate compared with diffusive cases. A change in sediment porosity can also have a substantial influence on the equilibration time owing to a change in the diffusional path length [36]. A diffusional equilibration sampler in a low porosity (0.5) medium, such as a saturated soil, takes approximately twice as long to equilibrate as one in sediment of porosity 0.9. These calculations assumed a molecular diffusion coefficient of the solute in porewater of $5 \times 10^{-6} cm^2 s^{-1}$.

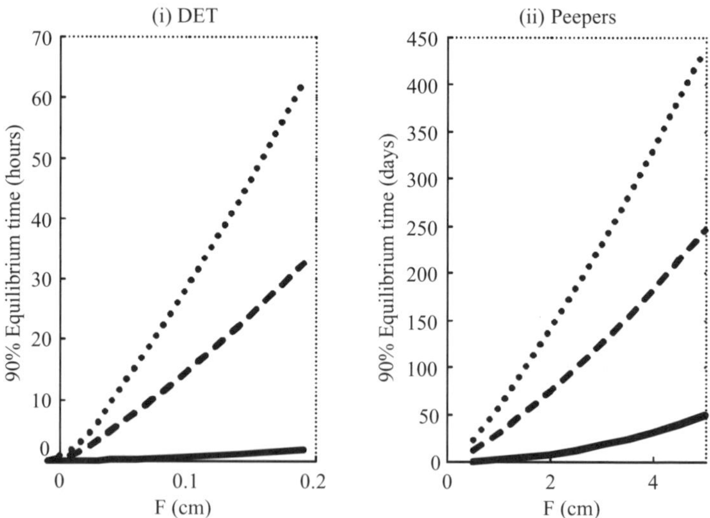

Figure 2. The effect of design factor and sediment resupply regimes on the equilibration time (90%) of (i) DET and (ii) dialysis samplers. Continuous line: sustained case. Dotted line: diffusional only. Dashed line: partially sustained.

Such a value is typical of transition metal ions, such as Mn at 14 °C. At lower temperatures equilibration times will be longer due to smaller values of D. Larger species, such as humic material, will equilibrate more slowly because (a) their intrinsic diffusion coefficients will be lower and (b) their effective diffusion coefficients may be even lower owing to the retarding effect of the gel or the filter.

The zones in the sediment where porewater concentrations become depleted are most pronounced for the diffusional case. Depletion to $< 99\%$ of the starting concentration extends up to 10 mm from the DET membrane and 80 mm from the dialysis membrane [33]. Adjacent samplers must be spaced at more than double these depletion distances to be certain that they are acting independently.

Webster *et al.* [37] recognised that, when dialysis samplers filled with deionised water are deployed in saline systems, salinity gradients can induce mixing within the SM. Their model assumed that the equilibration was potentially limited by three processes: (a) transport of solute through the external water (or sediment) to the membrane surface; they considered cases for either diffusional or well-mixed (sustained) resupply; (b) diffusion of solute across the dialysis membrane, and (c) diffusional or convective mixing of solute within the SM. Relationships were developed for estimating the time to 90 % equilibration in systems in which the external concentration was sustained or well mixed. These times were compared, for various designs of dialysis samplers and water salinities, to assess whether processes (b) or (c) were limiting [37].

For dialysis samplers deployed in very low ionic strength waters, diffusive mixing within the SM will be the limiting process, as modelled by Harper *et al.* [33]. However, with ionic strengths as low as 0.1 % of seawater ($I \approx 0.0007$), which is lower than many freshwaters, convective mixing, brought about by the instability of salinity (density) gradients, shortened the equilibration time [37]. At ionic strengths of 1 % of seawater and greater, convective mixing within the SM was sufficiently rapid that membrane transport was the limiting process [37]. This convective mixing within the SM was found to be most effective when the diffusional interface was upwards or sidewards, the latter being the most common orientation for dialysis samplers.

For deployment in sediments, diffusion of solutes within the porewaters will usually be the limiting process. However, in saline systems mixing within the SM should still lead to more rapid equilibration than described by the diffusional transport model used by Harper *et al.* [33] Good agreement ($< 5\%$ difference) was found between experimentally determined equilibration times for the conservative ions K, Na and Ca (and $< 10\%$ difference for Si, which was closer to the detection limit of the analytical technique) measured in laboratory microcosms using dialysis samplers where $F = 1.5$, 5.7 and 9.9 cm and the model predictions of Webster *et al.* [37]. For the diffusive case the 90 % equilibration time for K (the fastest diffusing ion measured, $D^0 = 19.6 \times 10^{-6}$ cm^2 s^{-1} at 25 °C) was 8 d for the $F = 1.5$ cm sampler and 134 d for the

$F = 9.9\,\text{cm}$ sampler. Sr ($D^0 = 7.93 \times 10^{-6}\,\text{cm}^2\,\text{s}^{-1}$ at 25 °C), which was the slowest diffusing species investigated, took about three times longer to attain the same extent of equilibration for each sampler.

For more reactive elements that have their porewater concentrations resupplied by the solid phase these times can be expected to be greatly reduced. The situation in which such resupply occurs along with convective mixing within the SM due to salinity gradients has not yet been modelled. However, the results are likely to be similar to those predicted by equation 1, reflecting very rapid equilibration.

Another effect that has been quantified [38], even more recently, also concerns the generation of density gradients. The initial exchange of solutes can reduce the density of the porewater adjacent to the membrane. In permeable sediment the resulting density gradients can produce a convective upward movement of the porewater, which distorts the concentration profile and extends the equilibration time. This effect is demonstrated in Figure 3, which shows daily

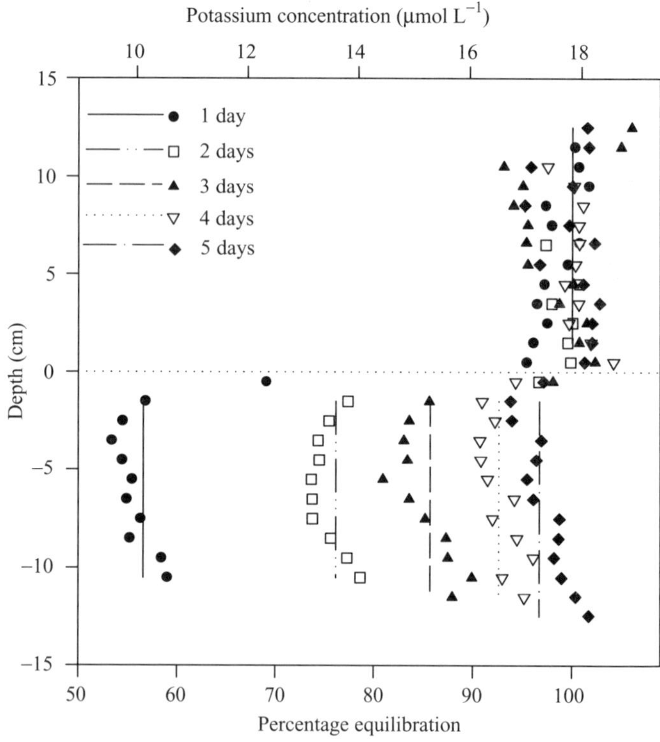

Figure 3. Concentration profiles and percentage equilibration of potassium obtained with peepers ($F = 1.0\,\text{cm}$) deployed for 1 to 5 d in a mesocosm of marine sediment and overlying water

concentration profiles for K obtained from peepers ($F = 1.0$ cm) filled with deionised water deployed in a marine mesocosm. The profile is expected to be close to vertical for the conservative K in sandy sediments, which was confirmed by allowing peepers to equilibrate fully. The overlying water chambers were more than 95 % equilibrated within 1 d. Only about 58 % equilibration was reached after 1 d for the chambers in the porewaters, however. This profile also showed a pronounced increase in concentration towards the bottom. This is predicted by the upward convective model, as equilibration occurs more rapidly where the convection effect is smallest. After 5 d about 97 % equilibration was achieved and the profile approached a more vertical shape. Overall the equilibration rate was greater than for diffusional equilibration, but was slower than that predicted using Webster's model of convective mixing within the cells [37].

Clearly using deionised water in the SM of dialysis samplers to be deployed in saline waters, in particular, has the potential to produce processes with conflicting effects on the equilibration time, depending upon the sediment permeability. To avoid such problems we recommend that the SM of dialysis samplers be filled with a solution of ionic strength similar to that of the waters being sampled. Then the transport term of equilibration should be predominantly diffusional as described by Harper *et al* [33]. As convection within the porewaters decreases with the volume of the SM [38], such effects should be minimal in DET. Even so DET samplers have been sensibly pre-equilibrated with saline solutions prior to deployment in marine sediments [39].

Other factors may affect equilibration times in the field. Bioirrigation and bioturbation processes [40,41], or physical pressure gradients within sediments arising from wave action or objects that obstruct the current [42,43] can move interstitial waters and consequently transfer solutes at rates far greater than molecular diffusion. The effect of such processes on measured pore water profiles is still largely unquantified [44].

2.1.2 PROFILE FIDELITY

The fidelity of diffusional samplers depends upon two factors relating to the non-continuity (*i.e.* discretisation) of sample measurement, profile averaging and phase shifts. In addition profile relaxation, both during and after sampling, may occur with unconstrained DET [33].

Sample discretisation arises because it is not possible to measure truly continuous analyte concentrations in the SM. In the case of dialysis and constrained DET samplers the SM is not continuous. Individual measurements are made on the sample in each compartment. With unconstrained DET the SM is continuous. However, for most measurements the gel needs to be sliced first and the sampled species eluted. In each case the single measurement obtained represents the average analyte concentration over the sampled interval. Spatially continuous measurements can be more closely approximated by using DET with a surface analytical technique, but even then the instrument has a finite resolution and some profile averaging consequently arises. Profile averaging is, however, less significant for high resolution samplers.

Another important consideration is whether the porewater peak is coincident with the sample compartment (*i.e.* in phase) or between compartments (out of phase) [33]. Fidelity is at a maximum for the in phase case. Once again smaller sample intervals are an advantage. Phase shifts are also more of a problem with dialysis and constrained DET because of the non-continuity of the SM.

For unconstrained DET, even if continuous measurements were possible, there would still be some loss of peak fidelity because the processes that sustain the peak in the porewaters are absent from the gel. Thus the peak shape undergoes relaxation (spreading) within the gel during deployment. The change in profile shape associated with profile averaging and relaxation is shown schematically in Figure 4. Once the DET sampler is removed from contact with the sediment pore waters, diffusional relaxation occurs to a greater extent as there is no mechanism which can sustain the profile shape within the gel. Thus it is essential that the solute profile is either chemically fixed in the gel or the gel is sliced immediately after retrieval (see section 2.4.1.4). Having a continuous SM therefore limits the effective resolution of unconstrained DET. Post-sampling relaxation was overcome by using the constrained DET sampler, with its isolated compartments. There is a trade-off: fidelity is lost as a result of either sample discretisation or diffusional relaxation.

The main consequence of phase shifts, profile averaging and relaxation is that profile fidelity is lost [33]. Narrower porewater peaks suffer a greater loss of

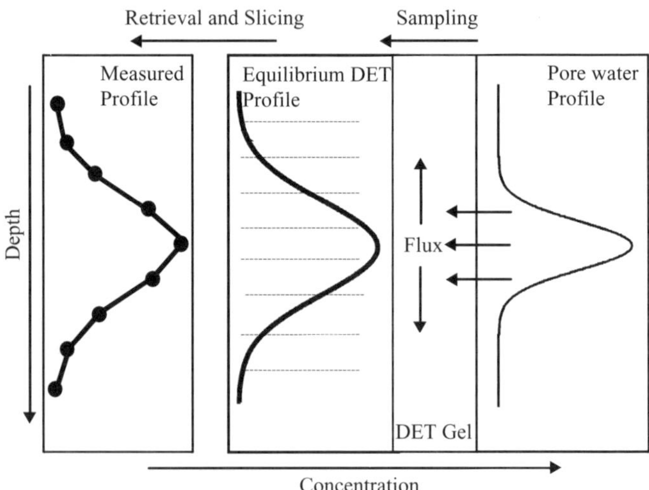

Figure 4. Lateral diffusion and profile averaging: processes resulting in loss of information in porewater profiles measured using DET

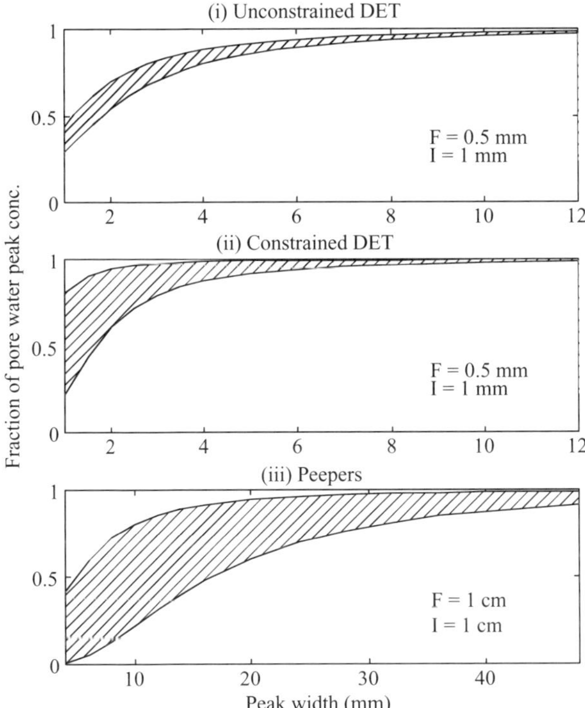

Figure 5. Range of possible measured peak concentrations as a proportion of porewater peak concentrations for (i) unconstrained DET, (ii) constrained DET and (iii) dialysis samplers

fidelity. This is demonstrated in Figure 5, which shows how the fraction of measured porewater peak concentrations compared with the actual concentrations (*i.e.* the fidelity) varies with the porewater peak width (defined by σ mm assuming a normal distribution). No sampler provides a perfect fidelity (*i.e.* a value of 1) under all conditions. The top line in each plot represents the highest achievable fidelity, limited by profile averaging effects and, in the case of unconstrained DET, diffusional relaxation. It also represents the case of the peak being in phase. The bottom line represents the lowest possible fidelity, as a result of a maximum phase shift. The shaded area represents the range of fidelities possible between these two cases. Constrained DET clearly has the potential to provide the highest fidelity profiles, but unconstrained DET is proportionately less affected by phase shift as there is no gap between the samples.

Both forms of DET are capable of reproducing 5–6 mm wide peaks at better than 90 % of their true peak height and 2 mm wide peaks at better than 54 %.

They are much better if by chance the peak is in phase. These fidelities would also be appreciably improved by using thinner gels than the 0.5 mm thick modelled case and by making measurements at more closely spaced intervals than 1 mm. A dialysis sampler with 1 cm sampling interval reproduces 30 mm wide peaks at about 80 % or more of their true values and 15 mm wide peaks at about 44 % or better. Peaks less then 5 mm in width could be missed altogether. The higher fidelity of DET samplers at low porewater peak widths is clear. For this exercise a constrained DET sampler of relatively low resolution (1 mm) was modelled. Even so the upper range of fidelity is better than that attained with unconstrained DET and the lower range is similar. The lower range fidelity would be improved significantly with a sampler of higher resolution.

Clearly dialysis samplers generally provide more averaged profiles than those obtained by DET. Consequently, fluxes calculated from concentration gradients may be underestimated, especially if the gradients are sharp. DET provides a more faithful sample of the porewater profile, providing that the gel thickness is not too great and there is minimal relaxation after retrieval. The coarser sampling with dialysis may yield smoother profiles in heterogeneous sediments, but fluxes calculated by averaging several differing DET concentration profiles should be more accurate. A high resolution constrained DET sampler should most accurately reproduce porewater profiles and therefore provide the best estimate of benthic fluxes[§]. Dialysis samplers may still produce a reasonable estimate of fluxes across the sediment–water interface because they measure at a spatial scale that observes more widespread and longer lasting trends. The sharper peaks measured more accurately by DET probably reflect short-lived and spatially localised microniche influences superimposed upon a profile determined by more regionalised influences, such as the redox conditions. However, this underlying trend should still be apparent with DET measurements.

2.2 THE EFFECT OF OXYGEN ON DIFFUSIONAL EQUILIBRATION SAMPLERS

Systematic errors may arise if traces of oxygen are present in diffusional equilibration samplers when they are deployed in anoxic sediments and waters. Iron, and to a lesser extent manganese, can be oxidised within the sampler to insoluble oxyhydroxides which may be present as colloidal-sized particles. These colloids will be present in addition to the equilibrium solute concentrations. As these colloids will probably dissolve when the samples are acidified, porewater Fe and Mn concentrations will consequently be overestimated. Any species that adsorb onto oxyhydroxide surfaces, such as trace metals and some nutrients, may also be overestimated.

§ For all solute flux measurements based on diffusional gradients, convective processes induced by, for example, bioturbation or bioirrigation, must be minimal.

Carignan and co-workers [45,46] demonstrated that oxygen could also diffuse out of the plastic construction materials of the sampler unless the whole assembly was deoxygenated for long times (weeks). Concentrations of total dissolved iron, manganese, reactive phosphorus and sulfate decreased while Fe(II) increased when dialysis assemblies were deoxygenated for several weeks compared with 48 h. This indicated that the total concentrations had been overestimated previously because of the oxidation artifacts described above and that reduced species were underestimated.

Generally, the procedure described by Carignan *et al.* [46] is recommended as it is unlikely that long term deoxygenation will cause any problems and it may improve the measurement. There is also a need to deoxygenate materials for DET samplers, although generally shorter times can be used as the thickness of plastic is much less than for dialysis samplers. A 48 h deoxygenation should be sufficient for the entire assembly, although more exhaustive testing is still required.

Most importantly, there is little point in deoxygenating samplers if they are then exposed to air before deployment. Exposure to air or oxygenated water for any time exceeding 20 s is likely to introduce errors in DET measurements [35]. Dialysis samplers, with their greater diffusional pathways should be less critical, but exposure should be minimised and restricted at most to about 1–2 min, including the positioning in the sediment.

A single dialysis cell (or several in some sediment types) may embrace a region where oxidants are diffusing downwards and reduced species are diffusing upwards [47]. This phenomenon may encompass a significant portion of a DET sampler. Alternatively reduced species may be supplied from microniches within a generally more oxidising region or there may be horizontal inhomogeneity within the typically 5 mm width of the dialysis cell. In any of these cases, because of the independent transport of oxidant and reductant along concentration gradients extending into the sampler, oxyhydroxides may form within the SM. It is advisable to make deployments for different times, in excess of the equilibration times established for conservative species, to try to assess whether such phenomena might be occurring.

2.3 DIALYSIS

2.3.1 Practical Considerations

2.3.1.1 Fabrication and preparation The main considerations for materials used in the construction of dialysis samplers are contamination of the chemical species of interest, introduction of oxygen into samplers being deployed in anoxic systems, and degradation of the membranes during the deployment.

Perspex is the most popular material for construction of peeper-type samplers. It is inexpensive, mechanically strong and easily machined. Although polycarbonate has also been used [45], oxygen is more soluble in it [46] and therefore without thorough deoxygenation it is more likely to induce oxyhydroxide precipitation [48]. Polycarbonate has been used as a sleeve for dialysis samplers of slightly different design without apparent interference [49]. PVC and stainless steel samplers have also been used to measure dissolved gases and major ions [50] where trace contamination and the effect of oxygen are less important.

A greater variety of materials have been used for soil and groundwater samplers. As soils tend to be oxic they are less susceptible to oxidation artifacts. Groundwater samplers have tended to use individual dialysis cells comprising a polyethylene bottle with membranes at one or either end, which can be stacked together or inserted within a PVC holder [12,29]. Glass bottles have been used for sampling organic compounds [29]. Stainless steel and Teflon devices have also been used [30].

Various membranes have been evaluated for use with dialysis samplers [27,45] with polysulfone and cellulose acetate eventually becoming the most popular. Polysulfone membranes (Gelman HT-450) were most resistant to microbiological degradation [27] and can be cleaned in acid to remove contaminants. Cellulose-based membranes did not degrade appreciably when deployed in low temperature (5 °C), anoxic environments, although they were prone to biodegradation at higher temperatures (28 C) [27,45]. Teflon membranes (75 μm) have been used to sample gases [51].

Membranes with pore sizes of 0.20 or 0.45 μm have most commonly been used. Carignan *et al.* [26] obtained similar concentrations when using 0.45 μm pore size membranes alone and combined with membranes with 0.03 μm pore sizes. They concluded that no colloids entered the SM through the 0.45 μm membrane. Some ions (Cd and Cu) were found to equilibrate more slowly with membranes that excluded 1000–10 000 Da molecular weight molecules (0.002 and 0.001 μm pores respectively) and were presumed to be appreciably complexed by organic material [26]. Cellulose acetate membranes (0.004 μm pore size) have been used for sampling inorganic ions in groundwater and nylon-coated Versapore membranes (0.2 μm pore size) for dissolved organic substances [29].

2.3.1.2 Deployment and retrieval Practical considerations for sampler deployment in rivers and estuaries include firm fixing to cope with strong currents and changing water levels, and sensible choice of site to avoid debris accumulation around the samplers, or periodic exposure to air in tidal systems and rapidly draining water bodies. Sediment samplers used in lakes have usually been deployed and retrieved by hand using divers. Deep water deployments of peeper samplers have been carried out using a remotely operated vehicle (ROV) [52].

During insertion of the sampler some sediment may be subducted. Usually, however, dialysis samplers are deployed for sufficient time (>4 d) to allow the prevailing redox conditions to re-establish after such a disturbance [53]. It can be quite difficult to insert a sampler into consolidated or sandy sediments. Sediment disturbance is likely and a gap may persist throughout the deployment, possibly causing elevated oxyhydroxide formation. Repacking sediment around the sampler removes the gap and produces more consistent results, but adds to the disturbance. When using a lander, such manipulations are not possible and so the likelihood of such artifacts arising should be assessed.

In highly bioturbated sediments it is inevitable that natural burrows will be occasionally encountered. This will often be indicated by the presence of oxyhydroxide deposits on the membranes. Although results from such samplers will probably be quite different from others at the same site, they will reflect genuine chemical changes that arise in heavily bioturbated sediment. Again the possibility that such results may not represent true dissolved concentrations needs to be considered.

Dialysis samplers for soil groundwater are usually suspended from a rope in screened wells with diameters only slightly larger than the sampler. Ronen *et al.* [12] have considered the possibility of using a floating sampler that moves automatically with rising and falling water levels.

Upon retrieval diffusional samplers should be exposed to different waters (*e.g.* overlying waters) for only a minimal period of time as solutes within the SM will begin to equilibrate with the other medium waters. This is more important for DET which has a more rapid response time, but can be important for dialysis samplers when deployed in deep water where rapid retrieval to the surface is impossible. In deep-sea deployments with an ROV, the samplers were inserted into a sealed 'scabbard' after retrieval from the sediment [52].

2.3.1.3 Sample preservation and analysis Before sub-sampling commences the sampler must first be cleaned thoroughly in order to prevent particles attached to the outside of the sampler from contaminating the samples. A high pressure water pistol has been used effectively for this purpose [32].

Brandl and Hanselmann [27] list appropriate preservation procedures after retrieval for measurement of gases, general water quality parameters, redox dependent species, nutrients and metabolites, and major ions. However, injection of acid directly into the sampler chamber is not recommended as this may lead to remobilisation of adsorbed or precipitated ions [32]. Oxygenation of the sample after retrieval can cause problems. Turbid white sulfur, as well as iron oxyhydroxides, have been observed in the sampling chamber [27]. Samplers previously exposed to anoxic conditions were found to accumulate oxygen at 4 μ M min^{-1} after exposing the device to air [45]. As iron (II) is oxidised rapidly, $t_{1/2} = 2.34$ min at pH 8 and 234 min at pH 7, with 1 mol of O_2 producing 4 mol of iron (III) [54], chambers should be sub-sampled within minutes, with anoxic chambers sampled first. The sampler can be kept in anoxic conditions prior to sub-sampling, but there are potential problems due to gas (H_2S, CO_2) volatilisation. When post-sampling oxidation occurs, acid can be used to re-dissolve oxyhydroxides and re-establish trace metals in solution, but such procedures may introduce artifacts and should be avoided wherever possible.

An advantage of dialysis and DET is that inadvertent contamination during preparation will decrease with deployment time as contaminants diffuse out of the assemblies.

2.3.2 Applications and Novel Developments

Table 2 provides details of applications using dialysis samplers. For brevity only field deployments have been included. They have been grouped according to analytes: gases; water quality parameters; major ions, nutrients and non-metals; trace elements and transition metals; and organic compounds. The stated concentration ranges apply to all deployments for similar systems; actual ranges for individual studies will usually be smaller. It is apparent that dialysis samplers offer great versatility in both the range of analytes and the natural water systems that have been investigated. Only recent, novel developments are discussed further.

Table 2A. Applications and features of studies using dialysis, DET and DGT: dissolved gases

Species	Sampling method	Analysis method	Concentrationrange*	System studied
$Ar_{(aq)}$	dial [51]	gas chromatography	5–10 μmol L^{-1}	marine sediment
$CH_4(aq)$	dial [8]	gas chromatography	0.45–2.4 mmol L^{-1}	riverine SWI
	dial [27, 45, 55–57]	gas chromatography	0–4.2 mmol L^{-1}	lacustrine SWI
	dial [58]	gas chromatography	0–1.2 mmol L^{-1}	saline wetlands
	dial [59]	gas chromatography	0–240 μmol L^{-1}	rice fields
	dial [51, 60]	gas chromatography	0–2.1 mmol L^{-1}	marine sed.
$\sum CO_2(aq)$	dial [50, 61]	CO_2 analyser/GC	2.8–50 mmol L^{-1}	marine SWI
	DET [39]	FIA conductometry	1–15 mmol L^{-1}	marine SWI
	dial [27, 55–57, 62]	FIA conductometry	0.01–7.2 mmol L^{-1}	lacustrine SWI
	dial [63]	ion chromatography	$0-7$ mmol L^{-1}	peat bogs
$N_2(aq)$	dial [51]	gas chromatography	150–500 μmol L^{-1}	marine sediment
$O_2(aq)$	dial [29, 64]	DO probe	0.7–0.5 mg L^{-1}	groundwater well

* For all studies cited in similar systems.

$\sum CO_2 = CO_2 + H_2CO_3 + HCO_3^- + CO_3^{2-}$.

DET = diffusive equilibration in thin films.
dial = dialysis.
DO = dissolved oxygen.
FIA = flow injection analysis.
GC = gas chromatography.

Table 2B. Applications and features of studies using dialysis, DET and DGT: water quality parameters

Species	Sampling method	Analysis method	Concentration range*	System studied
alkalinity	dial [27†]		1.8–4.5 meq L^{-1}	lacustrine SWI
	dial [65]	potentiometric titration	4–22 meq L^{-1}	salt marsh
	DET [39,66]	potentiometric titration	2–20 mmol L^{-1}	marine SWI
conductivity	dial [27†,67]	conductometry	300–450 μScm^{-1}	lacustrine SWI
	dial [12]	conductometry	0.8–100 μScm^{-1}	groundwater well
DIC	dial [45†,62,68]	gas chromatography	0.12–2.5 mmol L^{-1}	lacustrine SWI
DOC	dial [26,27,67,68]	carbon analyser	0.08–8.0 mmol L^{-1}	lacustrine SWI
	dial [69]	carbon analyser	0.525 μmol L^{-1}	riverine sed.
	dial [64]	carbon analyser	0.375–1.54 mmol L^{-1}	groundwater well
	dial [70]	carbon analyser	1.53–2.43 mmol L^{-1}	wetlands SWI
E_h	dial [70,71]	potentiometry	−192 to +90 mV	wetlands SWI
	dial [72–74]	potentiometry	−180 to +300 mV	marine sed.
pH	dial [27,56,57,62, 67,68,73,75–77]	potentiometry	3.95–8.2	lacustrine SWI
	dial [70,71]	potentiometry	6.54–7.46	wetlands SWI
	dial [29]	potentiometry	6.00–6.35	groundwater well
	dial [63]	potentiometry	4.1–6.5	peat bogs sed.
	dial [74]	potentiometry	7.56–8.08	marine SWI
salinity	dial [65]	titration	10–225	salt marsh SWI

* For all studies cited in similar systems.
†Not described in cited paper.
DIC = dissolved inorganic carbon.
DOC = dissolved organic carbon.

Table 2C. Applications and features of studies using dialysis, DET and DGT: major ions, nutrients and non-metals

Species	Sampling method	Analysis method	Concentration range*	System studied
$Br^-_{(aq)}$	dial [78]	ion chromatography	12.5–82.6 μmol L^{-1}	lacustrine sed.
	DET [79]	ion chromatography	0.02–2 mmol L^{-1}	lacustrine SWI
$Ca^{2+}_{(aq)}$	dial [27, 56, 57, 67, 80]	ICPAES/FAAS/ GFAAS	0.13–1.87 mmol L^{-1}	lacustrine SWI
	DET [81]	GFAAS	250–550 μmol L^{-1}	lacustrine SWI
	dial [70]	ICPAES	0.069–2.78 mmol L^{-1}	wetlands SWI
	dial [61,74]	FAAS	5.7–11.4 mmol L^{-1}	marine SWI
	DET [39]	HPLC	8–18 mmol L^{-1}	marine SWI
	dial [63]	ICPAES	0–2.5 mmol L^{-1}	peat bogs sed.
$Cl^-_{(aq)}$	dial [12, 29]	ion chromatography	0.28–8.17 mmol L^{-1}	groundwater well
	dial [73, 78]	ion chromatography	0.42–56.4 mmol L^{-1}	lacustrine SWI
	DET [39, 79, 82, 83]	ion chromatography	50–600 mmol L^{-1}	marine SWI
$HCO^-_{3(aq)}$	dial [64]	potentiometric titration	3.52–4.18 mmol L^{-1}	groundwater well
$K^+_{(aq)}$	dial [27, 55, 67, 80]	ICPAES/FAAS	0.01–51 μmol L^{-1}	lacustrine SWI
	DET [81]	GFAAS	20–70 μmol L^{-1}	lacustrine SWI
	dial [70]	ICPAES	32–276 μmol L^{-1}	wetlands SWI
	dial [64]	FAAS	†–640 μmol L^{-1}	groundwater well
$Mg^{2+}_{(aq)}$	dial [27, 55, 56, 67]	ICPAES/FAAS	2.6–412 μmol L^{-1}	lacustrine SWI
	DET [64]	GFAAS	80–160 μmol L^{-1}	lacustrine SWI
	dial [70]	ICPAES	0.069–2.36 mmol L^{-1}	wetlands SWI
	dial [84]	FAAS	493–572 μmol L^{-1}	spring fed fen

Table 2C. (*cont.*)

Species	Sampling method	Analysis method	Concentration range*	System studied
$Na^+_{(aq)}$	dial [27,67,80]	ICPAES	0.04–7.56 mmol L^{-1}	lacustrine SWI
	dial [64]	FAAS	52–349 μmol L^{-1}	groundwater well
$NH^+_{4(aq)}$	dial [27,45,46,55,56, 67,78,85]	colourimetry	0.001–1.5 mmol L^{-1}	lacustrine SWI
	dial [74]	colourimetry	0.004–1.53 μmol L^{-1}	marine SWI
	DET [39,79,83]	FIA colourimetry	0.02–1 mmol L^{-1}	marine SWI
	dial [86†]		0.18–1.17 mmol L^{-1}	riverine sed.
$NO^-_{2(aq)}$	dial [67]	colourimetry	0–0.5 μmol L^{-1}	lacustrine SWI
$NO^-_{3(aq)}$	dial [46,67]	Cd reduction /HPLC	0–45 μmol L^{-1}	lacustrine SWI
	DET [15]	HPLC	4–30 μmol L^{-1}	lacustrine SWI
	DET [39,79,83,87]	HPLC	0.01–1 mmol L^{-1}	marine SWI
	dial [12, 64, 88]	IC/Cd reduction	0.98–725 μmol L^{-1}	groundwater well
$PO^{3-}_{4(aq)}$	dial [8]	colourimetry	10–380 μmol L^{-1}	riverine SWI
	dial [74]	colourimetry	0.2–109.7 μmol L^{-1}	marine SWI
SRP	dial [45,46,49,55, 56, 78,89]	colourimetry/IC/ ascorbic acid reduction	0.02–200 μmol L^{-1}	lacustrine SWI
	dial [84]	Colourimetry	0.084–0.189 μmol L^{-1}	spring-fed fen
	feDGT [90]	FIA colourimetry	1.94 ± 0.06 μmol L^{-1}	pond water
	feDGT [90]	FIA colourimetry	0.1–10 μmol L^{-1}	lacustrine SWI
	feDGT [90]	FIA colourimetry	0–3.16 μmol L^{-1}	lacustrine SWI
total P	dial [27]	ICPAES	0–194 μmol L^{-1}	lacustrine SWI
$SiO_{2(aq)}$	dial [23†]		250–714 μmol L^{-1}	reservoir SWI
	dial [70]	ICPAES	0.053–2.39 mmol L^{-1}	wetlands SWI
	dial [27,45,55,56†, 67,80]	ICPAES	0.11–1.0 mmol L^{-1}	lacustrine SWI
$S^{2-}_{(aq)}$	dial [73]	colourimetric	0–0.14 μmol L^{-1}	wetlands SWI
	dial [72]	CSV	0–7 μmol L^{-1}	marine sed.
	dial [63]	LC-amperometry	0.2–0.5 μmol L^{-1}	Peat bogs
	dial [84]	titration	0–880 μmol L^{-1}	flood plain mire
	dial [65,91]	potentiometric titration	0–2.7 mmol L^{-1}	salt marsh
$H_2S_{(aq)}$	dial [61]	titration	0–3.8 mmol L^{-1}	marine sed.
	dial [55,56,77,89]	colourimetry	0.1–70 μmol L^{-1}	lacustrine SWI
$SO^{2-}_{4(aq)}$	dial [50,61,72]	turbidimetry/IC	1.5–29 mmol L^{-1}	marine SWI
	DET [39,79,83]	HPLC	15–30 mmol L^{-1}	marine SWI
	dial [27,46,56,67,73, 77,89,92]	ion chromatography	0.2–520 μmol L^{-1}	lacustrine SWI
	DET [15,81,87]	HPLC	10–100 μmol L^{-1}	lacustrine SWI
	dial [84]	turbidimetry	1.27–1.42 mmol L^{-1}	spring-fed fen
	dial [65†]		4–300 mmol L^{-1}	salt marsh
	dial [63]	turbidimetry	0.8–1.6 μmol L^{-1}	peat bogs
thiosulfate	dial [12,29,64]	turbidimetry	0.22–729 μmol L^{-1}	groundwater well
	dial [72]	CSV	0–15 μmol L^{-1}	marine sed.

* For all studies cited in similar systems.
† Not described in cited paper.
CSV = cathodic stripping voltammetry.
FAAS = flame atomic absorption spectrometry.
feDGT = diffusive gradient in thin films with a ferrihydrite accumulating layer.
GFAAS = graphite furnace atomic absorption spectrometry.
HPLC = high performance liquid chromatography.
IC = ion chromatography.
ICPAES = inductively coupled plasma atomic emission spectrometry.
SRP = soluble reactive phosphate.

Table 2D. Applications and features of studies using dialysis, DET and DGT: trace elements and transition metals

Species	Sampling method	Analysis method	Concentration range*	System studied
$Al_{(aq)}$	dial [56]	GFAAS	1.0–10 μmol L^{-1}	lacustrine SWI
	dial [70]	ICPAES	0–8.9 μmol L^{-1}	wetlands SWI
$As_{(aq)}$	dial [93–95]	ETAAS	0.02–0.60 μmol L^{-1}	lacustrine SWI
	chDGT [19]	PIXE	5–100 nmol L^{-1}	stream sed. & microbial mat
$Ba_{(aq)}$	dial [70]	ICPAES	0–0.55 μmol L^{-1}	wetlands SWI
$Cd_{(aq)}$	dial [68, 95–97]	GFAAS	2.0–14.1 nmol L^{-1}	lacustrine SWI
	dial [98,99]	GFAAS	0.15–19.5 nmol L^{-1}	lacustrine water
	chDGT [100]	GFAAS	18 ± 1.8 pmol L^{-1}	stream water
	dial [61]	DPASV	0.27–0.89 nmol L^{-1}	marine SWI
	chDGT [101]	GFAAS	0.22–0.31 nmol L^{-1}	marine water
	chDGT [16]	GFAAS	< 0.178 μmol L^{-1}	saturated soil
	chDGT [17]	GFAAS	0.1–1.3 nmol L^{-1}	lacustrine SWI
$Co_{(aq)}$	dial [26, 95]	GFAAS	0.1–0.5 μmol L^{-1}	lacustrine SWI
	chDGT [100]	GFAAS	0.226 ± 0.024 nmol L^{-1}	stream water
	chDGT [101]	GFAAS	0.26–0.53 nmol L^{-1}	marine water
$Cr_{(aq)}$	dial [26]	GFAAS	25–75 nmol L^{-1}	lacustrine SWI
	dial [70]	ICPAES	0–7.81 μmol L^{-1}	wetlands SWI
	dial [61]	GFAAS	1.0–6.6 nmol L^{-1}	marine SWI
	chDGT [101]	GFAAS	2.53–6.69 nmol L^{-1}	marine water
$Cs_{(aq)}$	x8DGT [102]	ICPMS	0.242 ± 0.005 nmol L^{-1}	lacustrine water
$^{137}Cs_{(aq)}$	x8DGT [103]	gamma-counting	0.102 Bq L^{-1}	lacustrine water
$Cu_{(aq)}$	dial [26,56,68,73,95, 97]	GFAAS	0.05–1.58 μmol L^{-1}	lacustrine SWI
	dial [99]	GFAAS	0.019–1.97 μmol L^{-1}	lacustrine water
	chDGT [100]	GFAAS	2.09 ± 0.08 nmol L^{-1}	stream water
	dial [69]	GFAAS	5.03 μmol L^{-1}	riverine sed.
	dial [70]	ICPAES	0.047–4.17 μmol L^{-1}	wetlands SWI
	dial [61]	DPASV	4.7–90 nmol L^{-1}	marine SWI
	chDGT [101]	GFAAS	1.33–2.14 nmol L^{-1}	marine water
	chDGT [16]	GFAAS	16–315 nmol L^{-1}	saturated soil
	chDGT [17]	GFAAS	0.5–2 nmol L^{-1}	lacustrine SWI
$Fe_{(aq)}$	dial [26,27,45,46,55, 56,67,68,73,75,80,89, 92,93,95,97]	AAS/ICPAES/ ICPMS	0.001–1.8 mmol L^{-1}	lacustrine SWI
	DET [9,14,15,31,35, 81]	PIXE/GFAAS	1.82–896 μmol L^{-1}	lacustrine SWI
	dial [84]	AAS	154–409 μmol L^{-1}	spring-fed fen
	dial [63]	ICPAES	0–450 μmol L^{-1}	peat bogs
	dial [70,71]	ICPAES	0–45 mmol L^{-1}	wetlands SWI
	dial [29]	ICPAES	179–806 μmol L^{-1}	groundwater well
	dial [74]	AAS	3.8–54.6 μmol L^{-1}	marine SWI
	DET [104]	GFAAS	1–20 μmol L^{-1}	marine SWI
	chDGT [13,101]	GFAAS	3.6–3.74 nmol L^{-1}	marine water
	dial [91*]		0–2.45 mmol L^{-1}	salt marsh SWI
	chDGT [19]	PIXE	1–7 μmol L^{-1}	stream sediment & microbial mat
	chDGT [17]	GFAAS	< 0.25 mmol L^{-1}	lacustrine SWI
Fe $(II)_{(aq)}$	dial [46]	colourimetry	0–80 μmol L^{-1}	lacustrine SWI
	dial [63]	ion chromatography	0.5–24.5 μmol L^{-1}	peat bogs
Fe $(III)_{(aq)}$	dial [63]	ion chromatography	0.5–10 μmol L^{-1}	peat bogs

Table 2D. (*cont.*)

Species	Sampling method	Analysis method	Concentration range*	System studied
$Hg_{(aq)}$	dial [105]	CVAFS	0.025–0.075 nmol L^{-1}	lacustrine sed.
	dial [105]	CVAFS	0.01–0.03 nmol L^{-1}	flooded soil
	dial [105]	CVAFS	0.02–0.21 nmol L^{-1}	peat bog sed.
$Mn_{(aq)}$	dial [27,45,55,56,67, 68,73,75,80,95,97]	AAS/ICPAES/ ICPMS	0.004–625 μmol L^{-1}	lacustrine SWI
	DET [14,15,35,81]	GFAAS	1.8–72.9 μmol L^{-1}	lacustrine SWI
	dial [29]	GFAAS	102–200 μmol L^{-1}	groundwater well
	dial [70, 71]	ICPAES	0–42 mmol L^{-1}	wetlands SWI
	dial [61]	GFAAS	0.15–0.60 μmol L^{-1}	marine SWI
	DET [104]	GFAAS	0.5–5 μmol L^{-1}	marine SWI
	chDGT [13,101]	GFAAS	0.8–3.71 nmol L^{-1}	marine water
	chDGT [19]	PIXE	1–2.5 μmol L^{-1}	stream sediment & microbial mat
	chDGT [17]	GFAAS	< 5 μmol L^{-1}	lacustrine SWI
$Ni_{(aq)}$	dial [26,58,68,95,97]	GFAAS	0.25–5.2 μmol L^{-1}	lacustrine SWI
	dial [99]	GFAAS	0.015–4.91 μmol L^{-1}	lacustrine water
	chDGT [100]	GFAAS	1.29 ± 0.11 nmol L^{-1}	stream water
	dial [70]	ICPAES	1.31–229 μmol L^{-1}	wetlands SWI
	chDGT [101]	GFAAS	3.73–4.52 nmol L^{-1}	marine water
	chDGT [16]	GFAAS	< 5.9 μmol L^{-1}	saturated soils
	chDGT [17]	GFAAS	2–10 nmol L^{-1}	lacustrine SWI
$Pb_{(aq)}$	dial [68, 95]	GFAAS	12–28 nmol L^{-1}	lacustrine SWI
	dial [69]	GFAAS	207 nmol L^{-1}	riverine sed.
	chDGT [100]	GFAAS	0.274 ± 0.014 nmol L^{-1}	stream water
	dial [61]	DPASV	1.7–17.9 nmol L^{-1}	marine SWI
	chDGT [101]	GFAAS	0.25–0.51 nmol L^{-1}	marine water
$Sr_{(aq)}$	x8DGT [102]	ICPMS	190 ± 4 nmol L^{-1}	lacustrine water
$Zn_{(aq)}$	dial [26,68,73,92,95, 97]	GFAAS	0.02–2 μmol L^{-1}	lacustrine SWI
	dial [99]	GFAAS	0.076–4.76 μmol L^{-1}	lacustrine water
	chDGT [100]	GFAAS	43.7 ± 1.84 nmol L^{-1}	stream water
	dial [71]	GFAAS	0.918 μmol L^{-1}	riverine sed.
	dial [67]	AAS	2.75–6.12 μmol L^{-1}	groundwater well
	chDGT [13, 106]	GFAAS	11–31 nmol L^{-1}	marine water
	chDGT [19]	PIXE	0.4–1 μmol L^{-1}	stream sediment & microbial mat
	chDGT [14, 17]	AAS	15–180 nmol L^{-1}	lacustrine SWI
	chDGT [16]	AAS	< 30.5 μmol L^{-1}	saturated soil

* For all studies in similar systems.
CVAFS = cold vapour atomic fluorometry spectrometry.
chDGT = diffusive gradient in thin films with a chelex accumulating layer.
DPASV = differential pulse anodic stripping voltammetry.
ETAAS = electro-thermal atomic absorption spectrometry.
ICPMS = inductively coupled plasma mass spectrometry.
LC = liquid chromatography.
PIXE = photon-induced X-ray emission spectrometry.
SWI = sediment–water interface (and refers to measurements made in both the overlying water and porewater).
x8DGT =diffusive gradient in thin films with a AG 50W-X8 cationic exchange resin accumulating layer.

Table 2E. Applications and features of studies using dialysis, DET and DGT: organic compounds

Species	Sampling method	Analysis method	Concentration range*	System studied
acetate	dial [27]	gas chromatogr.	0–67 $\mu mol\,L^{-1}$	lacustrine SWI
dimethyl-aniline	dial [29]	HPLC	61.9–140 $\mu mol\,L^{-1}$	groundwater well
toluene	dial [29]	HPLC	81.4–185 $\mu mol\,L^{-1}$	groundwater well
propionate	dial [27, 69]	gas chromatogr.	0–250 $\mu mol\,L^{-1}$	lacustrine SWI
xylene	dial [29]	HPLC	9.4–75.3 $\mu mol\,L^{-1}$	groundwater well

* For all studies in similar systems.

2.3.2.1 Recent developments of dialysis samplers for sediment deployment A volume-enhanced sampler for sediment pore water was developed [53] and used to measure concentration profiles of major ions, iron and manganese in Lake Erie sediment. It enabled screw cap bottles to be fitted to the sides of a sampler based upon the original peeper design. A channel linked each sampling chamber to an adjacent bottle. After retrieval each bottle was removed and capped, minimising opportunities for contamination. This provided a sample volume of about 30 mL at a sampling interval of 2 cm. The high design factor of this sampler ($F = 3.5$), resulted in slow equilibration. Laboratory studies with Fe, Mn, Ca and Si showed that it took about 2 weeks to approach equilibrium in sediment at 20 °C and three weeks at 4 °C. Equilibration of the conservative ions Na and K appeared to take somewhat longer, consistent with modelling predictions for no resupply from the solid phase [33,37]. Nevertheless, this sampler does have useful features and illustrates how the design and use of dialysis samplers frequently requires compromises.

A suitable preservative can be added before capping the bottles which removes the need for transferring the sample and therefore speeds up sample handling and minimises opportunities for analyte contamination [80]. Another feature is that the bottles can be isolated by remote operation while still in the sediments, which would prevent back equilibration during retrieval of the device and be particularly useful for deployment in sediment in deep water. One drawback of this sampler might be the introduction of artifacts by oxy-hydroxide formation, due to the longer sampling time required, leading to overestimation of some analytes.

A modified version of the Hesslein sampler was previously fitted with a shutter that could be slid up and down to uncover and cover the interfaces when being deployed and retrieved respectively [50]. This was used to measure effective diffusion coefficients of dissolved sulfate and total dissolved CO_2 within sediments as a function of sediment porosity.

Another modified peeper design allows removal of the sample individually from each chamber while still in the sediment and measurement of the concentration using on-line or field-based techniques [72]. Each chamber has a silicon tube leading to the surface through which the contents can be withdrawn with a syringe. Air is purged from each chamber by suction with a hand pump. The chambers were 10 cm apart, had a volume of 30 mL and an interfacial area of 2.4 cm^2. Consequently, F is large at 12.5 cm and equilibration times should be long. However, preliminary studies indicated that equilibration times may be as short as 2 weeks. Such times would only be possible if there is mixing within the sampling chambers [37] and either convection in the pore waters or rapid resupply from the solid phase [33]. Concentration profiles of sulfate, thiosulfate, sulfide, E_h, Fe, Mn, Zn, Pb, Cd and Cu [72,107] have been measured in tidal mud flats in the Lagoon of Venice [72]. This approach promises to provide direct temporal changes of porewater concentrations arising through natural cycles. However, there remain unresolved questions about this type of sampler. For example, at what water depths can they be deployed? How does having a long length of tubing in diffusive equilibrium with the sampling chamber affect the equilibration time and efforts to keep the chamber free of oxygen during long term deployments? As water is presumably drawn into the chamber at the same time that the sample is drawn up the silicon tubing, are successive samples obtained by filtration rather than dialysis?

2.3.2.2 Novel applications of dialysis samplers for groundwater studies A multilayer dialysis sampler has been used to sample aquifer colloids under natural gradient flow conditions [30]. Versapore (10 μm) membranes (Gelman) were used in samplers, of similar design to those in Figure 1-iii, to obtain a vertical profile extending over many metres. Laboratory studies with synthetic colloids (homogenous 0.6 μm latex and 0.4–0.7 μm mean kaolinite suspensions) in a flow-through cell indicated that equilibration occurred within 180 h. There appeared to be no fractionation or concentration of kaolinite colloids within the individual chambers and no clogging of the membranes. Large variations in the colloidal compositions (mainly aluminosilicates, $CaCO_3$, silica and organic matter) were noted between profiles obtained from the same aquifer, consistent with previous observations.

A multilayer dialysis sampler has also been used to study bacterial activity in aquifers [108]. Bacterial cultures (*Pseudomonas* sp, an atrazine reducing species) were encapsulated in calcium alginate gel and placed within individual dialysis cells interspaced with control cells containing water. The sampler was then placed within an aquifer contaminated with atrazine ($< 450\,\mu g\,L^{-1}$). A 0.2 μm nylon membrane was used to prevent the dispersal of the bacteria throughout the aquifer. Atrazine and oxygen concentrations were measured in the culture and control cells. Atrazine concentrations decreased by 90 % and oxygen by about 45 % in the cells containing the culture. This approach, combining the use of a dialysis sampler with that of a mini-reactor cell, highlights the potential for adapting dialysis and gel equilibration techniques to perform *in situ* experiments in a controlled way, as in the case of DGT (sections 3 and 4).

2.4 DIFFUSIVE EQUILIBRATION IN THIN FILMS (DET)

Samplers which rely on the free exchange of solutes between the water in a hydrogel and the immersion solution have only recently been introduced [9]. As they rely on there being no reaction between the hydrogel and solutes, each gel composition must be fully tested before it is used for *in situ* deployments.

2.4.1 Methodology Verification

2.4.1.1 Gel Preparation The main type of gel used for DET measurements has been based on polyacrylamide cross-linked with a derivative of agarose [13–15,35]. The acrylamide monomer structure and an idealised representation of the polymer formed by following this procedure are shown in Figure 6. Note that the hydrogel polymer structure is more open and flexible than that formed by using *N*,*N*′-methylene-bis-acrylamide (bis-acrylamide) as a cross-linking agent [109]. This is an important feature of the hydrogel for use in the sampling techniques described here.

The detailed procedure for preparation of this gel is given in Appendix I. As with all hydrogel preparation, attention to detail is vital to ensure that the resultant gel has reproducible properties. Diffusional characteristics of ions in the gel have been found to be particularly sensitive to the setting temperature and the concentration of cross-linker. Gels used for electrophoresis are usually cross-linked with bis-acrylamide. The diffusional characteristics of these gels vary greatly depending on the concentration of monomer and cross-linker and may be quite different from those of the gels described in Appendix I.

2.4.1.2 Probe construction and preparation As for dialysis, perspex has been the most popular material for the DET probe. The unconstrained DET probes are a much simpler construction than for dialysis, comprising only two flat sheets, one with a window (Figure 1iv). Various designs of DET probes have been used [9,35,39]. In the original design [9] a 1 mm thick layer of gel was sandwiched between two perspex plates (15 cm × 5 cm × 0.2 cm), one of which had a 10 cm × 1 cm window. The two plates were held together using either plastic clips or screws. A 0.45 μm (pore size) cellulose nitrate membrane was placed between the gel and top plate to prevent damage to the gel surface during insertion and authigenic deposition of oxyhydroxides on the gel during deployment. As DET assemblies are mainly deployed for short times (less than a week) the possible microbial degradation of the filter membrane is less of a problem than for dialysis. Cellulose nitrate, cellulose acetate and polysulfone membranes

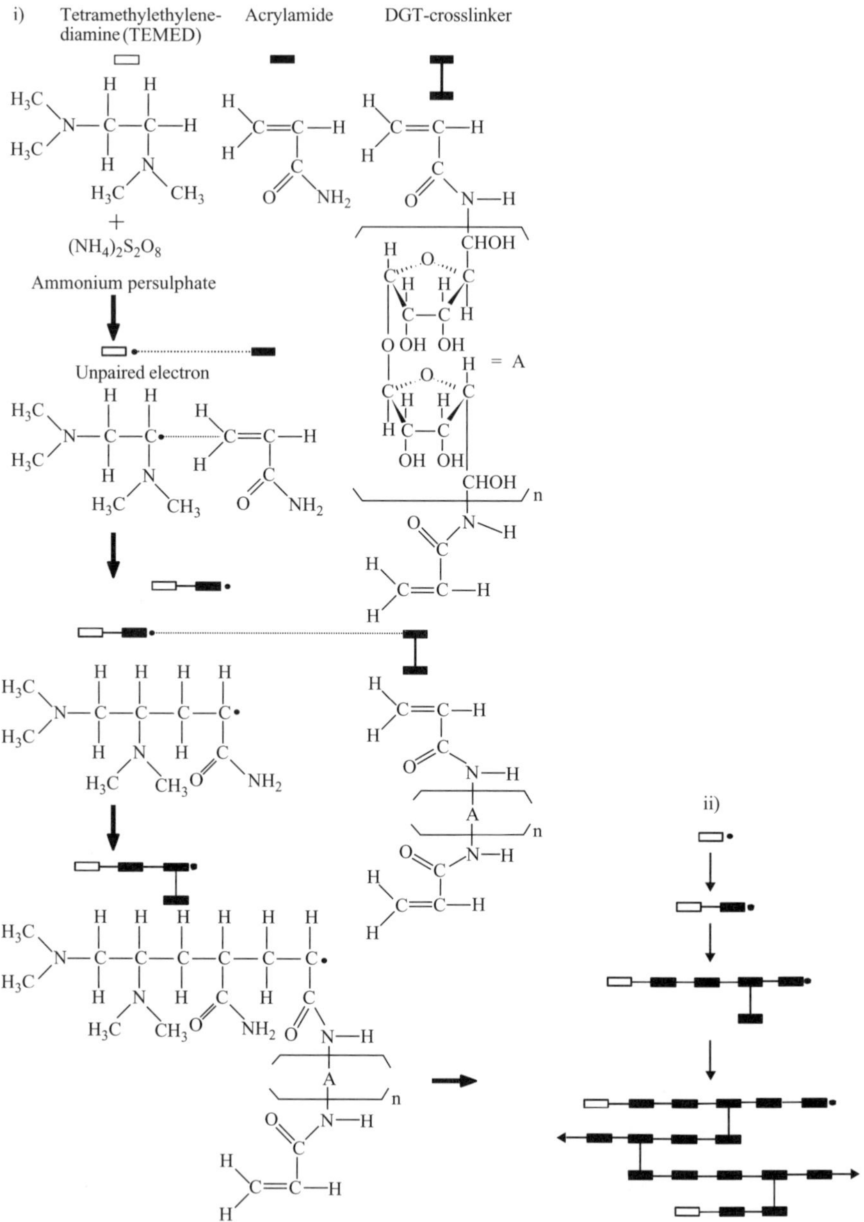

Figure 6. (i) Forms of the acrylamide monomer and agarose-derivative crosslinker; the structure and number of repeat saccharide units (*n*) of the cross-linker is actually unknown owing to commercial propriety, but it is obviously much bulkier than the bis-acrylamide cross-linker; (ii) idealised form of polyacrylamide hydrogel

have all been used successfully. As the gel has small pore sizes, it, rather than the filter (usually 0.45 μm), determines the size of the molecules that can equilibrate. Assemblies can be made to virtually any size. 50 cm long versions have been used for measuring Fe and Mn in lake bottom waters, while 4 cm wide versions have been used to measure two-dimensional distributions of Fe and Mn concentrations in sediment pore waters [82,110].

2.4.1.3 Porewater sampling Prior to deployment the entire gel assemblies need to be de-oxygenated (see section 2.2) by completely immersing them for at least 24 h in a container of similar ionic composition to the deployment matrix (see section 2.4.1.6) and vigorously bubbling with either argon or nitrogen. It is important that the container opening is almost closed to prevent back diffusion of oxygen. The probes must be transported in sealed oxygen-free containers of solution to the sampling site. Immersing the sealed container (not the probe itself) in a gas tight box filled with N_2 or in an Na_2SO_3 solution has been found to be a useful way of avoiding any O_2 ingress during transport [17]. Adding Na_2SO_3 to the degassing water is not advisable as it may diffuse into the gel and affect the results.

In shallow waters gels can be inserted directly into the sediment. Placement by divers has also been used in deeper waters [14,35,87]. Assemblies were transported in their sealed anoxic containers to the anoxic hypolimnion of the lake before opening the containers and inserting them into the sediment [35]. The assemblies had been previously fitted with a light nylon cord, which was attached to a small fishing float. This was used for retrieval. Landers have also been used to deploy gels *in situ* in freshwater and marine sediments [111].

Because DET probes are usually small compared to dialysis assemblies, they can be deployed in retrieved sediment cores [9,35,39,66,81,104], providing the cross-sectional area of the assembly is negligibly small compared with that of the core to minimise sediment displacement. Comparisons made between deployments *in situ* (divers) and in cores [35] (7 cm diameter) indicate that the porewater structure of the surface 1 cm of sediment may be lost when DET is used in cores.

2.4.1.4 Fixing/slicing Once unconstrained probes are removed from sediment, there is progressive relaxation of concentration differences by diffusional equilibration [33,35]. To retain the profile shape the gel must be sliced quickly or solutes must be chemically fixed. Fixing has been used for Fe and Mn. Upon retrieval, the probe is immersed in 0.01 M NaOH solution for several hours. In this strong base, Fe (II) and Mn (II) are oxidised and hydrolysed to immobile oxyhydroxides, which prohibits further relaxation [35]. Ideally the solution should be stirred to ensure rapid penetration of the gel within this time period. Fixing times longer than a few hours should be avoided as gel swelling occurs.

Measurements of the distribution of Fe and Mn with depth in the gel have shown that fixing is rapid so that the vertical relaxation extends to no more than the thickness of the gel [35]. Once retrieved and fixed, gels for the determination of Fe and Mn can then be sliced manually using a perspex cutter [14]. Early work used a ruler for the spacing, but more accurate slicing was achieved using a guillotine with a Teflon-coated stainless steel razor blade mounted on a Vernier stage [110]. Gel volumes have been successfully estimated by size (the cut volume) and by weight, [17,39,81]. Evaporative loss from the very small gel slices can introduce errors in weighing [110].

For solutes which cannot be fixed, such as Na^+, K^+, Ca^{2+}, Mg^{2+}, nitrate, phosphate, sulfate, chloride, ammonium, $\sum CO_2$ and alkalinity, gels have been rapidly sliced at intervals down to 1 mm. Upon retrieval there is insufficient time to dismantle probes that are either screwed or clipped together. Adhering sediment must be wiped away with a tissue, making sure the membrane is as clean as possible. Although washing has been successfully used [15], Mortimer *et al.* [39] recommended avoiding washing because re-equilibration potentially occurs. The central portion of the filter can then be cut out using a scalpel or a perspex cutter and removed. Mortimer *et al.* [39] recommended leaving 1 mm of gel either side of the window to act as a precaution against contamination. Sampled sections of gel were placed on a clean, dry cutting board, the filter was removed and the gel divided into 1 cm segments (within 20 s). These sections were then further subdivided to the required resolution. Relaxation of porewater structures will occur during this procedure and sharp sub-surface maxima may be diminished and boadened [33]. These factors must be taken into consideration when using this technique. However, broader centimetre scale structures will be little affected if the slicing is carried out immediately. Using thicker slices (e.g. 2.5–5 mm) automatically restricts the resolution of the measurement, but rapid slicing becomes less important. The individual pieces of gel were placed into pre-weighed 1.5 mL centrifuge tubes. The tubes were re-weighed to obtain the gel mass (and hence volume) by difference, and stored at 4 °C until analysis.

The above procedure can take up to 3 min for a complete probe by which time fine scale porewater structures (1–2 mm) will have diminished [33]. Harper *et al.* [33] concluded that the best possible spatial resolution without significant relaxation could be achieved by combining the desirable features of DET and dialysis. If gel is placed in compartments no relaxation can occur. Filling the compartments with gel rather than water allows manufacture of very small scale assemblies. Solutes can be analysed in the gel dried within the compartment using a beam technique. Fones *et al.* [31] developed such a constrained DET probe. It was manufactured from a 400 μm thick ceramic (Al_2O_3) sheet (10 cm × 3 cm). The probe contained a series of laser-cut compartments (200 μm × 3 mm), arranged in three columns, with the centre column offset by 200 μm. Analysis of the compartments provided three profiles each with a spatial resolution of

400 μm. Because of the high swelling nature of the gel usually used for DET, a polyacrylamide gel was used with a composition which did not swell [31]. Concentrations were measured by proton induced X-ray emission (PIXE) spectrometry using an internal standard of colloidal TiO_2 which was mixed into the gel solution.

The resolution limitations of this technique will be determined by the ability to manufacture probes with small compartments and to fill them with gel, as PIXE can be used on areas as small as 1 μm × 1 μm. Constrained probes with spatial resolutions of 0.4 mm have also been deployed *in situ* in deep-sea sediments (1000–4000 m) from a benthic lander and analysed by laser ablation ICPMS and PIXE [111]. Mortimer *et al.* [39] have also used a version of constrained DET using a ladder construction which was constructed from plastic with a spatial resolution of 2.5 mm. Gel slices were sufficiently large to be removed for conventional analysis.

2.4.1.5 Sample preparation and analysis The simplest way to analyse the concentration in the gel slices is to transfer the solute to solution. Gels fixed with NaOH are eluted with 2 M HNO_3 for 24 h and Fe and Mn measured using graphite furnace atomic absorption spectroscopy (GFAAS) [14,35]. If there has been no chemical treatment and there are no potential oxidation problems the solute can be extracted from the gel by back equilibrating it into a known volume of solution.

Appropriate back equilibration times depend on the volume of eluent used as it affects the diffusional path length and whether or not the tubes are well mixed [33]. Calculations show that for small volumes (0.3 mL) without mixing, 2 h ensures 95 % re-equilibration, but 1.5 mL would require 3 d [33]. Continuous mixing of these samples would reduce the times to 1 h in both cases.

Major cations were back equilibrated into 1 ml of deionised water for 24 h without mixing [81], while anions were back equilibrated into 1 ml of quiescent deionised water for 48–72 h prior to analysis [39]. For NH_4^+ and $\sum CO_2$ samples must be analysed within a few hours of equilibration to avoid volatility losses, even from quiescent solutions [39]. Haraldsson *et al.* [66] found that if samples from seawater deployments are back equilibrated into a 0.7 M solution of NaCl, instead of Milli-Q water, a more stable electrode response during the alkalinity titration can be obtained.

Various micro-analytical techniques have been used to measure solutes in the back equilibration solutions. NH_4^+ and $\sum CO_2$ have been analysed using the flow injection analysis method of Hall and Aller [83] which uses a gas-permeable membrane to transfer CO_2 or NH_3 from reagent streams into a receiving stream where they are measured using a conductivity detector. Anions have been analysed by ion chromatography and major cations by GFAAS [81]. Alkalinity has been determined by potentiometric titration (Gran) [66]. Phosphate concentrations were determined using colourimetric flow injection analysis [112].

Gel pieces in the 0.4 mm resolution constrained probe [31] were too small to remove. After retrieval of the probe from the sediment the protective filter was removed. The whole ceramic assembly was placed into a commercial gel dryer to dry out the compartments of gel prior to analysis by PIXE. The gel dryer uses a hot plate to heat up the gel while a vacuum draws away the water. This process is used because all gel samples must be free of any water prior to introduction into the PIXE chamber to avoid any cracking and blistering of the gel when the chamber is pumped down. The gel in each individual compartment was analysed using the 1 μm beam in 100μm × 100 μm rastors to provide an average concentration. In theory it should be possible to use PIXE in this way to determine many elements including Fe, Mn, Cl, Ca, Mg, Na, K and S. However, Fones *et al.* [31] could only measure Fe because of contamination from the ceramic affecting other determinands. The gel drier has also been used to dry down larger pieces of gel from unconstrained probes onto a filter membrane prior to PIXE analysis. This procedure maintains the dimensional integrity of the gel [14,35].

2.4.1.6 Gel–solute interactions and pre-conditioning When gels have been equilibrated in the laboratory with solutions of known solute concentration, the concentration of solute in the gel has generally been found to be the same as in solution [15,35,39,81] for a wide range of solutes including Fe(II), Mn(II), Cl^-, Ca^{2+}, Mg^{2+}, Na^+, K^+, $NH_4{}^+$, $\sum CO_2$, SO_4^{2-}, NO_3^-, Br^- and alkalinity. However, if the gel is transferred directly from its hydration solution of ultra-pure water to a low ionic strength ($< 10^{-3}$ M) medium, erratic results can be obtained, with concentrations in the gel varying from 10 % to 400 % of the concentration in solution [113]. If measurements were made in solutions with ionic strengths $\geqslant 10^{-3}$ M, corresponding to virtually all natural waters, no such problems were observed. It was generally found that conditioning gels in 0.01 M $NaNO_3$ solutions prior to deployment improved precision. After treatment with 0.1 M $NaNO_3$ the non-swelling gel used in constrained assemblies also gave 100 % recoveries [31]. Experiments with some bis-cross-linked gels, however, showed that some cations could accumulate in the gel [100]. This effect is greatly enhanced if gels are hydrolysed by treatment with ammonia.

Mortimer *et al.* [39] noticed erratic results for Cl^- and SO_4^{2-} when gels had simply been hydrated in deionised water and deployed in seawater. Storage in a medium of similar composition to seawater, after hydration in deionised water, gave reproducible results. These workers also reported that sulfate was not completely removed from their gels after a back equilibration time of 48 h. They attributed this observation to reaction of sulfate with the gel. However, as sulfate has a low diffusion coefficient and a relatively large volume of water (1.5 mL) was used for quiescent back diffusion, the results are consistent with incomplete back equilibration times predicted by the diffusion modelling of Harper *et al.* [33].

2.4.2 Applications

2.4.2.1 Freshwater Applications of DET in freshwater systems are listed in Table 2. DET was first used to determine vertical concentration profiles of Fe in the porewaters of cores retrieved from a productive lake (Esthwaite Water, UK) [9]. After fixing of Fe in NaOH, the gels were dried and Fe determined at sub-mm spatial resolution by PIXE. The measurements revealed steep concentration gradients and sub-surface maxima that demonstrated localised, reductive dissolution of fresh material. In a later study that systematically investigated the performance characteristics of DET, including response times, gel reactivity and spatial resolution, Fe and Mn were determined by slicing the gel followed by acid elution and measurement by AAS [35]. Back equilibration into water was used to measure major anions [15,39] and cations [81]. In all these cases good agreement was obtained between DET and alternative techniques, if the finer spatial resolution of DET was considered. Use of DET at a millimetre scale has shown that the concentrations of solutes in porewaters generally have more local structure than previously thought. Such structures can only be sustained if there is local resupply of solutes from solid phases coupled to efficient sinks. Sharp sub-surface maxima have been observed for Fe and Mn [14,35] (Figure 7), major cations [81] and anions [15], showing that there is a general source, which is probably the rapid decomposition of organic material. By using high resolution DET constrained probes, single point maxima observed using unconstrained DET at 1 mm resolution have been revealed as systematic peaks containing three or four data point [31].

Measurements of Fe and Mn by DET have been compared with DGT measurements [14] and used to assess the dynamics of re-supply from solid

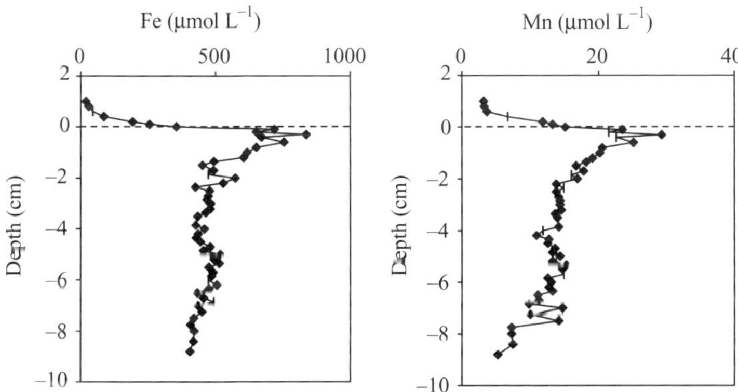

Figure 7. Concentration–depth profiles of Fe and Mn through the sediment–water interface at 14 m water depth in Esthwaite Water on July 8, 1993. Sampled *in situ* by the DET technique with fixing of Fe and Mn by NaOH, slicing and determination by AAS

phase to solution. Deployment of 4 cm wide DET gels that were subsequently sliced into 3 mm squares have provided contour maps of pore water concentrations [82]. These two dimensional images, along with multiple DET deployment, have shown that pore water structure in lake sediments is not usually horizontally uniform [82,110].

2.4.2.2 Marine DET was used in estuarine systems to determine nutrient fluxes [79,114], the effect of sediment mixing [44] and sulfate reduction. Comparison with suction samplers showed that the latter technique tends to overestimate concentrations [39]. Although a few deployments were made *in situ*, the majority involved insertion in collected cores (20 cm diameter). Mn determined in DET assemblies deployed in sediment cores collected from the Mediterranean agreed well with measurements made by slicing and squeezing sediment [104]. Profiles determined in sediments from the oxic region of the Black Sea showed closely spaced ($\sim$ 2 cm) sharp peaks for both Fe and Mn. The DET probes were deployed remotely from a lander which fixed the Fe and Mn by automatically releasing NaOH upon retrieval.

Similar sharp features for Fe and Mn have been observed in other marine sediments, suggesting that similar processes to those occurring in freshwater sediments may be operating. Such sharp-featured structures of solutes in pore waters revealed by the high resolution capability of DET is bringing to light previously unknown mechanisms of solute interactions. More systematic data at high resolution are required to elucidate them fully.

2.4.2.3 Future developments The popular use of DET is likely to continue to be the measurement of concentration profiles at 1 mm or greater intervals as the procedure is very straightforward. However, as the general heterogeneous nature of sediments becomes more generally accepted it is likely that greater attention will be paid to two-dimensional pore water structure. Therefore, wider probes sliced into grids, as used by Shuttleworth [82,110], are likely to find favour. Developments at higher resolution will be centred around the use of constrained probes. Laser ablation coupled to ICPMS may well become the dominant analysis tool, owing to its greater sensitivity and more general availability than PIXE. Constrained probes will probably be preferred for use in benthic landers [111] as they overcome relaxation problems. If remote fixing by NaOH is not used, the assembly must be covered immediately it is retrieved from the sediment to prevent any back re-equilibration. One lander application has designed a helical gel probe in a cylinder that encloses a core tube. A plastic sheath is automatically drawn over the assembly on retrieval [115]. Another lander application uses a flat perspex cover which is automatically positioned upon retrieval over the measurement window of a conventional flat probe [111].

3 DGT IN WATER

3.1 PRINCIPLES

The DGT technique is based on a simple device which accumulates solutes on a binding agent after passage through a hydrogel which acts as a well-defined diffusion layer (Figure 8) [13]. A binding agent such as a resin, selective to the target ions in solution, is immobilised in a thin layer of hydrogel (binding-gel). It is separated from solution by an ion-permeable hydrogel layer (diffusive gel) of thickness Δg. Between the diffusive gel and the bulk solution there is a diffusive boundary layer (DBL), of thickness δ where transport of ions is solely by molecular diffusion. Within a few minutes of immersion, a steady-state linear concentration gradient is established between the solution and the resin gel. By exploiting this simple steady-state condition the DGT technique can be used to measure concentrations *in situ*. The flux, J, of an ion through the gel is given by Fick's first law of diffusion (equation (3)), where D is the diffusion coefficient and $\mathrm{d}C/\mathrm{d}x$ is the concentration gradient.

$$J = D\,\mathrm{d}C/\mathrm{d}x \qquad (3)$$

If diffusion coefficients of ions in the diffusive gel are the same as in water, the flux is given by equation (4), where C is the bulk concentration of an ion and C' is the concentration at the boundary between the binding-gel and the diffusive gel.

$$J = D(C - C')/(\Delta g + \delta) \qquad (4)$$

If the free metal ions are in rapid equilibrium with the binding agent, with a large binding constant, C' is effectively zero, providing the binding agent is not

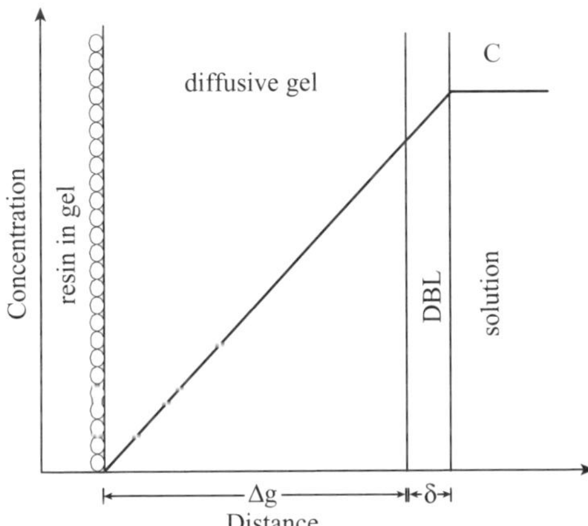

Figure 8. Conceptual view of the steady state concentration gradient of a solute through a DGT device deployed in a well stirred solution with solute concentration, C

saturated. In well-stirred solutions the boundary layer thickness, δ, is negligibly small compared with the thickness of the diffusive layer, Δg of ~ 1 mm. Equation (4) then simplifies to equation (5):

$$J = DC/\Delta g \tag{5}$$

In practice the DGT device is deployed for a fixed time, t. On retrieval the binding-gel layer is peeled off and the mass of the accumulated ions in this layer is measured. The mass can be measured directly in the binding-gel layer by drying it and using a beam technique such as PIXE [19] or in the case of radionuclides by direct counting [103]. More commonly, ions in the binding-layer are eluted with a known volume, V_e, of solution (1 or 2 M HNO_3 in the case of metals bound to Chelex resin) [13,14,17]. The concentration of ions in the eluent, C_e, is then measured by any suitable analytical technique after appropriate dilution. As the elution is in batch mode only a fraction of the bound ions are retrieved. The ratio of the eluted to bound metal is known as the elution factor, f_e. Values of f_e of 0.8 have been reported for Zn, Cd, Cu, Ni and Mn and 0.7 for Fe when using 1 or 2 M HNO_3 [116] to elute from Chelex resin. They were obtained by loading a known amount of metal onto a resin-gel by immersing it directly for 24 h in a stirred solution and measuring metal concentrations before and after immersing. Ideally values of f_e should be measured by each individual research group, especially if conditions such as acid strength are varied. Taking the elution factor into account the accumulated mass (M) of ions in the binding-layer can be calculated from equation (6) where V_g is the volume of gel in the binding-layer:

$$M = C_e(V_g + V_e)/f_e \tag{6}$$

M can be used to calculate the flux through the known area of the exposed diffusive layer, A (equation (7)):

$$J = M/At \tag{7}$$

Equating equations (5) and (7) and rearranging gives equation (8) which demonstrates that the concentration in the bulk solution can be calculated from the known values of Δg, D and A, the measured deployment time, t, and accumulated mass, M:

$$C = M\,\Delta g/DtA \tag{8}$$

This feature of DGT whereby concentration is calculated from the measured mass and deployment time make it ideal for *in situ* use. The relationship of external concentration to measured mass is determined by the values of Δg and A which are simple, fixed geometric quantities and the diffusion coefficient in the gel which can be measured for each temperature. It does not depend on the concentration of other components in the solution and therefore individual calibration in different media is unnecessary.

In most of the work reported to date the DGT devices have used gels and filters which permit free diffusion of ions with diffusion coefficients indistinguishable from their values in water. The diffusive boundary layer thickness in solution adjacent to the filter has been negligible because of effective mixing. These conditions facilitate the simple use of equation (8). For the case where the diffusive boundary layer thickness is not negligible and the gel (D_g), filter (D_f) and water (D_w) are all different, equation (9) applies, where Δg and Δf are the thicknesses of the gel and filter layers respectively [117].

$$\frac{1}{M} = \frac{1}{CAt}\left(\frac{\Delta g}{D_g} + \frac{\Delta f}{D_f} + \frac{\delta}{D_w}\right) \tag{9}$$

3.2 METHODOLOGY

3.2.1 Gel Preparation

The DGT technique relies on diffusion through the gel layer being reproducible. Depending on the proportions of cross-linker used and its composition, hydrogels based on acrylamide can have a wide range of pore sizes which determine effective diffusion coefficients. The hydrogel most commonly used for DGT is physically robust. When an appropriate proportion of agarose-derived cross-linker is used in its fabrication it permits free diffusion of simple solutes, with diffusion coefficients for metal ions being indistinguishable from those measured in water [113]. The diffusion coefficient can be less than in water and is sensitive to the amount and precise formulation of the cross-linker. The chemicals used and the preparation of the diffusive gel are identical to those given for DET (Appendix I).

A procedure for preparing a binding-gel which incorporates Chelex resin that can be used for the measurement of trace metals is given in Appendix II. Binding-gels for measuring phosphate and sulfide can be prepared using the same procedure as in Appendix II with substitution of ferrihydrite [90] and AgI [118] particles, respectively, for the Chelex resin. General cation exchange (AG 50W-X8, Bio-Rad) and anion exchange resins (AG 1-X8, Bio-Rad) have been used to measure Cs [102] and Sr [102], nitrate [119] and ammonia [119] in freshwaters. A selective resin (AMP) has been used for measuring Cs in seawater [103]. These resin-gels can only be set by using a higher temperature (60 °C) and higher concentrations of ammonium persulfate and TEMED than those used for the Chelex gel.

3.2.2 DGT Assembly

The gel assembly must be deployed in a holder which ensures that only the diffusive layer contacts solution. One design based on a simple tight fitting

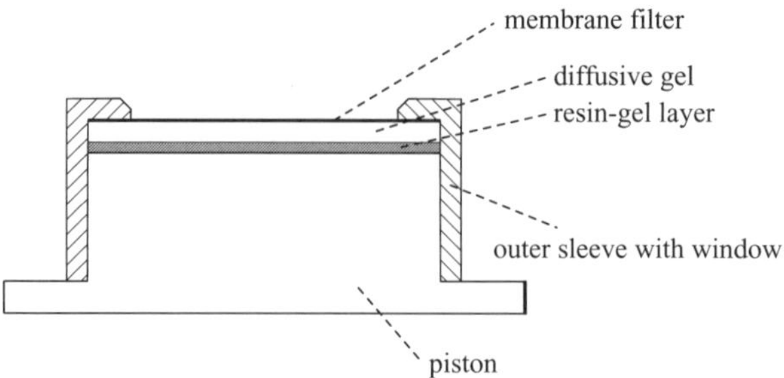

Figure 9. Schematic representation of a plastic DGT holder based on a simple piston design

piston (Figure 9) [13] is commercially available (DGT Research, Ltd, UK). It consists of a backing cylinder and front cap with a 2.0 cm diameter window. A layer of resin-gel is placed on top of the piston face, ensuring that the side with the settled resin faces upwards. A layer of diffusive gel is placed on top of it. For field applications a wet 100μm thick, 0.45μm Millipore cellulose nitrate filter or Gelman polysulfone filter is placed on top of the diffusive gel. The cap is then placed on top and pressed tight so that the layers are in good contact with one another and there is a good seal between filter and cap. It is essential that there are no trapped air bubbles to impede diffusion. The filter has been shown to behave as an extension of the diffusive gel layer, there being no significant difference in measured diffusion coefficients in a gel only and a gel plus filter combination [102,113]. However, when filter membrane alone is used a non-theoretical response has sometimes been observed. The filter generally protects the gel and minimises adhesion of particles to the surface of the device. Diffusive gels with typical thicknesses of 0.8 or 0.4 mm are used for DGT assemblies. It is essential that diffusive gels are conditioned in an electrolyte solution, such as $NaNO_3$ (0.01–0.1 M) prior to deployment. This prevents the possibility of a junction potential across the gel affecting diffusion [113].

3.2.3 Performance Test

A typical routine to test the performance of DGT assemblies prior to field deployment is given Appendix III. It is essential that each laboratory performs such a check. Agreement between the concentration measured by DGT and that calculated using equation (8) and the known solution concentration verifies that the correct experimental procedure is being used and that all components of the gel assembly are working correctly. The most likely causes of disparity are poor analysis, inadequate stirring and poor gel assembly allowing leakage or entraining air. Changes in the diffusion coefficients of solutes in the gel or filter are less likely causes, but would be detectable.

3.3 THEORETICAL RESPONSE

The basic principles of DGT have been verified in the laboratory. When gel assemblies with a single diffusive gel layer thickness were deployed in stirred $CdCl_2$ solutions for various times, their measured mass increased linearly with time (Figure 10i). Similarly when different diffusive layer thicknesses were used and deployment time was fixed, the measured mass was inversely proportional to gel layer thickness (Figure 10ii). In both cases the experimental data points agreed with the theoretical response calculated from the known concentration in solution, C, using equation (8). The diffusive layer thickness (gel layer plus filter), Δg, the exposed surface area, A, and the deployment time, t, were measured directly. The value of the diffusion coefficient in the gel and filter, D, was taken to be the same as that in water. Direct measurements using a diffusion cell with simple metal salt solutions have confirmed that the diffusion coefficients of metal ions in gel and filter assemblies are indistinguishable from values in water, indicating that there is no reaction between metal ions and the gel [35,113]. By contrast the diffusion of phosphate is slightly impeded by the gel (71 % of the value in water) [90]. Typical values for the diffusion coefficients of metal ions in water are given in Li and Gregory [120]. Corrections for temperature can be made using the Stokes–Einstein equation linking the temperature dependence to viscosity, as detailed in Zhang and Davison [13]. In freshwater the same value will apply, but in seawater they will be about 8 % lower. Clearly, as values of diffusion coefficients change with temperature, the DGT response is temperature dependent. If the structure, chemical reactivity and dimensions of the gel do not vary with temperature, the DGT temperature response should solely depend on the diffusion coefficient. This has been found to be the case. The measured accumulated mass was theoretically predicted [13] using equation (8) with the diffusion coefficients in the gel from 5 to 35 °C.

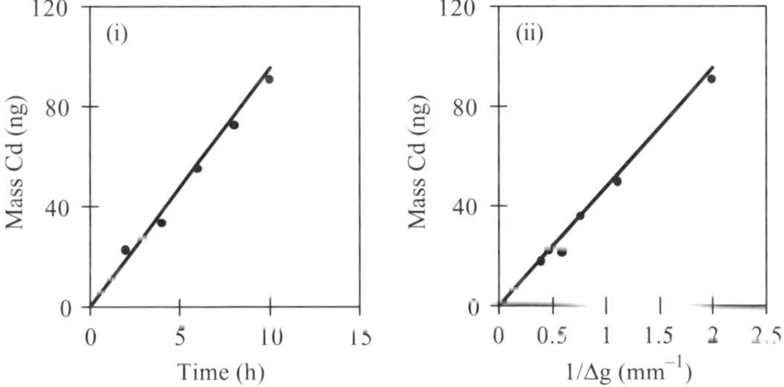

Figure 10. Measured mass of Cd in the resin layer for gel assemblies immersed in a stirred $CdCl_2$ solution (9 nmol L^{-1} Cd) at 22 °C (i) for different times; (ii) for different diffusive gel layer thicknesses exposed for 10 h. The solid lines are predicted by equation (8)

While metal ions diffuse freely through the gel described above, they may be impeded and have lower diffusion coefficients in gels with different proportions of cross-linker or if they are cross-linked with bis-acrylamide [100,113]. The diffusion properties depend on the precise gel composition. Providing the DGT units are well calibrated, by effectively determining the diffusion coefficient through the gel, they can still be used to measure metal ions accurately. With good temperature control and fixed stirring rates the accuracy and precision of replicate DGT measurements is limited only by the analytical method used and can be as low as $\pm$ 2 % [100].

3.4 EFFECT OF IONIC STRENGTH AND pH

Providing the diffusive gel of a DGT device is pre-conditioned in an electrolyte, the dependence of the DGT response on ionic strength reflects only changes in the diffusion coefficient. The 8 % lower diffusion coefficients in seawater than in pure water is attributed to the difference in viscosity [120]. Diffusion coefficients in the DGT gel in 0.7 M $NaNO_3$ solution have been measured to be about 10 % lower than in pure water [113]. Previous agreement between the DGT response assuming the D value for pure water and the direct measurement in 1 M $NaNO_3$ solution [13] is thought to be an artifact due to errors (signal suppression) associated with using ETAAS to measure metals in solution with high salt contents. Using the *D* value for pure water a 10 % lower value for DGT in 1 M $NaNO_3$ would be expected. If total cation concentrations are less than 2.10^{-4} M, diffusion coefficients of metal ions and the DGT response may be enhanced due to the diffusion of other ions and the requirement of electroneutrality (Tessier *et al*, pers. comm.).

The DGT response to solution pH depends on the properties of the binding agent determinand. Using Chelex-100 to bind cadmium, it is virtually independent of pH between pH 5 and 9. At pH 4 the response is still 90 % of theoretical, but at pH 3 or lower binding to Chelex is ineffective. At pH > 9 further swelling of the diffusive gel [113] may affect the DGT response.

3.5 EFFECT OF SOLUTION FLOW AND BIOFOULING

The flux of metal diffusing through the resin-gel of a DGT assembly will depend on the total thickness of the diffusive gel layer and the diffusive boundary layer (DBL) in solution. In well-stirred solutions the DBL must be negligibly small because the theoretical response is obtained. If there is no stirring in the solution or there is no flow during *in situ* deployment a lower response is obtained [13,117]. When DGT devices ($\Delta g = 0.8$ mm) were placed in a flume with their faces parallel to the flow the response was virtually independent of flow (within 5 %) for all measurable flow velocities (0.02–0.26 m. s^{-1}) [121]. It is expected, then, that the DBL will be negligibly small in rivers, estuaries and the well-mixed waters of seas and lakes. Deployment of devices with two or more different diffusion layer thicknesses permits calculation of concentration and DBL thickness irrespective of flow. When the DBL thickness, δ, is significant, equation (10) applies [13]. If measurements are made using several different gel layer thicknesses a plot of $1/M$ versus Δg should be linear with a slope $1/DCtA$ and an intercept $\Delta g/DCtA$ (equation (10)), enabling C and δ to be calculated. If the solution flow varies during deployment, δ will be an effective mean value. Clearly, where there is a significant flow all the devices with various Δg values should have a similar orientation

$$\frac{1}{M} = \frac{\Delta g}{DCtA} + \frac{\delta}{DCtA} \tag{10}$$

to ensure comparability.

Zhang *et al.* [90] used this procedure to calculate phosphate concentrations in a still pond water where the DBL was estimated to be 386 μm. Beaulieu [122] deployed DGT assemblies with three different gel layer thicknesses in different depths in several lakes. In each cases, plots of $1/M$ versus Δg were linear, enabling estimation of the DBL thickness in the range of 0.15 to 0.74 mm. With two gel layer thicknesses, Δg_1 and Δg_2, C can be calculated from the measured values of M_1 and M_2 (equation (11)) irrespective of the magnitude of the DBL thickness [90].

$$\frac{1}{M_1} - \frac{1}{M_2} = \frac{\Delta g_1 - \Delta g_2}{DCtA} \tag{11}$$

Simultaneous deployment of two or more devices, with different gel layer thicknesses, may also allow correction for bio-fouling although field testing is required. If the effects of bio-fouling are restricted to physically altering the surface they can be likened to a change in δ. Providing this surface modification is the same for all devices and that the effective surface area is unaltered, it will not affect the calculation of concentration. Equation (11) should not be used blindly. If concentrations calculated by equation (8) for the individual devices are different the system should be investigated to determine the likely cause before using equation (11). The accumulated mass measured by DGT may not in practice be very sensitive to bio-fouling for some metals. The mass of metal accumulated by DGT is very large compared with metal accumulated by biofilms and it is unlikely that a biofilm can significantly affect the steep diffusion gradients created by the device. The major effect might be enlargement of the diffusion layer thickness, but biofilms thicker than 100 μm would be required to alter the final DGT measurement appreciably.

3.6 CAPACITY

DGT has been shown to behave in accordance with the above theory provided the capacity of the resin to bind ions is not exceeded. If a non selective binding agent is used, other more common ions will quickly accumulate and saturate the binding sites. When a general cation exchange resin was used to measure caesium and strontium the performance was good in synthetic solutions, but in natural waters a linear response between measured mass and time was only found for a few hours [102]. After that time the resin became saturated with major cations such as calcium. Similar results were obtained when general ion exchange resins were used to measure nitrate and ammonia [119]. With a selective binding agent, however, very long deployment times can be used. Chelex is very selective for trace metals. DGT devices for trace metals are estimated to have maximum deployment times in the ocean of about 2 years and several months in more contaminated coastal waters [13]. Laboratory studies with synthetic lake water showed that the accumulated mass of cadmium increased linearly with time for at least 1 month [121].

3.7 SPECIATION

In principle three factors determine which species are measured by DGT, the binding agent, the diffusion layer thickness and the pore size of the gel. Chelex-100 has very strong binding groups, at high effective concentration, which will out-compete most other ligands for metal ions, especially at the low concentrations of organic components present in most natural waters. Metal is continuously removed from solution to the resin. If metal–ligand complexes rapidly dissociate they will contribute to this flux, but if they are kinetically inert they will not. The time available for this dissociation is roughly the time taken for a metal to diffuse through the gel layer which is determined by the gel layer thickness [13]. For a typically 0.4 mm thick gel it is about 2 min. Consequently only labile (dissociation time < 1 min) metal complexes are measured, in an analogous fashion to anodic stripping voltammetry [7,123] (for a detailed discussion of lability criteria see Chapter 8 of this book). If inert species can bind directly to the resin they will also be measured by DGT.

Any metal ion or metal complex measured by DGT has to diffuse through the pores of the gel. The agarose-derived cross-linked (DGT Research Ltd, UK) polyacrylamide gel composition used in the preliminary development of DGT has a very open structure which allows the free diffusion of simple metal ions; the diffusion coefficients of simple cations in the gel are indistinguishable from those in water [13,35,113]. Zhang and Davison [113] have shown that humic and fulvic substances extracted from both waters and soils diffuse through this original composition gel. However, as their molecular diffusion coefficients are appreciably less than that of simple metal ions, the transport of metal humic species to the resin layer in DGT will be retarded compared with simple metal ions. Therefore if DGT with an open gel as the diffusion layer was used to measure metal ions in an unknown natural water, the proportional contribution of any humic complexed metal would not be known. By varying the composition and cross linker of polyacrylamide gel, it is possible to restrict the pore size so that, while simple metal ions (tested for Cd^{2+} and Cu^{2+}) still diffuse freely, the diffusion of fulvic and humic species is very slow. In a DGT device which adopts such a restricted gel the contribution of the metal fulvic and humic complexes to the DGT accumulated metal may be negligible [113]. DGT then only measures small labile species in solution. Although simple organic species, such as acetate complexes, would be included in this small labile fraction, in practice it is likely to be dominated by simple inorganic species.

More accurate speciation measurements can be performed using two DGT devices, one fitted with an open-pore and the other with a partially restricted gel. The diffusion coefficients of metal ions and humic materials must be known in both gels. When the DGT units are deployed in the same solution, the mass of a metal accumulated in the resin layer of the DGT device with an open-pore gel, ${}^{o}M$, and with restricted gel, ${}^{r}M$, can be expressed as [117]:

$$ {}^{o}M = [{}^{o}D_{M}C_{i}) + ({}^{o}D_{H}C_{o})]At/\Delta g \tag{12} $$

$$ {}^{r}M = [({}^{r}D_{M}C_{i}) + ({}^{r}D_{H}C_{o})]At/\Delta g \tag{13} $$

where ${}^{o}D_{M}$ and ${}^{o}D_{H}$, ${}^{r}D_{M}$ and ${}^{r}D_{H}$ are the diffusion coefficients of the metal ion and humic substances in the open-pore and restricted gel, respectively. They can be independently determined using a diffusion cell [113]. If ${}^{o}M$ and ${}^{r}M$ are measured by DGT the concentration of labile inorganic metal, C_i, and the concentration of labile organic metal, C_o, can be calculated by solving equations (12) and (13) simultaneously. If C_i is known the free metal ion concentration can be easily calculated. C_i and C_o together account for the total concentration of labile metal. This may differ from the total dissolved concentration because there may be metal in solution that is inertly bound in complexes or present as colloids. Use of the above equations in natural waters depends on the labile large molecules being dominated by fulvic substances whose diffusion coefficient in the gels can be determined. Generally this is the case as all other large molecules are likely to bind sufficiently strongly that their complexes will not be labile [7]. Although metal fulvic complexes could be non-labile at extremely low metal concentrations when strong binding sites dominate, they are likely to be labile for most practical situations.

3.8 FIELD APPLICATIONS

Applications of DGT are listed in Table 2. Because there are few techniques that have DGT's capabilities of being able to measure labile species *in situ*, it is difficult to verify its performance in the field. When Zn was determined in a seawater sample using anodic stripping voltammetry (ASV) without any sample pretreatment there was good agreement with independent DGT measurements

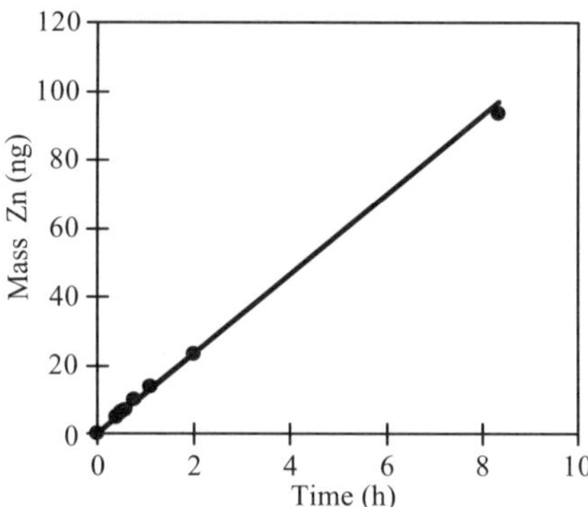

Figure 11. Measured mass of zinc in the resin layer for gel assemblies immersed in a stirred solution of natural seawater (pH 7.8) in the laboratory (28 °C) for different times. The concentration of labile Zn was measured directly by anodic stripping voltammetry (ASV) as 31 nmol L^{-1} and was used in equation (8) to calculate the theoretical line (the solid line)

(Figure 11) [10]. Taking the 10 % lower value of the diffusion coefficient in seawater into account, this result would have shown a 10 % higher figure for the DGT measured concentration. Such a result is consistent with the wider kinetic window of DGT which measures all species able to dissociate in < 1 min compared with < 100 ms for anodic stripping voltammetry [117].

Linear responses for the measured mass with respect to time and $1/\Delta g$ have also been obtained from DGT deployments made *in situ* in coastal water using different deployment times and/or gel layer thicknesses [10]. *In situ* DGT measurements have been made in estuaries where ASV and cathodic stripping voltammetry (CSV) were performed on samples collected from the same location [106]. Good agreement was obtained between free metal ion concentrations measured by CSV or by DGT, assuming DGT measures only labile inorganic species. DGT devices with different thicknesses have been deployed *in situ* in surface North Atlantic water and at 500 m depth [101]. In both cases the measured masses of Cu, Zn, Ni, Co, Pb and Mn increased linearly with $1/\Delta g$. Concentrations were consistent with other measurements of ocean metals.

Metals have also been measured *in situ* in lakes and rivers using DGT [100,121,124]. For some rivers, measurements of Cu and Cd by DGT agreed well with the total filtered (0.45 μm) fraction, indicating the absence of colloidal metal or metal complexed by large molecules (Figure 12). This latter fraction appeared to be present in other rivers where the DGT measured concentration was less [124]. Trace metal concentrations, measured in an unpolluted stream by deploying DGT with a bis-acrylamide cross-linked diffusive gel *in situ* for 10 d, were very low: Cd (0.018 nmol L^{-1}), Co (0.22 nmol L^{-1}), Pb (0.28 nmol L^{-1}), Ni (1.29 nmol L^{-1}), Cu (2.09 nmol L^{-1}) and Zn (0.044 nmol L^{-1}). The DGT

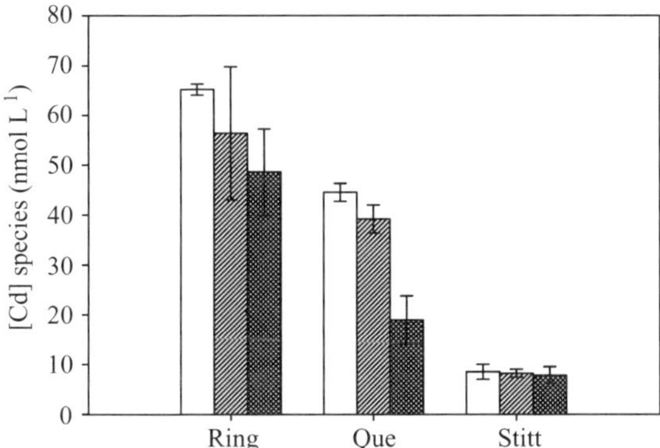

Figure 12. Speciation of Cd in tributary streams of the Pieman River in February 1998 (Mean ± 1/2 range of measured concentrations; No fill = total metal; striped = dissolved metal; hatched = DGT labile metal). Data from reference 124

measured concentrations were 97 % (Cd and Zn), 38 % (Co and Pb), 23 % (Ni) and 5 % (Cu) of the values obtained by measuring metals in a series of filtered samples [100]. Precision, of < 10%, was excellent at these low concentrations. The good agreement between DGT and direct measurement for Cd and Zn, indicates that all species were labile, with little complexation by large organic molecules. It also demonstrates the excellent accuracy of DGT. Suspension of DGT units at different depths in a stratified river enabled measurement of a concentration—depth profile of dissolved manganese (Figure 13) [124]. In such a dynamic system where Mn(II) is continually being produced by reductive dissolution and removed by oxidation the *in situ* capability of DGT is particularly important.

In situ measurements of phosphate by DGT in a small eutrophic pond were reproducible ($3.48 \pm 0.11\ \mu mol\ L^{-1}$, $n = 9$) and about 10 % higher than direct colourimetric measurements of phosphate on a series of filtered samples [90]. Ferrihydrite was used as the binding agent. Its precise preparation was critical to obtain reliable results, although it could be stored for several months at 4 °C.

A general cation exchange resin (AG50W-X8) has been used to measure Cs and Sr in a filtered sample from a softwater lake. The results compared well with direct measurement using ICP MS, providing deployment times were short (< 20 h) [102]. Competition from major cations would make this procedure impractical in hard waters or seawater. However, DGT has been used successfully to measure Cs in seawater using a selective resin AMP (ammonium molybdophosphate) [103]. DGT assemblies with this binding agent provided accurate measurements of ^{137}Cs when they were deployed *in situ* in a lake for 1 and 4 months. Concentrations measured using DGT did not appear to be affected by the development of a biofilm on the filter. After retrieval the

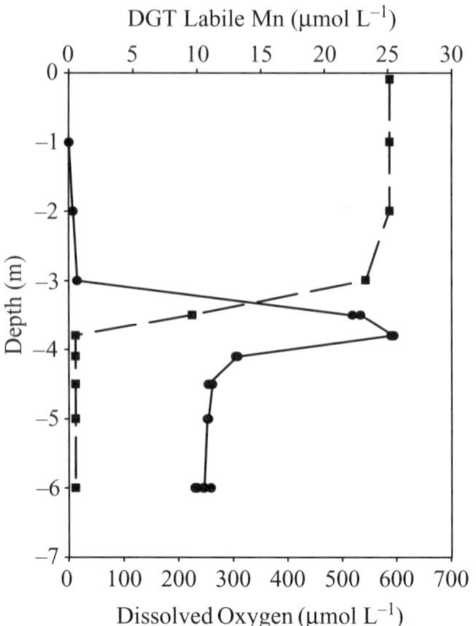

Figure 13. Concentration profile for Mn measured by DGT throughout the water column in the Hopkins River on 12 November 1997 (● = DGT labile Mn; ■ = dissolved oxygen)

resin-gel was simply air dried and placed directly on the gamma counter, eliminating the elution step. Full recoveries were obtained when DGT assemblies were deployed in known solutions, confirming the basic DGT theory (equation (8)) and the free diffusion of ions in the gel, without having to consider elution. Similar theoretical responses have been obtained when Chelex-based DGT assemblies were used to measure radioactive ^{60}Co and ^{65}Zn in synthetic lake water with direct counting of the resin gels.

One of the advantages of using DGT is to eliminate the matrix effect encountered when conventional analytical methods such as AAS and ICP MS are used directly. The only matrix involved in DGT for metal analysis is diluted HNO_3. Another attractive feature of DGT is its ability to pre-concentrate *in situ* which greatly aids the measurement of trace metals by overcoming contamination problems associated with sampling and handling. With the standard gel holder (Figure 9), the minimum practical elution volume is 0.3 ml. When a DGT device with a 0.5 mm thick diffusion layer (0.4 mm diffusive gel plus 0.1 mm filter membrane) is deployed for 24 h the concentration in the eluent is 60 times higher than in the solution. If electrothermal atomic absorption spectroscopy, with a typical detection limit of 1 nmol L^{-1} is used, the DGT detection limit for 24 h deployment will be about 16 pmol L^{-1}.

A novel use of DGT, which bridges applications in water and in sediments discussed in the next section, is its incorporation in sediment traps to provide an *in situ* measurement of remobilization from particles [125]. Deployment at different depths in a water column has provided direct measurements of remobilization fluxes of several trace metals.

4 DGT IN SEDIMENTS AND SOILS

This section discusses the application of DGT in sediments and saturated soils. For simplification the term sediments is used to include both media, unless stated otherwise. The underlying theory of DGT and its methodology are essentially the same as described in section 3. However, the interpretation of DGT measurements is not as straightforward in sediments as in solution. The well-mixed conditions that exist in solutions enable the interpretation of DGT measurements as concentrations (equation (8)). Pore waters are not well mixed and as a result the concentration adjacent to the DGT device may become depleted (cases 2 and 3 in Figure 14). Concentrations interpreted using equation (8) may therefore underestimate pore water concentrations. For this reason, DGT measurements are usually given as time-averaged fluxes from the pore water to the DGT device (equation (7)). Interpreted concentrations must be regarded as estimates unless supported by an independent technique (*e.g.* dialysis, DET, porewater extraction), or the use of multiple DGT deployments with different diffusion layer thicknesses (section 4.2.4).

The applications of DGT in sediments and soils can be subdivided into two types, characterised by different objectives and methodologies:

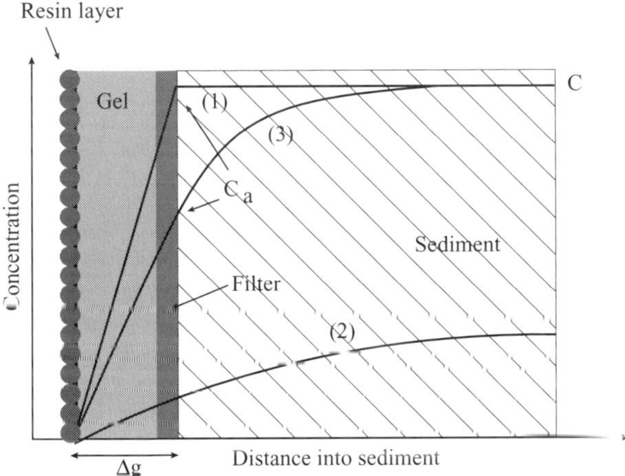

Figure 14. Cross-section through deployment illustrating the sustained (1), unsustained (2), and partially sustained (3) cases

(1) Homogeneous systems (or bulk deployments). The objective is to investigate quantitatively key sediment processes and properties. This is achieved by combining DGT measurements made at different times and gel layer thicknesses in an ideally well-mixed sediment, with independent porewater measurements and using a model of sediment–water interactions to fit the results (section 4.2.4).
(2) Heterogeneous systems (or high resolution deployments). Here the objective is usually to use *in situ* DGT measurements to provide vertical (and/or horizontal) pore water concentration profiles at very high resolution. This approach is discussed in section 4.3.

4.1 METHODOLOGY

4.1.1 Methodology in Homogeneous Systems

The piston-type DGT assemblies used in solution have also been deployed in well-mixed saturated soils [16] (Figure 9). Soils were initially well mixed to ensure homogeneity, placed in a plastic container, and sufficient water added to ensure saturation. The DGT device was pushed gently into the sediment. The dimensions of the container should be sufficient to ensure that the depletion of the porewater by DGT uptake does not progress to the edges of the container ($\geq$ 2 cm between the exposed filter and the container sides will be sufficient for a 24 h deployment, assuming the worst possible case of supply by diffusion alone). It is also important that no air pockets remain between the device and the sediment, and that the sediment does not dry out during the deployment.

The deployment time is usually 1 d. If deployment is significantly extended beyond 1 d, there are risks associated with exhausting the available pool of solute (which reduces the flux to the DGT device), saturating the resin if the concentration of metal is high, and changing the redox conditions due to the additional water. If less than 1 d is used, the DGT device may not attain a time-invariant response, although if resupply from the solid phase is rapid a pseudo-steady state can be attained within 1 h. As unsaturated soils are generally well aerated, they can be homogenised quite simply. Similar homogenisation of sediments would generally be difficult owing to their anoxic or sub-oxic conditions. The resin-gel is then analysed using the procedure described in section 3.1.

4.1.2 Methodology in Heterogeneous Systems

The most commonly used DGT device for obtaining high resolution (HR) vertical pore water concentration profiles was based on the design used for HR DET sampling (Figure 1iv), but with the addition of a resin-gel layer between the perspex back plate and the gel layer, which then acts as the diffusion gel [17].

DGT devices must be deoxygenated prior to deployment in most heterogeneous sediments owing to the potential presence of anoxic or sub-oxic horizons (section 2.2). During the deoxygenation process, any contamination of the deoxygenated water, particularly by the more ubiquitous trace elements such as zinc and copper, will result in their accumulation by the DGT device, and therefore significant contamination of the resin prior to deployment. Deoxygenation also needs to be carried out in an electrolyte similar to the sampled water (*e.g.* synthetic lake water or synthetic seawater) to ensure the correct diffusional properties for the gel [113]. This is normally incompatible with the objective of minimising contamination. The ideal deoxygenation solution would be a filtered sample of the water in which the DGT assembly is to be deployed with the species to be measured removed (*e.g.* trace metals can be removed by the addition of Chelex resin). Synthetic waters of similar ionic strength that have been rigorously cleaned can also be used.

The assembled, deoxygenated DGT device can be deployed in the same way as described earlier for DET devices: namely in cores [17], or *in situ* with insertion by hand from the surface [19], using divers [14], or by landers [111]. As with homogeneous systems, the device has usually been deployed for 1 d, but longer deployments can provide information about the size of the solid phase reservoir [18]. After retrieval the resin-gel is separated from the diffusion layer and sliced horizontally at appropriate intervals by similar methods to those described for DET. Slicing at intervals less than 1 mm may introduce heterogeneity associated with inadequate averaging of solutes attached to the ca. 100μm Chelex beads. A lower slicing limit of 1 mm is common [110,126]. Each slice is then weighed, eluted and analysed separately as described for solution DGT (section 3.1). As the volume of a 1 mm slice is usually less than 10 μL, a small elution volume (*e.g.* 200 μL) must be used to avoid the gain in DGT accumulation being cancelled by dilution.

The use of DGT in very high resolution (VHR) mode, where the spatial resolution may be as low as 100 μm, requires significant changes to the composition of the resin-layer gel and holder design. With HR DGT probes spatial resolution is limited to ~ 1 mm because of the relatively large size (100 μm diameter) of the Chelex 100 beads, and the relatively large probes (3–5 mm total thickness). These problems have been overcome by using a 200 μm thick resin layer [19] containing resin beads, ~ 0.2μm in diameter, overlain by a 200 μm thick diffusion layer. Relative to a conventional DGT device the reduced thickness of the diffusion layer acts to reduce the effect of lateral diffusion (sections 2.1.2 and 4.3), and the reduced resin bead size ensures a more homogeneous distribution on a sub-mm scale.

Details of the preparation of the diffusive and resin gels are given in Appendix IV. The sandwich of resin and diffusive gels and filter was held together by two 250 μm thick Al_2O_3 ceramic plates, one of which had a 1 cm × 7.5 cm window. The whole micro-DGT assembly had a total thickness of only 1 mm.

Accurate slicing of the resin-gel at a resolution of much less than 1 mm is technically very difficult. Instead the resin layer has been removed from the assembly, dried, and the concentrations in the resin layer determined by PIXE (see HR DET procedure, section 2.4.1). PIXE calibration was confirmed to within 10 % using DGT assemblies exposed to known concentrations of metals. Because of its more widespread use and increased sensitivity, laser ablation ICP MS, rather than PIXE, is likely to become the preferred method of VHR DGT analysis.

4.2 PRINCIPLES: HOMOGENEOUS SYSTEMS

As mentioned previously (section 4.1) the interpretation of DGT measurements in sediments is not as straightforward as for solutions. This section discusses the interaction between the sediment and DGT devices and illustrates the differences between interpretations of results from deployments made in solutions (detailed in section 3.1) and sediments. It then describes (section 4.2.3) a possible framework for interpreting DGT measurements in terms of fundamental kinetic and equilibrium parameters.

4.2.1 Sediments versus Solutions

When considering the operation of DGT in sediments it is instructive to review the principles of its use in solutions. The binding-layer binds solutes diffusing through the diffusive gel layer, removing them from solution. After a few minutes this sets up a steady-state linear concentration gradient through the diffusion layer, and therefore a steady-state flux to the binding-layer, enabling the calculation of the solute concentration external to the DGT device (equation (8)). This estimation of concentration requires that the solute concentration adjacent to the DGT device remains constant and effectively equal to that in the bulk solution beyond the influence of the DGT device. For deployments in solution there is a diffusive boundary layer where there is a concentration gradient between the well-mixed water and the device. The above condition is satisfied because this diffusive boundary is often negligibly thin owing to convection. In the porewaters of sediments there are usually no fast mixing processes and therefore the zone of solute depletion adjacent to the device is usually significant. Consequently the principles of using DGT in sediments are quite different from those in solution.

When developing their ideas on how DGT operates in sediments and what it measures, Zhang *et al.* [16,17], and Harper *et al.* [18] first considered, for simplicity, that the porewater and solid phase concentrations were uniform throughout the sediment. Such a situation has been created by homogenising soils [16,127]. Alternatively it might apply to sediments with little heterogeneity and local structure. To interpret DGT measurements in such sediments it is instructive to consider two significant differences from the use of DGT in water.

(1) Because of the lack of mixing it must be assumed that, in general, porewater concentrations adjacent to the DGT device become depleted (cases 2 and 3 in Figure 14). If there is no mechanism of supply of solutes other than diffusion, the zone of depletion adjacent to the DGT device becomes progressively larger with time. The flux to the DGT device, which is driven by the concentration gradient through the gel layer, therefore progressively decreases with time. It has been shown [18] that in the absence of any resupply of solutes to the porewater, after 24 h of deployment of the standard DGT sediment probe $C_a \approx 0.06C$, where C_a is the interfacial pore water concentration between the sediment and DGT device, and C is the labile pore water concentration in the bulk solution (see Figure 14). In practice C_a is usually much greater than this [13,16], implying that significant resupply of solutes is occurring. The local depletion in porewater concentrations is believed to induce remobilisation of solutes from the sediment solid phase.
(2) If it is assumed that C_a remains relatively constant during the deployment as a result of a constant resupply from the solid phase, the theory developed in section 3.1 can be adapted to deployments in sediments, using the interfacial porewater concentration between the DGT device and sediment, C_a, rather than the bulk solute concentration. Assuming steady state throughout the deployment, C_a can be calculated from the measured DGT mass (equation (14)) by analogy with equation (8). Therefore the concentration measured by DGT, C_{DGT}, is the porewater concentration adjacent to the DGT device, C_a, rather than the bulk porewater concentration, C.

$$C_{DGT} = C_a = \frac{M\,\Delta g}{DtA} \tag{14}$$

However in practice the steady-state condition does not hold throughout deployment (*i.e.* C_a changes with time) and the DGT measured concentration, C_{DGT}, effectively is a time-averaged value of C_a. If it is assumed that the flux to the resin layer at any given time is governed by C_a (as per equation (3)), then it is more correct to write:

$$C_{DGT} = \frac{1}{t}\int_{t_i=0}^{t} C_a(t_i)\mathrm{d}t \tag{15}$$

where $C_a(t_i)$ is C_a expressed as a function of time and t is the deployment time. In practice C_{DGT} will increase rapidly in the first few minutes as the steady state is established, and decrease gradually over hours, days or weeks as the capacity of the solid phase to resupply solutes becomes depleted. The initial increase in C_{DGT} has been shown to have negligible effect on the DGT measured concentration when deployments are for greater than 1 h [17] and there is usually negligible decline in C_{DGT} during the next 24 h. Therefore equation (14) usually holds in practice. The greatest difference between C_a and C_{DGT} is in the absence of any solute resupply when, after 24 h, $C_a = 0.06C$, but $C_{DGT} = 0.1C$ [18]. Hereafter the term C_{DGT} is used to represent DGT measured concentrations, but this can, in the great majority of cases, be thought of as the interfacial concentration, C_a.

4.2.2 Buffering

The DGT concentration obtained directly from the DGT measurement, C_{DGT}, will be less than or equal to the concentration of labile species in the pore water, C. Where independent measurements of C are available, DGT measurements can be interpreted in terms of the ratio R (equation (16)).

$$R = \frac{C_{DGT}}{C}, 0 < R < 1 \tag{16}$$

R may be obtained experimentally and used to characterise the deployment as one of the three cases described below. These cases are illustrated in Figure 14, which shows a cross-section through a DGT device during deployment.

(1) *Sustained Case ($R > 0.95$, curve 1 in Figure 14).* The porewater concentration adjacent to the DGT device is sustained at the bulk porewater concentration throughout the deployment. This can occur if (a) solute mixing rates in the sediment are fast (*e.g.* from tidal pumping, bioirrigation), or (b) the rate of resupply from the solid phase is fast compared with the rate of removal to the DGT device, and the capacity of the solid phase to resupply the porewater is large. Efficient porewater mixing will only occur exceptionally in surficial estuarine and riverine sediments so the sustained case will normally imply rapid resupply. The DGT measured concentration can be interpreted as the concentration of labile metal species in the porewaters. Note that the theoretical upper limit, of $R = 1$, is never fully achieved as it would require instantaneous resupply from an infinite capacity reservoir adjacent to the DGT device
(2) *Unsustained case ($R = R_{diff}$, curve 2 in Figure 14).* There is no resupply of solutes to the porewater. The DGT device is therefore supplied only by the diffusion of solutes through the porewaters, which become progressively depleted. R is then at its minimum possible value for the deployment time selected, and is termed R_{diff}. The exact value of R_{diff} depends on the solute diffusion coefficient in the sediment, the design of the DGT device, and the deployment time. Harper *et al.* [18] estimated that, for a typical sampler design in a high porosity sediment, $R_{diff} = 0.1$ (*i.e.* $C_{DGT} = 0.1C$) for a 24 h deployment.

(3) *Partially sustained case* ($R_{diff} < R < 0.95$, *curve 3 in Figure 14*). Significant resupply of the solute from the solid phase occurs, but it is insufficient to sustain fully porewater concentrations. The precise value of R is a quantitative measure of the ability of the solid phase to resupply the porewater in response to the depletion induced by the DGT sink.

4.2.3 *R* Values

The ratio, R, of DGT-measured to bulk concentration depends on a number of factors, including diffusion coefficients, sampler design, and the kinetics and capacity of solid phase resupply. If standardised methodologies and sampler designs are used, R can be interpreted in terms of both the capacity of the solid phase to resupply solutes to the porewater and the kinetics of this transfer. Such interpretation is possible only if simplifying assumptions are made about the interactions of solutes in the porewater with those sorbed to the solid phase. The response of the sediment and DGT device to the latter's deployment can then be modelled. A fuller discussion of the methodology and limitations of this approach is given in Harper *et al.* [18]

Two studies have interpreted R using such methods. Zhang *et al.* [16] proposed the relationship

$$R = K_D k_c \tag{17}$$

where K_D is a distribution coefficient that characterises the partitioning of labile metal between solid phase and pore water, and k_c is a composite constant incorporating the resupply rate constant from solid phase to pore water. The former is thus a capacity term, the latter a kinetic term. Note that k_c will also be dependent on factors such as sampler design and solute diffusion coefficients. Equation (17) yields qualitatively useful information when comparing DGT deployments made with the same equipment in similar media (see below for examples).

An extension of this method was developed by Harper *et al.* [18], based upon a simple model of solute exchange between solid phase and pore water, commonly used in fitting experimental sorption and desorption data [128,129]. The DIFS (DGT-induced fluxes in sediments or soils) model used by Harper *et al.* [18,130] is available as PC software from the authors. Solute exchange is considered to be a first order reversible process between a dissolved phase (porewater concentration, C_d) and a solid phase (sorbed concentration, C_s), described by equation 18 where k_f and k_b are the rate constants for sorption and desorption, respectively.

$$C_d \underset{k_b}{\overset{k_f}{\rightleftarrows}} C_s \tag{18}$$

The modelled response of the DGT device (R) was quantitatively related to K_{D} and the response time t_{c} of the (de)sorption process. t_{c} is defined (equation (19)) as the reciprocal of the sum of k_{f} and k_{b}.

$$t_{\mathrm{c}} = \frac{1}{k_{\mathrm{f}} + k_{\mathrm{b}}} \tag{19}$$

The relationship between R and K_{D} depends on t_{c}. Figure 15 shows this relationship for an assumed one-dimensional (1D) system (*i.e.* assuming an infinite planar surface for the DGT device). It illustrates two main points.

(i) For the sustained case to apply, when R must be greater than 0.9, K_{D} needs to be greater than or equal to $3 \times 10^4\,\mathrm{mL\,g^{-1}}$ and $t_{\mathrm{c}} \leqslant 4.2\,\mathrm{s}$. This quantifies the requirements of fast resupply from a solid phase, labile pool of large capacity.

(ii) As the speed of the resupply increases ($t_{\mathrm{c}} \rightarrow 0$) then R approaches a limiting curve best exemplified in Figure 15 by the $t_{\mathrm{c}} = 10^{-2}$ s line (A). The R values along this line can be related simply to $K_{\mathrm{D}}(\mathrm{mL\,g^{-1}})$ by a sigmoidal function (equation (20)):

$$K_{\mathrm{D}} = \left(\frac{1-R}{R}\right)^{a} \mathrm{e}^{b} \tag{20}$$

The R values for the 1D approximation presented in Figure 15 are, in most cases, within 1 % of those calculated for a more realistic 2D domain based on a

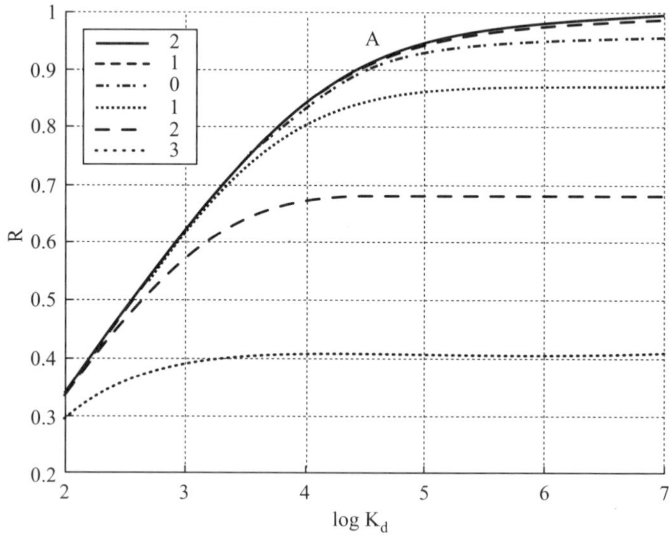

Figure 15. Plot of R against $K_{\mathrm{D}}(\mathrm{mL.g^{-1}}$ for different values of t_{c} (s). Log t_{c} is given in the inset. The line 'A' is the one fitted by equation (20)

Table 3. Coefficients a and b for equation (20)

	Porosity		
	0.6	0.75	0.9
$\Delta g = 0.53\,mm$			
$D_o = 1 \times 10^{-9}\,m^2\,s^{-1}$	$a = -1.940$	$a = -1.951$	$a = -1.959$
	$b = 5.357$	$b = 5.798$	$b = 6.639$
$D_o = 5 \times 10^{-10}\,m^2\,s^{-1}$	$a = -1.941$	$a = -1.954$	$a = -1.966$
	$b = 4.675$	$b = 5.111$	$b = 5.941$
$\Delta g = 0.33\,mm$			
$D_o = 1 \times 10^{-9}\,m^2\,s^{-1}$	$a = -1.909$	$a = -1.925$	$a = -1.946$
	$b = 6.302$	$b = 6.749$	$b = 7.591$
$D_o = 5 \times 10^{-10}\,m^2\,s^{-1}$	$a = -1.900$	$a = -1.921$	$a = -1.941$
	$b = 5.613$	$b = 6.058$	$b = 6.893$

standard sediment probe. The values of the constants a and b depend on geometric and transport terms including diffusion coefficients, diffusion layer thicknesses and porosities. A range of values for a and b are given in Table 3. D_o is the solute diffusion coefficient in the diffusion layer; t_c was taken to be 0.001 s.

R values obtained experimentally may not necessarily correspond to the theoretical definition given above. The species measured by DGT may not be exactly the same as those measured by the alternative technique for measuring porewater concentrations, leading to a degree of bias in the measured R values. However, the frequently observed clustering of data around $R = 1$ [16] suggests that such bias is often less significant than the random error, which effectively limits the precision of R value measurements to ~ 0.05. It is for this reason that the sustained case should practically be defined as $R \geq 0.9$ or 0.95. Figure 15 shows that estimates of K_D and t_c are very sensitive to small changes in R. Consequently the low precision associated with R value measurements translates into a large range of uncertainty for K_D and t_c.

Quantitative interpretations based on 1D numerical models, such as those used to generate Figure 15 and the values in Table 3, require that the deployment conforms as closely as possible to a 1D case. Porewater concentrations adjacent to the DGT assembly are depleted by the DGT device. Close to the edges of the exposure window this depletion may be less than that in the centre owing to enhanced resupply by diffusion from a greater area of sediment. At the centre, resupply will be essentially 1D, whereas it will be 2D at the edges. The distance from the edge of the exposure window over which supply will be 2D will approximately equal the distance, x, to which porewater concentrations are depleted by the DGT device. This has been estimated [18] to be $3\,\Delta g(1 - R)/R$. For typical DGT assemblies with $\Delta g = 0.5\,mm$, and $R = 0.3$, 0.6, and 0.9, x will be 3.5, 1 mm and 0.17 mm respectively. In order to ensure that 1D interpretations are appropriate, it is advisable to discard an appropriate amount of the resin layer next to the exposure window. However, in practice it is likely that there will be significant benefit only in the unsustained case (Figure 14, case 2).

It is important to appreciate that, in terms of the interpretations presented above, K_D is based on a reservoir of labile metal associated with the solid phase that is available *in situ*. This may differ from more conventional measurements of K_D [131].

4.2.4 Pore Water Concentrations

Although in general the DGT-measured concentration, C_{DGT}, is less than the bulk concentration, C, it is possible in principle to estimate C from several DGT measurements using devices with different diffusion layer thicknesses, Δg. Consider the partially sustained case (Figure 14, case 3) where there is some resupply, but the time-averaged flux to the DGT device, J, is less than the maximum possible flux, J_{max}, which occurs only in the sustained case, when the concentration gradient in the diffusion layer is maximal. J is less than J_{max} because of the inability of the sediment to resupply solute at a sufficient rate to satisfy the DGT demand. J is therefore limited to J_{sed}, the maximum flux from solid phase to porewater. Consider now what happens as Δg increases. J_{max} will decrease, but the potential rate of resupply from the solid phase remains the same. Eventually, when Δg reaches a critical value (Δg^*), J_{max} will be sufficiently low to be satisfied by J_{sed}, resulting in the sustained case. This reasoning can be expressed in a general form:

$$\begin{aligned} J &= J_{sed} < J_{max}, \quad \forall \Delta g < g^* \\ J &= J_{max} \leq J_{sed}, \quad \forall \Delta g \geq \Delta g^* \end{aligned} \tag{21}$$

If Δg is greater than or equal to some critical thickness, the deployment satisfies the sustained case, but otherwise it satisfies the partially sustained case. There are two drawbacks with this argument. Firstly, the sustained case can never truly be realised, so R can only approach 1. The concept of a critical Δg must be understood in light of this. Secondly, any critical value of Δg may fall outside the typical range of diffusive layer thicknesses used in DGT devices (0.3–3 mm). We can exploit the relationship between DGT-interpreted concentration and Δg to assess the extent of resupply and in some cases to estimate porewater concentrations. For example, consider Figure 16 which plots DGT-estimated concentration, C_{DGT}, against $1/\Delta g$. The solid line shows the theoretical result if the critical diffusion layer thickness, Δg^*, falls within the range of Δg in the deployed devices. It is clear that for $\Delta g > \Delta g^*$, C_{DGT} is equal to C and independent of Δg, but for $\Delta g < \Delta g^*$, $C_{DGT} < C$. The dashed line shows the result that may be expected if Δg^* falls outside the usual range of Δg (*i.e.* to the

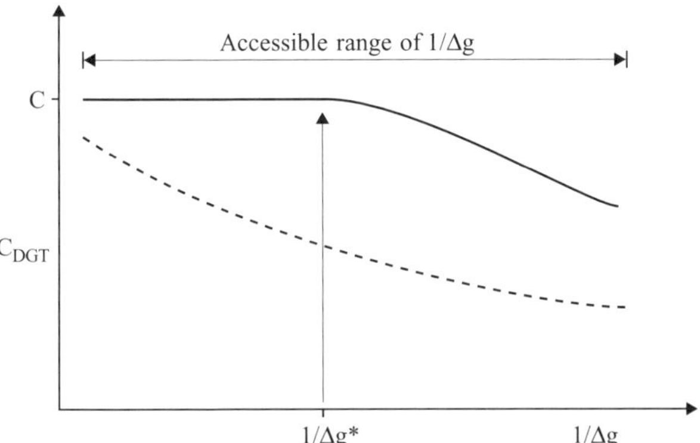

Figure 16. Schematic plot of C_{DGT} against $1/\Delta g$. Two cases are shown, where Δg^* is within the range of Δg used (solid line), and where the threshold diffusion layer thickness, Δg^* (see text for definition), is greater than the largest Δg used (typically 3 mm; dashed line)

left of the lines in Figure 16) in the DGT devices. C_{DGT} increases as Δg increases, but the sustained case is not attained. In this case we observe that, as $\Delta g \to \infty (1/\Delta g \to 0)$, $C_{DGT} \to C$ and $R \to 1$. The best estimate of C therefore comes from extrapolating the dashed line to the Y-axis (where $\Delta g = \infty$, and consequently $C_{DGT} = C$). These methods of estimating C must be used cautiously and are discussed in more detail by Harper *et al.* [18]

4.3 PRINCIPLES: HETEROGENEOUS SYSTEMS

The steep chemical gradients and fine scale structures frequently encountered close to the sediment–water interface [132] necessitate sampling at very fine spatial resolution. By slicing the resin-gel prior to analysis it is relatively easy to use DGT to make measurements of solutes at high vertical resolution. DGT has been most frequently used in sediments to investigate vertical structure. Typically, high resolution (HR) DGT measurements are reported as vertical flux profiles at a resolution of $\sim$ 1 mm, although measurements on the scale of 1–5 mm can be classed as HR when compared with conventional techniques, such as dialysis, which usually sample at 1 cm resolution. The interpretation of results in a system which exhibits structure in solute concentrations differs from that in a homogeneous system.

Fine scale structure in sediment porewater concentrations is created by the action of localised sources, such as the release of trace metals from decomposing organic matter. Persistent concentration gradients are created between these sources and sinks, which could include a well-mixed overlying water column, sorption to a solid phase or precipitation. As the principal objective in using

DGT in HR (or VHR) mode is usually to reproduce fine scale structure in vertical porewater concentration profiles, two aspects of DGT performance are of particular importance: its ability to represent accurately the shape of concentration profiles in the sediment; and its ability to measure accurately porewater concentrations. These can be termed 'profile shape fidelity' and 'concentration fidelity' respectively.

4.3.1 Profile Shape Fidelity

Profile shape fidelity refers to the extent to which DGT-measured profiles accurately reflect the shape of porewater concentration profiles. Fidelity as represented in Figure 17 is defined as the ratio of the actual porewater peak width to the DGT-measured peak width, expressed as a percentage. Thus a fidelity of 50 % indicates that the width of the DGT peak is twice that of the porewater peak. Relaxation of the profile shape is caused by the *in situ* process of *lateral diffusion* within the diffusion layer which has been described previously in the context of DET (section 2.1.2). Lateral diffusion occurs in the vicinity of steep concentration gradients. The effect is less important in DGT than DET owing to immobilisation of the solute by the resin layer. The smallest scale feature measurable using HR DGT depends on the thickness of the diffusion layer. A very sharp peak in porewater concentrations adjacent to the DGT device will spread as it diffuses through the diffusion layer, before being immobilised by the resin. Because of lateral diffusion then, the sharp pore water peak is ultimately represented in the resin layer by a broader peak.

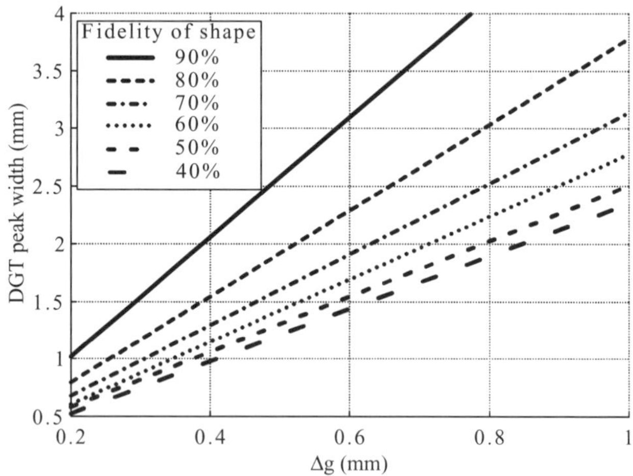

Figure 17. Profile shape fidelity, expressed as a percentage, as a function of Δg and width of the DGT profile peak (width equals 4σ of the assumed Gaussian shape of the peak). 'Good' fidelity is arbitrarily defined as being above the 80 % line

Experimental quantification of the profile shape fidelity at the millimetre scale is technically very difficult, as it would involve accurate measurement of *in situ* DGT and solute pore water profiles at a micrometre scale. An alternative approach has been to model numerically the interaction between DGT and sediment [133]. Although the use of a model inevitably oversimplifies the complexities of DGT–sediment interactions, it can serve to give a useful indication of the reliability of DGT measurements. The profile shape fidelity depends on the thickness of the diffusion layer (Δg), the vertical scale of porewater structures, and the characteristic time of the processes responsible for generating the vertical structure.

Harper *et al.* [133] assumed a relatively large K_D, a first order reversible exchange between sediment solid phase and porewater (as for the homogenised case, with characteristic time t_c), and a constant input flux which can be varied spatially to represent localised sources. This model was solved in a 2D domain representing the *in situ* deployment of a DGT device. The fidelity of the DGT profile shape relative to the pore water profile was determined for various values of the key parameters such as t_c, Δg and the vertical scale of the source term. A relationship between Δg, the width of a DGT measured peak, and fidelity of shape was then derived. Figure 17 shows the approximate fidelity of shape of DGT peaks, as a function of the width of the DGT peak and diffusion layer thickness, Δg. The peak was assumed to have a normal distribution with its width being expressed as four times the standard deviation.

Fidelity of shape is best for wide porewater peaks. If DGT-measured peak widths are greater than 2 mm, their shape will be reproduced well (fidelity $>$ 80 %) by DGT devices with gel layer thicknesses $\leqslant$ 0.5 mm.

Diffusion modelling of a point source adjacent to the DGT device has shown [133] that for a typical diffusion layer thickness ($\Delta g = 0.5$ mm) the minimum measurable peak width is $\sim$ 1 mm. Consequently spikes of a single data point in a 1 mm resolution profile may represent real features. This analysis was based on a 1D model of porewater structure; in reality diffusion in three dimensions may produce sharper features [134].

4.3.2 Concentration Fidelity

As discussed earlier, DGT-measured concentration profiles underestimate actual porewater concentrations by a ratio, R, which can vary between $\sim$ 0.1 and 1. One method of estimating porewater concentrations from DGT-measurements is to divide the DGT-measured concentration, C_{DGT}, by R. This is possible only if independent measurements of porewater concentrations, and therefore R values, are available at several different depths, and if those R values do not vary significantly with depth. While in homogeneous systems, C_{DGT}/R is by definition C (equation (16)), the action of lateral diffusion means that this is not necessarily the case in the presence of vertically structured porewater concentrations. Notwithstanding these limitations, Harper *et al.* [133] found that the ratio C_{DGT}/R was generally at least 90 % of C for peak widths >10 mm.

4.3.3 Very High Resolution

Very high resolution (VHR) DGT [19] resolves finer scale porewater features than is possible using HR DGT. It also better defines coarser features which are partially degraded at lower resolution, reproducing gradients more accurately and therefore providing more reliable estimates of vertical fluxes.

Because lateral diffusion within the diffusion layer acts to smooth out fine scale structure it is essential that the thickness of the diffusion layer is minimised. For example, with a diffusion layer of 0.1 mm (*i.e.* comprising just the filter) diffusion modelling suggests [133] that a sharp (10 μm) wide porewater peak 'relaxes' by lateral diffusion to become a 0.4 mm wide peak in the resin layer (it would become a 1 mm wide peak with a 0.5 mm diffusion layer). These calculations assume a 1D porewater structure. Variability in two or three dimensions would lead to sharper features in the DGT profile [134]. However, it is still unlikely that single spikes measured at 100μm resolution by VHR DGT represent real porewater features. They are more probably due to methodological problems.

4.4 APPLICATIONS

4.4.1 Homogeneous Systems

The measurement of trace metals in saturated soils by DGT [16] resulted in a range of R values between 0.3 and 1, depending on the specific metal and the treatment of the soil with sludge. It was possible to conclude that Zn and Cd are kinetically highly available in untreated soils. The kinetic availability of Zn and Cd was reduced in sludge-treated soils. No measurements of K_D were made and the R values were not interpreted directly as rate constants.

Similar interpretations have been made for *in situ* DGT deployments in undisturbed surficial lake sediments [17]. Average R values were calculated from vertical profiles of Cd, Cu, and Fe measured by DGT, DET and porewater extraction, and used to illustrate the use of the theory developed [133] for the kinetics and capacity of resupply. Cd was found to conform to the sustained case, whereas Cu and Fe were partially sustained. Generally this type of approach should be restricted to applications where sediments do not show significant vertical structure in porewater or DGT concentration profiles, or other sediment physico-chemical characteristics. Such structure is not consistent with the assumptions made in the development of the theory.

Deployments in unsaturated soils and soil slurries [127] have established the response of DGT to different soil moisture contents. At or above the maximum water holding capacity (MWHC) of the soil, the DGT response was largely independent of moisture content. Below MWHC, the flux to the DGT device decreased with decreasing moisture content. This was thought to reflect different diffusional regimes in the soil: decreasing soil moisture will lead to increased diffusional path lengths (tortuosity) and therefore lower measured fluxes. The flux of metals to the DGT device will therefore decrease. An alternative hypothesis is that low moisture contents may cause a slight dehydration of the diffusion gel, insufficient to change its dimensions, but enough to restrict diffusive supply to the resin. However, the dependence of metal uptake by cress on soil water content is similar to the response of DGT which argues against it being an artifact [135].

DGT assemblies have been deployed *in situ* in sediment traps in lakes [125]. Metal remobilised from the settled particulate material was accumulated by the DGT device. This additional mass could be subtracted from that accumulated from solution. The results were used to quantify the stoichiometric ratios of trace metals released during the reductive dissolution of manganese.

4.4.2 High Resolution Applications

DGT has been used *in situ* to measure high resolution concentration profiles of trace metals [14,17,110], phosphate [119], and sulfide [118]. Zhang *et al.* [17] measured concentrations and fluxes of Zn, Cd, Ni, Cu, Fe and Mn in the surface sediments of Esthwaite Water at 1 mm resolution. Those profiles for Cd, Fe and Mn are shown in Figure 18, together with alternative measurements of porewater concentrations, by porewater extraction and DET. Given the

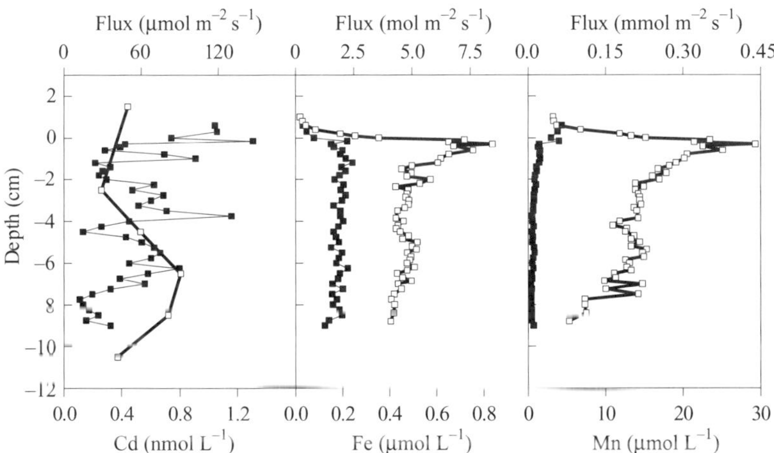

Figure 18. Concentration profiles for Cd, Fe and Mn measured by DGT (■) and an alternative technique (□, DET for Fe and Mn; porewater extraction for Cd)

apparent heterogeneity of Cd in the porewaters, there is reasonable agreement between measurements performed on two separate cores by DGT and extraction of porewaters, especially as the latter method averages over larger distances. Comparison of DGT results with the alternative measurements suggested that Cd is relatively rapidly resupplied from the solid phase, there is partial resupply of Fe and a diffusional only resupply for Mn. The sharp peaks, particularly in the profile of Cd, demonstrated the substantial local remobilisation of trace metals. Spatial separation of the peaks suggested that individual metals may have different mechanisms for remobilisation. While Figure 18 revealed structure suppressed by lower resolution measurements, profile maxima and minima were defined in some cases by only one or two points. It was clear from such work that higher resolution measurements would be needed to resolve these structures fully.

High resolution measurements have generally revealed considerable heterogeneity in the concentrations of solutes in pore waters. Quantitative interpretation in terms of the theory developed for homogeneous systems (Section 4.2) will not be strictly valid when there is substantial structure, although it might aid the conceptual appreciation of the system [133].

4.4.3 Very High Resolution Applications

Figure 19 illustrates VHR (100μm) profiles of Zn, Mn, Fe, and As in a surface sediment of a polluted stream overlain by a microbial mat. The profiles reveal geochemical processes acting on a sub-millimetre scale. They also illustrate the limitations of estimating vertical fluxes using conventional procedures for measuring solutes in porewaters. At a resolution of 1 cm the profiles would be represented by only two data points and flux estimates would most likely be grossly inaccurate. However, the concentration scales should be treated with caution without independent concentration measurements to establish the degree of buffering of porewater concentrations.

By rastering an area of resin with a 1μm PIXE beam, 2D maps of porewater concentration adjacent to the DGT device have been produced [19]. They have shown lateral concentration gradients on a millimetre scale, and the microniche remobilisation of trace metals. Preliminary results indicate that it should be possible to use laser ablation ICP-MS rather than PIXE for such measurements. The greater sensitivity of laser ablation should allow measurement of trace metals in unpolluted systems.

DGT has also been used to measure sulfide in pore waters at very high resolution [118]. Silver iodide rather than Chelex resin was used as the binding agent in the 'resin layer'. The silver iodide gel layer was dried and converted into a grey scale image using a flatbed scanner. Vertical sulfide concentration profiles were estimated by comparing the intensity of the sample image to that of prepared standards containing known concentrations of sulfide.

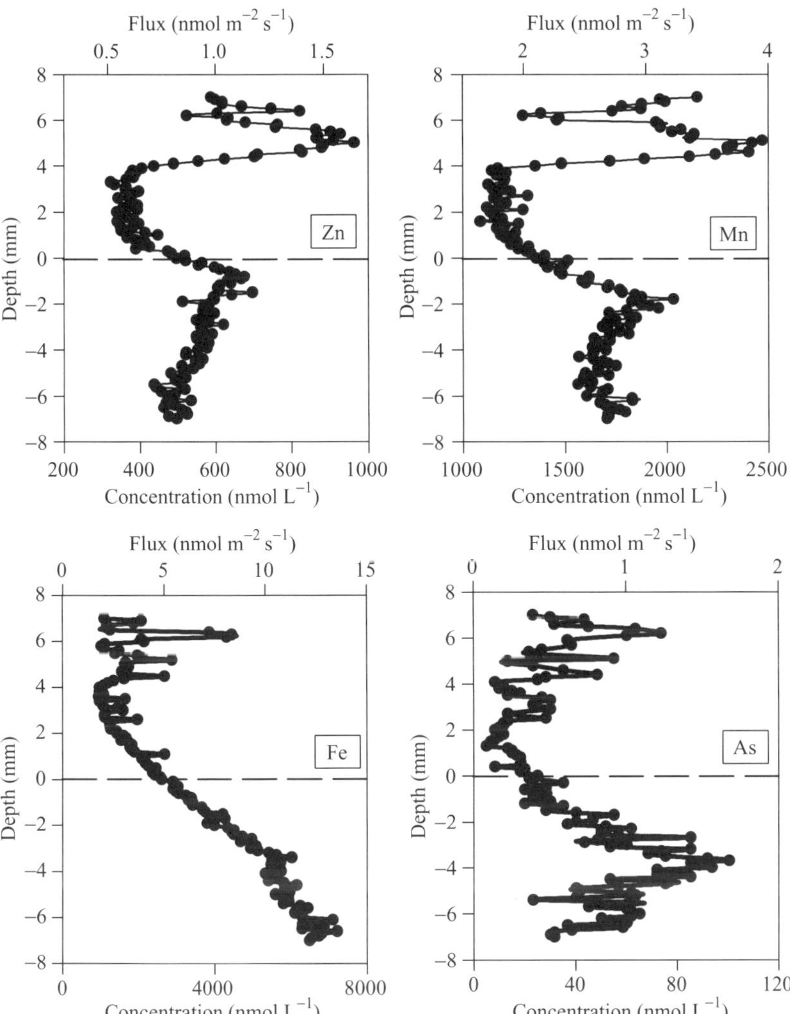

Figure 19. VHR vertical profiles of Zn, Mn, Fe, and As in surface sediment and an overlying microbial mat

5 CONCLUSIONS

The *in situ* use of dialysis has developed pragmatically, leading to many diverse applications, most notably in measuring solutes in pore waters of sediments and the saturated zone. Novel developments include the facility to remove or isolate samples periodically while leaving the dialysis unit in place [72,80] and the *in situ* assessment of microbial response by including organisms in dialysis cells [108]. This type of *in situ* experiment, which dialysis as well as DET and DGT can facilitate, is likely to be important in aiding our understanding of local processes. Only recently has it become clear that the time taken to reach equilibrium for a dialysis assembly placed in a sediment can provide information about the mechanism of solute supply within the system [33,37]. Comparison of results from deployments for different times or using samplers of different geometries may be a useful way to investigate the kinetics of solute resupply from solid phases.

There are two clear virtues of the DET technique: its relatively rapid response due to its thin layer of gel and the ease with which it can be used to perform measurements at high spatial resolution. Measurements at millimetre resolution have revealed a previously unrecognised spatial structure for many solutes in porewaters [15,35,81,82,110]. Such structure has implications for the nature of sediment–solute interactions, including the prevalence and temporal existence of microniches of supply and the rate of removal by sinks. Fine scale vertical and horizontal structure at the sediment–water interface calls into question the use of concentration–depth profiles to estimate fluxes of sediment–water exchange. Here the ease with which DET gels can be segmented into 2D arrays is likely to be particularly important [82]. The resulting 2D maps of porewater concentration are likely to provide (1) assessments of horizontal heterogeneity, (2) estimates of average fluxes at the sediment–water interface and (3) information on element coupling that can be used in mechanistic interpretation.

The highest sub-millimetre resolution attainable involves constraining gel to compartments, as for dialysis [31]. Such procedures are sufficiently time consuming, however, that constrained DET measurements are likely to remain a minority compared with those made with single sheets of gel at somewhat compromised spatial resolution.

DGT can be used to measure the concentrations of chemical species or fluxes from solid phase to solution in bulk media or at very high spatial resolution. Its in-built pre-concentration capabilities and provision of a simple analysis medium, even when used in seawater, make it useful as an analytical tool, but its chemical speciation capabilities are its most desirable features for use in bulk media. Laboratory measurements have indicated that it may be possible to exploit the range of porosities available for diffusive gels of different compositions to distinguish by size between organically and inorganically bound metal ions [117]. There is a clear need for a simple tool that can perform speciation measurements in natural waters. DGT has the potential to fill this gap and to complement alternative procedures, including electrochemical techniques and the use of speciation models.

The capability of DGT to measure fluxes from solid phase to solution in soils and sediments is likely to advance significantly our understanding of solute–solid interactions *in situ* [16,17]. These fluxes can be used to obtain kinetic and thermodynamic parameters, such as *in situ* dissociation rate constants from solid phase to solution and distribution coefficients [18]. However, such interpretations are dependent on the conceptual model of the soil or sediment system and on alternative measurements of porewater concentrations. There is a clear need to improve the data base to assess whether the interpreted quantities are truly constant or whether more realistic models need to be developed. At this early stage in the development of DGT, when there are few reported applications of its *in situ* use in sediments, it is best used in conjunction with alternative techniques, including dialysis and DET, that can provide complementary information.

The automatic fixing of solutes at a particular location on the resin-gel gives DGT the capability of being used to make measurements at high ($\sim$ 1 mm) and very high ($\sim$ 100 μm) spatial resolution in one or two dimensions [17,19]. These details of the spatial variability of the local fluxes of solutes in sediments can provide exciting new information on the structure and interactions of processes. However, it is important to realise that the observed structure may be relaxed from the true pore water profiles, that the true structure of solutes in pore waters will be 3D and that quantitative interpretation in terms of concentrations or fluxes may not be straightforward. Irrespective of these caveats, the potential for using DGT to investigate the structure of solute resupply is tremendous. There is an urgent need for more applications, particularly in soils where there have been no high resolution measurements to date.

Another area where DGT has great potential, but has yet to be applied and tested is its use as a simple surrogate for biologically available solutes. The continuous removal of solutes by DGT resembles the solute transport across a membrane that occurs with some plants and animals. In solution DGT measures labile species which are generally regarded as those that are bioavailable [136]. In sediments and soils, as well as assessing solute concentrations, DGT assesses their availability for resupply. Both these factors may affect availability to biota. The only study performed so far [135] indicated that the DGT response was better correlated to plant uptake of Zn than the Zn concentration in solution. Clearly further systematic uptake and toxicity studies are needed.

In their simplest forms, dialysis, DET and DGT, are all straightforward procedures which can be used in any laboratory using standard micro-analysis techniques. While dialysis has already been used widely to obtain *in situ* environmental information, DET and DGT have yet to be used extensively. Some aspects of DGT are still being developed, particularly with respect to the range of solutes it can be used to measure, although the list of trace metals, Fe, Mn, Cs, Sr, sulfide and phosphate has grown rapidly. With current complementary developments in micro-sensor technology, in appreciation of heterogeneity in soils and sediments and in understanding the role of chemical speciation, dialysis, DET and DGT will undoubtedly be used more widely and are likely to become standard techniques in the armoury of the environmental chemist.

REFERENCES

1. Glud, R. N., Gundersen, J. K., and Ramsing, N. B. (2000). Electrochemical and optical oxygen microsensors for *in situ* measurements. In *In Situ Monitoring of Aquatic Systems; Chemical Analysis and Speciation*, ed. Buffle, J. and Horvai, G., Wiley, Chichester, chapter 2, this volume.
2. Cai, W. J., and Reimers, C. (2000). Sensors for *in situ* pH and pCO_2 measurements in seawater and at the sediment–water interface. In *In Situ Monitoring of aquatic systems*; *chemical analysis and speciation*, ed. Buffle, J., and Horvai, G., Wiley, Chichester, Chapter 3, this volume.
3. Beer, De, D. (2000). Potentiometric microsensors for *in situ* measurements in aquatic environments. In *In Situ Monitoring of Aquatic Systems; Chemical Analysis and Speciation*, ed. Buffle, J., and Horvai, G., Wiley, Chichester, Chapter 5, this volume.
4. Klimant, I, Meyer, V., and Kuhl, M. (1995). Fibre optic oxygen microsensors, a new tool in aquatic biology, *Limnol. Oceanogr.*, **40**, 1159.
5. Davison, W., Zhang, H. and Miller, S. (1997). *In situ* procedures for measuring chemical speciation and availability of trace metals in soils and sediments. In *Contaminated Soils, Proceedings of the Third International Conference on the Biogeochemistry of Trace Elements*, ed. Prost, R., Les Colloques, p. 57.
6. Johnson, K. S., Elrod, V. A., Nowicki, J. L., Coale, K. H., and Zamzow, H. (2000). Continuous flow techniques for on-site and *in situ* measurements of metals and nutrients in sea water. In *In Situ Monitoring of Aquatic Systems; Chemical Analysis and Speciation*, ed. Buffle, J., and Horvai, G., Wiley, Chichester, Chapter 7, this volume.
7. Buffle, J., and Tercier, M-L. (2000). In situ voltammetry; concepts and practice for trace analysis and speciation. In *In Situ Monitoring of Aquatic Systems; Chemical Analysis and Speciation*, ed. Buffle, J., and Horvai, G., Wiley, Chichester, Chapter 9, this volume.
8. Hesslein, R. H. (1976). An in situ sampler for close interval pore water studies, *Limnol. Oceanogr.*, **21**, 912.
9. Davison, W., Grime, G. W., Morgan, J. W., and Clarke, K. (1991). Distribution of dissolved iron in sediment pore waters at submillimetre resolution. *Nature*, **352**, 323.
10. Davison, W. and Zhang, H. (1994). *In situ* speciation measurements of trace components in natural waters using thin-film gels, *Nature*, **367**, 545.
11. Benes, P. and Steinnes, E. (1974). *In situ* dialysis for the determination of the state of trace elements in natural waters, *Wat. Res.*, **8**, 947.
12. Ronen, D., Magaritz, M. and Levy, I. (1986). A multi-layer sampler for the study of detailed hydrochemical profiles in groundwater, *Wat. Res.*, **20**, 311.
13. Zhang, H. and Davison, W. (1995). Performance characteristics of the technique of diffusion gradients in thin-films (DGT) for the measurement of trace metals in aqueous solution, *Anal. Chem.*, **67**, 3391.
14. Zhang, H., Davison, W. and Grime, G. W. (1995). New in situ procedures for measuring trace metals in pore waters, In *Dredging, Remediation and Containment of Contaminated Sediments*, ed. Demars, K. R., Richardson, G. N., Yong, R. N. and Chaney, R. C. American Society for Testing and Materials, Philadelphia, PA, p. 170.
15. Krom, M. D., Davison, P., Zhang, H. and Davison, W. (1994). High-resolution pore-water sampling with a gel sampler, *Limnol. Oceanogr.*, **39**, 1967.
16. Zhang, H., Davison, W., Knight, B. and McGrath, S. (1998). *In situ* fluxes of trace metals in soils using DGT, *Environ. Sci. Technol.*, **32**, 704.
17. Zhang, H., Davison, W., Miller, S. and Tych, W. (1995). *In situ* high resolution measurements of fluxes of Ni, Cu, Fe and Mn, *Geochim. Cosmochim. Acta.*, **59**, 4181.

18. Harper, M. P., Davison, W., Tych, W. and Zhang, H. (1998). Kinetics of metal exchange between solids and solutions in sediments and soils interpreted from DGT measured fluxes, *Geochim. Cosmochim. Acta.*, **62**, 2757.
19. Davison, W., Fones, G. and Grime, G. W. (1997). Dissolved metals in surface sediments and a microbial mat at 100 μm resolution, *Nature* **387**, 885.
20. Prest, H. F., Huckins, J. N., Petty, J. D., Herve, S., Paasivirta, J. and Heinonen, P. (1995). A survey of recent results in passive sampling of water and air by semipermeable membrane devices, *Mar. Pollut. Bull.*, **30**, 543.
21. Buffle, J., Parthasarathy, N., Djane, N-K. and Mathiasson, L. (2000). Permeation liquid membranes for field analysis of trace compounds in aquatic media. In *In Situ Monitoring of Aquatic Systems; Chemical Analysis and Speciation*, ed. Buffle, J. and Horvai, G., Wiley, Chichester, Chapter 10, this volume.
22. Rudd, J. W. and Hamilton, R. (1975). Two samplers for monitoring dissolved gases in lake water and sediments, *Limnol. Oceanogr.*, **20**, 902.
23. Mayer, L. M. (1976). Chemical water sampling in lakes and sediments with dialysis bags, *Limnol. Oceanogr.*, **21**, 909.
24. Fanning, K. A. and Pilson, M. E. Q. (1971). Interstitial silica and pH in marine sediments: some effects of sampling procedures, *Science*, **173**, 1228.
25. Edmunds, W. M. and Bath, A. H. (1976). Centrifuge extraction chemical analysis of interstitial waters, *Environ. Sci. Technol.*, **10**, 467.
26. Carignan, R., Rapin, F. and Tessier, A. (1985). Sediment porewater sampling for metal analysis: A comparison of techniques, *Geochim. Cosmochim. Acta*, **49**, 2493.
27. Brandl, H. and Hanselmann, K. W. (1991). Evaluation and application of dialysis porewater samplers for microbiological studies at sediment water interfaces, *Aquat. Sci.*, **53**, 55.
28. De Lange, G. J., Cranston, R. E., Hydes, D. H. and Boust, D. (1992). Extraction of porewater from marine sediments: A review of possible artifacts with pertinent examples from the north Atlantic, *Mar. Geol.*, **109**, 53.
29. Kaplan, E., Banerjee, S., Ronen, D., Magaritz, M., Machlin, A., Sosnow, M. and Koglin, E. (1991). Multilayer sampling in the water-table region of a sandy aquifer, *Ground Wat.*, **29**, 191.
30. Weisbrod, N., Ronen, D. and Nativ, R. (1996). New method for sampling groundwater colloids under natural gradient flow conditions, *Environ. Sci. Technol.*, **30**, 3094.
31. Fones, G. R., Davison, W. and Grime, G. W. (1998). Development of constrained DET for measurements of dissolved iron in surface sediments at sub-mm resolution, *Sci. Tot. Environ.*, **221**, 127.
32. Teasdale, P. R., Batley, G. E., Apte, S. C. and Webster, I. T. (1995). Pore water sampling with sediment peepers, *TrAC*, **14**, 250.
33. Harper, M. P., Davison, W. and Tych, W. (1997). Temporal, spatial, and resolution constraints for *in situ* sampling devices using diffusional equilibration: dialysis and DET, *Environ. Sci. Technol.*, **31**, 3110.
34. Ronen, D., Magaritz, M., Gvirtzman, H. and Garner, W. (1987). Microscale chemical heterogeneity in groundwater, *J. Hydrol.*, **92**, 173.
35. Davison, W., Zhang, H. and Grime, G. W. (1994). Performance characteristics of gel probe used for measuring pore waters, *Environ. Sci. Technol.*, **28**, 1623.
36. Boudreau, B. P. (1996). The diffusive tortuosity of fine grained sediments, *Geochim. Cosmochim. Acta*, **60**, 3139.
37. Webster, I. T., Teasdale, P. R. and Grigg, N. J. (1998) A theoretical and experimental analysis of peeper equilibration dynamics, *Environ. Sci. Technol.*, **32**, 1727.
38. Grigg, N. J., Webster, I. T. and Ford, P. W. (1999). Porewater convection induced by peeper emplacement in saline sediment, *Limnol. Oceanogr.*, **44**, 425.

39. Mortimer, R. J. G., Krom, M. D., Hall, P. O. J., Hulth, S. and Stahl, H. (1998). Use of gel probes for the determination of high resolution solute distributions in marine sediments, *Mar. Chem.*, **63**, 119.
40. Aller, R. C. (1982). The effects of macrobenthos on chemical properties of marine sediment and overlying water. In *Animal–Sediment Relations*, ed. McCall, P. I. and Tevesz, M. J. S., Plenum, London, Chapter 2, p. 53.
41. Matisoff, G., Fisher, J. B. and Matis, S. (1985). Effects of benthic macroinvertebrates on the exchange of solutes between sediments and fresh-water, *Hydrobiol.*, **122**, 19.
42. Savant, S. A., Reible, D. D. and Thibodeaux, L. J. (1987). Convective transport within stable river sediments, *Wat. Resour. Res.*, **23**, 1763.
43. Webster, I. T. and Taylor, J. H. (1992). Rotational dispersion in porous media due to fluctuating flows, *Wat. Resour. Res.*, **28**, 109.
44. Mortimer, R.J.G., Davey, J. T., Krom, M. D., Watson, P. G., Frickers, P. E. and Clifton, R. C. (1990). The effect of macrofauna on pore-water profiles and nutrient fluxes in the interidal zone of the Humber Estuary. Estua. Coast. and Shelf. Sci., **48**, 683–699.
45. Carignan, R. (1984). Interstitial water sampling by dialysis: Methodological notes, *Limnol. Oceanogr.*, **29**, 667.
46. Carignan, R., St-Pierre, S. and Gächter, R. (1994). Use of diffusional samplers in oligotrophic lake sediments: Effects of free oxygen in sampler material, *Limnol. Oceanogr.*, **39**, 468.
47. Tessier, A., Carignan, R. and Belzile, N. (1994). Processes occurring at the sediment–water interface: emphasis on trace elements, In *Chemical and Biological Regulation of Aquatic Systems*, ed. Buffle, J. and De Vitre, R. R., Lewis, London, p. 139.
48. Teasdale, P. R., Allen, L., Apte, S. C., Batley, G. E. and Birch, G. (1998). *In situ* collection of diagenetic and induced oxyhydroxide precipitates from riverine and estuarine sediments, *Environ. Technol. Lett.*, **19**, 1191.
49. Bottomley, E. Z. and Bayly, I. L. (1984). A sediment porewater sampler used in root zone studies of the submerged macrophyte, Myriophyllum spicatum, *Limnol. Oceanogr.*, **29**, 671.
50. Kepkay, P. E., Cooke, R. C. and Bowen, A. J. (1981). Molecular diffusion and the sedimentary environment: results from the *in situ* determination of whole sediment diffusion coefficients, *Geochim. Cosmochim. Acta*, **45**, 1401.
51. Kipphut, G.W. and Martens, C. S. (1982). Biogeochemical cycling in an organic-rich coastal marine basin—3. Dissolved gas transport in methane saturated sediments, *Geochim. Cosmochim. Acta*, **46**, 2049.
52. Aller, R. C., Hall, P. O. J., Mackin, J. E., Rude, P. D. and Aller, J. Y. (1998). Biogeochemical heterogeneity and suboxic diagenesis in hemipelagic sediments of the Panama Basin, eastern tropical Pacific, *Deep-sea Res.*, **45**, 133.
53. Minnet, A. I. (1994). *Development of in situ measurements of sediment porewaters*, Honours thesis, University of Wollongong.
54. Davison, W. and De Vitre, R. (1992). Iron particles in freshwater. In *Environmental Particles, Vol. 1*, ed. Buffle, J. and van Leeuwen, H. P., Lewis London, p. 315.
55. Carignan, R. and Lean, D. R. S. (1991). Regeneration of dissolved substances in a seasonally anoxic lake: The relative importance of processes occurring in the water column and in the sediments, *Limnol. Oceanogr.*, **36**, 683.
56. Carignan, R. and Nriagu, J. O. (1985). Trace metal deposition and mobility in the sediments of two lakes near Sudbury, Ontario, *Geochim. Cosmochim. Acta*, **49**, 1753.
57. LaZerte, B. D. (1981). The relationship between total dissolved carbon dioxide and its stable carbon isotope ratio in aquatic sediments, *Geochim. Cosmochim. Acta*, **45**, 647.

58. Chanton, J. P., Martens, C. S. and Kelly, C. A. (1989). Gas-transport from methane-saturated, tidal, freshwater and wetland sediments, *Limnol. Oceanogr.*, **34**, 807.
59. Lindau, C. W. (1994). Methane emissions from Louisiana rice fields amended with nitrogen fertilizers, *Soil Biol. Biochem.*, **26**, 353.
60. Martens, C. S. and Klump, J. V. (1980). Biogeochemical cycling in an organic rich coastal marine basin. 1. Methane–water exchange processes, *Geochim. Cosmochim. Acta*, **44**, 471.
61. Gaillard, J.-F., Jeandel, C., Michard, G., Nicolas, E. and Renard, D. (1986). Interstitial water chemistry of Villefranche Bay sediments, *Mar. Chem.*, **18**, 233.
62. Boston, H. L. and Adams, M. S. (1987). Productivity, growth and photosynthesis of two small 'isoetid' plants, Littorella uniflora and Isoetes macrospora, *J. Ecol.*, **75**, 333.
63. Steinmann, P. and Shotyk, W. (1997). Chemical composition, pH, and redox state of sulfur and iron in complete vertical porewater profiles from two Sphagnum peat bogs, Jura Mountains, Switzerland, *Geochim. Cosmochim. Acta*, **61**, 1143.
64. Rettinger, S., Ronen, D., Amiel, A. J., Magaritz, M. and Bischofsberger, W. (1991). Tracing the influx of sewage from a leaky sewer into a very thin and fast-flowing aquifer, *Wat. Res.*, **25**, 75.
65. Casey, W. H. and Lasaga, A. C. (1987). Modelling solute transport and sulfate reduction in marsh sediments, *Geochim. Cosmochim. Acta*, **51**, 1109.
66. Haraldsson, C., Anderson, L. G., Hassellov, M. and Hulth, S. (1997). Rapid, high-precision potentiometric titration of alkalinity in ocean and sediment pore waters, *Deep-sea Res.*, **44**, 2031.
67. Bolliger, R., Brandl, H., Höhener, P., Hanselmann, K. W. and Bachofen, R. (1992). Squeeze-water analysis for the determination of microbial metabolites in lake sediments –Comparison of methods, *Limnol. Oceanogr.*, **37**, 448.
68. Tessier, A., Rapin, F. and Carignan, R. (1985). Trace metals in oxic lake sediments: possible adsorption onto iron oxyhydroxides, *Geochim. Cosmochim. Acta*, **49**, 183.
69. Ankley, G. T. and Schubauer-Berigan, M. K. (1994). Comparison of techniques for the isolation of sediment pore water for toxicity testing, *Arch. Environ. Contam. Toxicol.*, **27**, 507.
70. Mattuck, R. and Nikolaidis, N. P. (1996). Chromium mobility in freshwater wetlands, *J. Contam. Hydrol.*, **23**, 213.
71. Tarutis, W. J., Unz, R. F. and Brooks, R. P. (1992). Behavior of sedimentary Fe and Mn in a natural wetland receiving acidic mine drainage, Pennsylvania, U.S.A., *Appl. Geochem.*, **7**, 77.
72. Bertolin, A., Rudello, D. and Ugo, P. (1995). A new device for in-situ pore-water sampling, *Mar. Chem.*, **49**, 233.
73. Williamson, M. A. and Parnell, R. A. (1994). Partitioning of copper and zinc in the sediments and porewaters of a high-elevation alkaline lake, east-central Arizona, U. S. A. *Appl. Geochem.*, **9**, 597.
74. Viel, M., Barbanti, A., Langone, L., Buffoni, G., Paltrinieri, D. and Rosso, G. (1991). Nutrient profiles in the pore water of a deltaic lagoon: Methodological considerations and evaluation of benthic fluxes, *Estuar. Coast. Shelf Sci.*, **33**, 361.
75. Young, L. B. and Young, H. H. (1992). Geochemistry of Mn and Fe in lake sediments in relation to lake acidity, *Limnol. Oceanogr.*, **37**, 603.
76. Kelly, C. A., Rudd, J. W. M., Furutani, A. and Schindler, D. W. (1984). Effects of lake acidification on rates of organic matter deposition in sediments, *Limnol. Oceanogr.*, **29**, 687.
77. Holdren, G. R., Brunelle, T. M., Matisoff,. G. and Whalen, M. (1984). Timing the increase in atmospheric sulphate deposition in the Adirondack Mountains, *Nature*, **311**, 245.

78. Angelidis, T. N. (1997). Comparison of sediment pore water sampling for specific parameters using two techniques, *Wat. Air Soil Pollut.*, **99**, 179.
79. Mortimer, R. J. G., Krom, M. D., Watson, P. G. and Frickers, P. E. (1996). Sediment-water exchange of nutrients in the Humber Estuary. In *Proceedings of the Fourth International Symposium on the Geochemistry of the Earth's Surface*, ed. Bottrell, S. H., University of Leeds, p. 96.
80. Azcue, J. M., Rosa, F. and Lawson, G. (1996). An improved dialysis sampler for the *in situ* collection of larger volumes of sediment pore waters, *Environ. Technol. Lett.*, **17**, 95.
81. Zhang, H., Davison, W and Ottley, C. (1999). Remobilisation of major ions in freshly deposited lacustrine sediment at overturn, *Aquat. Sci.*, **61**, 354–361.
82. Shuttleworth, S. M., Davison, W. and Hamilton-Taylor, J. (1999). Two-dimensional and fine structure in the concentrations of iron and managanese in sediment pore waters, *Environ. Sci. Technol.*, **33**, 4169–4175.
83. Hall, P. O. J. and Aller, R. (1992). Rapid, small-volume, flow injection analysis for CO_2 and NH_4^+ in marine and freshwaters, *Limnol. Oceanogr.*, **37**, 1113.
84. Wheeler, B. D. and Giller, K. E. (1984). The use of dialysis cells for investigating pore water composition in wetland substrata, with particular reference to dissolved iron and sulphide, *Soil Sci. Plant Anal.*, **15**, 707.
85. Klingensmith, K. M. and Alexander, V. (1983). Sediment nitrification, denitrification and nitrous oxide production in a deep Arctic lake, *Appl. Environ. Microbiol.*, **46**, 1084.
86. Sarda, N. and Burton, G. A. (1995). Ammonia variation in sediments—spatial, temporal and method-related effects, *Environ. Toxicol. Chem.*, **14**, 1499.
87. Mortimer, R. J. G., Krom, M. D., Boyle, D. R. and Nishri, A. (1999). Use of a high resolution pore water gel profilers to measure groundwater fluxes at an underwater saline seepage site in Lake Kinneret, Israel, *Limnol. Oceanog.*, **44**(7) 1802–1809.
88. Dasika, R. and Atwater, J. (1995). Groundwater nitrate profiling by passive sampling over extended depth beneath the water-table, *Wat. Res.*, **29**, 2609.
89. Marnette, E. C. L., van Breemen, N., Hordijk, K. A. and Cappenberg, T. E. (1993). Pyrite formation in two freshwater systems in the Netherlands, *Geochim. Cosmochim. Acta*, **57**, 4165.
90. Zhang, H., Davison, W., Gadi, R. and Kobayashi, T. (1998) In situ measurement of dissolved phosphorus in natural waters using DGT, *Anal. Chim. Acta*, **370**, 29.
91. King, G. M, Klug, M. J., Wiegert, R. G. and Chalmers, A. G. (1982). Relation of soil water movement and sulfide concentration to Spartina alterniflora production in a Georgia salt marsh, *Science*, **218**, 61.
92. Tessier, A., Carignan, R., Dubreuil, B. and Rapin, F. (1989). Partitioning of zinc between the water column and the oxic sediments in lakes, *Geochim. Cosmochim. Acta*, **53**, 1511.
93. De Vitre, R., Belzile, N. and Tessier, A. (1991). Speciation and adsorption of arsenic on diagenetic iron oxyhydroxides, *Limnol. Oceanogr.*, **36**, 1480.
94. Belzile, N. and Tessier, A. (1990). Interactions between arsenic and iron oxyhydroxides in lacustrine sediments, *Geochim. Cosmochim. Acta*, **54**, 103.
95. Huerta-Diaz, M. A., Tessier, A. and Carignan, R. (1998). Geochemistry of trace metals associated with reduced sulfur in freshwater sediments, *Appl. Geochem.*, **13**, 1348.
96. Tessier, A., Couillard, Y., Campbell, P. G. C. and Auclair, J. C. (1993). Modeling Cd partitioning in oxic lake sediments and Cd concentrations in the freshwater bivalve *Anodonta grandis*, *Limnol. Oceanogr.*, **38**, 1.

97. Tessier, A., Fortin, D., Belzile, N., DeVitre and R. R., Leppard, G. G. (1996). Metal sorption to diagenetic iron and manganese oxyhydroxides and associated organic matter: Narrowing the gap between field and laboratory measurements, *Geochim. Cosmochim. Acta*, **60**, 387.
98. Hare, L. and Tessier, A. (1996). Predicting animal cadmium concentrations in lakes, *Nature*, **380**, 430.
99. Croteau, M.-N., Hare, L. and Tessier, A. (1998). Refining and testing a trace metal biomonitor (Chaoborus) in highly acidic lakes, *Environ. Sci. Technol.*, **32**, 1348.
100. Sangi, M. R. (1998). *Trace metal determination in river water by diffusion gradients in thin films, Ph.D. Thesis*, The University of Otago, New Zealand.
101. Zhang, H., Davison, W. and Statham, P. (1996). In situ measurements of trace metals in seawater using diffusive gradients in thin-films (DGT)., In *Proceedings of the Fourth International Symposium on the Geochemistry of the Earth's Surface*, ed. Bottrell, S. H., University of Leeds, p. 138.
102. Chang, L., Davison, W., Zhang, H. and Kelly, M. (1998). Performance characteristics for the measurement of Cs and Sr by diffusive gradients in thin-films (DGT), *Anal. Chim. Acta*, **368**, 243.
103. Chang, L. (1988). *Development of DGT for the measurement of radioactive and stable caesium and strontium in natural waters*, Ph.D. Thesis, Lancaster University, UK.
104. van Santwoort, P. J. M., deLange, G. J., Thomsom, J., Cussen, H., Wilson, T. R. S., Krom, M. D. and Strohle, K. (1996). Active post-depositional oxidation of the most recent sapropel (S1) in sediments of the Eastern Mediterranean, *Geochim. Cosmochim. Acta.*, **60**, 4007.
105. Montgomery, S., Mucci, A. and Lucotte, M. (1996). The application of in situ dialysis samplers for close interval investigations of total dissolved mercury in interstitial waters, *Wat. Air Soil Pollut.*, **87**, 219.
106. Twiss, M.R. and Moffett, J. (1998). Personal communication.
107. Bertolin, A., Mazzocchin, G. A., Rudello, D. and Ugo, P. (1997). Seasonal and depth variability of reduced sulphur species and metal ions in mud-flat pore waters of the Venice Lagoon, *Mar. Chem.*, **59**, 127.
108. Shati, M. R., Rönen, D. and Mandelbaum, R. (1996). Method for *in situ* study of bacterial activity in aquifers, *Environ. Sci. Technol.*, **30**, 2646.
109. Tanaka, T. (1981). Gels, *Sci. Am.*, **244**, 110.
110. Shuttleworth, S. (1999). *Heavy metal geochemistry in lakes: temporal and spatial variations measured with DET and DGT, Ph.D. thesis*, Lancaster University.
111. Fones, G., Black, K., Peppe, O., Davison, W., Hamilton-Taylor, J., Shimmield, G. and Christensen, K. (1998). DGT, DET & Microelectrodes: Techniques for in-situ pore water measurements at high spatial resolution of trace metals and O_2 using an autonomous benthic lander. In *Atlantic Frontier Environmental Conference*, University of Aberdeen.
112. Karlberg, B. and Pacey, G. E. (1989). *Techniques and instrumentation in Analytical Chemistry, Flow Injection Analysis—A Practical Guide*, Vol. 10, Elsevier, Amsterdam.
113. Zhang, H. and Davison, W (1999). Diffusional characteristics of hydrogels used in DGT and DET techniques, *Anal. Chim. Acta*, **398**, 329–340.
114. Mortimer, R. J. G., Krom, M. D., Watson, P. G., Frickers, P. E., Davey, J. T. and Clifton, R. C. (1998). Sediment–water exchange of nutrients in the intertidal zone of the Humber Estuary, *Mar. Poll. Bull.*, **37**, 261–279.
115. Tenberg, A. (1997). *Free-vehicle benthic lander technology for the study of biogeochemical processes in marine sediments*, Ph.D. Thesis, Goteborg University, Sweden.

116. Zhang, H, Teasdale, P. R. and Davison, W. (2000). Multi-element determination of metal species in natural waters using DGT with ICP-MS detection, *Analyst*, submitted.
117. Zhang, H. and Davison, W. (2000). Direct measurement of labile inorganic and organically bound metals in synthetic solutions and natural waters using DGT, *Anal. Chem*. submitted
118. Teasdale, P. R., Hayward, S. and Davison, W. (1999). In situ, high resolution measurement of dissolved sulfide using diffusive gradients in thin films with computer-imaging densitometry. *Anal. Chem.*, **71**, 2186–2191.
119. Kobayashi, T. (1999). *Development of DGT techniques for measuring nutrients in natural environments. Ph.D. Thesis*, Lancaster University, UK.
120. Li, Y. H. and Gregory, S. (1974). Diffusion of ions in sea water and in deep sea sediments, *Geochim. Cosmochim. Acta*, **38**, 703.
121. Davison, W. and Hutchinson, W. (1997). An assessment of the feasibility of using DGT procedures to measure trace metals and radionuclides in rivers, *Environment Agency*, R & D Technical Report, p. 92, EA.
122. Beaulieu, P. Y. (1999). *Verification d'une methode DGT pour mesurer les metaux traces et son utilisation in situ comparee a la dialyse in situ*, M.Sc thesis, Université du Quebec, INRS-Eau, Canada.
123. van Leeuwen H. P. (2000). Lability and mobility criteria relevant for enviromental chemical speciation measurements. In *In Situ Analytical Measurements in Aquatic Systems*, ed. Buffle, J. and Horvai, G, Wiley, Chichester, chapter 8, this volume.
124. Denny, S., Sherwood, J. and Leyden, J. (1999). In situ measurements of labile Cu, Cd and Mn in river waters using DGT. *Sci. Total Environ.*, **239**, 71–80.
125. Hamilton-Taylor, J., Smith E. J., Davison, W. and Zhang, H. (1999). A novel DGT-sediment trap device for the in situ measurement of element remobilisation from settling particles in water columns and its application to trace metal release from Mn and Fe oxides, *Limnol. Oceanogr.*, **44**, 1772–1790.
126. Soares, A. (1998). *An investigation of early diagenetic processes in marine coastal environments by the diffusive gradient in thin films (DGT) technique*, Ph.D. Thesis, Southampton University, UK.
127. Hooda, P. S., Zhang, H., Davison, W. and Edwards, A. C. (1999). Measuring bioavailable trace metals by diffusive gradients in thin films (DGT): soil moisture effects on its performance in soil. *Eur. J. Soil Sci.*, **50**, 1–10.
128. Nyffeler, U. P., Li, Y. H. and Santschi, P. (1984). A kinetic approach to describe trace element distribution between particles and solution in natural aquatic systems, *Geochim. Cosmochim. Acta*, **48**, 1513.
129. Jannasch, H. W., Honeyman, B. D., Balistrieri, L. S. and Murray, J. W. (1988). Kinetics of trace-element uptake by marine particles, *Geochim. Cosmochim. Acta*, **52**, 567.
130. Harper, M. P., Davison, W. and Tych, W. (2000). DIFS—a modelling and simulation tool for DGT induced trace metal remobilisation in sediments and soils, *Environ. Modell. Software*, **15**, 55–66.
131. Honeyman, B. D. and Santschi, P. H. (1988). Metals in aquatic systems, *Environ. Sci. Technol.*, **22**, 863.
132. Santschi, P., Hohener, P., Benoit, G. and Buchholtz-ten-Brink, M. (1990). Chemical processes at the sediment water interface, *Mar. Chem.*, **30**, 269.
133. Harper, M. P., Davison, W. and Tych, W. (1999). Estimation of pore water concentrations from DGT profiles: a modelling approach. *Aquat. Geochem.*, **5**, 337–355.

134. Harper, M. P., Davison, W. and Tych, W. (1999). One-dimensional views of three-dimensional sediments. *Environ. Sci. Technol.*, **33**, 2611–2616.
135. Davison, W., Hooda, P. S. Zhang., H. and Edwards, A. C. (2000). DGT measured fluxes as surrogates for uptake of metals by plants, *Adv. Environ. Res.*, **3**, 550–555.
136. Campbell, P. G. C. (1995). Interactions between trace metals and aquatic organisms: a critique of the free-ion activity model, *In Metal Speciation and Bioavailability in Aquatic Systems*, ed. Tessier, A. and Turner, D. R., Wiley, Chichester, p.45.

APPENDIX I

PROCEDURE FOR PREPARING A DIFFUSIVE GEL BASED ON POLYACRYLAMIDE CROSS-LINKED WITH AN AGAROSE-DERIVATIVE.

Chemicals:

Acrylamide solution (30 %)†, cross-linker solution (2 %)‡, freshly prepared ammonium persulphate (10 %, 1 g in 10 g of water), *N,N,N′N′*-tetramethyl-ethylene-diamine (TEMED, 99 %).

Gel Solution Preparation and Casting

Mix 25 ml of acrylamide solution, 17.5 ml of deonised water and 7.5 ml of cross-linker solution (measured by weighing 7.5 g) together in a clean plastic container. Make sure the resulting gel solution, referred to below as the gel solution, is well mixed by shaking or stirring. It can be stored in a refrigerator (4 °C) for at least 3 months. Pipette 10 ml of the gel solution into a container and add 70 μL of ammonium persulphate solution followed by 25 μ*L* of TEMED solution. Mix them well and cast immediately between two glass plates. All glass plates and plastic spacers have to be acid washed and rinsed with MilliQ water. Wipe dry with clean tissue papers rather than air-dry (air drying is preferred for resin gels to be used for low level trace metal studies, sections 3.8 and 4.4). Place a spacer round three edges and clip the glass plates together. To make pipetting of the gel solution easier offset the glass plates slightly, leaving a 1 mm overlap on the edge without spacer. Pipette the solution in a smooth controlled fashion. If air bubbles appear tilt plates to remove them before continuing pipetting. Maintain the assembly at about 42 °C for 45–60 min until the gel is completely set (no liquid remains). Thick gels require longer setting times. Gently prise open the assembly after the gel has set completely. Hydrate the gel in deionised water for 1 d (changing the water 2–3

† **Caution!** Acrylamide solution is a suspected human carcinogen. Handle with gloves.
‡ This is an olefinic agarose derivative of acrylamide (see secton 2.4.1). The solution is available from DGT Research Ltd., Skelmorlie, Bay Horse Road, Quernmore, Lancaster LA2 0QJ, UK.

times), then store the gel in 0.01–0.1 M $NaNO_3$§ solution. Spacers 0.25 mm thick produce approximately 0.4 mm thick gels when fully hydrated. Gels used for the analysis of ammonia/total CO_2 have been made using a similar procedure but use potassium peroxydisulphate as the initiator to minimise ammonia blanks [39]

APPENDIX II

PROCEDURE FOR PREPARING A RESIN GEL

For resin-gel preparation, first soak the resin in Milli Q (or deionised) water and then remove excessive water with clean tissues. Pipette 10 ml of the gel solution (acrylamide plus cross-linker) used for the diffusive gel into a container. Add 2 g (wet weight) of Chelex-100 (200–400 mesh) resin and mix well. Sufficient resin should be used to ensure that the resin density on the gel surface is maximal without affecting casting and setting of the gel. Then add 60 μL of ammonium persulphate solution and 20 μL of TEMED. Follow the casting procedures used for preparing a diffusive gel. Before pipetting make sure the resin particles are suspended in the gel solution by mixing it well. Resin will settle on one side of the gel by gravity if laid flat during setting. Use MilliQ (or deionised) water for both hydrating and storing the Chelex-gel. A typical resin-gel thickness is 0.4 mm after hydration.

APPENDIX III

PROCEDURE FOR TESTING THE PERFORMANCE OF DGT ASSEMBLIES

Prepare an immersion solution containing 0.01 M $NaNO_3$ with 10 ppb of Cd (or mixed metals) as follows. Into a 3 L plastic container, mix 2 L of MilliQ (or deionised) water and 20 mL of 1M $NaNO_3$ solution. Spike an appropriate amount of Cd standard solution (or mixed metal standard) to make up a 10 ppb solution (make sure that the pH of the solution is above 5). Synthetic lake water or synthetic seawater can be used rather than $NaNO_3$. Place three DGT units in the immersion solution. Make sure the solution is well stirred with a large magnetic follower rotating at a good speed but not cavitating to introduce bubbles. Take an aliquot of the immersion solution for analysis at the beginning and end of the experiment (typically 4 h). Record the deployment

§ This matrix is recommended for long term storage. However, for some applications in which back-equilibration of these ions might interfere with the system being studied the gels may have to be conditioned in the matrix in which it is to be deployed.

time and the solution temperature during exposure. Immediately rinse the surface of the retrieved DGT units with MilliQ water and dismantle the units. After peeling off the filter and diffusive gel, remove the Chelex gel and place it in a clean sample tube. Add 1 mL of 1 M HNO_3 solution and leave it for 24 h before analysis. Although the elution time can theoretically be reduced to 1 h with mixing, more consistent results are obtained with longer elution times. The concentration of Cd in the eluent acid should be about 20 ppb if you have followed the above procedures and assuming that the test was performed at about 25 °C. The concentration of Cd in the immersion solution can be calculated using equation (8).

APPENDIX IV

PREPARATION OF DIFFUSIVE AND RESIN GELS FOR ULTRA-HIGH RESOLUTION DGT

The resin used was a 10% SPR-IDA (Suspended particulate reagent – Iminodiacetate) solution with a bead size of 0.2 μm (Cetac Technologies Inc, Omaha, USA). The 200 μm thick resin layer was made by first adding 10 mL of acrylamide and 2.5 mL of DGT cross-linker together. The stock gel and 10% resin solution were then added together in a ratio of 1:1 to form a 5% IDA gel solution. For each 1 mL of solution 7 μL of persulphate and 2 μL of TEMED were added. The glass plates and plastic spacers were acid washed and rinsed with Milli-Q, and then air dried in a Class 100 laminar flow cabinet to ensure they were contamination free. A 0.1 mm plastic spacer was used to produce after hydration the 0.2 mm gel. 2 mL of IDA resin gel is pipetted evenly onto the glass plate, the other plate is then slid over the top ensuring no air bubbles are produced. The assembly was kept at 45 °C for 45 min. The gel was then hydrated in Milli-Q water cleaned with Chelex 100 resin. A 200μm thick layer of hydrogel of the same composition as that described in section 2.5.1, was made using the above method.

12 Microtechnology for the Development of *In Situ* Microanalytical Systems

G. C. FIACCABRINO, N. F. DE ROOIJ AND M. KOUDELKA-HEP
University of Neuchâtel, Neuchâtel, Switzerland

J. HENDRIKSE AND A. VAN DEN BERG
University of Twente, Enschede, the Netherlands

In Situ Monitoring of Aquatic Systems: Chemical Analysis and Speciation Edited by J. Buffle and G. Horvai.

1 INTRODUCTION

A microsystem generally comprises several functions such as sensing, signal processing and actuation [1]. In the particular case of miniaturised analytical systems, this definition is often extended to include large systems in their overall dimension that integrate any of the sensing, actuation or fluid handling parts in a miniaturised form [2–6]. The targeted attributes of a micro-analytical system range from low sample and/or reagent consumption to improved analytical performances (selectivity and speed of analysis), improved portability (low power consumption, reduced system dimensions), and added functionality (auto-calibration, self-testing and data transmission).

In spite of a limited number of reports in environmental applications [7–12], all these features could readily be exploited advantageously in field analysis. Whereas classical laboratory analyses provide intermittent data, microsystems can easily be decentralised to carry out unattended continuous or quasi-continuous analysis from remote sites. *In situ* sensing notably reduces sample manipulation, and consequently minimises handling costs and errors associated with possible sample degradation or contamination. By increasing temporal and spatial resolutions, microsystems may unveil short-term fluctuations of environmental relevance, and as a result become invaluable tools for the comprehension of biogeochemical processes. The ability to network autonomously and remotely controlled microsystems enables the collection of quasi-instantaneous analytical information over large areas, and thus provides additional information on the dynamics of these processes.

Microsystem technology (MST) encompasses various techniques, most of them originally adapted from the microelectronics industry, others specifically developed for the manufacture of microsystems. This chapter mainly focuses on the former, followed by a brief account of other fine engineering techniques such as LIGA. Although the overview is not complete, it is significantly reflective of the current MST activities. A critical discussion on different approaches for miniaturisation is given and illustrated with some examples of embodiment. For an in-depth review of the microfabrication field, the interested reader is referred to two recent publications [13,14]. A number of non-traditional techniques such as self-assembled monolayers, microcontact printing and micromolding techniques are reviewed elsewhere [15,16].

2 MICROSYSTEM TECHNOLOGY

Photolithography is still dominantly utilised for the manufacture of miniaturised components. Initially developed for the semiconductor industry, photolithography provides the means for a batch fabrication of microsystem parts at a micrometre scale [17]. Owing to the background of this technology, these parts are mostly fabricated on silicon substrates, although quartz or glass substrates are used likewise. In the following section, the different microfabrication techniques are presented as surface or bulk processes, depending on whether devices are constructed at the surface level or directly in the bulk of the substrate. Note that most of these techniques are often combined.

2.1 PHOTOLITHOGRAPHY

Although not unique, photolithography plays a key role in the patterning process, i.e. the transfer of a pattern onto a substrate [18]. The technique most simply consists in depositing a thin layer of UV sensitive material, namely photoresist, onto the substrate by spin-coating. The film is then selectively exposed to UV radiation through a quartz frame supporting the original pattern made in a thin chromium layer (Figure 1). As a result, exposed areas of photoresist are either made soluble (positive-tone photoresist) or insoluble (negative-tone photoresist). The soluble part of the photoresist can subsequently be removed by the appropriate developing solution, hence leaving on

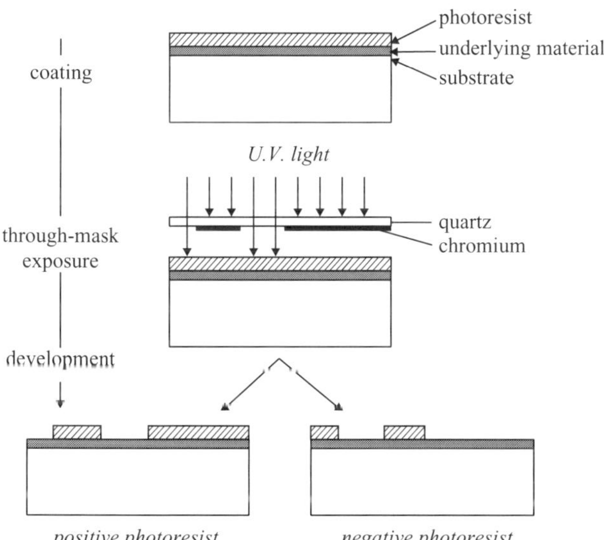

Figure 1. Postive-tone and negative-tone photoresist exposures

the substrate the copy image of the original pattern. Positive- and negative-tone photoresists used in UV lithography nowadays allow routinely 1 μm resolution to be achieved.

The ultimate role of photoresist in the patterning process depends on the patterning technique used. These can be divided in two categories, subtractive and additive processing. In subtractive processing, the patterning is performed by selectively protecting the material to be patterned, so that the unprotected parts can be etched away. Patterned photoresists therefore act as an etch mask. Different etching techniques are described in sections 2.2.2 and 2.3. Among the different possibilities, the preference is usually given to the technique preserving the photoresist integrity. Once the etch step is completed, the photoresist mask is removed from the surface in the appropriate solvent or in oxygen plasma. As an example, the technological sequence used for the realisation of the microdisc arrays described in Chapter 9 is illustrated in Figure 2.

Additive processing encompasses all techniques through which a selective deposition of material is accomplished. One of the dominant techniques nowadays is the so-called 'lift-off' technique. It is a simple alternative for the patterning of some thin films for which no etch process is compatible with

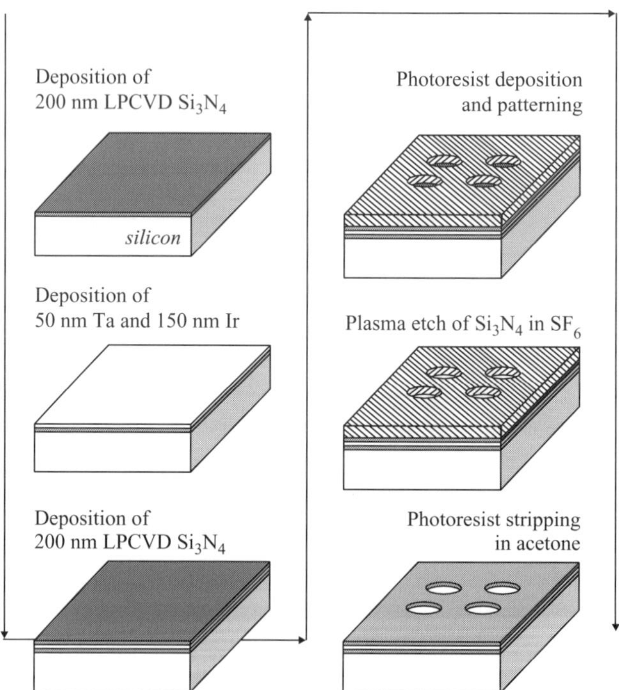

Figure 2. Technological sequence for the fabrication of the planar iridium microdisk arrays described in Chapter 9. (SU-8 containment ring patterning is not shown)

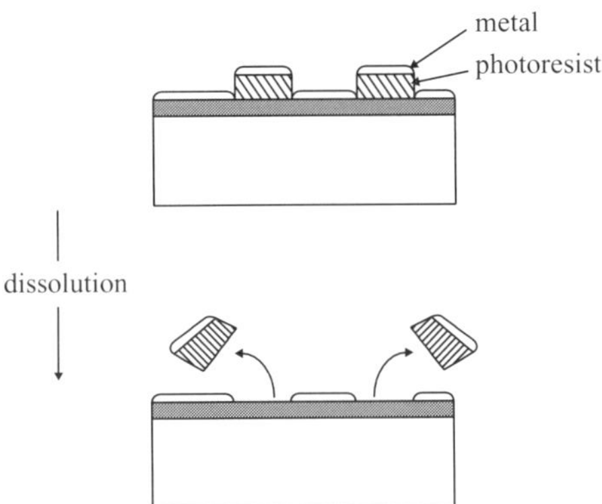

Figure 3. Example of a lift-off sequence for the patterning of noble metals

photoresist masking. For this reason it is frequently utilised for the patterning of noble metals, such as Pt or Ir. In the lift-off procedure the photolithographic step precedes the metal deposition, the patterned photoresist acting as a vertical stand-off evaporation mask (Figure 3). By dissolving the underlying photoresist in a suitable solvent, unwanted metallic parts are lifted off, thus leaving on the substrate the desired pattern.

The second most frequent technique, which also generally applies to metals, is electroplating or electroforming. The patterned photoresist then acts as a deposition mould. Metals are electrodeposited from solution onto a Cu, Ni, Au, or NiFe seeding layer. Patterning with high aspect ratio through-mask plating is addressed in the LIGA section.

2.2 SURFACE PROCESSES

2.2.1 Thin Film Technology

Thin film technology is of widespread use for the fabrication of planar devices, such as electrochemical and optical transducers. A selection of various materials and deposition techniques is presented hereafter and summarised in Table 1. Detailed description of these topics can be found in the suggested literature [19–22].

Thin metal films are ordinarily deposited by sputtering or electron gun evaporation (e-gun) from a metal target. Sputtering, however, is not only used for the deposition of metals but also for the deposition of various dielectric oxides and nitrides, and more recently carbon [23,24]. Noble metals, such as Pt, Au, Ag and Ir, often require the use of an intermediate layer to improve their adhesion to the substrate. Ti or Ta, deposited sequentially in a single pump-down process prior to the deposition of the noble metal, are frequently used as

Table 1. Classical thin film materials, deposition techniques and typical applications

Material	Deposition technique	Typical use
Pt, Ir, Au, Ag, C, Cu, Al, W, Ti, Ta, Ni	evaporation, sputtering, chemical vapour deposition, electrodeposition	electrodes, interconnection lines, adhesion layer, mechanical structures
Si (polycrystalline silicon)	chemical vapour deposition (low pressure or plasma enhanced), sputtering	sacrificial layer, interconnection lines, gate material, mechanical structures, resistors and piezoresistors
Si_3N_4, SiO_2, Al_2O_3, Ta_2O_5	chemical vapour deposition (atmospheric, low pressure or plasma enhanced)	encapsulation material, masking, capacitor dielectrics, gate material for ISFETs

adhesion promoters. The typical thickness of the adhesion layer is 0.05 μm and that of the noble metal layers is 0.15μm. These metals can be used to accomplish electrical, optical, mechanical and electrochemical functions. Microfabrication of thin film noble metal microelectrodes has been reviewed elsewhere [25].

As a dielectric material, silicon nitride (Si_3N_4) offers attractive characteristics such as a high dielectric breakdown strength, a high specific resistivity, and a near imperviousness to water and to the diffusion of alkali ions. For these reasons it is often used as a passivation material. The deposition of Si_3N_4 is obtained by chemical reaction of selected gases in a vapour phase. Thin dielectric films with controlled stress characteristics can be produced by adjusting the process variables, such as temperature, gas composition and pressure [19,20]. The built-in tensile stress is of critical importance when using Si_3N_4 films for the fabrication of suspended mechanical parts, such as membranes or levers.

Silicon oxide (SiO_2) is an excellent dielectric material, sometimes used as a sacrificial layer in surface micromachining, wave guide in optics, and gate oxide in field effect transistors (FETs) and ion-selective field effect transistors (ISFETs). It is less used as passivation layer owing to its permeability to alkali ions and its capacity to become hydrated.

Polycrystalline silicon, or polysilicon, is a well-known material used for the fabrication of gate contacts of FET devices, interconnection lines, capacitors, resistors and piezoresistors. Aside from its electric characteristics, polysilicon shows excellent mechanical properties suitable for the fabrication of free-standing structures. The crystalline structure of polysilicon, and thus its mechanical characteristics, is largely influenced by the deposition temperature and post-deposition thermal cycles [26]. A large variety of devices, e.g. accelerometers, cantilever tips for atomic force microscopy and micromotors, have been fabricated using the polysilicon surface micromachining technique described in section 2.2.3.

2.2.2 Etching Processes

Etching processes are subdivided into wet and dry (plasma) etching techniques [13,14]. Wet etching primarily involves a redox reaction for metals, and a displacement reaction for inorganic oxides (Table 2). The reaction product is a soluble salt or complex. The photoresist mask must be resistant to the chemicals involved in order to provide a good protection throughout the patterning process. Metal and dielectric masks can withstand in some cases harsher etch systems, such as those involving elevated temperatures or strong acids or bases.

Some of the solutions listed in Table 2 etch isotropically, that is, at an equal rate in all directions. As a consequence, the resulting pattern profile extends under the etch mask by an amount roughly equal to the etched thickness, giving rise to characteristic undercuts (Figure 4a). This effect irremediably limits the resolution and the aspect ratio that can be attained using these etching processes. Anisotropic etching of silicon in solution will be addressed in the bulk micromachining section.

Dry etching includes various techniques such as sputter etching, ion milling, plasma etching (PE) and reactive ion etching (RIE). All of those make use of a glow discharge. A number of halogen-containing gases, for example SF_6, CF_4,

Table 2. Typical wet etching systems

Etched material	Reagents	Etch type	Mask typically used
Al	Alu-etch (4:4:1:1 H_3PO_4/CH_3COOH / fuming HNO_3/H_2O) at 42°C	isotropic	PR*
Pt, Au	Aqua regia (3:1 fuming HNO_3 / HCl)	isotropic	PR limited
Au	Au-etch (4 g KI and 1 g I_2 in 40 ml H_2O)	isotropic	PR
SiO_2	BHF† (7:1 40% NH_4^+/HF)	isotropic	PR
Si_3N_4	85% H_3PO_4, at 160°C	isotropic	SiO_2
Si‡	KOH (40 wt%) at 60°C	anisotropic	SiO_2, Si_3N_4
	TMAH§ (typically 4–20 wt%, at 90°C	anisotropic	SiO_2, Si_3N_4
	EDP (ethylenediamine/pyrocathecol) (typically 750 ml ethylenediamine, 120 g pyrocatechol, 100 ml H_2O) at 115°C	anisotropic	SiO_2, Si_3N_4
	HNA¶ (typically 8:1:3 HNO_3, HF, CH_3COOH)	isotropic	SiO_2, Si_3N_4
Quartz	HF, NH_4F	anisotropic	Cr/Au
Borosilicate glass	HF	isotropic	Cr/Au

* PR, photoresist.
† BHF, buffered HF.
‡ Unlike Si, polysilicon is polycrystalline and wet etching is isotropic in all cases.
§ TMAH, tetramethyl ammonium hydroxide.
¶ HNA, HF/HNO_3/acetic acid.

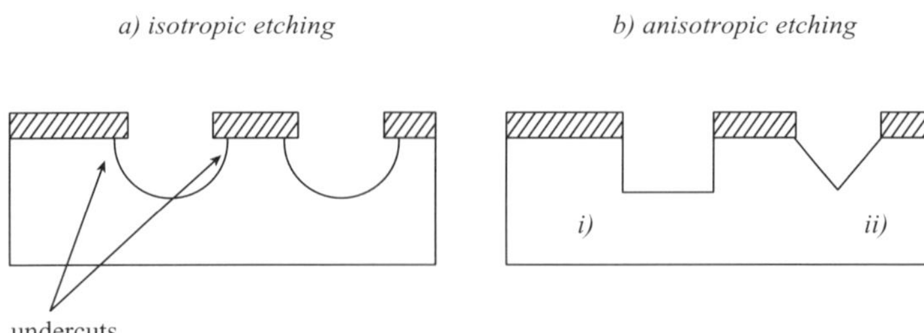

Figure 4. Characteristic etch profile resulting of (a) isotropic etching; (b) anisotropic dry (i) and wet (ii) etching

C_2ClF_5, CCl_4 and mixtures of these, with a possible addition of O_2, are regularly used for the patterning of Si, polysilicon, SiO_2, Si_3N_4 and Al [27]. The mechanism involved in the etching process can be classified as either purely physical (sputter etching), chemical (PE), or a combination of both (PE and RIE). Sputter etching is often depicted as the atomic scale version of sandblasting, with the virtues of removing any material. Its use is limited owing to the lack of a resistant etch mask, as erosion rate does not differ markedly from one material to another. In a PE process, the glow discharge generates active species such as atoms or free radicals that diffuse and chemically react with the surface to produce volatile compounds. Under high pressure conditions, the chemical etching process is controlled by the diffusion of the reactive species, and undercuts similar to those obtained in wet etching may also affect the resolution of the transferred pattern. Anisotropic etching with PE can be achieved by introducing a physical contribution to the etching process. The pressure is usually lowered and the substrate holder is polarised to increase ion bombardment. By placing the substrate normal to both the gas flow and the radio frequency field, anisotropic etching can be achieved as a result of the directional movement of the incident particles (Figure 4b). This combined effect of physical and chemical etching is found in RIE. Deep reactive ion etching (DRIE) will be addressed in section 2.3.

2.2.3 Surface Micromachining

Surface micromachined devices are free-standing structures built on the surface of a substrate, with a vertical dimension usually in the range 1–30 μm [28–30]. Surface micromachining is essentially based on the removal of a sacrificial layer. It can be summarised by the deposition and patterning of an SiO_2 sacrificial layer, the deposition and patterning of a polysilicon or a metallic film, and the removal of SiO_2 by lateral isotropic wet etching (Figure 5). This sacrificial layer technique has been applied for the fabrication of micromechanical devices, such as accelerometers [31,32], optical modulators [33] and relays [34].

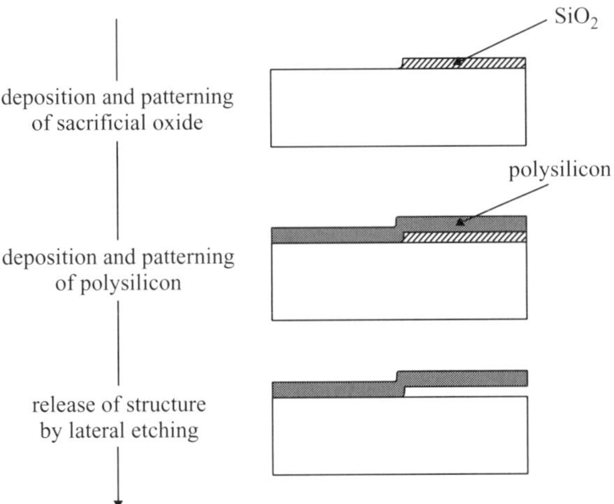

Figure 5. Basic polysilicon surface micromachining sequence

2.3 BULK PROCESSES

In this section bulk processes are associated with bulk micromachining, a technique by which three-dimensional devices are created by directly removing material from the bulk of the substrate itself. Note that silicon is an extremely attractive material for the manufacture of miniaturised actuators, as it exhibits excellent mechanical properties upon down scaling. Aside from the physical sensors and actuators field, bulk micromachining is a key step in the fabrication of various fluid handling devices [35,36], such as micro-pumps and valves [37–42], microchannels [43–45], filters [46–48], mixers [49,50], microreactors [51,52], or micro-cells [53,54]. Silicon wafer, borosilicate glass and quartz are usual substrates. Various techniques for the microfabrication of fluidic active components have been reviewed elsewhere [55].

The etching processes involved in bulk micromachining can again be classified into wet and dry etching categories. Bulk micromachining of crystalline silicon substrates by wet chemical etching is one of the earliest and simplest forms of micromachining. Different etch solutions have been investigated, the most popular and documented being KOH (Table 2). Unlike the case of polycrystalline silicon discussed in section 2.2.2, monocrystalline Si is etched anisotropically, as a consequence of the strong crystallographic dependence of the etching rate. The resulting micromachined parts often display characteristic shapes. Although the reasons for these etch rate differences are still unclear, the technique is well established [56–59]. As mentioned in section 2.2.2, masks cannot be made of photoresist, which is often replaced by SiO_2 or Si_3N_4.

Wet isotropic etching of bulk silicon is predominantly achieved in solutions composed of HNO_3, HF and CH_3COOH (Table 2). The well-accepted etch mechanism consists of the oxidation of the Si by nitric acid, followed by its dissolution by hydrofluoric acid. The etching proceeds at the same rate in all crystallographic directions, and thus gives rise to etch profiles characteristic of isotropic etching. Si_3N_4 masks are appropriate in this case.

In the anodic etching process, the oxidation of the Si is achieved by application of an anodic current. At low current density, typically 40 mA cm^{-2} at 0 V versus Ag/AgCl, porous silicon covered by SiO_2 is formed. Increasing the electrode potential increases the current density and the pore size. At high

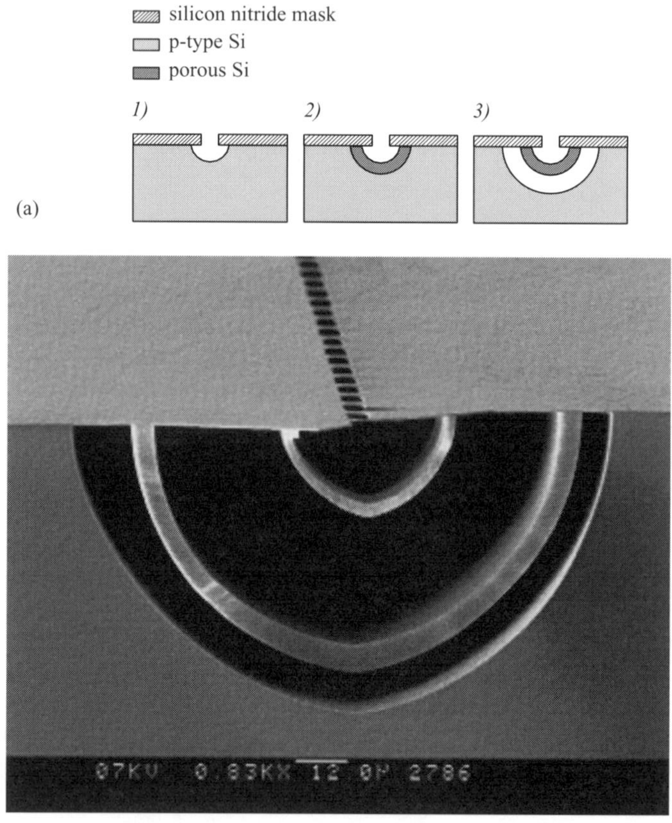

Figure 6. (a) Anodic etching of concentric microchannels in p-type Si. 1: A channel is isotropically etched at a high current density, starting from an opening in a silicon nitride mask. 2: The current density is lowered and the etching proceeds to form oxide covered porous silicon. 3: The current density is raised and the complete etching proceeds. During this step, the oxide layer inhibits the etching of the porous silicon. (b) Multi-wall microchannel containing two layers of free hanging porous silicon

Table 3. Typical dry etching systems

Etched material	Gases used for etching
SiO_2	CF_4, CHF_3, C_2F_6, C_3F_8
Si_3N_4	SF_6, CF_4/O_2, CF_4/H_2, CHF_3, C_2F_6
Polysilicon and Si	SF_6, CF_4, $SiCl_4/Cl_2$, CHF_3/Cl_2, $CHCl_3/Cl_2$, CCl_4/Cl_2
Al	Cl_2, CCl_4, $SiCl_4$
C, photoresist, polyimide	O_2, O_3

Note that chlorofluorocarbon gases (CFCs), known to be environmentally harmful, tend to be disregarded and are not listed above.

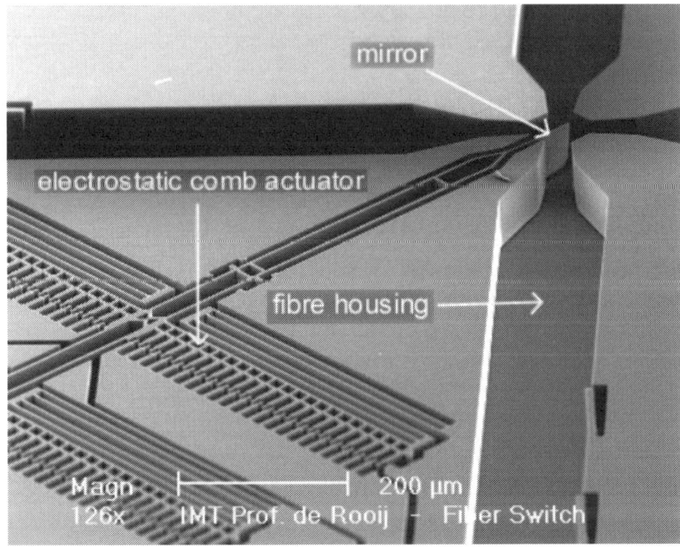

Figure 7. Scanning electron microscope image of a micromechanical optical switch, produced by ADRIE. By applying a potential difference between the two comb-like electrodes (comb actuators) an electrostatic force will develop and move the mirror in the optical path of the aligned fibres

current density, typically 180 mA cm^{-2} at 3 V versus Ag/AgCl, the etching is complete. By changing the electrode potential during the etching process, concentric regions of porous silicon separated by fully etched space can be created (Figure 6a). When the pores that are left in the Si wall are closed, for example by thermal nitridation in ammonia [60], concentric channels as those shown in Figure 6b, can be produced [61].

Bulk micromachining of borosilicate glass and quartz substrates is usually accomplished in 50 % HF. Masks of classical photoresist have a relatively short lifetime in solution, which limits the etch depth and the resolution. Chromium/gold masking layers are used instead of photoresist. Polysilicon has recently been proposed as a suitable etch mask, that can additionally be utilised in a subsequent anodic bonding step [62].

Until recently, bulk micromachining by the dry etching technique was disregarded as being a long and tedious task (Table 3). The technique has gained interest owing to recent developments in plasma etch equipment with higher plasma densities, suitable for the production of high aspect ratio features in silicon as well as in quartz [63,64]. The etching technique, known as advanced deep reactive ion etching (ADRIE), offers an aspect ratio of at least 15:1, and a good selectivity with respect to photoresist masking. Compared with anisotropic etching of Si, ADRIE additionally offer a higher density of integration and high design flexibility (e.g. Figure 7). A number of practical examples can be found in the suggested literature [65,66].

2.4 BONDING TECHNIQUES

The combination of various types of bonding techniques, i.e. art for completing a permanent and reliable joint between at least two parts, with surface or bulk micromachining, has been applied to the fabrication and packaging of various sensors and actuators. In the field of analytical chemistry, simple grooves and cavities micromachined in bulk silicon have been conveniently used as a variety of fluid handling parts [67]. One of the most popular bonding techniques is the anodic bonding [68,69], which allows silicon substrates to be permanently bonded to glass. Corning # 7740 borosilicate glass (Pyrex®) is widely used for this purpose as its thermal expansion coefficient matches almost perfectly that of silicon. Both substrates are brought into contact and heated at a temperature in the range 350–500 °C. At these temperatures the Pyrex® glass becomes slightly conductive. If a voltage between 200 and 2000 V is applied across the wafers, with the glass being polarised negatively, an electrostatic force will pull both substrates into intimate contact. At the same time, mobile Na^+ ions will leave the Si/Pyrex® interface, causing the depletion of positive charges, thus creating a fixed negatively charged region. Although the exact chemistry of the bonding is yet not thoroughly understood, it is assumed that the oxygen ions leave the glass and chemically bind with Si through the formation of Si–O bonds. Note that anodic bonding can be used on bare silicon wafers as well as on wafers covered with thermal SiO_2, sputtered glass, polysilicon or Si_3N_4.

The second most frequently used technique is called fusion bonding, which refers to the permanent joining of two substrates by thermal or chemical treatment. A number of applications of silicon fusion bonding, through Si–Si or Si-–SiO_2 bonds, have been proposed for Si solid-state physical sensors [70]. The Si substrates are first made hydrophilic by forming a high density of hydroxyl groups on their surfaces. Upon bringing both substrates together, a weak bonding due to hydrogen bonds occurs immediately. Thermal treatment strengthens the bonding, through a series of chemical reactions, presumably the transformation of silanol bonds to siloxane bonds. The fusion bonding of two quartz or Pyrex® substrates is intensively used for instance in the manufacture of capillary channels for electrophoresis [71–73].

2.5 LIGA AND ALLIED TECHNIQUES

LIGA†, as indicated by its German acronym, is the combination of lithographic, electroplating and moulding techniques [74,75]. As described in section 2.1, moulds with sharp vertical flanks are highly desirable in plating techniques. Sub-micrometre lithography with high aspect ratios surpassing by far that of classical photolithography is achieved by X-ray irradiation. Commonly used resists are copolymers of polymethyl methacrylate (PMMA), although poly (lactides) have also been proposed and seem more promising [13]. Upon exposure to X-rays, the long molecular chains in the exposed areas undergo a main-chain scission, thus greatly enhancing the solubility of these regions. As for classical negative-tone photoresists, selective developer dissolves the exposed areas, while leaving unexposed parts on the substrate. The patterned PMMA can either be used on its own, or utilised as a sacrificial mould for the electrodeposition of metallic micro-parts or mould inserts. The latter can be used to replicate the primary pattern by injection moulding or by hot embossing into polymeric materials. These secondary patterns can either be the final parts or again serve as sacrificial moulds. Because of its high performances and its versatility, the LIGA technique has received considerable attention [76,77]. Access to the technology is, however, fairly restrictive because of the need of a synchrotron to produce the X-rays.

When sub-micrometre resolution is no necessity, an interesting alternative technique can be used [78,79]. It is based on an thick epoxy negative-tone photoresist that can be readily patterned by classical UV sources. It is primarily composed of Shell SU-8 Epon resin, photosensitised with triaryl sulfonium salt. Its low optical absorption in the near UV spectrum favours good aspect ratio structures. Patterning of features up to 1200 μm in size with an aspect ratio of 18 has been demonstrated. In some cases, the use of SU-8 in microsystem technology can favourably replace the LIGA technology. A number of SU-8 applications have been reported [80–82]. It has been used for instance to produce a gel-integrated microelectrode array for *in situ* voltammetric measurements (Chaper 9, section 5.1.4). Containment rings made of SU 8 were formed at the surface of the sensor chip to enhance the mechanical stability of the anti-fouling gel membranes, and offer a better control of the membrane thickness (Figure 8).

---

†LIGA: Lithographie, Galvanoformung, Abformung.

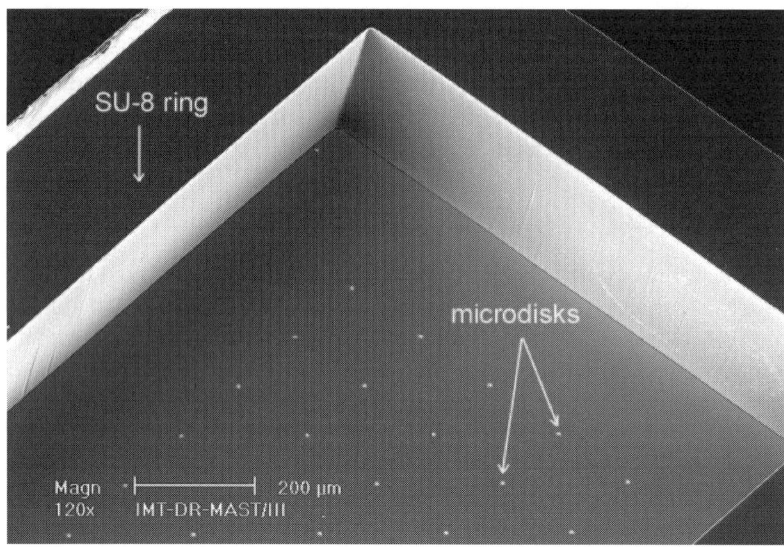

Figure 8. Scanning electron microscope image of the SU-8 containment ring, surrounding a microdisk array (top side view, tilted; see also Figure 13b in Chapter 9)

2.6 THICK FILM TECHNOLOGY

Thick film technology encompasses all techniques to deposit selectively on flat surface, coatings with typical thickness larger than several microns, e.g. silk-screening, screen printing. A paste or ink is generally pressed onto a substrate through openings in a screen. The paste consists of a mixture of the material of interest, an organic binder and a solvent. After printing, wet films are allowed to dry to remove solvents from the paste, and are subsequently fired to burn off the organic binder. Thick film technology is an interesting alternative to the thin film approach for manufacturing electrochemical sensors, with characteristic feature sizes of $\geq 50\mu m$ [83,84]. Electrodes can be silk-screened, screen printed or ink-jet printed on many different substrates, such as glasses, plastics or ceramics. Not only is this technique very inexpensive and versatile, but it also allows the pastes or inks to be blended with a number of specific modifiers, such as catalytic compounds or biomolecules [85,86]. Dielectric layers can also be laid onto the substrate, to cover precisely parts of the electrode and to define the active area of the device. Screen-printed carbon electrodes for instance have been proposed for the determination of lead, cadmium, copper and zinc [87,88].

2.7 LASER ABLATION AND LAMINATION OF POLYMERS

The use of laser techniques in micromachining is fairly wide and finds application in thermal treatment, lithography, as well as in different deposition and etching techniques. For the latter, focused beam milling or laser ablation can be utilised to machine a variety of materials, such plastics, glasses, ceramics and metals.

Laser ablation of selected polymers can be used for the production of flow channels at low cost. The photoablation process involves absorption of short-duration laser pulses in the UV region, with a concomitant bond breaking within the long-chain polymer molecules. The substrate material is expelled in the form of gas, polymer molecules and particulate matter, thus leaving behind a clean machined substrate. Such channels have been engraved in polyethylene terephthalate (PET) by UV laser ablation and closed on the top part by lamination with a PET/PE foil (Figure 9). The ability to generate electroosmotic flows in these channels makes this laser ablation/lamination technique very interesting for manufacturing microanalytical or micro-diagnostic systems [89]. Note that the lamination technique can be extended to seal channels in the millimetre size range, cut or machined in plastics by classical tools.

Combined with screen printing techniques, laser ablation has proven to be a good alternative to classical thin film technology for producing regular microdisc arrays [90]. In this case, an insulating layer entirely covering a screen-printed electrode is laser machined by ablating disc-shaped holes. Micro-machined membranes for liquid/liquid interface electrochemistry have been produced likewise [91].

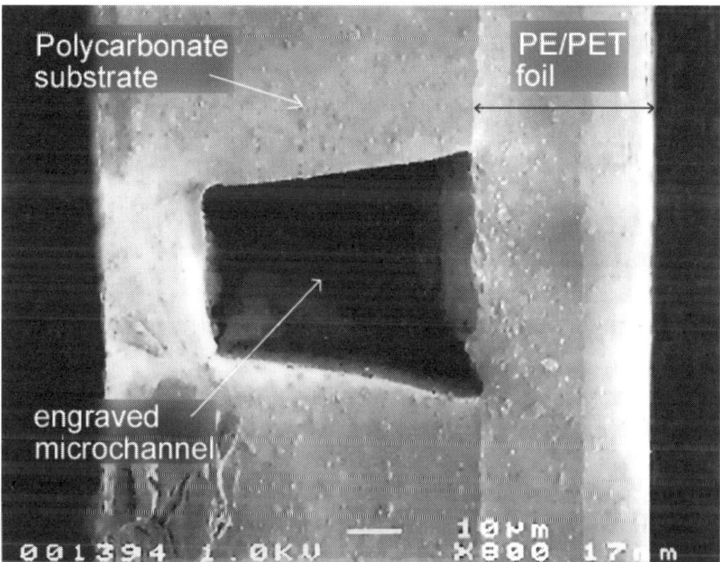

Figure 9. Scanning electron microscope image of a microchannel cross-section, produced in polycarbonate substrate by UV laser ablation and lamination. (Courtesy of J. Rossier and H.H. Girault, Laboratorie d'Electrochimie, EPFL, Switzerland)

2.8 HOT EMBOSSING

Hot embossing consists of pressing a master piece into a polymer at elevated temperature. This technique can be used to produce large numbers of replicate chips from a single master. In order to decrease either the cost or the possible structure size, masters can be produced by conventional machining or by the LIGA technique as described in section 2.5. In most cases, nickel is electroplated to form an embossing tool [92,93]. Channel diameters of 10–100 μm have been achieved by hot embossing in polymers such as polycarbonate (PC) and PMMA. As both the tool and the structure have to be cooled before the embossing cycle is completed, it takes a few minutes to produce a structure in this way. Because of the high cost of master fabrication, the method is less useful for prototyping.

3 SYSTEM MINIATURISATION

The selection of one of the various microsystem technologies described above for manufacturing a particular miniaturised device or system may *a priori* seem difficult. At first, one should evaluate carefully whether or not miniaturisation is appropriate. Indeed, some components can readily be purchased, if not as a miniature, at least in a fairly reduced format. Fluid handling components, such as small solenoid valves or pumps are currently available at moderate investment costs from a number of commercial sources. Microfabricated counterparts should ultimately be considered because of their enhanced performances, or when dictated by operational constraints such as size, weight, power consumption or ruggedness. This consideration also applies to sensors and data processing. If the need for miniaturisation is identified, the benefits of using any technology, including conventional engineering techniques, should be evaluated in terms of a device performance to fabrication cost ratio, as well as analytical constraints such as contamination, lost by adsorption etc.

Advantages and limitations in microsystem integration are discussed hereafter, together with specific aspects to consider for different microsystem components.

3.1 SMART SYSTEMS VERSUS HYBRID INTEGRATION

The attractive concept of 'smart system' consists in realising on the same Si chip all the sensing and data processing functions [14,94,95]. This idea has been applied with some success to physical sensors, but a few limiting factors should be mentioned. Because of the extreme cleanness that is needed to produce microelectronic devices, these have to be manufactured on a virgin wafer before any micromachining. The non-electronic parts can be realised as a second processing step on top of the electronic circuitry. On this basis, a fabrication sequence can be defined, which ensures both the integrity of the electronics and the characteristics of the micromachined parts. Defining this sequence, however, often proves to be a difficult task. Aside from the substantial efforts required to ensure process compatibility, the final cost per Si chip increases with integration steps for two reasons. Firstly, the fabrication yield decreases as the number of steps increases, and secondly, expensive silicon wafer surface area is uselessly carried over during the first manufacturing stages of the electronic part. Moreover, in the case of chemical sensors, the fact that only a fraction of the Si chip may be exposed to the sample media imposes complex and severe packaging constraints.

An alternative to the 'smart system' approach consists in a hybrid integration approach, where the various parts of the system are made separately and assembled together in a final step. By making use of conventional technology for some components, if necessary, this approach complies better with the complex nature of a full analytical system and the specific needs of the various applications. Discrete micro-components can be issued from optimised fabrication methods and assembled in final microsystems in a more cost efficient way. The difficulty lies in the art of making proper fluidic, electrical, optical and mechanical connections [96]. Platforms such as multi-chip module (MCM) and mixed, i.e. electronic and fluidic, circuit board (MCB) have been proposed to facilitate these interconnections. Examples of both platforms are shown in section 4.

3.2 FLUIDICS

The fluid handling part is one of the key parts of the microanalytical systems. Upon miniaturising, surface properties and dead volumes are critical issues that should not be overlooked to achieve representative and reproducible sampling and sample pretreatment, and to reduce response times. Table 4 compares general characteristics of macroscopic and micromachined fluidic devices.

3.2.1 Sample Pretreatment

It is not sufficiently recognised that proper design for the sampling part of the system is of the utmost importance in microanalytical systems. In particular, regardless of whether air or water samples are collected, dust and dirt particles form a threat to the proper functioning of microfluidic components. For that reason a separation module at the entry of the system is highly desirable. For example, the test compound can be brought into the system by pulling the water sample through a filter (filtration), or by letting it diffuse through a membrane into a carrier stream (dialysis).

Table 4. Comparison of performances of macroscopic vs micromachined fluidic devices

Issue	Macroscopic	Micromachined
Unwanted turbulent flow	yes	no
Very small dead volume	varies*	yes
Problems with purging bubbles	yes/no	yes
Efficient liquid pumps available	yes	not yet
Efficient liquid valves available	yes	not yet
Efficient gas pumps available	yes	not yet
Efficient gas valves available	yes	yes
Simple interconnection scheme	yes	no
Chemical resistant materials available	yes	varies
Low power consumption	no	varies†
Sub-mL volume	no	yes
High surface-area-to-volume ratio inside the device	no	yes
Batch fabrication	no	yes (not packaging)

*Usually larger than for microsystems.
†Usually lower than for macroscopic systems.
Source: Kovacs, G.T.A. [14].

With environmental samples, filtration can create a large number of artefactual results, and must be used with much caution. In correct conditions, however, in particular at low flow-rates, correct data can be obtained [97,98]. To increase the level of integration, filters can be micromachined in silicon [41]. As the filter thickness is usually smaller than the pore size, filters of this kind have very small dead volumes. Moreover, pore size distributions and pore shapes can be tailored to specific applications (Figure 10). For instance, filtration of yeast cells from beer was demonstrated, and pressure losses 40 times smaller than those achieved with comparable conventional filters were found [99]. Note that in filtration systems operated by a pump, the filtration flow will decrease upon fouling of the filter membrane. This is a well-known problem in large scale filtration processes. Incorporating a flow sensor close to the pump, to regulate the flow rate by means of a feedback loop can eventually solve the problem of decreasing flow rate [100]. This, however, is not applicable in analytical chemistry, as fouling of the filter usually results in wrong analytical data [97]. Fouling may be minimised by removing the particles that do not travel through the filter by a second fluid stream (cross-flow filtration). A better approach to separate the test compounds from colloids and particles of the sample is to make use of the fact that the diffusion coefficient of a chemical entity decreases when its size increases. Therefore, by flowing the sample stream and a carrier stream parallel to each other in a laminar flow through the same channel, small molecules diffuse from the sample to the carrier, while larger particles remain in the sample stream and are subsequently ejected from the sampling system [101]. Microsystems are particularly well suited for such applications,

Figure 10 Micrograph of a silicon microfilter, used for filtration or multi-lamination mixing

as perfect laminar flows are more easily obtained in micro- than in macro-systems.

With microdialysis, the test sample does not interact directly with the sensor since the test compound is separated in an internal solution. This allows better control of the physicochemical conditions of the solution analysed by the sensors. For example, possible disturbances by the sample pH in the case of enzymatic sensors are eliminated [102]. Microdialysis probes, commercially available and in use for years in neurochemistry, are starting to be used in environmental analysis such as gel integrated microelectrodes (chap. 9). However, in order to optimise the system with respect to response time and yield of recovery, other means of separation, such as PLM (Chapter 10), may be more useful.

3.2.2 Pumps

In most microanalytical systems, fluids are transported by one or several micromachined pumps. One such a pump is displayed in Figure 11, and selected examples are listed in Table 5. A number of additional references on silicon-based fluid handling parts are cited in section 2.3. Various actuation principles have been reported, notably electrostatic, piezoelectric, thermo-pneumatic, pneumatic for the most frequent ones. Polymeric membranes integrated in

Figure 11. Bottom view of a microfabricated silicon pump. The core of the pump is a micromachined silicon wafer, closed on both sides by two anodically bonded glass plates. The visible base plate contains the inlet and outlet channels of the micropump

plastic housings are an interesting alternative to silicon-based actuators [103–105]. Note that electroosmotic pumping, albeit a pumping mechanism only applicable to appropriate buffered electrolytes, is particularly efficient in handling down to picolitre volumes of samples. As the electroosmotic effect becomes stronger as the channel diameter decreases, pressure differences up to 55 bar can be generated in capillaries packed with micron-size silica beads [106]. Moreover, this way of pumping does not need any moving micromachined parts and enables the generation of an extremely steady flow. Electroosmotic pumping in miniaturised analysis systems is increasingly gaining interest [6,107].

Table 5. Characteristics of different micropumps

Actuation principle	Structure/material	Maximum flow rate ($\mu L\,min^{-1}$)	Maximum output pressure (10^3 Pa)	Reference
Electrostatic	silicon (four layers)	850	3.1	41
Piezoelectric	silicon–glass–silicon–glass	100	20	39
Thermo-pneumatic	silicon–glass–silicon	58	3	36
Pneumatic	glass–Au–Ti LIGA process	80	4.71	103
Electromagnetic	polymer–polymer	780	5.5	104

3.2.3 Flow Channels, Manifolds and Mixers

Passive components, such as flow-through cells, reaction chambers, or manifolds, can be realised in silicon, but they can also be machined, embossed, moulded or cut in polymeric materials. Extremely low cost channels can be realised by simple lamination of plastics. More complex geometries with finer feature sizes such as separation channels, mixers, reaction columns and filters can be fabricated by LIGA or bulk micromachining [2,13,14]. Anodic and fusion bonding of glass to silicon and glass to glass, respectively, offer a number of opportunities for producing microchannels, flow-through cells and separation channels for capillary electrophoresis. In principle the shape of the channel cross-section is severely limited because glass can only be etched isotropically. This problem can be overcome by first producing the desired channel shape in a silicon wafer and then depositing a thin silicon nitride film [108]. The wafer and nitride film are then anodically bonded to a Pyrex wafer, and the silicon is etched away. Free standing silicon nitride channels such as that shown in Figure 12 are the result.

Mixing is particularly difficult to achieve in small fluidic systems, as flow regimes in those systems are mostly laminar. The small typical size of channel

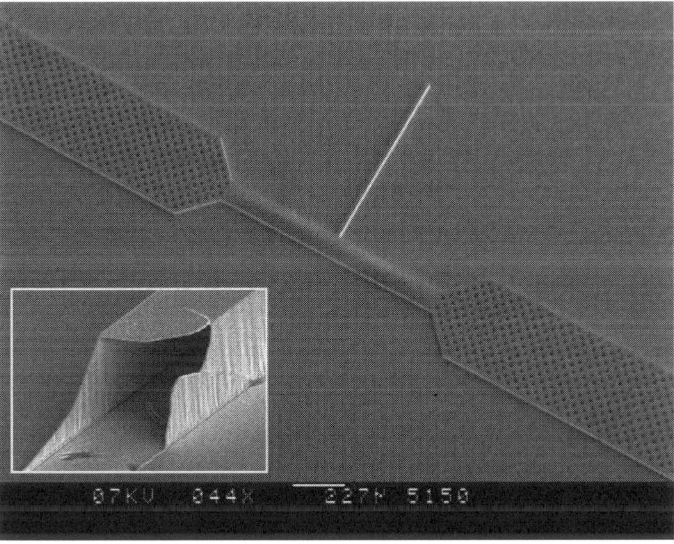

Figure 12. Scanning electron microscope picture of microflow channel connections. Low resistance web-like supply channels connected to each other including a junction are shown. Because the walls are only 390 nm there are pillars to support the structure. Normally channels wider than 100 μm are not possible and will collapse. Inset: Scanning electron microscope- picture of a channel junction cross-section showing the 390nm thick insulating wall. The channel is broken in order to show the wall thickness.

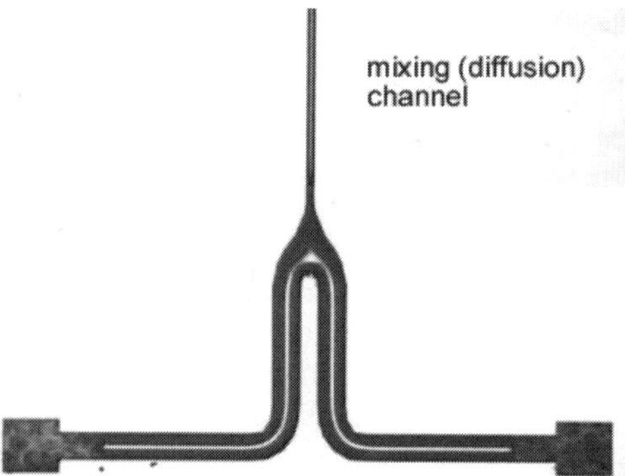

Figure 13. A simple mixer based upon the small diffusion length in a narrow channel

diameters makes it very difficult to induce a turbulent flow profile that is preferred for mixing. To circumvent this, an interesting approach consists in generating a turbulent flow regime by means of ultrasonically or piezoelectrically driven actuators [109,110]. In contrast, micro-channel diameters can be so small that diffusion becomes the dominant mixing mechanism in many cases. Thus, a very simple way of mixing is by flowing two laminar streams in parallel in a narrow channel [111,112], as shown in Figure 13. Faster mixing is usually achieved using a multi-lamination principle, which is obtained by splitting the input streams into alternating fluid laminates [49] or by injecting one input stream into the other via a micro nozzle-array [113]. The silicon microfilter that was shown in Figure 10 has been used for the same purpose [114].

3.2.4 Microreactors

When the sample has to undergo a reaction before analysis, all advantages of microreactors, such as a large surface to volume ratio, are available. If properly designed, the surface area of enzyme microreactors fabricated in silicon can be made large enough to reach the desired catalytic turnover rate by simple attachment of the enzyme to the reactor wall [51,52]. Moreover, if the reactor wall is made porous, e.g. by the anodic etching process described in section 2.3.1, its surface area can be further enlarged, hence resulting in even higher catalytic turnover rates [115]. An example of a tubular micromachined reactor is shown in Figure 14.

3.3 SENSORS

Over the past decades, active research works have been devoted at different sensing and transduction principles for various analytical applications [116–118].

Figure 14. Micromachined open tubular reactor (chip size 2.3 cm × 2.3 cm). The fabrication is based upon the anisotropic etching of {100} oriented Si to give v-grooves, and completed with an anodically bonded glass plate. This microreactor has been assembled with the micropump illustrated in Figure 11.

The sensing mechanism can be more or less straightforward (Chapters 2–6 and 9). Initially developed for clinical applications, biosensors and immunochemical methods show promising potential for the identification and quantitation of a number of compounds in environmental media [119,120]. Enzymes [121], antibodies [122], molecular imprinted polymers [123] and bacteria [124,125] have been used as sensing elements integrated in microanalytical systems for the analysis of a number of inorganic and organic species. However, for microsensor development, electrochemical and optical transduction appear to be the two most salient approaches for the monitoring of dissolved chemical species in natural waters and sediments.

A number of reports have addressed the microfabrication of electrochemical transducers and their properties [13,14,126,127]. Metallic micro-wires sealed in glass capillaries (Chapters 2 and 9), screen-printed or thin-film microelectrodes [25] (Chapter 9) have all their advantages and shortcomings. The selection of plastic or silicon as substrate material is based on similar considerations as for pumps and channels: silicon offers technological advantages compared with plastic, but is more expensive so should only be used when there are specific advantages. These include in particular the fabrication of almost perfectly smooth planar surfaces for electrodes and good geometric definition at small scale, and the possibility to use semiconductor devices such as FETs. In this way, ISFETs [128] and redox sensitive transistors [129] have been fabricated. The selection of material is also largely dependent on its chemical reactivity, possible release of interfering impurities or adsorption of the test compounds, in particular when trace analysis is concerned [130].

Optical transducers have benefited from improvements made in the field of telecommunications, such as fibre optics and integrated optics [131,132] (Chapter 2). Nowadays, laser diodes, modulators and detectors are compact and reasonably priced. Optical microsystems have been fabricated both in the hybrid fashion as well as using the integration approach (section 3.1). An example of the first approach is an elegant micro-analytical system for the detection of heavy metals [133]. The micro-spectrometer was fabricated by hot embossing, using mould inserts that were themselves produced by LIGA. An example of the second approach is a Mach–Zehnder interferometer [134] made using ZnO optical waveguides on a silicon substrate.

3.4 DATA PROCESSING

A number of options for data processing or system control are listed in this section. Commercial data acquisition plug-on cards for laptop computers with a PCMCIA‡ format already exhibit reduced overall dimensions compared with custom-made electronic boards. These acquisition cards are usually not application specific, but are sufficiently flexible to fit many applications. Note that some companies propose specific PCMCIA interfaces for their analytical sensors.

The multi-chip module (MCM) is a system composed of two or more electronic chips, closely mounted on a common substrate package. It offers a corresponding size reduction, but already requires some investment for a low to medium production volume. Some companies provide a full service, from design methodology to testing and quality assurance of MCM.

The ASIC, standing for application specific integrated circuit, is a customised and highly application specific integrated circuit. Reasonable costs are achieved by sharing the expenses with other users through multi-customer project, known as Multi-Project Wafer (MPW). Some silicon foundries and technical universities propose these services. The ASIC reaches by far the highest level of integration and specificity, and consequently requires a medium to high production volume.

4 SELECTED EXAMPLES

The development of integrated systems has reached a high degree of sophistication in the field of drug discovery and clinical monitoring. Some of those are commercially available. In contrast, systems for environmental monitoring are typically between the proof of principle phase and the prototype phase and

‡ Personal Computer Memory Card International Association.

quantitative specifications are not yet available. In section 3, interesting developments in various system parts have been discussed. In this section, the focus will be on a number of systems that have reached some level of integration, mostly based on parts that have a proven track record, rather than the highest possible level of sophistication. As there are yet only few environmental applications, general analytical examples are discussed here.

4.1 GAS PHASE ANALYSIS SYSTEMS

The gas chromatograph reported by Terry *et al.* [135] in the late 1970s is an early example of a miniaturised analysis system, in which a sample injection valve, a 1.5 m long separation capillary column and a thermal conductivity detector has been manufactured using some of the technologies described above [135]. Based on this work, a more recent publication reports the microfabrication of a precision injector and detector, integrated in a commercial micro-gas chromatograph [136]. The micromachined injector combines silicon etching and anodic bonding for the generation of two precision injection and sample microvalves (220 nL of internal volume), a 10 μL sample loop, and pressure matched sample and reference columns. The microvalves are pneumatically driven and exhibit a response time of about 15 ms, with an adjustable injection volume from 0.5 to 15 μL. The detector consists of four suspended hot wires positioned in two parallel flow channels, and arranged in a Wheatstone bridge configuration. The detector response mechanism is based on a difference in thermal conductivity between the eluting gas containing the analyte molecules and the reference gas stream. Whereas classical thermal conductivity detectors are considered to be poorly sensitive and responsive when used in microbore capillary, the extremely small dead volumes that can be achieved by micromachining have demonstrated that microfabricated counterparts are well suited for this purpose. Moreover, as a result of the low thermal mass of the device and the good heat transfer properties of silicon, thermal equilibrium is rapidly achieved and efficiently controlled. The reduced response time of the detector enables chromatographic peaks to be detected with temporal resolution of 10 ms. To reduce the manufacturing costs, the separation column in this example is not fabricated on chip but consists of a 4–10 m long microbore fused silica capillary, to which the micromachined injector, on the one end, and the detector, on the other end, are interfaced. The miniaturisation of selected parts of the gas chromatograph and the use of silicon technology has enabled the realisation of a high speed, portable, rugged, and moderate cost micro-gas chromatograph. Fast separation in less than 1 min was demonstrated for mixtures of ten common solvents and priority pollutants (benzene, trichloroethylene, toluene, ethylbenzene, *m*- and *p*-xylene, *o*-xylene).

A second example of gas detection system is that described by Rapp *et al.* [137], in which mass changes of an array of nine polymer films are measured upon exposure to various organic gases. The polymers are deposited on surface acoustic wave (SAW) devices. Upon adsorbing a gas, the mass change in the polymer is detected by corresponding frequency changes in the SAW devices. The analyser is mainly directed at the qualitative recognition of different gases, especially aliphatic hydocarbons (*n*-hexane, *n*-heptane and *n*-octane), methanol, chloroform and aromatic compounds (toluene, xylente, trimethylbenzene) by comparison of the various frequency shifts. The system includes integrated an RS232 computer interface, dust filtering, gas handling parts and temperature correction and uses 6 W of power, of which 4 W are used for temperature control. Internal zero calibration is achieved by pumping the gas in a closed loop and cleaning it by multiple passages through a built-in active carbon absorber. With a comparable modular approach, the MOSES (modular gas sensor system) [138] has been applied to samples from food and beverage industry (olive oils, coffee mixtures, tobacco, whisky), car industry (plastics, textile materials). The major advantage of this approach consists in the fact that sensors based on different detection mechanisms are grouped in optimised modules that can easily be interchanged to adapt the instrument over a varieties of samples.

4.2 LIQUID PHASE ANALYSIS SYSTEMS

The system shown in Figure 15a and b utilises the MCB approach described in section 3.1. The longest dimension of the system is approximately 4 cm. It comprises three inlets and outlets, two micro-pumps, two flow sensors and an optical absorption detector module. It can be used to measure chemical reaction products by detection of the (spectral) absorption intensity. Sample and reagent liquids are mixed in appropriate proportions on board (the actual mixing takes place during the solution flow in the channels) and the absorption of light produced by a number of light emitting devices is measured by a 64 pixel CCD detector.

The electronic control circuitry is located two levels below the MCB layer. It is based on a microcontroller system for the control of the liquid handling and for data processing. Implemented in the electrical circuitry are driving circuits for the micro-pumps, sensing circuits for the flow sensors, optical absorption measurement circuitry, power management and communications using an RS232 interface. The proof of the MCB concept has been successfully demonstrated and the device thoroughly characterised with different dyes.

Another embodiment of microsystems applied to environmental analyses is a hand-held instrument for direct potentiometric detection of potassium and nitrate [139]. The instrument is operated with a 12 V battery, controlled by a microprocessor, and enables on-spot analyses to be carried out with sufficient accuracy for preliminary investigations. Disposable sensor strips, including an ion-selective working electrode combined with a silver chloride reference electrode, are produced by screen-printing techniques.

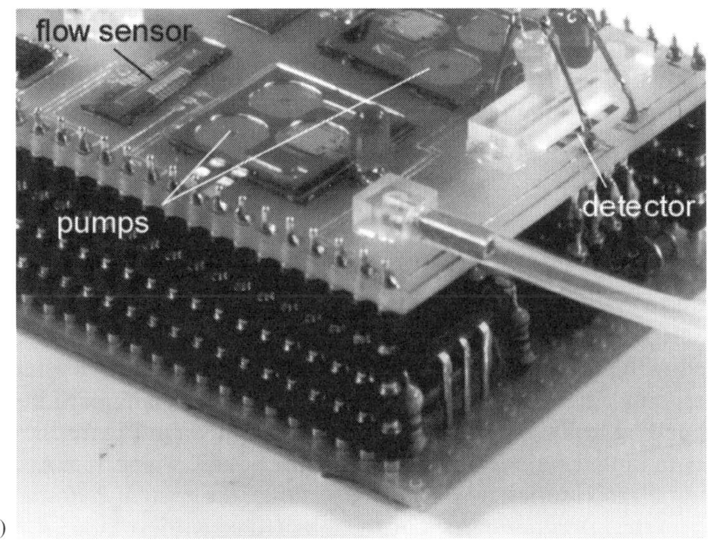

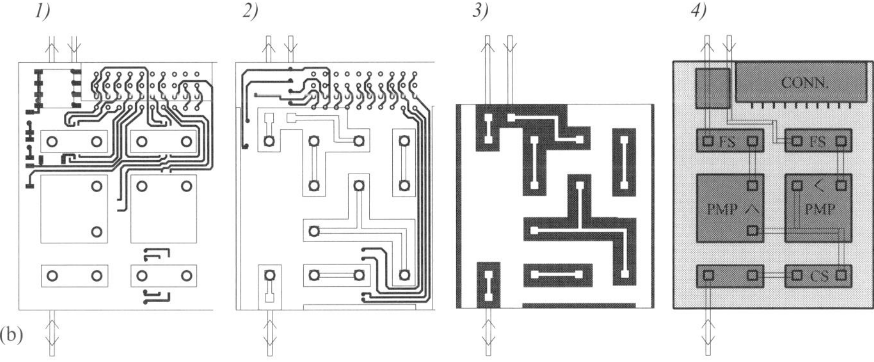

Figure 15. (a) Demonstrator microanalytical system modules mounted on an MCB. (b) Four levels of the MCB demonstrator shown in (a), from bottom to top. 1 and 2, printed circuit board levels with control and readout electronics; 3, fluidic connection level; 4, active component level. PMP = pump, FS = flow sensor, CS = chemical sensor

Similarly, a low power and hand-held system for voltammetric analysis of trace metals has been described [140]. It consists of a thin film Hg-plated iridium microelectrode array, an ASIC potentiostat and a microcontroller. The potentiostat includes two D/A and A/D converters§, controlling the electrode potential, and digitising the electrode current, respectively. Offset and gain errors are compensated with a differential switch capacitor circuitry

§ D/A, digital to analogue; A/D, analogue to digital.

embedded in the chip design. This integrated potentiostat is driven by the microcontroller, setting the timing and the parameters of the experiments, and storing the acquired data for a later retrieval. Detection limits are on the part par billion level (ca 10^{-8} mol L^{-1}). Other similar results in this application area have been reported [141,142]. In general, these systems with integrated potentiostat have a significantly higher noise than microfabricated electrodes arrays combined with classical electronics (Chapter 9, section 5.2), which presently limits their application to screening analysis in significantly polluted waters [143].

A system for automatic groundwater analysis that is to be part of a monitoring network using radio communication has been realised [144]. The system uses ion selective electrodes to measure nitrate and potassium while a micro-spectrometer combined with photometric reagents is used to determine iron levels. Microfluidics, including a mixer, were realised by anisotropic etching of Si in KOH and anodic bonding to Pyrex. The optofluidic cell is shown in Figure 16. The whole system is mounted onto a 75 mm × 500 mm board, which is not very small. However, all accessories needed for operating the system automatically are included on the board, in particular the tanks filled with calibration fluids.

As a final example, a commercially available microfabricated chlorine detector is presented. Chlorine is a commonly used sterilising compound in water. The residual concentration of chlorine is indicative of the potential resistance of the test water to organic or bacterial contamination. The continuous monitoring

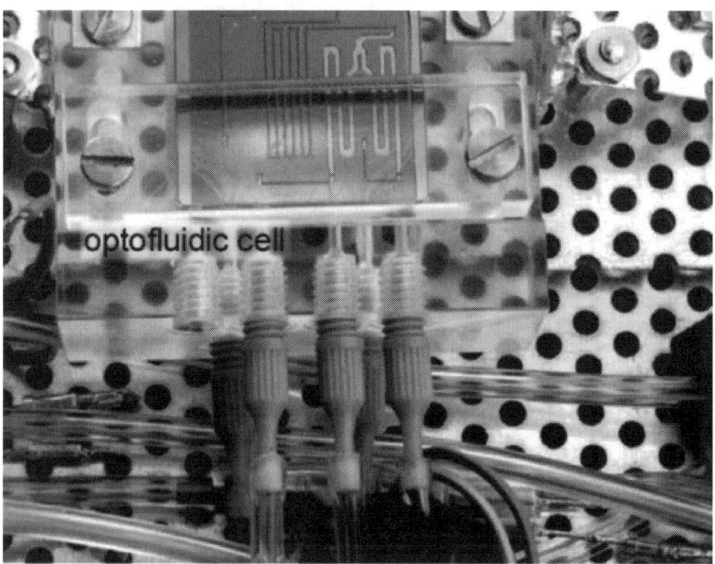

Figure 16. Optofluidic cell made by anodic bonding of a Pyrex glass cover to an anisotropically etched p{100} Si wafer. Iron concentrations are determined by mixing with appropriate reagents and measurement of the extinction of light

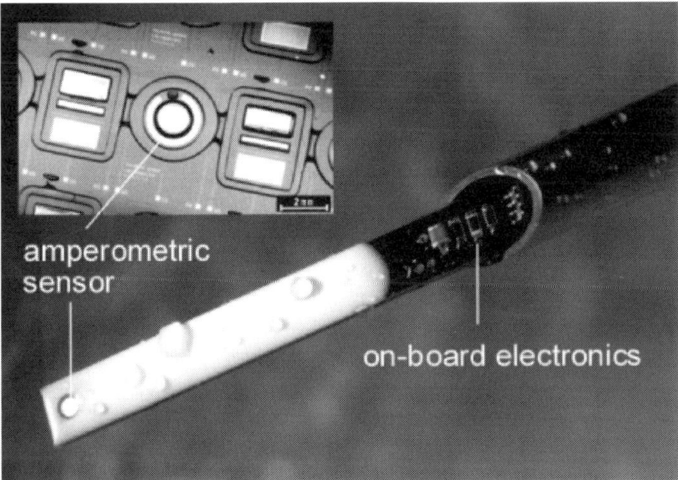

Figure 17. Commercial chlorine detector for drinking water monitoring. Inset: amperometric sensors produced on 4 in silicon wafer by thin film technology. (Courtesy of Alain Grisel, Microsens SA, Neuchâtel, Switzerland)

of chlorine is thus an asset for water quality control. At pH < 7.5, chlorine is present in solution as hypochlorous acid, which is detected amperometrically with a miniaturised three-electrode cell, composed of planar Pt working and counter-electrodes, and Ag/AgCl reference [145]. As illustrated in the inset to Figure 17, the amperometric sensors are fabricated on 4 in silicon wafer by thin film technology. A diffusion-limiting polyHEMA membrane, greatly enhancing the stability and reproducibility of the sensor response, is also photo-polymerised at wafer level. For specific applications, the sensor chip can be mounted on a custom-made board, containing a dedicated electronic interface. These systems have lifetimes of over 9 months of continuous use, excellent linearity over a dynamic range of $2 \times 10^{-8} - 2 \times 10^{-4}$ mol L^{-1} (1–10^4 ppb).

5 CONCLUSION

Microtechnology offers a number of possibilities for manufacturing selected parts of microanalytical systems. Current developments in the field of microanalytical systems include pumps, valves, channels, electrochemical and optical sensors. It is important to bear in mind that miniaturisation ought to be motivated by a clear improvement of the system functionality, and/or a substantial increase of the performance to cost ratio. This implies a good understanding of the system operating principles and of the microscale phenomena that can be exploited.

Additional aspects have to be considered when working continually for long periods of time with real samples. Issues such as sampling, reagent stability, prevention of clogging or biofouling of the system, automatic calibration at regular intervals, consumption of power and reagents, or minimum system size may become critical for *in situ* applications and should be addressed carefully.

It is reasonable to presume that microanalytical systems applied to the environment will occupy niche markets where their specificity is of real added value. Such markets could rise from the enforcement of environmental regulations on the basis of a 'polluter-pays' principle, driving bench top analysis to field analysis and urging the development of portable instrumentation, simple to use and free of artefact.

Alternatively, in laboratory analysis there is a tendency to move away from end-of-pipe measurements towards in-line process monitoring. In that case too, there are interesting opportunities for small rapid analysers. Presently the most important drive for developments in analytical microsystems clearly comes from non-environmental fields, in particular from the drug discovery and biotechnology areas with the rapid and sustained development of micro total analysis systems (μTAS). The major developments in these fields will eventually spin off to environmental monitoring systems as well.

ACKNOWLEDGEMENTS

S. Verpoorte, H. Girault, J. Rossier, A. Grisel, R. Schasfoort and A. Ehlert are kindly acknowledged for their contribution to the present chapter.

GLOSSARY

AD/DA Analogue to digital / digital to analogue conversions of an electric signal.

Anisotropic Exhibiting different values of properties in different crystallographic directions.

ASIC Application specific integrated circuit, an integrated circuit designed for a custom requirement.

Aspect ratio Ratio of the depth to the width of an etched hole or trench, often used as figure of merit to compare different processes.

Bulk micromachining Realisation of microstructures by removing material from the bulk of a substrate.

CCD Charge coupled device, semiconductor device used as sensitive element for the acquisition of images, e.g. video camera.

Chip An unpackaged device cut from a substrate, e.g. semiconductor device incorporating sensors, actuators, electronic circuitry.

CVD Chemical vapour deposition, deposition process of thin films onto a substrate obtained by the chemical reaction of selected gases, at atmospheric pressure and typical temperatures of 300–400 °C.

Dielectric A material with electrically insulating properties.

Dielectric breakdown (strength) Magnitude of an electric field that is necessary to cause passage of current across a dielectric material.

FET Field effect transistor, semiconductor device.

Gate Circuit input whose value affects that of the circuit output, e.g. in a FET, electrode controlling the current flow across the device.

Hybrid Assembly of several elements issued of distinct manufacturing techniques to produce a system with added functionality.

IC Integrated circuit, microfabricated semiconductor circuit containing electronic functions.

ISFET Ion-selective field effect transistor, semiconductor device combining the working principles of a field effect transistor with an ion selective electrode. The best known ISFET application is as pH sensor, although polymeric membranes incorporating ionophores can be deposited on the gate to make the ISFET specific to other ions.

Isotropic Having identical values of a property in all crystallographic directions.

LIGA Lithographie, Galvanoformung, Abformung, lithography, electrodeposition and moulding. The process involves the patterning of a thick layer of resist and the subsequent electrodeposition of a thick metal layer to produce a mould insert for precision injection moulding or a tool for hot embossing of polymers.

LPCVD Low pressure chemical vapour deposition, CVD carried out in a reduced pressure environment, typically 10^{-4} bar, and fairly elevated temperatures, typically 800 °C.

Mask A glass frame containing opaque pattern generally made of chromium and that is used in photolithography.

MCB Mixed circuit board, printed circuit board combining electronics and fluidics.

MCM Multi-chip module, assembly of several semiconductor chips on common base.

MPW Multi-project wafer, service offered by some IC manufacturers to lower the cost of technology, by sharing the expenses among different customers.

PCMCIA Personal Computer Memory Card International Association, format of plug-in extension cards mostly used for laptop personal computers.

PECVD Plasma-enhanced chemical vapour deposition, a glow discharge of selected gases is utilised to enhance the deposition process and reduce the deposition temperature 400 °C.

Packaging Protective enclosure of a chip (IC, sensor, transducer) or microsystem.

Photodiode Semiconductor circuit that produce an electrical signal (current/voltage) in response to illumination.

Polysilicon Polycrystalline silicon used as a sacrificial layer in surface micromachining, and as a conductor in integrated circuits, e.g. gate electrode in FET devices.

Printed circuit board (PCB) Insulating board supporting and offering electrical interconnections between electronic circuit components by means of metal tracks, typically copper lines.

Sacrificial layer A thin film that is later removed to release a microstructure from its substrate.

SAW Surface acoustic wave.

Signal-to-noise ratio Ratio of the output signal in presence of analyte to the output signal in absence of analyte..

Smart sensor A sensor in which the electronics that process the output signal of the sensor is partially or entirely integrated on the same substrate

Surface micromachining The realisation of microstructures is carried out on top of a substrate, i.e. the substrate is used as a base to build upon.

REFERENCES

1. Klein Lebbink, G. C. (1994). *Microsystem Technology—Exploring Opportunities*, STT Netherlands Study Centre for Technology Trends, The Hague.
2. Manz, A. and Becker, H. (ed.) (1998). *Microsystem Technology in Chemistry and Life Sciences*, Springer-Verlag, Berlin.
3. Van den Berg, A. and Bergveld, P. (ed.) (1995). *Micro Total Analysis Systems*, Kluwer, Dordrecht.
4. Rapp, R., Hoffmann, W., Süss, W., Ache, H. J. and Gölz, H. (1997). Performance of an electrochemical microanalysis system, *Electrochim. Acta*, **42**, 3391.
5. Ache, H. J. (1996). Chemical microanalytical systems: objectives and latest developments, *Fresenius' J. Anal. Chem.*, **335**, 467.
6. Harrison, D. J. and van den Berg, A. (Ed.) (1998). *Proc. Micro Total Analysis Systems '98*, Banff, Kluwer Dordrecht.
7. Blundell, N. J., Hopkins, A., Worsfold, P. J. and Casey, H. (1993). A portable battery-powered flow injection monitor for the *in situ* analysis of nitrate in natural waters, *J. Autom. Chem.*, **15**, 159.
8. Mason, A., Yazdi, N., Najafi, K. and Wise, K. D. (1995). A low-power wireless microinstrumentation system for environmental monitoring. In *Techical Digest of Transducers*, **Vol 1**, Royal Swedish Academy of Engineering Sciences, Stockholm, p. 107.
9. Ross, B., Steiner, G., Kiesshauer, M., Bradter, M. and Cammann, K. (1995). Instrument with integrated sensors for a rapid determination of inorganic ions, *Sens. Actuators B*, **26–27**, 380.
10. Cambiaso, A., Chiarugi, S., Grattarola, M., Lorenzelli, L., Lui, A., Margesin, B., Matinoia S., Zanini V. and Zen, M., (1996). An H^+-FET-based system for on-line detection of microorganisms in water, *Sens. Actuators B*, **34**, 245.

11. Ross, B., Chemnitius, G., Drost, S., Wörmann, W., Richter, M., Köster, O., Konz, W., Frimmet, F. H., Schuhmann, W., Ferretti, R. and Meixner, L. (1997). Microanalytical system with integrated chemical and biochemical sensors for environmental control. In *Proc. Eurosensors XI*, Warsaw, Poland, Vol. 2, p. 607.
12. Brecht, A. (1997) Integrated sensor systems for environmental monitoring, *Sens. Rev.*, **17**, 327.
13. Madou, M. (1997). *Fundamentals of Microfabrication*, CRC Press, New York.
14. Kovacs, G. T. A. (1998). *Micromachined Transducers Sourcebook*, WCB/McGraw Hill, New York.
15. Qin, D., Xia, Y., Rogers, J. A., Jackman, R. J., Zhao, X. M. and Whitesides, G. M. (1998). Microfabrication, microstructures and microsystems. In *Microsystem Technology in Chemistry and Life Sciences*, ed. Manz A. and Becker, H., Springer-Verlag, Berlin, Chapter 2.
16. Dario, P., Carrozza, M. C., Croce, N., Montesi, M. C. and Coccom M. (1995). Non-traditional technologies for microfabrication, *J. Micromech. Microeng.*, **5**, 64.
17. Ohlckers, P., Hanneborg, A. and Nese, N. (1995). Batch processing for micromachined devices, *J. Micromech. Microeng.*, **5**, 47.
18. Moreau, W. M. (1988). Semiconductor lithography: principles, practice and materials. In *Microdevices: Physics and Fabrication Technologies*, ed. Muray, J. J. and Brodie, I., Plenum, New York.
19. Bunshah, R. F. (1994). *Handbook of Deposition Technologies for Films and Coatings: Science, Technology and Applications*, Noyes, New York.
20. Schuegraf, K. K. (1988). *Handbook of Thin-film Deposition Processes and Techniques: Principles, Methods, Equipment*, Noyes, New York.
21. Sze, S. M. (1983). *VLSI Technology*, in Series in Electrical Engineering, McGraw-Hill, New York.
22. Ghandhi, S. K. (1983). *VLSI Fabrication Principles: Silicon and Gallium Arsenide*, Wiley, New York.
23. Fiaccabrino, G. C., Tang, X. M., Skinner, N., de Rooij, N. F. and Koudelka-Hep, M. (1996). Interdigitated microelectrode arrays based on sputtered carbon thin-films, *Sens. Actuators B*, **35–36**, 247.
24. Sreenivas G., Ang, S. S., Fritsh, I., Brown, W. D., Gerhardt, G. A. and Woodward, D. J. (1996). Fabrication and characterization of sputtered-carbon microelectrode arrays, *Anal. Chem.*, **68**, 1858.
25. Fiaccabrino, G. C. and Koudelka-Hep, M. (1998). Thin-film microfabrication of electrochemical transducers, *Electroanalysis*, **10**, 217.
26. Howe, R. T. and Muller, R. S. (1983). Polycrystalline and amorphous silicon micromechanical beams: annealing and mechanical properties, *Sens. Actuators*, **4**, 447.
27. Rossnagel S. M., Cuomo J. J. and Westwood W. D. (1990). *Handbook of Plasma Processing Technology, Fundamentals, Etching, Deposition, and Surface Interactions*, in Materials Sciences and Process Technology Series, Noyes, NJ.
28. Nathanson, H. C., Newell, W. E., Wickstrom, R. A. and Davis, Jr., J. R. (1967). The resonant gate transistor, *IEEE Trans. Electron Devices*, **14**, 117
29. Petersen, K. E. (1979). Micromachined membrane switches on silicon, *IBM J. Res. Dev.*, **23**, 376.
30. Linder, C., Paratte, L., Grétaillat, M.-A., Jeacklin, V. P. and de Rooij, N. (1992). Surface micromachining, *J. Micromech. Microeng.*, **2**, 122
31. Kuehnel, W. and Sherman, S. (1994). Surface micromachined silicon accelerometer with on-chip detection circuitry, *Sens. Actuators A*, **45**, 7.
32. Core, T. A., Tsang, W. K. and Sherman, S. J. (1993). Fabrication technology for an integrated surface-micromachined sensor, *Solid State Technol.*, **36**, 39

33. Jaecklin, V. P. (1994). Surface micromachined electrostatic actuators, Ph. D. dissertation, IMT, University of Neuchâtel, Switzerland, and references cited therein.
34. Gretillat, M.-A., Thiebaud, P., Linder, C. and de Rooij, N. F. (1995). Gretillat, M.-A. (1997). Electrostatic polysilicon microrelays, Ph. D. dissertation, IMT, University of Neuchâtel, Switzerland, and references cited therein.
35. Gravesen, P., Branebjerg, J. and Jensen O. S. (1993). Microfluidics—a review, *J. Micromech. Microeng.*, **3**, 168.
36. Elwenspoek, M., Lammerink, T. S. J., Miyake, R. and Fluitman, J. H. J. (1994). Towards integrated microliquid handling systems, *J. Micromech. Microeng.*, **4**, 227.
37. van Lintel, H. T. G., van de Pol, F. C. M. and Bouwstra, S. (1988). A piezoelectric pump based on micromachining of silicon, *Sens. Actuators*, **15**, 153.
38. Van de Pol, F. C. M., van Lintel, H. T. G., Elwenspoek, M. and Fluitman, J. H. J. (1990). A thermopneumatic micropump based on micro-engineering techniques, *Sens. Actuators A*, **21–23**, 198.
39. Gass, V., van der Schoot, B. H., Jeanneret, S. and de Rooij, N. F. (1993). Micro liquid handling using a flow-regulated silicon micropump, *J. Micromech. Microeng.*, **3**, 214.
40. Richtner, M., LinnemannR. and Woias, P. (1998). Robust design of gas and liquid micropumps, *Sens. Actuators A*, **68**, 480.
41. Maillefer, D., van Lintel, H., Rey-Mermet, G. and Hirshi, R. (1999) A high-performance silicon micropump for implantable drug delivery system. In *Proc. Micro Electro Mechanical Systems '99 (MEMS'99)*, FL, IEEE, New York, p. 541.
42. Neagu, C. R., Gardeniers, J. G. E., Elwenspeok, M. and Kelly, J. J. (1997). An electrochemical active valve, *Electrochim. Acta*, **42**, 3367.
43. Manz, A., Miyahara, Y., Miura, J., Watanabe, Y., Miyagi, H. and Sato, K. (1990). Design of an open-tubular liquid chromatography using silicon chip technology, *Sens. Actuators B*, **1**, 249.
44. Harrison, J. D., Manz, A., Fan, Z., Lüdi, H. and Widmer, H. M., (1992). Capillary Electrophoresis and sample injection systems integrated on a planar glass chip, *Anal. Chem.*, **64**, 1926.
45. Tjerkstra, W. R., de Boer, M., Berenschot, E., Gardeniers, J. G. E., van den Berg, A. and Elwenspoek, M. C. (1997). Etching technology for chromatography microchannels, *Electrochim. Acta*, **42**, 3399.
46. Van Rijn, J. C. M., Elwenspoek, M. C. and Fluitman, J. H. J. (1995) Micro filtration membrane with silicon micro machining for industrial and biomedical applications. In *Proc. Micro Electro Mechanical Systems (MEMS'95)*, Amsterdam, IEEE, New York, p. 83.
47. Brody, J. P., Osborn, T. D., Foster, F. K. and Yager, P. (1996). A planar microfabricated fluid filter, *Sens. Actuators*, **54**, 704.
48. He, B., Tan, L. and Regnier, F. (1999) Microfabricated filter for microfluidic analytical systems, *Anal. Chem.*, **71**, 1464.
49. Brenebjerg, J., Gravesen, P., Krog, J. P. and Nielsen, C. R. (1996). Fast mixing by lamination. In *Proc. Micro Electro Mechanical Systems (MEMS'96)*, San Diego, CA, IEEE, New York, p. 441.
50. Verpoorte, E. M. J., van der Schoot, B. H., Jeanneret, S., Manz, A., Widmer, H. M. and de Rooij, N. F. (1994). Three-dimensional micro flow manifolds for miniaturized chemical analysis systems, *J. Micromech. Microeng.*, **4**, 246.
51. Laurell, T., Drott, J. and Rosengren, L. (1995). Silicon wafer integrated enzyme reactors, *Biosens. Bioelectron.*, **10**, 289

52. Strike, D. J., Thiebaud, P., van der Sluis, A. C., Koudelka-Hep, M. and de Rooij, N. F. (1994). Glucose measurement using a micromachined open tubular heterogeneous enzyme reactor (MOTHER), *Microsystem Technol.*, **1**, 48.
53. Conrath, N., Czupor, N., Steinkuhl, Sundermeier, C., Trau, D., Wittkampf, M., Hinkers, H., Meusel, M., Chemnitius, G., Knoll, M., Spener, F. and Camman, C. (1997). Three-dimensionally structured transducers for chemical and biochemical sensors. In *Technical Digest of Transducers '97*, Chicago, IL, Vol. 1, IEEE, New York, p. 481.
54. Hsueh, Y.-T., Smith, R. L. and Northrup, M. A. (1996). A microfabricated, electrochemiluminescence cell for the detection of amplified DNA, *Sens. Actuators B*, **33**, 110.
55. van den Berg, A. and Lammerink, T. S. J. (1998). Micro Total Analysis Systems: microfluidic aspects, integration concept and applications, *Top. Curr. Chem.*, **194**, 21.
56. Kendal, D. L. and Shoultz, R. A. (1997). Wet chemical etching of silicon and SiO_2 and ten challenges for micromachiners,. In *Handbook of Microlithography, Micromachining and Microfabrication* Vol. 2, ed. Choudury, P. R. SPIE Press monograph, Washington, DC, Chapter 2.
57. Bohg, A. (1971). Ethylene-diamine–pyrocatechol–water mixture shows etching anomally in boron-doped silicon, *J. Electrochem. Soc.*, **118**, 401.
58. Reisman, A., Bekenbilt, M., Chan, S. A., Kaufman, F. B. and Green D. C. (1979). The controlled etching of silicon in catalyzed ethylene-diamine–pyrocatechol–water solutions, *J. Electrochem. Soc.*, **126**, 1406.
59. Kloek, B. (1989). Design, fabrication and characterization of piezoresistive pressure sensors, including the study of electrochemical etch-stop, Ph. D. dissertation, IMT, University of Neuchâtel, Switzerland, and reference cited therein.
60. Morazzani, Cantin, J. L., Ortega, C. Pajot, B., Rahbi, R., Rosenbauer, M., Bardeleben, H. J. and Vazsonyi, E., (1996). Thermal nitridation of p-type porous silicon in ammonia, *Thin Solid Films*, **276**, 32.
61. Tjerkstra, R. W., Gardeniers, J. G. E., Elwenspoek, M. C. and van den Berg, A., (1998). Electrochemical fabrication of multi walled micro channels. In *Proc. Micro Total Analysis Systems '98*, Banff, Kluwer, Dordrecht, p. 133.
62. Grétillat, M., Paoletti, F., Thiébaud, P., Roth, S., Koudelka-Hep, M. and de Rooij, N. F. (1997). A new fabrication method for borosilicate glass capillary tubes with lateral inlets and outlets, *Sens. Actuators A*, **60**, 219.
63. Coward, S. and Matthews, D. (1997). ZASE advances MEMS technology, *MST News Int. Activities Microsyst. Technol.*, **20**, 23.
64. Bartha, J. W., Greschner, J., Puech and M., Maquin, P. (1995). Low temperature etching of Si in high density plasma using SF_6/O_2, *J. Microelectron. Eng.*, **27**, 453.
65. Klaassen, E., Petersen, K., Noworolski J. M., Logan, J., Maluf, N., Brown, J., Sorment, C., McCulley, W. and Kovacs, G. T. A. (1996). Silicon fusion bonding and deep reactive etching: a new technology for microstructures, *Sens. Actuactors A*, **52**, 132.
66. Clerc, P.-A., Dellmann, L., Grétillat, F., Grétillat, M.-A., Indermühle, P.-F., Jeanneret, S., Luginbhul, P., Marxer, C. Pfeffer, T., Racine, G.-A., Roth, S., Stauffer, U., Stebler, C., Thiébaud, P. and de Rooij, N. F. (1998). Advanced deep reactive ions etching—a versatile tool for microelectromechanical systems, *J. Micromech. Microeng.*, **8**, 272.
67. Shoji, S. and Esashi, M. (1995). Bonding and assembling methods for realising a μTAS. In *Micro Total Analysis Systems*, ed. van den Berg, A. and Bergreld, P., Kluwer, Dordrecht, p. 165.

68. Wallis, G. and Pomerantz, D. I. (1969). Field assisted glass-metal sealing, *J. Appl. Phys.*, **40**, 10.
69. Ko, W. H., Suminto, J. T. and Yeh, G. J. (1985). Bonding techniques for microsensors. In *Micromachining and Micro-packaging for Transducers*, Elesevier, Amsterdam.
70. Barth, P. W. (1990). Silicon fusion bonding for fabrication of sensors, actuators and microstructures, *Sens. Actuators* A, **21–23**, 919.
71. Harrison, D. J., Fluri, K., Seiler, K., Fan, Z., Effenhauser, C. S. and Manz, A. (1993). Micromachining a miniaturized capillary electrophoresis-based chemical analysis system on a chip, *Sciences*, **261**, 895.
72. Fluri, K., Fitzpatrick, G., Chiem, N. and Harrison, D. J. (1996). Integrated capillary electrophoresis devices with an efficient postcolumn reaction in planar quartz and glass chips, *Anal. Chem.*, **68**, 4285.
73. Stjernström, M. and Roeraade, J. (1998). Methods for fabrication of microfluidic systems in glass, *J. Micromech. Microeng.*, **8**, 33.
74. Bley, P. (1993). The LIGA process for fabrication of three-dimensional microscale structures, *Interdiscipl. Sci. Rev.*, **18**, 267.
75. Romankiw, L. T., (1997). A path: from electroplating through lithograhic masks in electronics to LIGA in MEMS, *Electrochim. Acta*, **42**, 2985.
76. Ehrfeld, W. (1990). The LIGA process for microsystems. In *Proc. Micro System Technologies '90*, Berlin, p. 521.
77. Menz, W. (1996). LIGA and related technologies for industrial application, *Sens. Actuatorsy A*, **54**, 785.
78. Lee, K. Y., LaBianca N., Rishton, S. A., Zolgharnain, Gelorme, J. D. and Shaw, J., Chang, T. H.-P. (1995). Micromachining application for a high resolution ultrathick photoresist, *J. Vac. Sci. Technol. B*, **13**, 3012.
79. Lorenz, H., Despont, M., Fahrni, N., LaBianca, N., Renaud, P. and Vettiger, P. (1997). SU-8: a low-cost negative resist for MEMS, *J. Micromech. Microeng.*, **7**, 121.
80. Belmont-Hebert, C., Tercier, M. L., Buffle, J., Fiaccabrino, G. C., de Rooij, N. F. and Koudelka-Hep, M. (1998). Gel integrated microelectrode arrays for direct voltammetric measurements of heavy metals in natural waters and other complex media, *Anal. Chem.*, **70**, 2949.
81. Heuschkel, M. O., Guérin, L., Buisson, B., Bertrand, D. and Renaud, Ph. (1998). Buried microchannels in photopolymer for delivering of solutions to neurons in in a network, *Sens. Actuators B*, **48**, 356.
82. Strike, D. J., Michel, Ph. E., L'Hostis, E., Fiaccabrino, G. C., de Rooij N. F. and Koudelka-Hep, (1999). Microreactors and electrochemical detectors fabricated using Si and EPON SU-8. In *Digest of Technical Papers of Transducers'99*, Sendai, Japan, p. 130.
83. Holmes, P. J. and Loasby, R. G. (1976). *Handbook of Thick Film Technology, Electrochemical Publications.*
84. Prudenziati, M., ed. (1994). Thick film sensors. In *Handbook of Sensors and Actuators*, Vol. 1, Elsevier, Amsterdam.
85. Kalcher, K. (1990). Chemically modified carbon paste electrodes in voltammetric analysis, *Electroanalysis*, **2**, 419.
86. Kalcher, K., Kauffmann, J.-M., Wang, J., Svancara, I., Vytras, K., Neuhold, C. and Yang, Z. (1995). Sensors based on carbon paste in electrochemical analysis: a review with particular emphasis on the period 1990–1993, *Electroanalysis*, **7**, 5.
87. Desmond, D., Lane, B., Alderman, J., Hill, M., Arrigan, D. W. M. and Glennon, J. D. (1998). An environmental monitoring system for trace metals using stripping voltammetry, *Sens. Actuators B*, **48**, 409.

88. Wang, J. and Tian B. (1992). Screen-printed stripping voltammetric/potentiometric electrodes for decentralized testing of trace lead, *Anal. Chem.*, **64**, 1706.
89. Roberts, M. A., Rossier, J. S., Bercier, P. and Girault, H. H. (1997). UV laser machined polymer substrates for the development of microdiagnostic systems, *Anal. Chem.*, **69**, 2035.
90. Seddon, B. J., Shao, Y., Fost, J. and Girault, H. H. (1994). The application of excimer laser micromachining for the fabrication of disc microelectrodes, *Electrochim. Acta*, **39**, 783.
91. Osborne, M. C., Shao, Y., Pereira, C. M. and Girault, H. H. (1994). Micro-hole interface for the amperometric determination of ionic species in aqueous solutions, *J. Electroanal. Chem.*, **364**, 155.
92. Elders, J. Jansen and H. V. Elwenspoek, M. C. (1995). DEEMO: A new technology for the fabrication of microstructures. In *Proc. Micro Electro Mechanical Systems (MEMS'95)*, Amsterdam, IEEE, New York, p. 238.
93. Becker, H. Dietz, W. and Dannberg, P., (1998). Microfuidic manifolds by polymer hot embossing for μTAS applications. In *Proc. Micro Total Analysis Systems '98*, Banff, ed. Harrison, D. J. and van den Berg, A., Kluwer, Dordrecht, p. 253.
94. Gardner, J. (1994). Smart sensors. In *Microsensors: Principles and Applications*, ed. Gardner, J., Wiley, New York, p. 269.
95. Giachino, J. M. (1986). Smart sensors, *Sens. Actuators*, **10**, 239.
96. Schomburg, W. K., Maas, D., Bacher, W., Büstgens, B., Fahrenberg, J., Menz, W., Seidel, D. (1995). Assembly for micromechanics and LIGA, *J. Micromech. Microeng.*, **5**, 57
97. Buffle, J., Perret, D. and Newman, M. (1992). The use of filtration and ultrafiltration for size fractionation of aquatic particles, colloids and macromolecules, in *Environmental particles*. In *IUPAC Series in, Environmental Analytical and Physical Chemistry*, Vol. 1, ed. Buffle, J. and van Leeuwen, H.-P., Lewis, Boca Raton, FL, Chapter, 5, p. 171.
98. Kaptein, W. A., Zwaagstra, J. J., Venema, K. and Korf, J. (1998). Continuous ultraslow microdialysis and ultrafiltration for subcutaneous sampling as demonstrated by glucose and lactate measurements in rats, *Anal. Chem.*, **70**, 4696.
99. Kuiper, S., van Rijn, C. J. M., Nijdam, W. and Elwenspoek, M. C. (1998). Development of very high flux microfiltration membranes, *J. Membrane Sci.*, **150**, 1.
100. Gass, V., van der Schoot, B. H., Jeanneret, S. and de Rooij, N. F. (1994). Integrated flow-regulated silicon micropump, *Sens. Actuators B*, **43**, 335.
101. Yager, P. Bell, D., Brody, J. P., Qin, D., Cabrera, C., Kamholz, A. and Weigl, B., (1998). Applying microfluidic chemical analytical systems to imperfect samples. In *Proc. Micro Total Analysis Systems '98*, Banff, ed. Harrison, D. J. and van den Berg, A., Kluwer, Dordrecht, p. 207.
102. Böhm, S. Olthuis, W. and Bergveld, P. (1998). A flow-through cell with integrated coulometric pH actuator, *Sens. Actuators B*, **47**, 48.
103. Rapp, R., Schomburg, W. K., Maas, D., Schulz, J. and Stark, W. (1993). LIGA micropump for gases and liquids, *Sens. Actuators A*, **40**, 57.
104. Dario, P., Croce, N., Carrozza, M. C. and Varallo, G. (1996). A fluid handling system for a chemical microanalyzer, *J. Micromech. Microeng.*, **6**, 95–98.
105. Büstgens, B., Bacher, W., Menz, W. and Schomburg, W. (1994). Micropump manufactured by thermoplastic molding, In *Proc. Micro Electro Mechanical Systems (MEMS'94)*, Oiso, Japan, IEEE, New York, p. 18.
106. Paul, P. H., Arnold, D. W. and Rakestraw, D. J. (1998). Electrokinetic generation of high pressures using porous microstructures. In *Proc. Micro Total Analytical*

Systems '98, Banff, ed. Harrison, D. J. and van den Berg, A., Kluwer, Dordrecht, p. 49.
107. Manz, A., Effenhauser, C. S., Burggraf, N., Harrion D. J., Seiler, K. and Fluri, K. (1994). Electroosmotic pumping and electrophoretic separations for miniaturized chemical analysis systems, *J. Micromech. Microeng.*, **4**, 257.
108. Fintschenko, Y. Fowler, P., Spiering, V., Burger, G-J. and van den Berg, A. (1998). Characterisation of silicon-based insulated channels for capillary electrophoresis. In *Proc. Micro Total Analytical Systems '986*, Banff, ed. Harrison, D. J. and van den Berg, A., Kluwer, Dordrecht, p. 327.
109. Moroney, R. M., White, R. M. and Howe, R. T. (1991). Ultrasonically induced microtransport, *Proc. Micro Electro Mechanical Systems (MEMS'91)*, Nara, Japan, IEEE, New York, p. 227.
110. Woias, P., Hauser, K., Yacoub-George, E. and Hillerich, B. (1999). A silicon-based microreaction system for analytical applications, *Proc. Int. Conf. on Microreaction Technology IMRET 3*, Frankfurt, in press.
111. Erbacher, C., Bessoth, F., Busch, M., Verpporte, E. and Manz, A. (1999). Towards integrated continuous-flow chemical reactors, *Mikrochim. Acta*, in press.
112. Veenstra, T. T., Lammerink, T. S. J., Elwenspoek, M. C. and van den Berg, A. (1999) Characterisation method for a new diffusion-mixer applicable in micro Flow Injection Analysis systems, *J. Micromech. Microeng.*, **9**, (2), 199.
113. Yotsumoto, A., Nakamura, R., Shoji, S. and Wada, T. (1998). Fabrication of an integrated mixing/reaction micro flow cell for μTAS. In *Proc. Micro Total Analytical System '98*, Banff, ed. Harrison, D. J. and van den Berg, A., Kluwer, Dordrecht, p. 185.
114. Miyake, R., Lammerink, T. S. J., Elwenspoek, M. C. and Fluitman, J. H. J. (1993). *Proc. Micro Electro Mechanical Systems (MEMS'93)*, Fort Lauderdale, USA, IEEE, New York, p. 248.
115. Drott, J., Lindström, K., M., Rosengren, L. and Laurell, T. (1998). Porous silicon as the carrier matrix in a micro enzyme reactor to achieve a high efficient and long-term stable glucose sensor. In *Microreaction Technology: Proc. 1st International Conference on Microreaction Technology*, Springer, Berlin, p. 175.
116. Göpel, H. (ed) (1991). *Sensors, A Comprehensive Survey*, Vol. 2, VCH-Verlag, Weinheim.
117. Colombo, C., van den Berg, M. G. and Daniel, A. (1997). A flow cell for on-line monitoring of metals in natural waters by voltammetry with a mercury drop electrode, *Anal. Chem.*, **346**, 101.
118. Wang, J., Larson, D., Foster, N., Armalis, S., Lu, J., Rongrong, X. and Olsen, K., Zirino, A. (1995). Remote electrochemical sensor for trace metal contaminants, *Anal. Chem.*, **67**, 1481.
119. Rogers, K. R. and Lin, J. N. (1992). Biosensors for environmental monitoring, *Biosens. Bioelectron.*, **7**, 317.
120. Van Emon, J. M. and Lopez-Avila, V. (1992). Immunochemical methods for environmental analysis. *Anal. Chem.*, **64**, 79A–88A.
121. Ikebukuro, K., Nishida, R., Yamamoto, H., Arikawa, Y., Nakamure, H., Suzuki, M., Kubo, Takeuchi, T. and Karube, I. (1996). A novel biosensor system for the determination of phosphate, *J. Biotechnol.*, **48**, 67.
122. Meulenberg, E. P., Mulder, W. H. and Stoks, P. G. (1995). Immunoassays for pesticides, *Environ. Sci. Technol.*, **29**, 553.
123. Kriz, D., Ramström, O. and Mosbach, K. (1997). Molecular imprinting, new possibilities for sensor technology, *Anal. Chem.*, **69**, 345 A.

124. Larsen, L. H., T. Kjær and N. P. Revsbech. (1997). A microscale NO_3- biosensor for environmental applications, *Anal. Chem.*, **69**, 3527.
125. Damgaard, L. R., and N. P. Revsbech. (1997). A microscale biosensor for methane, *Anal. Chem.*, **69**, 2262.
126. Fleischmann, M., Pons, S., Rolison D. R., Schmidt, P. P. (1987). *Utramicroelectrodes*, Datatech Systems, Morganton, NC.
127. Montenegro, M. I., Queiros, A. A., Daschbach, J. D. (1990). *Microelectrodes: Theory and Applications*, Kluwer, Dordrecht.
128. Bergveld, P. and Sibbald, A., (1988). *Comprehensive Analytical Chemistry*, Vol. XXIII: *Analytical and Biomedical Application of Ion-sensitive Field Effect Transistors*, Elsevier, Amsterdam.
129. Hendrikse, J., Olthuis, W. and Bergveld, P. (1998). Characterization of the ^{E}MOSFET, a novel one electrode chemical transducer for redox measurements, *J. Electroanal. Chem.*, **458** (1–2) 23.
130. C. Belmont, C., Tercier, M. L., Buffle, J., Fiaccabrino, G. C. and M. Koudelka-Hep (1996). Mercury-plated iridium-based microelectrode arrays for trace metals detection by voltammetry: optimum conditions and reliability, *Anal. Chim. Acta*, **329**, 203.
131. Lambeck, P. (1992). Integrated opto-chemical sensors, *Sens. Actuators B*, **8**, 103.
132. Niessner, R. (1991). Chemical sensors for environmental analysis, *Trends Anal. Chem.*, **10**, 310.
133. Bier, W., Büstgens, B., Mohr, J., Müller, C., Schomburg, W. K., Fromhein, O., Kühner, O., Radloff, D. and Reichert, J. (1995). Ein Mikrosystem zur optischen Schadstoffanalyse, *FZK-Nachrichten*, **27**, 11.
134. Heideman, R. G. Veldhuis, G. J., Jager, E. W. H. and Lambeck, P. V. (1996). Fabrication and packaging of integrated chemo-optical sensors, *Sens. Actuators B*, **35–36**, 234.
135. Terry, S., Jerman, J. H. and Angell, J. B. (1979). A gas chromatographic air analyzer fabricated on a silicon wafer, *IEEE Trans. Electron Devices*, **26**, 1880.
136. Bruns, M. W. (1992). Silicon micromachining and high speed gas chromatography. In *Proc. International Conference on Industrial Electronics, Control, Instrumentation and Automation (IECON'92)*, San Diego, CA, IEEE, New York, p. 1640.
137. Rapp, M., Böss, B., Voigt, A., Gemmeke, H. and Ache, H. J., (1995). Development of an analytical microsystem for organic gas detection based on surface acoustic wave resonators, *Fresenius' J. Anal. Chem.*, **352**, 699.
138. Ulmer H., Mitrovics, J., Noetzel, G., Weimar, U. and Göpel, W. (1996). Odours and flavours identified with hybrid modular sensor systems, *Sens. Actuators B*, **43**, 24.
139. Ross, B., Steiner, G., Kiesshauer, M., Bradter, M. and Cammann, K. (1995). Instrument with integrated sensors for a rapid determination of inorganic ions, *Sens. Actuators B*, **26–27**, 380.
140. Reay, R. J., Flannery, A. F., Storment, C. W., Kounaves, S. P. and Kovacs, G. T. A. (1996). Microfabricated electrochemical analysis system for heavy metal detection, *Sens. Actuators B*, **34**, 450.
141. Desmond, D., Lane, B., Alderman, J., Hall, G., Alvarez-Icaza, M., Garde, A., Ryan J., Barry, L., Svehla, G., Arrigan D. W. M. and Schnittner, L. (1996). An ASIC-based system for stripping voltammetric determination of trace metals, *Sens. Actuators B* **34**, 466.
142. Uhlig, A., Paeschke, M., Schnakenberg, U., Hintsche, R., Diedich H. J. and Scholz, F. (1995). Chip-array electrodes for simultaneous stripping analysis of trace metals, *Sens. Actuators B*, **24–25**, 899.

143. Herdan, J., Feeney, R., Kounaves, S. P., Flannery, A. F., Storment, C. W. and Koracs, G. T. A. (1998). Field evaluation of an electrochemical probe for in situ screening of heavy metals in groundwater, *Environ. Sci. Technol.*, **32**, 131–136.
144. Ehlert, A. and Büttgenbach, S., (1998). Automatic sensor system for water analysis. In *Proc. Micro Total Analytical Systems '98*, Banff, ed. Harrison, D. J. and van den Berg, A., Kluwer, Dordrecht, p. 427.
145. van den Berg, A., Grisel, A., Verney-Norberg, E., van der Schoot, B. H., Koudelka-Hep, M. and de Rooij, N. F. (1993). On-wafer fabricated free-chlorine sensor with ppb detection limit for drinking-water monitoring, *Sens. Actuators B*, **13–14**, 396.

Index